Nano-Surface Chemistry

Nano-Surface Chemistry

edited by
Morton Rosoff
Long Island University
Brooklyn, New York

CRC Press
Taylor & Francis Group
Boca Raton London New York

CRC Press is an imprint of the
Taylor & Francis Group, an **informa** business

CRC Press
Taylor & Francis Group
6000 Broken Sound Parkway NW, Suite 300
Boca Raton, FL 33487-2742

First issued in paperback 2019

© 2002 by Taylor & Francis Group, LLC
CRC Press is an imprint of Taylor & Francis Group, an Informa business

ISBN-13: 978-0-8247-0254-0 (hbk)
ISBN-13: 978-0-367-39691-6 (pbk)

Visit the Taylor & Francis Web site at
http://www.taylorandfrancis.com

and the CRC Press Web site at
http://www.crcpress.com

Preface

Tools shape how we think; when the only tool you have is an axe, everything resembles a tree or a log. The rapid advances in instrumentation in the last decade, which allow us to measure and manipulate individual molecules and structures on the nanoscale, have caused a paradigm shift in the way we view molecular behavior and surfaces. The microscopic details underlying interfacial phenomena have customarily been inferred from in situ measurements of macroscopic quantities. Now we can see and "finger" physical and chemical processes at interfaces.

The reviews collected in this book convey some of the themes recurrent in nano-colloid science: self-assembly, construction of supramolecular architecture, nanoconfinement and compartmentalization, measurement and control of interfacial forces, novel synthetic materials, and computer simulation. They also reveal the interaction of a spectrum of disciplines in which physics, chemistry, biology, and materials science intersect. Not only is the vast range of industrial and technological applications depicted, but it is also shown how this new way of thinking has generated exciting developments in fundamental science. Some of the chapters also skirt the frontiers, where there are still unanswered questions.

The book should be of value to scientific readers who wish to become acquainted with the field as well as to experienced researchers in the many areas, both basic and technological, of nanoscience.

The lengthy maturation of a multiauthored book of this nature is subject to life's contingencies. Hopefully, its structure is sound and has survived the bumps of "outrageous fortune." I wish to thank all the contributors for their courage in writing. It is their work and commitment that have made this book possible.

Morton Rosoff

Contents

Contributors

Alain Carré   Fontainebleau Research Center, Corning S.A., Avon, France

Frank Caruso   Max-Planck-Institute of Colloids and Interfaces, Potsdam, Germany

Kwong-Yu Chan   Department of Chemistry, The University of Hong Kong, Hong Kong SAR, China

Noa Cohen   Department of Molecular Microbiology and Biotechnology, Faculty of Life Sciences, Tel Aviv University, Tel Aviv, Israel

Nir Dotan   Glycominds Ltd., Maccabim, Israel

V. Erokhin   Department of Biophysical M&O Science and Technologies, University of Genoa, Genoa, Italy

Amihay Freeman   Department of Molecular Microbiology and Biotechnology, Faculty of Life Sciences, Tel Aviv University, Tel Aviv, Israel

Xiaoan Fu   Department of Chemical Engineering, Case Western Reserve University, Cleveland, Ohio

Karl-Ulrich Fulda   Institute of Physical Chemistry, University of Cologne, Cologne, Germany

Suzanne P. Jarvis   Nanotechnology Research Institute, National Institute of Advanced Industrial Science and Technology, Ibaraki, Japan

Ori Kalid   Department of Molecular Microbiology and Biotechnology, Faculty of Life Sciences, Tel Aviv University, Tel Aviv, Israel

Achim Kampes   Institute for Physical Chemistry, University of Cologne, Cologne, Germany

Kazue Kurihara   Institute for Chemical Reaction Science, Tohoku University, Sendai, Japan

Bruce R. Locke   Department of Chemical Engineering, Florida State University, Tallahassee, Florida

Brian G. Moore   School of Science, Penn State Erie–The Behrend College, Erie, Pennsylvania

Claudio Nicolini   Department of Biophysical M&O Science and Technologies, University of Genoa, Genoa, Italy

Christof M. Niemeyer   Department of Biotechnology, University of Bremen, Bremen, Germany

Marie-Paule Pileni   Université Pierre et Marie Curie, LM2N, Paris, France

Dietmar Pum   Center for Ultrastructure Research, Universität für Bodenkultur Wien, Vienna, Austria

Syed Qutubuddin   Department of Chemical Engineering, Case Western Reserve University, Cleveland, Ohio

M. K. Ram   Department of Biophysical M&O Science and Technologies, University of Genoa, Genoa, Italy

Miquel Salmeron   Materials Sciences Division, Lawrence Berkeley National Laboratory, Berkeley, California

Margit Sára   Center for Ultrastructure Research, Universität für Bodenkultur Wien, Vienna, Austria

Bernhard Schuster   Center for Ultrastructure Research, Universität für Bodenkultur Wien, Vienna, Austria

Martin E. R. Shanahan   Adhesion, Wetting, and Bonding, National Centre for Scientific Research/School of Mines Paris, Evry, France

Uwe B. Sleytr   Center for Ultrastructure Research, Universität für Bodenkultur Wien, Vienna, Austria

Keith J. Stine   Department of Chemistry and Center for Molecular Electronics, University of Missouri–St. Louis, St. Louis, Missouri

Bernd Tieke   Institute for Physical Chemistry, University of Cologne, Cologne, Germany

Vladimir S. Trubetskoy   Mirus Corporation, Madison, Wisconsin

Vincenzo Turco Liveri   Department of Physical Chemistry, University of Palermo, Palermo, Italy

Jon A. Wolff   Departments of Pediatrics and Medical Genetics, University of Wisconsin–Madison, Madison, Wisconsin

Lei Xu   Materials Sciences Division, Lawrence Berkeley National Laboratory, Berkeley, California

Introduction

The problems of chemistry and biology can be greatly helped if our ability to see what we are doing, and to do things on an atomic level is ultimately developed—a development which I think can't be avoided.

Richard Feynman

God created all matter—but the surfaces are the work of the Devil.

Wolfgang Pauli

The prefix *nano-*, derived from the Greek word meaning "dwarf," has been applied most often to systems whose functions and characteristics are determined by their tiny size. Structures less than 100 nanometers in length (i.e., one-ten-millionth of a meter) are typical in nano-technology, which emphasizes the approach of building up from molecules and nano-structures ("bottom-up") versus the "top-down," or miniaturization, approach. *Nano-* actually refers not so much to the size of the object as to the resolution at the molecular scale. At such small scales, about half of the atoms are in the surface layer, the surface energy dominates, and the surface layer can be considered a new material with properties different from those of bulk. The hierarchy of scales, both spatial and temporal, is represented in the following table:

	Scale			
	Quantum	Atom/nano	Mesoscopic	Macroscopic
Length (meters)	10^{-11}–10^{-8}	10^{-9}–10^{-6}	10^{-6}–10^{-3}	$>10^{-3}$
Time (seconds)	10^{-16}–10^{-12}	10^{-13}–10^{-10}	10^{-10}–10^{-6}	$>10^{-6}$

Classical surface and colloid chemistry generally treats systems experimentally in a statistical fashion, with phenomenological theories that are applicable only to building simplified microstructural models. In recent years scientists have learned not only to observe individual atoms or molecules but also to manipulate them with subangstrom precision. The characterization of surfaces and interfaces on nanoscopic and mesoscopic length scales is important both for a basic understanding of colloidal phenomena and for the creation and mastery of a multitude of industrial applications.

The self-organization or assembly of units at the nanoscale to form supramolecular ensembles on mesoscopic length scales comprises the range of colloidal systems. There is a need to understand the connection between structure and properties, the evolution and dynamics of these structures at the different levels—supramolecular, molecular, and submolecular—by "learning from below."

When interaction and physical phenomena length scales become comparable to or larger than the size of the structure, as, for example, with polymer contour chain length, the system may exhibit unusual behavior and generate novel arrangements not accessible in bulk.

It is also at these levels (10–500 nm) that nature utilizes hierarchical assemblies in biology, and biological processes almost invariably take place at the nanoscale, across membranes and at interfaces. Biomolecular materials with unique properties may be developed by mimicking biological processes or modifying them. There is still much to discover about improving periodic arrays of biomolecules, biological templating, and how to exploit the differences between biological and nonbiological self-assembly.

The linkage of microscopic and macroscopic properties is not without challenges, both theoretical and experimental. Statistical mechanics and thermodynamics provide the connection between molecular properties and the behavior of macroscopic matter. Coupled with statistical mechanics, computer simulation of the structure, properties, and dynamics of mesoscale models is now feasible and can handle the increase in length and time scales.

Scanning proble techniques (SPM)—i.e., scanning tunneling microscopy (STM) and atomic force microscopy (AFM), as well as their variations—have the power to visualize nanoscale surface phenomena in three dimensions, manipulate and modify individual molecules, and measure such properties as adhesion, stiffness, and friction as well as magnetic and electric fields. The use of chemically modified tips extends the technique to include chemical imaging and measurement of specific molecular interactions. Improved optical methods complement probe images and are capable of imaging films a single molecule thick. Optical traps, laser tweezers, and "nano-pokers" have been developed to measure forces and manipulate single molecules. In addition, there is a vast range of experimental tools that cross different length and time scales and provide important information (x-ray, neutrons, surface plasmon resonance). Nevertheless, there is a further need for instrumentation of higher resolution, for example, in the decreased ranged of space and time encountered when exploring the dynamics and kinetics of surface films.

Chapter 1 is a view of the potential of surface forces apparatus (SFA) measurements of two-dimensional organized ensembles at solid–liquid interfaces. At this level, information is acquired that is not available at the scale of single molecules. Chapter 2 describes the measurement of surface interactions that occur between and within nanosized surface structures—interfacial forces responsible for adhesion, friction, and recognition.

In Chapter 3, Langmuir–Blodgett films of varying organizational complexity are discussed, as well as nanoparticles and fullerenes. Molecular dynamic simulation of monolayers and multilayers of surfactants is also reviewed. Chapter 4 presents those aspects of supramolecular layer assemblies related to the development of nanotechnological applications. Problems of preparing particle films with long-range two-dimensional and three-dimensional order by Langmuir–Blodgett and self-assembly techniques are dealt with in Chapter 5.

The next two chapters are concerned with wetting and capillarity. Wetting phenomena are still poorly understood; contact angles, for example, are simply an empirical parameter to quantify wettability. Chapter 6 reviews the use of scanning polarization force microscopy (SPFM), a new application of AFM using electrostatic forces, to study the nanostructure of liquid films and droplets. The effect of solid nanometric deformation on the kinetics of wetting and dewetting and capillary flow in soft materials, such as some polymers and gels, is treated in Chapter 7.

Chapter 8 presents evidence on how the physical properties of colloidal crystals organized by self-assembly in two-dimensional and three-dimensional superlattices differ from those of the free nanoparticles in dispersion.

A biomolecular system of glycoproteins derived from bacterial cell envelopes that spontaneously aggregates to form crystalline arrays in the mesoscopic range is reviewed in Chapter 9. The structure and features of these S-layers that can be applied in biotechnology, membrane biomimetics, sensors, and vaccine development are discussed.

DNA is ideally suited as a structural material in supramolecular chemistry. It has sticky ends and simple rules of assembly, arbitrary sequences can be obtained, and there is a profusion of enzymes for modification. The molecule is stiff and stable and encodes information. Chapter 10 surveys its varied applications in nanobiotechnology. The emphasis of Chapter 11 is on DNA nanoensembles, condensed by polymer interactions and electrostatic forces for gene transfer. Chapter 12 focuses on proteins as building blocks for nanostructures.

The next two chapters concern nanostructured core particles. Chapter 13 provides examples of nano-fabrication of cored colloidal particles and hollow capsules. These systems and the synthetic methods used to prepare them are exceptionally adaptable for applications in physical and biological fields. Chapter 14, discusses reversed micelles from the theoretical viewpoint, as well as their use as nano-hosts for solvents and drugs and as carriers and reactors.

Chapter 15 gives an extensive and detailed review of theoretical and practical aspects of macromolecular transport in nanostructured media. Chapter 16 examines the change in transport properties of electrolytes confined in nanostructures, such as pores of membranes. The confinement effect is also analyzed by molecular dynamic simulation.

Nanolayers of clay interacting with polymers to form nanocomposites with improved material properties relative to the untreated polymer are discussed in Chapter 17.

Morton Rosoff

1

Molecular Architectures at Solid–Liquid Interfaces Studied by Surface Forces Measurement

KAZUE KURIHARA   Tohoku University, Sendai, Japan

I. INTRODUCTION

Molecular and surface interactions are ubiquitous in molecular science, including biology. Surface forces measurement and atomic force microscopy (AFM) have made it possible to directly measure, with high sensitivity, molecular and surface interactions in liquids as a function of the surface separation. Naturally, they have become powerful tools for studying the origins of forces (van der Waals, electrostatic, steric, etc.) operating between molecules and/or surfaces of interest [1–4]. They also offer a unique, novel surface characterization method that "monitors surface properties changing from the surface to the bulk (depth profiles)" and provides new insights into surface phenomena. This method is direct and simple. It is difficult to obtain a similar depth profile by other methods; x-ray and neutron scattering measurements can provide similar information but require extensive instrumentation and appropriate analytical models [4].

Molecular architectures are self-organized polymolecular systems where molecular interactions play important roles [5]. They exhibit specific and unique functions that could not be afforded by single molecules. Molecular architecture chemistry beyond molecules is not only gaining a central position in chemistry but becoming an important interdisciplinary field of science. Investigations of molecular architectures by surface forces measurement is important for the following reasons.

1. It is essential to elucidate intermolecular interactions involved in self-organization, whose significance is not limited to material science but extends to the ingenuity of biological systems [5].
2. The importance of surface characterization in molecular architecture chemistry and engineering is obvious. Solid surfaces are becoming essential building blocks for constructing molecular architectures, as demonstrated in self-assembled monolayer formation [6] and alternate layer-by-layer adsorption [7]. Surface-induced structuring of liquids is also well-known [8,9], which has implications for micro- and nano-technologies (i.e., liquid crystal displays and micromachines). The virtue of the force measurement has been demonstrated, for example, in our report on novel molecular architectures (alcohol clusters) at solid–liquid interfaces [10].
3. Two-dimensionally organized molecular architectures can be used to simplify the complexities of three-dimensional solutions and allow surface forces measurement. By

1

employing this approach, we can study complex systems such as polypeptides and polyelectrolytes in solutions. For example, it is possible to obtain essential information such as the length and the compressibility of these polymers in solutions by systematically varying their chemical structures and the solution conditions [11].

Earlier studies of surface forces measurement were concerned mainly with surface interactions determining the colloidal stability, including surfactant assemblies. It has been demonstrated, however, that a "force–distance" curve can provide much richer information on surface molecules; thus it should be utilized for studying a wider range of phenomena [12]. Practically, the preparation of well-defined surfaces, mostly modified by two-dimensional organized molecules, and the characterization of the surfaces by complementary techniques are keys to this approach. A similar concept is "force spectroscopy" [13], coined to address force as a new parameter for monitoring the properties of materials. A major interest in force spectroscopy is the single molecular measurement generally employing an atomic force microscope. This measurement treats relatively strong forces, such as adhesion, and discusses the binding of biotin-streptavidin [14] and complementary strands of DNA [15] as well as the unfolding and folding of proteins [16]. On the other hand, the forces measurement of two-dimensionally organized molecules has advantages complementary to those of single molecule force spectroscopy. It can monitor many molecules at the same time and thus is better suited for studying long-range weaker forces. The measurement should bear a close relevance to real systems that consist of many molecules, because interactions between multiple molecules and/or macroscopic surfaces in solvents may exhibit characteristics different from those between single molecules.

The aim of this review is to demonstrate the potential of surface forces measurement as a novel means for investigating surfaces and complex soft systems by describing our recent studies, which include cluster formation of alcohol, polyion adsorption, and polyelectrolyte brushes.

II. SURFACE FORCES MEASUREMENT

Surface forces measurement directly determines interaction forces between two surfaces as a function of the surface separation (D) using a simple spring balance. Instruments employed are a surface forces apparatus (SFA), developed by Israelachivili and Tabor [17], and a colloidal probe atomic force microscope introduced by Ducker et al. [18] (Fig. 1). The former utilizes crossed cylinder geometry, and the latter uses the sphere-plate geometry. For both geometries, the measured force (F) normalized by the mean radius (R) of cylinders or a sphere, F/R, is known to be proportional to the interaction energy, G_f, between flat plates (Derjaguin approximation),

$$\frac{F}{R} = 2\pi G_f \qquad (1)$$

This enables us to quantitatively evaluate the measured forces, e.g., by comparing them with a theoretical model.

Sample surfaces are atomically smooth surfaces of cleaved mica sheets for SFA, and various colloidal spheres and plates for a colloidal probe AFM. These surfaces can be modified using various chemical modification techniques, such as Langmuir–Blodgett (LB) deposition [12,19] and silanization reactions [20,21]. For more detailed information, see the original papers and references texts.

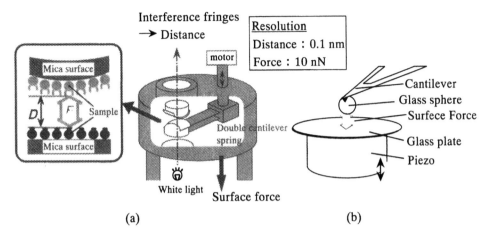

FIG. 1 Schematic drawings of (a) the surface forces apparatus and (b) the colloidal probe atomic force microscope.

III. ALCOHOL CLUSTER FORMATION ON SILICA SURFACES IN CYCLOHEXANE

Surface forces measurement is a unique tool for surface characterization. It can directly monitor the distance (D) dependence of surface properties, which is difficult to obtain by other techniques. One of the simplest examples is the case of the electric double-layer force. The repulsion observed between charged surfaces describes the counterion distribution in the vicinity of surfaces and is known as the electric double-layer force (repulsion). In a similar manner, we should be able to study various, more complex surface phenomena and obtain new insight into them. Indeed, based on observation by surface forces measurement and Fourier transform infrared (FTIR) spectroscopy, we have found the formation of a novel molecular architecture, an alcohol macrocluster, at the solid–liquid interface.

Adsorption phenomena from solutions onto solid surfaces have been one of the important subjects in colloid and surface chemistry. Sophisticated application of adsorption has been demonstrated recently in the formation of self-assembling monolayers and multilayers on various substrates [4,7]. However, only a limited number of researchers have been devoted to the study of adsorption in binary liquid systems. The adsorption isotherm and colloidal stability measurement have been the main tools for these studies. The molecular level of characterization is needed to elucidate the phenomenon. We have employed the combination of surface forces measurement and Fourier transform infrared spectroscopy in attenuated total reflection (FTIR-ATR) to study the preferential (selective) adsorption of alcohol (methanol, ethanol, and propanol) onto glass surfaces from their binary mixtures with cyclohexane. Our studies have demonstrated the cluster formation of alcohol adsorbed on the surfaces and the long-range attraction associated with such adsorption. We may call these clusters *macroclusters*, because the thickness of the adsorbed alcohol layer is about 15 nm, which is quite large compared to the size of the alcohol. The following describes the results for the ethanol–cycohexane mixtures [10].

Typical forces profiles measured between glass surfaces in ethanol–cyclohexane mixtures are shown in Fig. 2. Colloidal probe atomic force microscopy has been employed. In pure cyclohexane, the observed force agrees well with the conventional van der Waals attraction calculated with the nonretarded Hamaker constant for glass/cyclohexane/glass,

3.1×10^{-21} J. At an ethanol concentration of 0.1 mol%, the interaction changes remarkably: The long-range attraction appears at a distance of 35 nm, shows a maximum around 10 nm, and turns into repulsion at distances shorter than 5 nm. The pull-off force of the contacting surfaces is 140 ± 19 mN/m, which is much higher than that in pure cyclohexane, 10 ± 7 mN/m. Similar force profiles have been obtained on increasing the ethanol concentration to 0.4 mol%. A further increase in the concentration results in a decrease in the long-range attraction. At an ethanol concentration of 1.4 mol%, the interaction becomes identical to that in pure cyclohexane. When the ethanol concentration is increased, the range where the long-range attraction extends changes in parallel to the value of the pull-off force, indicating that both forces are associated with the identical phenomenon, most likely the adsorption of ethanol. Separation force profiles after the surfaces are in contact shows the presence of a concentrated ethanol layer near and on the surfaces (see Ref. 10a). The short-range repulsion is ascribable to steric force due to structure formation of ethanol molecules adjacent to the glass surfaces.

In order to understand the conditions better, we determined the adsorption isotherm by measuring the concentration changes in the alcohol upon adsorption onto glass particles using a differential refractometer. Figure 3 plots the range of the attraction vs. the ethanol concentration, together with the apparent adsorption layer thickness estimated from the adsorption isotherm, assuming that only ethanol is present in the adsorption layer [22]. For 0.1 mol% ethanol, half the distance where the long-range attraction appears, 18 ± 2 nm, is close to the apparent layer thickness of the adsorbed ethanol, 13 ± 1 nm. This supports our interpretation that the attraction is caused by contact of opposed ethanol adsorption layers. Half the attraction range is constant up to ~0.4 mol% ethanol and decreases with increasing ethanol concentration, while the apparent adsorption layer thickness remains constant at all concentration ranges studied. The discrepancy between the two quantities indicates a change in the structure of the ethanol adsorption layer at concentrations higher than ~0.4

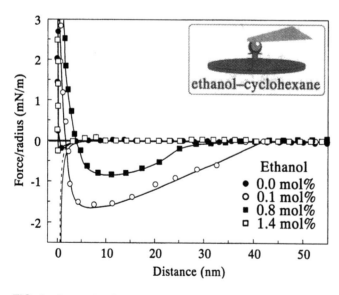

FIG. 2 Interaction forces between glass surfaces upon compression in ethanol–cyclohexane mixtures. The dashed and solid lines represent the van der Waals force calculated using the nonretarded Hamaker constants of 3×10^{-21} J for glass/cyclohexane/glass and 6×10^{-21} J for glass/ethanol glass, respectively.

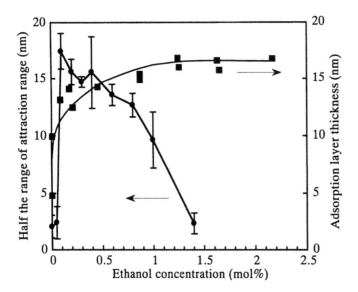

FIG. 3 Plots of half the range of attraction (see Fig. 2) and the apparent thickness of the ethanol adsorption layer vs. the ethanol concentration.

mol%. The structures of the adsorbed ethanol turned out to be hydrogen-bonded clusters, via the study employing FTIR-ATR spectroscopy.

FTIR-ATR spectra were recorded on a Perkin Elmer FTIR system 2000 using a TGS detector and the ATR attachment from Grasby Specac. The ATR prism made of an oxidized silicon crystal was used as a solid adsorbent surface because of its similarity to glass surfaces. Immediately prior to each experiment, the silicon crystal was treated with water vapor plasma in order to ensure the formation of silanol groups on the surfaces. Obtained spectra have been examined by referring to well-established, general spectral characteristics of hydrogen-bonded alcohols in the fundamental OH stretching region, because ethanol is known to form hydrogen-bonded dimers and polymers (clusters) in nonpolar liquids [23]. We have also experimentally examined hydrogen-bonded ethanol cluster formation in bulk cyclohexane–ethanol mixtures using transmission infrared spectroscopy.

FTIR-ATR spectra of ethanol in cyclohexane at various ethanol concentrations (0.0–3.0 mol%) are presented in Figure 4. At 0.1 mol% ethanol, a narrow negative band at 3680 cm^{-1}, a weak absorption at 3640 cm^{-1} (free OH), and a broad strong absorption (3600–3000 cm^{-1}) with shoulders at 3530 cm^{-1} (cyclic dimer or donor end OH), 3450, and 3180 cm^{-1} are observed. It is known that the isolated silanol group exhibits an absorption band at 3675–3690 cm^{-1} in a nonpolar liquid, e.g., CCl_4 and when the silanol groups hydrogen bond with esters, the absorption band shifts to a lower wavenumber (3425–3440 cm^{-1}) [24]. Thus, the negative absorption at 3680 cm^{-1} and the positive shoulder at 3450 cm^{-1} should correspond to the decrease in the isolated silanol groups and the appearance of the silanol groups hydrogen bonded with the adsorbed ethanol, respectively. The strong broad band ascribed to the polymer OH appeared at 3600–3000 cm^{-1} together with the relatively weak monomer OH band at 3640 cm^{-1}. This demonstrated the cluster formation of ethanol adsorbed on the silicon oxide surface even at 0.1 mol% ethanol, where no polymer peak appeared in the spectrum of the bulk solution at 0.1 mol% ethanol. With increasing ethanol concentration, the free monomer OH (3640 cm^{-1}) and the polymer OH peak (3330 cm^{-1}) increased, while the peaks at 3530, 3450, and 3180 cm^{-1} remained the same.

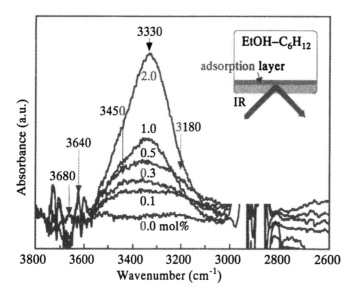

FIG. 4 FTIR-ATR spectra of ethanol on a silicon oxide surface in ethanol–cyclohexane binary liquids at various ethanol concentrations: 0.0, 0.1, 0.3, 0.5, 1.0, and 2.0 mol%.

At higher ethanol concentrations, ATR spectra should contain the contribution from
bulk species, because of the long penetration depth of the evanescent wave, 250 nm. To examine the bulk contribution, the integrated peak intensities of polymer OH peaks of transmission (A_{TS}) and ATR (A_{ATR}) spectra are plotted as a function of the ethanol concentration
in Figure 5. The former monitors cluster formation in the bulk liquid, and the latter contains
contributions of clusters both on the surface and in the bulk. A sharp increase is seen in A_{ATR}

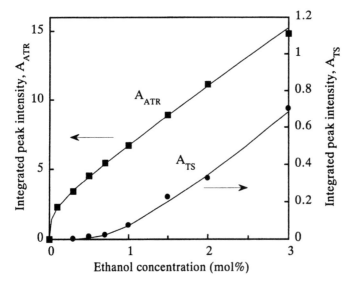

FIG. 5 Plots of integrated peak intensities of polymer OH (3600–3000 cm^{-1}) as a function of the
ethanol concentration. Filled circles represent the value obtained from the transmission spectra (A_{TS}),
while filled squares represent those from ATR (A_{ATR}).

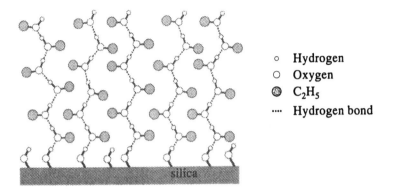

○ Hydrogen
○ Oxygen
⬢ C₂H₅
···· Hydrogen bond

FIG. 6 Plausible structure of the adsorption layer composed of ethanol clusters.

even at 0.1 mol% ethanol, but no significant increase is seen in A_{TS} at ethanol concentrations lower than 0.5 mol%. A comparison of A_{TS} and A_{ATR} clearly indicated that ethanol clusters formed locally on the surface at concentrations of ethanol lower than ~0.5 mol%, where practically only a negligible number of clusters exist in the bulk. The thick adsorption layer of ethanol most likely consists of ethanol clusters formed through hydrogen bonding of surface silanol groups and ethanol as well as those between ethanol molecules. A plausible structure of the ethanol adsorption layer is presented in Figure 6.

The contact of adsorbed ethanol layers should bring about the long-range attraction observed between glass surfaces in ethanol–cyclohexane mixtures. The attraction starts to decrease at ~0.5 mol% ethanol, where ethanol starts to form clusters in the bulk phase. It is conceivable that the cluster formation in the bulk influences the structure of the adsorbed alcohol cluster layer, thus modulating the attraction. We think that the decrease in the attraction is due to the exchange of alcohol molecules between the surface and the bulk clusters.

A similar long-range attraction associated with cluster formation has been found for cyclohexane–carboxylic acid mixtures and is under active investigation in our laboratory. Such knowledge should be important for understanding various surface-treatment processes performed in solvent mixtures and for designing new materials with the use of molecular assembling at the solid–liquid interfaces. For the latter, we have prepared polymer thin films by in situ polymerization of acrylic acid preferentially adsorbed on glass surfaces [25].

IV. ADSORPTION OF POLYELECTROLYTES ONTO OPPOSITELY CHARGED SURFACES

The process of adsorption of polyelectrolytes on solid surfaces has been intensively studied because of its importance in technology, including steric stabilization of colloid particles [3,4]. This process has attracted increasing attention because of the recently developed, sophisticated use of polyelectrolyte adsorption: alternate layer-by-layer adsorption [7] and stabilization of surfactant monolayers at the air–water interface [26]. Surface forces measurement has been performed to study the adsorption process of a negatively charged polymer, poly(styrene sulfonate) (PSS), on a cationic monolayer of fluorocarbon ammonium amphiphille **1** (Fig. 7) [27].

A force–distance curve between layers of the ammonium amphiphiles in water is shown in Figure 8. The interaction is repulsive and is attributed to the electric double-layer

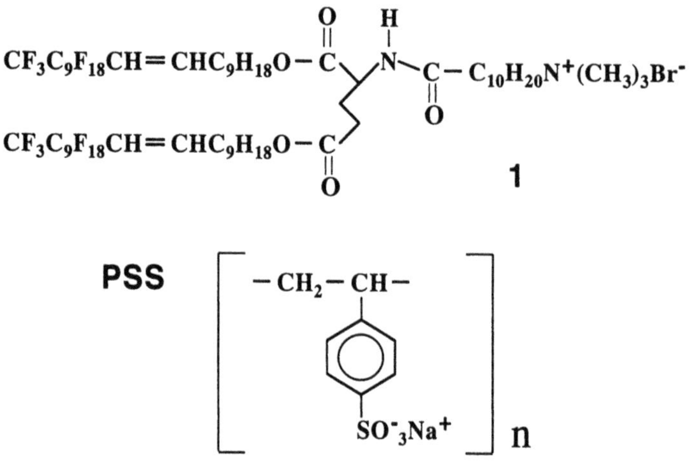

$$CF_3C_9F_{18}CH=CHC_9H_{18}O-\overset{\overset{O}{\|}}{C}\diagdown \underset{}{\overset{\overset{H}{\|}}{N}}-\overset{}{\underset{\overset{\|}{O}}{C}}-C_{10}H_{20}N^+(CH_3)_3Br^-$$

$$CF_3C_9F_{18}CH=CHC_9H_{18}O-\underset{\overset{\|}{O}}{C}\diagup$$

$$1$$

PSS $\left[\begin{array}{c} -CH_2-CH- \\ \\ \bigcirc \\ \\ SO_3^-Na^+ \end{array}\right]_n$

FIG. 7 Chemical structures of fluorocarbon ammonium amphiphile **1** and poly(styrene sulfonate) (PSS).

force. The addition of 0.7 mg/L PSS (1.4×10^{-9} M, equivalent to the addition of 0.7 nmol of PSS, which is close to the amount of the amphiphile on the surface) into the aqueous phase drastically alters the interaction. Here, the molecular weight (Mw) of PSS is 5×10^5. Over the whole range of separations from 5 to 100 nm, the force decreases more than one order of magnitude and does not exceed 100 μN/m. The analysis of the force profile has shown that more than 99% of the initial surface charges are shielded by PSAS binding. The

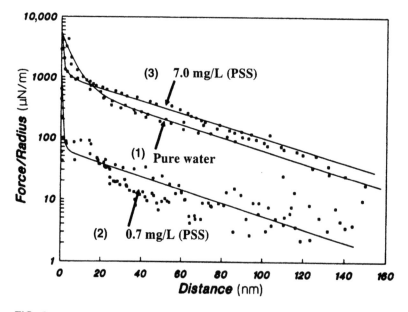

FIG. 8 Force–distance dependence for surfaces covered with fluorocarbon amphiphile **1** in pure water (1) and in aqueous solutions containing 0.7 mg/L poly (styrenesulfonate) (2) and 7.0 g/L poly (styrenesulfonate) (3). The molecular weight of the polymer is 5×10^5. Lines are drawn as a visual guide.

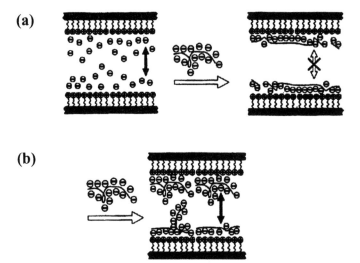

FIG. 9 Schematic illustration of adsorption of poly(styrenesulfonate) on an oppositely charged surface. For an amphiphile surface in pure water or in simple electrolyte solutions, dissociation of charged groups leads to buildup of a classical double layer. (a) In the initial stage of adsorption, the polymer forms stoichiometric ion pairs and the layer becomes electroneutral. (b) At higher polyion concentrations, a process of restructuring of the adsorbed polymer builds a new double layer by additional binding of the polymer.

thickness of the adsorbed layer of PSS is in the range of 1.5–2.5 nm (it is less than 1 nm in the case of PSS of 1×10^4 Mw). These data indicate flat and stoichiometric adsorption of the polyelectrolytes onto the monolayer surface (Fig. 9a).

Increased concentration of PSS at 7.0 g/L (1.4×10^{-5} M) leads to an increase in the force to value seven higher than that between the surfaces of fluorocarbon monolayers alone. The origin of this force is electrostatic in nature. Recharging of the surface by additional adsorption of PSS should occur as shown in Figure 9b.

Our results demonstrate well the complexities of polyelectrolyte adsorption and provide a basis for various surface treatments utilizing polyelectrolytes. They especially afford physical-chemical support for alternate layer-by-layer film formation of polyelectrolytes, which is becoming a standard tool for building composite polymer nano-films in advanced materials science.

V. POLYPEPTIDE AND POLYELECTROLYTE BRUSHES

Polypeptides and polyelectrolytes are essential classes of substances because of their importance in such areas as advanced materials science (functionalized gel) and biology (proteins, living cells, and DNA). Being polymers with charges and counterions and/or hydrogen bonding, they exhibit interesting, albeit complicated, properties. Two-dimensionally organized brush structures of polymers can simplify the complexities of the polyelectrolyte solutions. Attempts to investigate polyelectrolyte brushes have been carried out experimentally [11,28–32] and theoretically [33,34]. Direct measurement of surface forces has been proven useful in obtaining information about the concrete structures of polypeptide and polyelectrolyte brush layers. Taking advantage of the LB method, we prepared well-defined brush layers of chain-end-anchored polypeptides and polyelectrolytes [11,28–30].

We then investigated them based on the force profiles, together with FTIR spectra and surface pressure–area isotherms by systematically varying the polymer chain length, chemical structure, brush density, and solution conditions (pH, salt concentrations, etc). When the surfaces of the opposed polymer layers approach to a separation distance of molecular dimensions, the steric repulsion becomes predominant and hence measurable. By analyzing them, it is possible to obtain key parameters, such as thickness (length) and compressibility of polyelectrolyte layers, which are difficult to obtain by other methods, and to correlate them with polymer structures. Obtained information should form a basis for elucidating their properties and developing physical models. Moreover, it is more likely to discover new phenomena via a novel approach: We have found the density-dependent transition of polyelectrolyte brushes, which we have accounted for in terms of the change in the binding modes of counterions to polyelectrolytes [30].

A.   Brush Layers of Poly(glutamic acid) and Poly(lysine)

Polypeptides form various secondary structures (α-helix, β-sheet, etc.), depending on solution pHs. We have investigated end-anchored poly(L-glutamic acid) and poly(L-lysine) in various secondary structures [11,29,35,36], using the analytical method for the steric force

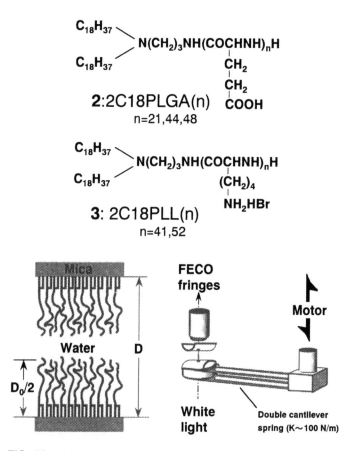

FIG. 10   Schematic drawing of surface forces measurement on charged polypeptide brushes prepared by LB deposition of amphiphiles **2** and **3**.

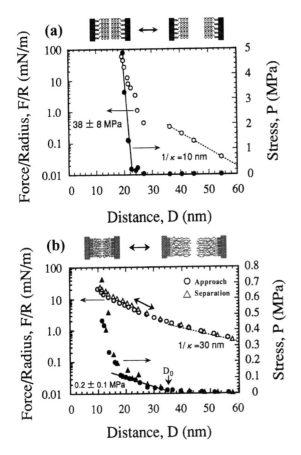

FIG. 11 Force profiles between poly(glutamic acid), 2C18PLGA(44), brushes in water (a) at pH = 3.0 (HNO₃), (b) at pH 10 (KOH) 1/κ represents the decay length of the double-layer force. The brush layers were deposited at π = 40 mN/m from the water subphase at pH = 3.0 and 10, respectively.

in order to examine more quantitatively the structures and structural changes in polyelectrolyte layers. The elastic compressibility modulus of polypeptide brushes was obtained, to our knowledge, as the first quantitative determination of the mechanical modulus of an oriented, monomolecular polymer layer in solvents.

Poly(L-glutamic acid) and poly(L-lysine) brush layers were prepared using amphiphiles **2** and **3** carrying the poly(L-glutamic acid) (2C18PLGA(n), degree of polymerization, n = 21, 44, 48) and the ply(L-lysine) segment (2C18PLL(n), n = 41), respectively (Fig. 10). They formed a stable monolayer at the air–water interface in which different secondary structures, such as α-helix and β-structures, were formed through intra- and intermolecular hydrogen bonding, depending on surface pressure and subphase pH. They were deposited onto mica surfaces and subjected to surface forces measurement. We used FTIR spectroscopy to study the formation and orientation of their secondary structures.

Figure 11a shows a force–distance profile measured for poly(L-glutamic acid) brushes (2C18PLGA(44)) in water (pH = 3.0, 10^{-3} M HNO₃) deposited at 40 mN/m from the water subphase at pH = 3.0. The majority of peptides are in the forms of an α-helix (38% determined from the amide I band) and a random coil. Two major regions are clearly seen in the force–distance profiles. At surface separations longer than 35 nm, the interaction is a typical double-layer electrostatic force, with a decay length of 10 ± 1 nm, which agrees well with the Debye length (9.6 nm) for 10^{-3} M HNO_3, due to ionized carboxyl groups. At separations shorter than ~20 nm, the repulsion is steric in origin and varies depending on the secondary structures existing in the surface layer. In order to examine detailed changes in the interactions, a force–distance profile is converted to a stress–distance (P-D) profile by differentiating the free energy of interaction G_f [Eq. (1)] between two flat surfaces as

$$P = -\frac{dGf}{dD} = -\left(\frac{1}{2\pi}\right)\left(\frac{d(F/R)}{dD}\right) \tag{2}$$

The stress curve sharply increases when the steric component appears upon compression. The initial thickness of a deformed layer is equal to be half the distance D_0 obtained by extrapolating the sharpest initial increase to stress zero. The value D_0 is 21 ± 1 nm, which is close the thickness of two molecular layers (19.2 nm) of the α-helix brush, calculated using the CPK model and the orientation angles obtained by FTIR analysis. We have calculated the elastic compressibility modulus Y,

$$Y = -\frac{dP}{dD/D_0} \tag{3}$$

to be 38 ± 8 MPa from the steepest slope of the stress–distance curve. This value is one to two orders of magnitude larger than the elasticity measured for a typical rubber (1 MPa).

Figure 11b shows a profile at pH 10, measured between the 2C18PLGA(44) LB surfaces prepared at 40 mN/m from the aqueous KOH subphase (pH 10). In this sample, two-thirds of the carboxylic acid groups dissociate; therefore, it behaves as a simple polyelectrolyte. The initial thickness of the deformed layer is 35 ± 2 nm, which is close to twice the length of 2C18PLGA(44) in the extended form, 37 nm. The elastic compressibility modulus is 0.2 ± 0.1, which is even smaller than the value for a typical rubber. Unexpectedly, the ionized layers are easily compressed. Counterion binding to the ionized chain should play an important role in decreasing the stress for compression by reducing the effective charges through the shielding and charge-recombination mechanisms.

Similar measurements have been done on poly(L-lysine) brushes. Table 1 lists a part of our data, which display specific features: (1) The value D_0 depends on the polymer chain

TABLE 1   Effective Length and Compressibility Modulus of Polypeptide Brushes Determined by SFA in Water

Peptide	pH	α-Helix content R_α (%)	D_0 (nm)	Compressibility modulus, Y (MPa)
PLL ($n = 41$)	10	34	16 ± 1	1.2 ± 0.6
(ionized chain)	11	47	19 ± 1	3.1 ± 0.8
	12	54	14 ± 1	3.3 ± 0.8
	4	0	32 ± 1	0.14 ± 0.05
PLGA ($n = 44$)	3	38	21 ± 1	38 ± 8
(ionized chain)	5.6	32	22 ± 1	22 ± 5
	10	0	35 ± 2	0.2 ± 0.1
PLGA ($n = 21$)	9.6	0	25 ± 2	0.2 ± 0.1
(ionized chain)				

The length D_0 corresponds to twice the thickness of the brush layers.

length as well as the secondary structures; thus it is a good measure for determining the thickness (length) of the polypeptide (or polyelectrolyte brush); (2) the compressibility modulus is sensitive to changes in the kind of secondary structures; (3) the moduli of α-helix brushes are one order of magnitude larger for poly(L-glutamic acid) than for poly(L-lysine), which is likely due to interchain hydrogen bonding between the carboxylic acid groups of neighboring poly(glutamic acid) chains; (4) the moduli of ionized chains are identical for poly(L-glutamic acid) and poly(L-lysine).

The stress–distance profile measured by the surface forces apparatus thus provides information on structural changes in polymers and polyelectrolytes in solvents. One advantage of our approach is that a model calculation is not necessary to extract physical parameters involved in the structural changes. One may note that the mechanical properties discussed here reflect not only the intrinsic flexibility of polypeptide chains but also other effects, such as the osmotic pressure of counterions present within charged brush layers. Such knowledge is essential for the theoretical understanding of polyelectrolytes and polypeptides. Our work employing surface forces measurement opens the door to studies on a wide range of structural changes of polymers in solvents, including proteins and polyelectrolyte networks in water. The complexities of their solution properties can be reduced by aligning them in a two-dimensional manner. Very recently, polyelectrolyte brushes have also begun to attract attention as a novel molecular architecture for nanotechnology [37]. The forces measurement should also provide valuable information for effectively designing such materials.

B. Density-Dependent Transition of Polyelectrolyte Layers

The ionized forms of polypeptides exhibit many characteristics in common; therefore, we have studied them under various conditions. The most interesting observation is the transition of a polyelectrolyte brush found by changing the polyelectrolyte chain density. The brush layers have been prepared by means of the LB film deposition of an amphiphile, 2C18PLGA(48), at pH 10. Mixed monolayers of 2C18PLGA(48) and dioctadecylphosphoric acid, DOP, were used in order to vary the 2C18PLGA(48) content in the monolayer.

Surface force profiles between these polyelectrolyte brush layers have consisted of a long-range electrostatic repulsion and a short-range steric repulsion, as described earlier. Short-range steric repulsion has been analyzed quantitatively to provide the compressibility modulus per unit area (Y) of the polyelectrolyte brushes as a function of chain density (Γ) (Fig. 12a). The modulus Y decreases linearly with a decrease in the chain density Γ, and suddenly increases beyond the critical density. The maximum value lies at $\Gamma = 0.13$ chain/nm^2. When we have decreased the chain density further, the modulus again linearly decreased relative to the chain density, which is natural for chains in the same state. The linear dependence of Y on Γ in both the low- and the high-density regions indicates that the jump in the compressibility modulus should be correlated with a kind of transition between the two different states.

To examine this peculiar behavior, we have converted the elastic compressibility modulus, *per unit area*, Y (Fig. 12a), to the modulus *per chain*, $Y' = Y/10^{18}\Gamma$ (Fig. 12b). The elastic compressibility modulus *per chain* is practically constant, 0.6 ± 0.1 pN/chain, at high densities and jumps to another constant value, 4.4 ± 0.7 pN/chain, when the density decreases below the critical value. The ionization degree, α, of the carboxylic acid determined by FTIR spectroscopy gradually decreases with increasing chain density due to the charge regulation mechanism (also plotted in Fig. 12b). This shows that α does not account for the abrupt change in the elastic compressibility modulus.

The density-dependent jump in the properties of polyelectrolyte brushes has also been found in the transfer ratio and the surface potential of the brushes [38], establishing the existence of the density (interchain distance)-dependent transition of polyelectrolytes in solutions.

The transition of the compressibility, and other properties of the polyelectrolyte brushes, is most likely accounted for in terms of the transition in the binding mode of the counterion to the polyelectrolytes, from the loosely bound state to the tightly bound one, which reduces inter- and intrachain repulsive interactions. The following supports this account: (1) At the critical density, $\Gamma_c = 0.20$ chain/nm^2, the separation distance between polyelectrolyte chains, d, is 2.4 nm. This distance is close to the sum, 2.6 nm, of the chain diameter, 1.3 nm, and the size of two hydrated K$^+$ counterions, 1.32 nm, indicating that the abrupt change in the compressibility modulus should be closely related to the counterion binding mode. (2) The critical distance satisfies the energy requirement for the tight binding of counterions (coulombic interaction between two unscreened elementary charges is equal to the thermal energy). (3) The stress profiles can be fitted to the theoretical equation derived based on the assumption that the stress of deformation arises from the osmotic pressure of the counterions. The analysis revealed that the osmotic coefficient in the high-brush-density region is one order of magnitude larger than that in the low-density region. (4) At the critical chain density $\Gamma_c = 0.2$ chain/nm^2, we have found that the distance between the ionized charges becomes close to twice the Bjerrum length [39]. Therefore, counterions must bind strongly to the polyelectrolytes at densities greater than the critical density.

In polyelectrolyte solutions, the counterion condensation on linear polyelectrolyte chains is known to occur when the charge density along the chain exceeds the critical value [40]. Our work indicates the existence of a critical value for the separation distance between chains, where the interchain interaction changes drastically, most likely due to the transition in the binding mode of the counterions (see Fig. 13). Many peculiar forms of behavior, which are often interpreted by the cluster formation or the interchain organization of polyelectrolytes, have been reported for high concentrations of aqueous polyelectrolytes

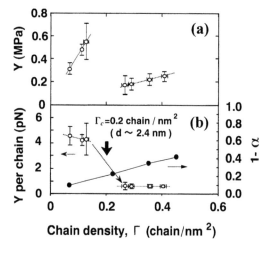

FIG. 12 Plots of elastic compressibility modulus (a) per unit area, Y; and (b) that per chain, Y', of 2C18PLGA(48) brushes as a function of chain density Γ. The ionization degree of the carboxylic acid group, α, is also plotted in part b.

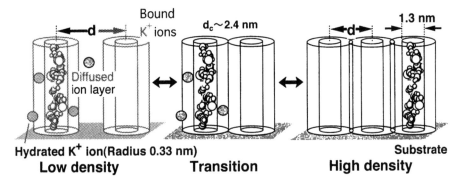

FIG. 13 Schematic drawing of possible binding modes of counterions to polyelectrolyte chains. Counterions loosely bind and form a cloud around the polyelectrolyte chains when the interchain distance (d) is greater than 2.4 ± 0.5 nm, while they strongly bind to form nearly neutral polyelectrolytes at smaller distances ($d < 2.4 \pm 0.5$ nm).

[41]. Our observation should be important in understanding these properties of polyelectrolytes in solutions and perhaps in gels.

VI.   CONCLUDING REMARKS

The nanometer level of characterization is necessary for nanochemistry. We have learned from the history of once-new disciplines such as polymer science that progress in synthesis (production method) and in physical and chemical characterization methods are essential to establish a new chemistry. They should be made simultaneously by exchanging developments in the two areas. Surface forces measurement is certainly unique and powerful and will make a great contribution to nanochemistry, especially as a technique for the characterization of solid–liquid interfaces, though its potential has not yet been fully exploited. Another important application of measurement in nanochemistry should be the characterization of liquids confined in a nanometer-level gap between two solid surfaces, for which this review cites only Refs. 42–43.

REFERENCES

1. JN Israelachivili. Intermolecular and Surface Forces. 2nd ed. London: Academic Press, 1992.
2. F Ohnesorge, G Binnig. Science 260:1451, 1993.
3. PC Heimenz, R Rajagopalan. Principles of Colloid and Surface Chemistry. New York: Marcel Dekker, 1994.
4. AW Adamson, AP Gast. Physical Chemistry of Surfaces. 6th ed. New York: Wiley, 1997.
5. J-M Lehn. Supramolecular Chemistry. Weinheim, Germany: VCH, 1995.
6. B Alberts, D Bray, J Lewis, M Raff, K Roberts, JD Watson. Molecular Biology of the Cell. 3rd ed. New York: Garlaud, 1994.
7. G Decher. Science 277:1232, 1997.
8. J Israelachivili, H Wennerstrom. Nature 379:219, 1996.
9. G Reiter, AL Demiral, S Granick. Science 263:1741, 1994.
10. (a) M Mizukami, K Kurihara. Chem Lett: 1005–1006, 1999; (b) M Mizukami, K Kurihara. Chem Lett 248, 2000.
11. T Abe, K Kurihara, N Higashi, M Niwa. J Phys Chem 99:1820, 1995.
12. K Kurihara. Adv Colloid Sci 71–72:243, 1997.

13. NA Burnham, RJ Colton. In: DA Bonnel, ed. Scanning Tunneling Microscopy and Spectroscopy. New York: VCH, 1993, p 191.
14. VT Moy, EL Florin, HE Gaub. Science 266:257, 1994.
15. GV Lee, LA Chrisey, RJ Colton. Science 266:771, 1994.
16. M Rief, J Pascual, M Saraste, HE Gaub. J Mol Biol 186:553, 1999.
17. JN Israelachivili, GE Adams. J Chem Soc: Faraday Trans 74:975, 1978.
18. WA Ducker, TJ Senden, RM Pashley. Langmuir 8:1831, 1992.
19. K. Kurihara, T Kunitake, N Higashi, M. Niwa. Thin Solid Films 210/211:681, 1992.
20. JL Parker, P Claesson, P Attard. J Phys Chem 98:8468, 1994.
21. H Okusa, K Kurihara, T Kunitake. Langmuir 10:3577, 1994.
22. M Mizukami, K Kurihara, manuscript in preparation.
23. U Liddel, ED Becker. Spectrochim Acta 10:70, 1957.
24. AK Mills, JA Hockey. J Chem Soc, Faraday Trans 71:2398, 1975.
25. S Nakasone, M Mizukami, K Kurihara. 78th JCS Spring Annual Meeting: 2PA175, 2000.
26. M Shimomura, T Kunitake. Thin Solid Films 132:243, 1985.
27. P Berndt, K Kurihara, T Kunitake. Langmuir 8:2486, 1992.
28. K Kurihara, T Kunitake, N Higashi, M Niwa. Langmuir 8:2087, 1992.
29. K Kurihara, T Abe, N Higashi, M Niwa. Colloids Surfaces A 103:265, 1995.
30. T Abe, N Higashi, M Niwa, K Kurihara. Langmuir 15:7725, 1999.
31. Y Mir, P Auroy, L Auvray. Phys Rev Lett 75:2863, 1995.
32. P Goenoum, A Schlachli, D Sentenac, JW Mays, J Benattar. Phys Rev Lett 75:3628, 1995.
33. SJ Miklavic, SJ Marcelja. J Phys Chem 92:6718, 1988.
34. P Pincus. Macromolecules 24:2912, 1991.
35. T Abe. PhD thesis, Nagoya University, Nagoya, 1997.
36. S Hayashi, T Abe, K Kurihara. manuscript in preparation.
37. RR Shah, D Merreceyos, M Husemann, I Rees, NL Abbott, CJ Hawker, JL Hedrick. Macromolecules 33:597, 2000.
38. S Hayashi, T Abe, N Higashi, M Niwa, K Kurihara. MCLC: in press.
39. T Abe, S Hayashi, N Higashi, M Niwa, K Kurihara. Colloids Surfaces A 169:351, 2000.
40. GS Manning, Ber Bunsen-Ges. Phys Chem 100:909, 1996.
41. H Dautzenberg, W Jaeger, J Kotz, B Philipp, Ch Seidel, D Stscherbina. Polyelectrolytes. New York: Hanser, 1994.
42. JN Israelachvili, PM McGuiggan, AM Homola. Science 240:189, 1988.
43. H-W Hu, S Granick. Science 258:1339, 1992.
44. C Dushkin, K Kurihara. Rev Sci Inst 69:2095, 1998.

2
Adhesion on the Nanoscale

SUZANNE P. JARVIS   Nanotechnology Research Institute, National Institute of Advanced Industrial Science and Technology, Ibaraki, Japan

I. ADHESION OVERVIEW

Sticky, one of our earliest childhood experiences and probably one of the first words to enter our vocabulary, is familiar to scientists and nonscientists alike. However, does our direct experience of stickiness, or scientifically speaking, adhesion, have any relevance at the nanoscale? How can adhesion be measured, how can it be manipulated, and what role does it play both in technological applications and intrinsically in nature? These are the questions that I will try to address in this chapter.

The adhesion of surfaces on a macroscopic scale is usually associated with specially designed glues or tapes, which are a prerequisite for holding two dry, solid surfaces together. Exceptions to this tend to be very smooth surfaces with small amounts of moisture between them, such as two sheets of glass or a rubber sucker on a bathroom tile. Intuitively then, even on a macroscopic scale, it is apparent that surface roughness and environment play a critical role in adhesion. Similarly, for many years scientists have realized that as the surfaces approach nanoscale dimensions, the surface roughness and the area of contact reach comparable dimensions, such that the apparent and true contact areas become approximately equal, as shown in Figure 1. This significantly increases the importance of adhesion in the interaction between the two surfaces. Ultimately it becomes necessary to consider the materials as ideal systems with properties no longer limited by defects, impurities, and contamination, which dominate for bulk materials. When considering the material on a near-atomic level, there are a number of attractive forces that can act between two surfaces brought into contact that can cause them to adhere to each other. The force required to separate the two surfaces is then known as the *adhesive force* or *pull-off force*. The magnitude of this force depends on the true contact area and the nature of the attractive forces holding the surfaces together. These forces could include, for example, van der Waals, capillary, or electrostatic forces. An excellent text explaining intermolecular and surface forces in detail is that of Israelachvili [1].

There are a number of industrial and technological areas in which nanoscale adhesion is important. One of the earliest fields concerned with adhesion on this scale was colloid science. Colloid particles lie in the intermediate region between macro and nano, with dimensions typically of the order of hundreds of nanometers up to a few microns. This means that their true contact areas lie well within the nano-domain and are influenced by interactions on this length scale. Adhesion between such particles is important, due to its influence on mineral separation processes and on the aggregation of powders, for example, on the walls of machinery or in the forming of medical tablets. In an extraterrestrial context, such

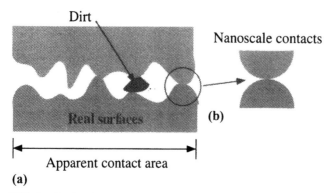

FIG. 1 Real and apparent contact areas. (a) macroscopic surfaces; (b) at the nanoscale.

processes are also important, because cosmic dust aggregation plays a role in planet formation.

As physical structures used in technological applications have been reduced in size, there has been an increasing need to understand the limiting processes of adhesion and to try to minimize them. For example, adhesion due to humidity is known to have a major effect on the durability and friction forces experienced at the recording head/disk interface. Microelectromechanical systems (MEMS) are also detrimentally affected by nanoscale adhesion, with their motion being perturbed or prevented.

On a molecular level there are a number of aspects of adhesion that are important. Preventing the infection of biocompatible materials by preventing bacterial adhesion is very important in the medical industry, particularly for artificial heart valves as well as the more commonly used contact lenses and dentures. A wider understanding of adhesion will be required to support the current boom in biotechnology, with particular regard to molecular motors and drug delivery systems. Adhesion and its manipulation may also lie at the heart of many biological functions and recognition processes.

II. THEORETICAL PERSPECTIVE

A. Continuum Mechanics

Various continuum models have been developed to describe contact phenomena between solids. Over the years there has been much disagreement as to the appropriateness of these models (Derjaguin et al. [2–4] and Tabor [5–7]). Experimental verification can be complex due to uncertainties over the effects of contaminants and asperities dominating the contact. Models trying to include these effects are no longer solvable analytically. A range of models describing contact between both nondeformable and deformable solids in various environments are discussed in more detail later. In all cases, the system of a sphere on a plane is considered, for this is the most relevant to the experimental techniques used to measure nanoscale adhesion.

1. Nondeformable Solids

(a) In Vacuum. For smooth, ideal, rigid solids, the Derjaguin approximation [8] relating the force law between a sphere of radius R and a flat surface to the energy per unit area $W(D)$ between two planar surfaces separated by a distance D gives:

$$F(D)_{\text{sphere}} = 2\pi R W(D)_{\text{plane}} \tag{1}$$

This equation is useful in that it is applicable to any type of force law so long as the range of interaction and the separation are much less than the radius of the sphere. Thus the force to overcome the work of adhesion between a rigid sphere and a flat surface written in terms of the surface energy $\Delta\gamma$ is:

$$F_{\text{pull-off}} = 2\pi R\,\Delta\gamma \qquad \text{where} \qquad \Delta\gamma = \gamma_{\text{sphere}} + \gamma_{\text{flat}} - \gamma_{\text{interface}} \qquad \text{(Dupré)} \quad (2)$$

If the sphere and the flat surface are the same material, then:

$$\gamma_{\text{interface}} = 0 \qquad \text{and} \qquad \gamma_{\text{sphere}} = \gamma_{\text{flat}} = \gamma_s \Rightarrow F_{\text{pull-off}} = 4\pi R\gamma_s \qquad (3)$$

This assumes that the only source of adhesion is the solid–solid contact.

(b) Forces Due to Capillary Condensation. For experiments conducted in air, the adhesive force acting between the two bodies may be dominated by the presence of capillary condensed water. These additional forces due to capillary condensation may be calculated for smooth, ideal, rigid solids. For a sphere and a flat surface joined by a liquid bridge the force F due to the Laplace pressure within the meniscus is given by [9]:

$$F = 4\pi R\gamma_{LV}\cos\theta \qquad (4)$$

where γ_{LV} is the surface tension of the liquid in the condensate and θ is the contact angle of this liquid on the solid.

(c) Nondeformable Solids in Condensable Vapor. The capillary forces just discussed act as an additional force; thus the force needed to separate a rigid sphere and a flat surface of the same material joined by a liquid bridge is given by:

$$F_{\text{pull-off}} = 4\pi R\gamma_{LV}\cos\theta + 4\pi R\gamma_{SL} \qquad (5)$$

2. Deformable Solids

The foregoing models considered incompressible bodies; however, this is never the case in practice. The following section discusses models that specifically consider contact between deformable solids.

(a) Hertz. For deforming solids, Hertzian analysis [10] gives the simplest approximation, for adhesive forces are ignored, i.e., no pull-off force and zero contact area for zero applied load. Given an applied force, P, and a tip radius, R, the contact diameter, $2a$, is

$$2a = 2\left(\frac{4RP}{3E^*}\right)^{1/3} \qquad (6)$$

where

$$E^* = \left(\frac{1-\nu_1^2}{E_1} + \frac{1-\nu_2^2}{E_2}\right)^{-1}$$

and E_1, ν_1 and E_2, ν_2 are Young's modulus and Poisson's ratio of the sphere and plane, respectively. Because surface forces are neglected in this model, it is not possible to apply it to adhesion measurements. However, it is included here because it is used as the basis for some of the other models.

(b) Johnson, Kendall, and Roberts (JKR). The theory of Johnson, Kendall, and Roberts [11] incorporates adhesion via the change in surface energy only where the surfaces are in contact. This gives:

$$2a = 2\left\{\frac{3R}{4E^*}\left(P + 3\pi R\,\Delta\gamma + \sqrt{6\pi RP\,\Delta\gamma + (3\pi R\,\Delta\gamma)^2}\right)\right\} \qquad (7)$$

where $\Delta\gamma$ is the work of adhesion. This is related to the force required to separate the sphere and the flat surface after contact (known as the adhesive force) $F_{\text{pull-off}}$ by the following equation:

$$F_{\text{pull-off}} = \frac{3}{2} \pi R \, \Delta\gamma \tag{8}$$

Note from Eq. [7] that even at zero applied load there is a finite contact area due to the adhesive forces alone, which is given by:

$$a_0 = \left(\frac{9\pi R^2 \, \Delta\gamma}{2E^*}\right)^{1/3} \tag{9}$$

Although this model seems to reflect well some experimental observations of contact and separation [6,7] the assumptions made in its formulation are in fact unphysical. They assume that the solids do not interact outside the contact region, whereas in reality electrostatic and van der Waals forces are nonzero at separations of several nanometers. The assumptions made by JKR lead to infinite values of stress around the perimeter of the connecting neck between sphere and plane.

The model is found to be most appropriate for contact between low-elastic-moduli materials with large radii when the work of adhesion is high. In comparison, the following model assumes that the surface forces extend over a finite range and act in the region just outside the contact. It is found to be more appropriate for systems with small radii of curvature, low work of adhesion, and high modulus [12].

(c) Derjaguin, Muller, and Toporov (DMT).   Derjaguin, Muller, and Toporov [2] assume that under the influence of surface forces, the sphere will deform in the contact region in accordance with the Hertzian model. Since the deformation is taken as Hertzian, the surfaces do not separate until the contact area is reduced to zero. At this instant, the pull-off force is predicted to be:

$$F_{\text{pull-off}} = 2\pi R \, \Delta\gamma \tag{10}$$

This is the same as Eq. [2], the value for nondeformable solids in vacuum. However, in the case of deformable solids, DMT theory gives a finite contact radius at zero applied load:

$$a_0 = \left(\frac{3\pi R^2 \, \Delta\gamma}{2E^*}\right)^{1/3} \tag{11}$$

Although the DMT theory attempts to incorporate distance-dependent surface interactions into the adhesion problem, it does not take into account the effect surface forces have on the elastic deformation. In other words, it does not predict the neck formation predicted by JKR.

(d) Maugis, Dugdale.   Maugis [13] included a surface force, which acts both inside and outside the contact perimeter. The attractive interaction is assumed to be constant up to a separation of h_o, at which point it falls to zero abruptly. The value of h_o is defined such that the maximum attractive force and the work of adhesion correspond to a Lennard–Jones potential, the Dugdale approximation [14]. The error of this somewhat arbitrary approximation is only apparent at low values of the elastic parameter lambda, λ,

$$\lambda = \sigma_o\left(\frac{9R}{2\pi w E^{*2}}\right)^{1/3} \tag{12}$$

where $\sigma_o h_o = w$, the work of adhesion, R is the radius, and E^* is the combined modulus as defined in Eq. (6). Even for low values of λ the discrepancy in the elastic compression is less than the atomic spacing. The net force acting on the contact is composed of a Hertzian pressure associated with the contact radius and an adhesive Dugdale component extending beyond the contact region up to a second radius. This radius can be found by solving two simultaneous equations.

(e)   Muller, Yushchenko, and Derjaguin (MYD)/Burgess, Hughes, and White (BHW). Muller et al. [12] and Burgess et al. [in 15] have formulated more complete descriptions for the adhesion between a sphere and a plane by allowing the solid–solid interaction to be a prescribed function of the local separation between the surfaces. The complete solution is mathematically complex and cannot be solved analytically. As a consequence it is rarely applied to experimental data. In any case, the uncertainties inherent in experiments of this type mean that a more precise model often gives very little advantage over the JKR and DMT models.

From the previous sections it is clear that there are a number of different possible models that can be applied to the contact of an elastic sphere and a flat surface. Depending on the scale of the objects, their elasticity and the load to which they are subjected, one particular model can be more suitably applied than the others. The evaluation of the combination of relevant parameters can be made via two nondimensional coordinates λ and P [16]. The former can be interpreted as the ratio of elastic deformation resulting from adhesion to the effective range of the surface forces. The second parameter, P, is the load parameter and corresponds to the ratio of the applied load to the adhesive pull-off force. An adhesion map of model zones can be seen in Figure 2.

As discussed by Johnson and Greenwood [16], in principle both a full numerical analysis using the Lennard–Jones potential and the Maugis analysis using the Dugdale approximation apply throughout the map with the Hertz, JKR, DMT and rigid zones being regions where some simplification is possible. However, in practice both analyses become less appropriate at high values of λ (in the JKR zone).

Unfortunately, to determine the appropriate zone of an adhesion experiment requires knowledge of a number of parameters, which may not be easily accessible, particularly for

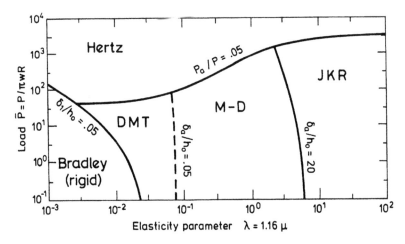

FIG. 2   Map of the elastic behavior of bodies. (Reprinted from Ref. 16.) Copyright 1997 by Academic Press.

nanoscale experiments. Such parameters include the radius of the probe or the combined modulus, which can be affected even by traces of contamination. In the literature, the most popular models tend to be DMT and JKR. The induced errors from the incorrect use of such models are usually less than the errors encountered in the experimental determination of the relevant parameters on the nanoscale.

(f) Liquid Bridge. When working in the presence of condensable vapors (i.e., water under ambient conditions), it is possible for an annulus of capillary condensate to form as the tip approaches the surface of the sample, as shown in Figure 3. Capillary forces can then dominate the interaction between the sphere and plane. The general formula for the adhesive force between the two due to capillary condensation is given by [17]:

$$F = F(\Delta P) + F_{\text{solid–solid}} + F(\gamma_{LV}) \tag{13}$$

where $F(\Delta P)$ is the force due to the Laplace pressure ΔP within the meniscus, $F_{\text{solid–solid}}$ is the force due to direct solid–solid interaction, and $F(\gamma_{LV})$ is the resolved force due to the liquid/vapor surface tension. Theory gives [5,18]:

$$F(\Delta P) = 4\pi R\gamma_{LV}\cos\theta \tag{14}$$

$$F_{\text{solid–solid}} = 4\pi R\gamma_{SL} \tag{15}$$

$$F(\gamma_{LV}) = 2\pi R\gamma_{LV}\sin\phi\sin(\theta + \phi) \tag{16}$$

where R is the radius of the sphere, θ is the solid/liquid contact angle, γ_{SL} is the solid/liquid interfacial free energy, γ_{LV} is the liquid/vapor interfacial free energy, and ϕ is as shown in Figure 3. In the case of deformable solids, Eq. [15] may be replaced by $F_{\text{solid–solid}} = 3\pi R\gamma_{SL}$. The preceding equations rely on the liquid's acting in a macroscopic, hydrodynamic manner. If the contact angle is assumed to be small and the radius of the sphere greatly exceeds the neck radius of the liquid condensate such that ϕ is also small, then $\sin\phi\sin(\theta + \phi) \ll 1$. Thus the direct surface tension contribution is negligible when compared to the Laplace pressure contribution.

When the sphere and plane are separated by a small distance D, as shown in Figure 4, then the force due to the Laplace pressure in the liquid bridge may be calculated by considering how the total surface free energy of the system changes with separation [1]:

$$U_{\text{total}} = \gamma_{SV}(A_1 + A_2) - \gamma_{SL}(A_1 + A_2) - \gamma_{LV}A_3 + \gamma_{SV}A_{\text{solids}}$$

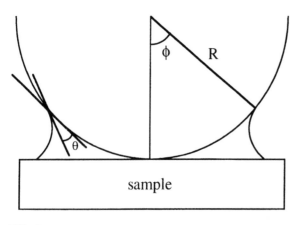

FIG. 3 Schematic of a liquid meniscus around the sphere.

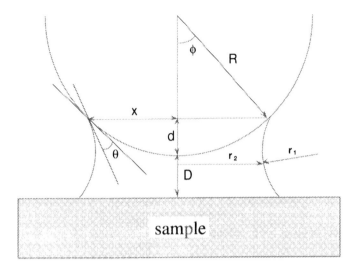

FIG. 4 Schematic of a liquid meniscus between the sphere and plane at finite separation.

where γ_{SV} is the solid/vapor interfacial free energy and A corresponds to the area of each interface as defined in Figure 3.

If $A_1 \approx A_2 \approx \pi R^2 \sin^2 \phi$, then for small ϕ,

$$U_{\text{total}} = 2\pi R^2 \phi^2 (\gamma_{SV} - \gamma_{SL}) + \text{constant} + \text{smaller terms}$$

But $\gamma_{SV} - \gamma_{SL} = -\gamma_L \cos \theta$ (Young equation), so

$$U_{\text{total}} \approx -2\pi R^2 \phi^2 \gamma_L \cos \theta + \text{constant}$$

giving

$$F = -\frac{dU_{\text{total}}}{dD} = 4\pi R^2 \phi \gamma_L \cos \theta \frac{d\phi}{dD} \tag{17}$$

Calculating the volume by considering a straight-sided meniscus and assuming that the liquid volume remains constant, Eq. [17] then becomes:

$$F = \frac{4\pi R \gamma_L \cos \theta}{1 + D/d} \tag{18}$$

Use of the condition of constant meniscus volume is most appropriate when growth and dissolution of the meniscus is comparatively slow. An alternative is to consider the Kelvin equilibrium condition. The Kelvin equation relates the equilibrium meniscus curvature (also known as the Kelvin radius) to the relative vapor pressure; and if Kelvin equilibrium is maintained during the separation process, then the adhesive force becomes [19]:

$$F = -4\pi R \gamma_{\text{LV}} \cos \theta \left(1 - \frac{D}{2r_{\text{m}} \cos \theta} \right) \qquad \text{(Kelvin condition of constant } r_{\text{m}}) \tag{19}$$

where the meniscus curvature,

$$r_m = \left(\frac{1}{r_1} + \frac{1}{r_2} \right)^1$$

and r_1 and r_2 are as defined in Figure 4.

In the limit of small surface separations, the adhesive force and its gradient tend to the same value for both the constant-meniscus-volume and Kelvin-equilibrium conditions.

The derivations of the foregoing equations have been based on the principles of thermodynamics and the macroscopic concepts of density, surface tension, and radius of curvature. They may therefore cease to be appropriate as the mean radius of curvature approaches molecular dimensions.

Including capillary condensation with the Hertz approximation, as considered by Fogden and White [20], introduces pressure outside the contact area; i.e., adhesion enters the problem nonenergetically through the tensile normal stress exerted by the condensate in an annulus around the contact circle. The resulting equations cannot be solved analytically; however, their asymptotic analysis may be summarized as follows.

For nonadhering bodies in contact in the presence of capillary condensation, the previous result for rigid solids is found to apply more generally to systems of small, hard, but deformable spheres in contact in vapor near saturation:

$$F_{\text{pull-off}} = 4\pi R \gamma_{\text{LV}} \qquad \text{for} \qquad \theta = 0 \tag{20}$$

In the limit of large, softer solids in vapor pressure closer to the value marking the onset of capillary condensation, the generalized Hertz and the original JKR theories are found to be qualitatively identical. However, the contact area for zero applied load will in general be different, since it is dependent upon the nature of the source of adhesion:

$$F_{\text{pull-off}} = 3\pi R \gamma_{\text{LV}} \qquad \text{and} \qquad a_0 = \left[9\pi \gamma_{\text{LV}} \frac{R^2}{E^*} \right]^{1/3} \tag{21}$$

Thus $(1/2)\Delta\gamma$ in the JKR approximation is replaced by γ_{LV}. The elastic modulus affects the contact area but not the adhesion force.

B. Molecular Dynamics and First-Principles Calculations

Molecular dynamics (MD) permits the nature of contact formation, indentation, and adhesion to be examined on the nanometer scale. These are computer experiments in which the equations of motion of each constituent particle are considered. The evolution of the system of interacting particles can thus be tracked with high spatial and temporal resolution. As computer speeds increase, so do the number of constituent particles that can be considered within realistic time frames. To enable experimental comparison, many MD simulations take the form of a tip-substrate geometry corresponding to scanning probe methods of investigating single-asperity contacts (see Section III.A).

One of the earliest molecular dynamics simulations to be compared directly to atomic-scale measurements was the work of Landman et al. [21]. They used MD simulations to investigate the intermetallic adhesive interactions of a nickel tip with a gold surface. The interatomic interactions of the nickel–gold system were described by many-body potentials, which were obtained using the embedded-atom method [22,23]. Long-range interactions, such as van der Waals forces, were neglected in the calculations. The theoretically calculated force versus tip–sample separation curves showed hysteresis, which was related to the formation, stretching, and breaking of bonds due to adhesion, cohesion, and decohesion. This was characterized by adhesive wetting of the nickel tip by the gold atoms. The wetting was instigated by the spontaneous jump to contact of the sample and the tip across a distance of approximately 4 Å. In this case, mechanical instabilities of the materials themselves caused the jump (dependent on the cohesive strength of the material), and it should not be confused with the jump-to-contact phenomenon often observed experimentally due to compliant measurement systems. Subsequent indentation resulted in plastic yielding, adhesion-induced atomic flow, and the generation of slip in the surface region of the gold substrate. The whole cycle of approach and retraction can be seen in Figure 5. A more detailed evaluation of the wetting phenomenon was obtained by considering pressure contours. In this way the atoms at the periphery of the contact were found to be under an extreme tensile stress of 10 GPa. In fact, both the structural deformation profile and the pressure distribution found in the MD simulations were similar to those described by contact theories that include adhesion (see Section II.A). On retraction of the tip, a conductive bridge of gold atoms was formed, the elongation of which involved a series of elastic and plastic yielding events, accompanied by atomic structural rearrangements. The eventual fracture of the neck resulted in gold atoms remaining on the bottom of the nickel tip.

Subsequently, Landman and Luedtke inverted the system to look at a gold (001) tip adhering to a nickel (001) surface [24] and noted a number of differences. In this case the tip atoms jumped to the surface, instead of the surface atoms coming up to meet the tip. Continuing the approach beyond this point resulted in the compression of the tip, wetting of the surface, and the formation and annealing of an interstitial layer dislocation in the core of the tip. Also, on retraction of the tip, a much more extended neck was drawn out between tip and sample, in this case, consisting solely of gold-tip atoms. This occurred regardless of whether or not additional force was applied to the contact region after the initial jump to contact.

The study of adhesive contacts was further extended by Landman et al. to include interionic CaF_2 interactions and intermetallic interactions mediated by thin alkane films [25]. In the case of CaF_2 the bonding is significantly different from that in the metallic systems studied previously. The authors treated the long-range coulombic interactions via the Ewald summation method and controlled the temperature at 300 K. As the tip and surface were brought together, the tip elongated to contact the surface in an analogous way to the

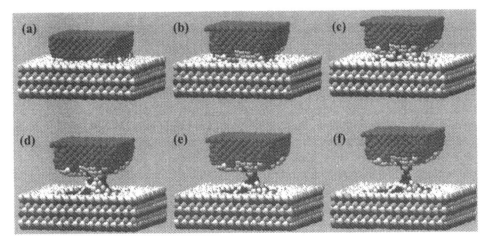

FIG. 5 Atomic configurations generated by the MD simulation. (Courtesy David Luedtke.)

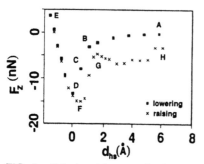

FIG. 6 Calculated force on the tip atoms for a CaF_2 tip approaching (■) and subsequently retracting (×) from a CaF_2 (111) surface. The distance from the bottom layer of the tip to the topmost surface layer for the points marked by letters is: A (8.6 Å), B (3.8 Å), C (3.0 Å), D (2.3 Å), E (1.43 Å), F (2.54 Å), G (2.7 Å), and H (3.3 Å). (Reprinted from Ref. 25, copyright 1992, with permission from Elsevier Science.)

jump to contact observed in metallic systems but involving smaller displacements of approximately 0.35 Å. Retracting the tip after a small net repulsive interaction resulted in clear hysteresis in the force–distance curve, as seen in Figure 6. This reflected the plastic deformation of the crystalline tip leading to eventual fracture. After the final pull-off, part of the tip remained fixed to the substrate surface. For their investigation of the effect of thin films, the authors considered n-hexadecane ($C_{16}H_{34}$), which they modeled using the interaction potentials developed by Ryckaert and Bellmans [26]. This was used as the mediating layer between a nickel tip and a gold (001) substrate. Before the onset of the interaction, the molecular film was layered on the gold substrate, with the layer closest to the surface exhibiting a high degree of orientational order parallel to the surface plane. The initial stages of the interaction involved some of the alkane molecules adhering to the tip, followed by flattening of the film and partial wetting of the sides of the tip. Continued reduction of the tip–sample distance induced drainage of the second layer of molecules under the tip and the pinning of the final layer. Subsequently, drainage of these molecules from under the tip was assisted by transient local inward deformations of the gold, which would seem to lower for the relaxation of unfavorable conformations of the alkane molecules. Finally, true tip–sample contact occured via displacement of gold-surface atoms toward the tip.

Harrison et al. use MD [27] simulations to investigate diamond–diamond adhesion, which occurs due to covalent bond formation (they neglect long-range interactions such as van der Waals). Typical MD interaction times are orders of magnitude faster than those typically used experimentally. However, they are still sufficiently slow to permit equilibration of the system. In the case of hydrogen-terminated diamond (111) tip and surface, a nonadhesive interaction was observed for an indentation with an applied load of 200 nN. This ceased to be the case if the applied load was increased to 250 nN, for plastic yielding of the diamond occurred and, consequently, adhesion between the tip and sample. For a non-hydrogen-terminated diamond surface and a hydrogen-terminated tip, adhesion was observed at both applied loads. Further, precise positions of the tip and surface atoms could be evaluated during the transition to contact and subsequent retraction, exposing different indentation and fracture mechanisms in the two cases. The main predictions of the simulation were that breaking the symmetry of the system and increasing the applied load both had the effect of facilitating adhesion. In addition, removing the hydrogen termination from the surface caused adhesive behavior to set in at smaller values of applied load, a feature they attributed to the increased surface energy of the diamond crystal when the hydrogen was removed. So far, it has proved extremely difficult to mimic such precision experimentally.

However, it has been observed experimentally that even low levels of impurities can have a strong influence on the adhesion of free surfaces and that they usually have the effect of lowering adhesion. Zhang and Smith [28] performed fully self-consistent, first-principles calculations to investigate the cause of such impurity influence on an Fe–Al interface. They found an asymmetry in the adhesion modification, depending on whether the impurities originated on the Fe or the Al surface, as shown in Figure 7. Clearly, the effects on the work of adhesion were much larger when the impurity was added to the Fe surface, due to the much larger adhesive energies of the impurity monolayers to the Fe surface. The effect of the impurity atoms were twofold: On one hand they pushed the surfaces apart, thus weakening the Fe/Al bonds; on the other, they formed bonds across the interface. Of these two competing effects, the former was always found to overwhelm the latter for their chosen systems of five different contaminant layers. This impurity lowering of adhesion has been found to be typical, both experimentally and theoretically.

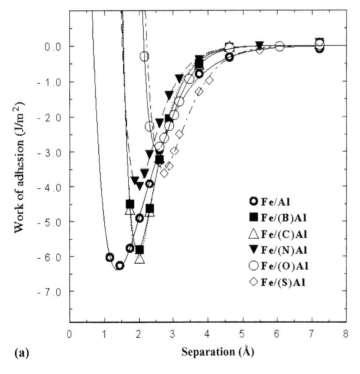

(a)

FIG. 7 Total energy per cross-sectional area as a function of interfacial separation between Fe and Al surfaces for the clean interface and for monolayer interfacial impurity concentrations of B, C, N, O, and S. Graph (a) is for the case where the impurity monolayer is applied to the free Al surface prior to adhesion, while graph (b) has the impurity monolayer applied to the free Fe surface prior to adhesion. The curves fitted to the computed points are from the universal binding energy relation. (From Ref. 28. Copyright 1999 by the American Physical Society.)

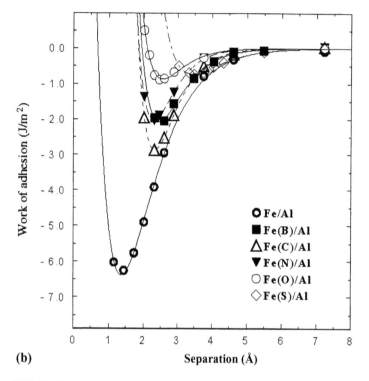

(b)

FIG. 7 Continued.

III. EXPERIMENTAL TECHNIQUES

A. Atomic Force Microscopy

One of the most popular methods of measuring adhesion on the nanoscale is to operate an atomic force microscope in "spectroscopy mode." The microscope consists of a small lever, of submillimeter dimensions, that is usually microfabricated. At the end of this cantilever is a sharp tip used to probe the interactions with a given sample. The lever on which the tip is mounted will change its position and also its apparent resonant frequency and stiffness according to the interaction with the surface. These changes can be sensitively measured by detecting either the static or dynamic motion of the lever with subangstrom sensitivity. This is usually done via either optical interferometry or the more popular method of optical beam deflection. The latter method, although inferior to the former in terms of sensitivity and ease of calibration, dominates, due to its ease of use and its widespread popularity among commercial AFM manufacturers. With this system, light from a laser diode or light-emitting diode is reflected from the back of the cantilever onto a four-segment photodiode. As the cantilever bends, the reflected beam moves across the photodiode, thus altering the relative voltages from each segment. It is then possible to analyze this to give normal and torsional motion of the lever, as seen in Figure 8.

The final component of the AFM is the tip–sample approach mechanism. Bringing the probe tip and the sample within interaction range in a nondestructive manner involves mounting one surface on an approach mechanism with subangstrom sensitivity. Most approach devices normally combine some form of stepper or slider with single step sizes of the order of microns, with a piezo device providing displacements of the order of one tenth of an angstrom up to a maximum of several microns. In microscopy mode, the piezo device

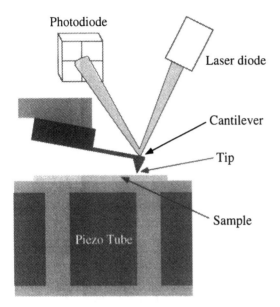

FIG. 8 Schematic of an atomic force microscope with optical beam deflection detection showing a typical angle of 10° between lever and sample.

makes relative lateral displacements between the probe and the sample in order to build up an image of the surface as a function of a particular value of applied force. However, it is the "spectroscopy mode" that is of most interest for measuring adhesion. In this case the tip and the sample are brought into intimate contact and the surfaces separated by applying small, known displacements to one of the surfaces. Simultaneously, the motion of the lever is recorded, resulting in a curve of the form shown in Figure 9. The attractive force in the

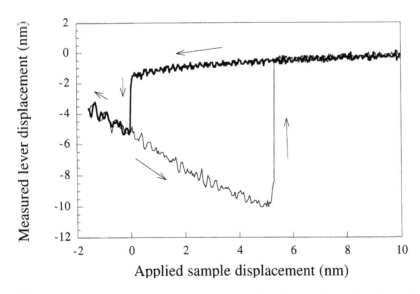

FIG. 9 Curve of sample motion versus lever motion. The experimental starting point for generating a force versus distance curve.

unloading part of the force–distance curve reflects the adhesion between the tip and the sample. From this it is possible to extract the adhesion force from the pull-off distance using Hooke's law. Thus, the force is derived by multiplying together the displacement and the cantilever stiffness.

There are some exceptions to this method. For example, it is possible to apply forces directly to the end of the cantilever rather than displacements to the sample in order to control the approach and separation of the two surfaces [29,30]. This more direct method reduces unwanted relative lateral motion between the tip and the surface. The application of direct forces in this way also has alternative uses, such as enabling sensitive dynamic measurements to be made [29].

One point, which is often disregarded when using AFM, is that accurate cantilever stiffness calibration is essential, in order to calculate accurate pull-off forces from measured displacements. Although many researchers take values quoted by cantilever manufacturers, which are usually calculated from approximate dimensions, more accurate methods include direct measurement with known springs [31], thermal resonant frequency curve fitting [32], temporary addition of known masses [33], and finite element analysis [34].

The lever deflection must also be calibrated. In the case of interferometric detection, the calibration is readily at hand from the wavelength of the light used. In the case of the optical beam method, the best calibration technique is less obvious. The most usual method is to apply a large displacement to the sample via the piezo, with the tip and sample in contact. If the lever is very compliant and the contact very stiff, then the lever and sample can be assumed to be moving together, and thus the deflection sensor can be calibrated from the known sample displacement. There are clearly a number of errors with this method, including deformation in the contact region and unintentional lateral sliding of the tip on the sample. An alternative calibration method for small displacements relies on the high-resolution imaging of materials exhibiting a stepped structure of known height.

1. Relationship Between Pull-Off and Adhesion

The magnitude of the pull-off force depends on the nature of the tip–sample interaction during contact. Adhesion depends on the deformation of the tip and the sample, because attractive forces are proportional to the contact area. Quantifying the work of adhesion is difficult. The measured magnitude of $\Delta\gamma$ is strongly dependent on environment, surface roughness, the rate of pull-off, and inelastic deformation surrounding the contact.

An important consideration for the direct physical measurement of adhesion via pull-off measurements is the influence of the precise direction of the applied force. In AFM the cantilever does not usually lie parallel to the surface, due to the risk that another part of the cantilever chip or chip holder will make contact with the surface before the tip. Another problem relates to the fact that the spot size in the optical beam deflection method is usually larger than the width of the lever. This can result in an interference effect between the reflection from the sample and the reflection from the cantilever. This is reduced if the cantilever and sample are not parallel. Most commercial AFM systems use an angle in the range of 10°–15° between the sample and the cantilever. Depending on this angle and the extent to which the cantilever is bent away from its equilibrium position, there can be a significant fraction of unintentional lateral forces applied to the contact.

Chang and Hammer [35] investigated this issue theoretically by simulating the detachment of receptor-coated hard spheres from ligand-coated surfaces using normal, tangential, and shear forces after allowing the particles to bind to steady state. They point out that we should not expect different types of force to be equally effective at removing the spherical particles and attempt to quantify this by an adhesive dynamics computer simulation method. The interacting molecules were regarded as linear springs, with their rate of reaction dependent on the distance from the substrate. They found that tangential forces were transmitted to the bonds at the interface much more efficiently than normal forces, due to the small ratio of bond length to bead radius, causing a large axial strain in the former case. Tangential forces were found to be 20 times more effective for detaching a single bond and 56 times more effective when there were many bonds, as demonstrated in Figure 10. Were the radius of the spherical particle to be reduced, the difference between the normal and tangential forces to break the adhesive bonds would also become smaller. The authors make some harsh conclusions related to AFM. They show that if the force is to be exerted normal to the substrate, then it must be controlled precisely, for deviations as small as 10° can significantly alter the measured value of the adhesion. They also recommend that for systems where the sphere–substrate separation is small, the calculation of average force per bond made by dividing the pull-off force by the number of bonds in the interface should be abolished, because the method produces very misleading results. Actually, a related problem was also observed with surface forces apparatus (see Section III.B). Christenson [36] found that the use of double cantilevers to reduce relative torsional movement of the two surfaces resulted in larger measured adhesion values than with a single cantilever structure.

Stuart and Hlady [37] found that unintentional lateral forces influenced their measurements of adhesion between surface-bound protein molecules and colloid probe–bound ligands. They noted a greatly exaggerated separation distance and a stick-slip behavior in their adhesion curves, which they attribute to rolling and buckling of the cantilever under the influence of lateral forces as the sample was retracted with the probe stuck to it.

A similar concern due to lateral motion also applies to purely qualitative imaging. Sugisaki et al. [38] observed an artifact in their simultaneously topographic and adhesion

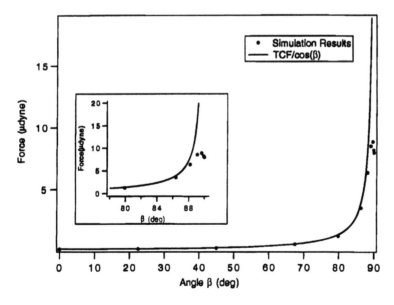

FIG. 10 Critical force needed to rupture all the bonds as a function of β, the angle at which the force is exerted. Simulation results are given by solid circles. Tangential critical force TCF/cos β is also plotted for comparison (solid line). (Reprinted with permission from Ref. 35. Copyright 1996 American Chemical Society.)

images as a consequence of lateral sliding of the tip on the surface during the retraction of the surface. They showed schematically how the two images, although obtained almost instantaneously, actually correspond to different points on the surface due to this lateral sliding of the tip. The shift between the two points depended on the amount of cantilever deflection. The deflection depends on the stiffness of the cantilever and the strength of the adhesion. Thus, for compliant levers on highly adhesive samples, the lateral sliding during retraction is likely to be large. They do not discuss how this lateral sliding could also influence the quantitative measurement of adhesion.

2. Adhesive Force Measurements with AFM

Within the first five years after its invention, many adhesion measurements with AFM were motivated by the fact that adhesion was seen to be limiting the imaging resolution of the instrument. High adhesive forces were often observed that totally dominated the interaction between the tip and the sample. The additional load from these surface forces was sufficient in some cases to prevent the study of delicate biological samples. Weisenhorn et al. [39] investigated adhesion between a microfabricated tip and a mica surface in air and water and found adhesive forces larger than 100 nN while operating in air. These could be reduced by two orders of magnitude by operating in water. They also saw a two-stage pull-off in air, which they attributed to tip–sample separation followed by tip–meniscus separation.

A subsequent and more detailed study by Weisenhorn et al. [40] investigated a Si_3N_4 tip on mica and a tungsten carbide tip on metallic foils in various liquids. While in ethanol the adhesion force of both systems was sub-nN and reproducible, results in water were much more erratic. The metallic surfaces showed high adhesion in water, more than an order of magnitude larger than that measured in ethanol, which could be due to the adsorption of alkane contaminants in air prior to the experiment. These would be removed by the solvent action of ethanol but not by water. The influence of applied load on the adhesion force was also investigated for a Si_3N_4 tip on mica in water. Increasing the applied load from 0.35 nN to 5.3 nN increased the measured pull-off force from 0.2 nN to 0.7 nN. Due to the compliance of the levers in these measurements ($k = 0.035$ N/m), applying such loads involved large displacements of the sample in contact with the tip. Due to the cantilever's lying at an angle to the surface (typically 10°–15° in most commercial systems), it is likely that a significant displacement would also have occurred in the lateral direction in the form of sliding of the tip across the surface. Such effects were not usually considered in early AFM measurements.

Operating in dry nitrogen, Burnham et al. [41] measured the adhesive forces between a large tungsten tip ($R = 2000$ nm) and a variety of surfaces, including mica, graphite, and two thin films of $CH_3(CH_2)_{16}COOH$ and $CF_3(CH_2)_{16}COOH$. Using a double cross-cantilever designed to limit the motion of the tip to the direction perpendicular to the sample, they compared measured adhesive forces with sample surface energies as determined from advancing contact angle measurements. The high-surface-energy samples showed the expected square-root dependence on surface energy, but with lower magnitude, probably due to the influence of microasperities on the tip. The lower-surface-energy materials, such as the stearic acid films, exhibited significant deviation from the square-root dependence, which may be related to mechanical-property differences. Interestingly, a large difference in adhesive force was reported between the –CH$_3$ and –CF$_3$ terminated thin films, even without any special treatment of the tip, suggesting chemical sensitivity without chemical modification of the probe (see Section III.A.6).

In a similar environment, Blackman et al. [42] measured adhesive forces between a smaller tungsten tip ($R = 50$–100 nm) and various molecularly thin organic films on a silicon substrate. They found three distinctly different types of adhesive behavior, corresponding to a liquid for unbound films of perfluoropolyether, intermediate for the bonded films, and soft-solid for multilayers of cadmium arachidate. By comparison of their results, they considered that contaminants and surface roughness of the tip may have had a significant effect on the measurements of Burnham et al. [41] and that forces in addition to van der Waals, such as capillary and chemical bonding, were wrongly neglected in their interpretation.

Mizes et al. [43] used AFM to investigate the dependence of adhesion on surface roughness and material inhomogeneities. The former basically amounts to a dependence on contact area, which was already well understood. If surface roughness and probe size are of similar magnitude, then the contact area, and hence the adhesion, naturally depends on whether the probe sits at the top of a ridge or the bottom of a ridge. To investigate the spatial variation in adhesion, they created an adhesion map from pull-off measurements made systematically in a 2D raster scan across the surface. Although not measured simultaneously with topography, the thermal drift was sufficiently low that adhesion and surface topography could be compared. For doped polymers, the adhesion map showed structure that was not seen in the topography, indicating that adhesion could be used to identify regions of varying material property on the sample. The authors suggested extending this by coating the cantilever to facilitate adhesion measurements between any two materials of interest.

Weihs et al. [44] made a conservative estimate of the work of adhesion in air between an LB film and a tungsten tip in order to investigate the resolution limit for contact-mode AFM imaging of soft samples. Pull-off measurements were performed with a compliant cantilever ($k \approx 1\ \mathrm{Nm^{-1}}$), no applied force, and a large tungsten tip ($R \approx 2000$ nm). The use of a compliant cantilever allowed easy detection of contact and prevented the application of forces, which can lead to inelastic deformation and greater adhesion. Their large tip would tend to underestimate $\Delta\gamma$ because of small asperities on the tip, which have smaller adhesive forces. The average pull-off force for several tests was 510 ± 100 nN, which combined with the measured macroscopic tip radius of 2000 nm gave a work of adhesion equal to 54 mJ-m^{-2}. The authors used this value of $\Delta\gamma$ to predict the adhesive force for a 200-nm and a 20-nm tungsten tip and found it to be 51 nN and 5.1 nN, respectively. They then used a JKR model to find the contact diameter, and hence resolution, as a function of small applied loads, the important aspect being that adhesive forces appeared to limit the best possible imaging resolution. Even at zero applied load, the contact diameter for the 20-nm tip was calculated to be 6.3 nm. Further, from consideration of the yielding force, it was shown that tip sharpening could be detrimental to the imaging resolution because it increased the risk of sample damage through inelastic deformation. All measurements were made in air and clearly showed that to avoid damage to compliant materials during contact of the probe, it is necessary either to carefully consider the mechanics of the interaction or to move to liquid environments, where the adhesive force can be more readily manipulated.

With the application of AFM to the nondestructive imaging of biological samples in mind, Hoh et al. [45] investigated adhesion based on the formation of hydrogen bonds between a silicon nitride tip and a glass sample in aqueous environments. They found that the adhesive force could be manipulated by altering pH, ion type, and concentration of buffer solutions. Adhesion of the Si_3N_4-glass system was found to be very sensitive to pH, with adhesion decreasing to undetectable levels above a pH of 9. This indicated that a significant fraction of the adhesion was due to hydrogen bonding. The authors note that consistent with this hypothesis was their observation of discrete steps in the adhesive interaction. However, later work by the same group confirmed that this was not in fact due to hydrogen bonding (see Section III.A.4). Phosphate-buffered saline (PBS) was also found to reduce adhesion, even at a near neutral pH of 7.2, primarily due to the phosphate, although this effect was also concentration dependent. Also, PBS was compared with three other buffers of tris(hydroxymethyl) aminomethane (Tres), glycene, and N-2-hydroxyethylpiperazine N'-2-ethanesulfonic acid (Hepes). The PBS was found to be the most effective at reducing adhesion at low concentration.

In order to be able to more rigorously define the nature of the tip, Feldman et al. [46] used either oxygen plasma–treated cantilevers or gold-coated cantilevers, with the former providing a polar tip and the latter a nonpolar tip. Interactions with various polymer surfaces were investigated in a liquid medium of perfluorodecalin. This nonpolar liquid has both a low dielectric constant and a low refractive index, as well as being inert toward most polymers. Thus, it was a very appropriate medium in which to measure the dispersion component of the van der Waals forces between the tip and surface. The authors found that for a given series of polymers, their adhesive ranking varied greatly between polar and nonpolar tips regardless of the precise composition. For nonpolar polymers they found a correlation between adhesion force and calculations based on the Lifshitz theory of van der Waals interactions, while with polar polymers a reasonable correlation with water contact angle was found.

Ralich et al. [47] investigated the affect of chain length for adhesion measurements in primary alcohols ranging from methanol to 1-nonanol. Measurements were made between a layer of phosphonic acid adsorbed on an aluminium substrate and an aluminium-coated silicon nitride tip, chosen for their stability. Adhesion forces were found to decrease with increasing chain length and to increase with increasing dielectric constant of the medium. The latter observation rules out the possibility of electrostatic shielding, which would serve to weaken the tip–sample interaction as the solvent permittivity increased. The use of deionized water as a solvent in the same tip–sample system resulted in an adhesive force at least an order of magnitude larger than that of methanol, the medium which had the highest adhesive force of the alcohols. The authors could not explain their results in terms of a single model and thus concluded that a combination of mechanisms must be evoked in order to describe the observed adhesion.

One of the few experimental measurements of adhesion in ultrahigh vacuum (UHV) was made by Carpick et al. [48]. Operating in UHV significantly reduces the contamination present, and, provided that the tip and sample are heated in UHV, it is also possible to remove any water layer on the surfaces. Carpick et al. investigated the adhesion of a platinum-coated tip to a mica surface cleaved in vacuum and used JKR theory (see Section II.A.2.b) to determine the interfacial adhesion energy. During the experiment a sudden large decrease in adhesion was observed after scanning the tip in contact with the mica. The same decrease in adhesion could not be observed simply by performing repeated force spectroscopy measurements at the same spot. Thus the scanning must have produced a change in the interfacial chemistry or structure. The stability of the platinum coating was checked after the experiment by measuring the resistance of the tip in contact with a conducting sample. Because the tip material was insulating apart from the platinum film, this gave a good indication that the platinum remained on the tip after adhesive contact and scanning on the mica. One possible explanation for the reduction in adhesion energy suggested by the authors was the scanning-induced transfer of potassium from the mica to the tip, which would lower the surface free energy of the platinum. The measured decrease in the work of adhesion was 0.39 J/m^2, which corresponds quite well with the energy of adsorption of potassium on Pt(111) at one-monolayer coverage (0.94 J/m^2). Similar experiments were performed with silicon and silicon nitride tips, but no such variation in adhesion was observed, lending further weight to the possibility that the chemistry of the tip plays a role. However, the material properties are also different, as the authors point out, so structural changes resulting in adhesion reduction could not be ruled out completely without further experiments.

3. Adhesion Due to Capillary Forces

Adhesion due to capillary forces can depend on humidity, the wettability of the tip and substrate, and the tip radius. Results of experiments on mica with a Si$_3$N$_4$ tip, published by Thundat et al. [49], correlated adhesive force with both humidity and imaging resolution. They retracted the tip after making contact with the sample and measured the force acting at the pull-off point. Although they do not show a pull-off curve in their paper, from their description it appears that the pull-off occurs as a single jump rather than in two parts consisting of tip–mica separation and tip–meniscus separation as described in the work of Weisenhorn et al. [39]. By altering the humidity, Thundat et al. managed to change the measured adhesion force by over a factor of 2. They also managed to match the experimentally measured adhesion approximately with a theoretical force derived from a solid–solid component obtained using a Lennard–Jones potential and a capillary force (see Section II.A.2.f). Experimentally, the adhesion force no longer decreased as the relative humidity was reduced below 10%, and atomic periodicity in their images was abruptly lost as adhesion reached its minimum value. The authors account for their observation of resolution dependence on humidity in terms of the tip's jumping between force minima created by the solvation shell structure of the water next to the mica. However, for mica it is also important to consider that the periodic images may be formed due to the periodic sliding of the mica sheets over each other, with one small mica sheet attached to the tip. In this case it is not possible to easily evaluate the imaging resolution based on the observation of atomic periodicity alone. Changes in relative adhesion between the mica sheets and between mica and the tip as a function of humidity could also account for the sudden changes in imaging characteristics.

Binggeli and Mate [50] investigated the influence of capillary condensation for three surfaces exhibiting very different contact angles with water: clean silicon oxide with a contact angle of 2°, amorphous carbon with 45°, and perfluoropolyether with 80–90°. All surfaces were investigated with an etched tungsten tip approximately 100 nm in radius. Due to the high spring constants of their probes (30–70 N/m) they did not see a jump from contact, rather, the attractive force from the meniscus smoothly decreased to zero. They thus measured the length of this transition as a function of humidity. Probably due to the stiffness of their force sensors, they observed very little change in the adhesive force as a function of relative humidity below around 60%. Above this, the dependence on humidity decreased with increasing hydrophobicity, as shown in Figure 11. One important aspect not discussed in the paper is the wettability of the tip. It is possible that as a metal in air, their tip would have adsorbed some hydrophobic alkanes. Perhaps this could explain the lack of humidity dependence of the tungsten–perfluoropolyether interaction, where both surfaces are probably resistant to the formation of a liquid meniscus between them.

Jarvis and Pethica [51] also investigated adhesion as a function of surface wettability but used a diamond tip for its known hydrophobic properties. They also used magnetic force–controlled AFM and applied forces to magnetic material behind the tip to bring the

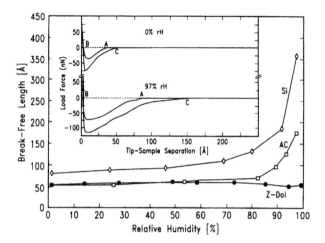

FIG. 11 Break-free length (distance B–C on inset) as a function of relative humidity for the clean Si(100) wafer (Si), the amorphous carbon film (AC), and the bonded perfluoropolyether film (Z-Dol). The inset depicts two dipping experiments on amorphous carbon, at 0% and 97% relative humidity (rH). (From Ref. 50. Copyright 1994 American Institute of Physics.)

tip and the surface into contact. They investigated hydrophilic silicon oxide as well as two different hydrophobic modifying layers of hexamethyldisilazane (HMDS) and dichlorodimethylsilane (DDS) at a humidity of 40%. The same tip was used in all the experiments and cleaned in between by dipping it in N-methyl-2-pyrolidone and heptane. Resulting curves are shown in Figure 12. On the silicon oxide they saw evidence of a soft intermediate layer, which tended not to wet the tip and was thus assumed to be water adsorbed onto the hydrophilic surface. No adhesion was resolved with their 48-N/m can-

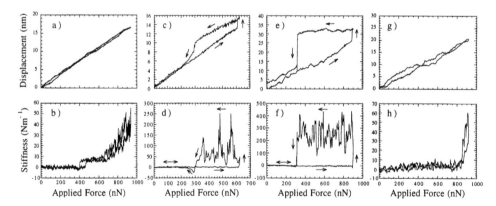

FIG. 12 Approaches of a diamond tip to the three silicon surfaces. Approach on hydrophilic silicon showing an instability in the stiffness curve (b) followed by a plateau. Note the change in gradient for the last 2 nm in the force–displacement curve (a). Graphs (c) and (d) show a typical approach of a diamond tip to an HMDS-treated silicon wafer. Large instabilities may be observed in both graphs. Graphs (e) and (f) show a commonly seen approach for a DDS treated surface. These figures have similar characteristics to those seen in (c) and (d) and are attributed to hydrophobic regions on the surface. Another commonly seen approach on the DDS treated surface is shown in graphs (g) and (h). In this case the characteristics are more similar to graphs (a) and (b) and are attributed to patches of bare silicon. (Reprinted from Ref. 51, copyright 1996, with permission from Elsevier Science.)

tilevers. Interactions with the HMDS were found to start with an instantaneous jump to contact, and a large adhesive force of approximately 400 nN had to be overcome in order to separate the tip and the sample. Evidence of the hydrophobic monolayer being stretched out between the tip and the sample on retraction was also seen in the pull-off curve. On the DDS, higher adhesive forces of approximately 550 nN were observed. However, there were also regions on the surface that showed the same interaction characteristics as the silicon oxide surface, suggesting that, unlike the HMDS modifying layer, the DDS exhibited pinhole in the monolayer, where the underlying hydrophilic substrate was exposed.

Eastman and Zhu [52] performed a study of Si_3N_4 tips, subsequently coated with gold and finally with paraffin thin films, adhering to mica. The study was purely qualitative, because they did not have accurate values for either their lever stiffnesses or their tip radii. By enclosing their AFM in a sealed container and controlling the humidity they found that hydrophobic tips show a lower adhesion with a surface expected to have a thin water film, as also observed by Jarvis and Pethica [51]. Surprisingly they did not see any dependence of adhesion on humidity for any of their tips.

4. Quantized Adhesion

Quantized adhesion was observed by Hoh et al. [53] for a Si_3N_4 tip breaking contact with a glass surface in water, which had been NaOH adjusted to pH 8.5. As described at the beginning of Section III.A, a more accurate description of the measurement would be quantized displacement, which can be evaluated as quantized adhesion by multiplying the observed displacement by the lever stiffness. The authors speculate that their observations could be explained either by the breaking of discrete numbers of hydrogen bonds between the tip and surface or by the breakdown of the continuum properties of water in close proximity to a solid surface.

Similar measurements were subsequently made by Cleveland et al. [54], using an improved low-noise AFM. These measurements provided conclusive evidence for the latter interpretation, for the tip remained in the vicinity of the sample without detection of any contact that would lead to bond formation. Measurements were made in milli-Q water on freshly cleaved calcite and barite crystals with silicon tips with a UV-cleaned, native, hydrophilic silicon oxide. Due to the dissolving, of the crystals, divalent ions would be expected in the water. By approaching the tip close to the surface and monitoring the thermally induced motion of the cantilever, a number of preferred tip positions were found that corresponded approximately to the diameter of a water molecule. The two main forces acting between surfaces in liquids are the attractive van der Waals forces and repulsive electric double-layer forces if there are charged surface groups. However, at small separation, continuum models break down due to the discrete nature of the liquid itself. Thus, in the Cleveland experiment, when the water is confined by the sample surface or between a blunt AFM tip and surface, it can form ordered layers. These layers will have increasing stiffness with decreasing separation, which will influence the interaction between the tip and the sample.

This structuring of liquids into discrete layers when confined by a solid surface has been more readily observable in liquid systems other than water [1,55]. In fact, such solvation forces in water, also known as hydration forces, have been notoriously difficult to measure due to the small size of the water molecule and the ease with which trace amounts of contamination can affect the ordering. However, hydration forces are thought to be influential in many adhesive processes. In colloidal and biological systems, the idea that the hydration layer must be overcome before two molecules, colloidal particles, or membranes can adhere to each other is prevalent. This implies that factors affecting the water structure, such as the presence of salts, can also control adhesive processes.

Recently a continuous measurement of solvation shells in pure water was made by Jarvis et al. [56]. They detected six layers next to a hydrophilic self-assembled monolayer (SAM) in pure water. The SAM had first been imaged with nanometer resolution using the same tip, thus opening up the possibility of making site-specific measurements of solvation structure with high lateral resolution. This technique could prove extremely useful for investigating the role of water structure in a variety of processes, including biological function. The authors could make a reasonable estimate of their tip radius because they used a multiwalled carbon nano-tube as a probe. Such tubes often have well-defined and asperity-free capped ends and are also extremely robust during the experiment [57]. The use of carbon nano-tubes as tips may provide a working compromise between full atomic-scale characterization of the probe with field ion microscopy (see Section III.A.8) and the poorly characterized microfabricated probes in common use. The additional possibility of chemical modification of the tubes [58] opens up a new area of highly localized investigation of chemically specific adhesion, where unintentional changes in the probe size or shape or surface roughness can be minimized.

The alternative interpretation of Hoh et al. [53] of quantized adhesion in terms of a specific number of bonds breaking has also been extended experimentally, although usually a single specific interaction is deliberately isolated between tip and sample. This type of measurement of single-bond energy will be dealt with in more detail in the following section. Because adhesion measurements are often performed in a liquid environment, it is clear that as AFM is developed to operate routinely at high sensitivity, it will become important to distinguish between quantized adhesion due to bond breaking and quantized displacement due to the structure of the solvent medium.

5. Single Molecule Force Spectroscopy

As measurements with AFM became more precise in terms of measurement sensitivity, control of the environment, and control of contact area, it became feasible to measure the adhesion of a discrete number of specific bonds. Florin et al. [59] investigated adhesion forces between ligand–receptor pairs by bringing an avidin-derivatized AFM tip into contact with agrose beads functionalized with biotin, desthiobiotin, or iminobiotin. In order to limit the number of binding events during contact, most of the biotin on the beads was blocked by avidin. With such systems, steps were observed in the retraction curve. By considering only the last step in the retraction curve, because other steps could be nonlinear convolutions of multiple unbinding processes, a histogram could be built up showing quantization of the adhesive force in units of 160 pN at a scan rate of 1 Hz. This value was assumed to correspond to a single avidin–biotin pair. The authors consider that the observation of quantized adhesion was highly dependent on their use of an agarose bead as the biotin substrate. The bead could facilitate the buildup of force at parallel binding sites via the tension of soft agarose springs, thus making the lateral positions of such sites of negligible importance. Its lateral mobility could also release the lateral stresses, which may build up during retraction of the sample (see also Section III.A.1). In the case of desthiobiotin, no clear quantization could be seen. But with iminobiotin, quantization in units of 85 pN was observed. This is consistent with the lower value of the binding enthalpy compared to biotin.

Lee et al. [60] investigated the adhesion of a single pair of DNA strands. They identified two types of forces: interchain forces associated with Watson–Crick base pairing between complementary strands, and intrachain forces associated with the elasticity of single strands. For studying interchain interactions, complementary oligomers $(ACTG)_5$ and

$(CAGT)_5$ were covalently attached to a silica tip and surface. A histogram of adhesive forces between the two, recorded at a retraction speed of 10 to 0.1 nm/s, showed four peaks, one of which could be identified as the nonspecific adhesive interaction by comparison with the adhesive force for noncomplementary oligonucleotides. The three other peaks, at 0.83, 1.11, and 1.52 nN, were associated with the interchain interactions of single pairs of DNA oligonucleotides 12, 16, and 20 base pairs in length. Interchain adhesion was studied by immobilizing C_{20} oligomers on both surfaces and then hybridizing a homopolymer of inosine on one side. The homopolymer was thought to have an average length of 160 bases but was actually composed of a wide distribution of molecular weights, resulting in a broad distribution of rupture lengths. The magnitude of the rupture force was typically 0.46 nN.

Dammer et al. [61] studied the specific interaction between polyclonal biotin–directed antibodies and biotinylated bovine serum albumin. To ensure that their curves corresponded to specific antibody–antigen interactions, the authors executed a number of control experiments, including change of buffer pH and using a nonspecific antibody. For the specific interaction, typical curves showed a number of adhesion peaks of the order of piconewtons. Only the final jump out was used for the histogram data, which exhibited peaks at interger multiples of 60 pN. The authors express a number of concerns regarding the experimental method, including the influence of unintentionally applied lateral forces and nonspecific background forces.

Clearly there are a number of uncertainties in the largely statistically derived results just outlined. Histogram peaks are somewhat dependent on bin size, and changes in tip or surface characteristics during a series of many hundreds of approaches can be hard to account for. It can also be difficult to compare results from different groups on the same system due to differences in rates of pull-off and direction of pull-off (the extent to which lateral forces are involved), which can affect the measured adhesion. Lo et al. [62] repeated the measurements on the biotin–avidin and biotin–streptavidin systems in order to demonstrate the use of the Poisson statistical analysis method. The main advantages of the method are that it does not require any assumptions about surface energies or contact areas, a large number of force measurements are not necessary, and it is not limited by the force resolution of the AFM. Using this analysis it is possible to extract a value for a single unbinding force from the slope of the linear regression fit of the variance-vs.-mean plot. Before the experiment the authors paid special attention to evaluating the tip and substrate coating using x-ray photoelectron spectroscopy, time-of-flight secondary ion mass spectrometry, and contact-angle measurements to optimize the conditions for coverage. For the biotin–avidin and biotin–streptavidin systems, the unbinding forces were found to be 173 and 326 pN, respectively, at a pull-off rate of 5 μm/s. The authors also tried the histogram approach but found that for their signal-to-noise level they were unable to observe quantized adhesion as reported by others [59].

Harada et al. [63] addressed the problems of lack of control over the orientation of the probing molecule and the conformational changes that proteins may undergo during adsorption to the tip and surface. They used direct immobilization of the molecules onto an inorganic substrate without the commonly used spacer, in order to more easily discriminate between specific and nonspecific adhesive forces. With oriented Fab fragments of an anti-ferritin antibody on their AFM probes, they investigated a ferritin-coated gold substrate. Displaying their adhesion measurements as histograms with three different bin sizes, they then applied Fourier analysis to each distribution. Such analyses always revealed dominant peaks corresponding to 63 and 130 pN for a retraction speed of either 0.1 or 1 μm/s, giving a quantized force of approximately 63 pN/molecular pair.

Ideally one would wish to remove the need for statistics by directly and reproduce-ably measuring a single bond only. One problem with the measurement of specific individual bond energies is that it is extremely difficult, even with a tip of small radius, to isolate a single bond species between the tip and the sample. To form a single bond in a controlled way requires the cantilever to be stiffer than the maximum force gradient experienced during the approach, but stiffer levers exhibit less sensitivity. If multiple bonds are formed, then it can be difficult to make an independent calculation of the contact area and hence the number of bonds involved.

Jarvis et al. [64] attempted to measure a single bond energy directly, between a silicon tip and sample, in an ultrahigh-vacuum environment. Using a compliant lever stabilized by magnetically applied forces to compensate for its bending toward the sample, they mapped out the interatomic force in a reproducible and reversible manner. They note that even if the interaction is limited to a single atom each on the tip and the sample, it is unlikely that a single bond potential is measured. At the very least, back bonds should also stretch, meaning that the total measured displacement will be the sum of contributions, due to a number of bonds in series and parallel. From their force curve the maximum tensile force was 0.3 nN, which is somewhat lower than the expected 1.5 nN for a single Si–Si bond but larger than that expected for a purely van der Waals interaction.

One method used by Lantz et al. [65] was to form a small number of bonds between the tip and the sample by isolating them perpendicular to the surface normal in a synthetic peptide molecule of cysteine$_3$–lysine$_{30}$–cysteine. The experimental conditions were selected such that the peptide was in an alpha-helix structure resulting from the formation of hydrogen bonds between every forth residue in the peptide chain. This structure was confirmed by circular dichroism spectroscopy. In order to isolate the peptides under the tip in such a way that a single stretching event could be guaranteed, the active peptides were self-assembled in a low concentration alongside nonreactive peptides of cysteine–lysine$_{30}$. Thus, by bringing a gold tip into contact with the surface and retracting, it was possible to measure adhesive forces characteristic of the breaking of 31 internal hydrogen bonds and the sulphur–gold bond with the tip or sample. As an additional aid to analyzing the stretching process of the peptide, the authors made a continuous measurement of its compliance by oscillating the lever with a known force and measuring the resulting oscillation amplitude [29]. All measurements were performed in MOPS buffer adjusted to pH 11 using NaOH. The nonspecific interaction was investigated on films of purely unreactive peptides. From their applied force-vs.-elongation of the pepetide curve, the authors calculated the work done on the peptide. By subtracting a value for the sulphur–gold bond from the literature and dividing by the 31 hydrogen bonds broken, they calculated a value of 20.2 kJ/mol, which is surprisingly close, considering the environment of the experiment, to the ab initio calculated value for the hydrogen-bond energy in vacuum.

Grandbois et al. [66] measured the rupture force of a single covalent bond by anchoring single polysaccharide molecules to a glass or gold surface using different surface chemistries. Polysaccharides, upon stretching, go through a reversible conformational transition, which can be easily detected as a plateau in the pull-off force. In the case of amylose, this plateau lies at 275 pN, and the authors used this characteristic feature to determine in which retraction curves they had only a single molecule stretched between tip and surface. In this way, by altering the attachment chemistry, they found that a silicon–carbon bond ruptured at 2 nN, while the sulphur–gold bond ruptured at 1.4 nN, both measured at a retraction rate of 10 nN/s. In order to identify the ruptured bond in the case of carbodiimide chemistry, bond rupture probability densities were theoretically derived. In the case of sulphur–gold attachment it was not possible to confirm whether the bond was truly ruptured or whether the gold atom was extracted from the metal surface. All measurements were made in phosphate-buffered solution at pH 7.4.

Recently, Oesterhelt et al. [67] combined imaging and spectroscopy to locate and measure the unfolding pathway of specific bacteriorhodopsin molecules within a purple membrane, including extraction from the membrane. From prior imaging of the surface, a single molecule was chosen and a load of 1 nN applied for 1 s, which had a 15% chance of causing the protein molecule to adsorb onto the tip. After measuring the force vs. extension characteristics of the protein at a retraction rate of 40 nm/s, the surface was again imaged to confirm the anticipated presence of a vacancy site left after extraction of the protein. Before-and-after images, together with the extension characteristics, are shown in Figure 13.

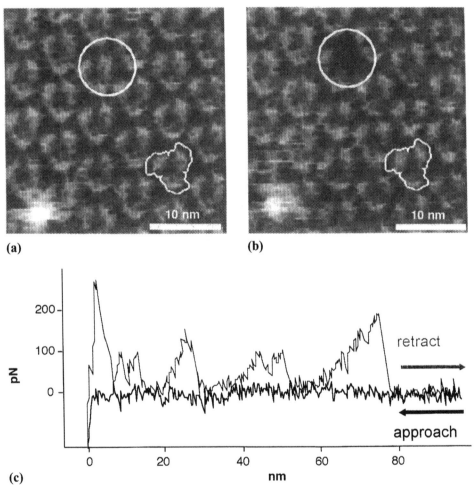

FIG. 13 Controlled extraction of an individual bacteriorhodopsin from native purple membrane: (a) Typical high-resolution AFM topograph of the cytoplasmic surface of a wild-type purple membrane showing the trimer-assembled bacteriorhodopsin. (b) After the adhesive force peaks were recorded, a topograph of the same surface was taken to show structural changes. (c) Pull-off characteristics at a separation velocity of 40 nm/s. Discontinuous changes in the force indicate a molecular bridge between tip and sample that is found to extend up to separations of 75 nm, which corresponds to the length of one totally unfolded protein. (Courtesy of Filipp Oesterhelt.)

Although it was possible to accurately locate and manipulate an individual molecule, it was not possible to control which part of the protein attached to the tip, this was thought to be the cause of significant variations in the force-vs.-extension curves. The authors identified a subset within their data that, from the extension length and characteristics, they believed corresponded to proteins attached as the cytoplasmic COOH-terminus. From analysis of these curves they determined that the protein was extracted two helices at a time. This technique combining high-resolution imaging with spectroscopy is clearly very powerful for investigating specific molecular interactions and adhesive processes. It will be interesting to see if it can be applied to smaller molecules in the future.

6. Chemical Force Microscopy

A significant proportion of adhesion studies on the nanoscale come under the heading of chemical force microscopy (CFM). Technically there is no clear distinction between CFM and the method of specific interaction measurement detailed in the previous section. Instead of measuring a particular interaction such as ligand–receptor or antibody–antigen, which usually involves multiple interactions of large complex molecules, CFM tends to involve much smaller molecules, where a single specific interaction is anticipated, such as van der Waals or hydrogen bonding. Often the sample will be designed such that there is a variation in the interaction at different locations on the surface. Frequently, microcontact printing is used for this purpose [68]. An image of the surface can then be built up due to the absence or presence of a particular interaction.

The most common method for functionalizing AFM probes for CFM is to make use of the strong, covalent sulphur–gold bond, in order to form robust and highly ordered crystalline-like self-assembled monolayers (SAMs). One of the drawbacks of this technique is that the tip must be coated first with gold and then with the SAM, resulting in a tip of significantly larger radius than the original tip. This broadening of the radius subsequently reduces the imaging resolution when operated as a microscope. There is also the problem of limiting the experiment to a known number of interacting molecules so that a specific interaction can be quantified in terms of the force or energy involved.

Often, JKR is used to calculate the spherical contact area at pull-off, and hence the number of interacting molecules can be calculated. One inconsistency with this method is that little attention is paid to the molecular arrangement on tip and surface. Calculations, for example, giving the area of interaction to cover two molecules, which is not physically possible for a spherical contact. A further inconsistency is the assumption that the pull-off represents all bonds breaking simultaneously, rather than as a discretely observable series of ruptures indicative of the variation in bond extension, which must occur under the tip.

Another interesting angle on pull-off force measurements as made by CFM is the potential to control the environment. One obvious application is the study of pH-dependent adhesion force measurements. This type of study enables the determination of pKa values of surface-bound ionizable groups and the investigation of acid–base behavior in a way analogous to prior studies with surface forces apparatus. The understanding of such interactions is highly relevant to phenomena such as protein folding and enzymatic catalysis.

Amongst the earliest measurements involving chemical functionality of the probe were those of Nakagawa et al. [69]. They investigated octadecyltrichlorosilane (OTS) chemically modified tips against chemically adsorbed monolayers of different alkyltrichlorosilanes in ethanol, as shown schematically in Figure 14. When both tip and surface were modified by OTS, a large adhesive force was observed that was not present for the case of an unmodified silicon nitride tip on an OTS-modified surface. Additionally there

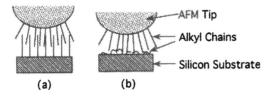

FIG. 14 Schematic drawing of entwining between the alkyl chains of CA molecules on a tip and those on a silicon substrate for (a) $n \geq 8$ and (b) $n < 8$. (Reprinted from Ref. 69. Copyright 1993 by the Publication Board, Japanese Journal of Applied Physics.)

was found to be a strong correlation between the molecular chain length of the molecules on the substrate and adhesion. They speculate that this is due to an entwining of the longer molecules on the tip and the sample, thus increasing the interaction area of the alkyl chains and increasing the adhesion. Very little adhesion was observed if the chain length was reduced below $n = 8$ for $CH_3(CH)_nSiCl_3$, which they associated with the shorter films' not being uniformly self-assembled but having a large and random tilt angle. Even in this early paper there is a significant indication that the mechanics of the contact and the effective interaction area are the critical parameters, rather than chemical interactions, in many systems.

The mapping of spatial arrangements of chemical functional groups was performed by Frisbie et al. [70]. Samples of lithographically patterned organic SAMs, including CH_3-(hydrophobic) and COOH- (hydrophilic) terminated end groups were investigated with similarly functionalized tips. Working on the hypothesis that the magnitude of the friction force signal in AFM is directly related to the magnitude of the adhesive force between tip and sample, the authors used the lateral signal to form the image of the patterned surface with resolution comparable to an optical condensation image. They also found from force spectroscopy measurements that the adhesive interaction between the functionalized tip and the sample exhibited the following trend: COOH/COOH > CH_3/CH_3 > COOH/CH_3. Thus, the interaction between hydrophilic groups, which can form hydrogen bonds, was stronger than between hydrophobic groups or mixed groups in the experimental environment of complete immersion in ethanol.

A similar experiment was performed some years later by the same group, but this time utilizing carbon chemistry to improve lateral resolution and durability of the functionalized probes [58]. Wong et al. used multiwalled carbon nanotube tips attached to the end of a standard AFM tip. The tubes were produced with open ends by shortening them in an oxidizing environment. Carboxyl groups are expected at these open ends based on spectroscopic studies of bulk samples. This was tested directly by measuring the adhesion force as a function of pH between unmodified and modified nanotube tips and a hydroxy-terminated SAM, as shown in Figure 15. In this way the authors determined that carboxyl groups were indeed present, and, from the similarity of their measured pKa value to the bulk solution value for benzoic acid, they concluded that the groups are well solvated and accessible to reaction. Tips were subsequently modified with benzylamine and ethylenediamine and adhesion investigated in the same manner. To form an image of a patterned surface of CH_3- and COOH-terminated groups as a function of adhesion, the authors utilized their prior observation [71] that phase-lag differences of an oscillating tip making intermittent contact with a sample can be qualitatively related to differences in the adhesion force. Although high resolution was not demonstrated, it would seem that the potential is there, given the small probe radius of a typical carbon nanotube.

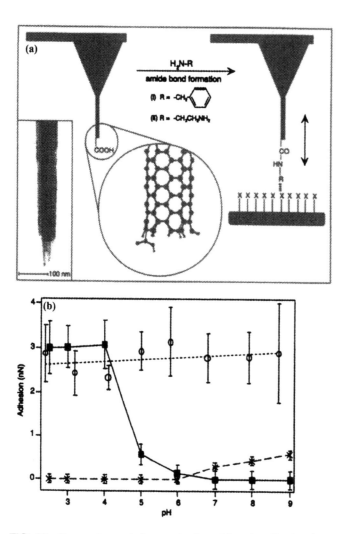

FIG. 15 Preparation and characterization of functionalized carbon nano-tube tips. (a) Diagram illustrating the modification of a nanotube tip, by coupling an amine (RNH$_2$) to a pendant carboxyl group, and the application of this probe to sense specific interactions with functional groups (X) of a substrate. The circular inset is a molecular model of a single nanotube wall with one carboxyl group at the end. Inset, TEM image showing the open end of a shortened nanotube tip. (b) Adhesion force as a function of pH between the nanotube tips and a hydroxy-terminated SAM (11-thioundecanol on gold-coated mica): filled squares, carboxyl (unmodified); open circles, phenyl (modified with benzylamine); and crosses, amine (modified with ethylenediamine). Each data point corresponds to the mean of 50–100 adhesion measurements, and the error bars represent one standard deviation. (Reprinted with permission from Ref. 58. Copyright 1998 Macmillan Magazines Limited.)

Thomas et al. [72] used a modified force microscope in which a compensatory force was applied to the probe to keep its displacement at zero. With this system they studied interactions between organomercaptan molecules with CH$_3$, NH$_2$, or COOH end groups. All measurements were performed in dry nitrogen. From SEM-measured tip radii and pull-off force they calculated the work of adhesion using the DMT model. They found that the work of adhesion values qualitatively scaled as expected for van der Waals, hydrogen bonding, and acid–base interactions, but the interaction length scales were much longer than expected. The effect of tip radius was also investigated, and the normalized adhesive interaction was seen to fall above a tip radius of 400 nm, due to the difference between the real and apparent contact areas.

Van der Vegte and Hadziioannou [73] have presented a detailed cross-reference study of five different termination groups (Table 1). Qualitatively the data showed the expected trend of small adhesive forces for tip and substrate whose interactions were pure van der Waals compared to the tips capable of hydrogen bonding. Of all possible combinations it was found that dissimilar pairs of van der Waals interacting tips exhibited the weakest adhesive forces. However, their data showed that by exchanging the terminating groups on the tip and substrate they did not always see the same adhesive force. If the measurement is purely an indication of the bond energy, why should this be so? In fact, as the authors note, the measurements will be influenced by the number of bonds formed, which relates to the tip radius and contact area prior to pull-off. The latter parameter is not easy to control to within the area of a single molecule (approximately 20 Å^2 for an alkanethiol) in AFM, particularly with larger tip dimensions and poorly controlled transitions to contact.

For the investigation of bacterial adhesion important for manmade implants and contact lenses, Ong et al. [74] developed a technique for attaching a layer of bacteria to Si_3N_4 AFM tips, shown in Figure 16. First the tips were coated with polyethyleneimmine, then the cells were added, and finally the tips were treated with glutaraldehyde to firmly anchor the cells. The lipopolysaccharide molecules coating the cell surface were found to greatly influence the bacterial adhesion properties of the cell. This was investigated using two genetically similar *E. coli* strains, which differed only in lipopolysaccharide composition, or, in other words, only in surface charge and hydrophobicity. The strain with stronger net negative charge was strongly repelled by the hydrophilic, negatively charged surfaces of mica and glass. In contrast, the same strain was attracted to polystyrene and strongly attracted to Teflon, which corresponds to the increased hydrophobicity of the Teflon surface. Conversely, the other strain was attracted to both mica and glass, probably due to van der Waals interactions or bridging effects, and repelled by polystyrene and Teflon. Further experiments involved the addition of 100 mM NaCl to the 1mM Tris buffer. This was found to reduce the repulsive interaction, thus identifying the effect as being electrostatic in nature

TABLE 1 Single Chemical Bond Forces (in pN) for Every Tip–Substrate Combination, Calculated on the Basis of the JKR Theory of Adhesion Mechanics[a]

	Substrate				
Tip	CH_3[b]	OH	NH_2	COOH	$CONH_2$
CH_3[b]	81	57	59	61	601
OH	50	101	113	112	117
NH_2	54	88	98	95	100
COOH	95	109	105	114	137
$CONH_2$	62	110	102	125	120

[a] All bond strengths apply to measurements in ethanol.
[b] Although the van der Waals interaction is not a "two-center" bond interaction in this experimental setup, the calculated values for F_{single} represent an effective binding force experienced by the molecular pairs and are listed for comparison reasons.
Source: Ref. 73. Copyright 1997 American Chemical Society.

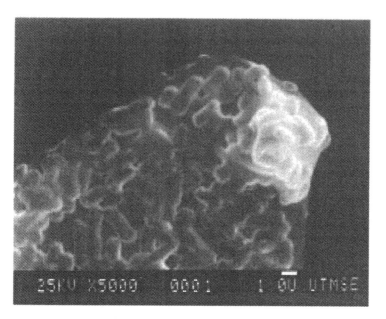

FIG. 16 SEM micrograph of a microfabricated Si_3N_4 tip coated with *E. coli* D21 cells. (Reprinted with permission from Ref. 74. Copyright 1999 American Chemical Society.)

while having no effect on the attractive interaction of the other strain. Investigations with hydrophobic glass found that both bacterial strains formed strong adhesive bonds. The authors conclude that cell adhesion appears to result from a combination of van der Waals, electrostatic, and hydrophobic interactions but that a complete picture must include steric interactions, bridging effects, and/or receptor–ligand interactions.

Kidoaki and Matsuda [75] used SAMs with carboxyl terminal groups to attach proteins of albumin, immunoglobulin, and fibrinogen to their AFM tips. Using long-chain alkanethiolates self-assembled on gold as well-defined model substrates they found that, irrespective of the type of protein, exceptionally large adhesive hysteresis and multiple jumps in the retraction curves were obtained on the CH_3-terminated SAM in phosphate-buffered saline. This was in stark contrast to the curves on NH_2-, OH-, and COOH-terminated SAMs, leading to the conclusion that the adhesion force due to hydrophobic interactions is dominant in aqueous systems. They observed that for low loading forces of less than 2 nN they could not always establish adhesion between the protein and the SAM. This, together with a correspondence between protein size and the range and number of jumps in the retraction curves, implied that the approach first involved induced deformation or denaturing of the protein under applied load and that then, on retraction, the protein reformed toward the original conformation. The detachment of the protein from the sample surface then commenced, together with deformation and stretching of the protein, with possible unfolding. Finally the protein–sample interface was broken and the protein refolded toward the native state. Clearly, with so many processes occurring simultaneously, it is difficult to attribute specific regions of the force curve to specific processes.

It has been observed by Koleske et al. [76] that factors other than specific chemical interaction can influence adhesion measurements. They studied a film of mixed-chain-length fatty acid and found that the adhesion of the tip over the shorter chain was 20% larger than over the longer chain. despite the fact that both had the same CH_3 terminating group, shown in Figure 17A. This increased adhesion they attribute to the tip's penetrating more deeply into the shorter-chain-length LB film due to compliance differences between the two chains, seen schematically in Figure 17B. This being the case, it is clear that when interpreting chemical force microscopy data it is essential to consider both mechanical and chemical effects.

One group performed an elegant experiment to sidestep the problems associated with mixed-chain-length films. McKendry et al. [77] used chemical force microscopy to investigate chirality. The importance of chirality in pharmacology was brought to the fore by thalidamide in the 1960s. McKendry and coworkers distinguished between the two enantiomers of mandelic acid arrayed on a surface via differences in both the adhesion forces and the frictional forces measured by their S-enantiomer functionalized probe. Adhesion forces were essentially found to double for different entomers on tip and surface as compared to values for the same entomer on both surfaces. The most important aspect of the work was that the mechanical differences between the different chemical interactions was minimized.

In summary, chemical force microscopy appears to be a powerful tool for the study of adhesion on the nanoscale provided that issues such as the detrimental influence of mechanical effects, substrate roughness, packing density, unknown tip radius, and reliance on statistical analysis can somehow be resolved in a consistent and logical way.

7. Adhesion Between Atomically Characterized Surfaces

One significant drawback in all the previously described experiments in the lack of precise knowledge about the tip, particularly with regard to shape and chemical nature on the

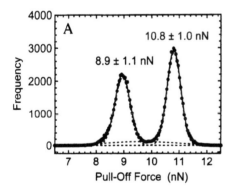

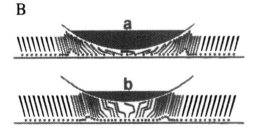

FIG. 17 (A) Histogram of pull-off force from 86,000 force curves. The heavy line is the sum of three Gaussians used to fit the data. (B) Schematic representation of the compression of the LB film by the SPM tip over the C_{16} region (a) and the C_{24} region (b). The shaded area of the tip indicates the contact region. (Reprinted from Ref. 76.)

atomic scale. Even this is not always sufficient to explain specific interactions, and additional information regarding the form of the dangling bond is required [78]. One group has made a significant effort to characterize their tips on an atomic scale, both before and after bringing the tip and sample into contact, so as to study adhesion of metals on the atomic scale. This was done by combining scanning tunneling microscopy, atomic force microscopy, and field ion microscopy [79]. In this setup an STM tip was electrochemically etched from a single-crystal tungsten wire oriented in the (111) direction and sharpened by field evaporation. The tip was characterized on an atomic scale at 150 K using field ion microscopy. The trimer-terminated W(111) tip (see Figs. 18a and b) was then pressed into a Au(111) surface on which the Au(111) 22 × $\sqrt{3}$ reconstruction had previously been imaged in STM mode. In order to measure the forces acting during the approach and retrac-

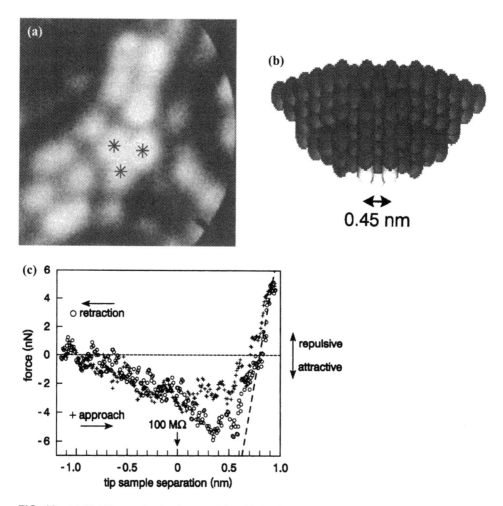

FIG. 18   (a) Field ion projection image of the (111)-oriented W tip. Trimer atoms are marked with an asterisk (*). (b) Hard-sphere model of the tip apex reconstructed from FIM images (apex trimer highlighted in bright tones, vertical scale is expanded by a factor of 1.8). (c) Force versus tip-sample separation measured on a flat terrace using a W-trimer tip. (From Ref. 79. Copyright 1998 by the American Physical Society.)

tion of tip and sample, the sample was mounted on a small, stiff cantilever beam ($k = 116$ N/m) whose deflection was accurately measured using differential optical interferometry.

Surprisingly, attractive forces were found to extend over almost 2 nm, and there was no spontaneous jump to contact (see Fig. 18c). Thus, in contrast to a Ni tip on a Au substrate [21], it would appear that spontaneous wetting of the W-trimer tip by the Au substrate does not occur, even at contact pressures of 25 GPa. They also observed hysteresis corresponding to an energy dissipation of 7 eV in the range of maximum adhesive interaction. However, they found that adhesion hysteresis, due to wetting, does occur for polycrystalline W tips and results in a hillock's remaining on the surface, signifying plastic deformation [80].

8. Colloid Probe

One of the inherent problems in AFM is uncertainty regarding the effective tip radius. This was particularly bad before the advent of sharp microfabricated probes, when tip radii could be as large as several microns. On this scale it is not reasonable to assume that the tip has an atomically smooth surface or even a good approximation to one. In this case it is no longer valid to assume a single-asperity contact, which causes considerable complications in evaluating the resulting data. An additional problem in early experiments was that the interaction with the sample was not fully controlled, and "tip crashes" often occurred, resulting in actual change of tip shape during the experiment.

These problems have been improved in recent years by the microfabrication of sharp tips with radii less than 10 nm, the observation in an SEM or STEM of the exact radius before and after the experiment, the use of robust carbon-nanotube probes, and general improvements in control electronics. However, another method used initially was the attachment of a small colloid particle in place of the AFM tip. These particles were considered a reasonably good approximation to a single-asperity contact; their radii were accurately known and remained the same for the duration of the experiment. Such probes have also been used to investigate colloids where surface roughness is an important aspect of the colloid interaction.

Ducker et al. [81] first used colloid particles attached to microfabricated AFM cantilevers as a means of defining the geometry and material of both surfaces. The spheres used had a radius of 3.5 μm and a surface roughness of 3 nm_{p-p} over 0.45 μm², which was reduced still further by scanning. Their initial investigation was into the approach characteristics of the colloid and sample in solutions of sodium chloride. Subsequent work involved studying the stability of various colloids after surface treatment [82]. Hydrophobic silica surfaces exhibited very large adhesion of 0.4 N/m when normalized for tip radius. In the case of gold surfaces in weak sodium chloride solution, the adhesive force was found to be highly variable. Due to the strength of the adhesive force, the authors consider that the contact region may be changed during the series of measurements.

The colloid probe technique was first applied to the investigation of surfactant adsorption by Rutland and Senden [83]. They investigated the effect of a nonionic surfactant petakis(oxyethylene) dodecyl ether at various concentrations for a silica–silica system. In the absence of surfactant they observed a repulsive interaction at small separation, which inhibited adhesive contact. For a concentration of 2×10^{-5} M they found a normalized adhesive force of 19 mN/m, which is small compared to similar measurements with SFA and is probably caused by sufactant adsorption's disrupting the hydration force. The adhesive force decreased with time, suggesting that the hydrophobic attraction was being screened by further surfactant adsorption. Thus the authors concluded that adsorption occurs through small aggregates rather than through growth of a monolayer. Increasing the concentration still further to 3×10^{-5} M and then to 4×10^{-5} M decreased the adhesion until a concentration above the critical micelle concentration was reached. At this point the aggregates appeared to form bilayer structures, and repulsive steric forces were observed.

Preuss and Butt [84] also investigated the influence of surfactant, in their case between a silica colloid probe and an air bubble. Hydrophilic probes were repelled by the bubble and did not form an adhesive contact. Silanized hydrophobic probes did form an adhesive contact with the bubble in aqueous electrolyte, and a normalized force of 0.64 N/m was required to separate them. The effect of two different surfactants was investigated, one that strongly adsorbs to silica and is positively charged and one that does not adsorb to silica and is negatively charged. With the addition of the negatively charged surfactant, the force between the hydrophobic probe and the bubble became repulsive, requiring greater applied force to instigate an adhesive contact. In the case of the positively charged surfactant and the hydrophilic probe, a strong adhesive contact was formed that increased slightly with concentration of the surfactant. This change in the type of interaction from repulsion to strong adhesive contact formation on the addition of surfactant is probably because this surfactant exhibits strong adhesion to the silica, making the formerly hydrophilic probe become hydrophobic. When the concentration was increased above 6 mM, the nature of the interaction again changed to being predominantly repulsive, probably due to positively charged surfactant's being adsorbed at the air/water interface. The small adhesive interaction that was observed was found to increase monotonically with applied load. The authors could not explain this observation but speculate that it could be due to a change of the surfactant structure at the interfaces as the particle and air bubble approach.

Fuji et al. [85] used porous and nonporous silica particles to measure adhesion as a function of humidity. They altered the pore size distribution of the particles by using different conditions of hydrothermal treatment. One particle was glued to the end of an AFM cantilever, and the other particle was glued to the substrate. Measurements were made across a range of relative humidities from 40% to 90%. All adhesion forces were normalized relative to the radii of the two particles. From the varying humidity dependence of the different particles they could assume that the generation mechanism of the adhesive force was closely related to the geometric structure of the surface. For the rough spheres, capillary condensation was hindered by surface asperities unless the humidity was very high, in which case the water filled in the gaps, allowing the capillary force to act over the entire area. The critical humidity could be calculated by setting the thickness of the adsorbed water layer plus the Kelvin radius (see Section II.A.2.f) equal to the critical pore radius. The calculated value of critical humidity and the observed correspondence between relative humidity and adhesive force showed a good match, assuming that capillary condensation occurred after filling the pores and that this would cause a marked increase in the adhesion. At lower humidities the hydrogen-bonding force was considered to dominate instead of the capillary force, and thus the adhesion should depend on the hydrophilicity and the contact area. This was indeed seen to be the case, with the adhesion force increasing with decreasing pore size and decreasing with decreasing surface density of silanol groups.

Sagvolden et al. [86] also combined the use of colloids with AFM force sensors to study adhesion. In their case, instead of attaching the colloid to the end of the AFM probe and applying a normal force, they approached the free colloids from side on, with the AFM cantilevered at an angle of approximately 30° to the surface normal. Thus, they applied a predominantly lateral force to the colloid particles. The colloids were coated with protein molecules, and their adhesion was studied against three nonbiological surfaces, consisting of glass, hydrophobic, and hydrophilic polystyrene. By not having the colloids permanently attached to the force sensor it was possible for the protein adsorption to progress over hours without mechanical interference. Measurements were made in Tris-buffered saline, pH 7.4, and various proteins were investigated with different net charge, size, and structural rigidity. They found that the strength of protein adhesion increased with the hydrophobicity of the surfaces and that loosely folded proteins adhered more strongly than rigid ones to the hydrophobic polystyrene. However, adhesion to hydrophilic glass was charge controlled. Despite the dependence of the absolute value of adhesion force on the direction in which the bonds are broken [35], it seems reasonable to use the method for comparative studies of this type.

Heim et al. [87] used silica particles of various radii between 500 nm and 2.5 μm (determined by scanning electron microscopy) to investigate the dependence of adhesion force on radius. Such particles were used for both tip and sample, and positioning was done optically. At an effectively quasistatic separation velocity of 0.5 μm/s, the adhesive force was found to be a linear function of the particle size, as predicted by both DMT and JKR continuum models (see Sections II.A.2.b and c). Their pull-off data show considerable scatter, although they maintain that their measured adhesive force did not depend on the previously applied load (0–600 nN), ambient air pressure (10^2–10^5 Pa), or humidity (10%–40%). Results from repeated measurements with the same particles were highly reproducible, while different pairs of particles could have a standard deviation from the mean value of up to 25%. This could indicate that surface roughness was playing an important role; however, direct measurement of the roughness found it to be considerably lower than the expected deformation over an area of the order of the contact area, which would imply that its effect would be negligible.

Vakarelski et al. [88] also investigated the adhesive forces between a colloid particle and a flat surface in solution. In their case they investigated a silica sphere and a mica surface in chloride solutions of monovalent cations CsCl, KCl, NaCl, and LiCl. The pH was kept at 5.6 for all the experiments. To obtain the adhesive force in the presence of an electrostatic interaction, they summed the repulsive force and the pull-off force (coined F_{off} by the authors!) to obtain a value for the adhesive force that is independent of the electrostatic component.

In their preliminary experiments, in pure water, they checked for any time or load dependence in the measurements and found a significant time dependence, with the measured adhesion increasing by nearly a factor of 5 as the contact time was increased from 0.1 s to 100 s. As for load, they could find no discernible difference in adhesion when the load was increased from 1 to 50 mN/m. Results were also found to be highly repeatable for a single probe tip but not for different tips, probably due to variations in surface roughness. The authors propose an explanation of these effects based on the presence of solvation shells between the probe and the sample that are breached during longer contact times. A rough calculation of the minimum and maximum gap at pull-off as a function of time correspond to 0.32 nm and 0.55 nm, which are approximately equal to solvation shells in water measured by others [1,54,56]. However, the surface roughness of the sphere is of the same order, so the calculation does not prove conclusive.

In KCl, the time dependence showed a marked decrease as the concentration was increased, becoming practically independent of time at a concentration of 1.0 M. This could be explained by adsorbed layers of cations on the surfaces, which reduce the attractive force between them. Further, this reduction with concentration became greater as the hydration enthalpy of the cations increased. This was interpreted in terms of the observations of Higashitani and Ishimura [89], who reported that the thickness of the adsorbed layer of cations increases with the absolute hydration enthalpy. This is because cations with larger values of hydration enthalpy, such as Li^+ and Na^+, form thicker but weakly adsorbed layers, unlike Cs^+ and K^+, which absorb directly onto the surfaces, forming thin but strongly adsorbed layers, as shown schematically in Figure 19.

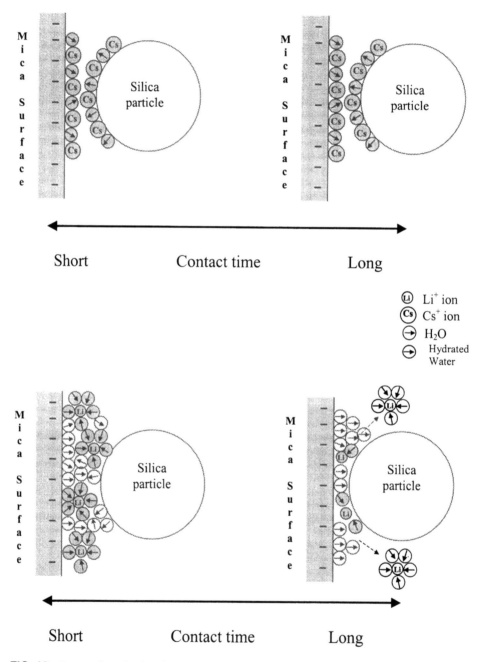

FIG. 19 Proposed mechanism for the difference of adhesive force between cations of low and high hydration enthalpies. (Reprinted from Ref. 88. Copyright 2000 by Academic Press.)

B. Surface Forces Apparatus

The surface forces apparatus (SFA) predates the AFM by approximately two decades [90]. It was the first apparatus that successfully measured forces between hard surfaces as a function of their separation below 100 nm; it consists of crossed mica cylinders with radii of the order of centimeters. In a very similar way to the AFM, one surface is fixed while the other is mounted on a flexible force-sensing beam, the main difference being that the interacting surfaces are 10^4–10^6 bigger in the case of SFA. The distance between the two cylinders is usually measured by white light interferometry using fringes of equal chromatic order (FECO). This direct measurement of separation is one of the major advantages over AFM where separation must be calculated. Various improvements and alterations have been made to the original SFA, and the equipment is now capable of measuring forces in the range 10^{-8}–1 N with a vertical distance resolution of ± 0.5–1 Å [1,91]. The SFA has subsequently been used to study various forces, including van der Waals, double-layer, viscous, and hydration forces, as well as adhesion.

Although the SFA and its derivatives have produced many interesting results, they are subject to several limitations. One of the biggest limitations is that there are very few solid surfaces that may be studied in this way, the most common being mica, although sapphire [92] and silica [93] have also been used, with certain limitations due to surface roughness. In addition the instrument does not provide any mapping or imaging of the surface.

Other, similar methods for measuring surface–surface interactions, which come under the generic heading of surface force apparatus, include the crossed-filament method. This utilizes a beam deflection technique similar to that now being used in some AFMs for the measurement of surface displacement [94]. Another technique for displacement measurement used in a similar SFA is that of a capacitance transducer. Both techniques suffer the criticism that separation is not measured at the point of interest, i.e., the gap between the two surfaces as measured in the FECO technique.

More details on SFA and its applications can be found in Chapter 1.

C. Quartz Crystal Microbalance

The quartz crystal microbalance (QCM) is usually used to monitor the deposition of thin films or to study gas adsorption and reactions on surfaces, although it can also be applied to the study of cell adhesion. The instrument consists of a small disk (radius typically of the order of a centimeter) cut from a quartz crystal with two electrodes, one of which is used as the substrate for deposition or adsorption. Measurements are usually implemented by measuring the shift in the resonant frequency of the QCM due to the mass change involved in adsorption of material onto the microbalance [95]. Additional information can be obtained by also measuring the change in dissipation or Q value of the QCM due to viscoelastic energy loss. This method has recently been used to study living cell adhesion processes [96]. By analyzing both the frequency shift and the dissipation it was possible to obtain a fingerprint of the cell adhesion process, which included the properties of the surface and the type of cell. This method tends to operate on the scale of tens of cells.

IV. ADHESION IN NATURE

Nature takes various approaches to adhesion, the most common being to control contact area through deformation of a multitude of hairlike fibrils or alternatively to modify the surface property in a way analogous to chemical force microscopy. These processes have been refined for eons via processes of natural selection, thus making their study of potential benefit not only to improve our understanding of evolutionary processes but also as model reference systems to the new breed of microelectromechanical systems and similar nano-technological applications.

Dammer et al. [97] investigated the cell adhesion proteoglycan of the marine sponge *Microciona prolifera*, which consists of rings with diameter 200 nm and approximately 20 irradiating arms, each 180 nm in length. The adhesion proteoglycan was attached to an AFM tip and surface, which were brought together in seawater Tris buffer. On retraction of the surface at a rate of 0.1 Hz, interactions were observed up to a separation of 200 nm, which were interpreted as the lifting and extension of the stringlike arms. Such extensions were often accompanied by multiple jumps of the AFM lever, indicating polyvalent binding, and with an average adhesive force of 40 pN per jump corresponding to a pair of adhesion proteoglycan arms. The average adhesion force of 125 pN and the maximum adhesive force of 400 pN, both measured in a physiological Ca^{2+} concentration of 10 mM, could thus be interpreted as the binding between 3 and 10 pairs of arms, respectively. From analysis of the slope of the force curves it was found that the arms did not behave like ideal springs but that their stiffness increased with extension. This, together with the 400-pN noncovalent binding force, indicated how cell dissociation could occur without destroying the adhesion proteoglycans while maintaining the ability to hold the weight of approximately 1600 cells in physiological solution.

A prime example of the impressive use of adhesion in nature is the gecko, which can hang from the ceiling and climb with ease up smooth vertical surfaces (Fig. 20a). Its ability to do this was known for some time to be associated with the many thousands of hairs (or setae) comprising each foot (Fig. 20b). In addition, each seta was found to con-

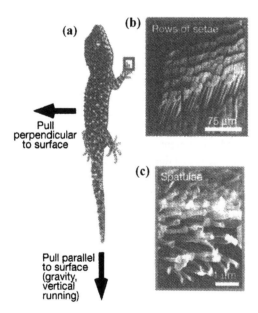

FIG. 20 (a) Tokay gecko with toe outlined. (b) SEM of rows of setae. (c) The finest terminal branches of a seta, called *spatulae*. (Reprinted with permission from Ref. 99. Copyright 2000 Macmillan Magazines Limited.)

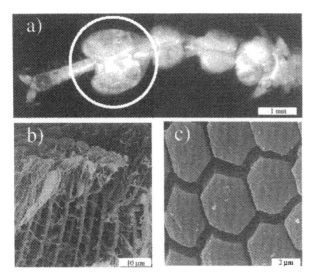

FIG. 21 Microstructure of the attachment pads. (a) Complete leg section containing the attachment pads. (b) Fracture of a shock-frozen pad. (c) Top view of the pad surface. (From Ref. 100. IOP Publishing Limited.)

sist of hundreds of spatula-shaped projections approximately 200–500 nm in diameter (Fig. 20c). Using a specially designed AFM force sensor capable of simultaneous normal and lateral force measurements with high sensitivity [98], Autumn et al. [99] investigated the adhesion characteristics of a single seta. By analysis of the uncurling of the gecko's toe and attendance at some interesting parties, the authors determined that orientation and loading were crucial parameters for the setal force capacity. For example, setal contact with its many spatulae not projected toward the surface resulted in an adhesive force of less than 300 nN. Further, even with the spatulae projected toward the surface, both perpendicular loading and lateral forces were necessary for effective attachment. The maximum adhesive forces of 194 ± 25 μN were observed after a lateral displacement of approximately 5 μm, a distance imperceptibly small relative to the size of the gecko. Direct measurements determined that the adhesion mechanism was not related to suction or friction but was more likely due to intermolecular forces. However, the role of adsorbed water in the adhesion mechanism was not determined. Assuming that the tip of each spatula could be approximated by the curved segment of a sphere of radius 2 μm and that it is separated from the surface by 3 Å, the associated van der Waals force would be 400 nN. This would give a setal force between 40 and 400 μN, which corresponds to the measured value.

Scherge and Gorb [100] used a microfabricated silicon force sensor very similar to an AFM cantilever but without the tip to investigate adhesion of the pads of the great green bush cricket (Fig. 21a). Motion of the force sensor was measured using a fiber-optic interferometer. They found the insect's attachment pads to be flexible in such a way that they could self-adjust to different scales of roughness (Fig. 21b). Additionally, to reduce the effect of capillary action due to adsorbed water, the insect appeared to secrete a hydrophobic layer of long-chain hydrocarbons onto the pads. For this reason it was found to be essential to keep their insects alive for the duration of the experiment in order to fully understand the adhesion mechanisms.

Specifically, the pads themselves have a hemispherical cross section, the upper layer of which is a 180-nm-thick film. The microstructure consists of hexagons separated by hemolymph-filled trenches (Fig. 21c). The inner structure is made up of branches within branches, the finest of which are only 80 nm in diameter, thus permitting the insect to keep a constant contact area regardless of substrate roughness. Due to the flexibility of the pad, the contact area can be controlled. The pads also exhibit viscoelastic properties and have higher stiffness for quick compression. Adhesion remains low below a critical load of about 800 μN, above which the pad and substrate snap into contact. This corresponds well to the distribution of the weight of the insect between its pads. What is less well understood is the insect's detachment mechanism. The authors speculate that detachment could be via a rolling process or by inflating its pad via its heartbeat.

V. SUMMARY

Many years have passed since the early days of AFM, when adhesion was seen as a hindrance, and it is now regarded as a useful parameter for identification of material as well as a key to understanding many important processes in biological function. In this area, the ability of AFM to map spatial variations of adhesion has not yet been fully exploited but in future could prove to be particularly useful. At present, the chemical nature and interaction area of the AFM probe are still rarely characterized to a desirable level. This may be improved dramatically by the use of nanotubes, carbon or otherwise, with functionalized end groups. However, reliance on other measurement techniques, such as transmission electron microscopy and field ion microscopy, will probably be essential in order to fully evaluate the tip–sample systems under investigation.

The volume of work measuring, or relating to, adhesion on the nanoscale appears to be increasing exponentially, and regrettably in this chapter it has only been possible to give the reader a taste of the many excellent papers on the subject. The growth of the field is clearly, in part, due to the advent of atomic force microscopy and new methods for functionalizing and characterizing nanoscale structures. It is also undoubtedly driven by the expanding areas of nano- and biotechnology, which demand a full understanding of interaction processes on the nanoscale. Beyond measurement lies the important possibility of controlling adhesion on the nanoscale, particularly with regard to biological systems, an area that will likely constitute one of the main targets of the field in the coming years.

REFERENCES

1. JN Israelachvili. Intermolecular and surface forces. Academic Press, London, 1992.
2. BV Derjaguin, VM Muller, YP Toporov. J Colloid Interface Sci 53:314–326, 1975.
3. BV Derjaguin, VM Muller, YP Toporov. J Colloid Interface Sci 67:378–379, 1978.
4. BV Derjaguin, VM Muller, YP Toporov. J Colloid Interface Sci 73:293, 1980.
5. D Tabor. J Colloid Interface Sci 58:2–13, 1977.
6. D Tabor. J Colloid Interface Sci 67:380, 1978.
7. D Tabor. J Colloid Interface Sci 73:294, 1980.
8. BV Derjaguin. Zh Fiz Khim 6:1306, 1935.
9. JS McFarlane, D Tabor. Proc Roy Soc Lond A 202:224–243, 1950.
10. H Hertz. J Reine. Angew Math 92:156, 1881.
11. KL Johnson, K Kendall, AD Roberts. Proc Roy Soc Lond A 324:301–313, 1971.
12. VM Muller, VS Yushchenko, BV Derjaguin. J Colloid Interface Sci 77:91–101, 1980.
13. D Maugis. J Colloid Interface Sci 150:243–269, 1992.

14. DS Dugdale. J Mech Phys Solids 8:100–104, 1960.
15. RG Horn, JN Israelachvili, F Pribac. J Colloid Interface Sci 115:480–492, 1987.
16. KL Johnson, JA Greenwood. J Colloid Interface Sci 192:326–333, 1997.
17. LR Fisher, JN Israelachvili. Colloids Surfaces 3:303–319, 1981.
18. J Visser. Surface Colloid Sci 8:3, 1976.
19. HK Christenson. J Colloid Interface Sci 104:234–249, 1985.
20. A Fogden, LR White. J Colloid Interface Sci 138:414–430, 1990.
21. U Landman, WD Luedke, NA Burnham, RJ Colton. Science 248:454–461, 1990.
22. SM Foiles, MI Baskes, MS Daw. Phys Rev B 33:7983–7991, 1986.
23. JB Adams, SM Foiles, WG Wolfer. J Mater Res Soc 4:102, 1989.
24. U Landman, WD Luedtke. J Vac Sci Technol B 9:414–423, 1991.
25. U Landman, WD Luedtke, EM Ringer. Wear 153:3–30, 1992.
26. JP Ryckaert, A Bellmans. Discuss Faraday Soc 66:96, 1978.
27. JA Harrison, CT White, RJ Colton, DW Brenner. Surf Sci 271:57–67, 1992.
28. W Zhang, JR Smith. Phys Rev Lett 82:3105–3107, 1999.
29. SP Jarvis, A Oral, TP Weihs, JB Pethica. Rev Sci Instrum 64:3515–3520, 1993.
30. A Schemmel, HE Gaub. Rev Sci Instrum 70:1313–1317, 1999.
31. TP Weihs, S Hong, JC Bravman, WD Nix. J Mater Res 3:931, 1988.
32. JL Hutter, J Bechhoefer. Rev Sci Instrum 64:1868–1873, 1993.
33. JP Cleveland, S Manne, D Bocek, PK Hansma. Rev Sci Instrum 64:403–405, 1993.
34. JE Sader, I Larson, P Mulvaney, LR White. Rev Sci Instrum 66:3789–3798, 1995.
35. K-C Chang, DA Hammer. Langmuir 12:2271–2282, 1996.
36. HK Christenson. J Colloid Interface Sci 121:170–177, 1988.
37. JK Stuart, V Hlady. Langmuir 11:1368–1374, 1995.
38. K Sugisaki, N Nakagiri, Y Kinjo. Langmuir 15:5093–5097, 1999.
39. AL Weisenhorn, PK Hansma, TR Albrecht, CF Quate. Appl Phys Lett 54:2651–2653, 1989.
40. AL Weisenhorn, P Maivald, H-J Butt, PK Hansma. Phys Rev B 45:11226–11232, 1992.
41. NA Burnham, DD Dominguez, RL Mowery, RJ Colton. Phys Rev Lett 64:1931–1934, 1990.
42. GS Blackman, CM Mate, MR Philpott. Phys Rev Lett 65:2270–2273, 1990.
43. HA Mizes, K-G Loh, RJD Miller, SK Ahuja, EF Grabowski. Appl Phys Lett 59:2901–2903, 1991.
44. TP Weihs, Z Nawaz, SP Jarvis, JB Pethica. Appl Phys Lett 59:3536–3538, 1991.
45. JH Hoh, J-P Revel, PK Hansma. Nanotechnology 2:119–122, 1991.
46. K Feldman, T Tervoort, P Smith, ND Spencer. Langmuir 14:372–378, 1998.
47. RM Ralich, Y Wu, RD Ramsier, PN Henriksen. J Vac Sci Technol A 18:1345–1348, 2000.
48. RW Carpick, N Agraït, DF Ogletree, M Salmeron. Langmuir 12:3334–3340, 1996.
49. T Thundat, X-Y Zheng, GY Chen, RJ Warmack. Surf Sci Lett 294:L939–L943, 1993.
50. M Binggeli, CM Mate. Appl Phys Lett 65:415–417, 1994.
51. SP Jarvis, JB Pethica. Thin Solid Films 273:284–288, 1996.
52. T Eastman, D-M Zhu. Langmuir 12:2859–2862, 1996.
53. JH Hoh, JP Cleveland, CB Prater, J-P Revel, PK Hansma. J Am Chem Soc 114:4917–4918, 1992.
54. JP Cleveland, TE Schäffer, PK Hansma. Phys Rev B 52:R8692–R8695, 1995.
55. SJ O'Shea, ME Welland, JB Pethica. Chem Phys Lett 223:336–340, 1994.
56. SP Jarvis, T Uchihashi, T Ishida, H Tokumoto, Y Nakayama. J Phys Chem B 104:6091–6094, 2000.
57. H Dai, JH Hafner, AG Rinzler, DT Colbert, RE Smalley. Nature 384:147–150, 1996.
58. SS Wong, E Joselevich, AT Woolley, CL Cheung, CM Lieber. Nature 394:52–55, 1998.
59. E-L Florin, VT Moy, HE Gaub. Science 264:415–417, 1994.
60. GU Lee, LA Chrisey, RJ Colton. Science 266:771–773, 1994.
61. U Dammer, M Hegner, D Anselmetti, P Wagner, M Dreier, W Huber, H-J Güntherodt. Biophys J 70:2437–2441, 1996.

62. Y-S Lo, ND Huefner, WS Chan, F Stevens, JM Harris, TP Beebe. Langmuir 15:1373–1382, 1999.
63. Y Harada, M Kuroda, A Ishida. Langmuir 16:708–715, 2000.
64. SP Jarvis, H Yamada, S-I Yamamoto, H Tokumoto, JB Pethica. Nature 384:247–249, 1996.
65. MA Lantz, SP Jarvis, H Tokumoto, T Martynski, T Kusumi, C Nakamura, J Miyake. Chem Phys Lett 315:61–68, 1999.
66. M Grandbois, M Beyer, Matthias Reif, H Clausen-Schaumann, HE Gaub. Science 283:1727–1730, 1999.
67. F Oesterhelt, D Oesterhelt, M Pfeiffer, A Engel, HE Gaub, DJ Müller. Science 288:143–146, 2000.
68. A Kumar, GM Whitesides. Appl Phys Lett 63:2002–2004, 1993.
69. T Nakagawa, K Ogawa, T Kurumizawa, S Ozaki. Jpn J Appl Phys 32:L294–L296, 1993.
70. CD Frisbie, LF Rozsnyai, A Noy, MS Wrighton, CM Lieber. Science 265:2071–2074, 1994.
71. A Noy, CH Sanders, DV Vezenov, SS Wong, CM Lieber. Langmuir 14:1508–1511, 1998.
72. RC Thomas, JE Houston, RM Crooks, T Kim, TA Michalske. J Am Chem Soc 117:3830–3834, 1995.
73. EW Van der Vegte, G Hadziioannou. Langmuir 13:4357–4368, 1997.
74. Y-L Ong, A Razatos, G Georgiou, MM Sharma. Langmuir 15:2719–2725, 1999.
75. S Kidoaki, T Matsuda. Langmuir 15:7639–7646, 1999.
76. DD Koleske, WR Barger, GU Lee, RJ Colton. Mat Res Soc Symp 464:377–383, 1997.
77. R McKendry, M-E Theoclitou, T Rayment, C Abell. Nature 391:566–568, 1998.
78. FJ Giessibl, S Hembacher, H Bielefeldt, J Mannhart. Science, 289:422–425, 2000.
79. G Cross, A Schirmeisen, A Stalder, P Grütter, M Tschudy, U Dürig. Phys Rev Lett 80:4685–4688, 1998.
80. A Schirmeisen, G Cross, A Stalder, P Grütter, U Dürig. Appl Surf Sci 157:274–279, 2000.
81. WA Ducker, TJ Senden, RM Pashley. Nature 353:239–241, 1991.
82. WA Ducker, TJ Senden, RM Pashley. Langmuir 8:1831–1836, 1992.
83. MW Rutland, TJ Senden. Langmuir 9:412–418, 1993.
84. M Preuss, H-J Butt. Langmuir 14:3164–3174, 1998.
85. M Fuji, K Machida, T Takei, T. Watanabe, M Chikazawa. J Phys Chem B 102:8782–8787, 1998.
86. G Sagvolden, I Giaever, J Feder. Langmuir 14:5984–5987, 1998.
87. L-O Heim, J Blum, M Preuss, H-J Butt. Phys Rev Lett 83:3328–3331, 1999.
88. IU Vakarelski, K Ishimra, K Higashitani. J Colloid Interface Sci 227:111–118, 2000.
89. K Higashitani, K Ishimura. J Chem Eng Jpn 30:52–58, 1997.
90. D Tabor, RHS Winterton. Nature 219:1120–1121, 1968.
91. PF Luckham, BA de L Costello. Adv Colloid Int Sci 44:183–240, 1993.
92. RG Horn, DR Clarke, MT Clarkson. J Mater Res 3:413, 1988.
93. RG Horn, DT Smith. J Non-Crystall Solids 120:72–81, 1990.
94. Ya I Rabinovich, BV Derjaguin, NV Churaev. Adv Colloid Int Sci 16:63–78, 1982.
95. V Marchi-Artzner, J-M Lehn, T Kuntake. Langmuir 14:6470–6478, 1998.
96. C Fredriksson, S Kihlman, M Rodahl, B Kasemo. Langmuir 14:248–251, 1998.
97. U Dammer, O Popescu, P Wagner, D Anselmetti, HJ Güntherodt, GN Misevic. Science 267:1173–1175, 1995.
98. BW Chui, TW Kenny, HJ Mamin, BD Terrisand D Rugar. Appl Phys Lett 72:1388–1390, 1998.
99. K Autumn, YA Liang, ST Hsieh, W Zesch, WP Chan, TW Kenny, R Fearing, RJ Full. Nature 405:681–684, 2000.
100. M Scherge, SN Gorb. J Micromech Microeng 10:359–364, 2000.

3

Langmuir Monolayers: Fundamentals and Relevance to Nanotechnology

KEITH J. STINE University of Missouri–St. Louis, St. Louis, Missouri

BRIAN G. MOORE Penn State Erie–The Behrend College, Erie, Pennsylvania

I. INTRODUCTION TO LANGMUIR MONOLAYERS AND LANGMUIR–BLODGETT FILMS

The manipulation of single-molecule-thick films at the water–air interface and their transfer onto solid substrates are among the earliest efforts setting the foundation for nanotechnology. The investigation of Langmuir–Blodgett (LB) films dates back to the work of Langmuir and Blodgett at General Electric [1–3]. A renewed interest in LB films began with the work on energy transfer between dye molecules incorporated in fatty acid LB films by Kuhn and coworkers [4] and has continued unabated ever since. The range of materials studied at one time or another in LB films encompasses most of chemistry.

This chapter will begin with a brief introduction to the preparation of LB films and a survey of the methods for their study. Nearly 7000 papers have been published on LB films since 1990. Research on LB films has been summarized in the books by Roberts [5], Tredgold [6], and Petty [7]. The book by Gaines [8] remains an excellent introduction to the field of Langmuir films, as is the more recent text by MacRitchie [9]. The main portion of this chapter will focus on the application of LB film techniques to some recent topics associated with nanotechnology, namely, films of nanoparticles and of fullerene materials. Recent efforts to produce LB films with patterning within the plane of the film will also be surveyed. The application of computer simulations, especially those at the molecular level, to understanding the organization of the monolayers from which LB films are formed will be reviewed.

A. Basic Features of Langmuir–Blodgett Film Formation and Study

The first step in the preparation of an LB film is the successful spreading of a monolayer of the material of interest, which may be molecular, polymeric, or particulate and need not be amphiphilic in the traditional sense. This is accomplished by depositing drops of a dilute solution of the material in an appropriate "spreading solvent" onto the water surface. The concentration is generally millimolar or less, and the solvent selected should be one that will spread across the surface rapidly and evaporate without remaining at the surface or dissolving into the subphase. Common spreading solvents include chloroform, benzene, and hexane, sometimes with a small fraction of ethanol or methanol added to the latter two to alter the polarity. For vertical LB transfer, the monolayer must be able to be compressed to a surface pressure (Π) that remains stable during transfer to the substrate.

A schematic of a LB trough is shown in Figure 1. The main components are the Teflon trough, barrier(s) that can be used to smoothly compress the monolayer at a controlled rate, a surface pressure transducer, and the dipper capable of raising and lowering a substrate smoothly at a controlled rate. The simplest LB troughs have one barrier, with a dipping well at one end of the trough, while others have two opposing barriers for symmetrical compression of the monolayer and the dipping well in the middle of the trough. Most LB troughs allow for temperature control of the subphase. The area of the water surface is important; larger surface areas are needed for deposition of LB films of many layers. Typical LB trough areas are in the 200–1500-cm^2 range. The trough does not need to be more than 2–3 mm deep. The dipping well must be deep enough and wide enough to accommodate the desired substrate. Smaller-area troughs are desirable for scarce biological samples. Also available are "alternate-layer" LB troughs, with two separate subphase compartments allowing for transfer of layers of two different materials in controlled sequences.

The majority of commercial LB troughs use the Wilhelmy plate method for measurement of surface pressure (Π), although some use the alternate Langmuir float method. The plate material most commonly used is cut pieces of filter paper, of negligible cost and completely wetted by water. The other type of plate used is a piece of high-purity platinum metal, which can be cleaned in a flame and gives a reproducible contact angle with water of 60°.

Langmuir–Blodgett films have been deposited on many different substrates. The substrates used include different types of glass (such as quartz for UV-visible spectroscopy); CaF$_2$ plates for transmission infrared spectroscopy; silicon, germanium, and ZnSe plates for internal reflection infrared spectroscopy. For electrochemical applications, LB films

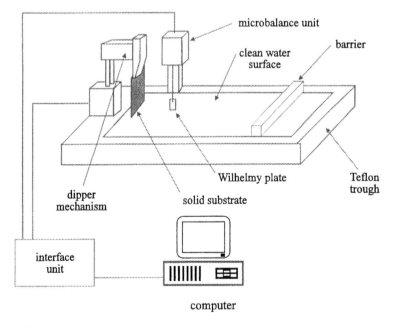

FIG. 1   Major components of a Langmuir–Blodgett trough.

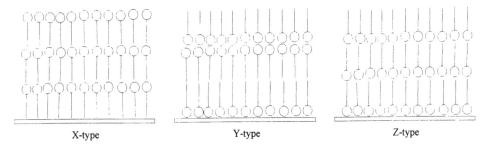

FIG. 2   Common modes of Langmuir–Blodgett film transfer: X-type, Y-type, and Z-type.

have been transferred to evaporated metal films, highly oriented pyrolytic graphite (HOPG), glassy carbon, glass coated with a transparent layer of indium-tin oxide (ITO glass), semiconductor, and metal surfaces. For examination by scanning probe microscopy, freshly cleaved mica and HOPG have been frequently used because of their flatness at the molecular level.

The deposition of the LB film can be conducted in a vertical or horizontal (Langmuir–Schaefer) fashion. In the traditional vertical dipping mode, the substrate is immersed (lowered) and emersed (raised) through the interface. The transfer of a single layer onto a hydrophilic substrate is often performed by immersing the substrate prior to spreading of the monolayer. The film is then compressed to the desired target surface pressure, and the substrate is drawn up through the interface while the trough feedback system adjusts the barrier position to maintain a constant surface pressure. If multilayer deposition is desired, then immersion followed by emersion is repeated for the desired number of cycles. A pause is needed before reimmersion of the substrate to allow for evaporation of the entrained water. If a layer is transferred on every upward and downward pass through the interface, then the transfer is known as Y-type, the most common. It can also be the case that a layer is transferred only on the downward passes (X-type) or only on the upward passes (Z-type). The LB films resulting from these three types of transfer are depicted in Figure 2. The character of the transfer for each pass is described by the *transfer ratio,* defined as the decrease in area of the monolayer divided by the area of the substrate. A transfer ratio near unity nominally indicates uniform transfer. Transfer ratios below unity indicate partial transfer, and those much greater than unity indicate significant monolayer collapse, structural relaxation, or dissolution during the transfer process. Dipping speeds of ~0.1 to several millimeters per second are commonly used. In the horizontal mode of transfer, the substrate is held parallel to the water surface and lowered until it touches the compressed monolayer; it is then gently lifted up. Horizontal transfer has been applied to films that cannot maintain a constant surface pressure or are unsuitable for vertical dipping.

B.   Langmuir Monolayers and Langmuir–Blodgett Films of Fatty Acids

Traditional amphiphiles contain a hydrophilic head group and the hydrophobic hydrocarbon chain(s). The molecules are spread at molecular areas greater (~2–10 times) than that to which they will be compressed. The record of surface pressure (Π) versus molecular area (A) at constant temperature as the barrier is moved forward to compress the monolayer is known as an *isotherm*, which is analogous to P-V isotherms for bulk substances. Π-A isotherm data provide information on the molecular packing, the monolayer stability as determined by the collapse pressure (Π_c), the occurrence of two-dimensional phase transitions, and the compressibility of the different phases [8]. Π-A data as a function of temperature allow mapping of phase boundaries and determination of transition enthalpies for first-order phase transitions [8–10]. Plateaus, usually not precisely flat, in Π-A curves of single-component monolayers are indicative of first-order phase transitions, with two-phase coexistence occurring over the range of molecular area demarked by the plateau. Other phase transitions, especially those between more ordered phases at higher values of Π, appear as changes in slope and may be either first order or second order [10–12]. The "limiting molecular area" of a monolayer is estimated by linear extrapolation of the steepest portion of the isotherm to zero surface pressure, representing a close packing of molecules in the same structural arrangement at zero surface pressure. Such extrapolated values provide a comparison between different monolayers and different conditions.

The most common two-dimensional phases in monolayers are the gaseous, liquid-expanded, liquid-condensed, and solid phases. A schematic Π-A isotherm is shown in Figure 3 for a fatty acid for the phase sequence: gas (G) → G + liquid-expanded (LE) → LE →

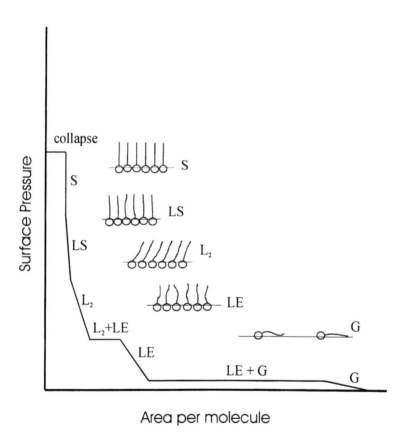

FIG. 3 An isotherm is depicted for a Langmuir monolayer of an amphiphile showing the Π-A variation for the phase sequence; gas (G) → G + liquid-expanded (LE) → LE → LE + tilted condensed phase (L_2) → L_2 → vertical condensed phase (LS) → S (solid). Schematic depictions of the molecular organization in the phases are shown above the isotherm.

LE + L_2 → L_2 → LS → S (solid), followed by collapse. The L_2 and LS phases are both examples of condensed phases [11,12]. The LS phase is a liquid-crystalline phase with vertical molecular orientation and has a lower surface viscosity than the L_2 phase, in which the molecules are tilted away from vertical; hence the origin of the term *superliquid* [13]. The drop in surface viscosity is attributed to the greater ease of molecules sliding past each other in the LS phase. The molecules in the gas phase are several to tens of angstroms apart and weakly interacting, behaving much like a two-dimensional gas. The hydrocarbon tails are lying flat on the water in the gas phase. At very low surface pressures, a long plateau is shown that is a coexistence region between the gas phase and the liquid-expanded (LE) phase [14]. The LE phase is a two-dimensional liquid, the hydrocarbon chains are off the water surface and have a significant degree of conformational freedom, and there is only short-range correlation in the positions of the head groups. The coexistence region between the G and LE phases indicates that this is a first-order phase transition, similar to the coexistence of bulk gas and liquid phases below the critical point [10,14,15]. Studies of the LE-G coexistence region are difficult due to the low surface pressures and the very stringent requirements on cleanliness required for meaningful isotherms [14]. It has been shown that the width of the LE-G plateau in the Π-*A* isotherms decreases with temperature [15]. The surface pressure of the plateau increases with temperature, but it remains low and the plateau narrows as the expected critical point is approached. It has been confirmed using fluorescence microscopy that the thermodynamic lever rule governs the area fractions of the G and LE across the plateau region near room temperature [15]. Compression to the point where a significant rise in Π is first observed in this particular isotherm indicates that the monolayer is now in the LE phase. The features present on the surface in the LE-G region can be imaged by fluorescence or Brewster-angle microscopy (BAM). The features observed include "droplets," "gas bubbles," and foamlike patterns, with polygonal cells [15,16] that can undergo complex structural instabilities on compression or heating [17] and that can coarsen over time [18,19].

Further compression along the isotherm shown first results in conversion of the monolayer to the LE phase. The monolayer is then in a one-phase region and is homogeneous across the water surface. The surface pressure rises until the second plateau is encountered; at this point in the isotherm the film enters the coexistence region with the L_2 phase, one of the liquid-condensed (LC) phases. As the film is compressed across the plateau, the area fraction of the L_2 phase increases in accordance with the thermodynamic lever rule [15]. On exiting the plateau, which may be defined as occurring at the point where the slopes of the immediate pre- and postplateau regions intersect, the film will be in the L_2 phase. The L_2 phase is a two-dimensional liquid-crystalline phase with bond orientational order [27] in the positions of the head groups. The hydrocarbon chains are tilted at a consistent angle and are tilted toward their nearest neighbors. Polarized fluorescence microscopy [20–24] and BAM [25,26] reveal micron-sized regions in which molecules have the same azimuthal tilt direction, as well as many fascinating details concerning the domain morphologies. The L_2 phase is analogous to the smectic-I phase of liquid crystals [11,27]. Further compression gives the LS phase, analogous to the smectic-B (hexatic) liquid-crystal phase. In this phase, the molecules are oriented vertically, the bond orientational order remains, and the positional ordering is short range. Transition to the solid phase (S) gives positional ordering over larger but finite distances, analogous to the smectic-E phase.

Monolayer phase sequences are temperature dependent. For chain lengths greater than that of pentadecanoic acid, the LE phase will not be observed at room temperature. For chain lengths shorter than myristic acid (C14), condensed phases will not be observed at room temperature, and loss of molecules due to dissolution into the subphase will be significant. At high molecular areas and very low surface pressures, the longer-chain fatty acids exhibit coexistence of the gas phase and a condensed phase [15]. This is in contradiction to the often-cited description of the isotherm region of very low pressure as being solely the gas phase. Visualization by BAM clearly shows floating islands of condensed phase separated from each other by the gas phase. The point at which the surface pressure first rises is that at which these domains contact each other. Further transitions to different condensed phases then occur along the rising isotherm. The initial rising portion of the isotherm of stearic acid on an acidified subphase has often been erroneously assigned as the disordered liquid phase, whereas it is actually the L_2 phase.

The structure of the two-dimensional liquid-condensed and solid phases of fatty acids, esters, and alcohols have come be understood as related to the phases of smectic liquid crystals [11,27]. There are other condensed and solidlike phases beyond those already mentioned, sometimes described by different workers using different nomenclature. The ordering of the phases can be described using the same language of "order parameters" used to described smectic liquid crystals, such as positional order, tilt order, bond orientational order, tilt-bond coupling, and herringbone order. The description draws upon Π-A isotherm data, synchrotron x-ray diffraction studies of monolayers at the water–air interface, and observation of domain patterns and orientational patterns using polarized fluorescence microscopy and BAM. The earliest discussion of the possible structures of these phases was

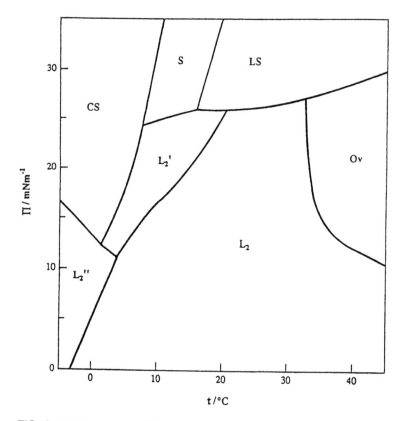

FIG. 4 Surface pressure (Π)–temperature (T) phase diagram for heneicosanoic acid. (Reproduced with permission from Ref. 31a. Copyright 1997 American Institute of Physics.)

made solely based on π-A data by Stenhagen and Lundquist [28–31]. The relation between these monolayer phases and the liquid-crystal phases has been described by Knobler and Desai [11]. A phase diagram for heneicosanoic acid is shown in Figure 4; a phase not discerned from the isotherms was discovered by optical microscopy, the Ov phase, which is similar to the smectic-L phase and is a tilted phase [32]. The phase sequences are dependent on pH and the presence of multivalent subphase cations. The addition of divalent cations to the subphase results in a shift of the isotherm toward greater molecular area and generally a loss of the liquid-crystalline phases [33]. A change of pH from 5.5 to 6.5 removes the L_2 phase of arachidic acid [34].

II. METHODS OF STUDY

A. Classical Methods

The primary classical method of study of Langmuir monolayers is clearly that of recording Π-A isotherms. Another classical method applied to the study of Langmuir monolayers is the measurement of surface potential [8,9], which is sensitive to changes in the orientation and density of the molecular dipoles of the monolayer. In addition, surface potential fluctuations were clearly observed in the coexistence region of palmitic acid [35].

The surface viscosity varies significantly along the isotherm and across monolayer phase boundaries. Addition of subphase metal ions increases the surface viscosity drastically, as was recently reinvestigated [36]. Recently, microscopy methods have been used to image velocity profiles of different monolayer phases flowing through a narrow channel, such as used in the canal viscometer [37]. The two main methods used to study monolayer viscosity are the canal viscometer and the oscillating disc method [8,9].

B. Microscopy Methods

Langmuir monolayers have a complex and widely variable domain structure. Over the past two decades, much insight into film structure and phase behavior has been gained by observation of these domains using two optical microscopy methods. The first method employed was fluorescence microscopy [38–40], in which a very small amount of a surfactant covalently bonded to a fluorescent group is spread along with the monolayer. The amount incorporated is generally 0.1–2.0 mol%. An appropriately chosen probe partitions itself such that it is excluded from condensed-phase regions and is preferentially soluble in the LE phase. In this case, the LE phase will appear bright, and the condensed-phase regions will appear dark. Similarly, the LE phase will appear brighter than the gas phase due to its higher surface density. Within condensed phases with tilt order, it is possible to observe regions of different tilt orientation using excitation by polarized light, provided the probe is oriented by the surrounding amphiphiles and is present in the condensed phase [20–24,38]. The method has been applied to monolayers of fatty acids, phospholipids, and related compounds. An image of a pentadecanoic acid monolayer in the LE + G coexistence region is shown in Figure 5a, and an image in the LE + liquid-condensed phase (L_2 in this case) coexistence region is shown in Figure 5b.

Brewster-angle microscopy dispensed with the need for a probe molecule [41,42]. Brewster-angle microscopy is based solely on the reflectivity properties of p-polarized light. The reflectance of p-polarized light at the water–air interface vanishes at 53.1° (using refractive indices at 20°C) if the interface is perfectly sharp; in reality there is a deep minimum near zero. The presence of a monolayer gives an intervening layer of different re-

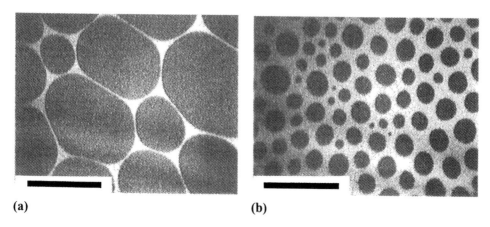

(a)   (b)

FIG. 5 Fluorescence micrographs of a monolayer of pentadecanoic acid at 22°C: (a) at 0.80 nm^2 molecule^{-1} in the coexistence region between the liquid-expanded (LE) and the gas (G) phases. The LE phase forms the bright walls of the foamlike pattern, and the darker, interior regions are the G phase; (b) at 0.29 nm^2 molecule^{-1} in the coexistence region between the LE phase and the liquid-condensed phase ("LC" or LS). The LC domains are the dark, round regions. The subphase is milli-Q water acidified to pH 1.8 with HCl. The fluorescent probe (NBD-hexadecylamine) is present at a mole fraction of 0.005. The scale bar in the lower left of each image is 200 μm.

fractive index, and light will be reflected. The reflected light is conveniently collected with an objective and focused onto a charge-coupled device (CCD) camera. The reflected light intensity is sensitive to the density and refractive index of the monolayer, and the polarization is sensitive to any in-plane anisotropy in the refractive index. The method can be applied to systems in which the behavior of a probe molecule would be problematic, such as spread polymers, spread films of particles, and adsorbed films. An image of a pentadecanoic acid monolayer in the LE + condensed-phase (L$_2$) coexistence region is shown in

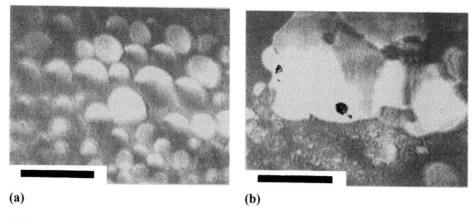

(a)   (b)

FIG. 6 (a) Brewster-angle microscopy (BAM) image of a monolayer of pentadecanoic acid at 22°C, 0.25 nm^2 molecule^{-1} in the coexistence region between the liquid-expanded and the liquid-condensed (L$_2$) phases (b) BAM image of stearic acid at 22°C, 0.60 nm^2 molecule^{-1} in the coexistence region between the gas (G) and the liquid-condensed (L$_2$) phases. In each of these images, the polarizer angle has been set to 60°. The subphase is milli-Q water acidified to pH 1.8 with HCl. The scale bar in the lower left of each image is 450 μm.

Figure 6a. In Figure 6b, an image of a stearic acid monolayer in the L_2 + G coexistence region is shown, clearly indicating the presence of ordered islands of molecules even at very low surface pressure. The sections of different contrast represent regions of different molecular tilt orientation. Both fluorescence microscopy [44] and BAM [45] can be applied to LB monolayer films on glass substrates and have been used to examine the LB transfer process. For multilayer LB films, polarized optical microscopy can provide images of defects such as disclinations and other orientational patterns [46].

Scanning force microscopy has been widely applied to the study of the molecular-scale structure of LB films. Molecular-scale images of the two-dimensional lattices of a range of LB films, including fatty acid salts and phospholipids, have been resolved [47–53]. An image of a three-layer cadmium arachidate LB film is shown in Figure 7a [54]. High-resolution images of the molecular lattices are shown, as well as Fourier transforms that show distinct spots in reciprocal space. A lower-resolution image shows that the three-layer LB film reorganizes on exposure to water, yielding one-, three-, and five-layer regions. Figure 7b shows an image of three-layer LB films that are a mixture of cadmium arachidate and arachidic acid, having been prepared at pH 5.0, where the fatty acid head groups are partially ionized [34]. The films have been subjected to a process known as *skeletonization*—immersion in benzene to remove the more soluble arachidic acid.

More recently, the method of scanning near-field optical microscopy (SNOM) has been applied to LB films of phospholipids and has revealed submicron-domain structures [55–59]. The method involves scanning a fiber-optic tip over a surface in much the same way an AFM tip is scanned over a surface. In principle, other optical experiments could be combined with the SNOM, such as resonance energy transfer, time-resolved fluorescence, and surface plasmon resonance. It is likely that spectroscopic investigation of submicron domains in LB films using these principles will be pursued extensively.

Microscopy methods based on nonlinear optical phenomena that provide chemical information are a recent development. Infrared sum-frequency microscopy has been demonstrated for LB films of arachidic acid, allowing for surface-specific imaging of the lateral distribution of a selected vibrational mode, the asymmetric methyl stretch [60]. The method is sensitive to the surface distribution of the functional group as well as to lateral variations in the group environmental and conformation. Second-harmonic generation (SHG) microscopy has also been demonstrated for both spread monolayers and LB films of dye molecules [61,62]. The method images the molecular density and orientation field with optical resolution, and local quantitative information can be extracted.

C. Other Methods

Langmuir–Blodgett films can be studied by the full range of optical spectroscopic methods. Kuhn and coworkers dispersed donor and acceptor molecules in fatty acid LB films and studied resonance energy transfer using fluorescence spectroscopy [4]. External-reflection infrared spectroscopy can be used to determine molecular orientation in LB films [63]. While nonresonance Raman scattering is too weak for directly recording the spectrum of a monolayer at the water–air interface, surface-enhanced Raman scattering (SERS) can be achieved for LB layers on SERS-active substrates [64]. Growing silver nanoparticles beneath the head groups of a fatty acid monolayer enabled the recording of SERS spectra for the surfactant film [65]. Resonance Raman spectroscopy has been used to study monolayers of the amphiphilic dye cetyl orange at the water surface [66].

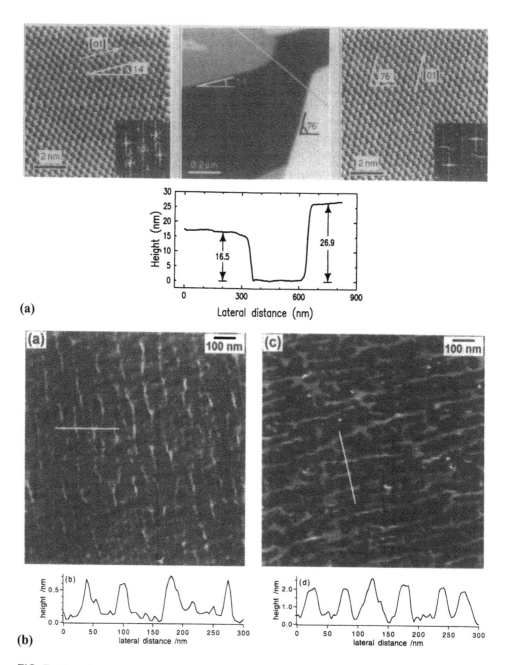

FIG. 7 Scanning probe microscopy images of Langmuir–Blodgett (LB) films. (a) LB film of cadmium arachidate transferred onto mica at 30.0 ± 0.1 mN m^{-1}. The center image is of a 1-μm $\times$ 1-μm scan of a five-layer film aged under water for 25 hours and shows parts of three-bilayer (left of image) and 5-bilayer (right of image) regions. The image on the left is a high-resolution scan showing the orientation of the molecular lattice on the three-bilayer-high island on the left of the center image. The image on the right is a high-resolution scan showing the molecular lattice on the five-bilayer-high island on the right of the central image. (Reprinted with permission from Ref. 54. Copyright 1992 American Chemical Society.) (b) Cadmium arachidate/arachidic acid three-layer LB film deposited onto mica at pH 5.0 before (a, left) and after (c, right) skeletonization by exposure to benzene. Height profiles (b, d) along the straight lines drawn in the images are shown beneath. (Reprinted with permission from Ref. 34. Copyright 1996 American Chemical Society.)

X-ray diffraction has been applied to spread monolayers as reviewed by Dutta [67] and Als-Nielsen et al. [68]. The structure of heneicosanoic acid on Cu^{2+} and Ca^{2+} containing subphases as a function of pH has been reported [69], as well as a detailed study of the ordered phases of behenic acid [70], along with many other studies. Langmuir–Blodgett films have also been studied by x-ray diffraction. Some recent studies include LB film structure just after transfer [71], variations in the structure of cadmium stearate LB films with temperature [72], and characterization of the structure of cadmium arachidate LB films [73]. X-ray [74,75] and neutron reflectivity [76,77] data on LB films can be used to model the density profile normal to the interface and to obtain values of layer thickness and roughness.

III. NANOPARTICLES AND LANGMUIR–BLODGETT FILMS

A. Introduction

Nanoparticles, or "quantum dots," in the ~2–10-nm size range represent an intermediate state of matter, with tunable electronic and optical properties between those of molecules or small clusters and those of the bulk substance [78–80]. A potential application of nanoparticles is the construction of devices based on single-electron phenomena [81–84]. A quantum dot is placed between the source and the drain of a transistor device. Charge quantization effects can be observed for quantum dots when the tunneling resistance is high enough and the capacitance is low enough. In the "Coulomb staircase" effect, the current versus source–drain voltage (at a fixed gate voltage) will show steplike variations as individual electrons tunnel onto the quantum dot. Sweeping the gate voltage at a fixed source–drain voltage will result in conductance spikes as individual electrons tunnel onto the quantum dot; the spikes are referred to as *Coulomb oscillations*. The current steps and conductance spikes are determined by the quantized energy levels of the nanoparticle. For example, a single-electron transistor (SET) based on a CdSe nanoparticle bound between two gold leads on a Si substrate exhibits Coulomb effects associated with the particle bound inside the ~5-nm gap between the electrodes [85]. Applications in the development of nanoelectronic devices and quantum computing based on single-electron phenomena are being pursued. The LB film method facilitates the study of properties of thin films of nanoparticles, but does not provide the ability to place quantum dots onto surfaces in a controlled arrangement. It is possible to use the LB film transfer method to create a "quantum-well" structure, such as recently achieved using a layer of CdSe nanocrystals. Quantum-well structures have been constructed from LB films of pthalocyanines and other organic molecules [86]. Controlled placement of nanoparticles onto surfaces is currently being pursued using lithography to create patterns of functional groups capable of selectively binding the nanoparticles [87]. Single-electron oxidation/reduction has been observed using cyclic voltammetry in self-assembled monolayers (SAMs) of gold nanoparticles modified by organic ligands [88].

Generation of nanoparticles under Langmuir monolayers and within LB films arose from earlier efforts to form nanoparticles within reverse micelles, microemulsions, and vesicles [89]. Semiconductor nanoparticles formed in surfactant media have been explored as photocatalytic systems [90]. One motivation for placing nanoparticles within the organic matrix of a LB film is to construct a superlattice of nanoparticles such that the optical properties of the nanoparticles associated with quantum confinement are preserved. If monolayers of capped nanoparticles are transferred, a nanoparticle superlattice can be constructed. Methods for the construction of LB films containing nanoparticles include transfer of monolayers of capped particles, *in-situ* growth of nanoparticles under monolayers, the reduction of LB films of fatty acid metal salts, and binding of nanoparticles to LB films [91–95].

B. Langmuir–Blodgett Films Containing Metallic Nanoparticles

1. Formation of Metallic Nanoparticles Beneath Langmuir Monolayers

Efforts to form metallic particles beneath monolayers were initiated by Fendler (96). Silver-particulate films were formed by electrocrystallization at the water–air interface. Monolayers of a dialkylphosphate surfactant were spread in a dish containing a solution of silver nitrate. A silver electrode was immersed in the subphase, and a platinum electrode was floated on the water surface. A potential of 1.8–1.9 V was applied. Observation of the water surface showed a film of silver particles starting at the Pt cathode and growing concentrically, covering the water surface at a rate of 1–2 cm^2 hour^{-1}. Silver-particle films also formed under arachidate monolayers and those of another dialkylphosphate, but not under those of cationic surfactants. No silver-particle film formation was observed on bare water. Scanning tunnel microscopy on horizontally transferred films showed interconnected silver spheroids with 20-nm short axes and 30-nm long axes and 10–20 nm in height. A monolayer of surfactant-capped silver particles, prepared by photoreduction of silver nitrate in reverse micelles, adsorbed onto highly oriented pyrolytic graphite (HOPG) slowly removed from the solution (97).

The electrocrystallization of silver beneath monolayers of 13 different surfactants was studied [98]. The monolayers were spread in a trough containing an saturated calomel electrode (SCE) reference electrode, Ag wire counterelectrode, and a silver (or Pt/Ir) working electrode brought into contact with the surface. The spread monolayer was compressed to the desired area prior to application of the reducing potential. The application of a potential below −150 mV vs. SCE resulted in the formation of shiny silver-particle films. Silver-particle film formation was observed under arachidic acid monolayers, two long-chain dialkylphosphates, and three phospholipids. Transmission electron microscopy (TEM) observation of horizontally transferred films showed interconnected 40–90-nm-size Ag particles, as seen in Figure 8. The resistivity in the plane of the film was 5–20 Ω cm^{-1}. The mass of silver reduced during a given period of film growth was determined coulometrically from the total charge passed. As the silver-particle films grew radially outward from the working electrode, it was possible to estimate a film thickness of 60 nm using the mass of the reduced silver, the film radius, and the density of silver.

It was later demonstrated that silver-particle films could be formed beneath monolayers using a reducing agent [99]. Monolayers of four different surfactants were examined as templates for silver-particle film formation, the most successful being *N,N′*-bis(2-aminoethyl)-2-hexadecyl-1,3-propanediamide . The monolayers were spread on silver nitrate subphases in a circular trough covered by a jar with a dish of formaldehyde placed next to the trough. The formaldehyde vapor caused slow growth of silver-particle films. *In-situ* optical reflectivity was used to monitor the film growth and indicated film thickness in the range 8–50 nm. The films could not be transferred by vertical dipping; however, horizontal transfer was successful. The optical micrographs of the films growing beneath monolayers of dihexadecyl phosphate for 8 hours showed separated fractal treelike aggregates. After longer periods, the growth patterns gradually took a dendritic appearance. The ab-

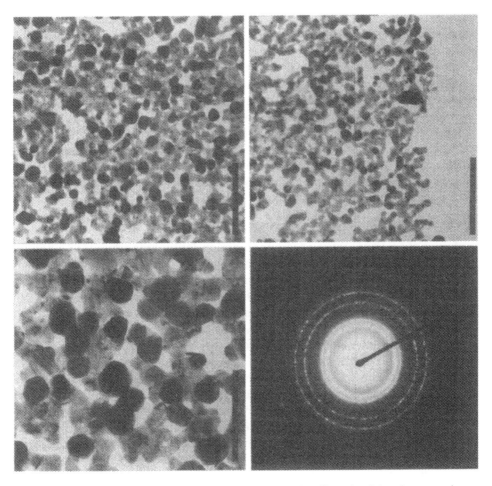

FIG. 8 TEM images of silver particulate film formed under dihexadecylphosphate monolayers (0.45 nm^2 molecule^{-1}, 51 mN m^{-1}) by electrocrystallization on a 0.01 M silver nitrate subphase at −0.200 V vs SCE. The top left shows a representative image from the middle of the film; the top right shows a representative image from the edge of the film. The bottom left shows an image at higher magnification. The bottom right shows the diffraction pattern. (Reprinted with permission from Ref. 98. Copyright 1996 American Chemical Society.)

sorbance spectra of these films transferred onto quartz after 12 hours showed a peak at 270 nm and a growing plasmon absorbance at 440 nm that became dominant for films exposed for up to 21 days. Silver films grown beneath monolayers of the propanediamide surfactant looked different. After 12 hours of exposure, well-separated circular islands were seen that subsequently enlarged into irregularly shaped islands. The silver ion reduction was faster under these monolayers. The absorbance spectra after 12 hours showed maxima at 270 nm and 380 nm, associated with nonmetallic Ag$_4^{2+}$ clusters and with 5-nm-diameter silver particles. After one week, "interband" transitions, arising from contact of silver particles, were observed. The lateral resistivity of the films was dependent upon the thickness and connectivity of the silver clusters and islands.

Monolayers of octadecylmercaptan, dioctadecyldithioethylammonium bromide, and dioctadecyldimethylammonium bromide were used as templates for the growth of gold nanoparticles [100]. The monolayers were spread on 0.5 mM HAuCl$_4$. Reduction by carbon monoxide or photoreduction initiated by light from a 150-W xenon lamp resulted in gold particle formation. After exposure to CO for 30 minutes, a color change at the water surface was observed. Absorbance maxima from 564 to 580 nm were recorded for the three systems for films horizontally transferred onto quartz slides. Gold particles grown under the mercaptan had a range of sizes and often appeared as coalesced aggregates. Gold particles grown under the dithioethylammonium head groups were numerous and smaller. Gold particles grown under dioctadecyldimethylammonium bromide showed some highly regular shapes, including triangular, isohedral, and decahedral. The nucleation density and regularity of the particle growth was sensitive to the environment around the surfactant head groups. It proved difficult to fit the plasmon absorption bands to the Mie theory, due to the range of sizes and morphologies.

2. Binding of Metal Nanoparticles to Langmuir Monolayers

Production of LB films containing metallic nanoparticles has been achieved by exploiting the electrostatic attraction of charged nanoparticles in the subphase to the oppositely charged head groups of surfactant monolayers [101]. Gold colloids prepared by reduction of chloroauric acid (HAuCl$_4$) by sodium citrate (12.6 $\pm$ 2.9 nm) were capped with carboxythiophenol. The hydrosols of capped, anionic gold particles were used as subphases for octadecylamine monolayers. The subphase was adjusted to pH 8 so that the carboxyl groups on the particles would be fully ionized and the amine head groups fully protonated. The isotherm of the spread monolayer was observed to expand with time (shift toward higher area). The substrates were made hydrophobic by depositing one monolayer of lead arachidate. AT-cut quartz substrates were used for quartz crystal microbalance (QCM) experiments. Mass uptake per LB transfer cycle was linear for the eight cycles studied. A 15-minute drying time was required between each cycle. The mass uptake corresponded to a surface coverage of 25% Au particles per layer. The plasmon absorption of the Au particles near 650 nm was clearly seen in the LB films and was stable for many weeks. In a follow-up study, data at pH 8, pH 9, and pH 11.5 were compared [102]. The mass uptake versus number of transfers was linear at all pH values, with the slope larger at pH 8 than at pH 9. At pH 11.5, the slope was consistent with transfer of only the octadecylamine monolayer. The UV absorbance at 650 nm versus number of layers was also linear; films transferred at pH 11.5 show no absorbance at 650 nm, the wavelength of the particle plasmon absorbance.

The general method was also applied to the assembly of LB films containing capped silver nanoparticles [103]. Silver nanoparticles (0.73 $\pm$ 0.12 nm) were prepared by the reduction of silver sulfate by borohydride followed by capping with carboxythiophenol. The capping shifted the plasmon absorption from 386 nm to 405 nm in the hydrosol. The hydrosol was used as the subphase for octadecylamine monolayers at pH 9. The Π-A isotherm shifted from a limiting area of 0.20 nm^2 on water to 0.45 nm^2 on the hydrosol subphase. Using AT-quartz resonators as substrates, the mass increase per layer was determined up to 20 layers (Fig. 9a). The slope was greater at pH 9 than at pH 12, where deprotonation of the ammonium head groups resulted in loss of particle binding. Optical absorbance and interferometry were also used to follow the stepwise buildup of the particle films (Fig. 9b). A particle surface coverage of 30% per layer was obtained. The surface coverage could also be controlled using mixed monolayers of octadecylamine and octadecanol [104]. For 9:1 octadecanol:octadecylamine films, the surface coverage of colloidal particles was reduced to 19.5%. The equilibration time for the surface cluster density at 25 mN m^{-1} was reported as near 15 hours. The silver particles could also be coated with a bilayer of lauric acid, and

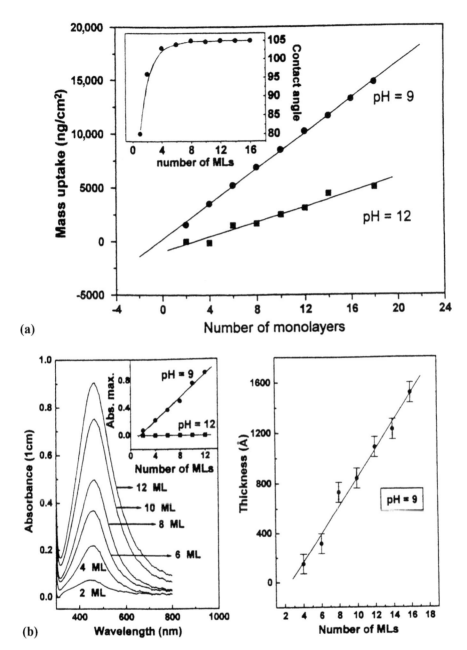

FIG. 9 Silver nanoparticles capped by 4-carboxythiophenol electrostatically adsorbed to positively charged octadecylamine monolayers. (a) Mass uptake versus number of layers at subphase pH 12 and pH 9; the inset shows the contact angle of water versus the number of layers. (b) Absorbance spectra as a function of the number of layers transferred (left), with the inset showing the plasmon absorbance at 460 nm versus the number of layers. Thickness versus number of layers as determined by optical interferometry is shown on the right. (Reprinted with permission from Ref. 103. Copyright 1996 American Chemical Society.)

these particles electrostatically adsorbed onto octadecylamine monolayers [105]. Examination of lauric acid adsorption onto octadecanethiol SAMs using QCM gave a surface coverage of ~25% at pH 3.5. Above pH 5, no adsorption of a lauric acid monolayer onto the SAM was detected. The formation of the bilayer of lauric acid on the highly curved particle surface was attributed to the ability of the chains to interdigitate [106]. The adsorption of octadecanethiol-capped silver nanoparticles covered with a monolayer of octadecylamine onto arachidic acid monolayers at pH 9 was also successful and gave a particle surface coverage of 15%. The octadecylamine secondary monolayer was adsorbed onto the silver particles in ethanol after they had been capped with octadecanethiol. The adsorption of octadecanethiol-capped silver particles covered by lauric acid onto octadecylamine monolayers at pH 9 gave a surface coverage of 27%.

Using a mixed hydrosol subphase containing both silver (70 ± 12 nm) and gold (130 ± 30 nm) nanoparticles, the production of LB films containing controlled ratios of the two was possible [107]. The gold particles adsorbed preferentially at the interface relative to their fraction in the subphase, with an adsorbed gold particle fraction of 0.7 obtained from a 1:1 mixed hydrosol. This was attributed to the smaller distortion in the octadecylamine film adsorbed on the less curved gold-particle surfaces. The cluster density at the interface took up to 12 hours to stabilize. Gold particles were also adsorbed to octadecylamine monolayers from mixed hydrosols of carboxythiophenol-capped silver particles and uncapped gold particles [108]. This indicated that the capping molecule was transferred from the capped silver particles to the uncapped gold particles; it was proposed that this occurred during encounters between the particles. Transfer onto a gold colloid film improved the photocurrent response of LB films of a 9-*cis*-retinal Schiff base derivative [109]. The ITO substrates were immersed in gold sols (1, 3, and 5 nm) prior to transfer of the LB monolayer.

3. Direct Spreading of Metallic Nanoparticles

Spread monolayers of silver nanoparticles capped by octadecanethiol were studied [110]. The silver nanoparticles were generated in sodium bis(2-ethylhexyl) sulfosuccine (AOT) reverse micelles in isooctane with added aqueous silver nitrate by sodium borohydride reduction. Capping by addition of octadecanethiol resulted in the precipitation of the particles, which were collected and redispersed in chloroform. The mean particle diameters varied from 3.0 to 6.8 nm, increasing with the ratio $w = [H_2O]/[AOT]$, which is related to the size of the aqueous cores of the reverse micelles. The spread monolayers of the capped particles were too rigid for vertical deposition, so horizontal deposition was used. The areas per particle in the close-packed monolayers of the three capped particle preparations were 20 ($w = 2.5$), 58 ($w = 5$), and 130 ($w = 10$) nm^2 particle^{-1}, clearly showing the dependence on micelle core size, as confirmed by TEM. Brewster-angle microscopy of spread uncapped particles showed smaller aggregates than for spread monolayers of the capped particles, which showed large islands. The islands merged to a uniform film on compression that fractured irreversibly upon expansion. The attraction between the hydrophobic capping molecules favored extensive aggregation at the water surface.

An extraction method was also employed for the preparation of stabilized silver nanoparticles [111,112]. A silver-particle sol was added to oleic acid in hexane and emulsified. On standing, the silver particles transferred into the organic layer and the oleic acid adsorbed onto them. Dilution of the oleic acid–stabilized silver particles ($d = 10.3 ± 0.4$ nm) in hexane was used to prepare the spreading solution. The isotherms (Fig. 10a) showed five distinct regions and indicated a strong degree of immiscibility between oleic acid regions and regions of oleic acid–stabilized silver particles. Regions 1 and 2 are the gaseous and liquid states of the film. The first plateau (3) along the compression, near 30 mN m^{-1}, was interpreted as collapse within the oleic acid regions. Along stage 4, further compression occurs. The second collapse (5), near 60 mN m^{-1}, was related to collapse of the particle monolayer. The area per particle, determined from the isotherm regime prior to the collapse of the particle layer, was twice that for close-packed bare silver particles. The

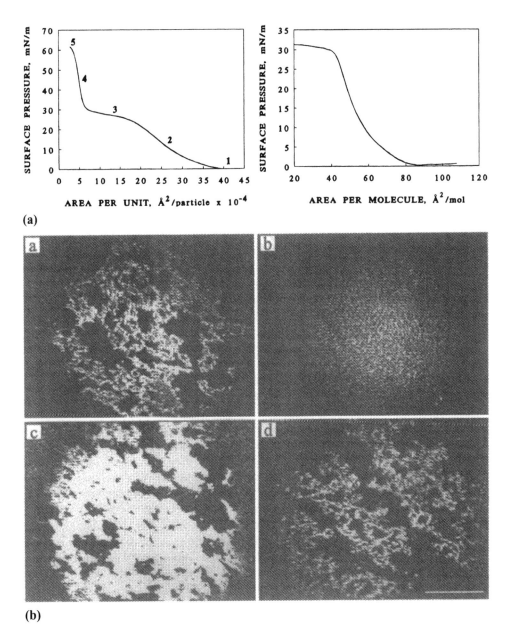

(a)

(b)

FIG. 10 (a) Π-A isotherms for oleic acid–capped silver particles; the labels refer to the different phases as described in the text. (b) BAM micrographs at surface areas of (a) 5000-nm^2 particle^{-1}, (b) 3200-nm^2 particle^{-1} during compression, (c) 1500-nm^2 particle^{-1} during compression, and (d) 5000-nm^2 particle^{-1} after re-expansion. The scale bar represents 1 mm. The micrographs are identified by the letters in the upper left corner of each image. (Reprinted with permission from Ref. 111. Copyright 1996 American Chemical Society.)

BAM showed bright regions due to the Ag particles and dark regions occupied primarily by oleic acid (Fig. 10b). TEM showed monolayer domains of surfactant-coated silver particles within regions of oleic acid. Vertical transfer onto quartz plates of up to eight layers with good linearity was reported.

The extraction method was applied to the preparation of surfactant-stabilized nanoparticles of platinum, palladium, and platinum/palladium alloys [113] using distearyldimethylammonium chloride. Platinum particles coated by the surfactant were prepared by mixing the surfactant in chloroform with an aqueous solution of H_2PtCl_6. The metal salt was extracted into the reverse micelles and reduced by formaldehyde. Surfactant-coated palladium particles were prepared using $PdCl_2$ reduced by sodium borohydride. Monolayers were spread from chloroform solutions of the surfactant-stabilized metal particles. The isotherms showed plateaulike features related to collapse of the surfactant regions of the films followed by a rise to higher pressure until collapse of the particle monolayers. Vertical transfer was reported as successful for compressed films if a slow dipping rate was used. Brewster-angle microscopy showed islands, and TEM showed separate regions of surfactant-coated particles and regions of surfactant. Reflectivity data for the films on water gave thickness values consistent with monolayer coverage.

Platinum nanoparticles formed by the reduction of H_2PtCl_6 by sodium borohydride were capped by polyvinylpyrrolidone [114]. The hydrosol was centrifuged, dried, and redispersed in chloroform prior to spreading at the water–air interface. The average size of the polymer-stabilized Pt particles was determined using TEM to be 29 nm. The TEM micrographs of the transferred films clearly showed the Pt clusters. It was not possible to image the metal core and polymer shell separately. The Pt core diameter was estimated as 3.8 nm. The Π-A isotherms showed a liftoff area of 5.0×10^4 nm^2 per cluster. The isotherms showed little hysteresis, an essentially linear rise up to 18 mN m^{-1} and back, and a relatively high compressibility. The large liftoff area was attributed to partial detachment of the polymer from the particles. Quartz Crystal Microbalance (QCM) analysis of the mass change gave a surface coverage of 50% for the coated particles. X-Ray Photoelectron Spectroscopy (XPS) analysis of the Pt 4f electrons showed the presence of some PtO_2.

The production of fatty acid–capped silver nanoparticles by a heating method has been reported [115]. Heating of the silver salts of fatty acids (tetradecanoic, stearic, and oleic) under a nitrogen atmosphere at 250°C resulted in the formation of 5–20-nm-diameter silver particles. Monolayers of the capped particles were spread from toluene and transferred onto TEM grids. An ordered two-dimensional array of particles was observed. The oleic acid–capped particle arrays had some void regions not present for the other two fatty acids.

The spreading of Au_{55} clusters capped by triphenylphosphine has been described [116]. The cluster $Au_{55}(PPh_3)_{12}Cl_6$ has a metallic core of 1.4 nm and a 0.7-nm-thick shell [117]. Monolayers of the clusters were spread from freshly prepared solutions in methylene chloride; the Π-A isotherm showed a compact area of 3.7 nm^2 per cluster. The films were compressed above 30 mN m^{-1} without evidence of collapse. And BAM observation showed large islands, hundreds of microns in size, merging into a uniform film under compression. The monolayers were transferred onto mica, silicon, and gold. Atomic force microscopy observation of the boundary of the film edge and bare mica showed a 1.9-nm step. Scanning tunneling microscopy observation of the film on Au showed some regions of locally ordered clusters. The cluster $Au_{55}(PPh_3)_{12}Cl_6$ was also spread mixed with two different polymers and the lateral arrangements of the clusters compared using TEM [118]. For the case of a polymer with ionizable carboxylic acid groups and incorporation of cysteamine in the subphase, a homogeneous distribution of particles was observed.

Two-dimensional arrays of metallic nanoparticles separated by electron tunneling barriers could potentially exhibit single-electron conductivity effects. Monolayers of dodecanethiol-capped 3-nm gold clusters were compressed and examined with optical microscopy [119]. The large extinction coefficient of the particles made direct observation possible. The surface pressure began to rise near 18 nm^2 particle^{-1}, and jagged islands were seen that did not fully merge until a surface pressure above 5 mN m^{-1} was reached. Langmuir–Blodgett monolayers were transferred onto mica at 12 mN m^{-1}. Observation by AFM confirmed the 3.5–3.7-nm height expected for a monolayer, although bilayers covered about 10% of the surface. Monolayers transferred onto Si/SiO$_2$ substrates with interdigitated gold electrodes were treated with a bifunctional capping compound, 2,5″-bis(acetylthio)-5,2′,5′,2″-terthienyl. The room-temperature in-plane conductivity was 5×10^{-6} S cm^{-1} before exposure to the bifunctional capping compound, and increased to 5×10^{-3} S cm^{-1} after exposure. The bifunctional capping compound displaced dodecanethiol to some extent and lowered the tunneling barrier between neighboring particles. And STM measurements of current from the tip through a single particle for cluster monolayers on gold/mica showed nonlinear features in the *I-V* curve, suggesting single-electron tunneling. Clusters of Au$_{55}$(Ph$_2$PC$_6$H$_4$SO$_3$H)Cl$_6$ self-assembled onto LB monolayers of eicosamine showed single-electron tunneling in *I-V* curves measured at 90 K using STM [120].

Efforts to determine and explain the surface pressure versus temperature phase diagram of capped metallic nanoparticle monolayers have been reported [121]. In this work, the point is made that V_e the excess volume, is a factor determining the interactions between capped particles. V_e is the difference between the size of the capping molecule and the cone of space projecting away from the nanoparticle surface that it occupies. In this model, the particle is assumed spherical and the capping agent taken as the cylinder of space occupied by the stretched hydrocarbon chain. A geometrical calculation can be used to determine V_e in nm^3, which is a function of particle radius and chain length of the capping molecule. Dodecanethiol- and nonanethiol-capped gold particles, dodecanethiol-capped silver particles, and oleylamine-capped gold particles of narrow size distribution were examined. Distinctly different phase diagrams were obtained comparing the regime $V_e > 0.35$ nm^3 (case I) and 0.15 nm$^3 < V_e < 0.35$ nm^3 (case II). The phases were assigned by horizontal transfer of the monolayers at different temperatures and surface pressures followed by TEM examination. Case I (large free volume) monolayers showed two distinct regions at nonzero surface pressure: foamlike phases and "one-dimensional" phases. The one-dimensional phases were primarily strings of particles. The foamlike phases, resembling monolayer foams [15–18,122] or monolayers of latex particles [123,124] consisted of lines of particles forming the cell walls of a two-dimensional foam. Very low coverages were assigned as the two-dimensional gas phase. For case II (smaller V_e) particle monolayers, the phase diagram consisted of a condensed, hexagonally packed phase at lower temperature and higher surface pressure and a collapsed, less ordered three-dimensional phase at higher temperature and surface pressure. The TEM images of horizontally transferred films of 2.8-nm-diameter silver particles capped by dodecanethiol are reproduced in Figure 11; a hexagonal lattice is evident, with the occasional vacancy or defect. For very low free volume, the particle monolayers exist as aggregates. The more closely packed particle monolayers show red-shifted absorbance spectra.

The cospreading of hexanethiol-capped gold particles with the cross-linking agent 4,4′-thiobisbenzenethiol (TBBT) has been reported as a means of producing cross-linked monolayers of hexanethiol-capped Au nanoparticles [125]. In the presence of TBBT, the

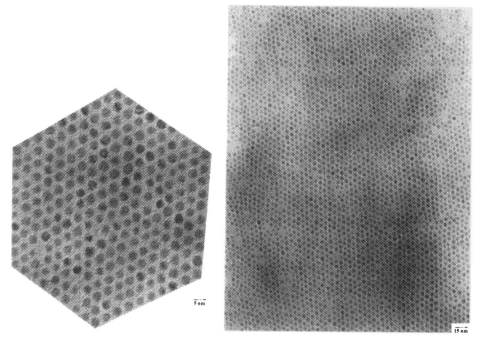

FIG. 11 TEM images of 2.8-nm-diameter silver particles capped by dodecanethiol that were horizontally transferred from the water surface at a surface pressure just below that at which the film would collapse. The top figure is a higher-resolution image of this phase of particles. (Reprinted with permission from Ref. 121. Copyright 1997 American Chemical Society.)

surface pressure rose earlier in the compression of the capped gold particles. The exchange/cross-linking reaction was allowed to occur for 6 hours with the film held in a compressed state. Upon reopening the barriers, particle patches as large as 1.0 cm² were visible by eye. The UV spectra of the transferred films showed a red shift of the surface plasmon band, and TEM images showed a close-packed particle structure.

For monolayers of capped silver nanoparticles in the condensed, close-packed regime, an insulator–metal transition upon compression was observed by monitoring both the linear (reflectivity) and nonlinear optical (SHG) properties of the film [126]. The monolayers were prepared from 2.7-nm silver particles capped by decanethiol or hexanethiol and 4.0-nm-size particles capped by propanethiol. As the film was compressed, the SHG signal enhancement rose to almost 500 and then fell sharply at the collapse point, remaining constant thereafter near 300. The SHG signal enhancement was normalized to the SHG of the uncompressed monolayer. Since the SHG is sensitive only to the particles at the water–air interface, there was essentially no background signal. Compression increased the wavefunction overlap between the metal cores of the capped nanoparticles. Quantum coupling between the particles was demonstrated by the exponential decrease of $\Delta\chi^{(2)}$ with increasing center-to-center particle separation. The reflectivity, measured using a white light source and a diode array detector, dropped, and the plasmon peak flattened at the same point in the compression. The reflectivity of the monolayer became that expected for a thin metallic film. The insulator–metal transition was determined to occur when the separation between the surfaces of neighboring metal cores dropped below 0.5 nm. The transition was reversible upon expansion. Calculation of the second-harmonic response of a hexagonal lattice of metal particles as a function of the interparticle separation also gave a peak, attributed to the transition from localized states to a bandlike structure [127]. The metal–insulator transition in the monolayer of capped silver particles was also observed using an admittance method [128]. Two platinum wires were contacted with the water surface, and admittance was measured over the frequency range 10 kHz to 1 MHz using an inductance capacitance resistance (LCR) meter. The change in the frequency dependence on compression was consistent with a transition to metallic conductivity. Monolayers of capped silver nanoparticles were transferred to HOPG and studied with STM [129], using decanethiol, hexanethiol, and pentanethiol as the capping agents. The coulomb blockade effect was observed for the LB films of the decanethiol-capped 2.6-nm silver particles but not for those capped by the two shorter alkanethiols. For the two shorter capping agents, strong quantum mechanical exchange between particles was inferred.

Monolayers of size-selected 3–5-nm octanethiol-capped silver particles spread and transferred to TEM grids horizontally at low surface pressure showed wirelike structures [130]. Strands of nanoparticles, 8–25 nm wide and one particle thick, were observed, as shown in Figure 12. The "wires" of capped silver particles were microns long and tended to align roughly parallel with their neighbors. The wires survive LB transfer well. Such patterns could also be produced in computer simulations and were argued as arising from the same competing forces that result in stripe patterns in Langmuir films of dipolar surfactants [131].

4. Production of Metallic Nanoparticles Inside Langmuir–Blodgett Films

Gold nanoparticles have been produced photochemically in LB films [132]. Octadecylamine, 4-hexadecylaniline, benzyldimethylstearylammonium chloride, and two phospholipids, dipalmitoyl-L-α-phospatidylcholine (DPPC) and dipalmitoyl-DL-a-phosphatidyl-L-serine (DPPS), were used to form monolayers on pH 3.7 subphases containing 0.1 mM $HAuCl_4$. Langmuir–Blodgett transfer onto Si wafers for XPS and AFM, quartz plates for UV-visible spectroscopy, and carbon-coated Formvar–covered titanium grids for TEM observation was carried out. The cationic head groups of these monolayers bind $AuCl_4^-$ anions, and illumination of these LB films by UV light leads to the formation of gold particles. Exposure of a 13-layer octadecylamine LB film to UV light (750 mW cm^{-2}) followed by examination by UV spectroscopy showed the gold particle plasmon band at 550 nm grow as the color of the films changed to deep purple. The gold particles formed by a photoreduction reaction. The most intense plasmon bands were observed for the octadecylamine films. No particle growth was detected for DPPS LB films, and only a monolayer of DPPC could be transferred. And XPS confirmed that the phospholipids bound very little $AuCl_4^-$. Different gold particle morphologies were observed for the three ammonium head-group surfactants, and it was determined that the most oriented particles formed for the octadecylamine and benzyldimethylstearylammonium chloride LB films, which are fully protonated at pH 3.7. The hexadecylaniline was about 50% protonated at this pH. In the first two cases, well-formed 20–300-nm Au platelets formed, including twinned crystals and neat triangular single crystals. Irregular-looking gold clusters formed in the hexadecylaniline films. This group has also reported the formation of oriented CdI_2 particles inside LB multilayers of cadmium arachidate exposed to HI vapor [133].

(a)

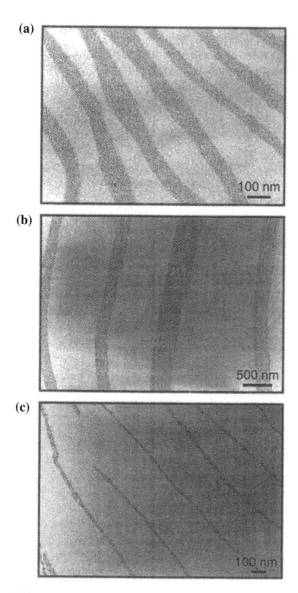

FIG. 12   TEM micrographs of wirelike assemblies of silver nanoparticles. (a) Octanethiol-capped silver nanoparticles of average diameter 3.4 nm deposited from hexane solution. (b) The same particles deposited from heptane solution. (c) Octanethiol-capped silver particles of average diameter 4.4 nm deposited from heptane solution. (Reproduced with permission from Ref. 130. Copyright 1998 American Chemical Society.)

C.   Langmuir–Blodgett Films Containing Semiconductor Nanoparticles

1.   Formation of Semiconductor Nanoparticles Underneath Langmuir Monolayers

The generation of semiconductor nanoparticles below spread monolayers was first reported by Fendler [134]. Poly(styrenephosphonate diethyl ester) and its 1:1 copolymer with poly(methyl methacrylate) were spread on subphases containing $Cd(NO_3)_2$, $Pb(NO_3)_2$, $ZnCl_2$, and $InCl_3$. Infusion of H_2S gas into the subphase resulted in nucleation and growth of nanoparticles of CdS, PbS, ZnS, or In_2S_3 at the interface. Using reflectivity data, fractional coverages of 0.22 (In_2S_3 under polymer) to 0.69 (CdS under copolymer) were determined. For CdS and ZnS, particle sizes near 7.5–8.0 nm were dominant; the other two materials formed larger structures when examined by TEM on horizontally transferred films.

An effort to use the pH dependence of metal ion complexation to control the size of CdS particles grown beneath a monolayer was reported [135]. Monolayers of $C_{16}H_{33}C(H)[CON(H)(CH_2)_2NH_2]_2$ were spread over 0.25 mM $CdCl_2$ at pH values between 6.1 and 9.1, with increased pH resulting in stronger binding of Cd^{2+}. The limiting area determined from the isotherms shifts from 0.24 nm^2 $molecule^{-1}$ at pH 6.1 to 0.44 nm^2 $molecule^{-1}$ at pH 9.1. Infusion of H_2S gas resulted in CdS-particle growth beneath the monolayers. The growth of the particle layers was monitored by optical reflectivity. Films were transferred horizontally onto quartz plates after 3-, 5-, and 10- minute exposures to H_2S. The UV absorbance spectra gave absorption edges corresponding to bandgaps of 2.82, 2.78, and 2.74 eV respectively. The bandgaps correspond to 3–5 nm particle diameters, respectively. Longer exposures resulted in porous films, 2–3 nm thick, of disc-shaped particles. Higher pH values favored the formation of smaller particles. It was found that if the quartz-supported films were illuminated at $\lambda > 400$ nm in the presence of methylviologen and benzyl alcohol, electrons excited into the CdS bandgap would reduce methylviologen.

A further study of CdS- and ZnS-nanoparticle films included electrical and photoelectrical characterization [136]. It was noted that vertical LB transfer could not be employed due to the fragility of the films; thus horizontal transfer was employed. Films of CdS particles were grown of 20–30 nm thickness, while those of ZnS particles were 20–200 nm thick. Reflectivity data fit to Fresnel equations was used to determine the layer thickness and effective refractive index of the layer of semiconductor particles using a four-layer model air/monolayer/particle film/substrate or air/monolayer/particle film/subphase. The equivalency of the optical parameters obtained for the particle film in the two cases was taken as good evidence for complete transfer. Optical reflectivity versus time was also useful for monitoring the film growth and determining when the film thickness had stopped increasing. The bandgap for bulk CdS is 2.40 eV; for the 6.3-nm-thick films, the bandgap determined from the absorbance spectrum was 2.54 eV. The bandgap of 2.54 eV corresponds to an average particle diameter of 5 nm.

CdS and ZnS particulate films formed under arachidate monolayers were investigated [137]. X-ray diffraction was carried out on the films transferred onto Teflon, spanning 2.0 mm pinholes. The x-ray diffraction patterns indicated that the hexagonal phase was dominant. For ZnS, the particles were found to be of the face-centered cubic crystalline phase (sphalerite). TEM was used to analyze the morphology of the particle film and the particle sizes after H_2S exposure for 3, 15, and 30 minutes. After 3 minutes, CdS particles were 3.0 ± 0.5 nm in size. There were also some conical clusters only 0.6 nm high and 2–3 nm in diameter. As time progressed, the particles aggregated, larger ones appeared, and chains and discs of particles formed. The bandgap was 2.83 eV for early-stage films with 1 nm "optical thickness," dropped to 2.6 eV for films of 3.5-nm optical thickness, and then slowly decreased toward 2.5 eV. The particle films appeared to have layers, the first containing 3–4-nm-thick discs that were 3–8 nm in diameter and were connected. The films were pictured as interacting arrangements of nanoparticles whose sizes and spatial distribution was governed by the interfacial growth process.

Cadmium sulphide–particle films grown beneath cadmium arachidate monolayers have been subjected to investigation by scanning electrochemical microscopy (SECM), photoelectrochemical measurements, and STM methods [138]. In the first configuration, the reference and working electrodes were in the subphase beneath the particle film. The tip of an STM was used as a working electrode. In the second configuration, the particle film was transferred to titanium foil placed at the bottom of a Teflon cell. The STM tip was immersed in the cell for scanning over the film surface. The reference electrodes and counterelectrodes were immersed in the electrolyte solution, and the light from the end of an optical fiber was used to illuminate the surface for photoelectrochemical measurements. The films were found to be subject to photocorrosion in the presence of oxygen. The main products of the photoreaction are Cd^{2+} and SO_4^{2-}. The flat-band potential was estimated as -1.3 V. Using STM to measure current versus voltage, curves showed clear evidence for current rectification.

The morphology of PbS particles was found to depend strongly on the nature of the monolayer under which they were grown [139,140]. Epitaxial growth of the particles beneath the monolayer head groups depended on how well the lattice of the semiconductor phase being formed matched the spacing preferred by the surfactant head groups. Monolayers of arachidic acid, octadecylamine, and mixtures were spread on subphases 1 mM in lead nitrate at pH 5.8. After compression to 20 mN m^{-1}, H$_2$S gas was introduced. The PbS particles formed on the lead arachidate monolayers were equilateral triangles, and electron diffraction showed that they grew epitaxially with the [111] plane parallel to the monolayer surface. Lead sulphide particles could not be grown under mixed monolayers of 1:5 arachidate: octadecylamine. Under 5:1 arachidate: octadecylamine monolayers, subtle changes in shape occurred, but the triangular particles still formed. The growth of PbS under lead arachidate monolayers even at zero surface pressure indicated the presence of islands of the solid phase at very low surface pressure. The Pb–Pb distance was 0.420 nm in the [111] plane, and thus had only a very slight mismatch with the 0.416-nm spacing of arachidate head groups. This close match allowed epitaxial growth of crystallites with well-defined morphologies and orientations. The addition of octadecylamine altered the growth preference in a complex manner for 2:1 arachidate: octadecylamine monolayer templates, growth was oriented along the [100] face at 30 mN m^{-1}, but along the [110] face at 0 mN m^{-1}. Lead sulphide particles grown under 1:2 arachidate:octadecylamine monolayers were rectangular, randomly oriented, and of variable size.

The photoelectrochemical behavior of PbS-nanoparticle films was morphology dependent [141–143]. The PbS nanoparticles grown under arachidate monolayers and under mixed monolayers of arachidic acid and octdecylamine had regular triangular shapes (equilateral or right-angled), while those grown under hexadecylphosphate had irregular, nonepitaxial, disclike shapes. The absorbances of the three forms had maxima near 380 nm and absorption edges near 800 nm. The small differences in the absorbance spectra were amplified when potential-dependent spectra were recorded for films transferred to ITO-coated glass electrodes. The photocurrent versus potential onset was different for the three forms: The onset of the photocurrent was near 0 V for the equilateral triangular particles but decreased to about -0.35 V for the right-angled triangular particles. The photocurrent arose from the generation of electron-hole pairs, and these studies showed that this process is sensitive to particle morphology.

The formation of CdS particles beneath monolayers of N-methyl-p-(p-tetradecyloxystyryl)pyridinium iodide, a cationic surfactant with a hemicyanine portion was studied [144]. The monolayers were spread on a circular glass trough coated with paraffin, and the trough was covered by a jar. The CdCl$_2$ subphase also contained EDTA, so the Cd^{2+} was complexed and in the form of Cd-EDTA with an overall charge of -2. The monolayer was spread to 0.25 nm^2 molecule^{-1} and H$_2$S gas was injected. The TEM images of transferred particulate films showed that increasing the concentration of Cd-EDTA^{2-} (0.05–1 mM) or the reaction time (5–60 min) increased the particle size (15–100 nm).

The formation of CdS and PbS particles under stearic acid has been studied using BAM and TEM [145,146]. Some H$_2$S gas was injected beneath monolayers compressed to either 0.20 nm^2 molecule^{-1} or 0.40 nm^2 molecule^{-1} on CdCl$_2$ and observed with BAM. At 0.20 nm^2 molecule^{-1}, the initially smooth film broke into floating islands. At 0.40 nm^2 molecule^{-1}, separated floating domains were present upon spreading. Injection of H$_2$S caused the appearance of small bright dots in the image, but did not significantly alter the domain structures. And TEM showed that the particle films grown under films at 0.20 nm^2 molecule^{-1} covered most of the surface and the electron diffraction pattern had sixfold spots. In contrast, particle films grown under films at 0.40 nm^2 molecule^{-1} were loosely aggregated in a random manner. A stearic acid film on Pb(NO$_3$)$_2$ was also treated with H$_2$S in a similar fashion. And TEM showed the PbS particles shaped like equilateral triangles. The BAM showed formation of straight ridges at 0.20 nm^2 molecule^{-1}, suggesting that particle formation had induced monolayer collapse. The degree of commensurability between the lattice spacing of the solid being formed and that of the head groups of the surfactants was concluded as the main factor determining the changes in the monolayer morphology upon particle formation. For stearic acid monolayers compressed to 0.20 nm^2 molecule^{-1}, the surface pressure of 37.5 mN m^{-1} was observed to drop to 20 mN m^{-1} upon H$_2$S gas injection. Rodlike (100–150 nm long, 5–15 nm wide) as well as dotlike particles (8–15 nm) were observed by TEM. The rodlike particles were oriented in three essentially distinct directions 120° apart. The arcual diffraction spots were found to fall into two distinct rings of six spots each, rotated by 30° with respect to each other, and were attributed to two different sets of particles. A type of lattice structure not observed for bulk CdS (cubic or hexagonal) was identified as consisting of one-dimensionally ordered planes of atoms that were packed in threefold symmetrical orientations (the rodlike particles). The interaction between the surfactant and the growing particles is possibly complex enough to alter the lattice structure of the materials being formed.

Cadmium sulphide nanoparticles also grew epitaxially beneath arachidic acid monolayers. At room temperature, the nanoparticles formed appeared in two crystalline types, rodlike crystallites 50–300 nm long and 5–15 nm wide that grew in three directions separated by 120°, and irregular disclike crystals. Selected area diffraction indicated that the hexagonal form of CdS had formed that possesses a wurtzite type of structure. The temperature dependence of the head-group packing of monolayers was found to influence the growth of CdS beneath monolayers of arachidic acid. When CdS particle growth was induced beneath cadmium arachidate monolayers at 3–4°C, the lowered temperature reduced the growth rate and resulted in the formation of narrower crystallites. These exhibited an absorption edge at 460 nm versus 500 nm for the CdS crystallites grown at 20°C [147]. The absorption edge for bulk CdS is 500 nm. The CdS nanocrystals grown at lower temperature also tended to cover the entire surface, whereas those grown at room temperature were clustered.

2. Binding of Semiconductor Nanoparticles to Langmuir Monolayers

The binding of semiconductor nanoparticles to LB films was introduced by Fendler [148]. A silanized quartz slide was first coated by a two-layer LB film of arachidic acid. A layer of dioctadecyldimethylammonium chloride was deposited at 40 mN m^{-1} on the down-stroke such that the cationic head groups remained exposed. The substrate was immersed into a Teflon beaker placed in the trough beneath the dipper containing the hydrosol of CdS nanoparticles (2.8 nm). The CdS nanoparticle were stabilized by adsorbed hexametaphos-phate and were thus anionic and bound to the exposed cationic head groups of the LB film. Complete coverage took 60–80 minutes. Upon emersion, a second layer of dioctade-cyldimethylammonium chloride was deposited, creating a sandwich structure. The process was repeated up to 15 times, creating alternating layers of cationic surfactant and anionic nanoparticles. The STM images on highly oriented pyrolytic graphite (HOPG) revealed two-dimensional patches of interconnected particles.

It is also possible to allow nanoparticle colloidal solutions to bind to spread mono-layers by electrostatic attraction and then to transfer the film and bound particles to the sub-strate by LB dipping [149]. Langmuir monolayers of a 3:1 mixture of stearylamine and be-henic acid were spread and compressed to 35 mN m^{-1}. A colloidal solution of 5.2-nm CdS particles was placed beneath the subphase in a beaker. Three hours were allowed for the CdS particles to bind onto the LB film. Transfer onto a glass plate for measurement of ab-sorbance and fluorescence spectra or a copper grid covered with a carbon film for TEM fol-lowed. The presence of the CdS particles in the transferred film was confirmed. The fluo-rescence emission spectra indicated that binding to the amine head groups passivated certain defect states, for the emission from these was eliminated and the quantum yield for the bandgap emission increased 10-fold. The binding of PbI$_2$ particles was pursued in a somewhat different manner. A stearic acid bilayer LB film was placed in a solution of a di-cationic, rigid bipolar pyridinium compound that electrostatically adsorbed on top of the stearic acid bilayer. Subsequent immersion in the colloidal suspension of PbI$_2$, anionic due to the slight excess of I$^-$ used in the preparation, resulted in particle binding, as confirmed by TEM [150]. Alternating layers of the PbI$_2$ particles and the cationic bipolar spacer com-pound were also formed on top of SAMs [151].

Cadmium sulphide particles stabilized by hexametaphosphate were used as the sub-phase for monolayers of dioctadecyldimethylammonium bromide. The anionic stabilized particles were bound to the cationic surfactant head groups and transferred by vertical dip-ping to create surfactant monolayer–anionic particle–surfactant monolayer alternating sandwich structures [152]. The CdS nanoparticles were prepared from aqueous solutions of Cd(ClO$_4$)$_2$ and (NaPO$_3$)$_6$ adjusted to pH between 8.0 and 10.0; the particle size produced by the subsequent reaction with H$_2$S decreased with pH (6 nm at pH 8.4, 4 nm at pH 9.4). The fluorescence wavelength depended upon particle size, and maximal fluorescence was obtained by adjusting the pH to 10.5 after the particle formation was complete. The ab-sorption and fluorescence spectra versus number of layers are reproduced in Figure 13. The absorption edge of 495 nm corresponds to a bandgap of 2.50 eV (Fig. 13a). Time-resolved fluorescence emission of the CdS nanoparticles was measured on the LB films after exci-tation by a pulsed, frequency-tripled Nd:YAG laser at 355 nm. The fluorescence decay curves were fit to a stretched exponential, $I = I_0 \exp[-(t/\tau)^\beta]$, where $0 < \beta < 1$ is related to the distribution of lifetimes and τ is the peak of the lifetime distribution. A narrow fluo-rescence emission band that peaked at 495 nm was assigned as the excitonic fluorescence and that at 670 nm was assigned to emission due to decay of trapped holes (Fig. 13b). Rel-ative to the sol, the emission peaked at 495 nm was sharpened in the LB film and the trap-emission band was suppressed. The trap states on the particle surfaces in solution are due primarily to solvation. Addition of Cd^{2+}, Sr^{2+}, Ag$^+$, triethylamine, or aqueous hydroxide enhanced the fluorescence intensity from the LB films by factors of up to 3.8 relative to the

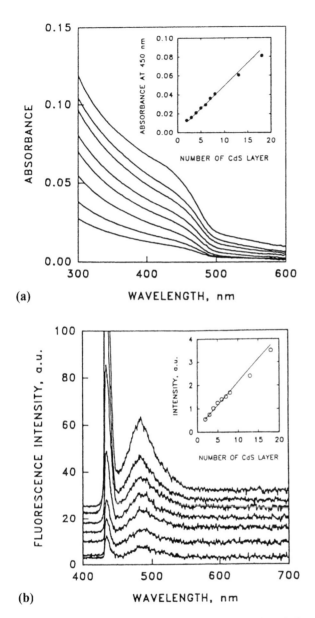

(a)

(b)

FIG. 13 Langmuir–Blodgett films of dioctadecyldimethylammonium bromide incorporating 6.0-nm-diameter CdS nanoparticles stabilized by 0.2 mM sodium hexametaphosphate. (a) Absorbance spectra for 2, 3, 4, 5, 6, 7, 8, and 13 layers (bottom to top); the inset shows the linearity of the absorbance at 450 nm versus the number of layers. (b) Fluorescence spectra (excitation wavelength = 400 nm) for 2, 3, 4, 6, 8, 13, and 18 layers (bottom to top); the inset shows the emission intensity at 480 nm versus the number of layers. (Reproduced with permission from Ref. 152. Copyright 1994 American Chemical Society.)

dry state. The fluorescence lifetime parameter τ increased from 10 ns in the sol to 15 ns for the LB film in water and then decreased to 12 ns for the LB film in aqueous triethylamine.

The binding of carboxylic acid–derivatized CdS nanoparticles to monolayers of octadecylamine was studied [153]. The CdS hydrosol was prepared by reaction of $CdSO_4$ with Na_2S in the presence of a capping molecule, 4-carboxythiophenol. The average particle size, deduced from the absorption edge and the previously reported absorption edge versus particle size data, was 3.5 nm. Octadecylamine was spread on the subphase of the CdS hydrosol. The subphase was kept at pH 9 or pH 12. At pH 9, the amines were protonated (pKa = 10.6 for the ammonium head group) and the carboxylates unprotonated (pKa = 4.5). At pH 12, the amine head groups were unprotonated, and thus the strength of the attraction of the anionic coated CdS particles for the monolayer was adjusted by varying the pH. After an hour of equilibration at 25 mN m^{-1}, multilayer LB films were formed by transfer onto quartz, Si, or AT-cut quartz. A monolayer of lead arachidate was deposited first to make the substrates hydrophobic. Isotherms recorded at pH 9 gave liftoff areas (areas at which the pressure rises began) of 0.20 nm^2 for octadecylamine and 0.55 nm^2 after equilibration with the capped CdS hydrosol. Little effect of the hydrosol on the isotherm liftoff area was reported at pH 12. The mass uptake (ng cm^{-2}) vs. number of layers was linear for the 12 layers deposited. From the mass per transfer, a surface coverage of 30% of CdS particles was determined.

Hydrosols of 6-nm TiO_2 particles were used as subphases for monolayers of stearic acid [154,155]. The pH of the subphase was adjusted to 3.12, and Y-type LB films were obtained by transfer at 15 mN m^{-1}. The LB films were examined by TEM and showed closely packed TiO_2 particles. FTIR spectra (on CaF_2 plates) showed that the hydrocarbon chains remained ordered, and x-ray diffraction on Si gave a spacing of 12 nm, assigned to both the particle and the stearate bilayer. In a subsequent study [156], LB films of TiO_2–stearate were prepared on Si made hydrophobic by exposure to HF and on Si made hydrophilic by UV irradiation. Good-quality films were obtained on the hydrophilic Si, as evidenced by AFM examination. The surface photovoltage obtained for the TiO_2-stearate on n-type silicon increased, while deposition on p-type silicon decreased the photovoltage. This demonstrates the potential application of these films in photovoltaic devices.

3.   Direct Spreading of Semiconductor Nanoparticle Monolayers

Direct spreading of nanoparticle films at the water–air interface for LB transfer has been successful. Nearly monodisperse CdSe nanoparticles capped with trioctylphosphine oxide [157] were spread as Langmuir monolayers [158]. The CdSe particles were prepared by a pyrolysis procedure in five sizes, from 0.25 to 0.53 nm. The Π-A isotherms measured on these nanoparticles are shown in Figure 14a. The LB films were deposited onto glass slides coated with a layer of sulfonated polysytrene by a single upstroke, with the coated slide being immersed prior to spreading of the film. Transfer for TEM examination was performed using magnetic nickel grids coated with amorphous carbon held onto a thin magnet. The Π-A isotherms shifted with particle size, and the areas per particle in the fully compressed monolayer were within 15% of the cross-sectional area of the particles. The monolayers could withstand Π up to 65 mN m^{-1} and were stable over several hours, although the isotherms were subject to hysteresis. The measured optical absorption and photoluminescence spectra (excited at 430, 490, and 540 nm) showed the expected blue shift with decreasing particle size (Fig. 14b). Relative to the solution spectra, an additional blue shift attributed to some degradative effect of the water surface was observed. The TEM images showed a local hexagonal symmetry. Image analysis to obtain a radial distribution function

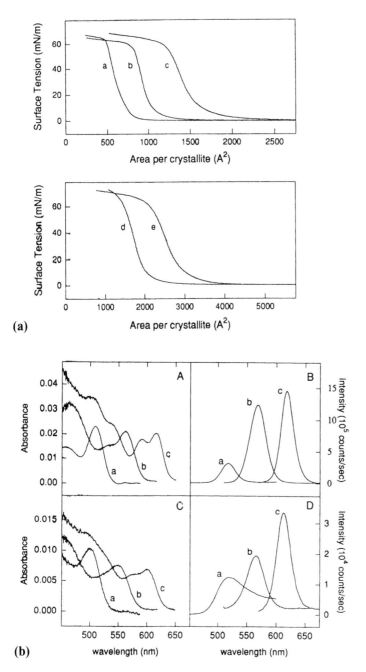

FIG. 14 Measurements on monolayers and LB films of CdSe nanoparticles of narrow size distribution (a) Π-*A* isotherms for Langmuir monolayers of CdSe nanoparticles of diameter 2.5 nm (curve a), 3.0 nm (curve b), 3.6 nm (curve c), 4.3 nm (curve d), and 5.3 nm (curve e). The area per nanoparticle was determined by dividing the trough area by the estimated number of particles deposited on the surface. (b) Absorbance and photoluminescence spectra of the nanoparticles in solution (A, B) and in monolayers on sulfonated polystyrene–coated glass slides (C. D). The nanoparticle diameters are 2.5 nm (curves labeled *a*), 3.6 nm (curves labeled *b*), and 5.3 nm (curves labeled *c*). The excitation wavelengths are (a) 430 nm, (b) 490 nm, and (c) 540 nm. (Reproduced with permission from Ref. 158. Copyright 1994 American Chemical Society.)

gave four clearly visible nearest-neighbor shells. An average of 5.7 nearest neighbor particles per particle was obtained.

Cadmium sulphide particles capped with a copolymer have been prepared as monolayers [159]. The polymer used was polymaleic anhydride octadecyl ester. The degree of esterification was such that the ratio of carboxyl:ester groups was 3:2. An emulsion was formed by mixing aqueous $CdCl_2$ and the polymer dissolved in chloroform. The lower organic phase was extracted and gradual injection of thioacetamide resulted in the generation of H_2S. The Cd^{2+} was complexed by the polymer in the form of reverse polymer micelles, and the reaction with H_2S produced CdS particles capped by the polymer and dissolved in chloroform. The polymer-capped CdS particles were spread, compressed to 30 mN m^{-1}, and transferred onto CaF_2 or Si. The Π-A isotherm for the spread polymer-capped particles was condensed relative to that of the monolayer of the polymer, due to the complexation of the polymer around the particles. X-ray diffraction gave a single Bragg peak corresponding to a spacing of 6.3 nm. Photoluminescence with excitation at 308 nm indicated energy transfer between the polymer and the CdS particles. Changing the ratio of carboxylic acid:ester to 7:1 gave particles quoted as nearly the same size, less than 2 nm [160].

The photoelectrochemical properties of CdS nanoparticles modified with 2-aminoethanethiol (2-AET) was investigated [161]. The CdS particles were prepared in reverse micelles of the surfactant AOT in heptane by reaction of Cd^{2+} with Na_2S, followed by capping with 2-AET. The capped particles ($\langle d \rangle$ = 2.7 nm) were redispersed in benzene/DMSO. Monolayers were spread on subphases that were 10 mM in the cross-linking agent glutaraldehyde at 0°C. The subphase temperature was then raised to 25°C, and the cross-linking reaction was allowed to occur for 1 hour. The cross-linking was carried out at a ratios of 1.0–6.0 of A_{link}/A_{CdS}, where A_{link} was the area to which the particle layer was compressed before raising the temperature and A_{CdS} = 5.7 nm^2, the average area of a CdS particle. The cross-linked particle monolayers were transferred onto Au/mica modified by 2-AET at 17 mN m^{-1}. Collapse was observed near 19 mN m^{-1} for all A_{link}/A_{CdS} ratios; the limiting areas varied from 8.6 nm^2 particle^{-1} (no glutaraldehyde) to 50 nm^2 particle^{-1} (A_{link}/A_{CdS} = 6.0). The STM images of LB monolayers prepared from non-cross-linked CdS particles showed that the particles clumped into islands; in contrast, those prepared from particle monolayers subjected to cross-linking were well dispersed. Photoelectrochemical measurements were conducted in aqueous KCl, with triethanolamine as the electron donor. The photocurrent onset was at −1.15 V vs. SCE and was zero for more negative potentials; at zero V it was near 1.4 μA cm^{-2} for a single-layer film. The photocurrent was proportional to the number of particle layers transferred with one, two, and three layers having been examined. The action spectrum had essentially the same wavelength dependence as the absorbance spectrum.

Success in the transfer of multilayer LB films of CdS nanoparticles 2.65–3.40 nm in diameter stabilized by dodecylbenzenesulfonic acid spread from chloroform was achieved [162]. The surfactant stabilized the sol against Ostwald ripening, the growth of larger particles at the expense of smaller ones. In this system, unbound surfactant was not a significant part of the spread monolayer. The particle monolayers could be spread in a gaseous regime and compressed to a densely packed multilayer film four to five particles thick. Close-packed structures were seen using TEM on the transferred films, and the close-packed areas from the monolayer regime of the isotherms were consistent with hexagonal packing. The fluorescence emission (380-nm excitation) of the LB films had a maximum at 455 nm, compared to 510 nm in the sol. Cadmium sulphide particles capped by hexametaphosphate and by dioctadecyldimethylammonium bromide or hexadecyltrimethylammonium bromide have been spread as monolayers [163]. DODAB was present in excess and formed monolayer regions between the particles. On the other hand, the CTAB-capped particles were nearly close packed. The fluorescence emission from the particle multilayers was studied down to 8 K. The emission intensity increased down to 30 K and then decreased again, due to the presence of shallow surface traps. Scanning tunneling spectroscopy was used to measure *I-V* curves (current versus tip bias voltage) and n-type semiconductor behavior was observed.

Nanoparticles of the semiconductor titanium dioxide have also been spread as monolayers [164]. Nanoparticles of TiO_2 were formed by the arrested hydrolysis of titanium isopropoxide. A very small amount of water was mixed with a chloroform/isopropanol solution of titanium isopropoxide with the surfactant hexadecyltrimethylammonium bromide (CTAB) and a catalyst. The particles produced were 1.8–2.2 nm in diameter. The stabilized particles were spread as monolayers. Successive cycles of II-*A* isotherms exhibited smaller areas for the initial pressure rise, attributed to dissolution of excess surfactant into the subphase. And BAM observation showed the solid state of the films at 50 mN m^{-1} was featureless and bright; collapse then appeared as a series of stripes across the image. The area per particle determined from the isotherms decreased when sols were subjected to a heat treatment prior to spreading. This effect was believed to arise from a modification to the particle surface that made surfactant adsorption less favorable.

Monolayers have also been spread from a mixture of an organosol of SnO_2 nanoparticles capped by dodecylbenzene sulfonic acid (DBS) and arachidic acid [165]. SnO_2 particles of 4.5 nm average diameter, formed by mixing $SnCl_2$ and aqueous hydroxide, were peptized with 2.0 M HCl, capped, and then extracted into chloroform to which arachidic acid was added. The measured II-*A* isotherms show limiting areas of 22 nm^2 (with no arachidic acid) and 23 nm^2 (with molar ratio SnO_2:arachidic acid of 10:1). From the absence of sulfur in the XPS spectra of the LB films and absence of S=O in the infrared spectra, it was concluded that DBS had dissolved in the subphase during spreading and compression. Scanning electron microscopy of a 151-layer LB film on Si showed good uniformity and domains of about 100-nm size of clusters of particles.

Coulomb blockade effects have been observed in a tunnel diode architecture consisting of an aluminum electrode covered by a six-layer LB film of eicosanoic acid, a layer of 3.8-nm CdSe nanoparticles capped with hexanethiol, and a gold electrode [166]. The LB film serves as a tunneling barrier between aluminum and the conduction band of the CdSe particles. The conductance versus applied voltage showed an onset of current flow near 0.7 V. The curve shows some small peaks as the current first rises that were attributed to surface states. The data could be fit using a tunneling model integrated between the bottom of the conduction band of the particles and the Fermi level of the aluminum electrode.

4. Formation of Semiconductor Nanoparticles Inside Langmuir–Blodgett Films

The formation of semiconductor nanoparticles and related structures exhibiting quantum confinement within LB films has been pursued vigorously. In 1986, the use of the metal ions in LB films as reactants for the synthesis of nanoscale phases of materials was described [167]. Silver particles, 1–2 nm in size, were produced by the treatment of silver behenate LB films with hydrazine vapor. The reaction of LB films of metal salts (Cd^{2+}, Ag^+, Cu^{2+}, Zn^{2+}, Ni^{2+}, and Pb^{2+}) of behenic acid with H_2S was mentioned. The use of HCl, HBr, or HI was noted as a route to metal halide particles. In 1988, nanoparticles of CdS in the "Q-state" size range (below 5 nm) were prepared inside LB films of cadmium arachidate [168]. The work was motivated by the desire to produce superlattices of Q-state semi-conductor particles with good lateral contact between neighboring particles. Arachidic acid was spread on a subphase of 0.3 mM $CdCl_2$. Because the pH was greater than 5, the carboxylate groups were complexed with the cadmium ions. The monolayers were transferred onto glass at 30 mN m^{-1}. The resulting LB films of 1, 2, 3, and 7 layers were exposed to H_2S gas. The bandgap for bulk CdS is 2.8 eV. The absorption spectrum for the gas-treated LB films was 2.4 eV, suggesting a particle size of 5 nm. The size was estimated using the results from the electron/hole pair in a spherical box calculation, assuming that the effective mass of the electron/hole pair was the same as the bulk value. The particle size was independent of the number of layers. An average thickness increase of 0.3 nm per layer after CdS particle formation was noted. It was pointed out that either spherical or disclike particles were possible. In a subsequent x-ray diffraction study of CdS particle growth in cadmium stearate, it was found that the layer structure was preserved after successive cycles of H_2S gasing and immersion in aqueous $CdCl_2$ [169]. The absorbance onset was found to increase over six successive gassing and immersion cycles, from 440 nm after the first cycle to 486 nm after the sixth cycle, still below the bulk onset of 520 nm. X-ray diffraction showed that the layer structure was weakened during the gassing and then partly restored during the immersion in aqueous $CdCl_2$. Grazing-angle FTIR showed that particle formation caused some tilting of the hydrocarbon chains away from the vertical. The lead stearate LB films were deposited at 30 mN m^{-1} from the vertical solidlike S phase.

It is not always the case that exposure to H_2S gas results in the formation of spherical nanoparticles and disruption of the LB film layer structure. The exposure of LB films of lead stearate to H_2S gas at 1 torr did not change the layer spacing of 5.05 ± 0.02 nm as determined by x-ray diffraction [170]. The spacing determined was essentially that defined by the planes of Pb^{2+} ions coordinated to the stearate head groups. Infrared spectroscopy further confirmed that the highly ordered structure of stearate bilayers with Pb^{2+} coordinated to two stearate molecules was not disrupted. The UV-visible absorbance spectrum of the LB films after gas treatment showed an absorption edge at 690 nm. This corresponded to a 1.4-eV shift relative to bulk PbS. It was concluded that quantum confinement arose due to the formation of sheets or lines of PbS within the LB film. An attempt to generate sulfides of platinum and palladium in LB films of dimethyldioctadecylammonium bromide containing complex ions of these species resulted in observation of the metal sulfides with blue-shifted absorbance onsets; however, the H_2S treatment also resulted in side reactions [171]. Treatment with hydrazine resulted in the formation of palladium particles of size 3.6 nm.

Production of nanoparticles of CdS, CdSe, CdTe, CdSSe, CdSTe, and CdSeTe was reported [172, 173]. The particles were produced inside LB films of cadmium arachidate, a diacetylenic fatty acid, nonacosa-10,12-diynoic acid, and a diacetylenic long-chain fatty acid with a benzoic acid head group. The H_2Se gas for reaction with the LB films was produced by reaction of FeS and HCl, and the H_2Te gas was produced by reaction of Al_2Te_3 and HCl. The LB films of 2–20 layers were exposed to these gases, and then the absorbance spectra were obtained. A bandgap of 2.1 eV was obtained for the CdSe particles, 0.4 eV greater than that for the bulk material. The bandgap for the CdS particles was 2.8 eV versus 2.4 eV for bulk CdS and was 1.7 eV for CdTe versus 1.4 eV for bulk CdTe. The particles could be generated inside the LB films of the diacetylenic fatty acids. These films could then be polymerized by UV irradiation to produce the blue form of the polydiacetylene, which could be converted to the red form by heating.

The use of a polymer monolayer to form LB films to be gas treated to form CdS nanoparticles has been reported [174]. The polymer use was poly(maleic anhydride) esterified with octadecanol such that the ratio of carboxylic acid groups to octadecyl ester groups was 2:1. The polymer was spread on $CdCl_2$ or $PbCl_2$ containing subphase, transferred at 18–20 mN m^{-1}, and then exposed to H_2S gas; Y-type multilayers were obtained. The substrates were Si, quartz, CaF_2, and glass. X-ray diffraction gave a layer spacing of 5.7 nm. It was claimed that each nanoparticle formed was contained within a single polymer molecule. Very much smaller particles than those produced in stearic acid films were reported, with the bandgap of PbS produced in the polymer LB films being 3.18 eV as compared to 1.96 eV for stearic acid. It was hypothesized that the particle size could be controlled by varying the degree of esterification of the polymer. It was demonstrated that CuS nanoparticles could be formed in LB films of this polymer [175]. From the intensity decrease of the carboxylate stretching bands during the H_2S treatment, it was concluded that the percent conversion to the carboxylic acid form was 70%, 50%, and 40%, for the cadmium-, copper-, and lead-containing polymer LB films, respectively. In another example of use of a polymer LB film to form CdS nanoparticles, a cross-linked ethylenediamine-epichlorohydrin "microgel" with stearate side chains was used. CdS particles of 4 nm average diameter formed on H_2S exposure [176].

The treatment of LB films of copper behenate (10–50 layers) with H_2S gas resulted in formation of the semiconductor Cu_2S [177]. In this case, the LB films of behenic acid alone were formed and then exposed to solutions of copper chloride. Conversion of the carboxyl groups to carboxylate groups upon copper complexation was confirmed by infrared spectroscopy. Resistivity measurements versus temperature confirmed the formation of semiconducting Cu_2S in one case, and showed a linear increase in log(R) versus $1/T(K)$. All of the samples became insulators on exposure to air; maintaining the conductivity required storage under vacuum. The formation of Cu_2S sheets in some of the sample was concluded from optical microscopy and resistivity data.

Langmuir–Blodgett films of arachidic acid transferred from subphases containing $CuSO_4$ and NH_3 were used to generate nanoparticles of CuS [178]. The added ammonia suppressed the formation of a dinuclear complex $Cu_2(OOCR)_4$ and promoted formation of a mononuclear complex $Cu(NH_3)_2(OOCR)_2$ in the monolayers. The dinuclear species had poor LB transfer characteristics. For 10 dip cycles, the average transfer ratio was 0.94 for the mononuclear complex. The UV-visible absorbance curve for the transferred films showed a peak at 241 nm due to a carboxylate to Cu^{2+} ligand-to-metal charge transfer band. CuS particles were formed by exposure to H_2S gas. The LB film environment stabilized the CuS particles against oxidation. Tapping mode AFM after washing away the surfactant with chloroform gave an average CuS particle size of 4.1 nm. And LB films of behenic acid spread on subphases containing Cd^{2+}, Hg^{2+}, and mixed subphases of the two were exposed to H_2S gas [179]. Cd^{2+} was not observed in the XPS spectra of LB films transferred from subphases below 5.3. The LB films of the mixed metal behenate salts exposed to H_2S resulted in separate CdS and HgS particles, as deduced by modeling the UV spectra. Immersion of a film containing CdS particles into a Hg^{2+} solution resulted in exchange of Cd^{2+} for Hg^{2+} at the surface, resulting in a HgS shell.

Changes in thermal stability and mass due to the formation of CdS nanoparticles in LB films were examined [180]. The LB films were formed onto gold-coated quartz oscillators from monolayers of arachidic acid or nonacosa-10,12-diynoic acid on $CdCl_2$ containing subphases. The films were exposed to H_2S gas until the mass change indicated complete conversion of Cd^{2+} to CdS. The thermal stability of the H_2S-treated films was reduced, with significant mass loss initiating at ~55°C, compared to minimal mass loss in the untreated films up to at least ~80°C under mild vacuum. The average CdS-particle size from TEM was 2.5 nm. A more thermally stable film was obtained if films of just arachidic acid exposed to H_2S gas were then exposed to aqueous $CdCl_2$; however, the mass change suggested one-third less particle formation. The LB films of the diacetylenic fatty acid treated with H_2S gas and polymerized with UV irradiation were more thermally stable. The formation of CdS particles inside cadmium arachidate LB films was studied in more detail using QCM, UV-visible spectroscopy, and electron microscopy [181]. The mass uptake for 10-, 20-, and 40-layer films was studied and was consistent with complete conversion of Cd^{2+} to CdS. The UV-visible spectra showed an onset near 440 nm, corresponding to 2–3-nm-size particles. Examination by TEM showed a broader range of sizes, the largest being 14 nm and 2–4 nm being the most common size. Immersion of the films in $CdCl_2$ caused conversion of arachidic acid back to cadmium arachidate, and further gassing increased the number of CdS particles present.

The relation between LB film structure and CdS particle formation was studied using x-ray reflectivity [182]. Langmuir–Blodgett films of nine layers of cadmium arachidate were deposited on quartz at 30 mN m^{-1} and exposed to H_2S gas for 10, 30, and 60 minutes; the absorption edge was 455 ± 5 nm. The reflectivity data were fit to a model of alternating stacks of different electron density. It was concluded that the CdS particles retained a layered arrangement, restricted to 1.4 nm either side of the original planes of metal atoms. The effect of CdS particle formation on the structure of LB films of cadmium stearate was studied by AFM imaging of the surface of the LB film [183]. Prior to H_2S exposure, the LB film surface was smooth and the ordered packing of the tail groups of the molecules could be clearly seen. A rectangular herringbone lattice with $a = 0.842$ nm and $b = 0.510$ nm was evident. After particle formation, the surface became bumpy and the ordered lattice structure could no longer be resolved. Grazing-angle FTIR was used to follow the effect of CdS particle formation on the structure of 20-layer LB films of cadmium arachidate [184]. The LB films were subjected to cycles of H_2S gassing followed by immersion in aqueous $CdCl_2$. Analysis of the carboxyl/carboxylate and methyl/methylene infrared modes revealed that CdS particle formation caused significant tilting, but not disordering, of the alkyl chains. A persistent fraction of arachidate not converted back to arachidic acid was believed to be capping the CdS particles. Curiously, a study of particle formation of CdS in cadmium stearate LB films concluded that all of the stearate was converted to stearic acid after gas treatment [185]. The effect of the transfer pressure on the production of CdS particles in cadmium stearate LB films was examined [186]. It was found from x-ray diffraction that the layer spacing was retained when CdS particles were generated in cadmium stearate films transferred at 37.5 mN m^{-1}, while gas treatment destroyed the periodicity in films transferred at 25 mN m^{-1}. CdS nanoparticles (3–6.4 nm) formed by H_2S gas treatment of LB films of cadmium behenate were exposed to hexane to dissolve away the organic matrix [187]. The remaining particle film was studied by Raman, infrared, and UV spectroscopy. In the infrared spectra, it was observed that the surface optical phonon frequencies shifted from 273 cm^{-1} to 280 cm^{-1} after the removal of behenic acid. The longitudinal optical phonon mode frequencies seen in the Raman spectra of the CdS particles were lower than those for bulk CdS.

The photoelectrochemical properties of CdS nanoparticles formed in LB films of cadmium arachidate on ITO glass (indium tin oxide–coated glass) were investigated [188]. The CdS particles were formed by exposure to H_2S gas, and then the cadmium arachidate structure was "regenerated" by exposing the gas-treated films with aqueous solutions of $CdCl_2$. Gassing/immersion cycling increased the particle size from 2.3 ± 0.7 nm after one cycle to 9.8 ± 2.4 nm after five cycles. The 9.8-nm particles showed UV-visible absorbance characteristic of the bulk semiconductor. The photoelectrochemical cell consisted of a Pt counterelectrode, a SCE reference electrode, and a 150-W Xe lamp. Open-circuit photovoltages as large as -800 mV were observed. The supporting electrolyte was sodium sulfite. The mechanism for the photoresponse involved creation of electron-hole pairs in the nanoparticles, followed by oxidation of aqueous sulfite (SO_3^{2-}) to dithionite ($S_2O_6^{2-}$). The long-term degradation of the photovoltage was due to the photodissociation of CdS back to Cd^{2+} and S. Electron/hole pairs moved from the initial particle in the LB film to the electrode surface. Treatment of the CdS-particle-containing LB films with H_2Se gas resulted in formation of $CdS_xSe_{(1-x)}$ core-shell particles [189]. These core-shell particles contained within LB films exhibited a lower open-circuit voltage and much greater short-circuit currents under illumination, for the CdSe shell has a lower bandgap.

One of the problems with nanoparticles of CdS and other materials is the presence of defect states on the particle surface that trap electron-hole pairs. This is similar to the problem of surface passivation of bulk semiconductors. These states reduce the quantum yield for radiative electron-hole recombination; passivation can be achieved by binding of amines to CdS [190,191]. CdS particles prepared from LB multilayers of cadmium arachidate by exposure to H_2S gas were subjected to 1.0-MeV H^+ ions [192,193]. The UV-visible absorption spectra of films gas treated at 80°C had absorption threshold of 430–450 nm (3.1–3.6-nm particle size) versus 370–390 nm (2.2–2.4-nm particle size). From the in situ ion-induced emission spectra and photoluminescence (325-nm excitation, He-Cd laser) after the ion treatment, it was concluded that the ion irradiation removed mid-gap states. The defect states included those due to Cd^0 (formed from Cd^{2+} in $CdSO_4$ generated by heating-induced oxidation), those due to sulfur atoms, and those due to interstitial sulfur atoms. The emission due to the trapping states associated with Cd^0 and interstitial sulfur were eliminated after irradiation with H^+. The emission from the sulfur-related defects eliminated as well for the particles prepared by gassing at 80°C, the absorption threshold did not change after irradiation. Time-resolved ion-induced emission showed that the lower-energy emissions (510 nm and 600 nm) had multiexponential decay related to the distribution of photogenerated electrons recombining with trapped holes at the particle surface.

Nanoparticles of CdSe can be formed in LB films by exposure to H_2Se gas [194]. The gas was produced by heating a suspension of FeSe in concentrated HCl. The absorption edge increased with gassing time from 540 nm after 20 minutes to 578 nm after 125 minutes (Fig. 15). Using absorption onset versus particle size data, this corresponds to particle size increasing from 1.1-nm radius to 1.3-nm radius. The absorption edge for bulk CdSe was given as 743 nm. By comparing the absorbance versus gassing time for 4-layer and 20-layer films, it was concluded that not all the Cd^{2+} was converted to CdSe in the 20-layer film. FTIR confirmed conversion of the arachidate to the acid form upon gassing and a tilting of the alkyl chains upon particle formation. XPS spectra of the Se 3d-peak showed the presence of some elemental selenium in the LB films. It was noted that, due to the inelastic mean free path of the photoelectrons, XPS probed approximately the outer 5–6 layers.

In an effort to restrict the location of semiconductor nanoparticles in LB films and inhibit aggregation, the formation of CdS in LB films of calixarenes was investigated [195]. Limiting areas of 3.0 nm^2 and 1.8 nm^2 were obtained on 0.5 mM $CdCl_2$, compatible with the cross-sectional areas of the calixarenes. Y-type LB films were prepared at 25 mN m^{-1} on glass, quartz, and silicon. The substrates had been made hydrophobic by treatment with a silane vapor. After H_2S treatment overnight in sealed jars, UV absorbance spectra and XPS data were obtained. The absorption edge for the CdS particles formed in the calixarene LB films transferred at pH 5.5 was 3.3 eV as compared with 2.7 eV for films formed in cad-

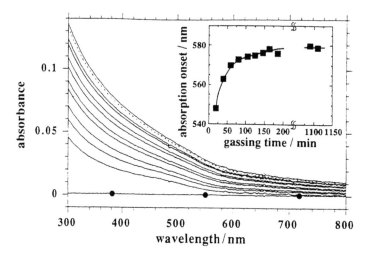

FIG. 15 UV-Visible absorbance spectra of CdSe nanoparticles forming in a 20-layer cadmium arachidate Langmuir–Blodgett film by repeated H_2Se gas treatments followed by recording the spectra. The flat line with the circles (•) is the spectra obtained by gassing a 20-layer arachidic acid film with H_2Se. The lines above this were recorded after gassing the 20-layer cadmium arachidate film for periods of 20, 40, 60, 80, 105, 125, 145, 185, 1089, and 1109 minutes. The contributions due to the quartz plate and the LB film were subtracted. The inset shows the absorption onset wavelength versus gassing time. (Reproduced with permission from Ref. 194. Copyright 1995 American Chemical Society.)

mium stearate LB films. Smaller-size particles formed in the calixarene LB films than in the cadmium stearate LB films. The value of the absorption edge decreased with pH, which corresponded to larger CdS particles formed from films transferred at higher pH, the range 5.5–7.5 being examined. The pH effect was explained as due to greater binding of Cd^{2+} by the monolayers at higher pH. XPS confirmed the presence of the CdS particles, and the greater Cd^{2+} content in films transferred at higher pH. The smaller particle size was attributed to restriction of the growth of the CdS particles, to some extent, by the calixarene cavities. In the case of 1.5-nm CdS particles formed in calixarene LB films [196], the UV absorbance spectrum showed the presence of three overlapping transitions due to transitions between discrete energy levels of the nanoparticles. The x-ray reflectivity on LB films before and after H_2S treatment showed that the periodicity of the calixarene LB films was hardly affected. In contrast, the regular periodicity of cadmium stearate LB films was reported as severely diminished. While the CdS particles were too large to be included inside the calixarene cavities, it was proposed that partial inclusion reduced particle aggregation.

The goal of constructing superlattices has motivated efforts to remove the fatty acid molecules after nanoparticle formation. CdS-nanoparticle films formed from cadmium arachidate LB films followed by dissolution of the fatty acid were investigated by QCM and STM [197]. These films were subsequently investigated by AFM and SEM [198]. The mass changes detected by QCM were consistent with the increase expected during formation of the nanoparticles and the decrease associated with dissolving-out of the surfactant molecules. Removal of the surfactant matrix was achieved by immersion in chloroform. X-ray reflectivity showed a loss of all the Bragg peaks after the chloroform treatment. The size of the CdS particles was 4.4 nm when prepared in 20-bilayer LB films and was 2–2.5 nm when prepared in a single LB bilayer. The nanoparticles arranged themselves in 100-nm triangular clusters that were often further arranged in needlelike forms. The surface of some regions of the film was flat enough for atomic-scale AFM or STM imaging, and both methods showed the lattice periodicity of 0.42 nm expected for bulk CdS. The formation of superlattices of nanoparticles of different types of particles has also been reported [199]. The superlattices were formed by first exposing LB films of the first metal salt to H_2S and then washing out the arachidic acid with chloroform. This was followed by deposition of a LB multilayer from a subphase containing a second metal salt on top of the layer of nanoparticles of the first metal sulfide. Exposure to H_2S followed by chloroform washing produced a layer of nanoparticles of the second metal sulfide on top of the first layer. A layered film of CdS-MgS-CdS-MgS-CdS was prepared and SEM images, viewed edge on, showed a 60-nm thickness for each layer. And LB films of up to 200 layers of yttrium arachidate were reported to yield a uniform thin film of Y_2O_3 upon heating at 550°C [200]. The layers were characterized by SEM, optical reflectance, grazing-angle x-ray diffraction, and Rutherford backscattering spectroscopy. It was mentioned that other trivalent metal salt LB films produced good oxide films, and this was a possible route to thin layers of high-temperature superconductors.

The observation of monoelectron conductivity through CdS nanoparticles formed in single-bilayer LB films of cadmium arachidate using STM has been reported [201]. Steplike variations in the current versus bias voltage corresponded to individual electrons moving onto the nanoparticle. A double-tunneling barrier structure was required, and this was achieved by positioning an STM tip over a single nanoparticle in the single layer of nanoparticles formed in a bilayer LB film on graphite. More reproducible *I-V* curves were obtained in a modified procedure that involved forming the CdS particle directly on a tungsten STM tip [202]. The tip was immersed and removed from cadmium arachidate monolayer to form a single bilayer on the end of the tip. The CdS particles were then formed on the tip. Highly oriented pyrolytic graphite (HOPG) was used as the second tunneling junction. Symmetrical steplike behavior and sometimes unsymmetrical steplike behavior were seen in the *I-V* curves, as shown in Figure 16. The unsymmetrical behavior was attributed to charge trapping by an impurity. In 40% of the samples, no steplike behavior was observed, indicating that a CdS nanoparticle had not formed at the tip during the H_2S gas treatment.

D. Langmuir–Blodgett Films Containing Magnetic Nanoparticles

Magnetic nanoparticles are of interest for their potential application in magnetic recording media and magneto-optical devices. The initial accomplishments in this area were also from Fendler [203]. Two layers of cadmium arachidate on silicon were formed by LB transfer of arachidic acid spread on 0.25 mM $CdCl_2$ at pH 5.6. The silicon substrate was raised through the water surface once and then immersed into a beaker placed in the trough that contained a dispersion of colloidal magnetic particles. The colloidal magnetic particles were cationic Fe_3O_4, 8.1 ± 0.5 nm in diameter, previously found to adhere strongly to lipid bilayers [204]. A third layer of arachidate was deposited over the nanoparticle layer to create a sandwich structure. The adsorption of the Fe_3O_4 particles onto the two-layer Cd arachidate LB film was followed by optical reflectivity; fitting of the reflectivity versus angle data to the Fresnel matrix equations gave a thickness consistent with the proposed structure. Repeating the dipping and particle adsorption process resulted in the layered structures, up to seven uniformly deposited alternating cadmium arachidate bilayer + Fe_3O_4-particle layers. These films had magneto-optical properties and showed a Kerr ef-

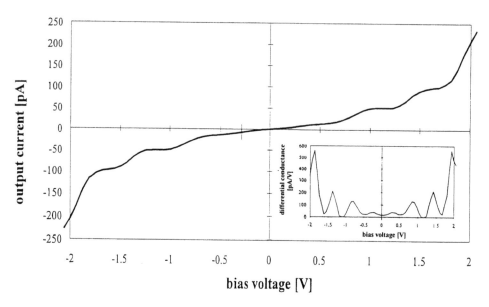

FIG. 16 Current versus bias voltage for a CdS nanoparticle on the end of an STM tip. The CdS particles were formed by exposing a bilayer of cadmium arachidate on the STM tip to H_2S gas. The other conducting surface is a highly oriented pyrolytic graphite electrode. The inset is a plot of differential conductance versus the bias voltage. (Reproduced with permission from Ref. 202. Copyright 1996 National Academy of Sciences, U. S. A.)

fect, the rotation of polarization of reflected light during application of a magnetic field parallel to the film surface. The Kerr effect also provided a contrast mechanism for imaging micron-scale domains of magnetic-particle orientation; some rod-shaped domains parallel to the dipping direction were observed. Fe_3O_4 particles stabilized by surfactants were later studied at the water–air interface using BAM and as transferred films using TEM [205]. The TEM images showed ~100–500-nm circular islands of magnetite particles, and further compression pushed these islands close together. The lauric acid was segregated from the magnetite clusters. Reflectivity data for the collapsed films gave a thickness of 13.8 nm. The diameter of the prepared particles was 13.0 ± 0.3 nm from TEM analysis.

Another approach to forming LB films containing magnetic nanoparticles used a suspension of Fe_3O_4 particles (average size = 8.5 nm) as the subphase for monolayers of poly(octadecene-co-maleic anhydride) [206]. The negative charge of the carboxylate groups in the polymer monolayer attracted the positively charged Fe_3O_4 particles. The Π-A isotherm of the polymer on the nanoparticle suspension was significantly expanded, the limiting area increased from 0.24 nm^2 to 0.29 nm^2. And BAM observation showed a homogeneous surface for the polymer monolayer up to the collapse. On the nanoparticle subphase, compression of the polymer monolayer results in the reversible formation of island-like domains. And TEM revealed a random distribution of the Fe_3O_4 particles, which were bound to the monolayer by electrostatic attraction. The LB films of γ-Fe_2O_3 nanoparticles were prepared by spreading stearic acid monolayers on a dispersion of the nanoparticles [207]. As for the polymer monolayer, the binding of the nanoparticles expanded the films. And FTIR confirmed the complexation of the stearic acid carboxylate head groups to the particles. In an earlier study, a dispersion of α-Fe_2O_3 nanoparticles was used as the subphase for stearic acid monolayers [208]. The stearate–Fe_2O_3 complexes were transferred onto a variety of substrates (up to 21 layers), with transfer ratios near unity. Transmission FTIR showed hydrocarbon chains in trans conformations. Small-angle x-ray data gave a spacing of 16 nm, consistent with a multilayer Y-type structure with two layers of α-Fe_2O_3 particles between the head groups of the stearate bilayers. The absorption spectra of the films were similar to those of the hydrosol, the absorption edge being near 600 nm. A subsequent study [209] used a polymaleic anhydride polymer with a nitroazobenzene derivative side chain. And TEM of the transferred films showed domains of closely packed particles separated by low-density regions. It was noted in this study that a single monolayer of these particles enhanced the photovoltage response of silicon substrates. In another study, monolayers of stearic acid were spread on a subphase of a commercial ferrofluid containing both Fe_3O_4 (8 nm) and Fe_2O_3 (17 nm) nanoparticles [210]. The Fe_3O_4 particles adsorbed preferentially onto the stearic acid monolayers, which were transferred onto Si or HOPG for analysis by x-ray diffraction, tapping-mode AFM, and STM. One-dimensional 0.3–0.5-μm chains of particles were found in the LB films, as were 0.1–0.2-μm oval domains having an oblique lattice. The spontaneous organization of these particles was attributed to the magnetic dipole interactions.

A study of the effect of magnetic nanoparticle size on the monolayer behavior aimed to examine the balance between magnetic dipole forces and van der Waals interactions [211]. Nanoparticles of γ-Fe_2O_3 in three size ranges were examined: 7.5 nm ($\sigma = 0.1$ nm), 11 nm ($\sigma = 0.15$ nm), and 15.5 m ($\sigma = 0.2$), where σ is for a log-normal distribution. The cationic particles were coated with lauric acid, yielding a precipitate that was redispersed into hexane for spreading onto the water surface. The Π-A isotherms were reproducible after the first compression–expansion cycle. After normalizing the x-axis to A/A_o, where A_o is the area per particle in a close-packed configuration, it could clearly be seen that the 7.5-nm particles attained a much closer packed structure on compression than the larger particles. For the 7.5-nm-size particles, A/A_o approached unity under compression, while for the 15.5-nm particles, a limit near 1.5 was approached, and $\Pi \sim 5$ mN m^{-1} was recorded at A/A_o ~ 4. The particle monolayers on water could be imaged by reflection microscopy. The 15.5-nm-diameter particles showed stringy, open aggregates, while the 7.5-nm particles formed a uniformly bright layer that fractured on expansion. And TEM of transferred films showed that the larger particles were arranged in chains while the smaller particles formed dense circular islands. The effect of a magnetic field (30 mT) applied across the water–air interface on the isotherm was small but reproducible. The observation of denser aggregates for the larger particles was viewed as consistent with theoretical expectation for systems with both dipolar interactions and isotropic interactions.

Langmuir–Blodgett films of iron(III) arachidate have been used as precursors for the production of ultrathin layers of iron oxide for potential use as magnetic storage media [212]. Such LB films were transferred from subphases of 0.05 mM $FeCl_3$ onto Si wafers and glass or quartz slides, the number of layers ranging from 29 to as many as 305. A certain set of transfer conditions was necessary to produce optimally smooth films. The films were subjected to three treatments: (1) heating in ambient air up to 350–390°C for 15 minutes, (2) heating in ambient air up to 900°C for 1–2 days, (3) heating at 500°C in a tube furnace under a flow of H_2 gas. The films heated in air consisted mainly of α-Fe_2O_3 (antiferromagnetic), and those heated under H_2 consisted of Fe_3O_4 (ferrimagnetic), FeO, and metallic α-Fe (ferromagnetic). The types of iron oxides present were determined by analysis of the Fe 2p peak in the XPS spectra, x-ray diffraction, and analysis of the Fe and O content by ERDA (elastic recoil detection analysis). Only the LB films heated under H_2 had magnetizations potentially suitable for use as magnetic storage media.

If a stearic acid monolayer is spread on a subphase containing $CdCl_2$, $MnCl_2$, and $NaHCO_3$, LB films containing Mn^{2+} and Cd^{2+} can be obtained and used to produce magnetic semiconductor nanoparticles [213]. Such particles have tunable magneto-optical properties. It was noted that the Cd/Mn ratio obtained in the LB films differed from the Cd/Mn ratio in the subphase due to the higher binding affinity of the carboxylate groups for Cd^{2+}. Exposure to $H_2S + NH_3$ mixed gas resulted in the formation of nanoparticles of $Cd_{1-x}Mn_xS$, a magnetic semiconductor. The formation of the magnetic semiconductor particles within 51-layer LB films was confirmed by analysis of the electron spin resonance (ESR) spectra. From UV spectra of gas-treated films with $x = 0.20$, 0.27, and 0.45, particle diameters near 3 nm (with a wide size distribution) were deduced. And ESR on the gas-treated film showed a narrow single line, while the untreated films showed six hyperfine lines. The ESR spectra lost their angular dependence after the gas treatment, indicating that the particle formation disrupted the layer structure of the LB film. Irradiation of the LB films with 1.0-MeV H^+ passivated the surfaces of the particles [214].

Ferroelectric lead zirconium titanate (PZT) particles coated by lauric acid of a size at most 1 μm were successfully spread at the water–air interface [215]. The particles were obtained by grinding PZT and removing the larger particles by precipitation. A dispersion of the particles could be spread on the water surface, giving a stable film that could be transferred. "Fixing" of the transferred layer of particles by applying drops of a solution of lead acetate, titanium isopropoxide, and zirconium ethoxide was necessary for successful transfer of the next layer. The II-A isotherm reached a pressure near 80 mN m^{-1} on the first cycle and showed a transition attributed to squeezing out of the lauric acid. And AFM confirmed that the interparticle distances were smaller for films transferred from the second cycle. The firing of the LB films at 250°C for 2 hours burned off the organic components. A gold electrode was evaporated on top of the films, and graphite tips or STM tips were used as second electrode to measure the polarization. The polarization versus electric field showed ferroelectric hysteresis loops.

IV. LANGMUIR–BLODGETT FILMS OF FULLERENES

A. Introduction

The fullerenes and related carbon nanostructures are central to the development of nanotechnology. The fullerenes include the well-known members, C60 and C70, and other closed-cage compounds of carbon, such as C76, C78, C82, and C84. Fullerenes are widely promoted as fundamental building blocks for nanoscale devices and systems. An active part of the effort in fullerene research concerns thin films containing C60 and C70 and synthetically modified derivatives of C60, including the efforts reviewed in this section to study Langmuir–Blodgett films containing C60 and C70. Due to the enormity of the literature in the field, the discussion is restricted to LB films. A number of excellent recent books have reviewed the physical properties of fullerenes [216–219] and nanotubes [220]. The synthetic chemistry of fullerene modification has also been reviewed [221].

The fullerenes were discovered during the pulsed laser ablation of graphite in a helium gas environment [222]. The prominent C60 and other carbon clusters were identified by time-of-flight mass spectroscopy. In 1990, a method for the production of milligrams-grams of C60, C70, and other fullerenes was introduced [223,224]. The method involves resistive heating of a graphite rod in an inert gas atmosphere, helium giving the highest fullerene yields, to produce soot from which the fullerenes are extracted and separated. The commercial availability of C60 and C70 and the relative simplicity of the soot production gave numerous research groups access to C60 and C70.

C60 has the geometry of a truncated icosahedron formed by hexagons and pentagons, each pentagon being surrounded by hexagons to satisfy the "isolated pentagon rule." The C—C bond length is 0.146 nm on the pentagons and 0.140 nm on the hexagons. C60 is the smallest fullerene to satisfy this rule, C70 being next smallest. The inner diameter of C60 is 0.709 nm, and the outer diameter, including the outer electron cloud, is 1.034 nm. The bonding would be similar to the sp^2 bonding of graphite, but the curvature results in some sp^3 character. C60 crystallizes in a cubic structure with a nearest-neighbor distance of 1.002 nm. While C60 has the well-known soccer-ball shape, C70 is slightly longer along one axis (inner diameters = 0.796 nm, 0.712 nm). The solid-state phase behavior of C70 is more complex, the room-temperature phase is hexagonal close packed. The fullerenes C78 and larger exhibit progressively more geometric isomers, some of which are chiral [225,226]

C60 and C70 are both excellent electron acceptors, and C60 can accept up to six electrons into its $3t_{1u}$ LUMOs. A Huckel calculation for C60 shows that the HOMOs are 5 h_u orbitals holding 10 electrons. The related π orbitals of C70 are nondegenerate but close in energy. The ionization energy of an isolated C60 molecule is 7.6 eV, and the electron affinity is 2.65 eV; thus the molecule is expected to be an electron acceptor. Six consecutive and reversible reduction/oxidation waves were seen using cyclic voltammetry in acetonitrile/toluene solvent at $-10°C$ with tetrabutylammonium phosphoroushexafluoride as the supporting electrolyte. The reduction potentials (vs. Fc/Fc$^+$, ferrocene standard) were -0.98 V, -1.37 V, -1.87 V, -2.35 V, -2.85 V, and -3.26 V; C70 also exhibits six reversible reductions at a series of potentials fairly close to those of C60 [227].

The doping of C60 by alkali metals results in stable crystalline phases for M_1C60, M_3C60, M_4C60, and M_6C60, and is achieved by exposing C60 to metal vapors. The observation of superconductivity at 18 K for potassium-doped C60 (K_3C60) resulted in widespread interest in doped fullerene films [228]. Cs_3C60 enters a superconducting state at $T < T_c = 40$ K [229]. In exohedral doping, the dopant atom resides outside the C60 cages. Undoped fullerenes exhibit semiconductivity. Doping with alkali metals lowers the resistivity, with the lowest values reached for M_3C60; metallic conductivity is observable near this stoichiometry [230].

B. Langmuir–Blodgett Films Containing C60 and C70

1. General Properties of Monolayers and Langmuir–Blodgett Films of C60 and C70

Most studies of spread films of C60 and C70 on water include II-A isotherms, with limiting areas near 0.2–0.3 nm^2 molecule^{-1} indicating the formation of multilayers. The affinity of C60 and C70 for the water surface is low, for they are uniformly hydrophobic. However, spreading of monolayers giving isotherms with limiting areas on compression near 0.95 nm^2 molecule^{-1}, consistent with the size of C60, can be achieved if the concentration of the spreading solution is low enough. The first report of Langmuir films of C60 [231] included the isotherm reproduced in Figure 17; the films were produced by spreading of 0.05–0.1 mM solutions of C60 in benzene. The isotherms were run on a trough using the Langmuir float method to measure II. The films were very rigid, sustaining surface pressures II > 65 mN m^{-1}. The limiting areas derived from the II-A isotherms of films spread from solutions of greater concentration were reported to give radii near 0.35 nm, compared with 0.56 ± 0.07 nm for the monolayer films. The multilayer films sustained values of II

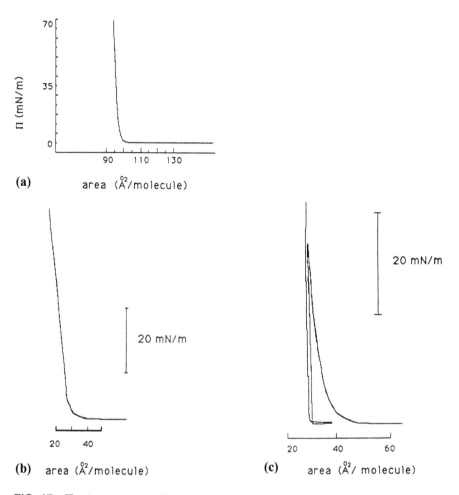

FIG. 17 Π-*A* isotherms for C60 on a pure aqueous subphase at 21°C: (a) Monolayer prepared by spreading 50 μL of a 0.1 mM solution in benzene; (b) monolayer prepared by spreading 200 μL; and (c) a compression–expansion cycle of a monolayer prepared by spreading 100 μL of a 0.5 mM solution. (Reproduced with permission from Ref. 232. Copyright 1993 American Chemical Society.)

near 100 mN m^{-1}. The formation of needle crystals under compression of multilayer films was sensitive to the presence of residual solvent [232]. The transfer of C60 films compressed to 15 mN m^{-1} onto silver-coated quartz oscillators raised and lowered vertically through the water–air interface was reported, and the mass change derived from the frequency change per transfer cycle [233]. The oscillator was lowered and raised five times over 150 minutes. After each raising up through the interface, approximately 15 minutes was required for the evaporation of entrapped water. The final frequency decrease per cycle for all but the first was 3.1–3.3 times that expected for a C60 monolayer. The surface-enhanced Raman spectrum of C60 monolayer deposited on glass covered by silver island films was reported and the expected C60 Raman peaks, as well as some peaks assigned to higher-order Raman modes, were assigned [234]. The formation of multilayer films by C60, C70, and a C60/C70 mixed film was reported in the Π-*A* measurements of Back and Lennox [238], who also observed a kink in the Π-*A* isotherm of C70, suggesting a reorientation transition.

Mixed films of C60 and arachidic acid gave limiting molecular areas near 0.28 nm^2 molecule^{-1}, suggesting that the C60 units were embedded in the hydrocarbon matrix. The molecular areas obtained deviated from that expected for ideal mixing for mole fractions of C60 above ~0.3 [231]. In a related study, a limiting area near 0.17 nm^2 molecule^{-1} was reported for C60 films spread from benzene [236], consistent with formation of multilayers. Equimolar mixed films of C60 with arachidic acid and with dioctadecyldimethylammonium perchlorate (DODMAP) gave reproducible isotherms. The contribution of C60 to the limiting area per molecule was 0.07 nm^2 molecule^{-1} for the 1:1 C60/arachidic acid films and 0.06 nm^2 molecule^{-1} for the 1:1 C60/DODMAP films. These observations are consistent with the C60 units residing primarily in the hydrocarbon matrix of the mixed films. The UV absorbance spectra of the mixed transferred films showed three main absorbances at 340, 265, and 220 nm, red-shifted by about 10 nm relative to those of the solutions in benzene. The peak absorbances were proportional to the number of layers transferred. The AFM images of the C60 LB film on highly oriented pyrolitic graphite (HOPG) showed crystallites similar to those obtained from casting of solutions of C60 in benzene (Fig. 18a). In contrast the AFM images of the mixed LB films of C60 and arachidic acid were smooth to better than 5 nm (Fig. 18b). The AFM images of DODMAP/C60 LB films on HOPG showed doughnutlike structures attributed to DODMAP vesicles for monolayer LB films and the 20 layer films had a rough surface with 200 nm undulations [237]. In a study of C60 films spread from benzene, limiting areas below 0.20 nm^2 molecule^{-1} were observed, and the films would not transfer to fused-silica slides [238]. Equimolar mixed films with stearyl alcohol could be transferred to fused silica.

Studies of LB deposition of C60, C70, and mixtures with arachidic acid onto glass, single-crystal silicon, quartz, and aluminum-coated glass were reported [239,240]. The transfer ratios for C70 films spread from more concentrated solutions were 0.7 ± 0.1; ellipsometry gave a thickness per transferred layer of 9.5 nm up to eight layers of Z-type deposition onto silicon. Conductivity measurements, both parallel and perpendicular to the plane of the film on Al-coated glass and with Al evaporated top electrodes, were performed. The in-plane measurements were all short circuits, and the perpendicular measurements gave 10^{-11}–10^{-12} S cm^{-1}. Alternate 35-layer LB films of C60 and copper pthalocycanine were reported, and the UV spectra contained the unchanged peaks from the two components [241]. The LB films of 1:4.2 C60 and stearic acid were subjected to four successive compression cycles; the isotherm shifted closer to that of stearic acid with each compression cycle. This suggests that the C60 molecules were squeezed out of the hydrocarbon matrix with successive compressions to form a sheet on top of the stearic acid chains [242]. X-ray diffraction on 31-layer films gave a layer thickness of 2.9 nm, slightly greater than that for stearic acid. Using ellipsometry and reflectivity, a refractive index of 1.75 was determined for the mixed LB films.

BAM observations on C60 and C70 monolayers [243] showed the coexistence of a two-dimensional gas phase with a two-dimensional condensed phase at high molecular areas. BAM was used to distinguish monolayer regions from bilayer or multilayer regions, at least on the microns size range. Upon spreading from dilute solutions (~0.01 mM) at very large molecular areas, circular domains and foamlike patterns were observed. The formation and growth of brighter, multilayer "discs" occurred frequently during the compression, until by 0.80–1.20-nm^2 molecule^{-1} the films contained many multilayer regions. Spreading from concentrated solutions (above 0.1 mM) clearly showed sheetlike regions of different thickness, some monolayer and others multilayers.

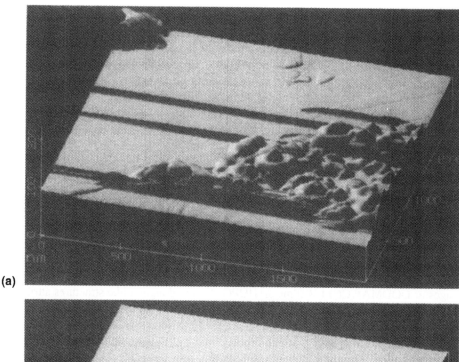

(a)

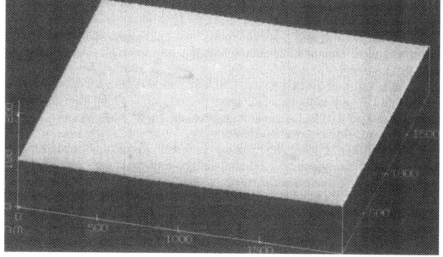

(b)

FIG. 18 Scanning force microscopy images. (a) C60 transferred horizontally onto highly oriented pyrolytic graphite (HOPG) at 25 mN m^{-1}. (b) 1:1 mixed film of C60 and arachidic acid transferred horizontally onto HOPG at 25 mN m^{-1}. (Reproduced with permission from Ref. 235. Copyright 1996 American Chemical Society.)

High-resolution transmission electron microscopy (HRTEM) was applied to LB films of C60. Selected area diffraction of the C60 LB film showed a hexagonal symmetry with a 1.0-nm unit cell. In addition to well-ordered regions, there were amorphous regions where lattice fringes could not be observed [244]. HRTEM was applied to image C60 films transferred from a subphase that contained dissolved phenol [245]. The Π-A isotherms showed multilayer formation; however, optical micrographs of LB films on hydrophobic substrates revealed a smooth texture for the films transferred from the phenol containing subphase, versus a patchy texture for the films transferred from pure water.

The results suggested overall trilayer formation. The HRTEM images of the LB films transferred from phenol containing subphase onto holey carbon films showed regions of hexagonal symmetry, regions of distorted hexagonal packing, and amorphous regions. And HRTEM examination of Langmuir–Schaefer films of C60 on copper grids showed variations in morphology from area to area; the most ordered regions were consistent with a reorganization into a face-centered cubic structure [246]. STM images of 20-layer LB films on HOPG were also shown to have a lattice structure consistent with fcc packing. X-ray diffraction on the 20-layer LB films of C60 showed no periodicity. An SEM image of C60:arachidic acid LB films of molar ratio 1:4.2 showed some occasional agglomerates of C60 [247]. Forty-layer LB films prepared from equimolar mixtures of C60 and C70 with arachidic acid showed periodic fringes in small-angle x-ray diffraction, giving a periodicity of 5.57 nm [248]. This periodicity is similar to that of LB films of pure arachidic acid (5.62 nm) and represents a bilayer periodicity. It was conjectured that the film structure is determined by the arachidic acid molecules and that the fullerenes are immersed in the hydrocarbon chain matrix. Transfer ratios of C60 and C70 of 0.4–0.6 were reported, while those for the mixed films were near 1 for C60 and 0.8 for C70. C60 films examined by STM showed some regions of ordered molecules and some regions of aggregated molecules. HRTEM images of C60 showed crystalline regions, as well as defects and amorphous regions, while the images for C70 films had smaller crystalline regions. STM of the mixed film of arachidic acid and C60 were mostly smooth with some small bumps, indicating some C60 aggregation but that most of the C60 was dispersed in the hydrocarbon chains.

The preparation of good-quality LB films of C60 mixed with poly(p-methyl)phenyl methacrylate was reported; on the basis of ellipsometry, an average thickness per transferred layer of 1.6 nm was reported [249]. An AFM study of C60 LB films spread in a monolayer regime (0.05 mg mL^{-1} in toluene) and a concentration in a multilayer regime (0.10 mg mL^{-1} in toluene) and transferred at different surface pressures was carried out [250]. The LB films prepared on freshly cleaved mica at 1.0 mN m^{-1} showed fractal-like collections of aggregated C60 particles. Transfers at higher pressure showed denser aggregates of similar-size C60 particles at 7 mM m^{-1}; larger particles nearly 2 mm in size were also found. The smaller size C60 particles were 130 nm in diameter and about 15 nm in height. Images of LB films transferred at 0 mN m^{-1} (before the pressure rise) were relatively smooth. In these authors' experiments, the Π-A isotherm appears less steep, and C60 aggregation occurred upon spreading or early in the compression. The size of the C60 particles depended upon the spreading solvent, with smaller particles seen upon spreading from methylene chloride.

2. Mixed Monolayers of Fullerenes and Amphiphilic Macrocycles

One of the methods for separating and purifying C60 utilizes their complexation with calixarenes [251,252]. Fullerene complexation by macrocycles produces water-soluble complexes, as can be achieved by refluxing γ-cyclodextrin with C60 [253]. Studies of Langmuir monolayers of fullerenes spread with amphiphilic derivatives of macrocycles demonstrate encapsulation of the fullerene as a host–guest complex and at least partial inhibition of fullerene aggregation. Mixed monolayers of C60 and C70 with an amphiphilic hexa-aza-crown and an amphiphilic octa-azacrown were studied at the water–air interface [254]. The azacrowns were functionalized on the ring nitrogens by amide linkages to 3-methoxystyryl, 4-dodecyl ether groups. Mixed films of C60 and the amphiphilic hexa-azacrown showed a plateaulike feature under compression whose width increased as the C60 content was in-

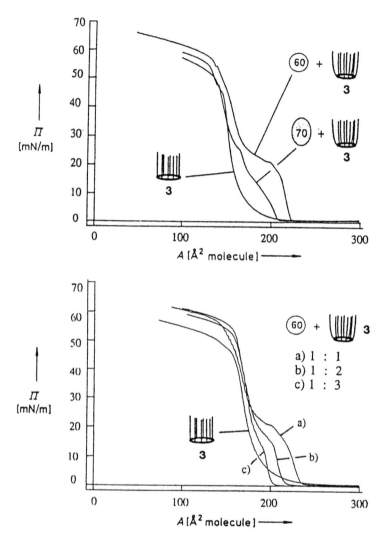

FIG. 19 Π-*A* isotherms of amphiphilic hexa-azacrown (noted as 3 in the figure) and 1:1 mixtures with C60 and C70 (top) and for 1:1, 1:2, and 1:3 mixtures of C60 and the amphiphilic hexa-azacrown (bottom). (Reproduced from Ref. 254 with permission of the author. Copyright 1991 Wiley-VCH.)

creased toward 50% (Fig. 19), the plateau was also observed for mixed monolayers with C70. The plateau was not observed for equimolar mixtures of C60 or C70 with the octa-aza-crown. The collapse pressure was ~5 mN m^{-1} higher in the 1:1 mixed films with C70 and ~10 mN m^{-1} higher for the 1:1 mixed films with C60 than for the monolayer of the pure aza-crown. The structure proposed for the mixed films has the C60 occupying the azacrown cav-ities. It was proposed that decomplexation followed by C60 crystallization occurs upon com-pression for the C60 + hexa-azacrown films; this did not occur upon compression of the C60 + octa-azacrown films. Well-defined LB multilayers could be deposited on either hy-drophobic or hydrophilic supports. The regularity of the multilayers was confirmed by ob-servation of Kiesig fringes in small-angle X-ray scattering (SAXS) data. AFM micrographs provided support for decomplexation of the complexes upon compression. Smooth images were obtained for 1:1 mixed films transferred at 5 mN m^{-1}, while C60 crystallites were seen in films transferred at 25 mN m^{-1}, a pressure above the plateau in the isotherm.

A subsequent study examined mixed monolayers of *p-tert*-butylcalix[8]arene with C60 and C70 using II-*A* isotherm measurements combined with BAM [255]. These authors prepared the spreading solutions (in chloroform) for monolayer formation from the solid calix[8]arene:C60 (or C70) complexes, prepared by precipitation from refluxing solutions. In both the calixarene:C60 and calixarene:C70 monolayers, the isotherms and BAM observations indicated a transition from a two-dimensional gas phase coexisting with a two-dimensional condensed phase near zero surface pressure to a condensed phase as the surface pressure increased. The BAM images of the calixarene:C60 monolayers clearly show large islands (>100 μm) in a dark background (the gas phase). The calixarene:C70 monolayers also showed islands, but with inner regions of distinctly different brightness interpreted as due to the presence of C70 bilayers, trilayers, and multilayers. Upon compression, the islands "crashed" into each other, merging and healing, although regions of overlap became visible. The compressed films were reported as fragile and prone to fracture if the trough was disturbed. The BAM observations provide important information on the morphology of the films.

In a study of mixed monolayers of C60 and *p-tert*-butylcalix[8]arene, different isotherm behavior was obtained [256]. The surface pressure was observed to rise at a lower molecular area (1.00 nm^2 molecule^{-1} vs. 2.30 nm^2 molecule^{-1} in the prior study). Similar isotherms were observed whether a 1:1 mixture or a solution prepared by dissolving the preformed 1:1 complex was spread. The UV spectra of the transferred LB films appeared different than that of bulk C60. It was concluded that a stable 1:1 complex could be formed by spreading the solution either of the mixture or of the complex. This was confirmed in a later study by the same group that included separate spreading of the calixarene and the C60 [257]. This is clearly a complex system, for it is pointed out that the calixarene-limiting area depends on the spreading solvent and is subject to a "memory effect." In another study, mixed monolayers and the resulting LB films of C60 with two derivatized calix[4]resorcinolarenes and with *p-tert*-butylcalix[6]arene and *p-tert*-butylcalix[8]arene were examined [258]. The C60 in the mixed monolayers with the calix[4]resorcinolarenes was either phase separated or within the hydrocarbon matrix of the macrocycles. With *p-tert*-butylcalix[6]arene, C60 caused an expansion of the monolayer and was believed to be in the calixarene and stabilizing the cone configuration. C60 also clearly expanded the *p-tert*-butylcalix[8]arene monolayers and was concluded to be fully encapsulated. The expansion of calixarene monolayers upon C60 inclusion was attributed to stabilization of the open-cone configuration and prevention of the calixarene from reorienting on compression from perpendicular to parallel to the water surface. Even layer-by-layer LB film deposition was confirmed using ellipsometry. SEM images showed that only the *p-tert*-butylcalix[8]arene LB films were smooth and free of patches indicative of C60 aggregation.

The suppression of C60 crystallite formation in mixed LB films was attempted by mixing C60 and amphiphilic electron donor compounds [259]. Observation of the C60 LB film transferred horizontally by TEM clearly showed 10–40-nm-size crystallites. The diffraction pattern gave an fcc lattice with unit cell length 1.410 nm. Examination of the mixed films with arachidic acid by TEM showed extensive crystallite formation. Mixed LB films of three different amphiphilic derivatives of electron donors with C60 were examined. One particular derivative showed very little formation of C60 crystallites when LB films were formed from monolayers of it mixed with C60 in a 1:2 ratio, while two others reduced C60 crystallite formation but did not eliminate it.

3. Langmuir–Blodgett Films of Synthetic Derivatives of C60

While spreading of C60 and C70 in monolayer states can be accomplished, both are clearly subject to a tendency to aggregate. Much work has been concerned with the synthetic modification of fullerenes [221], and the addition of hydrophilic functional groups improves the prospects for monolayer spreading. One of the first studies of a modified fullerene concerned 1-*tert-butyl*-9-hydro-fullerene-60, C60 with a *t*-butyl group and a hydrogen added across a C60 double bond [260]. The use of a dilute spreading solution gave isotherms consistent with a monolayer, while a more concentrated spreading solution gave thicker films. This appears to be the first successful spreading of a fullerene as monolayer in a trough using the Wilhelmy plate method rather than the Langmuir float method. The monolayer films did not transfer well, but the thicker films gave transfer ratios near unity for transfer onto single-crystal silicon, quartz, and aluminum-coated glass. Ellipsometry gave a thickness per transferred layer of 5.56 nm [261]. The electrical properties perpendicular to the plane of the film were studied and a dc conductivity of 10^{-10} S cm^{-1} was found, indicating that the films were insulators.

A comparison of the monolayer behavior of C60, the methylene-bridged $C61H_2$, and the fullerene epoxide C60O was reported [262]. The behavior of $C61H_2$ and C60O was similar to that of C60 (monolayer spreading only from dilute solutions); however, C60O was easier to spread as a monolayer, due to the modest hydrophilicity introduced by the epoxide group.

A set of C60 derivatives formed by attachment of a *N*-acylpyrrolidine group was studied [263]. Derivatives with hydrocarbon acyl groups (methyl, hexadecyl, arachidyl) required spreading from dilute solutions to attain monolayer rather than multilayer behavior, a derivative where the acyl group was perfluoroheptyl spread as monolayers regardless of the concentration [264]. Being more hydrophobic than C60, the fluorocarbon chains preferentially orient away from the water surface and inhibit fullerene aggregation.

C60 modified by nucleophilic addition of dodecylamine forms C60-(NH-$(CH_2)_{11}$-$CH_3)_x$ where $x = 5 \pm 2$ by elemental analysis and the dodecyl chains are randomly distributed [265]. In situ neutron reflectivity data on the C60 films and of the C60-dodecylamine adduct at the water–air interface were obtained. These authors report obtaining limiting molecular areas of 0.10–0.40 nm^2 molecule^{-1}. The neutron reflectivity data were fit to a model consisting of two slabs and yielded a total film thickness near 10.0 nm. Within this total thickness, the film is rough, with many areas 8–9 molecules thick and other, thinner regions 2–3 molecules thick. In the model, the lower slab consists of C60 units and water, and the upper slab consists of C60 units and air. The C60-dodecylamine adduct exhibited reproducible monolayer spreading with Π rising near 1.50 nm^2 molecule^{-1}. Both neutron and x-ray reflectivity data were obtained on the films of the C60-dodecylamine adduct. Fitting to a three-box model (chains + water, C60 units, chains in air) gave a thickness for the fullerene slab of 1.1 nm, larger than that for a completely smooth C60 monolayer of 0.82 nm. The surface roughness of the C60-dodecylamine films was 0.4 nm, as compared to nearly 10.0 nm for the C60 films.

A C60 derivative with an attached fluorinated chain gave a limiting area of 0.78 nm^2 molecule^{-1} [266]. It was reported that this film was so mechanically rigid that it pushed the Wilhelmy plate out of the water at Π ~ 14 mN m^{-1}. The monolayer spreading of this compound arises from the even greater hydrophobicity of the fluorocarbon chains and their orientation away from the water surface. The LB films with a fluorinated tetrathiafulvalene derivative did not show evidence of charge transfer in their UV spectra.

Matsumoto et al. reported the first C60 derivative made hydrophilic in a manner such that it spread nicely with a hydrophilic group immersed in the subphase [267]. This C60 derivative had a single chain with two carboxy groups terminated by a carboxyl group. The limiting area obtained was 0.78 nm^2. The films successfully transferred onto hydrophilic substrates by vertical dipping to form LB monolayers or multilayers. The red shift of the 260-nm and 320-nm C60 peaks in the UV spectra indicated close packing of the C60 units. Reflection and transmission infrared spectra were used to determine the molecular orientation in the LB films. Analysis of the relative intensities of the C—O—C and CH$_2$ stretching modes was consistent with a molecular orientation in which the attached chain was at an angle to the surface. Hydrogen bonding between the carboxyl groups of neighboring molecules within a single layer was observed, because the absorption due to carboxylic acid dimers was observed. And AFM of a single monolayer transferred onto mica showed a relatively smooth image, although holes and some bilayer regions were clearly observed. The area fraction of the monolayer regions was estimated as 88%.

The characterization of the monolayer behavior of eight related C60 derivatives was reported [268–270]. The isotherms of a set of three monofunctionalized ester derivatives (ethyl, propyl, and dodecyl) gave limiting areas consistent with multilayer behavior: 0.28 nm^2, 0.32 nm^2 and 0.40 nm^2, respectively. Functionalization of C60 solely by alkyl chains was not sufficient to inhibit aggregation of C60. More hydrophilic derivative bearing a bis(triethyleneglycol monomethyl ether group) spread as a monolayer with a broadly rising isotherm and a limiting area of 0.94 nm^2. The introduction of bis(ethoxycarbonyl)methylene groups onto C60 was found to give monolayer spreading. The limiting area was lower for the equatorial form than for the two trans forms of these compounds. An equatorial tris isomer spread as a monolayer and showed transitions from gaseous to liquidlike to a solid-like phase coexisting with bilayers. Monolayers of this C60 derivative compressed above 35 mN m^{-1} and expanded did not show monolayer behavior upon recompression, suggesting that the formation of bilayers at higher Π was irreversible. The films were examined on water with BAM, on LB films on mica with AFM, and on quartz using UV spectroscopy.

A number of additional monosubstituted C60 derivatives were investigated [271]. Derivatives bearing two methoxyphenol groups, a pentanoic acid group, and a methyl pentanoate group were all subject to varying degrees of multilayer formation. A derivative bearing a bis phenol group was sufficiently hydrophilic to spread as a monolayer. Esterification of the pentanoic acid group with hydrocarbon and fluorocarbon alcohols gave derivatives that spread as monolayers. The derivative esterified with the fluorinated alcohol could form good Z-type LB films. An absorbance band characteristic of aggregated C60 [272] was not observed in the LB films. A Diels–Alder adduct of C60 and 8-(9-anthryl)-7-oxaoctanoic acid was pursued as a compound that could form good monolayers and ordered LB films [273]. Heating of the LB films to reverse the Diels–Alder reaction and drive out the anthracene derivative was proposed as a way to produce ordered multilayers of pure C60. The limiting areas of this derivative were 0.7–0.9 nm^2, depending on the conditions, and the collapse pressures were 50–60 mN m^{-1}. FTIR confirmed dimerization of the carboxyl head groups within a layer by hydrogen bonding. While the transfer ratios indicated formation of Z-type films, after the transfer x-ray diffraction indicated transformation to a bilayer Y-type film structure. The heating of multilayer LB films immersed in ethanol/water at 70°C for two days confirmed the removal of 40% of the anthracene derivative. UV and FTIR spectral data confirmed the formation of C60 in the LB films, although complete conversion to C60 multilayers did not occur. C60 covalently attached to a cryptate has been prepared and found to form monolayers; it was impossible to separate the bound sodium from the cryptate and study the effect of adding NaCl to the subphase on the isotherm [274]. Four different C60-glycosides were also studied, which exhibited varying degrees of monolayer stability. It was found that with benzyl-protected glucose hydroxyls monolayer spreading was not achieved. Monolayers of benzo-18-crown-6-methanofullerene were studied by BAM and UV-visible spectroscopy [275]. It was found that the limiting area of 0.84 nm^2 was increased to 1.01 nm^2 on 1.0 M KCl subphase. BAM showed smooth films under compression. It was proposed that the K$^+$ cations complexed to the crown ether increased the hydrophilicity and expanded the films. Monolayers of C60 attached to azacrowns showed an increase in limiting area from 0.90 nm^2 to 1.10 nm^2 on going from pure water to a 1 M KCl subphase [276].

The first report of spread films containing C60 side-chain polymers concerns spreading of a C60-containing polyamic acid amine salt that formed a LB film that could be converted to a polyimide upon heating [277]. Another report involved a tri-arm polyethyleneoxide, each arm terminated by an azide group reacted with C60 to attach a C60 unit to the end of each of the three chains [278]. This proved to be an effective way to prevent the aggregation of C60 on the water surface, for monolayers of these tri-arm C60 copolymers exhibited a surface pressure rise starting at 13 nm^2 arm^{-1}. The monolayers became rigid and compact if compressed above 10 mN m^{-1}.

4. Electrochemical Properties of Langmuir–Blodgett Films of C60 and Its Derivatives

Fullerenes are electron acceptors; C60 is capable of accepting up to five electrons reversibly in benzene solutions. In principle, it can accept up to six electrons in its LUMO. The electrochemical behavior of LB films containing C60 derivatives was first studied by Bard and coworkers [279] using cyclic voltammetry (CV), scanning tunneling microscopy (STM), and scanning electrochemical microscopy (SECM). Solvent-cast films were studied [280] and compared with LB films. The CV scans for films solvent-cast onto Pt and glassy carbon electrodes were conducted in acetonitrile at a series of potential scan rates and with different negative potential limits. By varying the negative potential limit, it was possible to subject the film to only the first reduction, to the first and second reductions, to three consecutive reductions, or to four consecutive reductions. The first cathodic peak potential (E_{pc}) was -1.16 V, and the corresponding anodic peak potential (E_{pa}) was -0.72 V, measured at 20 mV sec^{-1} using a silver quasi-reference electrode. The large separation between the cathodic and anodic peak potentials, found to increase if the scan rate was increased, indicated structural reorganization of the film during the reduction/oxidation cycle. The peak separation of 445 mV could be compared to that of 60 mV for C60 dissolved in benzene, the expected value for a diffusing, reversible one-electron redox species. In principle, reversible immobilized noninteracting redox species should exhibit no separation between their reduction and oxidation peaks. The solvent-cast films were ~0.2 μm thick, equivalent to ~160 layers and thus much thicker than the LB films. The amounts of charge passed during the first reduction and reoxidation waves were slightly less than that expected for one-electron reduction of all the C60 in the film. The structural change occurring during the reduction process involved rearrangement of the C60$^-$ anions to accommodate the tetrabutylammonium (TBA$^+$) counterions. The initial C60 film consisted of crystallites in the face-centered cubic structure known for C60. Accommodation of the large counterions required rearrangement of the film structure. As the TBA$^+$ counterions left during the reoxidation, a more resistive and compact film of diminished solvent accessibility formed, an effect confirmed by observing the loss of peak current during continuous cycling. Extending the negative potential limit revealed a smaller peak splitting of 160 mV for the second reduction and reoxidation waves ($E_{pa} = -1.22$ V, $E_{pc} = -1.38$ V at 20 mV sec^{-1}). The amount of charge passed was equivalent to 50–70% of the C60$^-$ present. The product of the third reduction was inferred to be chemically unstable, and the fourth reduction was an irreversible multielectron process. The behavior in the presence of electrolytes containing Li$^+$, Cs$^+$, and K$^+$ cations was studied; for Cs$^+$ and K$^+$ an overall three-electron reduction appeared to occur in one wave. The structural rearrangements and overall redox processes for C60 films were dependent on the nature of the cation. The STM observations of the initial C60 films could distinguish the individual C60 molecules; however, good STM images after reduction could not be obtained. And SECM observations indicated that the initial films were nonconductive and that after reduction cycling became less conductive but more uniform in their topography. Electrochemical quartz crystal microbalance (EQCM) experiments, combined with evidence from laser desorption mass spectrometry (LDMS), indicated that there was some loss of C60$^-$ from the film upon reduction, in addition to uptake of the TBA$^+$ counterions [281]. The TBA$^+$ counterions remained trapped inside the film after reoxidation.

The LB films of C60 were successfully transferred to alkylthiol-modified gold and to iodine-modified polycrystalline Pt and Au, which are all hydrophobic surfaces. The LB films were prepared having 1–5 layers. The observed peak splitting for the first reduction/oxidation process was smaller, 150 mV, as compared with 580 mV at 200 mV s^{-1} scan rate (Fig. 20). The transfer of monolayers by vertical dipping was successful if hydrophobic substrates were used at surface pressures of 5–10 mN m^{-1}. Films transferred at II $>$ 30 mN m^{-1} were reported as visibly patchy. The peak splitting for the first reduction/oxidation process was near 400 mV for the LB films formed from multilayer films spread from more concentrated solutions; this value is closer to that of the solvent-cast films. LB monolayers and drop-cast films of ethylenediamine-modified C60 (C60EA) and its tetramethyl ester (C60EM) were studied by cyclic voltammetry in acetonitrile [282]. The CV scans for the LB film of C60EM were less reversible than for drop-coated films on ITO electrodes. The LB films and drop-coated films of C60EA gave similar CV behavior.

The electrochemical behavior of the C70 solvent-cast films was similar to that of the C60 films, in that four reduction waves were observed, but some significant differences were also evident. The peak splitting for the first reduction/oxidation cycle was larger, and only about 25% of the C70 was reduced on the first cycle. The prolate spheroidal shape of C70 is manifested in the II-A isotherm of C70 monolayers. Two transitions were observed that gave limiting radii consistent with a transition upon compression from a state with the long molecular axes parallel to the water surface to a state with the long molecular axes perpendicular to the water surface.

Monolayers of 1-*tert*-butyl-1,9-dihydrofullerene-60 on hydrophobized ITO glass exhibited three well-defined reduction waves at -0.55 V, -0.94 V, and -1.37 V (vs. saturated calomel electrode, SCE), with the first two stable to cycling [283]. Improved transfer ratios near unity were reported. The peak splitting for the first two waves was 65–70 mV, much less than reported for the pure C60-modified electrodes. The reduction and oxidation peak currents were equal; however, the peak currents were observed to be proportional to the square root of the scan rate instead of being linear with the scan rate as normally expected for surface-confined redox species.

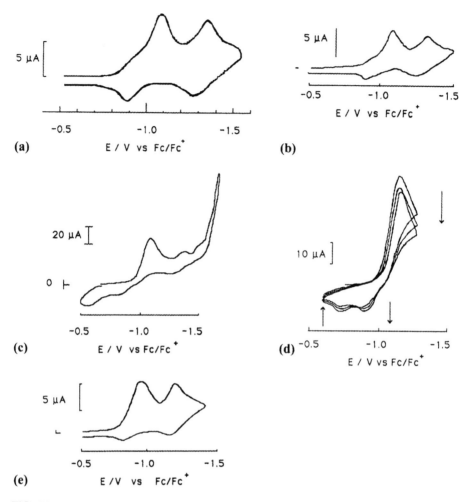

FIG. 20 Cyclic voltammograms of Langmuir–Blodgett films of C60 on a gold electrode made hydrophobic by exposure to octadecylmercaptan in ethanol. The measurements were performed in 0.1 M tetrabutylammonium tetrafluoroborate in acetonitrile. The LB films were prepared by vertical dipping. (a) Monolayer formed by spreading 50 μL of 10^{-4} M C60, transferred to substrate at $\Pi = 5$ mN m^{-1}; (b) same as (a) but transferred at $\Pi = 20$ mN m^{-1}; (c) monolayer formed by spreading 100 μL of 5×10^{-4} M C60, transferred at $\Pi = 50$ mN m^{-1}; (d) repetitive scans under the same conditions as in (c); and (e) C60 + arachidic acid 1:1 mixture transferred at $\Pi = 5$ mN m^{-1}. The potential scan rate was 0.2 V s^{-1}. (Reproduced with permission from Ref. 232. Copyright 1993 American Chemical Society.)

C60 has been used to produce solvent-cast and LB films with interesting photoelectrochemical behavior. A study of solvent-cast films of C60 on Pt rotating disc electrodes (RDEs) under various illumination conditions was reported [284]. Iodide was used as the solution-phase reductant. The open-circuit potential shifted by 74 mV per decade of illumination intensity from a continuous wave (cw) argon-ion laser. The photocurrent versus power was measured at −0.26 V under chopped illumination (14-Hz frequency, vs. SCE) up to 30 mW cm^{-2} and was close to linear. The photoexcitation spectrum (photocurrent versus wavelength) was measured at 0.02 V (vs. SCE) from 400 to 800 nm and found to be similar to the absorption spectrum of a sublimed film. Photoactivity extended to 720–740 nm, corresponding to a bandgap of 1.7 eV. The dark current of the films was quite large.

Some LB monolayer films of a series of pyrrolidine derivatives of C60 were studied on SnO_2 electrodes in 1 M KCl electrolyte solution [285]. The C60 derivatives themselves exhibit three successive reductions at potentials slightly different from those of the parent C60 when studied dissolved in acetonitrile with 0.1 M (n-butyl)$_4$NClO$_4$. To avoid the electroactivity of oxygen, the measurements were performed under a nitrogen atmosphere. Oxygen quenches the triplet excited state of C60. Using an excitation wavelength of 355 nm, anodic photocurrents near -120 nA were observed that rose to -19 nA to -38 nA (-65 nA to -92 nA if N_2 bubbling was maintained). The anodic photocurrent increased with bias voltage over the range studied. The effect of adding ascorbic acid as an electron donor or methylviologen as an electron acceptor into the electrolyte solution was examined. Addition of ascorbic acid caused an increase in photocurrent up to 1.9 mmol with a slope of 120–200 nA mM^{-1}, followed by a nonlinear relation that leveled off at higher concentration. The photocurrents could be switched on and off tens of times with little change. Methylviologen was found to reduce the photocurrents. A positive bias and a soluble electron donor resulted in optimal photocurrent generation for these LB films. The triplet state of the C60 generated on illumination is a much stronger electron acceptor, having a reduction potential near 1.14 V vs. SCE. The intersystem crossing rate from the singlet to the triplet state is high. Illumination in the absence of an electron donor resulted in the reduction of the fullerenes by water, with the generation of O_2 quenching the triplet state unless it was removed by N_2 bubbling. In the presence of ascorbic acid, the excited triplet state fullerenes were reduced by the ascorbic acid and O_2 was not generated. Quantum yields of 1.2%–8.2% were reported.

In a subsequent study, a C60 derivative formed by reacting glycine methyl ester and C60 under photochemical conditions was used to prepare monolayer LB films on the SnO_2 electrodes [286]. A limiting area of 0.95 nm^2 and a transfer ratio of 0.90 were reported. Similar photoelectrochemical behavior as the C60-pyrrolidine derivatives was reported. A quantum yield of 2.5% was reported. The photocurrents appeared higher for those derivatives with more electron-donating groups. For an aminodicarboxylate derivative [287], the photocurrent exhibited a linear increase versus bias voltage from -0.35 V to 0.40 V vs. SCE of 0.1 nA mV^{-1}. Photocurrent saturation was not observed for illumination intensities up to 2.10 mW cm^{-2}. The maximum quantum yield was 3.0% in this case. The photocurrent spectrum tracked the absorbance spectrum very closely. In a study of C60 iminodiacetic acid ester, it was observed that the photocurrent increased with pH, an effect attributed to more effective oxidation of water, whose redox potential drops with pH [288]. Another derivative studied was also a pyrrolidine derivative, a quantum yield of 1.9% being reported [290].

A set of dicarboxylic acid derivatives of C60 was prepared [291]. The monolayers were stabilized by dissolved cations, and the II-A isotherms were sensitive to pH. For these derivatives, the photocurrent was reported as cathodic, increased in the presence of O_2, and was favored by a negative bias. The photocurrent changed from cathodic to anodic when the pH increased above ~6 [292]. In this case, C60 was derivatized by the tetramethyl ester of ethylenediaminetetraacetic acid (EDTA) and then hydrolyzed to the tetraacid form. The limiting areas from the II-A isotherms were 1.25 nm^2 and 1.02 nm^2, respectively. The lower limiting area of the tetraacid form was attributed to hydrogen bonding between the carboxyl groups. The LB transfers were carried out at 15–20 mN m^{-1}, below the collapse pressures, which were 18–25 mN m^{-1} for these compounds. Above the collapse pressure, the isotherms indicated a monolayer-to-bilayer transition. Transfer ratios of 0.95 onto quartz or ITO plates were reported. On quartz, two of the UV peaks underwent a red shift relative to the solution-phase spectrum. The derivatives exhibited four well-defined reversible redox waves at potentials not more than 15 mV different from those of C60. For the tetramethylester derivative, illumination of the LB films resulted in an anodic photocurrent that increased upon addition of ascorbic acid or quinone as electron donors up to a limiting value. Addition of large amounts of the electron donor methylviologen was able to reverse the direction of the photocurrent. The photocurrent of the LB films of the tetra-acid was strongly pH dependent, being cathodic below pH ~6 and anodic above. The photocurrent of the tetramethylester form was not pH dependent. In the proposed mechanism for the cathodic photocurrent, the excited C60 accepts an electron from the conduction band of the ITO electrode that is then accepted by the viologen from the C60 radical anion. In the mechanism for the anodic photocurrent, the excited C60 accepts an electron from the electron donor in solution, and an electron is transferred from the C60 radical anion to the conduction band of the ITO electrode. The pH regulates which process is preferred.

In two other studies, it was observed that C60 in LB films can quench the fluorescence of pyrene [293] and of 16-(9-anthroyloxy)palmitic acid [294] by photoinduced electron transfer. In these studies, both C60 and the electron-donating fluorophore were incorporated into a tricosanoic acid LB film in different ratios.

5. Superconductivity in Langmuir–Blodgett Films of C60 and Derivatives

Superconductivity was observed in LB films of C60 doped with potassium using ESR spectroscopy [295,296]. LB films of 50 layers were prepared on sheets of poly(ethylene terephthalate) and placed in quartz ESR tubes. Potassium metal was introduced into the other end of the tube, which was the placed in a double furnace and heated in vacuum. The temperature near the end with the LB films was 200°C, that near the side with potassium was 150°C, and the doping time was varied up to 700 minutes. The optimal doping time was near 500 minutes. For C60 LB films doped for near this amount of time, a low-field ESR signal was observed below 8.1 K, indicating the transition to a superconducting state. The low-field signal ESR method does not require electrical contact with the sample, allows the sample to be kept away from air, and is sensitive to very small sample sizes and to the formation of small superconducting regions within a sample. This value of T_c is lower than that for the bulk C60 doped to K_3C60 of 18.6 K. Intermediately doped species can undergo the superconducting transition, while both C60 and K_6C60 are insulators. Horizontally transferred LB films of C60 doped by heating with K metal, Rb metal, RbN_3 (rubidium azide), and KN_3 were studied using ESR [297]. There was no evidence of charge transfer between C60 and potassium, although the ESR signal was observed to decrease. It is interesting to note that the spin density in the LB films was 10 times that of C60 powder (10^{-3} per molecule vs. 10^{-4} per molecule). No low-field signal was observed from films doped by heating in the presence of K metal, Rb metal, or KN_3. The films doped by heating in the presence of RbN_3 had spin densities of 0.16 per molecule, indicating that a large number of $C60^-$ anions formed. The temperature dependence of the spin susceptibility suggested the emergence of metallic regions (estimated at 20% volume fraction) in the films on cooling. The metallic phase of the film was assigned as o-Rb_1C60. ESR low-field signal (LFS) measurements showed some irreproducibility, although a transition to a superconducting state was clearly observed near 23 K in some of the RbN_3-doped films. In situ LFS and ESR studies of doping of C60 LB films by potassium were described in fuller detail [298]. At an optimal doping time of 480 min, a transition temperature of 12.9 K was observed. In over-doped samples, no LFS was observed and the ESR data showed g values associated with K_3C60, K_4C60, and K_6C60; only the K_3C60 phase can become superconducting. Resistivity versus temperature for a doped sample showed semiconducting behavior.

6. Electrical Behavior of Langmuir–Blodgett Films Containing C60 and Derivatives

The electrical properties of C60 LB films have been studied. The conductivity of bulk films of C60 increases in an activated fashion from 5×10^{-12} S cm^{-1} at 250 K to about 2×10^{-10} S cm^{-1} by 400 K [299]. The LB films doped with iodine vapor exhibited small shifts of about 4 nm in their UV spectra and the appearance of six hyperfine lines in their ESR spectra. The results suggested a partial charge transfer from iodine to C60, although the compound decomposed easily in air [300]. Field effect conductance measurements on C60 and C60/C70 mixed films were not successful for films up to 11 layers. The LB films of polyhexylthiophene/arachidic acid and C60/C70 + arachidic acid films did not show a bipolaron peak in their absorbance spectra. Significantly higher conductivity values near 10^{-3} were found for three-layer LB films of an octabromo derivative of C60 on glass between two silver paste electrodes [301]. It was thought that the Br atoms provided bridges for electron transport. The conductivity dropped on either exposure to UV light (375 nm) or exposure to iodine vapor by as much as two orders of magnitude. The octabromo-C60 compound spread as a monolayer from a 0.1 mM solution in benzene. The LB films of C60 mixed with the organic conductors tetrahexadecylthiotetrathiafulvalene (THT-TTF) and biethyldithiane-tetrathiafulvalene (BEDT-TTF) showed significant conductivity values of 10^{-4}–10^{-5} S cm^{-1} and 4–20 S cm^{-1} after doping with iodine vapor. It was reported that C60 contributed to the charge transport, for the conductivity was 2–4 orders of magnitude lower when stearic acid was used rather than C60 [302]. The LB films consisting of C60 and metallated (Pd, Fe, and Co) tetra-*tert*-butylpthalocyanines were formed on hydrophobized-glass slides. The electrical conductivity measured using the four-point probe technique was 10^{-9} S cm^{-1} for C60 + palladium tetra-*tert*-butylpthalocyanine films, 10^{-6} S cm^{-1} for the C60 + iron tetra-*tert*-butylpthalocyanine, and 10^{-4} S cm^{-1} for the C60 + cobalt tetra-*tert*-butylpthalocyanine 50-layer films [303]. Mixtures of fullerenols and of a copper pthalocyanine derivative were found to form phase-separated monolayers, as observed in AFM on the LB monolayers [304]. The electrical conductivity showed no effect of mixing these in LB films. The palladium-containing system exhibited a photocurrent larger than the dark current versus applied voltages up to 1 volt. Schottky photoelectric cells have been fabricated with LB films of C60 and arachidic acid (1:5) deposited on Ag-coated glass with thin semitransparent Al electrodes evaporated on top. The LB film photoelectric cells were subjected to a 420-nm, 300-ps, 1-μJ pulse and the photovoltage measured. The photovoltage rose within less than 15 ns to a value of 6 mV after incidence of the laser pulse, a much faster rise than observed for merocyanine LB films. The absorption spectrum for 11-layer LB films and the photocurrent action (photocurrent vs. illumination wavelength) spectrum were quite similar.

Conjugated polymers doped with C60 become p-type semiconductors [305,306] some LB films of two polyalkylthiophenes mixed with arachidic acid and doped with C60 have been prepared [307]. The films of polyalkylthiophene + arachidic acid + C60 (spread from mixtures of 1.0:0.33:0.1 ratio) on ITO glass had a well-defined layer structure, as confirmed by x-ray diffraction. The bilayer distance obtained from the Bragg equation was 5.6 nm, the same as for arachidic acid LB films. Since the films were spread on subphases containing

0.5 mM CdCl$_2$, the diffraction comes from the ordered cadmium planes. Schottky diodes were made by evaporating an aluminum counterelectrode on top of the LB film. The films clearly exhibited rectification in the current versus voltage curves. The turn-on voltage was 1.0 V for poly-3-octadecylthiophene and 0.3 V for poly-3-butylthiophene, indicating that this parameter might be tunable. A subsequent study included poly-3-octylthiophene, poly-3-dodecylthiophene, and poly-3-hexadecylthiophenes [308]. Schottky diodes made from the octylthiophene-containing LB films showed sharp onset of rectification near zero volts. The break in the behavior between the butyl and octyl derivatives and the longer chain lengths was not explained, although it was pointed out that the rectification ratio decreased with alkyl chain length. The longer-chain derivatives showed more gradual rectification curves, with the current rising near 1 V. It was pointed out that for electron donors strong enough to reduce C60, air sensitivity could be a problem. As a prelude to the goal of studying electron transfer from photoexcited nanoparticles of cadmium sulfide (CdS), the formation of LB films from mixed monolayers of arachidic acid and C60 spread on Cd(ClO$_4$)$_2$-containing subphases was studied [309]. The LB films were exposed to H$_2$S gas to form the nanoparticles, and this was reported to cause some disruption of the film structure.

7. Nonlinear Optical Properties of Langmuir–Blodgett Films Containing C60 and Derivatives

The nonlinear optical properties of fullerene-containing LB films have been investigated. Solvent-cast C60 films were found to exhibit second-harmonic generation [310,311]. The significant NLO response of C60 is attributed to the delocalized nature of its electronic structure and its large polarizability, giving electric-quadrupole and magnetic-dipole contributions. Degenerate four-wave mixing was used to generate forward phase-conjugate waves in C60/arachidic acid LB films [312]. In this study, it was concluded that aggregation of C60 accounted for the magnitude of the nonlinear optical effects and that pulsed laser irradiation tended to promote some conversion back to C60 monomers. Degenerate four-wave mixing demonstrated that the third-order susceptibility was large (7×10^{-12} esu) and that the NLO response was faster than 35 ps [313]. The linear absorption coefficient at 1064 nm was small (6 cm^{-1}), as measured using NIR spectra. The refractive index was 2.0, determined from the fringe pattern of the NIR spectrum. The hyperpolarizability of LB films of a piperazine derivative of C60 was measured as $(3.6 \pm 1.2) \times 10^{-29}$ esu [314], determined using measurements on LB films of one, three, and five layers. Spreading from a dilute solution in CH$_2$Cl$_2$/CS$_2$ 10:1 gave a well-behaved monolayer and could be transferred as Z-type LB films.

C60 LB films deposited on a glass waveguide were studied and found to give a nonlinear relation between output power and input power between 6 mW and 14 mW using the 514.5-nm line of an argon-ion laser [315]. A subsequent study by this group [316] concerned LB films of two C60-glycine ester derivatives (methyl and ethyl esters). The monolayers of these derivatives had collapse pressures slightly below 20 mN m^{-1}, and the slope of the Π-A isotherms indicated that these films were more compressible than those of C60. The monolayers could be transferred onto quartz, and the transfer ratio was consistent and near 0.9 for the methyl ester derivative; however, the transfer ratio dropped with layer number for the ethyl ester derivative. The second-order molecular hyperpolarizability was determined to be $(2.3 \pm 1.0) \times 10^{-29}$ esu, and the third-order molecular hyperpolarizability was determined to be 3.4×10^{-31} esu for the C60-glycine methyl ester; similar values were reported for the ethyl ester. The quadratic dependence of the second-harmonic signal on layer number was observed, at least upon comparing one- and three-layer films.

The third-order susceptibility of LB films of 1-benzyl-9-hydrofullerene-60 was reported as 2.1×10^{-11} esu; the second-order polarizability was reported to be below the detection limit [317]. A 4-acetalphenyl substituted C60-pyrolidine was found to form monolayers and gave a transfer ratio of 0.95 ± 0.05 and Z-type films at 20 mN m^{-1}. The second-order susceptibility was evaluated as 6×10^{-9} esu [318]. It has been observed that LB films of C60 and of a tetracyanoethyleneoxide derivative of C60 [319] generated a coherent SHG response for five-layer films, while an incoherent SHG response was generated by single-monolayer LB films [320]. The coherence of the response is assessed by measuring the polarization and direction of the SHG, incoherent response being diffuse and depolarized. The thickness of the C60 monolayers was 2.9 ± 0.1 nm and that of the C60-CN was 2.5 ± 0.1 nm, equivalent to about three times the C60 molecular diameter, and was determined by interference microscopy. The intensity of the SHG response of the C60-CN layers was about five times that of C60, this molecule producing electric-dipole SHG as well as the electric-quadrapole and magnetic-dipole SHG generated by C60.

The SHG response of LB films of ethoxycarbonyldecylene-fullerene-60 (ECDF) and bis(ethoxycarbonyldecylene)-fullerene-60 (BECDF) was studied [321]. The limiting area of ECDF was 0.52 nm^2, indicating greater than monolayer coverage, but is not as small as the values near 0.3 nm^2 for C60 multilayer spreading. For BECDF, Π was observed to rise near 1.4 nm^2, and a long plateau from 1.0 nm^2 to 0.5 nm^2 was observed with Π near 10 mN m^{-1}. The plateau was interpreted as due to a monolayer + bilayer coexistence ending in a bilayer state. High-resolution TEM revealed both well-ordered regions as large as 100 nm in size and distorted regions and stacking faults for LB films transferred onto electron microscope grids by horizontal lifting [322]. A related study concerned LB films of a dodecylamino-C60 and of dodecylamino polyhydroxy-C60 [323]. The limiting areas for these two compounds were 0.60 nm^2 and 0.65 nm^2, respectively. Well-ordered rows of molecules were observed; regions of a two-dimensional hexagonal lattice could be seen as well as regions where the lattice was distorted. The nearest-neighbor distance was 0.99 nm for the dodecylamino-C60 films and 0.95 nm for the polyhydroxy-C60 films. The values of the second-order susceptibility ($\chi^{(2)}$) and the second-order molecular hyperpolarizability (β) were determined to be 2.3×10^{-6} esu and 1.0×10^{-28} esu for ECDF and 4.9×10^{-6} esu and 2.0×10^{-28} esu for BECDF, respectively. An acetalphenyl-substituted C60-pyrrolidine spread as monolayers, with a limiting area of 0.96 nm^2 [324]. The film could be compressed to $\Pi = 20$ mN m^{-1} and then expanded with little hysteresis, with compression beyond this pressure resulting in partial conversion to bilayers or multilayers. The LB films could be transferred onto glass or quartz with transfer ratio of 0.90 ± 0.05; the measured value of the second-order susceptibility was 5.5×10^{-9} esu. It was pointed out that such a value is 100 times smaller than that for typical SHG materials, such as hemicyanine dyes. Mixed films with 22-tricosenoic acid were also a part of this study. A different experimental approach, namely, the Pockels response measured using electro-optically modulated attenuated total reflection spectroscopy, was used to measure the value of $\chi^{(2)}$ and gave 1.3×10^{-9} esu [325]. The experiment was conducted on C60 + arachidic acid LB films, and there was no effect of arachidic acid on the C60 NLO properties.

8. Friction, Wear, and Wetting Properties of Langmuir–Blodgett Films of C60 and Derivatives

The friction and wear properties of fullerene LB films have been investigated. The coefficient of kinetic friction was measured using a steel ball-on-glass disk method, with the LB films deposited onto the glass disk [326,327]. The friction coefficient dropped from ~0.8

for glass to ~0.1 for the LB film of 1:1 C60 + stearic acid. The friction coefficient decreased as the number of layers varied from 1 to 11 to 25, and was independent of the applied load between 0.5 and 5 N. The wear lifetime for a stearic acid:C60 LB film was about 20 times longer than for a behenic acid LB film or for a solvent-cast 1:1 stearic acid:C60 film. SEM on the LB films of stearic acid and C60 showed the C60 aggregated in superfine particles of ~100-nm size. The wear lifetime was greater for the C60 + stearic acid LB films than for pure C60 LB films, and it was proposed that the superfine C60 particles "rolled around" in the hydrocarbon matrix under the applied load. In a study of wetting behavior, a C60 derivative with a single succinic acid group attached was used to form LB films on glass and on polyethylene terephthalate [328]. The advancing contact angle of water was studied for C60 and the derivative transferred under different conditions. On the PET slides, transfer of C60 at higher pressures gave a contact angle near 95°, higher than the value of 80° for the bare substrate. The contact angle rose earlier as a function of transfer pressure for the films transferred from multilayer C60 films than for those transferred from monolayer films. Transfer of the amphiphilic derivative onto glass gave contact angles in the 80–90° range, depending on transfer pressure and subphase pH.

9.   Langmuir–Blodgett Films Containing Other Fullerene Materials

Langmuir–Blodgett films containing single-wall nanotubes (SWNTs) have been reported [329]. The soot produced by catalytic arc discharge [330] was added to an aqueous solution of lithium dodecyl sulfate. Ultrasonication of the mixture disentangled the SWNTs, resulting in stable dispersions of surfactant-coated SWNT bundles and single surfactant-coated SWNTs, as confirmed by AFM examination of cast films of the dispersions. The subphase upon which the dispersion was spread to form a Langmuir film contained poly(allyamine hydrochloride), which served to form a polyionic complex with the negatively charged LDS-coated SWNTs and thus prevent them from dissolving into the subphase. An isotherm with a collapse pressure near 32 dyne cm^{-1} was observed for the spread SWNT dispersion, and was the same as that for the LDS solution spread on the same subphase. Horizontal deposition by the Langmuir–Schaefer method from the monolayer of the SWNT dispersion onto Si/SiO$_2$ wafers was successful. AFM observation revealed well-separated SWNTs flat on the surface.

V.   LANGMUIR–BLODGETT FILMS WITH NANOSCALE PATTERNS

The observation that Langmuir monolayers often exist as phase-separated domains has begun to be exploited as a means to produce LB monolayers with two-dimensional patterns. This approach aims to form, transfer, and stabilize these domains on appropriate substrates by combining the Langmuir–Blodgett method with the covalent bonding aspect of self-assembly. There are a small number of studies, and the possible further functionalization of the domains and use of different combinations of materials make this area promising for the construction of patterned films.

The transfer of alkylsilane monolayers onto mica was explored by Knobler and coworkers [331]. Monolayers of octadecyltrichlorosilane (OTS) were spread at the water–air interface. The trichlorosilane head groups rapidly undergo hydrolysis and cross-linking reactions to produce two-dimensional polysiloxanes on acidic subphases. On less acidic subphases, the polymerization rate is slower and the monolayer can be spread and transferred before the polymerization is significant. The Π-A isotherm for OTS spread on a pH 5.7 subphase (unadjusted milli-Q water) shows a plateau signifying a liquid-expanded (LE) + liquid-condensed (LC) phase coexistence region. The plateau will diminish with time as the polymerization slowly occurs, but there is ample time to spread the film and transfer it to the substrate. The spread films were transferred at different points along the isotherm onto acid-treated mica, heated in an oven, and examined by AFM. By this procedure, the formation of siloxane bonds to the mica surface was achieved. When transferred along the rising portion of the isotherm associated with the LE phase, the cured films on mica showed small islands of uniform size. The formation of these islands was attributed to the dewetting of water during the transfer process, resulting in the breakup of the LE phase into islands. The size of the islands was 0.4 μm when transferred at 5 mN m^{-1} and 1 mm min^{-1}. Transfer at the onset of the coexistence plateau showed the formation of both the small islands and larger domains, 1–1.5 μm in diameter, attributed to the LC phase. The surface coverage of larger domains increased as the transfer position was moved further inward along the plateau. There was no difference in height between the two types of domains, and the mica surface could be clearly distinguished between the domains. Examination by AFM showed that the morphologies remained unchanged 45 days later.

In a subsequent study, mixed silanes were used to generate regions of different height or of different hydrophobicity [332]. The compounds used were OTS, dodecyltrichlorosilane (DTS), and a fluorinated silane, $1H,1H,2H,2H$-perfluorodecyltrichloro silane (FTS). The spread monolayers of these compounds were transferred to mica, cured, and then immersed into solutions of one of the other compounds so that it would self-assemble onto the remaining bare portions of the mica surface. An OTS monolayer transferred onto mica and cured and then immersed into FTS in toluene and cured produced a surface with higher hydrocarbon islands (OTS domains) in a background of fluorinated chains (FTS). An FTS monolayer transferred onto mica and cured and then exposed to OTS in solvent and cured produces a surface of fluorinated domains in a higher hydrocarbon background. The expected height difference between OTS and FTS regions of 0.9 nm was confirmed using the topographic AFM mode, and the expected greater friction on the FTS regions was observed using the friction force AFM mode. The density and size of islands could be controlled by varying the initial LB deposition conditions, surface pressure, and dipping speed. Using AFM, it was shown that bovine serum albumin preferentially adsorbed onto the hydrocarbon regions (Fig. 21). In another study, it was shown that bacteriophage T4 preferentially adsorbed onto the fluorocarbon regions [333].

Efforts to produce the two-dimensional analogs of polymer-dispersed liquid crystals have been reported [334,335]. In the preparation of these liquid-crystal "nanoparticles," the surfactant monomer 2-pentadecylaniline (2PDA) and a ferroelectric liquid crystal (FLC) were cospread on the water surface. The liquid crystal was phase-separated from the 2PDA and remained so after the 2PDA was polymerized by adding an oxidant to the subphase. The polymerizations were carried out at controlled surface pressures. By varying conditions, the size of the two-dimensional liquid-crystal domains could be varied from 50 to 5000 nm. If the polymerization was carried out below a certain critical pressure, the FLC domains remained two-dimensional above this pressure the polymerization induced their collapse into three-dimensional "nanoparticles" of size 5–25 nm. The polymerization of the 2PDA matrix surrounding the FLC domains freezes in the phase-separated morphology. The construction of alternating-layer LB films, with some layers containing the FLC particles, and polymer layers was envisioned as a route to optical devices such as switchable Bragg gratings.

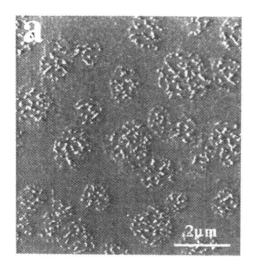
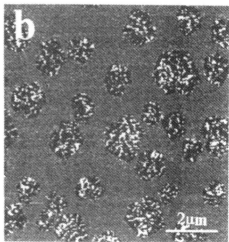

FIG. 21   Scanning force microscopy images of bovine serum albumin (BSA) adsorbed on a phase-separated monolayer on mica prepared from dodecyltrichlorosilane and 1*H*,1*H*,2*H*,2*H*-perflourode-cyltrichloro silane (FTS). The islands of DTS were prepared from monolayers transferred at 1.5 mN m^{-1} and 5.0 mm min^{-1}. The upper topographic image (a) reveals that the adsorbed BSA is not uniformly distributed. The lower friction force image (b) shows that the BSA is adsorbed only on the DTS islands, which appear darker than the surrounding FTS. (Reproduced with permission from Ref. 332. Copyright 1996 American Chemical Society.)

Monolayers of a thiolipid and palmitic acid were found to phase separate on compression and yielded separate, highly structured domains of the thiolipid [336,337]. The monolayer was transferred onto a gold substrate and the fatty acid washed off. The binding of the sulfur to gold, as in self-assembled monolayers, results in stable, immobilized domains. These bare regions were then used to self-assemble ω-mercaptocarboxylic acid complexed to copper and build up multilayers. The thiolipid consisted of two phospholipids linked near the head-group region by a disulfide bond. The domains of thiolipid produced were several tens of microns across. When deposited onto mica, the thiolipid domains had a flowerlike structure with six "petals," each petal having molecules oriented with different tilt azimuthal angles.

VI. MOLECULAR-LEVEL SIMULATIONS OF LANGMUIR MONOLAYERS AND LANGMUIR–BLODGETT FILMS

We review here results of computer simulations of monolayers, with an emphasis on those models that include significant molecular detail to the surfactant molecule. We start with a focus on hydrocarbon chains and simple head groups (typically a COOH group in either the neutral or the ionized state) and a historical focus. A less comprehensive review follows on simulations of surfactants of other types, either nonhydrocarbon chains or different head groups. More detailed descriptions of the general simulation techniques discussed here are available in a book dedicated to simulation techniques, for example, Allen and Tildesley [338] or Frenkel and Smit [339].

A. Simulations of Surfactants with Hydrocarbon Chains and Carboxyl Head Groups

Some of the first molecular dynamics simulations attempting a reasonably realistic three-dimensional representation of the hydrophobic chain were performed by Kox et al. [340]. Their model chain consisted of seven united atom units representing the methylene (CH_2) groups. Bond lengths were kept fixed, and a strong harmonic force was used to keep the next-nearest-neighbor distances constant in a way such that the tetrahedral bond angle was maintained. Energy differences between *trans* and *gauche* configurations were not included. Within the chain only strong repulsive forces were included, the attractions being included only among the head groups, which interacted through a Lennard–Jones 12-6 (L-J) potential,

$$V(r) = 4\varepsilon \left[\left(\frac{\sigma}{r} \right)^{12} - \left(\frac{\sigma}{r} \right)^{6} \right]$$

where r is the distance between groups, ε is a parameter describing the strength of the interaction, and σ is the diameter of the repulsive core of the group. The system contained 90 molecules with periodic boundary conditions, under conditions of constant temperature. Two temperatures were run, both at relatively low monolayer densities. The calculated isotherm at the lower temperature showed a loop, while the higher-temperature one did not, suggestive of a first-order phase transition, probably that between the gas and the LE phase.

Not long after the early simulation work by Kox et al., van der Ploeg and Berendsen [341] performed a simulation of a lipid bilayer that improved on several of the features of the chain model. While not a direct simulation of a monolayer, many of the techniques they employed were used in later simulations of monolayers. Guided by previous molecular dynamics simulations of alkane chains [342], they included dihedral angle potential functions and L-J interactions among the united atoms representing the methylene units on the chain, as well as bond angle potentials. The bond lengths were constrained using the SHAKE algorithm [343,344]. The L-J parameters for the head group were derived from polarizabilities and van der Waals radii. The head groups were held in the plane by a harmonic potential in the z-direction. Two layers of 16 decanoate molecules (placed with chains facing each other) were simulated at an area of 0.25 nm^2 molecule^{-1}. Bond order parameters were measured, showing good agreement with experimental (NMR) data. They also observed a significant fluctuating collective tilt to the chains. They suggested that the collective tilt could be an artifact of the small system size. A run on two layers of 64 molecules showed that the collective tilt did not extend over the whole unit cell.

The occurrence of kink defects in the hydrocarbon chains was the focus of Northrup and Curvin [345,346]. In this context, a kink is two *gauche* rotations in opposite directions, with a *trans* rotation in between. As compared to the all-*trans* configuration, a kink defect roughly preserves the overall linear extension of the chain, allowing it to remain densely packed even at a relatively high surface density. Part of the motivation for concentrating on *gauche* defects and kinks as a theme for interpreting the LE–LC transition (and also the main transition in bilayers) was previous theoretical work by Nagle [347,348]. In Northrup and Curvin's work, 36 hydrocarbon chains containing 15 carbon atoms were simulated by molecular dynamics at a high density of 0.183 nm^2 molecule^{-1}. As in the previous works, united atoms were used for each methylene group and the terminal methyl. The potential energy, derived from simulation work with proteins, included bond angle deformations, dihedral angle energies, and van der Waals nonbonded interactions. Potential energy, free energy, and entropy profiles for kink transitions were determined from the simulations using an umbrella sampling technique [349,350]. Among their conclusions were that cooperative tilt effects are important in the formation of kink defects as well as isolated *gauche* states.

One feature of molecular-level computer simulations is that structural features are relatively easier to extract compared to collective effects such as phase transitions. This aspect can be seen in the simulations of Harris and Rice [351] of a monolayer of pentadecanoic acid on water. Part of their stated motivation was to study the LE–LC transition, for which substantial experimental evidence exists [14,37] for a first-order transition; near room temperature (~300 K) the LE phase is shown by these experiments to exist in a range of molecular areas of approximately 0.45–0.32 nm^2 molecule^{-1}. Their model for the chain again used united atoms for methylene groups in a manner similar to the work by van der Ploeg and Berendsen described earlier. The interface was assumed infinitely sharp and interacted with the chain though a Lennard–Jones 9-3 potential as a function of height off the surface. The 9-3 potential comes from integration of the normal 12-6 potential over the surface. The head–surface interactions were made strong enough to keep the head groups from leaving the surface even in a low-surface-density state. Runs were made for both dense states (~0.22–0.23 nm^2 molecule^{-1}) and a more dilute surface density (0.64 nm^2 molecule^{-1}) as well as clusters. Their results showed *no* evidence of the LE phase. For the simulation run at 0.64 nm^2/molecule, a solid cluster spontaneously formed from the initial homogenous, low-density configuration. For the dense states simulated, a collective tilt was observed of approximately 30°, apparently in the direction of the nearest neighbor. Tilt was also observed in the clusters, although slightly smaller, about 20°, with molecules near the center nearly upright. The transverse structure factor was calculated for the dense states simulated. Sharp peaks at three unique nonzero wave vectors were observed (in a plot of the structure factor in the k_x,k_y plane this appears as one central peak with six side peaks). Chain tilt makes the peaks not all of the same height; ordering is better in the direction along the tilt. Density profiles normal to the interface and the number of *gauche* configurations as a function of the position along the chain were also measured. The absence of any evidence of the LE phase was discussed by Harris and Rice. They suggested that the most likely source for this discrepancy between simulations and experiment was the continuum model used for the interface.

Molecular dynamics simulations of monolayer chains with 20 carbon atoms were undertaken by Bareman et al. [352,353]. One of their models for the chain was similar, in terms of methodology, to the earlier van der Ploeg and Berendsen work and to the work of Harris and Rice. This was a united-atom approach, with potentials guided by work on bulk alkanes [342]. The head groups were not included explicitly; the first carbon on the chain was confined to, but allowed to move within, the anchoring plane. Initial work on molecular areas of 0.21, 0.26, and 0.35 nm^2 molecule^{-1} at 300 K showed an ordered and tilted monolayer at the smallest molecular areas and a much more disordered but still tilted monolayer at the higher areas. Some of the snapshots at the higher areas showed evidence of two regions of "phases," with areas of more and less order. Further work on the 20-carbon chain [353] employed both the united-atom model and an all-atom representation of the chain, with the head groups again not treated explicitly but anchored in the plane. The all-atom potentials included appropriate bond and angle constraints and interactions between atoms on different chains of the form

$$V(r) = Ae^{-Br} - \frac{C}{r^6}$$

with parameters developed previously by Williams [354]. These potentials had been used in computer simulation work on n-alkanes [355]. The Bareman et al. simulations concentrated on dense packings, in the range of 0.18–0.22 nm^2 molecule^{-1}, partially with an idea of comparing to experimental work on arachidic acid monolayers studied by x-ray diffraction and reflectivity. The average tilt angle as a function of molecular area for both the united-atom and all-atom models were compared to experimental data, with better agreement being seen with the all-atom model. The onset of tilt occurred between ~0.18 and 0.19 nm^2 and then rose quickly within a few nm^2/molecule to about 30°. The direction of tilt differed for the two models. In the united-atom model, the tilt was in the nearest-neighbor (NN) direction; for the all-atom model, collective tilt persisted for long periods of time in either the NN direction or toward the next-nearest neighbor (NNN). For both models, the chains were found to be in a rotator phase at 300 K.

Subsequent to the work already described, Bareman and Klein [356] simulated a 21-carbon alkanoic acid (heneicosanoic acid) on water, with an all-atom potential for the chains *and* with the water modeled in atomic detail as well—apparently one of the first molecular dynamics simulations to treat both the chains and the water in full atomic detail. The water molecules were modeled with the SPC (simple point charge) model [357]. A block of 1200 water molecules was equilibrated first; then the surfactants were placed on the water surfaces on opposing sides of the water slab. The simulation was then moved forward, with molecular dynamics at constant energy with periodic boundary conditions in all three directions, including the application of the Ewald summation [358] to treat the long-range interactions. A high density (0.1867 nm^2 molecule^{-1}) and relatively low temperature (274.2 K) were simulated, motivated partially by results of grazing incidence x-ray diffraction experiments that had been performed on heneicosanoic acid in the high-pressure region of the phase diagram at low temperature. The results of the simulation showed a substantial tilt of about 10°, even at this high density, in contrast to the experiments and the previous simulations of chains on a flat surface, which showed only a very small or zero tilt at this molecular area. Examination of the snapshots of the simulation indicated the likely reason for this difference—the head groups sat on the water in an uneven manner where adjacent molecules were offset from one another by one bend of the hydrocarbon chain. This feature also led to nearly structureless density profiles normal to the surface, in contrast to the previous simulations that assumed a completely flat interface.

Moller et al. [359] looked at both united-atom and all-atom representations in the model of the chain. The united-atom model was again of a form similar to that used by van der Ploeg and Berendsen [341]. The potential for the all-atom representation of the chain was in this case also based on the Williams [354] parameters, but refitted to a Lennard–Jones 12-6 form to help improve computational speed. The head group was modeled as a single center, with parameters derived from the polarizability and size of a carboxylate group and no inclusion of dipole moment. The surface was smooth continuum interacting with the molecules through a Lennard–Jones 9-3 potential, with parameters derived from graphite. Both energy minimizations and molecular dynamics simulations were performed on a stearic acid chain. Both the united-atom and all-atom models showed a sudden onset of tilt as the area was increased. In the united-atom model, this occurred at 17.5 nm^2 molecule^{-1}, the tilt rising quickly to about 19°. For the all-atom model, onset of tilt was at 20.9 nm^2 molecule^{-1}, rising to about 30°. In comparison, the x-ray diffraction work on arachidic acid monolayers on water had showed the onset at about 19.8 nm^2 molecule^{-1} [360]. The minimum-energy structures were interpreted in comparison with the B and C forms of bulk crystals of stearic acid, their conclusion being that the energy-minimized structures of the monolayers can be viewed as relaxed forms of the bulk crystal structure. In the molecular dynamics simulations, 64 molecules were simulated using both chain models. Two molecular areas were chosen, one on each side of the area observed for the onset of tilt in the energy minimizations. The united-atom model showed a substantial negative surface pressure and was not pursued further. The all-atom chain model, simulated at 0.2079 and 0.212 nm^2 molecule^{-1}, showed a near-zero tilt at the lower area and a tilt of about 9° at the higher area. The direction of tilt was in the most part NN, but they observed long-lived excursions from this situation.

An extension of the stearic acid simulations of Moller et al. was performed by Kim et al., [361] the most significant difference being the treatment of the head group. They treated the carboxylic acid group in an all-atom manner, with dipole–dipole interactions included via partial charges; the interaction of the head group with the surface included the effect of image charge contributions. The all-atom representation of the chain was used. As before, both energy-minimization calculations and molecular dynamics simulations were performed. The inclusion of the more realistic head group had only a small effect on the energy-minimization results: Sudden onset of tilt was seen at about 0.208 nm^2 molecule^{-1}, with the tilt rising quickly from a value of zero to near 30°. In the molecular dynamics simulations, which were performed at 0.206 and 0.212 nm^2 molecule^{-1}, the tilt transition could again be seen, but with a larger tilt at the higher area of 18.8° (as compared to about 9° for the united-atom head-group simulations). The tilt azimuthal angle remained essentially constant with time, as opposed to the long-lived excursions observed in the previous work, and the direction of tilt was viewed from the snapshot at 21.2 nm^2/molecule appears to be in the NN direction.

A modification of the united-atom approach, called the anisotropic united-atom (AUA) model was the focus of extensive work by Karaborni et al. [362–365]. As in the other models of hydrocarbon chains described so far, the AUA approach to monolayers was preceded by work on alkanes [367]. In the AUA model the interaction site is located at the geometrical mean of the valence electrons of the atoms it represents, while the pseudoatom itself is located at the carbon atom position. The movement of each interaction center depends on the conformation of the molecule as a whole.

Karaborni and Toxvaerd [362] simulated a monolayer consisting of chains with 15 methylene groups and a head group using molecular dynamics, in a simulation box containing 64 molecules. The head groups were dipolar, with the head–head dipolar repulsion also included. The interaction of the surfactant with the surface was developed using solubility data for head groups and the CH$_2$ groups as a means to develop an external potential for this interaction; thus the interface in this work is not a perfectly flat interface. The simulations were performed for a range of molecular areas from 0.18 to 30 nm^2 molecule^{-1}, at a controlled constant temperature of 300 K. Further work on this model was reported by Karaborni et al. [364], including the use of two different simulation box shapes, rectangular and a box with unequal x- and y-dimensions chosen to fit a hexagonal structure. Surface pressure vs. area isotherms were calculated, showing a possible first-order transition in the range between 0.21 and 0.23 nm^2 molecule^{-1}, more pronounced in the rectangular box simulations. Snapshots of the simulated chains at these two molecular areas near the transition are shown in Figure 22. Above 0.23 nm^2 the surface pressure decreased to near zero at about 27 nm^2 molecule^{-1}. The molecular tilt was analyzed; the tilt angle θ was defined as

$$\cos(\theta) = \frac{1}{N} \sum_{i=1}^{i=N} \frac{\mathbf{u}_i \cdot \mathbf{u}_n}{|\mathbf{u}_i| \, |\mathbf{u}_n|}$$

(a)

(b)

FIG. 22 Side view snapshots of a simulation of a 16-carbon hydrogenated surfactant chain with a carboxylate-like head group on a water surface at 300 K. The view in (a) (top) is an area of 0.21 nm^2 molecule^{-1}; (b) (bottom) is at 0.21 nm^2 molecule^{-1}. These two areas roughly bracket a first-order transition with some features of the LE–LC transition. See also Figure 23 for the corresponding pressure–area isotherm. (Reproduced with permission from Ref. 364. Copyright 1992 American Chemical Society.)

where $\mathbf{u}_i$ is the longest eigenvector in the moment of inertia tensor for the molecule, $\mathbf{u}_n$ is the surface normal, and N is the number of molecules. The tilt calculated in this fashion showed a nearly linear rise with molecular area from 2.5° at the highest density achievable in the simulations (0.185 nm^2 molecule^{-1}) increasing to about 23° at 0.30 nm^2 molecule^{-1}. Translational diffusion was measured for both head groups and the centers of mass of the molecules. By both measures, the diffusion was seen to be zero up to 0.22 nm^2 molecule^{-1}. At and above this area, measurable diffusion was seen, with the diffusion constant D increasing with molecular area from that point onward. The onset of diffusion was seen at just the same molecular area as the change in curvature in the pressure–area isotherm, leading to the conclusion that a liquid–solid transition occurs at this point. Chain conformations were studied in detail. At both 0.185 and 0.19 nm^2 molecule^{-1}, there are no *trans* bonds in the chain interior. From 0.20 nm^2 up, *gauche* defects begin to appear in the chain. This was interpreted as a continuous transition occurring within the solid state (before the onset of diffusion).

Karaborni and Toxvaerd [363] and Karaborni [365] also performed a set of simulations on a Langmuir monolayer with a longer chain, 19 methylene groups plus head group, using the AUA model and other details of the simulations very similar to the previous work on the shorter chain [362,364]. The simulations were performed for molecular areas ranging from 0.185 to 0.25 nm^2 molecule^{-1}, again at 300 K. The surface pressure isotherm calculated from simulation data showed a fairly good agreement as compared to an experimental isotherm for a 20-carbon surfactant (*n*-docosyl acetate); the pressure decreased quickly from very high values at small areas, going to zero near 0.24 nm^2 molecule^{-1}. Between 0.24 and 0.25 nm^2 molecule^{-1} there was evidence of a first-order phase transition. The tilt of the molecules was analyzed. The plot of θ vs. molecular area showed a sharp increase starting at 0.20 nm^2; between the runs made at 0.20 and 0.21 nm^2 the tilt jumps from

3.5° to 13.5°. On the same plot, the experimental results for the tilt angle measured by x-ray scattering [360] were presented, with fairly good agreement. At 0.23 and 0.24 nm^2 molecule^{-1} the direction of tilt was clearly NN; at 0.21 and 0.22 nm^2 the tilt direction was not well defined. Snapshots of the monolayer at 0.22 nm^2 molecule^{-1} showed precession as a function of time. The monolayer tilt was in the NN direction part of the time and NNN part of the time. By an area of 0.25 nm^2, the tilt distribution—probability of occurrence of a given tilt $P(\theta)$ as a function of angle θ—was very broad, which was interpreted as the monolayer's being in a melted state at that area, however, diffusion measurements indicated that the system was not in a true diffusive state at 0.25 nm^2. The different orderings and tilts were also seen by examining the structure factor. At an area of 0.20 nm^2, they showed six peaks (three independent peaks) of roughly equal-height peaks about the central peak; by 0.21 nm^2, this had changed qualitatively, showing the presence of tilt, in that the peaks are not all of equal height; at 0.23 nm^2, the three peaks were indicative of well-defined NN tilt. The chain conformations were studied in a similar manner to the work on the shorter chains, with a similar conclusion: At areas between 0.185 and 0.20 nm^2 there appears to be a continuous transition from all-*trans* state to a state with kink defects in the chains.

One can try and compare the isotherms from the Karaborni et al. simulations [362–365] with the available experimental data. Given the several approximations that enter into the potentials used and the simple model of the interface, one cannot expect quantitative comparisons. However, the work on the 16-carbon and 20-carbon chains at the same temperature allows us at least to attempt to deduce the trend with chain length. The calculated isotherms for both chains are shown in Figure 23. The loop between 0.21 and 0.23 nm^2 molecule^{-1} in the 16-carbon chain was associated with a first-order transition, possibly the LE–LC transition. For the 20-carbon chain, the first-order transition is starting to occur between 0.24 and 0.25 nm^2 molecule^{-1} and at a lower pressure. This trend with chain length is in fact exactly that seen for the LE–LC transition experimentally, [366] a remarkable result considering the relative simplicity of the model in this very complicated system. Especially to be noted is the data points for the shorter chain at areas higher than 0.25 nm^2 molecule^{-1}; in this region the monolayer is in a highly diffusive, chain-disordered state yet still with a measurable surface pressure, all characteristics that one associates with the LE state. The hint of the beginning of the transition at 0.25 nm^2 molecule^{-1} in the 20-carbon chain is harder to understand as a possible LE–LC transition. Experimentally, the 20-carbon carboxylic acid (arachidic acid) is far below its triple point at a temperature of 300 K. This combined with the facts that the pressure is nearly zero at the point of the transition and that high diffusion was not seen clearly at this point suggest that this might be just the beginning of the LC–G transition.

The important issue of size effects was addressed by Karaborni and Siepmann [368]. They used the same chain model and other details employed in the Karaborni et al. simulations described earlier [362–365] and the 20-carbon chain. System sizes of 16, 64, and 256 molecules were employed with areas of 0.23, 0.25 and 0.27 nm^2 molecule^{-1}; simulations with 64 molecules were also performed for areas ranging from 0.185 to 0.40 nm^2 molecule^{-1}. The temperature used was 275 K, as opposed to 300 K used in the previously discussed work by Karaborni et al. with the 20-carbon chain. At the smaller areas no significant system size dependence was found. However, the simulation at 0.27 nm^2 molecule^{-1} showed substantial differences between $N = 64$ and $N = 256$ in ordering and tilt angle. The 64-molecule system showed more order than the 256-molecule system and a slightly lower tilt angle. The pressure–area isotherm data for these simulations are not

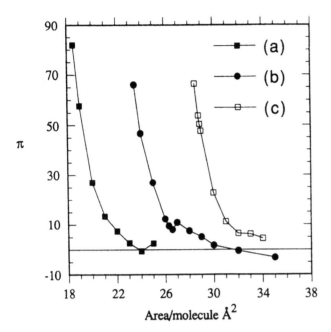

FIG. 23 Surface pressure vs. area/molecule isotherms at 300 K from molecular dynamics simulations of Karaborni et al. (Refs. 362–365). All are for hydrocarbon chains with carboxylate-like head groups. (a) (filled squares) A 20-carbon chain. (b) (filled circles) A 16-carbon chain with a square simulation box; the curve is shifted 5 Å^2 to the right. (c) (open squares) A 16-carbon chain with a non-square box with dimensions in the ratio $x/y = (3/4)^{1/2}$ to fit a hexagonal lattice; the curve is shifted 5 Å^2 to the right. (Reproduced with permission from Ref. 365. Copyright 1993 American Chemical Society.)

presented; but given the earlier work on the 20-carbon chain [362,364] at a higher temperature, we can assume that the pressure at the area of 0.27 nm^2 molecule^{-1} is near zero, suggestive of the monolayer being in a two-phase region. Thus in the interpretation of the size-dependence work, the issue of inhomogenous or possible domain structures is a natural one to consider. The authors suggest that their results indicate that when the system size is too small, perhaps a larger tilt angle is assumed instead of domains being formed.

The simulation of a first-order phase transition, especially one where the two phases have a significant difference in molecular area, can be difficult in the context of a molecular dynamics simulation; some of the works already described are examples of this problem. In a molecular dynamics simulation it can be hard to see coexistence of phases, especially when the molecules are fairly complicated so that a relatively small system size is necessary. One approach to this problem, described by Siepmann et al. [369] to model the LE–G transition, is to perform Monte Carlo simulations in the Gibbs ensemble. In this approach, the two phases are simulated in two separate but coupled boxes. One of the possible MC moves is to move a molecule from one box to the other; in this manner two coexisting phases may be simulated without an interface. Siepmann et al. used the chain and interface potentials described in the Karaborni et al. works [362–365] for a 15-carbon carboxylic acid (i.e. pentadecanoic acid) on water. They found reasonable coexistence conditions from their simulations, implying, among other things, the existence of a stable LE state in the Karaborni model, though the LE phase is substantially denser than that seen experimentally. The re-

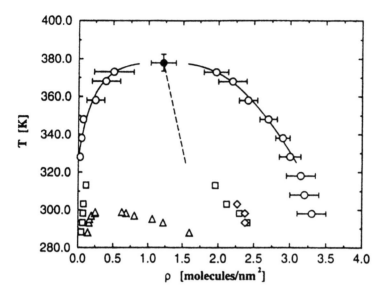

FIG. 24 Monolayer G–LE coexistence conditions from the simulations of Siepmann et al. (Ref. 369) on a pentadecanoic acid model using Gibbs ensemble Monte Carlo simulation. The filled circles are the simulation results. Experimental results are also shown from Ref. 370 (triangles), Ref. 14 (squares), and Ref. 15 (diamonds). (Reproduced with permission from Ref. 369. Copyright 1994 American Chemical Society.)

sults of their work are shown in Figure 24, with comparisons of the simulation results given to experimental measurements [14,15,370] of the G–LE phase boundaries of pentadecanoic acid. They estimate a critical point in the vicinity of 100°C, which is not necessarily at odds with the experimental data, given that the experimental measurements have been made at temperatures substantially below the critical point. Finally, it is interesting to note that the best fit to the simulation coexistence data was using a critical exponent of 0.32, which is what one would expect from a three-dimensional system.

B. Simulations of Other Surfactants

We have focused so far on single-chain surfactants with hydrocarbon chains, mostly with COOH or closely related head groups. Computer simulations have also been performed on a variety of other surfactants. We do not attempt here to exhaustively review all work, but describe some (hopefully) representative samples.

Extensive molecular dynamics simulations on surfactants with fluorinated or partially fluorinated chains have been performed by Shin et al. [371–373], Collazo et al. [374], Schmidt et al. [375,376], and Kim and Shin [377,378]. These works were motivated partially by experimental results on these systems, especially x-ray diffraction, and partly to address fundamental issues of the effect of molecular flexibility on the phase transitions and the microscopic structure of the phases in these systems. Because of the larger size of the fluorine atoms as compared to hydrogens, fluorinated chains are more rigid, with the all-*trans* configuration being more favored than for a hydrogenated chain. This important issue of flexibility, especially as it relates to phase transitions in monolayers, was addressed on a theoretical basis by Schmid and Schick [379], they found that chain flexibility (along with segment rigidity) is required to produce two liquid phases in a surfactant monolayer system.

The molecular dynamics model used by Shin et al. [371] bears some similarity to the previously described work of Harris and Rice [351], with appropriate modification of parameters to apply to the fluorinated chain, for example, the larger *trans–gauche* energy difference and barrier height. Fluorinated chains are known to assume a helical twist along the chain; in this initial work this effect was not included. The surface interactions and head group were treated in a similar manner as in the Harris and Rice [351]. Three different partially fluorinated possibilities for a 12-carbon carboxylic acid were studied: one having a sole CH_2 group near the head, one having a single CH_2 group at the C-5 position on the chain, and one having a "flexible spacer" of four methylenes adjacent to the head group. Both constant-area and constant-pressure runs were performed; for the constant-pressure simulations the surface pressure was a low value of of 0.16 mN/m. The most extensive analysis was done for the fluorinated chain with the single methylene near the head, $CF_3(CF_2)_9CH_2COOH$. At 300 K, constant pressure, and 0.297 nm^2 molecule^{-1}, they observed a fluctuating tilt near 15°, smaller than that of comparable hydrogenated surfactants, in qualitative agreement with experiment. Simulations on this monolayer were also undertaken at constant area for larger areas, up to 0.70 nm^2 molecule^{-1}. At the larger areas simulated, the monolayer was seen to break up into islands. This was seen for both periodic boundary condition runs and cluster simulations. Within an island, the average tilt was near the same as for the smaller-area simulations, again in qualitative agreement with experimental results. By contrast, the molecule with the single methylene group further up the chain (not adjacent the head group) showed a more regular tilt ordering, and the molecule with the 4-carbon flexible spacer near the head group was more disordered than the other two molecules studied.

Shin and Rice [373] applied these methods to a series of 15-carbon carboxylic acid surfactants possessing partial block fluorination—i.e., either just the seven carbons near the head fluorinated, the seven adjacent the tail, or a 5-4-5 arrangement of alternating fluorinated-hydrogenated-fluorinated. In the case of the first two, the monolayers are seen to be significantly disordered, with the disorder naturally concentrated in the hydrocarbon region of the chain. The 5-4-5 alternating arrangement, however, showed a well-ordererered structure.

Shin et al. [372] simulated perfluorinated and partially fluorinated 12-carbon surfactants, with two important differences from the previous work: First, they included a modification to the dihedral potential in order to mimic the helix structure; and second, they modified their method for calculating the tilt angle. In the work described, the inclusion of the helical component to the dihedral potential did not have a major effect on the monolayer structure. The change of method for calculating tilt did have a substantial effect, however. In their earlier work they had not taken into account the azimuthal distribution of tilts in calculating the average. For densities near close packing, this new means of calculating the tilt led to a much smaller average tilt angle (4–5°) in much closer agreement with the x-ray data. They also confirmed their earlier higher-area simulation results, where the fluorinated surfactant is seen to break up into dense islands at larger areas. Figure 25 shows side and top snapshots of simulations contrasting perfluorinated and hydrogenated chains. One can see the tilt is much smaller in the fluorinated chains and with a varying tilt direction; the hydrogenated chains of the same length show much more tilt and an NN tilt direction.

Further simulations on monolayers of perfluorinated chains were performed by Schmidt et al. [375,376]. Part of the stimulus for this work was x-ray results on these systems showing evidence for azimuthal ordering in these systems—in other words, the possibility of the monolayer analog of rotator transitions, which are well known in bulk alka-

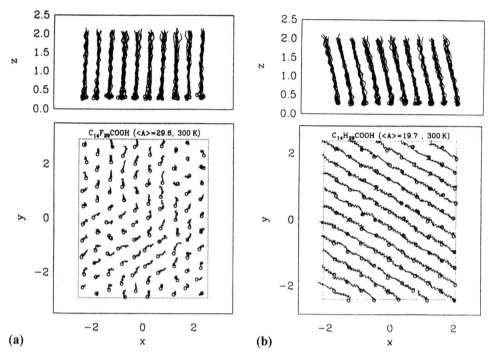

FIG. 25 Top view and side view snapshots comparing simulation results for perflourinated and 15-carbon hydrogenated chains from Shin and Rice (Ref. 373), both at 300 K and surface pressure of 8.0 mN/m. (a) F(CF$_2$)$_{14}$COOH; (b) H(CH$_2$)$_{14}$COOH. (Reproduced with permission from Ref. 373. Copyright 1994 American Chemical Society.)

nes and also have been studied in some bulk fluorocarbon long chains. Rotator transitions involve detailed interactions between chains at relatively high densities; thus the details of the chain interaction potentials become important. Perhaps for this reason, Schmidt et al. chose to adopt a model of the fluorinated chains similar to their previous work [351,371,372] but also with the incorporation of the anisotropic united-atom (AUA) representation for the chain—recall the AUA model for hydrocarbon chains in monolayers was used extensively by Karaborni and coworkers, as described previously in this review [362–365]. Two completely fluorinated chains were studied, a 12-carbon chain with a COOH head group [375] and a 20-carbon chain with no head group [376]. For the 12-carbon surfactant, their results for tilt and ordering were similar to their previous work: At high density the molecules were nearly normal to the interface, while at areas larger than this the monolayer broke up into islands that had much the same structure as the homogeneous monolayer at high density. In analyzing for azimuthal order, they looked at two widely differing temperatures, 150 K and 275 K, finding that at the higher temperature the molecules are azimuthally disordered (in a rotator phase), while at the lower temperature weak azimuthal order was seen; in this work the UA and AUA models were about the same in terms of their predicted structure. The case of the 20-carbon perfluorinated chain without a head group [376] was motivated by experiments [380] showing that this molecule forms an ordered monolayer on water even though it lacks a head group. The simulation methods employed to model this surfactant were similar to those used on the 12-carbon surfactant [375], again incorporating the AUA potential. They found that the monolayer was in a rotator phase at 275 K while being azimuthally ordered at 150 K. In this respect, the AUA model was seen to be superior to the UA model in generating predictions of a rotator transition between these two temperatures—this conclusion is partially based on comparison to the experimental data of lamellar crystals of this molecule.

Kim and Shin [377,378] have extended the simulations on monolayers of chain molecules with no head group by looking at partially fluorinated chains, as well as focusing on the dynamics of the process that occurs when the monolayer is formed from an initially disordered system. They looked primarily at the molecule $F(CF_2)_{12}(CH_2)_{18}H$, which has been studied experimentally [381], employing simulation techniques similar to the earlier study on the 20-carbon perflourinated chain [376]. Close-packed, initially ordered structures were simulated both in the hydrocarbon-up and hydrocarbon-down configuration; both were found to be stable. Interestingly, the tilt is different in the two blocks, leading to a kink in the molecules at the bond between the two blocks. At molecular areas larger than close packed (0.35 nm^2), more disordered, metastable (negative surface pressure) structures were seen. Of the two temperatures run here, 275 and 400 K, the more uniform structures were seen at 400 K. A series of simulations of temperature quenches were performed. For the quenches, the initial state was an essentially completely disordered state achieved by first running at 700 K for an extended period of time before starting the quench. The quenches were performed down to a variety of lower temperatures, ranging from 350 to 450 K. Surprisingly, highly ordered monolayer states developed spontaneously from the initial random configuration. All of the final states showed mixed configurations relative to the interface (a mixture of hydrocarbon up and hydrocarbon down). The degree of order in the final state was dependent on the final temperature; the most ordered final state was seen at an intermediate temperature of 400 K. Both higher and lower final temperatures led to more disordered final configurations. These results were interpreted as a combination of kinetic and thermodynamic effects. If the melting point of the system is near 400 K, then below the melting point we expect higher temperatures to favor a more rapid organization toward an ordered equilibrium; above the melting point, of course, the final equilibrium state is disordered.

Peters et al. have performed molecular dynamics simulations of monolayers formed from diglycerides, specifically 1,2-*sn*-dipalmitoylglycerol [382–384]. They described a model of the dual chains consisting of the AUA interaction type and full atomic detail for the rest of the molecule. The subphase was taken as a continuous medium with an external field of the form described by Karaborni and Toxvaerd [362]. Simulations were performed at a constant temperature of 298 K and molecular areas ranging from 0.362 nm^2 to 40.5 nm^2. The plot of surface pressure as a function of molecular area showed clearly three nearly linear regions, separated by two breaks in slope near 0.383 and 0.398 nm^2 molecule^{-1}, which agrees very well qualitatively with the experimental isotherm. The first transition was described as a swelling of the molecules, traced to a complicated mechanism involving the competition at high density between the packing of the chains and the attraction of the two coupled ester groups to the interface. The second transition was seen to be coincident with a large, discontinuous increase in tilt angle and was described as a first-order transition. The third paper of this series [384] describes not only simulations for this system but also parallel experimental measurements, including both pressure–area data as well as extensive x-ray diffraction data. This thorough parallel study largely confirmed the interpretation of the two transitions described in the two earlier papers.

Substantial effort has been extended in the direction of treating the water in detail, as opposed to a continuum. One approach has been to simulate individual surfactant molecules on a realistic water surface. Porohille and Benjamin [385] simulated a molecule of p-n-pentylphenol on or near a water surface modeled as individual water molecules. They found the phenol head group mostly immersed in the water but the hydrocarbon chain adopting configurations placing it most often parallel to the interface, as a result of the dispersion attractions between the chain and the water molecules. Their choice of this surfactant was motivated by second-harmonic generation experiments; they reported that their simulations led to calculated free energies of absorption that agreed with the experimental results. More recent molecular dynamics simulations on sodium dodecyl sulfate at a water liquid–vapor interface by Schweighofer et al. [386] led to similar results. The also modeled the interface in a detailed way using individual water molecules; they found that the hydrocarbon tail tends to preferentially lay down on the water surface, so the surfactant most often adopts a bent configuration in order to keep the head group inserted in the water while the hydrocarbon chain is on the water surface.

Bocker et al. [387] used molecular dynamics to simulate a dense-packed monolayer of n-hexadecyltrimethylammonium chloride on a water surface, with the water modeled in detail, again using the TIP4P/SPC model for water [357]. Their approach was similar to that described earlier by Bareman and Klein [353] in that they used a slab of water molecules and set up monolayers on both sides. In this case, after equilibrating the two-sided slab configuration they also then took one of the monolayers and about four layers of water from the two-sided slab and extended it by adding it to itself in the x- and y-directions to make a larger system that was then run further in simulations; a Lennard–Jones wall was put beneath the water layers to confine them. As in the work by Bareman and Klein, they found smoother density profiles than those seen in simulations with a featureless, flat interface. The chains adopted a tilt with a fairly broad probability distribution having a maximum at around 20°.

Tarek et al. [388] studied a system with some similarities to the work of Bocker et al. described earlier—a monolayer of n-tetradecyltrimethylammonium bromide. They also used explicit representations of the water molecules in a slab orientation, with the monolayer on either side, in a molecular dynamics simulation. Their goal was to model more disordered, liquid states, so they chose two larger molecular areas, 0.45 and 0.67 nm^2 molecule^{-1}. Density profiles normal to the interface were calculated and compared to neutron reflectivity data, with good agreement reported. The hydrocarbon chains were seen as highly disordered, and the diffusion was seen at both areas, with a factor of about 2.5 increase from the smaller molecular area to the larger area. They report no evidence of a tendency for the chains to aggregate into ordered islands, so perhaps this work can be seen as a realistic computer simulation depiction of a monolayer in an LE state.

Kuhn and Rehage [389,390] have performed simulations of a monododecyl pentathylene glycol monolayer at the air–water interface, with explicit inclusion of water in the molecular dynamics simulations. This is perhaps a surfactant particularly suited to the inclusion of explicit water molecules because of the extended length of the chain with polar groups, and in fact they find significant water presence all along the glycol parts of the chains. These waters apparently have an impact on the chain configurations, for the glycol ether part of the molecule was seen to be fairly well ordered and aligned nearly perpendicular to the plane. By contrast, the alkyl parts of the molecules were much more disordered and adopted a significant tilt angle of about 43° with respect to the surface normal. Even though the hydrocarbon parts of the chains were disordered, only very small diffusion was observed, presumably due to the close binding of the ether parts to the water substrate.

REFERENCES

1. KB Blodgett. J Am Chem Soc 56:495, 1934.
2. KB Blodgett. J Am Chem Soc 57:1007–1022, 1935.
3. KB Blodgett, I Langmuir. Phys Rev 51:964–982, 1937.
4. H Kuhn, D Möbius, H Bucher. In: A Weissberger and BW Rossiter, eds. Techniques of Chemistry. Vol 1, Part IIIB. New York: Wiley, 1972, pp 577–702.
5. G Roberts, ed. Langmuir–Blodgett Films. New York: Plenum Press, 1990.
6. RH Tredgold. Order in Organic Thin Films. Cambridge, UK: Cambridge University Press, 1994.
7. MC Petty. Langmuir–Blodgett Films: An Introduction. Cambridge, UK: Cambridge University Press, 1996.
8. GL Gaines. Insoluble Monolayers at Liquid–Gas Interfaces. New York: Wiley, 1966.
9. F MacRitchie. Chemistry at Interfaces. San Diego, CA: Academic Press, 1990.
10. CM Knobler. In: I Prigogine, SA Rice, eds. Advances in Chemical Physics. Vol 77. New York: Wiley, 1990, pp 387–425.
11. CM Knobler, RC Desai. Ann Rev Phys Chem 43:207–236, 1992.
12. AM Bibo, CM Knobler, IR Petersen. J Phys Chem 95:5591–5597, 1991.
13. WD Harkins, LE Copeland. J Chem Phys 10:272–286, 1942.
14. NR Pallas, BA Pethica. J Chem Soc Faraday Trans 1, 83:585–590, 1987.
15. B Moore, CM Knobler, S Akamatsu, F Rondelez. J Phys Chem 94:4588–4595, 1990.
16. B Moore, CM Knobler, D Broseta, F Rondelez. J Chem Soc, Faraday Trans 2, 86:1753–1761, 1986.
17. KJ Stine, CM Knobler, RC Desai. Phys Rev Lett 65:1004–1007, 1990.
18. KJ Stine, SA Rauseo, BG Moore, JA Wise, CM Knobler. Phys Rev A 41:6884–6892, 1990.
19. CM Knobler, K Stine, BG Moore. In: A Onuki, K Kawasaki, eds. Dynamics and Patterns in Complex Fluids (Springer Series in Physics, vol. 52). New York: Springer-Verlag, 1990, pp 130–140.
20. X Qiu, J Ruiz-Garcia, KJ Stine, CM Knobler, JV Selinger. Phys Rev Lett 67:703–706, 1991.
21. DK Schwartz, CM Knobler. J Phys Chem 97:8849–8851, 1993.
22. B Fischer, MW Tsao, J Ruiz-Garcia, TM Fischer, DK Schwartz, CM Knobler. J Phys Chem 98:7430–7435, 1994.
23. CM Knobler. Il Nuovo Cimento 16D:1367–1372, 1994.
24. DK Schwartz, J Ruiz-Garcia, Xia Qiu, JV Selinger, CM Knobler. Physica A 204:606–615, 1994.
25. MW Tsao, TM Fischer, CM Knobler. Langmuir 11:3184–3188, 1995.
26. S Rivieré, S Hénon, J Meunier, DK Schwartz, MW Tsao, CM Knobler. J Chem Phys 101:10045–10051, 1994.
27. PG de Gennes, JE Prost. The Physics of Liquid Crystals. 2nd ed. Oxford, UK: Clarendon Press, 1993.
28. E Stenhagen. In: EA Braude, FC Nachod, eds. Determination of Organic Structures by Physical Methods. New York: Academic Press, 1955, pp 325–371.
29. S Ställberg-Stenhagen, E Stenhagen. Nature 156:239–240, 1945.
30. M Lundquist. Chem Scr 1:5–20, 1971.
31. M Lundquist. Chem Scr 1:197–209, 1971.
31a. E Teer, CM Knobler, C Lautz, S Wurlitzer, J Kildae, TM Fischer. J Chem Phys 106:1913–1920, 1997.
32. GA Overbeck, D Möbius. J Phys Chem 97:7999–8004, 1993.
33. MC Shih, TM Bohanon, JM Mikrut, P Zschack, P Dutta. J Chem Phys 96:1556–1559, 1992.
34. ML Kurnaz, DK Schwartz. Langmuir 12:4971–4975, 1996.
35. NR Pallas, BR Pethica. Langmuir 1:509–513 (1985).
36. RS Ghaskadvi, S Carr, M Dennin. J Chem Phys, 111:3675–3678, 1999.
37. DK Schwartz, CM Knobler, R Bruinsma. Phys Rev Lett 73:2841–2844, 1994.

38. HM McConnell. Ann Rev Phys Chem 42:171–195, 1991.
39. KJ Stine. Microsc Res Tech 27:439–450, 1994.
40. KJ Stine. In: A Baskin, W Norde, eds. Physical Chemistry of Biological Interfaces. New York: Marcel Dekker, 1999, pp 749–768.
41. D Hönig, D Möbius. J Phys Chem 95:4590–4592, 1991.
42. D Hönig, D Möbius. Thin Solid Films 210/211:64–68, 1992.
44. H Riegler, K Spratte. Thin Solid Films 210–211:9–12, 1992.
45. D Hönig and D Möbius. Chem Phys Lett 195:50–52, 1992.
46. IR Petersen, JD Earls, IR Girling, GJ Russell. Mol Cryst Liq Cryst 147:141–147, 1987.
47. R Viswanathan, DK Schwartz, J Ganaes, JAN Zasadzinski. Langmuir 8:1603–1607, 1992.
48. J Garnes, DK Schwartz, R Viswanathan, JAN Zazadzinski. Nature 357:54–57, 1992.
49. DK Schwartz, J Garnaes, R Visnwanathan, S Chiruvolu, JAN Zasadzinski. Phys Rev E 47:452–460, 1993.
50. DK Schwartz, R Viswanathan, J Garnaes, JA Zasadzinski. J Am Chem Soc 115:7374–7380, 1993.
51. DK Schwartz, R Viswanathan, JAN Zasadzinski. Phys Rev Lett 70:1267–1270, 1993.
52. SW Hui, R Viswanathan, JA Zasadzinski, JN Israelachvili. Biophys J 68:171–178, 1995.
53. DK Schwartz. Surf Sci Rep 27:241–334, 1997.
54. DK Schwartz, R Viswanathan, JAN Zasadzinski. J Phys Chem 96:10444–10447, 1992.
55. Y Horiguchi, K Yagi, T Hosokawa, N Yamamoto, H Muramatsu, M Fujihara. J Microsc 194:467–471, 1999.
56. H. Shiku, RC Dunn. J Microsc 194:455–460, 1999.
57. AK Kirsch, V Subramaniam, A Jenei, TM Jovin. J Microsc 194:448–454, 1999.
58. CW Hollars, RC Dunn. J Phys Chem 101:6313–6317, 1997.
59. CW Hollars, RC Dunn. Biophys J 75:342–353, 1998.
60. M Florscheimer, C Brillet, H Fuchs. Langmuir 15:5437–5439, 1999.
61. M Florscheimer, H Salmen, M Bosch, C Brillet, M Wierscham, H Fuchs. Adv Mater 9:1056–1060, 1997.
62. M Florscheimer, M Bosch, C Brillet, M Wierscham, H Fuchs. Supramol Sci 4:255–263, 1997.
63. JF Rabolt, FC Burns, NE Schlotter, JD Swalen. J Chem Phys 78:946–952, 1983.
64. R Aroca. J Mol Struct 292:17–28, 1993.
65. J Chamberlain, JE Pemberton. Langmuir 13:3074–3079, 1997.
66. T Takenaka, H Fukuzaki. J Raman Spect 8:151–154, 1979.
67. P Dutta. Curr Opin Solid State Mater Sci 2:557–562, 1997.
68. J Als-Nielsen, D Jacquemain, K Kjaer, F Leveiller, M Lahav, L Leiserowitz. Phys Rep 246:251–313, 1994.
69. B Lin, TM Bohanon, MC Shih, P Dutta. Langmuir 6:1665–1687, 1990.
70. RM Kenn, C Bohm, AM Bibo, IR Petersen, H Mo¯hwald, J Als-Nielsen, K Kjaer. J Phys Chem 95:2092–2097, 1991.
71. MK Durbin, A Malik, AG Richter, KG Huang, P Dutta. Langmuir 13:6547–6549, 1997.
72. JB Peng, GJ Foran, GT Barnes, IR Gentle. Langmuir 13:1602–1606, 1997.
73. J Claudius, T Gerber, J Weigelt, M Kinzler. Thin Solid Films 287:225–231, 1996.
74. F Rieutord, JJ Benattar, L Bosio, P Robin, C Blot, R de Kouchkovsky. J Physique 48:679–687, 1987.
75. J Dalliant, L Bosio, JJ Benattar, C Blot. Langmuir 7:611–614, 1991.
76. MR Buhaenko, MJ Grundy, RM Richardson, SJ Roser. Thin Solid Films 159:253–265, 1988.
77. MJ Grundy, RJ Musgrove, RM Richardson, SJ Roser, J Penfold. Langmuir 6:519–521, 1990.
78. JH Fendler, ed. Nanoparticles in Solids and Solutions. New York: Kluwer Academic, 1996.
79. JH Fendler, ed. Nanoparticles and Nanostructured Films: Preparation, Characterization and Applications. New York: Wiley, 1998.
80. G Schmid, M Baumle, M Geerkens, I Helm, C Osemann, T Sawitowski. Chem Soc Rev 28:179–185, 1999.

81. LP Kouwenhoven, PL McEuen. In: G Timpe, ed. Nanotechnology. New York: Springer-Verlag, 1999, pp 471–536.
82. DL Feldheim, CD Keating. Chem Soc Rev 27:1–12, 1998.
83. G Schön, U Simon. Colloid Polym Sci 273:202–218, 1995.
84. R Turton. The Quantum Dot: A Journey into the Future of Microelectronics. New York: Oxford University Press, 1995.
85. DL Klein, R Roth, AKL Lim, AP Alivisatos, PL McEuen. Nature 389:699–701, 1997.
86. M Burghard, C Mueller-Schwanneke, G Philipp, S Roth. J Phys Condens Matter 11: 2993–3002, 1999.
87. T Vossmeyer, S Jia, E Delonno, MR Diehl, SH Kim, X Peng, AP Alavisatos, JR Health. J Appl Phys 84:3664–3670, 1998.
88. SW Chen, RW Murray. J Phys Chem B 103:9996–10000, 1999.
89. JH Fendler. Tetrahedron Lett 43:1689–1700, 1987.
90. JH Fendler. J Phys Chem 89:2730–2740, 1985.
91. JH Fendler. Colloids Surf A 71:309–315, 1993.
92. JH Fendler, FC Meldrum. Adv Mater 7:607–632, 1995.
93. JH Fendler. Curr Opinion Coll Interface Sci. 1:202–207, 1996.
94. JH Fendler. Chem Mater 8:1616–1624, 1996.
95. JH Fendler. Curr Opinion Solid State Mat Sci 2:365–369, 1997.
96. XK Zhao, JH Fendler. J Phys Chem 94:3384–3387, 1990.
97. C Dolan, Y Yuan, T Jao, JH Fendler. Chem Mater 3:215–218, 1991.
98. NA Kotov, MED Zaniquelli, FC Meldrum, JH Fendler. Langmuir 9:3709–3716, 1993.
99. KC Yi, Z Horvolgyi, JH Fendler. J Phys Chem 98:3872–3881, 1994.
100. KC Yi, VS Mendieta, RL Castanares, FC Meldrum, C Wu, JH Fendler. J Phys Chem 99: 9869–9875, 1995.
101. KS Mayya, V Patil, M Sastry. Langmuir 13:2575–2577, 1997.
102. KS Mayya, V Patil, M Sastry. J Chem Soc Faraday Trans 93:3377–3381, 1997.
103. M Sastry, KS Mayya, V Patil, DV Paranjape, SG Hedge. J Phys Chem B 101:4954–4958, 1997.
104. KS Mayya and M Sastry. Langmuir 14:74–78, 1998.
105. V Patil, KS Mayya, SD Pradham, M Sastry. J Am Chem Soc 119:9281–9282, 1997.
106. M Sastry, KS Mayya, V Patil. Langmuir 14:5921–5928, 1998.
107. KS Mayya, M Sastry. J Phys Chem B 101:9790–9793, 1997.
108. KS Mayya, M Sastry. Langmuir 14:6344–6346, 1998.
109. YH Sun, JR Li, BF Li, L Jiang. Langmuir 13:5799–5801, 1997.
110. FC Meldrum, NA Kotov, JH Fendler. J Chem Soc Faraday Trans 90:673–680, 1994.
111. FC Meldrum, NA Kotov, JH Fendler. Langmuir 10:2035–2040, 1994.
112. FC Meldrum, NA Kotov, JH Fendler. Mater Sci Eng C 3:149–152, 1995.
113. FC Meldrum, NA Kotov, JH Fendler. Chem Mater 7:1112–1116, 1995.
114. M Sastry, V Patil, KS Mayya, DV Paranjape, P Singh, SR Sainkar. Thin Solid Films 324:239–244, 1998.
115. K Abe, T Hanada, Y Yoshida, N Tanigaki, H Takiguchi, H Nagasawa, M Nakamoto, T Yamaguchi, K Yase. Thin Solid Films 327–329:524–527, 1998.
116. LF Chi, S Rakers, M Hartig, H Fuchs, G Schmid. Thin Solid Films 327–329:520–523, 1998.
117. G Schmid. Chem Rev 92:1709–1727, 1992.
118. M Burghard, G Philipp, S Roth, K von Klitzing, G Schmid. Opt Mater 9:401–405, 1998.
119. JP Bourgoin, C Kergueris, E Lefevre, S Palacin. Thin Solid Films 327–329:515–519, 1998.
120. LF Chi, M Hartig, T Dreschler, Th Schwaack, C Seidel, H Fuchs, G Schmid. Appl Phys A 66:S187–190, 1998.
121. JR Heath, CM Knobler, DV Leff. J Phys Chem B 101:189–197, 1997.
122. KJ Stine, MF Bono, JS Kretzer. J Coll Interface Sci 162:320–322, 1994.
123. J Ruiz-Garcia, R Gámez-Corrales, BI Ivlev. Physica A 236:97–104, 1997.

124. J Ruiz-Garcia, R Gámez-Corrales, BI Ivlev. Phys Rev E 58:660–663, 1998.
125. SH Chen. Adv Mater 12:186–189, 2000.
126. CP Collier, RJ Saykally, JJ Shiang, SE Henrichs, JR Heath. Science 277:1978–1981, 1997.
127. F Remacle, CP Collier, JR Heath, RD Levine. Chem Phys Lett 291:453–458, 1998.
128. G. Markovich, CP Collier, JR Heath. Phys Rev Lett 80:3807–3810, 1998.
129. G Medeiros-Ribeiro, DAA Ohlberg, RS Williams, JR Heath. Phys Rev B 59:1633–1636, 1999.
130. SW Chung, G Markovich, JR Heath. J Phys Chem B 102:6685–6687, 1998.
131. RP Sear, SW Chung, G Markovich, WM Gelbart, JR Heath. Phys Rev E 59:R6225–R6229, 1999.
132. S Ravaine, GE Fanucci, CT Seip, JH Adair, DR Talham. Langmuir 14:708–713, 1998.
133. JK Pike, H Byrd, AA Morrone, DR Talham. J Am Chem Soc 115:8497–8498, 1993.
134. Y Yuan, I Cabasso, JH Fendler. Chem Mater 2:226–229, 1990.
135. KC Yi, JH Fendler. Langmuir 6:1519–1521, 1990.
136. XK Zhao, JH Fendler. Chem Mater 3:168–174, 1991.
137. XK Zhao, JH Fendler. J Phys Chem 95:3716–3723, 1991.
138. XK Zhao, L McCormick, JH Fendler. Chem Mater 3:922–935, 1991.
139. J Yang, JH Fendler. J Phys Chem 99:5505–5511, 1995.
140. JH Fendler. Supramol Chem 6:209–216, 1995.
141. Y Tian, C Wu, N Kotov, JH Fendler. Adv Mater 6:959–962, 1994.
142. Y Tian, C Wu, N Kotov, JH Fendler. Mater Res Soc Symp 358:259–264, 1995.
143. Y Tian, JH Fendler. Chem Mater 8:969–974, 1996.
144. WL Yu, W Huang, BY Zhu, GX Zhao. Mater Lett 221–223, 1997.
145. L Zhang, J Liu, Z Pan, Z Lu. Supramol Sci 5:577–581, 1998.
146. L Zhang, G Shen, Z Pan, Z Lu. Mater Chem Phys 55:160–163, 1998.
147. J Yang, FC Meldrum, JH Fendler. J Phys Chem 99:5500–5504, 1995.
148. S Xu, XK Zhao, JH Fendler. Adv Mater 2:183–188, 1990.
149. Z Du, Z Zhang, W Zhao, Z Zhu, J Zhang, Z Jin, T Li. Thin Solid Films 210/211:404–406, 1992.
150. M Gao, X Zhang, B Yang, J Shen. J Chem Soc Chem Commun 2229–2230, 1994.
151. M Gao, M Gao, X Zhang, Y Yang, B Yang, J Shen. J Chem Soc Chem Commun 2777–2778, 1994.
152. Y Tian, C Wu, JH Fendler. J Phys Chem 98:4913–4918, 1994.
153. KS Mayya, V Patil, PM Kumar, M Sastry. Thin Solid Films 312:300–305, 1998.
154. LS Li, Y Chen, S Kan, X Zhang, X Peng, M Liu, T Li. Thin Solid Films 284–285:592–595, 1996.
155. LS Li, Z Hui, Y Chen, X Zhang, X Peng, Z Liu, TJ Li. J Coll Interface Sci 192:275–280, 1997.
156. LS Li, J Zhang, LJ Wang, Y Chen, Z Hui, TJ Li, LF Chi, H Fuchs. J Vac Sci Technol B 15:1618–1622, 1997.
157. CB Murray, DJ Norris, MG Bawendi. J Am Chem Soc 115:8706–8715, 1993.
158. BO Dabbousi, CB Murray, MF Rubner, MG Bawendi. Chem Mater 6:216–219, 1994.
159. LS Li, J Jin, YQ Tian, YY Zhao, TJ Li, SM Jiang, ZL Du, GH Ma, N Zheng. Supramol Sci 5:475–478, 1998.
160. J Jin, LS Li, YQ Tian, YJ Zhang, Y Liu, YY Zhao, TS Shi, TJ Li. Thin Solid Films 327–329:559–562, 1998.
161. T Torimoto, N Tsumura, M Miyake, M Nishizawa, T Sakata, H Mori, H Yoneyama. Langmuir 15:1853–1858, 1999.
162. NA Kotov, FC Meldrum, C Wu, JH Fendler. J Phys Chem 98:2735–2738, 1994.
163. Y Tian, JH Fendler. Chem Mater 8:969–974, 1996.
164. NA Kotov, FC Meldrum, JH Fendler. J Phys Chem 98:8827–8830, 1994.
165. L Cao, L Huo, G Ping, D Wang, G Zeng, S Xi. Thin Solid Films 347:258–262, 1999.
166. SH Kim, G Markovich, S Rezvani, SH Choi, KL Wang, JR Heath. Appl Phys Lett 74:317–319, 1999.

167. A Raudel-Teixier, J Leloup, A Barraud. Mol Cryst Liq Cryst 134:347–354, 1986.
168. ES Smotkin, C Lee, AJ Bard, A Campion, MA Fox, TS Mallouk, SE Webber, JM White. Chem Phys Lett 152:265–268, 1988.
169. I Moriguchi, I Tanaka, Y Teraoka, S Kagawa. J Chem Soc Chem Commun 1401–1402, 1991.
170. X Peng, S Guan, X Chai, Y Jiang, T Li. J Phys Chem 96:3170–3174, 1992.
171. DJ Elliot, DN Furlong, TR Gengenbach, F Grieser, RS Urquhart, CL Hoffmann, JF Rabolt. Colloids Surf A 103:207–219, 1995.
172. DJ Scoberg, F Grieser, DN Furlong. J Chem Soc Chem Commun 515–517, 1991.
173. F Grieser, DN Furlong, D Scoberg, I Ichinose, N Kimizuki, T Kunitake. J Chem Soc Faraday Trans 88:2207–2214, 1992.
174. X Peng, R Lu, Y Zhao, H Chen, T Li. J Phys Chem 98:7052–7055, 1994.
175. L Wang, L Li, L Qu, R Lu, X Peng, Y Zhao, T Li. Mol Cryst Liq Cryst 295:175–180, 1997.
176. H Xiong, H Li, Z Wang, X Zhang, J Shen, M Gleiche, L Chi, H Fuchs. J Coll Inter Sci 211:238–242, 1999.
177. J Leloup, A Ruaudel-Teixier, A Barraud. Thin Solid Films 210/211:407–409, 1992.
178. DJ Elliot, DN Furlong, F Greiser. Colloids Surf A 141:9–17, 1998.
179. DJ Elliot, DN Furlong, Franz Greiser. Colloids Surf A 155:101–110, 1999.
180. DN Furlong, R Urquhart, F Greiser, K Tanaka, Y Okahata. J Chem Soc Faraday Trans 89:2031–2035, 1993.
181. RS Urquhart, DN Furlong, H Mansur, F Greiser, K Tanaka, Y Okahata. Langmuir 10:899–904, 1994.
182. JK Basu, MK Sanyal. Phys Rev Lett 79:4617–4620, 1997.
183. XM Yang, GM Wang, ZH Lu. Supramol Sci 5:549–552, 1998.
184. RS Urquhart, CL Hoffmann, DN Furlong, NJ Geddes, JF Rabolt, F Greiser. J Phys Chem 99:15987–15992, 1995.
185. I Moriguchi, K Hosoi, H Nagaoka, I Tanaka, Y Teraoka, S Kagawa. J Chem Soc Faraday Trans 90:349–354, 1994.
186. Z Pan, J Liu, X Peng, T Li, Z Wu, M Zhu. Langmuir 12:851–853, 1996.
187. A Milekhin, M Friedrich, DRT Zahn, L Sveshnikova, S Repinsky. Appl Phys A 69:97–100, 1999.
188. HS Mansur, F Greiser, MS Marychurch, S Biggs, RS Urquhart, DN Furlong. J Chem Soc Faraday Trans 91:665–672, 1995.
189. HS Mansur, F Greiser, RS Urquhart, DN Furlong. J Chem Soc Faraday Trans 91:3399–3404, 1995.
190. T Dannhauser, M O'Neil, K Johansson, G McClendon. J Phys Chem 90:6074–6076, 1986.
191. A Eychmuller, A Hasselbarth, L Katsikas, H Weller. Ber Bunsenges Phys Chem 95:79–84, 1991.
192. K Asai, T Yamaki, S Seki, K Ishigure, H Shibata. Thin Solid Films 284–285:541–544, 1996.
193. T Yamaki, K Asai, K Ishigure, H Shibata. Radiat Phys Chem 50:199–205, 1997.
194. RS Urquhart, DN Furlong, T Gegenbach, NJ Geddes, F Grieser. Langmuir 11:1127–1133, 1995.
195. AV Nabok, T Richardson, F Davis, CJM Stirling. Langmuir 13:3198–3201, 1997.
196. AV Nabok, T Richardson, C McCartney, N Cowlam, F Davis, CJM Stirling, AK Ray, V Gacem, A Gibaud. Thin Solid Films. 327–329:510–514, 1998.
197. P Facci, V Erokhin, A Tronin, C Nicolini. J Phys Chem 98:13323–13327, 1994.
198. P Facci, A Diaspro, R Rolandi. Thin Solid Films. 327–329:532–535, 1998.
199. V Erokhin, P Facci, L Gobbi, S Dante, F Rustichelli, C Nicolini. Thin Solid Films. 327–329:503–505, 1998.
200. DT Amm, DJ Johnson, T Laursen, SK Gupta. Appl Phys Lett 61:522–524, 1992.
201. V Erokhin, P Facci, S Carrara, C Nicolini. J Phys D Appl Phys 28:2534–2538, 1995.
202. P Facci, V. Erokhin, S. Carrara, C Nicolini. Proc Natl Acad Sci 93:10556–10559, 1996.
203. XK Zhao, S Xu, JH Fendler. J Phys Chem 94:2573–2581, 1990.

204. XK Zhao, PJ Herve, JH Fendler. J Phys Chem 93:908–916, 1989.
205. FC Meldrum, NA Kotov, JH Fendler. J Phys Chem 98:4506–4510, 1994.
206. YS Kang, S Risbud, J Rabolt, P. Stroeve. Langmuir 12:4345–4349, 1996.
207. YS Kang, DK Lee, P Stroeve. Thin Solid Films 327–329:541–544, 1998.
208. X Peng, Y Zhang, J Yang, B Zou, L Xiao, T Li J Phys Chem 96:3412–3415, 1992.
209. X Peng, M Gao, Y Zhao, S Kang, Y Zhang, Y Zhang, D Wang, L Xiao, T Li. Chem Phys Lett 209:233–237, 1993.
210. SA Iakovenko, AS Trifonov, M Giersig, A Mamedov, DK Nagesha, VV Hanin, EC Soldatov, NA Kotov. Adv Mater 11:388–392, 1999.
211. S Lefebure, C Menager, V Cabuil, M Assenheimer, F Gallet, C Flament. J Phys B 102: 2733–2738, 1998.
212. A Brugger, Ch Schoppmann, M Schurr, M Seidl, G Sipos, CY Hahn, J Hassman, O Waldmann, H Voit. Thin Solid Films 338:231–242, 1999.
213. T Yamaki, T Yamada, K Asai, K Ishigure, H Shibata. Thin Solid Films. 327–329:581–585, 1998.
214. T Yamaki, T Yamada, K Asai, K Ishigure. Thin Solid Films. 327–329:586–590, 1998.
215. NA Kotov, G Zavala, JH Fendler. J Phys Chem 99:12375–12378, 1995.
216. MS Dresselhaus, G Dresselhaus, PC Eklund. Science of Fullerenes and Carbon Nanotubes. New York: Academic Press, 1996.
217. HW Kroto, DRM Walton, eds. The Fullerenes: New Horizons for the Chemistry, Physics, and Astrophysics of Carbon. Cambridge, UK: Cambridge University Press, 1993.
218. P Delhaes, H Kuzmany, eds. Fullerenes and Carbon-Based Materials. New York: Elsevier Science, 1998.
219. R Saito, G Dresselhaus, MS. Dresselhaus Physical Properties of Carbon Nanotubes. London: Imperial College Press, 1998.
220. R Taylor, ed. The Chemistry of the Fullerenes (Advanced Series in Fullerenes, Vol. 4). River Edge, NJ: World-Scientific, 1995.
221. H Aldersey-Williams. The Most Beautiful Molecule: The Discovery of the BuckyBall. New York: Wiley, 1995.
222. HW Kroto, JR Heath, SC O'Brien, RF Curl, RE Smalley. Nature 318:162–163, 1985.
223. W Krätschmer, LD Lamb, K Fostiropoulos, DR Huffman. Nature 347:354–358, 1990.
224. RE Haufler, J Conceicao, LPF Chibante, Y Chai, NE Byrne, S Flanagan, MM Haley, SC O'Brien, C Pan, Z Xiao, WE Billups, MA Ciufolini, RH Hauge, JL Margave, LJ Wilson, RF Curl, RE Smalley. J Phys Chem 94:8634–8636, 1990.
225. JM Hawkins, M Nambu, A Meyer. J Am Chem Soc 116:7642–7645, 1994.
226. R Ettl, I Chao, F Diederich, RL Whetten. Nature 353:149–153, 1991.
227. Q Xie, E Pérez-Cordero, L Echegoyen. J Am Chem Soc 114:3978–3980.
228. AF Hebard, MJ Rosseinsky, RC Haddon, DW Murphy, SH Glarum, TTM Palstra, AP Ramirez, AR Kortan. Nature 350:600–601, 1991.
229. TTM Palstra, O Zhou, Y Isawa, PE Sulewski, RM Fleming, BR Zegarski. Solid State Commun 93:327–330, 1995.
230. RC Haddon. Acc Chem Res 25:127–133, 1992.
231. YS Obeng, AJ Bard. J Am Chem Soc 113:6279–6280, 1991.
232. LOS Bulho˜es, WS Obeng, AJ Bard, Chem Mater 5:110–114, 1993.
233. B Seo, H Lee, J Chung, W Seo, Y Cho, KH Lee. Synth Met 86:2423–2424, 1997.
234. GG Siu, L Yulong, X Shishen, X Jingmei, L Tiankai, X Linge. Thin Solid Films 274:147–149, 1996.
235. T Nakamura, H Tachibana, M Yumura, M Matsumoto, R Azumi, M Tanaka, Y Kawabata. Langmuir 8:4–6, 1992.
236. T Nakamura, H Tachibana, M Yumura, M Matsumoto. Synth Met 55–57:3131–3136, 1993.
237. J Milliken, DD Dominguez, HH Nelson, WR Barger. Chem Mater 4:252–254, 1992.
238. R Back, RB Lennox. J Phys Chem 96:8149–8152, 1992.

239. G Williams, C Pearson, MR Bryce, MC Petty. Thin Solid Films 209:150–152, 1992.
240. G Williams, AJ Moore, MR Bryce, YM Lvov, MC Petty. Synth Met 55–57:2955–2960, 1993.
241. Y Li, R Sun, A Lu, Y Fan, D Jiang, L Zhang. Thin Solid Films 248:83–85, 1994.
242. J Cirák, P Tomcik, D Barancok, M Halus`ka, V Nadazdy. Synth Met 73:285–287, 1995.
243. R Castillo, S Ramos, J Ruiz-Garcia. J Phys Chem 100:15235–15241, 1996.
244. C Long, Y Xu, Y Li, D Xu, Y Yao, D Zhu. Solid State Commun 82:381–383, 1992.
245. Y Tomioka, M Ishibashi, H Kajiyama, Y Taniguchi. Langmuir 9:32–35, 1993.
246. P Wang, M Shamsuzzoha, W Lee, X Wu, RM Metzger. Synth Met 55–57:3104–3109, 1993.
247. R Rella, P Siciliano, L Valli. Phys Stat Solidi 143:K129–K131, 1994.
248. Y Xu, J Guo, C Long, Y Li, Y Yao, D Zhu. Thin Solid Films 242:45–49, 1994.
249. H Lee, SR Park, DW Kim, BY Seo. Synth Met 71:2065–2066, 1995.
250. T Imae, Y Ikeo. Supramol Sci 5:61–65, 1998.
251. Atwood, GA Koutsantonis, CL Raston. Nature 368:229–231, 1994.
252. T Suzuki, K Nakashima, S Shinkai. Chem Lett 699–702, 1994.
253. T Andersson, K Nilsson, M Sundahl, G Westman, O Wennerström. J Chem Soc Chem Commun 604–606, 1992.
254. F Diedrich, J Effing, U Jonas, L Jullien, T Plesnivy, H Ringsdorf, C Thigen, D Weinstein. Angew Chem Intl Ed Engl 31:1599–1602, 1992.
255. R Castillo, S Ramos, R Cruz, M Martinez, F Lara, J Ruiz-Garcia. J Phys Chem 100:709–713, 1996.
256. P Lo Nostro, A Casnati, L Bossoletti, L Dei, P Baglioni. Coll Surf A 116:203–209, 1996.
257. L Dei, P LoNostro, G Capuzzi, P Baglioni. Langmuir 14:4143–4147, 1998.
258. ZI Kazantseva, NV Lavrik, AV Nabok, OP Dimitriev, BA Nesterenko, VI Kalchenko, SV Vysotsky, LN Markovskiy, AA Marchenko. Supramol Sci 4:341–347, 1997.
259. TS Berzina, VI Troitsky, OY Neilands, IV Sudmale, C Nicolini. Thin Solid. Films. 256:186–191, 1995.
260. A Hirsch, A Soi, HR Karfunfel. Angew Chem Intl Ed Engl 31:766–767, 1992.
261. G Williams, A Soi, A Hirsch, MR Bryce, MC Petty. Thin Solid Films 230:73–77, 1993.
262. NC Maliszewskyj, PA Heiney, DR Jones, RM Strongin, MA Cichy, AB Smith. Langmuir 9:1439–1441, 1993.
263. M Maggini, A Karlsson, L Pasimeni, G Scorrano, M Prato, L Valli. Tet Lett 35:2985–2988, 1994.
264. M Maggini, L Pasimeni, M Prato, G Scorrano, L Valli. Langmuir 10:4164–4166, 1994.
265. JY Wang, D Vaknin, RA Uphaus, K Kjaer, M Losche. Thin Solid Films 242:40–44, 1994.
266. S Ravaine, B Agricole, C Mingotaud, J Cosseau, P Delhaès. Chem Phys Lett 242:478–482, 1995.
267. M Matsumoto, H Tachibana, R Azumi, M Tanaka, T Nakamura, G Yunome, M Abe, S Yamago, E Nakamura. Langmuir 11:660–665, 1995.
268. DM Guildi, Y Tian, JH Fendler, H Hungerbuhler, K Asmus. J Phys Chem. 99:17673–17676, 1995.
269. DM Guildi, T Tian, JH Fendler, H Hungerbuhler, K Asmus. J Phys Chem 100:2753–2758, 1996.
270. DM Guildi, K-D Asmus, Y Tian, JH Fendler. Proc Electrochem Soc 96–10:501–508, 1996.
271. S Ravaine, C Mingotaud, P Delhaes. Thin Solid Films. 284–285:76–79, 1996.
272. RV Bensasson, E Bienvenue, M Dellinger, S Leach, P Seta. J Phys Chem 98:3492–3500, 1994.
273. T Kawai, S Scheib, MP Cava, RM Metzger. Langmuir 13:5627–5633, 1997.
274. U Jonas, F Cardullo, P Belik, F Diedrich, A Gügel, E Harth, A Herrmann, L Isaacs, K Mullen, H Ringsdorf, C Thilgen, P Uhlmann, A Vasella, CAA Waldraff, M Walter. Chem: Eur J 1:243–251, 1995.
275. S Wang, RM LeBlanc, F Arias, L Echeygoyen. Langmuir 13:1672–1676, 1997.

276. DA Leigh, AE Moody, FA Wade, TA King, D West, GS Bahra. Langmuir 11:2334–2336, 1995.

277. Y Seo, C Jung, M Jikei, M Kakimoto. Thin Solid Films. 311:272–276, 1997.

278. D Taton, S Angot, Y Gnanou, E Wolert, S Setz, R Duran. Macromol 31:6030–6033, 1998.

279. C Jehoulet, YS Obeng, Y-T Kim, F Zhou, AJ Bard. J Am Chem Soc 114:4237–4247, 1991.

280. C Jehoulet, AJ Bard, F Wudl. J Am Chem Soc 113:5456–5457, 1991.

281. F Zhou, S Yau, C Jehoulet, DA Laude, Z Guan, AJ Bard. J Phys Chem 96:4160–4162, 1992.

282. H Luo, N Li, W Zhang, L Gan, C Huang. Electroanalysis 11:582–585, 1999.

283. LM Goldenberg, G Williams, MR Bryce, AP Monkman, MC Petty, A Hirsch, A Soi. J Chem Soc Chem Commun 1310–1312, 1993.

284. B Miller, JM Rosamilia, G Dabbagh, R Tycko, RC Haddon, AJ Muller, W Wilson, DW Murphy, AF Hebard. J Am Chem Soc 113:6291–6293, 1991.

285. C Luo, C Huang, L Gan, D Zhou, W Xia, Q Zhuang, Y Zhao, Y Huang. J Phys Chem 100:16685–16689, 1996.

286. C Luo, L Gan, D Zhou, C Huang. J Chem Soc Faraday Trans 93:3115–3117, 1997.

287. W Zhang, L Gan, C Huang. J Mater Chem 8:1731–1734, 1998.

288. Y Zhao, L Gan, D Zhou, C Huang, J Jiang, W Liu. Solid State Commun 106:43–48, 1998.

289. W Zhang, L Gan, C Huang. Synth Met 96:223–227, 1998.

290. Y Huang, L Gan, C Huang, F. Meng. Supramol Sci 5:457–460, 1998.

291. Y Huang, Y Zhao, L Gan, C Huang, N Wu. J Coll Interf Sci 204:277–283, 1998.

292. W Zhang, Y Shi, L Gan, C Huang, H Luo, D Wu, N Li. J Phys Chem B 103:675–681, 1999.

293. MI Sluch, IDW Samuel, MC Petty. Chem Phys Lett 280:315–320, 1997.

294. MI Sluch, IDW Samuel, A Beeby, MC Petty. Langmuir 14:3343–3346, 1998.

295. P Wang, RM Metzger, S Bandow, Y Maruyama. J Phys Chem 97:2926–2927, 1993.

296. RM Metzger, P Wang, X Wu, GV Tormos, D Lorcy, I Scherbakova, MV Lakshmikantham, MP Cava. Synth Met 70:1435–1438, 1995.

297. K Ikegami, S Kuroda, M Matsumoto, T Nakamura. Jpn J Appl Phys 34:L1227–L1229, 1995.

298. P Wang, Y Maruyama, RM Metzger. Langmuir 12:3932–3937, 1996.

299. J Paloheimo, H Isotalo, J Kastner, H Kuzmany. Synth Met 55–57:3185–3180, 1993.

300. Y Xiao, Z Yao, D Jin, F Yan, Q Xue. J Phys Chem 97:7072–7074, 1993.

301. YF Xiao, AQ Wang, Y Liu, XM Liu, ZQ Yao. Thin Solid Films 251:4–6, 1994.

302. YF Xiao, ZQ Yao, DS Jin. Thin Solid Films 251:94–95, 1994.

303. M Rikukawa, S Furumi, K Sanui, N Ogata. Synth Met 86:2281–2282, 1997.

304. H Ding, D Sun, H Cui, S Xi, F Tian. Supramol Sci 5:611–613, 1998.

305. NS Saricifiti, L Smilowitz, AJ Heeger, F Wudl. Science 258:1474–1476, 1992.

306. Y Wang. Nature 356:585–587, 1992.

307. Y Li, Y Xu, D Xu. Solid State Commun 95:695–699, 1995.

308. Y Liu, Y Xu, D Zhu. Synth Met 90:143–146, 1997.

309. CT Ewins, B Stewart. Thin Solid Films 284–285:49–52, 1996.

310. XK Wang, TG Zhang, WP Lin, SZ Liu, GK Wong, MM Kappes, RPH Chang, JB Ketterson. Appl Phys Lett 64:2785–2787, 1992.

311. TG Zhang, ZY Xu, PM Lundquist, WP Lin, JB Ketterson, GK Wong, XK Wang, RPH Chang. Opt Commun 111:517–520, 1994.

312. X Zhang, X Ye, K Chen. Opt Commun 113:519–522, 1995.

313. ZH Kafafi, JR Lindle, RGS Pong, FJ Bartoli, LJ Lingg, J Milliken. Chem Phys Lett 188:492–495, 1992.

314. LB Gan, DJ Zhou, CP Lou, CH Huang, TK Li, J Bai, XS Zhao, XH Xia. J Phys Chem 98:12459–12461, 1994.

315. R Sun, Y Li, J Zhang, D Li, Y Fan, A Lu. Thin Solid Films 248:100–101, 1994.

316. D Zhou, L Gan, C Luo, H Tan, C Huang, Z Liu, Z Wu, X Zhao, X Xia, S Zhang, F Sun, Z Xia, Y Zou. Chem Phys Lett 235:548–551, 1995.

317. S Ma, X Lu, J Chen, K Han, L Liu, Z Huang, R Cai, G Wang, W Wang, Y Li. J Phys Chem 100:16629–16632, 1996.

318. D Zhou, L Gan, C Luo, C Huang, Y Wu. Solid State Commun 12:891–894, 1997.
319. VR Novak, SL Vorob'eva, IV Myagkov. Mol Mater 7:175–178, 1996.
320. ED Mishina, AA Fedyanin, D Klimkin, AA Nikulin, OA Aktsipetrov, SL Vorob'eva, MAC Devillers, T Rasing. Surf Sci 382:L696–L699, 1997.
321. Y Xu, C Zhu, C Long, Y Liu, D Zhu, A Yu. Mol Cryst Liq Cryst 294:7–10, 1997.
322. C Long, Y Xu, C Zhu, D Zhu. Chin Sci Bull 43:649–652, 1998.
323. C Long, Y Xu, C Zhu, D Zhu. Solid State Commun 101:439–442, 1997.
324. D Zhou, GJ Ashwell, R Rajan, L Gan, C Luo, C Huang. J Chem Soc Faraday Trans 93:2077–2081, 1997.
325. G Wang, J Wen, Q Houng, S Qian, X Lu. J Phys D: Appl Phys 32:84–86, 1999.
326. J Zhang, Q Xue, Z Du, Z Zhu. Chin Sci Bull 39:1184–1187, 1994.
327. QJ Xue, J Zhang. Tribol Intl 28:287–291, 1995.
328. E Tronel-Peyroz, G Miquel-Mercier, P Seta. Synth Met 81:33–38, 1996.
329. V Krstic, GS Duesberg, J Muster, M Burghard, S Roth. Chem Mater 10:2338–2340, 1998.
330. C Journet, WK Maser, P Bernier, A Loiseau, M Lamy de la Chapelle, S Lefrant, P Deniard, R Lee, JE Fischer. Nature 388:756–759, 1997.
331. J Fang, CM Knobler. J Phys Chem 99:10425–10429, 1995.
332. J Fang, CM Knobler. Langmuir 12:1368–1374, 1996.
333. J Fang, CM Knobler, M Gingery, FA Eiserling. J Phys Chem B 101:8692–8695, 1997.
334. TE Herod, RS Duran. Langmuir 14:6606–6609, 1998.
335. TE Herod, RS Duran. Langmuir 14:6956–6968, 1998.
336. C Duschl, M Liley, H Vogel. Angew Chem Intl Ed Engl 33:1274–1276, 1994.
337. L Santesson, TMH Wong, M Taborelli, P Descouts, M Liley, C Duschl, H Vogel. J Phys Chem 99:1038–1045, 1995.
338. MP Allen, DJ Tildesley. Computer Simulation of Liquids. Oxford, UK: Oxford University Press, 1987.
339. D Frenkel, B Smit. Understanding Molecular Simulation. New York: Academic Press, 1996.
340. AJ Kox, JPJ Michels, FW Wiegel. Nature 287:317–319, 1980.
341. P van der Ploeg, HJC Berendsen. J Chem Phys 76:3271–3276, 1982.
342. JP Ryckaert, A Bellemans. Faraday Discuss Chem Soc 66:95–106, 1978.
343. JP Ryckaert, G Ciccotti, HJC Berendsen. J Comput Phys 23:327–341, 1977.
344. MP Allen, DJ Tildesley. Computer Simulation of Liquids. Oxford, UK: Oxford University Press, 1987, pp 92–98.
345. SH Northrup. J Phys Chem 88:3441–3446, 1984.
346. SH Northrup, MS Curvin. J Phys Chem 89:4707–4713, 1985.
347. JF Nagle. J Chem Phys 63:1255–1261, 1975.
348. JF Nagle. Ann Rev Phys Chem 31:157–195, 1980.
349. GM Torrie, JP Valleau. J Comp Phys 23:187–199, 1977.
350. D Frenkel, B Smit. Understanding Molecular Simulation. New York: Academic Press, pp 176–180, 1996.
351. J Harris, SA Rice. J Chem Phys 89:5898–5908, 1988.
352. JP Bareman, G Cardini, ML Klein. Phys Rev Lett 60:2152–2155, 1988.
353. JP Bareman, ML Klein. J Phys Chem 94:5202–5205, 1990.
354. DE Williams. J Chem Phys 47:4680, 1967.
355. T Yamamoto. J Chem Phys 89:2356–2365.
356. JP Bareman, ML Klein. In: KS Liang et al., eds. Interface Dynamics and Growth. Mat. Res. Soc. Symp. Vol. 237, MRS, pp 271–279, 1992.
357. WL Jorgensen, J Chandrasekhar, JD Madura, RW Impey, ML Klein. J Chem Phys 79:926, 1983.
358. D Frenkel, B Smit. Understanding Molecular Simulation. New York: Academic Press, pp 347–362, 1996.
359. MA Moller, DJ Tildesley, KS Kim, N Quirke. J Chem Phys 94:8390–8401, 1991.

360. K Kjaer, J Als-Nielsen, CA Helm, P Tippman-Krayer, H Möhwald. J Phys Chem 93:3200–3206, 1989.
361. KS Kim, MA Moller, DJ Tildesley, N Quirke. Mol Sim 13:77–99, 1994.
362. S Karaborni, S Toxvaerd. J Chem Phys 96:5505–5515, 1992.
363. S Karaborni, S Toxvaerd. J Chem Phys 97:5876–5883, 1992.
364. S Karaborni, S Toxvaerd, OH Olsen. J Phys Chem 96:4965–4973, 1992.
365. S Karaborni. Langmuir 9:1334–1343, 1993.
366. WD Harkins, E Boyd. J Chem Phys 45:20–43, 1941.
367. S Toxvaerd. J Chem Phys 93:4290–4295, 1990.
368. S Karaborni, JI Siepmann. Mol Phys 83:345–350, 1994.
369. JI Siepmann, S Karaborni, ML Klein. J Phys Chem 98:6675–6678, 1994.
370. MW Kim, DS Cannell. Phys Rev A 13:411–416, 1976.
371. S Shin, N Collazo, SA Rice. J Chem Phys 96:1352–1366, 1992.
372. S Shin, N Collazo, SA Rice. J Chem Phys 98:3469–3474, 1993.
373. S Shin, SA Rice. Langmuir 10:262–266, 1994.
374. N Collazo, S Shin, SA Rice. J Chem Phys 96:4735–42, 1992.
375. ME Schmidt, S Shin, SA Rice. J Chem Phys 104:2101–2113, 1996.
376. ME Schmidt, S Shin, SA Rice. J Chem Phys 104:2114–2123, 1996.
377. N Kim, S Shin. J Chem Phys 110:10239–10242, 1999.
378. N Kim, S Shin. J Chem Phys 111:6556–6564, 1999.
379. F Schmid, M Schick. J Chem Phys 102:2080–2091, 1995.
380. M Li, AA Acero, Z Huang and SA Rice. Nature 367:151, 1994.
381. Z Huang, AA Acero, N Lei, SA Rice, Z Zhang, ML Schlossman. J Chem Soc Faraday Trans 92:545, 1996.
382. GH Peters, S Toxvaerd. Tens Surf Det 30:264–268, 1993.
383. GH Peters, S Toxvaerd, A Svendsen, OH Olsen. J Chem Phys 100:5996–6010, 1994.
384. GH Peters, NB Larson, T Bjornholm, S Toxvaerd, K Schaumburg, K Kjaer. Phys Rev E 57:3153–3163, 1998.
385. A Porohille, I Benjamin. J Phys Chem 97:2664–2670, 1993.
386. KJ Schweighofer, U Essmann, M Berkowitz. J Phys Chem B 101:3793–3799, 1997.
387. J Bocker, M Schlenkrich, P Bopp, J Brickmann. J Phys Chem 96:9915–9922, 1992.
388. M Tarek, DJ Tobias, ML Klein. J Phys Chem 99:1393–1402, 1995.
389. H Kuhn, H Rehage. Tens Surf Det 35:448–453, 1998.
390. H Kuhn, H Rehage. J Phys Chem B 103:8493–8501, 1999.

4

Supramolecular Organic Layer Engineering for Industrial Nanotechnology

CLAUDIO NICOLINI, V. EROKHIN, and M. K. RAM   University of Genoa, Genoa, Italy

I. INTRODUCTION

This chapter reviews the present status of supramolecular layer engineering with respect to the development of a nanotechnology capable of yielding an industrial revolution along the routes foreseen by the programs launched some time ago by the National Science and Technology Council in Italy and more recently with heavy funding by President Bill Clinton starting beginning fiscal year 2000.

II. SUPRAMOLECULAR LAYER ENGINEERING

A. Langmuir Film Engineering

Salts of fatty acids are "classic" objects of LB technique. Being placed at the air/water interface, these molecules arrange themselves in such a way that its hydrophilic part (COOH) penetrates water due to its electrostatic interactions with water molecules, which can be considered electric dipoles. The hydrophobic part (aliphatic chain) orients itself to air, because it cannot penetrate water for entropy reasons. Therefore, if a few molecules of such type were placed at the water surface, they would form a two-dimensional system at the air/water interface. A compression isotherm of the stearic acid monolayer is presented in Figure 1. This curve shows the dependence of surface pressure upon area per molecule, obtained at constant temperature. Usually, this dependence is called a π-A isotherm.

Initially, the compression does not result in surface pressure variations. Molecules at the air/water interface are rather far from each other and do not interact. This state is referred to as a *two-dimensional gas*. Further compression results in an increase in surface pressure. Molecules begin to interact. This state of the monolayer is referred as *two-dimensional liquid*. For some compounds it is also possible to distinguish liquid-expanded and liquid-condensed phases. Continuation of the compression results in the appearance of a *two-dimensional solid-state* phase, characterized by a sharp increase in surface pressure, even with small decreases in area per molecule. Dense packing of molecules in the monolayer is reached. Further compression results in the collapse of the monolayer. Two-dimensional structure does not exist anymore, and the multilayers form themselves in a non-controllable way.

The floating monolayer can be transferred onto the surface of solid supports. Two main techniques are usually considered for the monolayer deposition, namely, Lang-

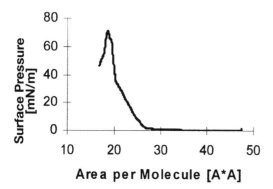

FIG. 1   Variation of pressure vs. area isotherm of stearic acid.

muir–Blodgett (or vertical lift) and Langmuir–Schaefer (or horizontal lift). Specially pre-pared substrate is passed vertically through the monolayer. The monolayer is transferred onto the substrate surface during this passage. The important point is to have the monolayer electrically neutral. If some charges in the monolayer molecule head groups are uncom-pensated, the deposition will not be performed and electrostatic interaction of this charge with water molecules will be higher than the hydrophobic interactions of chains with the hydrophobized substrate surface.

The other method of monolayer transfer from the air/water interface onto solid sub-strates is illustrated in Figure 2. This method is called the Langmuir–Schaefer technique, or horizontal lift. It was developed in 1938 by I. Langmuir and V. Schaefer for deposition of protein layers. Prepared substrate horizontally touches the monolayer, and the layer transfers itself onto the substrate surface. The method is often used for the deposition of rigid monolayers and for protein monolayers. In both cases the application of the Lang-muir–Blodgett method produces defective films.

Deposited films are usually divided into three types, schematically shown in Fig-ure 4, namely, X-, Y-, and Z-types. As it is clear from the figure, the Y-type is a cen-trosymmetrical one, while the X- and Z-types are polar ones, which differ only by the ori-entation of the head groups and hydrocarbon chains with respect to the substrate surface. Such a difference appears due to the fact that in some cases there is no monolayer transfer during upward or downward motion of the substrate in the case of LB deposition. In the case of the LS deposition, moreover, the layers always seem to be transferred in a polar

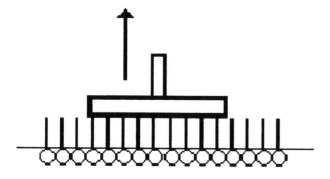

FIG. 2   Schematic of Langmuir–Schaefer deposition.

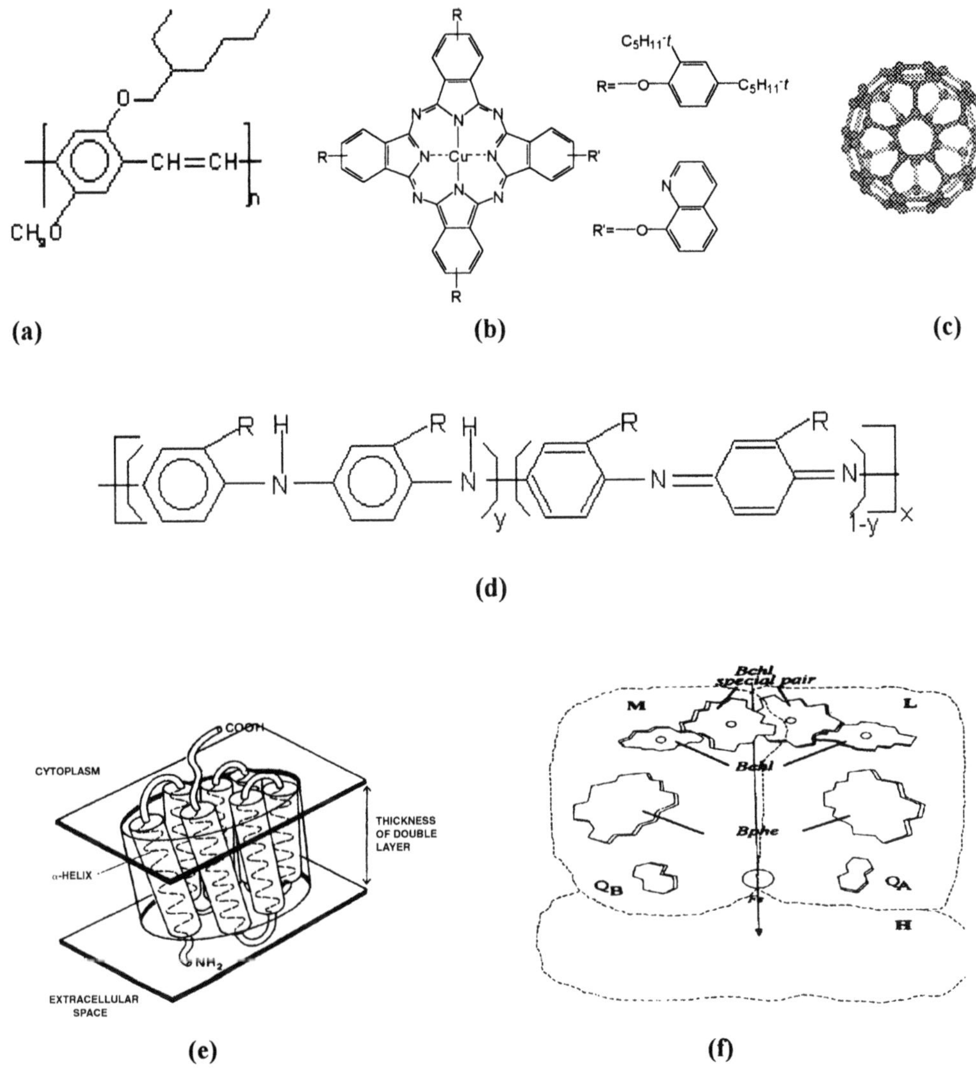

(a)   **(b)**   **(c)**

(d)

(e)   **(f)**

FIG. 3   Organic molecules.

manner. However, the X- and Z-types are almost never realized in practice. Even if some nonlinear properties, such as pyroelectricity, realizable only in polar structures, were observed, and the structures were considered as polar ones, detailed investigations have revealed that the films are of Y-type with unequal filling of odd and even layers. Moreover, in the case of LS deposition, it was shown that the last three transferred monolayers are involved in structural reorganization during pasage through the meniscus, in order to realize thermodynamically stable Y-type packing.

A large number of potential applications for organized protein monolayers have recently motivated considerable research activity in this field (Boussaad et al. 1998, Kiselyova et al. 1999). Construction of specific interaction-directed, self-assembled protein films has been performed at the air–water interface. The Langmuir–Blodgett (LB) technique has been extensively used to order and immobilize natural proteins on solid surfaces (Tronin et

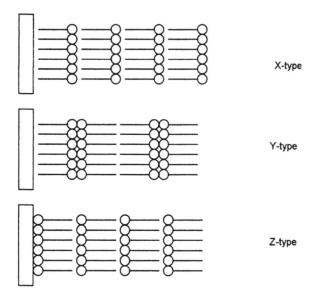

X-type

Y-type

Z-type

FIG. 4   Types of LB deposition.

al. 1994, 1995, and Facci et al. 1993). The technique has several advantages not available in solution studies: (1) A protein monolayer at an air–water interface can be transferred after establishing a molecular organization of the film selected at different surface pressure or surface density, (2) a protein monolayer formed at an air–water interface can be transferred to many different surfaces, including those prepared for electronic studies, (3) protein multilayers can be obtained by repeating the monolayer transfer process as many times as desired, and (4) the transferred film can be studied in different environments, such as air or solution (Guryev et al. 1997). The success of the LB technique draws on the property of amphiphilic molecules when spread and compressed at an air–water interface to form a compact monolayer, with hydrophobic and hydrophilic parts directed to air and water, respectively. When this technique is applied to nonamphiphilic water-soluble proteins, difficulties can be encountered. Simple water-soluble proteins at the air–water interface tend to expose their hydrophobic interior to the air, causing dramatic changes of their native structure, which may be detrimental to the function of the proteins or even cause the liganded cofactors to dissociate from the protein. In-plane order can be greatly enhanced by exploiting the combination of planar orientation and mobility, which this interface provides.

B.   Layer-by-Layer Film Engineering

It is most interesting and challenging to construct ultrathin films with a supramolecular architecture, in which the individual organic molecules are macroscopically oriented and where the molecules with different functionality can be incorporated into individual layers. Recently, layer-by-layer (LBL) assembly processes based on electrostatic or other molecular forces represent a unique technique that presents a new approach to the formation of supramolecular architectures by adsorption of consecutively alternating polyelectrolytes. The self-assembly of charged polyelectrolytes (i.e., proteins, conducting polymers, zirconium phosphate, optical dyes, metal nanoparticles, aluminosilcates, and clay) by LBL can be considered an alternative to the Langmuir–Blodgett, spin-coating, and chemical vapor deposition techniques. The most substantial advantage of LBL self-assembly is the highly accurately controlled average thickness of the polyelectrolytes layers, where the macroscopic properties of molecular film can be controlled by microscopic structure. In this approach, thin films are built up on a layer-by-layer basis by alternately exposing a substrate to positive and negative polyelectrolytes: polyanion and polycation. On each deposition cycle the charge on the exposed surface is compensated and reversed by adsorbed polymer. The amount of material deposited on each cycle approaches a constant and reproducible value, permitting any number of layers to be incorporated. The instruments required are minimized; the technique has proven to be a powerful way to produce thin structures with defined composition. The multilayer films in which both the anionic and the cationic layers were polyelectrolytes and the principle of the layer-by-layer adsorption of polyanions and polycations are depicted schematically in Figure 5. Decher and coworkers (1992 and 1996) applied this approach to linear polyanions and bipolar amphiphiles. This technique was then successfully applied to at least more than 20 water-soluble, linear polyanions, including conducting polymers, ceramics, dyes, porphyrin, DNA, and proteins (Decher et al. 1992, Keller et al. 1994, Cheung et al. 1994, Fou et al. 1996).

The chemical structures of the polyanions, such as poly(styrene sulfonate) (NaPSS), poly(sulfonated aniline), poly(allyamine hydrogen chloride) (PAH), and poly(etheneimine) (PEI), are shown in Figure 6. In most of the polyelectrolyte solutions the NaCl or the KCl etc. salts are found at a very high concentration in the depositing solution. The adsorption time varies from 5 minutes to 1 hour for various polyelectrolytes. The astonishing control of individual average-layer thicknesses is simply due to the well-known changes in the coil structure of polyions in solutions of different ionic strength. At low concentrations of the salt, the charged groups in the polyelectrolyte repel each other and the polymer adopts an extended conformation, and the chain adsorbs flat. At higher concentrations of the salt, the repulsion of charged groups is partly screened and the polyelectrolytes conformation approaches a random coil, and the chain adsorbs rather loopy (Decher et al. 1992). The two different layer thicknesses of the same materials can be maintained in a single film by adjusting the ionic strength of the adsorption solutions. The effect of the deposition process depends on the salt effect, "which does the fine-tuning of the layer pair thickness," the temperature effect, "which does the reversible change in the film thicknesses." The deposition process is more or less independent of the molecular weight of the polyelectrolytes. Using neutron reflectometry, Decher et al. (1992, 1996, 1997) resolved the internal structure of self-assembled polyelectrolyte multilayer films with high resolution, which showed that such surface films consisted of stratified structures in which polyanions and polycations of individual layers are interdigitated with one another. For alternating layers of poly(styrene sulphonate), PSS, and poly(allyamine hydrochloride), PAH, adsorbed onto atomically flat surfaces, a roughening of successively deposited layers leads to a progressively larger number of adsorption sites for consecutive generations of adsorbed polymer, and thus to an increase in layer thicknesses with an increasing number of deposited layers. The fine-tuning of the thicknesses of individual layers can indeed be used for the construction of complex film architectures, even in multilayers composed of only two materials, a single polyanion and a single polycation. A potential application of the LBL technique is the fabrication of homogenous ultrathin film of conjugated polymers. Molecular-level processing of conjugated polymers (i.e., polypyrrole, polyaniline, poly(phenylene vinylene), poly(o-anisidine)) by the LBL technique (Ferreira and Rubner 1995, Ram et al. 1999) was also shown in the literature.

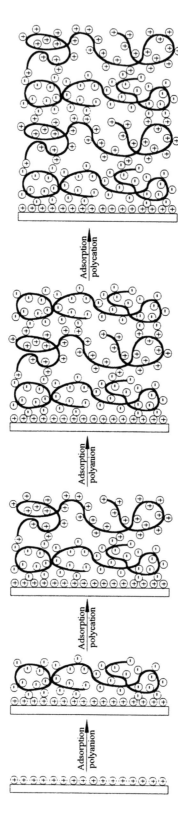

FIG. 5 Schematic of multilayer assembly by consecutive adsorption of polyanion and polycation.

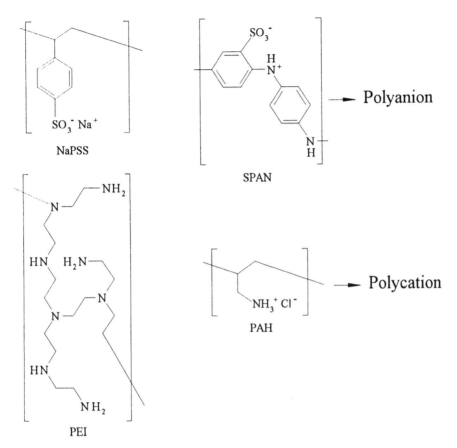

FIG. 6 Structures of an important class of polyelectrolytes used for deposition of layer-by-layer self-assembly.

C. Organic Materials

The structures of the new organic materials used are summarized in Figure 3. The major drawback of the polymeric materials is the multistep synthesis required for the functionalization and the stringent process requirements of the condensation polymerization. It seemed desirable therefore to find shorter synthetic routes to process the polymers with predictable absorption wavelengths of light. Several types of poly(p-phenylenevinylene) (PPV) polymer derivatives have been synthesized in our laboratory (Ram et al., 1997a), namely, poly(2-methoxy-5-(2′-ethyl)hexyloxy-p-phenylenevinylene) (MEHPPV). The Gilch route has been modified in order to increase the processability for this specific device application. Poly(phenylene vinylene) (PPVs) are main-chain conjugated polymers, which have very interesting electrical and photoconjugated properties, and that make them suitable for applications in optoelectronic and microelectronic devices, such as the PV cells described here.

Photosynthetic reaction centers from *Rhodobacter sphaeroides* and bacteriorhodopsin (BR) from purple membrane (PM) have been used for their unique optoelectronic properties and for their capability of providing light-induced proton and electron pumping. Once assembled they display extremely high thermal and temporal stability

(Nicolini 1997, 1998a,b). These features make them two of the most promising biological molecules for developing devices (see Nicolini 1996c). Inorganic nanoparticles, such as CdS, PbS, and TiO_2, and organic systems, namely, conducting polymer, dyes, and fullerene (C60) nanocrystals, were prepared and organized in thin films to fabricate donor (D)–acceptor (A) supramolecular assemblies (Nicolini 1997).

III. SUPRAMOLECULAR LAYER CHARACTERIZATION

A. Brewster-Angle Microscopy

The method is based on the fact that plane-polarized light does not reflect from the interface when it is incident at the Brewster angle, determined by the equation

$$tg\phi = \frac{n_2}{n_1}$$

where n_1 and n_2 are refractive indexes of the two media at the interface. The value of this angle for the air/water interface is 53.1°. Therefore, it is possible to adjust the analyzer position in such a way that it will be a dark field when imaging the air/water interface. Spreading of the monolayer varies the Brewster conditions for both air/monolayer and monolayer/water interfaces, making visible the morphology of the monolayer. A typical Brewster-angle image of the monolayer at the air/water interface is shown in Figure 7. To some extent, fluorescence and Brewster-angle microscopies coincide. However, Brewster-angle microscopy is preferred because it does not demand adding anything to the monolayer, leaving the situation unaffected by any disturbing factors.

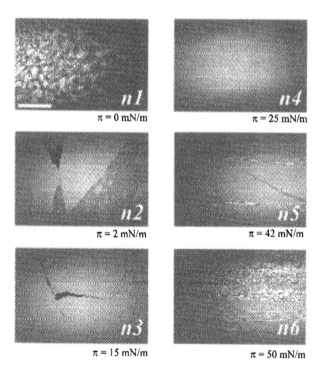

FIG. 7   Brewster-angle microscopy: Image of POAS monolayer at air/water interface at pH 1 and various pressures.

B. X-Ray Scattering

Several methods based on the utilization of x-rays are used to study LB film structure. Diffractometry and reflectometry are the usual ones. Diffractometry is used mainly when well-ordered periodic structure is under the investigation and several Bragg reflections are in the x-ray pattern. The position of these reflections is determined by the Bragg equation:

$$2D \sin \Theta = n\lambda$$

where D is the thickness of the periodic unit (period or spacing), λ is the wavelength of the incident x-ray beam, Θ is the incident angle, and n is the number of the reflection. It is therefore possible to obtain directly the thickness of the periodic unit (usually, bilayer) from the angular position of the Bragg reflection. The other information, which can be obtained directly from the x-ray pattern, is a correlation length (L). This parameter can be considered a thickness for which the film can still be considered an ordered one, and it is determined by the following formula:

$$L = \frac{\lambda}{2 \sin(\Delta\Theta)}$$

where $\Delta\Theta$ is a half-width of the Bragg reflection.

More information can be obtained if several reflections were registered. Each Bragg reflection can be considered a component in the Fourier row representing the electron density of the repeating unit of the film. Therefore, the electron density of a period of the film is proportional to the intensity of the Bragg reflection, but the phase of this harmonic must also be taken into account. Fortunately, most LB films have Y-type structure, and, therefore, the phases for them can be only 0 or π, thus transferring, the phase problem typical for crystallography into a more simple sign problem. This problem is usually solved by fitting the electron density in some known regions. Most often, this region is where close packing of hydrocarbon chains takes place.

Not so direct as diffractometry, reflectometry is the other approach for studying LB firms with x-ray techniques. It also allows, however, studying not very well-ordered films and even monolayers. The reflectometry approach considers all registered scattering curves. It can contain also Bragg reflections and Kiessig fringes. Kiessig fringes correspond to the interference of the x-ray beam reflected from the air/film and film/substrate interfaces; their angular position gives information about the total thickness of the LB film. The electron density profile in the case of reflectometry is calculated from models, fitting the experimental data.

C. Atomic Force Microscopy

Atomic force microscopy, a type of scanning probe microscopy, utilizes a sharp probe to scan across the surface of a sample. A laser is focused on the tip, and the beam is reflected to the split photodiode detector. As the cantilever is deflected over the sample surface, the photodiode monitors the changes of the laser beam. The changes are stored in the computer to produce a topographic image of the sample surface. There are three different modes in atomic force microscopy: contact mode, tapping mode, and noncontact mode. In *contact mode*, the tip physically comes in contact with the sample. Atomic resolution can be reached in contact mode; however, there is a risk of damage to soft samples. During *tapping mode*, the cantilever is oscillated at or near its resonance frequency. The scanner, in a pendulum-type motion, enables the tip to "tap" the sample surface as the scanner comes to

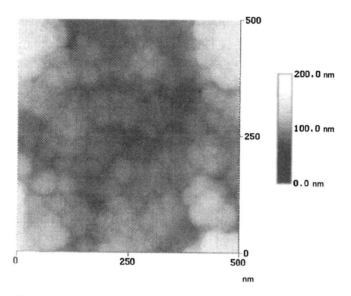

FIG. 8 SiO_2 AFM image of six bilayers in situ self-assembled layer-by-layer films of polypyrrole coated with SiO_2 and poly(styrene sulfonate).

the bottom of its swing. There is less damage in tapping mode and higher lateral resolution; however, there is a slightly slower scan speed than in contact mode. The cantilever is oscillated at a frequency slightly above the cantilever's resonance frequency during *noncontact* AFM. The tip oscillates above the adsorbed fluid layer on the surface. It does not come in contact with the sample surface. The sample is not damaged during noncontact mode; however, the scan speed is much slower, there is lower lateral resolution, and it may be used only with very hydrophobic samples. Noncontact mode is not used very often because of these disadvantages. In our case, the AFM used was a home-built instrument (Polo Nazionale Bioelettronica), (Sartore et al. 1999) working in contact mode. It operated in air, at constant deflection (i.e., vertical contact force) with triangular-shaped, gold-coated Si_3N_4 microlevers (commercially available Park Scientific Instrument chips) (Ram et al. 1999). The tips of the microlevers had a standard aspect ratio (about 1:1), and the levers had a nominal force constant of 0.03 N/m. The constant-force set point was about 0.1 nN, while the images acquired were 256×256-pixel maps. During acquisition, the row-scanning frequency was 4 Hz, i.e., a physical tip–sample motion speed of 8-4-2 μm/sec in the 2-1-0.5-μm scan size images, respectively. All images are standard top-view topographic maps, where the brightness is proportional to the quota of the features over the sample surface; i.e., light means mountain, dark means valley. The images shown are generally representative of the samples, since the same images appear in four different regions of the analyzed samples, positioned at the vertices of a 4-mm-side square, centered at the specimen. Figure 8 shows an AFM image of six bilayers in situ self-assembled layer-by-layer films of polypyrrole (PPY) coated with SiO_2 nanoparticles along with a sodium salt of poly(styrene sulfonate).

D.   Fourier Transform Infrared Spectroscopy

Fourier transform infrared (FTIR) spectroscopy is a powerful analytical tool for characterizing and identifying organic molecules. The IR spectrum of an organic compound serves as its fingerprint and provides specific information about chemical bonding and molecular structure. FTIR involves the vibrations of molecules. Molecules can bend or stretch at their bonds. They can also wiggle around in a wag, or twist. Molecules vibrate in certain modes, depending on the symmetry properties of the molecules' shape (Pepe et al. 1998). The energy it takes to excite a vibrational mode varies depending on the strength of the bond and the weight of the molecule. FTIR involves the conversion of energy to molecular vibrations. Infrared radiation (wavenumbers of 4800–400 cm^{-1}) can be converted to vibrations in the molecule, which causes the molecule to go from a ground vibrational state to an excited vibrational state. Samples are run either as pure substance or in KBr pellets. A beam of infrared radiation is passed through the sample. A detector generates a plot of percent transmission of radiation versus the wavenumber or wavelength of the transmitted radiation. When the percent transmission is below 100, some of the light is being absorbed by the sample. Each peak in the spectrum represents absorption of light energy and is called an *absorption band*. The amount of energy required to stretch a bond depends on many things, one of which is the strength of the bond and the masses of the bonded atoms. The higher wave numbers are achieved when the bonds are stronger and the atoms are found to be smaller.

E. Gravimetric Measurements

Gravimetric measurements were carried out by means of a homemade gauge with a sensitivity of 0.57 ± 0.18 ng/Hz using quartz oscillators with frequency of 10 MHz. Calibration of the quartz balance was performed according to Facci et al. 1993. Results of the gravimetric study of deposited monolayers are presented in Figure 9 for wild-type and recombinant proteins. Linear dependence of the frequency shift upon the number of deposited layers indicates the reproducibility and homogeneity of the deposition. Knowing the molecular weight and dimensions of the protein, it is possible to compare the area per molecule in the film, calculated from gravimetric measurements, with that for the closely packed system.

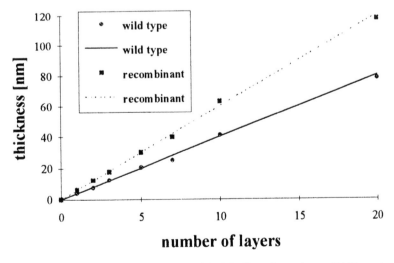

FIG. 9 Dependence of the thickness of the LB film of cytochrome P450 scc, both wild type (solid line) and recombinant (dashed line), upon the number of transferred layers.

F. Electrical Characterization

Electrical conductivity in general reflects the net charge motion brought about by an electric field E. The value of electrical conductivity, σ can be expressed as: $\sigma = J/E = nq\mu$, where J is the current density, E is the electric field; n is the concentration of charge carriers (solitons, polarons, and bipolarons), and μ is the mobility of the charge carriers. It is necessary for a conducting polymer to contain an overlapping set of molecular orbits to provide reasonable carrier mobility along the polymer chain. The dc and ac conductivity can be experimentally measured. The industrial application of this class of ultrathin electroconducting material is promising. So the electrical conductivity of each polyanilines LS films was studied by depositing on the interdigitated electrodes. Figure 10 shows the *I-V* characteristics of polyaniline films measured at a scan rate of 20 mV/sec. The current–voltage characteristics for PANI (curve 1), POT (curve 2), and POAS (curve 3) do not show ohmic behavior (Ram et al 1999). It can be related to the possible redox reaction of the interdigitated electrodes with HCl during the preparation of LB films, which could also be related to some potential barrier with the degenerately doped conducting polyanilines. The only curve showing a different behavior is that of PEOA (curve 4), with a linear relationship in *I-V* characteristics. The PANI LS films (curve 1) show the higher magnitude of current for each measured potential. The magnitude of current shows a minimum for the PEOA LS films. The decrease in the current magnitude in POT can be related to the decrease in interchain order, which is related to the conductivity value, whereas POAS leads to an increase in electronic localization attributable to the decrease in conductivity. Still the larger-substituent ethoxy group in PEOA shows another effect like charge localization along the polymer chains, which increases the orientation, thus diminishing the current (or conductivity) value.

G. Electrochemistry

Cyclic voltammetry and other electrochemical methods offer important and sometimes unique approaches to the electroactive species. Protein organization and kinetic approaches (Correia dos Santos et al. 1999, Schlereth 1999) can also be studied by electrochemical survey. The electron transfer reaction between cytochrome P450scc is an important system for

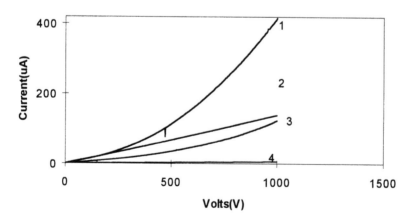

FIG. 10   Current–voltage characteristics of polyanilines LS films of 40 monolayers deposited on interdigitated electrodes: (1) PANI, (2) POT, (3) POAS, (4) PEOA.

investigating fundamentals regarding long-range electron transfer in biological systems; it has attracted considerable experimental and theoretical attention. Cyclic voltammetry, as an efficient method to investigate electron transfer reactions between proteins, is being exploited increasingly for obtaining electrochemical information on metalloproteins.

IV. ROLE OF MOLECULAR CLOSE PACKING ON PROTEIN THERMAL STABILITY

Langmuir–Blodgett (LB) films of proteins (Tiede 1985, Hwang et al. 1977, Furuno et al. 1988, Lvov et al. 1991) and lipid–protein complexes (Fromherz 1971, Phillips et al. 1975, Mecke et al. 1987, Kozarac et al. 1988) were intensively studied and characterized by different techniques.

Recent publications on the thermal stability of proteins organized in dense solid films, deposited by LB (Nicolini et al. 1993, Facci et al. 1994, Erokhin et al. 1995) and by self-assembling (Shen et al. 1993), leave several questions unanswered. In particular, it is still not completely clear which parameter is responsible for this phenomenon. Two main factors are discussed when speaking about induced thermal stability, namely, decreased water content and molecular close packing (Nicolini et al. 1993). It seems that both of them work in parallel, and unfortunately it is difficult to settle directly which one plays the dominant role.

It is interesting to compare the thermal-treatment effect on the secondary structure of two proteins, namely, bacteriorhodopsin (BR) and photosynthetic reaction centers from *Rhodopseudomonas viridis* (RC). The investigation was done for three types of samples for each object–solution, LB film, and self-assembled film. Both proteins are membrane ones and are objects of numerous studies, for they play a key role in photosynthesis, providing a light-induced charge transfer through membranes—electrons in the case of RC and protons in the case of BR.

It is necessary to note that the initial conditions of the samples in solution were absolutely different. RC was extracted from the membranes by detergent (lauryldimethy-lamineoxide LDAO); the solution contains the individual protein molecules surrounded by a detergent "belt" shielding the hydrophobic areas of the protein surface. In the case of BR the situation is different. BR is the main part of purple membranes (about 80%) and is already close packed in it. It is difficult to extract BR in the form of individual molecules, for they are very unstable (Okamura et al. 1974). Thus, the initial solution of BR was in reality the solution of sonicated membrane fragments.

Described differences in the initial protein solution conditions, of course, differentiates the processes of film formation, both in the case of the LB technique and in the case of self-assembling.

CD spectra of different samples of RC and BR after heating at different temperatures are presented in Figures 11 and 12, respectively. Comparison of the CD spectra of RC and BR in solution allows one to conclude that BR in solution is much more heat resistant with respect to RC. The other interesting point is that the temperature behaviors of the BR in LB and self-assembled films are absolutely identical. Both of them demonstrate high thermal stability, and significant differences in the CD spectra appear only after heating up to 200°C. The situation is absolutely different in the case of RC. In fact, LB films show high thermal stability, and significant differences appear, as in case of BR, only after heating to 200°C. Self-assembled films of RC, in contrast, are much more affected by thermal treatment.

RC Rps. viridis in solution

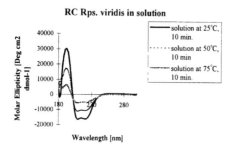

RC Rps. viridis in LB (20 layers)

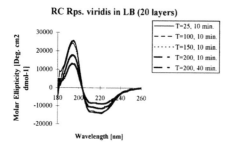

RC Rps viridis spread film after treatment
at different temperatures

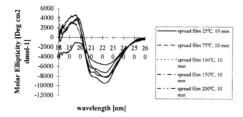

FIG. 11   CD spectrum of RC in solution and LB and spread films.

Such differences in the secondary structure behavior with respect to temperature can be explained by suggesting that molecular close packing of proteins in the film is the main parameter responsible for the thermal stability. In fact, as in the case of BR, we have close packing of molecules even in the solution (membrane fragments); there are practically no differences in the CD spectra of BR solution at least till 75°C (denaturation takes place only for the sample heated to 90°C). RC in solution begins to be affected even at 50°C and is completely denatured at 75°C, for the solution contains separated molecules.

The fact that CD spectra of BR in LB and self-assembled films show similar behavior with respect to temperature is also not strange. Because the basic block of the film in both cases is the membrane fragment, which is already closely packed, there is no principal difference between these samples with regard to packing. The difference in the distribution of these fragments cannot be critical for thermal stability.

In the case of RC there are big differences in the temperature dependencies of CD spectra of LB and self-assembled films. The differences are due to the different organization of molecular structures in the film. In the case of self-assembled films, the molecules are randomly distributed. Moreover, the presence of the detergent in the dried film can cause some damage of the protein structure, as is illustrated by the difference in the CD spectra of the film with respect to solution even without any heating. In contrast, the LB technique allows depositing dense films practically free of the detergent.

Molecular regular close packing of BR in membrane fragments is well known and was confirmed also by STM measurements (Fedorov et al. 1994). In the case of RC LB films, this was also reported for that from *Rhodobacter sphaeroides* (Facci et al. 1994). Because it was not reported for RC from *Rhodopseudomonas viridis*, STM imaging was performed in order to evaluate the molecular packing in the RC LB film. The typical image is presented in Figure 13. The periodicity of the image features is about 6.5 nm, which is comparable to the dimensions of the protein molecule. The image illustrates that a regular close-packed film was obtained.

This comparative study pointed out molecular close packing as a key parameter responsible for the thermal stability of proteins in films. In the case of BR, this close packing is reached due to the nature of the sample, while LB organization seems to be a more general procedure, for the same goal can be reached for practically any type of protein sample. The last statement was even confirmed by the comparison of the thermal behavior of extracted separated BR in self-assembled and LB films. It was found that BR in LB films is more stable for this kind of sample. The results will be reported in detail elsewhere.

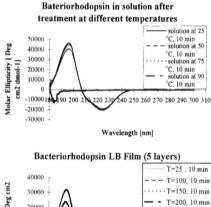

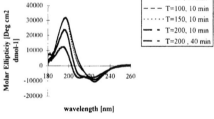

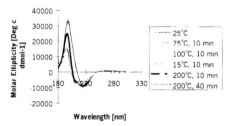

FIG. 12   CD spectrum of BR in solution and LB and spread films.

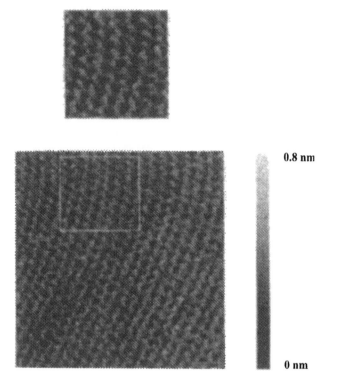

0.8 nm

0 nm

FIG. 13   STM image of RC monolayer.

V. INDUSTRIAL BIOCATALYSIS

The Langmuir–Blodgett (LB) technique was successfully applied for the deposition of thin protein layers (Langmuir and Schaefer 1938, Tiede 1985, Lvov et al. 1991). LB organization of protein molecules in film not only preserved the structure and functionality of the molecules, but also resulted in the appearance of new, useful properties, such as enhanced thermal stability (Nicolini et al. 1993; Erokhin et al. 1995).

Enhanced thermal stability enlarges the areas of application of protein films. In particular it might be possible to improve the yield of reactors in biotechnological processes based on enzymatic catalysis, by increasing the temperature of the reaction and using enzymes deposited by the LB technique. Nevertheless, a major technical difficulty is that enzyme films must be deposited on suitable supports, such as small spheres, in order to increase the number of enzyme molecules involved in the process, thus providing a better performance of the reactor. An increased surface-to-volume ratio in the case of spheres will increase the number of enzyme molecules in a fixed reactor volume. Moreover, since the major part of known enzymatic reactions is carried out in liquid phase, protein molecules must be attached chemically to the sphere surface in order to prevent their detachment during operation.

The aim of the work was the development of a technique to deposit enzyme LB films on the surface of small glass spheres and to test the enzymatic activity of such samples before and after thermal treatment.

The experiments were carried out with two enzymes, urease and glutathione-S-transferase (GST). Urease catalyses the hydrolysis of urea (Avrameas 1978), while GST is an enzyme that catalyzes the reduction of compounds such as alkylants with a nucleophilic addition of the thiol of glutathione to electrophilic acceptors (e.g., aryl and alkyl halides, quinones, organic peroxides) (Pickett and Lu 1989). Both enzymes were chosen, since their activity can easily be tested by spectrophotometric measurements.

The LB technique was chosen for covering the spheres because it was shown to provide enhanced thermal stability of many types of proteins in deposited layers (Nicolini et al. 1993, Erokhin et al. 1995, Antolini et al. 1995), which no other technique is able to achieve. Since only the upper protein layer is involved in the catalytic activity, no special attention was paid to check whether the deposited layer is a monolayer or multilayer. However, the samples were thoroughly washed to remove protein molecules not bound covalently to the sphere surface, since during the functional test these molecules could contribute to the measured apparent catalytic activity.

Borosilicate glass spheres with a diameter of 2 mm were used as substrates for the deposition.

The surface of the spheres was activated in the following way. Spheres were treated with boiling chloroform, rinsed on a glass filter, and dried under nitrogen, to be subsequently silanized with 3-glycidoxypropyl trimethoxysilane following the technique proposed by Malmquist (Malmquist and Olofsson 1989). Silanization of the spheres was performed in nitrogen flux in order to prevent reciprocal attachment of spheres and to activate their surface homogeneously.

The essential steps of the deposition procedure, which are the same for both enzymes, are illustrated in Figure 14. The protein solution was spread over the water subphase, and the monolayer was compressed up to 25 mN/m. Activated spheres were distributed over the monolayer in the following way: A plate with spheres over it was moved along the monolayer, while the weak nitrogen flow was used for transferring the spheres from the plate to the layer. It is important to have the plate in close vicinity to the water surface in order to keep all the particles floating over the monolayer. The spheres were kept at the surface for 30 minutes in order to provide chemical linking of the monolayer to activated surface. After this time the feedback system was switched off and the Wilhemy plate was removed from the water. The layer with particles was compressed until the minimum area (20 cm^2), which corresponds to the collapse of the monolayer, was reached. Even though this compression yields a multilayer film, such action seems to be necessary, since otherwise only half of the sphere surface was covered with protein monolayer, while compression induced the motion both of spheres and the monolayer, covering other regions of the spheres. The

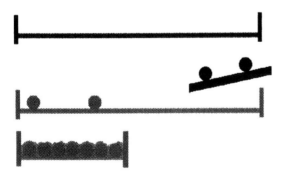

FIG. 14 Schematic of the deposition of LB films on to the spherical substrate.

spheres were collected, washed with substrate buffer in order to remove parts of the mono-layer not attached chemically to the sphere surface, and dried.

The activity of enzymes in the film was estimated in the following way: In order to test the activity of urease, we utilized a calorimetric assay based on urea hydrolysis; the enzymatic reaction was followed at 590 nm, the suitable wavelength for bromcresol purple (Chandler 1982). Urea concentration was 1.67 ; ts 10^{-2} M.

The enzymatic activity of GST was evaluated spectrophotometrically following the conjugation of glutathione (GSH) thiol group to 1-chloro2,4-dinitrobenzene (CDNB) at a wavelength of 340 nm (Habig et al. 1974) by a double-beam spectrophotometer (Jasco 7800). The GSH and CDNB concentrations were 2.5 mM and 0.5 mM, respectively. Ten covered spheres were placed into the cuvette. The diffusion effects (Antolini et al. 1995) were avoided by carrying out the measurements under continuous stirring with a magnetic microstirrer (Bioblock scientific) at a speed of 600 rpm.

The deposition procedure described earlier allows one to obtain protein films chemically bound to the activated surface of spherical glass particles. Subsequent compression of preformed protein monolayer with these particles permitted to coverage of the particle area that initially has not come in contact with the monolayer, as schematically shown in Figure 14. Even if such a procedure does not initially result in deposition of strictly one monolayer, this fact does not seem to be critical, because only the monolayer chemically attached to the surface remains after washing.

From the results of the urease activity test summarized in Figure 15, it is clear that the deposition procedure preserved to a certain extent the enzyme catalytic activity. Heating the sample before testing decreased the enzyme in the film by about 30% but did not eliminate it completely. The results of the activity test of two samples are summarized in Table 1 together with reference values for a spontaneous reaction without enzyme. It is necessary to underline that enzymatic activity on spherical supports was higher than the respective value in "flat" films, which could indicate that apparent catalytic efficiency was improved due to an increased area-to-volume ratio.

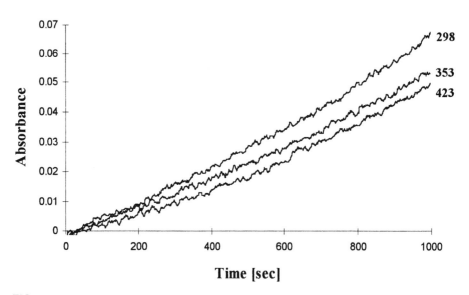

FIG. 15 Urease activity test at different temperatures (K).

TABLE 1 Slopes of Activity for Two
Samples of Urease and for Spontaneous
Reaction

Sample 1	7.01×10^{-4}
Sample 2	5.97×10^{-4}
Average	6.49×10^{-4}
Mean error	7.4×10^{-5}
Spontaneous reaction	2.49×10^{-6}

As evident from results of the GST activity test presented in Figure 16, the enzyme is active in LB films. Comparison of these catalytic activity values with the balance values of GST LB film deposited onto flat silanized surfaces (Antolini et al. 1995), and presuming a linear dependence of activity on enzyme concentration, gives an effective amount of 2.44 pmol in the sample. Knowing the total area of spheres in the reaction medium (10 spheres with total area of about 125 mm^2) and taking into account that only the upper layer is involved in the reaction, we can estimate the surface density of the enzyme in the layer. The area per molecule was found to be 83 nm^2. On the other hand, the area per molecule can be estimated from strictly geometric considerations of protein sizes approximations taken from the Protein Data Bank, and such calculations yield a value of 34 nm^2. Thus, comparing these values the conclusion can be reached that the amount of active enzyme in the film is about 41% of a maximum possible in the close-packed layer. The resultant activity is reported in Table 2 together with the value for a spontaneous reaction.

The comparison of the results on enhanced thermal stability of proteins in LB films reported here and those already published again underlines that molecular close packing is a critical parameter responsible for this phenomenon.

The described procedure allows one to deposit protein, in particular, enzyme, LB films onto the surface of small spheres. Deposited multilayer film was washed in order to leave at the surface only a layer covalently attached to the activated surface. The enzyme

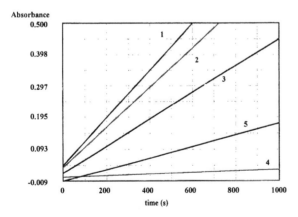

FIG. 16 GST activity test. (1) Solution (slope 106×10^{-5} 1/s), (2) proposed method (the best, slope 90×10^{-5} 1/s), (3) proposed method (average, slope 53×10^{-5} 1/s), (4) Langmuir–Schaefer method (the best, slope 6×10^{-5} 1/s), (5) spontaneous reaction (average, slope 16.4×10^{-5} 1/s). Curves 1–4 show the difference of absorbance for catalyzed and spontaneous reactions.

TABLE 2 Slopes of Activity Graphs for Two
Samples of GST and for Spontaneous Reaction

Sample 1	4.7×10^{-5}
Sample 2	5.4×10^{-5}
Average	5.05×10^{-5}
Mean error	4.95×10^{-6}
Spontaneous reaction	3.1×10^{-5}

in the film preserves its catalytic activity and demonstrates highly increased thermal stability. The procedure can be useful for the fabrication of heat-proof active elements for bioreactors based on enzymatic catalysis.

VI. REACTION-CENTER-BASED PHOTOCELLS

The recent progress in research indicates that the disentanglement of photosynthesis will give strong impulse to the application of solar cells. Moreover, as far as the bioconversion of sunlight is concerned, it is known that photosynthesis starts with a charge separation process in the photosynthetic reaction centers (RCs): a photoactive bacterial protein. Therefore, several experiments have been carried out to test the possibility of the conversion of light to electrical energy by photochemical cells (or simply photocells) containing such photosynthetic bacterial proteins. The photocells developed have different geometries, and therefore it is very hard to classify the reliability of such devices. Nevertheless, the results indicated that the efficiency of quantum energy conversion is practically 100% (all the light energy was converted into the charge separation). Nevertheless, the energy conversion efficiency of the realized photosensitive units crucially depends on molecular orientation and is very difficult to estimate, because articles usually contain incomplete information.

Usually, such proteins are immobilized by thin-film fabrication techniques, such as self-assembly, electrical sedimentation, Langmuir–Blodgett, polymer matrix, or gel entrapment. Among such techniques, the most promising seems to be Langmuir–Blodgett. In fact, this technique allows the protein molecules to be organized in an ordered 2D array. There are many publications on films of RCs. The films were dually characterized from different points of view. The discovery of heat and temporal stability had opened big possibilities in using the protein in devices (Nicolini et al. 1993).

In the literature, Japanese authors (Yasuda et al. 1993) suggested the making of a photodevice, using an RC LB film sandwiched between two electrodes (one of them transparent). Photovoltage registered in this work was 4–5 mV for the film, which contained 44 layers. In our work (Facci et al. 1998) we have shown that it is possible to adjust a tilt of RC molecules in the layer by controlling the surface pressure. On the other hand, for BR the possibility of increasing anisotropy by electric field application was shown. We can hope that by applying our technique of electric field–assisted deposition we will be able to increase this number (according to existing estimations, there is only about 12% of prevalent orientation with respect to the opposite one; with our technique, we can hope to increase this anisotropy). The other possible way of increasing the film anisotropy was suggested by Miyake in LB8 (Kazuyuki et al. 1998). They voltage-biased the substrate during deposition with respect to the water subphase. He reported that even in this case the anisotropy of the film was improved. Thus, the films can be useful for the construction of devices for converting light to electron energy, working in the range of visible–near-IR spectrum (Nicolini, 1998a).

VII. PURPLE-MEMBRANE-BASED OPTOELECTRONIC APPLICATIONS

In recent years our understanding of the structure and function of several biological systems has grown rapidly. The study of bacteriorhodopsin (BR) protein and the elucidation of its function as a light-driven proton pump represents one of the most interesting examples (Oesterhelt et al. 1991, Brächle et al. 1991, Birge 1990). BR is a light-transducing protein in the purple membrane (PM) of *Halobacterium Halobium*. Its features allow one to identify and design several potential bioelectronics applications aimed at interfacing, integrating, or substituting for the silicon-based microelectronics systems, as well as developing molecular devices (Birge 1992). BR is a notable exception as compared to the usual biological molecules, being mechanically robust and chemically and functionally stable in extreme conditions, such as high temperatures (Hampp 1993, Shen et al. 1993, Zeisel and Hampp 1996), which usually represents one of the key parameters of working conditions. Furthermore, it possesses remarkable photonic and photovoltaic properties, which have been exploited for molecular device construction. For these reasons BR has been adopted as a building block for a number of experimental prototypes, such as filters, photocells, artificial photoreceptors, optical memories, image sensors, and biosensors (Birge 1990, Oesterhelt et al. 1991, Fukuzawa et al. 1996, Fukuzawa 1994, Miyasaka et al. 1992, Storrs et al. 1996, Maccioni et al. 1996, Chen and Birge 1993).

In addition, thin-film technologies (Ulman 1991) allow the assembly of biological materials in a 2D system, usually required for device development. Among these technologies, Langmuir–Blodgett (LB) (Ulman 1991, Roberts 1990, Zasadzinski et al. 1994) seems to be one of the most promising, due to its ability to form molecular systems having high packing degree and molecular order. Moreover, it has been possible to assess that such a technique allows the fabrication of 2D protein close-packed structures, showing an enhancement of some chemical-physical properties or an induction of new properties commonly known for proteins in solution or even in membranes (Shen et al. 1993; Nicolini et al. 1993, Pepe and Nicolini 1996, Maxia et al. 1995). These properties include long-term stability against thermal and functional (photochemical) degradation.

Therefore, investigators have shown considerable interest in the adoption of the Langmuir–Blodgett technique, or its modifications, to make molecular electronic devices using, in particular, as an active component, a light-transducing protein, such as BR. In fact, the ability of BR to form thin films with excellent optical properties and the intrinsic properties themselves make it an outstanding candidate for use in optically coupled devices.

BR thin layers have been widely studied (Hwang et al. 1977, Ikonen et al. 1993, Shibata et al. 1994, Méthot et al. 1996, Sugiyama et al. 1997) because they exhibit bistability in optical absorbance and they provide light-induced electron transport of protons through the membrane. Furthermore, their extremely high thermal and temporal stability also allows considering them as sensitive elements for electrooptical devices (Erokhin et al. 1996, Miyasaka et al. 1991). However, in order to use BR properties to provide photovoltage and photocurrent, it is necessary to orient all the molecules in such a way that all the proton pathways are oriented in the same direction. The LB technique in its usual version does not allow this. When BR-containing membrane fragments are spread at the air/water interface, they orient themselves rather randomly in such a way that the proton pathway vectors are oriented in opposite directions in different fragments. Nevertheless, a technique of electrochemical sedimentation is known that allows the deposition of highly oriented BR layers.

However, the layers deposited with this technique are rather thick and not well controllable in thickness.

The Langmuir–Blodgett (LB) technique allows one to form a monolayer at the water surface and to transfer it to the surface of supports. Formation of the BR monolayer at the air/water interface, however, is not a trivial task, for it exists in the form of membrane fragments. These fragments are rather hydrophilic and can easily penetrate the subphase volume. In order to decrease the solubility, the subphase usually contains a concentrated salt solution. The efficiency of the film deposition by this approach (Sukhorukov et al. 1992) was already shown. Nevertheless, it does not allow one to orient the membrane fragments. Because the hydrophilic properties of the membrane sides are practically the same, fragments are randomly oriented in opposite ways at the air/water interface. Such a film cannot be useful for this work, because the proton pumping in the transferred film will be automatically compensated; i.e., the net proton flux from one side of the film to the other side is balanced by a statistically equal flux in the opposite direction.

On the other hand, the technique of electrochemical sedimentation is known allowing the formation of rather thick BR films by orienting them in the electric field.

Therefore, the following method was suggested and realized (the scheme is shown in Fig. 17). A 1.5 M solution of KCl or NaCl (the effect of preventing BR solubility of these salts is practically the same) was used as a subphase. A platinum electrode was placed in the subphase. A flat metal electrode, with an area of about 70% of the open barriered area, was placed about 1.5–2 mm above the subphase surface. A positive potential of $+50 \div -60$ V was applied to this electrode with respect to the platinum one. Then BR solution was injected with a syringe into the water subphase in dark conditions. The system was left in the same conditions for electric field–induced self-assembly of the membrane fragments for 1 hour. After this, the monolayer was compressed to 25 mN/m surface pressure and transferred onto the substrate (porous membrane). The residual salt was washed with water. The water was removed with a nitrogen jet.

The dependence of the surface pressure upon the time with and without applied electric field is shown in Figure 18. It is clear that the electric field strongly improves the ability of the membrane fragments to form a monolayer at the water surface.

X-ray measurements of the deposited multilayers revealed practically the same structure in films prepared with the usual LB technique and electric field–assisted monolayer

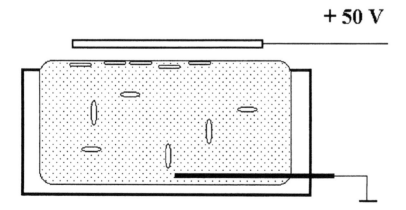

+ 50 V

FIG. 17 Schematic of electric field–assisted BR monolayer formation.

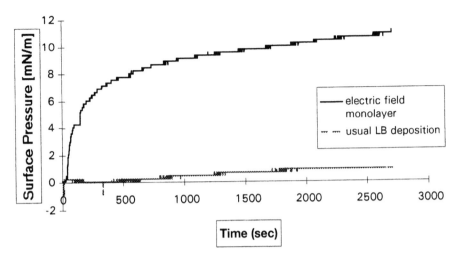

FIG. 18 Dependence of the surface pressure of BR monolayer upon the time in the presence and absence of the electric field.

formation. This finding does not seem strange. In fact, an electric field only aligns the fragments at the air/water interface, providing equal orientation of the proton pathways. The layered structure in this case remains the same. X-ray curves from both types of samples revealed Bragg reflections corresponding to a spacing of 46 Å, which is in a good correspondence with the membrane thickness.

In order to control the degree of BR orientation, photoinduced current was measured. Photosignal was also measured, as a function of the illumination wavelength (Fig. 19). Moreover, one monolayer of BR was deposited onto the porous membrane. The results are summarized in Table 3. It is clear that the photoresponse in the case of electric field–assisted monolayer formation is much higher compared to that after a normal LB deposition (in the last case the signal value is comparable with the noise, indicating a mutually compensating orientation of the membrane fragments in the film). The observed results allow the conclusion that the suggested method of electrically assisted monolayer formation is suitable for the formation of BR LB films, where the membrane fragments have preferential orientation.

Since electric field–assisted monolayer formation at the air/water interface turned out to provide the possibility of highly oriented BR LB film formation, it is possible to suggest another application of BR films for transducing purposes. The principles of device realization are described next. The scheme of the proposed device is presented in Figure 20. Porous membrane with deposited BR film separates two chambers with electrolytes and two platinum electrodes. A light fiber is attached to the X–Y mover, which allows illuminating desirable parts of the membrane. Illumination of the membrane part will result in the proton pumping through it, carried out by BR. Therefore, a current between the electrodes will appear. This current must depend upon several factors, such as light intensity, pH of the electrolytes, and gradient of the pH on the membrane. One of the possible applications of the suggested device is mapping of 2D pH distribution in the measuring chamber, which can result from the working of enzymes immobilized in this chamber. By scanning the light over the membrane it will be possible to obtain the current proportional to the pH gradient at the illuminated point, and by maintaining the pH value fixed in the reference chamber it

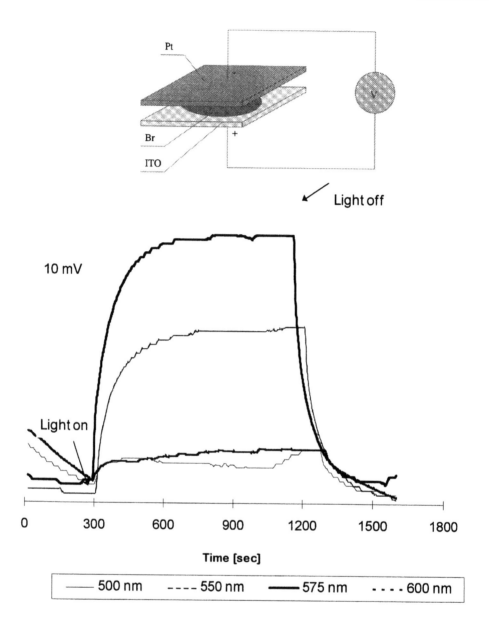

FIG. 19 Photosignal measured for BR photoinduced current as a function of time in the structure shown on top.

TABLE 3 Photocurent Observed in a System Using Porous Membranes Covered with BR Film Deposited by the Usual LB Technique and Electric Field–Assisted Technique

	Light-on current [pA]	Light-off current [pA]
Usual LB technique	15	10
Electric field–assisted monolayer formation	820	10

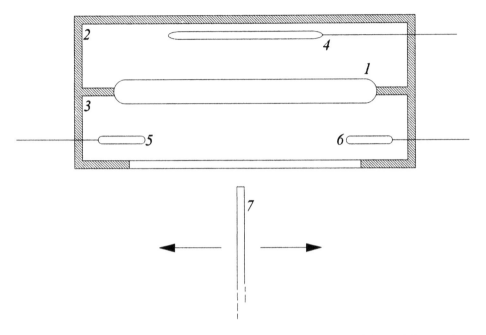

FIG. 20 Schematic of device for light-addressable proton pumping. (1) BR containing membrane, (2,3) working sections, (4–6) electrodes, and (7) light source.

will be possible to calculate absolute pH values at different points over the whole membrane surface. If different types of enzymes, producing or consuming protons during their functioning, are distributed over the area close to the membrane, the device will allow one to determine the presence of different substrates in the measured volume, performing, therefore, as a multiple enzymatic biosensor.

Space resolution of the transducer, in principle, is extremely high. Because each BR molecule performs proton pumping, it will be comparable with the protein size (about 2 nm). In practice, however, it will be limited by the possibility of focusing the light beam. But, in any case, it will be more than in existing transducers, subject to technological problems.

VIII. METALLOPROTEIN MONOLAYERS FOR HEALTH SENSORS

All known cytochromes are heme proteins, which take part in the electron transport process. Their classification is based on their number, referring to the wavelength of maximum absorption of the reduced form. Indeed it is no exaggeration to state that P450 is the most versatile biological catalyst known. Cytochrome (Cyt) P450 is known to catalyze hydroxylation, dehalogenations, N-, S-, and O-dealkylations, N-oxidations, sulfdoxidations, and other reactions. The Cyt P450 catalyst reacts with xenobiotic drugs, including antibiotics, carcinogens, antioxidants, solvents, anaesthetics, dyes, pesticides, petroleum products, alcohol, and odorants, and physiologically occurring compounds such as steroids, eicosanoids, fatty acids, lipid hydroperoxides, retinoids, acetone, and acetol. The reactive site of all of these enzymes is extraordinarily simple, containing only an iron protoporphyrin IX with cysteinnate as the fifth ligand, leaving the sixth coordination site to bind the activated molecular oxygen. The local environment of oxygen binding and activation is also very simple, with mostly hydrophobic protein residues and a single threonine hydroxyl, which is essential for catalysis for some but not all P450s, as shown in Table 4 (Albertus et al. 1998). Other enzymes add negatively charged groups to the hydroxy group, thus rendering soluble a medicine or harmful metabolite that otherwise would remain confined in the plasma membrane. Fabrication of ultrathin cytochromes by self-assembly or Langmuir–Blodgett technique enhances the stability against temperature and environmental effects.

A. Metalloproteins and Metal-Binding Sites

Regardless of the category, each metal cofactor–binding site is defined by functional groups, generally amino acid side chains, which serve as ligands to the metal. In some cases, the functional group can be generated by posttransitional modification of an amino acid that occurs for example, during the biosynthesis of urease, where the nickel-binding ligand is generated by CO_2^--dependent modification of a lysine residue, or during the activation of clotting factors, where a Glu residue is carboxylated to generate a bidentate ligand for phospholipid bound Ca^{2+}. Metals (Lewis acids) bind to these functional groups (Lewis bases) according to preferences determined by the "hard–soft" theory of acids and bases, which states that hard acids bind to hard bases and soft acids to soft bases (Michel et al. 1998). Thus, Ca^{2+}, Mg^{2+}, Mn^{2+}, and Fe^{3+} are generally found bound to oxygen ligands (carboxylates, phenolates, carbonates, and phosphates), while Zn^{2+}, Ni^{2+}, and Fe^{2+} have an affinity for imidazolyl nitrogen, and Cu^+ has a strong preference for sulphur ligands (thiols, thiolates, thioethers). The indicated ligand preferences are according to the hard–soft then which is intended to serve only as a guideline for thinking about metal-binding sites in proteins. The nature of the ligand affects dramatically the chemical properties of the metal, as shown in Table 5.

The primary physiological role of the P450 family is that of a mono-oxygenase. The catalytic reaction can be summarized as

$$RH + O_2 + 2H^+ \, 2e^- \rightarrow ROH + H_2O$$

where RH can be one of a large range of possible substrates. The specificity of a given P450 is determined by the contact residues, which define the active site of the enzyme. These can vary widely between different P450s; however, the principal component of the active site of all P450s is a heme moiety. The iron ion of the heme moiety is the site of the catalytic reaction, and it is also responsible for the strong 450-nm absorption peak in combination with CO. The accepted catalytic cycle can be seen in Figure 21. This begins when the substrate binds to the active site. If the reaction is to proceed further, this displaces a water molecule

TABLE 4 Types of Cytochromes and Their Functionality

Type of protein	Axial ligand	Function	Mechanism
1. Hemoglobin	His	Transport of O_2	$HB + 4O_2 = Hb(O_2)$
2. Myoglobin	His	Storage of O_2	$Mb + O_2 = Mb(O_2)$
3. Cytochrome P450	Cys	Oxidation of substrate	$RH + O_2 + 2e^- + 4H^+ \rightarrow ROH + H_2O$
4. Cytochrome c oxidase	His	Reduction of O_2	$O2 + 4e^- + 4H^+ \rightarrow 2H_2O$
5. Metallothioneins	Cys	Zn and Cu binding	

TABLE 5 Ligand Preferences for Metals Commonly Found in Metalloproteins

Metals, hard	Ligands	
Mn^{2+}	H_2O	OH^-
Fe^{3+}	ROH	RO^-
Mg^{2+}	PO_4^{3-}	$ROPO_3^{2-}$
Ca^{2+}	$R\!-\!CH_2COO^-$	CO_3^{2-}
	$RNH\!-\!COO^-$	RNH_2
Metal, intermediate		
Fe^{2+}	Imidazolyl nitrogen	
Ni^{2+}		
Zn^{2+}		
Co^{2+}		
Cu^{2+}		
Metal, soft		
Cu^+	R_2S	RS^-
	RSH	

that forms a ligand to the heme iron atom in unbound P450. This is accompanied by a change in the spin of the ion from a low spin (1/2) state, in which the $5/3d$ electrons are maximally paired, to a high spin (5/2) state, in which the electrons are maximally unpaired. This in turn causes a change in the redox potential of the iron from approximately -300 mV to approximately -170 mV. This is sufficient to make the reduction of the iron by the redox partner

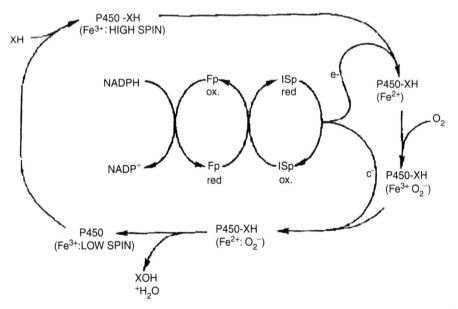

FIG. 21 Schematic diagram of cytochrome P450 mono-oxygenase reaction with substrate RH and product ROH.

of the cytochrome, usually NADPH or NADH, thermodynamically favorable (Shaik et al. 1998). This is followed by the binding of a molecule to a separate site adjacent to the Fe^{+3} ion. This state is not stable and is easily autooxidized, releasing O_2^-; however, if the transfer of a second electron occurs, the catalytic reaction continues. It then reacts with protons from the surrounding solvent to form H_2O, which is released, leaving an activated oxygen atom that then may react with the substrate molecule, resulting in an hydroxylated from of the substrate. This entire reaction cycle usually takes between 1 and 10 seconds.

B. Cytochrome P450 Side-Chain Cleavage and Cholesterol Monitoring

Encoding gene: CYP11A; hemeprotein (481 residues); location: inner mitochondrial membranes; and molecular weight: 57 kDa. A cytochrome P450 of particular interest is indeed the cytochrome P450scc (side-chain cleavage), a redox responsible for cholesterol oxidation. Mitochondrial cytochrome P450scc (CYP11A1) is the mixed-function mono-oxygenase that catalyzes the initial step of the steroid hormone biosynthesis–cholesterol side-chain cleavage reaction to form the pregnenolone. It is an integral mitochondrial membrane-bound enzyme, which catalyzes the first reaction of the steroidogenic pathway, i.e., the conversion of cholesterol to pregnenolone. This reaction requires the input of six electrons that are transferred to the hemeprotein from NADPH via the flavoprotein NADPH-adrenodoxin Reductase and the iron-sulfur protein [2Fe-2S] adrenodoxin. The electrons used for the mono-oxygenase reaction are received from NADPH through an electron transfer chain consisting of two proteins: the FAD-containing adrenodoxin reductase and [2Fe-2S] ferredoxin called adrenodoxin (AD). Cytochrome P450scc forms a stable complex with AD, and the interaction of the P450scc and AD has been reported to be mainly electrostatic. Cytochrome P450scc is a globular metalloprotein, which is formed by a heme group and a long polypeptide chain, 70% of which is α-helix. The molecular weight is about 56.4 kDa. The cytochrome P450 scc in intact human cells is localized in mitochondria on internal membrane. The transcription of the gene of cytochrome P450scc proceeds with higher activity in mitochondria of cells of subrenal glands but takes place in liver. The main function of cytochrome P450scc in subrennal glands is the biosynthesis of steroids whereas in the liver it is the metabolism of xenobiotics.

In order for the cytochrome P450scc to function as mono-oxygenase, two binding sites are necessary. The first binding site is the oxygen packet that binds the O_2 molecule, whereas the second site is the substrate pocket. Under normal conditions, the necessary requirements for the mono-oxygenation process to proceed is the cyclic electron transfer according to the scheme that utilizes Adx [adrenodoxin (Fe-S protein)] and AdR [adrenodoxin reductase (flavoprotein)]. The reduced oxidized states of the iron atom in cytochrome P450scc can be determined spectrophotometrically in the model system with sodium dithionate and CO by the appearance of the peak at 450 nm in the oxidized state.

The monolayer film consisting of cytochrome P450scc molecules exhibits well-defined surface-pressure dependence. This is reflected in a change of the molecular orientation of the proteins at the surface. It has been shown that monolayers of P450scc can be transferred from the air/water interface onto a solid substrate and covalently immobilized without damage to the hemeprotein structure. The P450scc films can be deposited on solid supports by Langmuir–Schaefer techniques using either hydrophobic or hydrophilic adsorption. Moreover, the cytochrome most likely will not denature on the water surface because after being transferred from the interface, it still exhibits specific electron transfer ac-

TABLE 6   Surface Density and Area per Molecule of Cytochrome P450scc wild type
and Recombinant in LB Film Deposited onto Solid Substrate

	Surface density [ng/mm^2]	Area per molecule [nm^2]
Wild type	2.32	40.23
Recombinant	3.16	29.53

tivity. However, up to now the orientation of the hemeprotein molecules in LB films has been unknown. Langmuir monolayer formation was also optimized by the selection of the proper subphase composition. It was possible to assess that the P450 formed stable Langmuir films and that the molecules were close-packed, overcoming the surface pressure of 20 mN/m. Comparison of these results with the protein sizes allowed one to make a conclusion about the close packing of the protein molecules in the layer. The increased thermal stability of the secondary structure of this protein in the film was also demonstrated.

Comparison of the areas of wild-type and recombinant proteins in deposited layers is presented in Table 6. As is clear from the table, we have a denser layer in the case of the recombinant protein. The area per molecule in this case (29.53 nm^2) is of the same order of magnitude as obtained by Turko et al. (1992) and corresponds well to the calculated area per molecule assuming the close packing of molecules and taking into consideration its sizes from the Protein Data Bank. The protein molecule can be estimated as a block with the following sizes: 5 nm $\times$ 6 nm $\times$ 4 nm, thus giving in one cross section an area of about 30 nm^2. The value obtained for the wild-type protein monolayer is higher, indicating that closest possible packing was not reached.

X-ray measurements were carried out with a small-angle diffractometer with a linear position–sensitive detector. Cu Kα radiation ($\lambda = 0.154$ nm) was used (Mogilevski et al. 1984). The samples were rotated with respect to the incident beam, while the intensity was registered by linear position–sensitive detector. The angular resolution of the detector was 0.01°. The curves were acquired in the 2Θ range of 0.3–2.0°. X-ray reflection curves are presented in Figures 22 and 23 for wild-type and recombinant proteins, respectively. The

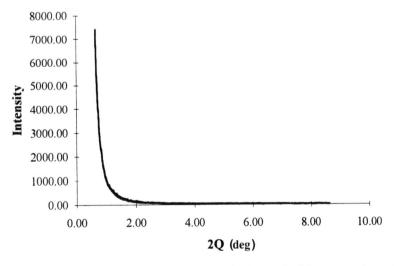

FIG. 22   X-ray pattern of LB film containing 20 layers of wild-type cytochrome P450 scc.

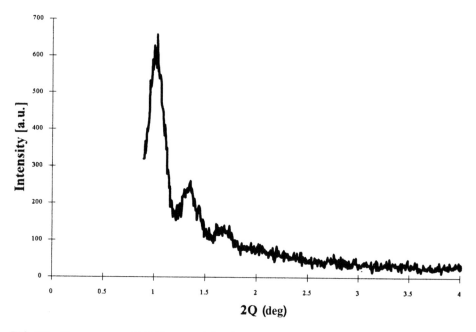

FIG. 23   X-ray pattern of LB film containing 20 layers of recombinant cytochrome P450 scc.

curve obtained from the LB film of wild-type protein (Fig. 22) presents neither Bragg reflections nor Kissig fringes. Such a result means that the film is not ordered and that there
is not uniformity of the thickness along the sample area. In the case of recombinant protein
(Fig. 23), we see Kiessig fringes, whose angular position depends upon the number of deposited layers. The average monolayer thickness calculated from these data is about 6 nm,
corresponding well to both the ellipsometric data and the protein sizes from the Protein
Data Bank.

It has been suggested that the electrochemical behavior of metalloproteins is dependent on several factors, such as the nature of the proteins, the solution environments, the
sensitivity of the electrodes, the pH of the electrolyte solution. The cyclic voltammogram
was performed at 20 mV/s between 0.4 and −0.6 V vs. Ag/AgCl. Various numbers of
monolayers displayed a cathodic peak potential (−472 to −480 mV), with other peaks
varying from −114 to −120, with a small anodic counter varying from −144 to −120 mV
and a small peak at −268, indicative of an irreversible process. There is a shift in the peak
potential as a function of monolayer, as shown in Figure 24. Figure 25 shows the dependence on scan rate for the 30 monolayers of cytochrome P450scc films on ITO glass plate
in 0.1M KCl containing 10 mM phosphate buffer. It shows the cathodic peak potential at
−470 and −134 mV, with the feeble anodic peak at −312 and −76 mV. The cathodic peak
current at −470 mV was plotted with the scan rate and also suggests an irreversible redox
process (figure not shown). It was interesting to see that two cathodic peaks are visible. The
change in the redox peak potential can be observed as the pH of the system varies from 6.8
to 7.4 of phosphate buffer. Table 7 shows the redox peak potential of cytochrome P450scc
at different pH of phosphate buffer.

Cholesterol is found in many biological membrane and is the main sterol of animal organisms. It is equimolar with phospholipids in membranes of liver cell, erythrocytes, and
myelin, whereas in human stratum corneum it lies in the outermost layer of the epidermis

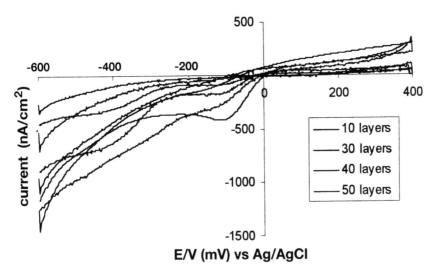

FIG. 24 CVs of Langmuir–Schaefer films of cytochrome P450scc on indium-tin oxide glass plate (ITO) in 10 mM phosphate buffer containing 0.1 M KCl at a scan rate of 20 mV/s between 0.4 and −0.6 V vs. Ag/AgCl.

(which represents 20 wt% of the lipid fraction). Cholesterol has been extensively studied as a known regulator of membrane ordering, but it needs good regulation of its relative concentration in the human body. If found in large quantity, it causes death to the human being by blocking the blood pumping to the heart. In this context, a sensor based on the cholesterol oxidase enzyme has been developed, but was found unsuitable for commercial exploitation due to the short lifetime of cholesterol oxidase in various immobilized matrices. At present we are investigating enzyme or protein, which carries on hydroxylation to cholesterol. In this regard, we see the best possibility based on cytochrome P450scc. A cytochrome

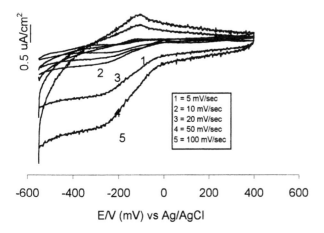

FIG. 25 Cyclic voltammograms of Langmuir–Schaefer films of cytochrome P450scc on indium-tin oxide glass plate (ITO) in 10 mM phosphate buffer containing 0.1 M KCl as a function of scan rate between 0.4 and −0.6 V vs. Ag/AgCl? LS films on ITO functional as the working electrode, platinum as the counter, and Ag/AgCl as the reference electrode.

TABLE 7   Redox Peak Potential of Cytochrome P450scc at Different pH of
Phosphate Buffer

Phosphate buffer	Anodic peak (mV)	Cathodic peak
pH 6.8	−315, −159	−306, −206.7
pH 6.8 with KCl	−316, 78.01	−224.4, 126.6
pH 7.0	−367.7, 189, 238	−44.5, 65.64
pH 7.0 with KCl	−342.0, −74.0	−234, 30.0
pH 7.1 with KCl	−470, −106	42
pH 7.4 with KCl	−470, −115.5	276.0, −17

P450 of particular interest is indeed cytochrome P450scc (scc means side-chain cleavage), the component of the redox chain responsible for cholesterol oxidation. The components of this chain are cytochrome P450scc, adrenodoxin, and adrenodoxin reductase, which are used together to create the chain of electron transfer necessary for the hydroxylation of cholesterol. One of the reasons this mettaloprotein is of particular interest for Bioelectronics is based on its oxidizing action and not mere electron transfer. In this context the transfer of electrons through the metalloprotein will be performed by electrochemical technique. The sensor will be developed using the Langmuir–Blodgett films of cytochrome P450 along with conducting polymer. Conducting polymers such as polypyrroles and polyanilines will be used as electron transfer carriers in the hydroxylation of cholesterol.

Cytochrome P450scc of bovine adrenocortical mitochondria catalyzes the side-chain cleavage of cholesterol to produce pregnenolone, the first and second rate-limiting steps in the biosynthesis of steroid hormones. The reaction occurs in three sequential mono-oxygenation steps, and requires two electrons and one molecule of oxygen in each step via NADPH-adrenodoxin reductase and adrenodoxin. Adrenodoxin (ferredoxin) forms a 1:1 complex with P450scc and functions as a mobile electron shuttle. Electrostatic interactions are important for binding of the complex. Steady-state and transient-state kinetic studies have demonstrated that cholesterol is hydroxylated initially at the 22R position and then at the 20R position, followed by C20–C22 bond cleavage to yield pregnenolone; both 22R-hydroxycholesterol and 20R, 22R-dihydroxycholesterol have been deduced to be natural intermediates in the P450scc-catalyzed side-chain cleavage of cholesterol. It has been reported that 20R, 22R-dihydroxycholesterol accumulates following its production in the incubation mixture of 22R-hydroxycholesterol with mitochondria from bovine adrenal cortex or from human placenta. However, since the reaction proceeds successively without the release of intermediates, they do not usually accumulate in the reconstituted system; therefore it lacks direct evidence as to their structure. The structure of the intermediate 20R, 22R-dihydroxycholesterol offered valuable information for elucidation of the mechanism of the cleavage of C20–C22 bonds. The cholesterol is metabolized to a hormone precursor, "pregnenolone," by the way of 22R-hydroxycholesterol and 20R, 22R-dihydroxycholesterol by P450scc. The flavoprotein NADPH-adrenodoxin reductase and iron sulfur protein adrenodoxin functions as a short electron transport chain, which donates electrons one at a time to adrenal cortex mitochondrial cytochrome P450scc. The soluble adrenodoxin acts as a mobile one-electron shuttle, forming a complex first with NADPH-reduced adrenodoxin reductase, from which it accepts an electron, and then dissociating and finally reassociating by donating an electron to the membrane-bound cytochrome P450. Dissociating and reassociating with flavoprotein then allows a second cycle of electron transfers. Complex factors govern the sequential protein–protein interactions, which compromise this adreno-doxin shuttle mechanism: among these factors, reduction of the iron sulfur by flavin weakens the adrenodoxin–adrenodoxin reductase interaction, thus promoting dissociation of this complex to yield free reduced adrenodoxin. Substrate (cholesterol) binding to cytochrome scc both promotes the binding of the free adrenodoxin to the cytochrome and alters the oxidation-reduction potential of the heme so far to favor reduction by adrenodoxin. The cholesterol-binding sites on cytochrome P450scc appear to be in specific effects of both phopholipid head groups headways direct communication with the hydrophobic phospholipids in which this substrate is dissolved.

Figure 26 shows the redox potential of 40 monolayers of cytochrome P450scc on ITO glass plate in 0.1 KCl containing 10 mM phosphate buffer. It can be seen that when the cholesterol dissolved in X-triton 100 was added 50 μl at a time, the redox peaks were well distinguishable, and the cathodic peak at −90 mV was developed in addition to the anodic peak at 16 mV. When the potential was scanned from 400 to −400 mV, there could have been reaction of cholesterol. It is possible that the electrochemical process donated electrons to the cytochrome P450scc that reacted with the cholesterol. The kinetics of adsorption and the reduction process could have been the ion-diffusion-controlled process.

Comparative study of LB films of cytochrome P450 wild type and recombinant revealed similar surface-active properties of the samples. CD spectra have shown that the secondary structure of these proteins is practically identical. Improved thermal stability is also similar for LB films built up from these proteins. Marked differences for LB films of wild type and recombinant protein were observed in surface density and the thickness of the deposited layer. These differences can be explained by improved purity of the recombinant sample. In fact, impurity can disturb layer formation, preventing closest packing and diminishing the surface density and the average monolayer thickness. Decreased purity of

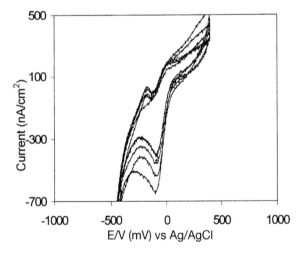

FIG. 26 Cyclic voltammograms of 40 monolayers of Langmuir–Schaefer films of cytochrome P450scc on indium-tin oxide glass plate (ITO) in 10 mM phosphate buffer at a scan rate of 20 mV/s between 0.4 and −0.4 V vs. Ag/AgCl. LS films on ITO worked as the working electrode, platinum as the counter, and Ag/AgCl as the reference electrode. Cholesterol dissolved in X-triton 100 was added 50 μl at a time: (1) with cholesterol, (2) 50 μl of cholesterol, (3) 100 μl cholesterol, and (4) 150 μl of cholesterol.

the sample can be also the reason for pure homogeneity of the deposited layers in the case of wild-type protein. The cathodic peak current of cytochrome P450scc was found to be pH dependent from 6.8 to 7.4, with the exchange of two protons. The kinetics of adsorption and the reduction process could have been the ion-diffusion-controlled process.

IX. SINGLE-ELECTRON AND QUANTUM PHENOMENA IN ULTRASMALL PARTICLES

Currently, single-electron phenomena attract the attention of many research groups. Arising from early attempts at explanation on the basis of a very simple model (Giaever and Zeller 1968, Zeller and Giaever 1969), a formal theory of single-charge phenomena was outlined in 1986 (Averin and Likharev 1986a).

D. V. Averim and K. K. Likharev developed a theory for describing the behavior of small tunneling junctions based on electron interactions. They had started from previous work on Josephson junctions (Likharev and Zorin 1985, Ben-Jacob 1985, Averin and Likharev 1986b) and established the fundamental features of the single-charging phenomena. Their work is based on a quantization theory and handles the tunneling phenomenon as a perturbation, described by annihilation and creation operators of a Hamiltonian.

The theory foresees the possibility of *coulomb blockade* phenomenon in such junctions. Averim and Likharev had investigated the conditions of vanishing for the Josephson tunneling and demonstrated the possibility of having normal electrodes in the junction. That is, no superconducting electrodes are necessary, and, therefore, coulomb blockade is possible to observe, in principle, even at room temperature.

The work of Averim and Likharev had suggested considering two-junction systems trapping the electron inside it, based on the ideas emerging from the theory.

Following this idea, two-junction systems were investigated, and steplike behavior was observed. It was related to the exclusion of the next incoming electron into the intermediate granule due to the electric field created by the previously entering ones (Mullen et al. 1988).

In the traditional lithography approach, researchers continued to consider the idea that modern STM (scanning tunnel Microscopy) could be the proper tool for the formation of two-junction systems when working with very small particles. This consideration had related the studies of single-electron phenomena to the concept of quantum dots (Glazman and Shechter 1989).

In particular, considering a ballistic model for the charge transport through a dot, it was possible to demonstrate that the current through it should be represented as a series of equidistant peaks whose positions correspond to the steps in the coulomb staircase.

Moreover, the possibility of considering single-electron phenomena in a frame of a dot-based system theory allows consideration of even semiconductor nanoparticles as quantum dots, useful for single-electron junctions (Averin et al. 1991).

The modeling of junctions based on these semiconductor quantum dots reveals that their behavior in terms of single-electron phenomena can result in current–voltage characteristics with differential negative resistance regions (Gritsenko and Lazarev 1989). This fact was connected to the possibility of resonating tunneling through quantized energy levels inside the dot (Guinea and García 1990, Beenakker et al. 1991, Sumetskii 1993, Groshev et al. 1991).

On the other hand, some work on the topic considers the presence of negative differential resistance in the current–voltage characteristics and the possibility of a coulomb staircase as different output of the very same phenomenon and, therefore, has tried to consider both of them in the very same conceptual frame (Beenakker 1991, Stone et al. 1992, Prigodin et al. 1993). This approach seems to be successful; in fact, it was possible to see models describing current–voltage curves presenting both stairs and negative resistance along them (He and Das Sarma 1993, Carrara et al. 1996).

All of the theoretical work proposed during the past 10 years forces experimental researchers to develop real systems to observe the described phenomena. In reality, only a year after the very first work of Averim and Likharev, the first measurements of the coulomb blockade and the coulomb staircase phenomena were published.

The Likharev group performed the first measurements of an oscillation of the right periodicity in respect to the single-electron blockade in an indium granule–based junction (Kuzmin and Likharev 1987).

A few months later, the observation of the coulomb blockade appeared (Fulton and Dolan 1987). In the case of a coulomb staircase, only one additional mouth was required (Barner and Ruggiero 1987).

On the other hand, even in particle systems the coulomb blockade (Van Bentum et al. 1988a) and the coulomb staircase (Van Bentum et al. 1988b) were observed, some nonlinear effects were observed in the current–voltage characteristics (Wilkins et al. 1989), and behavior related to the quantized energy levels inside the particles was described (Crommie et al. 1993, Dubois et al. 1996).

However, the main research result from those years was the discovery of the room-temperature single-electron phenomenon. In the 1990s, STM experiments on liquid crystal had shown a very weak staircase (Nejoh 1991); only one year later, the clear observations of the coulomb blockade and the coulomb staircase were demonstrated on gold nanoparticles (Shönenberger et al. 1992a) and the role of system symmetry on the appearance of these two phenomena was outlined (Shönenberger et al. 1992b).

The effect of structure parameters on the features of the junctions was studied. For example, different materials were tested as insulator for spacing the two-junction systems (Shönenberger et al. 1992b, Dorogi et al. 1995), the role of the barrier width in such systems was tested as a possible parameter for creating regions with differential negative resistance in the current–voltage characteristics (Erokhin et al. 1995a), and the role of different semiconducting materials in constructing the particles useful for the experimental system was investigated (Erokhin et al. 1996).

Finally, it was possible to build up the first stand-alone room-temperature single-electron junction by depositing a semiconducting particle directly onto the tip of a very sharp electrode, avoiding in this case the use of an STM microscope, and it was possible to observe the coulomb staircase in such a system (Facci et al. 1996).

In order to understand the basic principles, let us consider the following scheme (Fig. 27). A small granule is placed between two electrodes and is separated from them by tunneling gaps. The structure is asymmetric. Tunneling barriers are characterized by their heights and widths. These parameters determine the tunneling probabilities for each barrier. Therefore, it is possible to attribute to each of them the times τ_1 and τ_2 (the times for electrons to pass through them), reversibly proportional to tunneling probability. The time during which the electron can stay in the granule is τ_g. The following equation must be valid:

$$\tau_1 < \tau_g < \tau_2$$

On the other hand, these junctions can be also characterized by their capacities. Let us consider what will happen when the voltage is applied to this system. When the voltage value

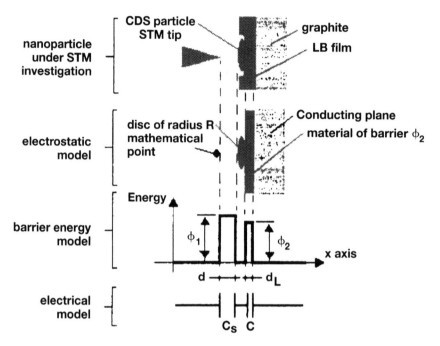

FIG. 27 Schematic of the single-electron junction. (Graphite—electrodes; disc—granule; STM tip—electrodes.)

is rather small, the electron will be tunneled from electrode 1 to the granule after time τ_1. The arrival of the electron will result in the appearance of the electric field equal to e/C, where C is the capacity of the junction. This electric field will be directed opposite to the bias voltage and will prevent the arrival of new electrons before the bias voltage become more then e/C. Therefore, in time τ_1 the electron will be tunneled back to electrode 1. Thus, there will be no current in the circuit (coulomb blockade). When the bias voltage overcome the value of e/C, one electron will always stay at the granule; therefore, after time τ_2 it will be able to tunnel to the second electrode, providing a constant current, whose amplitude will be reversibly proportional to τ_2. When the bias voltage overcomes the value $2e/C$, two electrons will stay in the granule and after τ_2 will be tunneled independently to electrode 2, providing a constant current of doubled amplitude. By increasing further the bias voltage it will be possible to come to the situation when three, four, and more electrons will stay in the granule, producing staircase behavior of the voltage–current characteristics (coulomb staircase).

The described behavior can be observed in systems when thermal excitation is less then electrostatic energy, namely:

$$\frac{e^2}{2C} > kT$$

Therefore, there always exists a temperature where single-electron phenomena are no longer observable for each geometry of the junction. An increase in this temperature demands a decrease in the capacity value and, therefore, granule size. In order to make possible the observation of such phenomena at room temperature, granule size must be less than 3 nm. Therefore, single-electron conductivity in the circuits, fabricated with the traditional microelectronic technology, was observed at very low temperatures, for the granule sizes are much more then 3 nm. However, the approach of continuous reduction of the element sizes with the development of traditional technology approaches is still the mainstream for fabrication of single-electron devices.

In addition to the mainstream of element formation, several nontraditional technological approaches were carried out for the formation of elements with nanometer sizes and their utilization for construction of single-electron elements (Wilkins et al. 1989, Shönenberger et al. 1992a, Dorogi et al. 1995, Erokhin et al. 1995a).

In 1988 a method for the formation of CdS particles in Langmuir–Blodgett (LB) matrix was suggested (Smotkin et al. 1988). LB film of cadmium arachidate was exposed to an atmosphere of H_2S. During the reaction, the head groups of arachidic acid were protonated, and CdS was produced according to the following reaction:

$$[CH_3 (CH_2)_{18} COO]_2 Cd + H_2S \rightarrow 2CH_3 (CH_2)_{18} COOH + CdS$$

Several experimental techniques were applied to characterize these objects. It was found that CdS was formed as small particles inside the LB film with sizes in the nanometer range. Similar work was carried out that resulted in the formation of PbS, CuS, HgS, etc. The sizes of the particles produced by such approaches turned out to be rather similar to that of CdS. The observed sizes suggest that the objects could be useful for the formation of nanogranules for room-temperature single-electron junctions.

However, the formation of the junction is not a trivial task, even if the granule is formed. It is necessary to provide contacts to it.

Scanning tunnel microscopy (STM) was chosen as a tool for realization of this task (Wilkins et al. 1989). CdS nanoparticles were formed in a bilayer of cadmium arachidate deposited onto the surface of freshly cleaved graphite (Erokhin et al. 1995a). The graphite was used as the first electrode. Initially, STM was used for localizing the position of the particles. Figure 28 shows the images of different areas of the sample. The particles are vis-

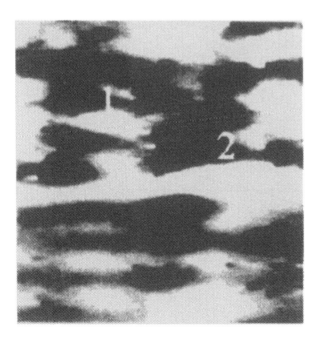

FIG. 28 STM image of CdA LB film after reaction with H_2S.

ible as wells in the corrugated matrix. Their sizes correspond well to the sizes estimated for these objects with other techniques. It suggests that the wells in the images could be related to CdS particles formed in the LB film of the arachidic acid. The reaction process strongly disturbs the structure of the film, resulting in an increased corrugation of the arachidic acid matrix, at least in some regions surrounding the particles. Taking into account the hydrophilic properties of CdS, we can suppose that the matter in the bilayer is redistributed after the reaction in such a way that the CdS particles are not covered by the arachidic acid molecules. This consideration can account for the negative features visible in the STM images and therefore could be related to the CdS particles.

As the second step, the STM tip was locked over the desired particle, feedback was temporally switched off, and voltage–current (*V-I*) characteristics were measured. The typical trend of the *V-I* characteristics is shown in Figure 29. Current steps are clearly observable in the presented curve, indicating that the single-electron junction was formed. It is worth mentioning that the characteristics observed in areas without particles demonstrate a normal tunneling behavior (see Fig. 30).

Part from the characteristics similar to those shown in Figure 29, other characteristics with periodic current oscillations were observed. *V-I* characteristics measured by placing the STM tip over particles 1 and 2 in Figure 28 are shown in Figure 31. Regions with negative resistance, visible in these characteristics, are not usual for a coulomb staircase. However, several features of these curves are similar for single-electron junctions, namely, equal steps in voltage corresponding to the current steps. Moreover, there is a dependence of the voltage step width on the particle size over which the tip was locked. It is clearly visible that the voltage step in the curve, obtained over the object with smaller sizes (smaller capacity), is bigger, corresponding well qualitatively to a value of *e/C* as the voltage step.

Nevertheless, the appearance of regions with negative resistance is generally not typical for the coulomb staircase phenomenon. However, several articles reported similar features both theoretically (Beenakker 1991, Stone et al. 1992, Prigodin et al. 1993) and experimentally (Reed et al. 1988).

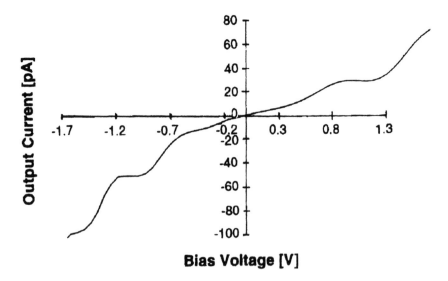

FIG. 29 Voltage–current characteristics with single-electron conductivity.

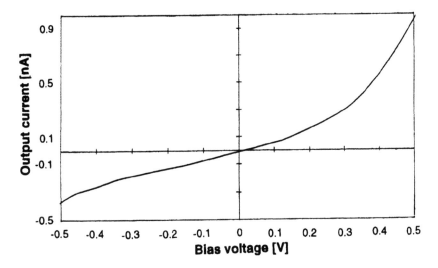

FIG. 30 *V-I* characteristics acquired outside the nanoparticle region.

The observed phenomena can be explained if we consider that different behaviors in the *V-I* characteristics of the same granule (staircase and negative resistance) are measured when different values of current are locked by the STM feedback. This fact implies, of course, that different tip–granule distances are attained in the two cases. By considering the structure as a two-barrier system, we can suggest that one barrier, namely, that between the

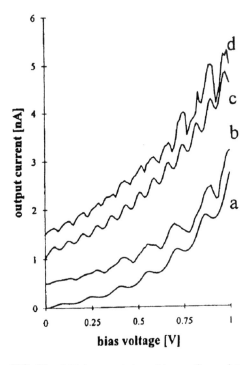

FIG. 31 *I-V* characteristics with negative resistance obtained with CdS particles.

graphite and the granule, is fixed (cannot be varied), while the second can be adjusted by displacing the vertical position of the tip. From these considerations, we can conclude that by varying the locked current, we can change the asymmetry of the structure.

Asymmetry of the structure is one of the basic parameters responsible for the coulomb staircase phenomenon (Shönenberger et al. 1992b). As we *have* said previously, it is absolutely necessary to have an asymmetrical system in order to observe steplike *V-I* characteristics.

In the case of a rather symmetrical system, tunneling probabilities become of the same order of magnitude for both barriers. Therefore, the total current through the system is controlled by both barriers. Before reaching the threshold value e/C for a bias voltage, a normal tunneling current through both junctions is present. This is why a marked suppression of conductivity around zero bias voltage was not observed. When the bias voltage matches e/C, the charge existing in the granule unbalances the voltage distribution through the junctions, preventing the new ingress of electrons to the granule and facilitating their drain. This behavior is the possible reason for the appearance of regions with negative resistance in the measured *V-I* characteristics. More detailed consideration of the model, describing both single-electron phenomena and differential negative resistance, is presented in (Carrara et al., 1996), so here we will use only the results obtained from application of the model.

Starting from this consideration, the fitting was performed, taking into account a two-barrier system where the voltage is redistributed when successive electrons enter the granule. The results of the fitting are presented in the same figure (Fig. 31) and describe well the observed behavior.

The results just presented clearly show the possibility of the organization of single-electron junctions. Nevertheless, it still cannot be considered as staying along the junction, for it is necessary to find the particle with STM. Moreover, to go out of tunneling and then to land the tip once more, the probability of coming to the same place will be practically zero due to several factors, such as thermal drift and mechanical vibrations.

A good solution for this problem was performed by synthesizing the particle directly on the tip of a metal stylus (Facci et al. 1996). Junction preparation involves several steps. The first step, Figure 32a, consists of stylus etching. Styli were prepared from a 0.5-mm tungsten wire by electrochemical etching in 1 M KOH (pH 12.4) with a special etcher, designed for the STM tip preparation, by applying a 20-V peak-to-peak ac voltage between the wire and a toroidal graphite electrode and setting a shutoff current of 0.5 A. After etching, tips were washed in pure water and in isopropanol. The tip surface turned out to be hydrophobic, as indicated by the downward bending of the water meniscus when the stylus was immersed into the trough, and, hence, provided a suitable surface for depositing an even number of monolayers.

The second step, Figure 32b, consists of the covering of the styli with cadmium arachidate LB films. Monolayers of arachidic acid (in principle, it is also possible to use stearic or behenic acids with practically the same results) were spread over the surface of 10^{-4} M $CdCl_2$ water subphase in a Langmuir trough. The monolayer was compressed to a surface pressure of 27 mN/m and transfered onto styli by a vertical dipping technique. Up to six monolayers were deposited.

The third step, Figure 32c, consists of the formation of CdS nanoparticles inside the LB film by exposing precoated tips to the atmosphere of H_2S for a time, sufficient for completing the reaction according to the film thickness (Facci et al. 1994a).

The treated tips were connected to the experimental setup (Fig. 33), which is able to bring them into close proximity with the atomically flat electrode (plates of freshly cleaved,

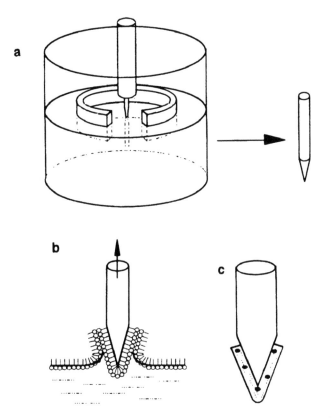

FIG. 32 Procedure for tip preparation for single-electron conductivity.

highly oriented pyrolytic graphite in our case). Practically, the tips were connected to the one-dimensional piezo mover, controlled by a feedback circuit.

After landing the tip in a random place and typically locking a current of 0.1 nA at the bias voltage of 1 V to avoid tip–substrate impact, voltage–current characteristics were measured at room temperature by switching off the feedback system, sweeping the bias voltage, and recording the current flowing through the system.

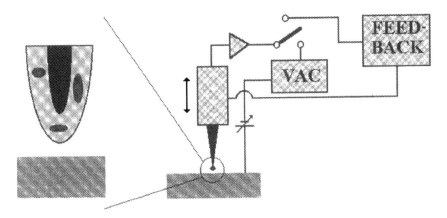

FIG. 33 Experimental setup for single-electron-phenomena measurements.

182

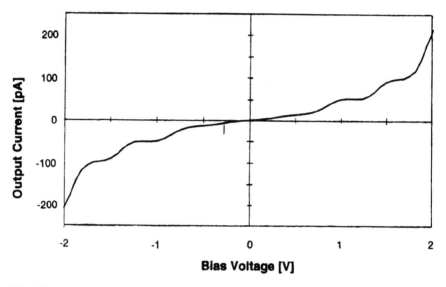

FIG. 34 *V-I* characteristics obtained with the setup shown in Figure 33.

Two typical characteristics obtained in such prepared junctions are shown in Figures 34 and 35 (Facci et al. 1996). These curves are representative of the overall behavior of the set of studied junctions. It is clear that their behavior is highly nonlinear, displaying several interesting features. In both cases it is possible to see a marked depression of the current around the zero-bias voltage, which is typical for single-electron junctions, revealing the so-called coulomb blockade. Moreover, in both types of characteristics, the current revealed the steplike behavior on the bias voltage. Figure 34 shows rather ideal symmetrical characteristics. The steps in voltage are equidistant, as in coulomb staircase phenomena, allowing estimation of the junction capacitance, which turned out to be 1.5×10^{-19} F (estimated from the voltage step of 0.54 V). This value is consistent with these allowing obser-

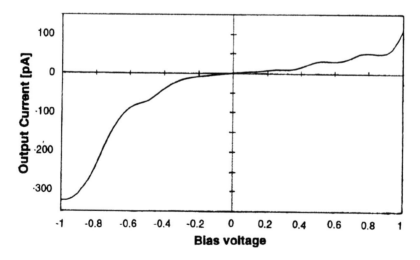

FIG. 35 Voltage–current characteristics obtained with the setup shown in Figure 33.

vations such phenomena at room temperature. In fact, the electrostatic energy of the junction calculated from these values is about 540 meV, which is much higher then the thermal excitation energy at room temperature (26 meV).

Another typical behavior of voltage–current characteristics is reported in Figure 35. These data are different from those in the previous case mainly because of the fact that the curve is asymmetrical in the value of the offset voltage. This behavior, together with the possibility of observing a wider depression of the conductivity around the zero-bias voltage, has been reported earlier (Kuzmin and Likharev 1987, Wilkins et al. 1989, Shönenberger et al. 1992b). It is likely due to charge trapping in the particle by means of impurities. In fact, being initially nonneutral, the particle will shift the voltage offset around zero, due to the variation in the electrostatic energy of the granule itself.

Reported trends (steplike characteristics of both types) were registered in about 60% of all prepared samples, a percentage that is consistent with the yield of good tip production (a good tip allows one to obtain atomic resolution when used in STM). The other 40% of samples displayed the usual tunneling characteristics without any steps.

The approach described represents one more step toward the realization of a completely stand-alone single-electron junction based on nanoparticles and produced in organic matrix. Quantum dot synthesis directly on the tip of a metal stylus does not require the use of STM for localizing the particle position and requires only the use of atomically flat electrodes and a feedback system for maintaining a desirable double-barrier structure.

Furthermore, the reproducibility of the results in this case is much better than that obtained by measuring voltage–current characteristics on CdS nanocrystals formed on a flat electrode and localized by the STM tip. Indeed, the present approach allows for a good repeatability of voltage–current characteristics, since the mutual position of the tip and the granule is fixed and the only adjustable gap is that between the granule and the flat electrode, which is feedback controlled. Instead, in the usual STM approach, even disregarding the problem of localizing the particle position, thermal drift and external noise can affect the stability and reproducibility of the data, since the positioning stage (X–Y mover) is not equipped with a feedback system; e.g., it is practically impossible to achieve tip repositioning in different tip–substrate approaches.

Thus, previously described experiments had demonstrated the possibility of realization of single-electron junctions based on CdS nanoparticles. Nevertheless, because only one type of particle was tested, the question about the role of the material's properties for successful single-electron junction formation was still open.

The aim of the next stage of the work was to study the possibility of the realization of structures that would exhibit single-electron phenomena by forming the nanoparticles with the same technique but from different materials. Comparison of the properties of such structures with those built up with CdS particles will make clearer the role of a material's characteristics on the final properties of the structure.

PbS was chosen as the object of this study (Erokhin et al. 1997). In fact, CdS is a broadband semiconductor (2.42 eV), while PbS is a narrow-band semiconductor (0.37 eV); thus, the differences in properties of the structures based on them, if any, should be easily distinguishable.

Structures were prepared in a way similar to that used for preparing CdS nanoparticles. The only difference was that lead arachidate LB films were deposited after spreading and transferring arachidic acid monolayers at the surface of 10^{-4} M $Pb(NO_3)_2$.

As in the case of CdS granules, two types of $V\text{-}I$ characteristics were registered, namely, one with negative resistance regions (Fig. 36) and the other with steplike behavior

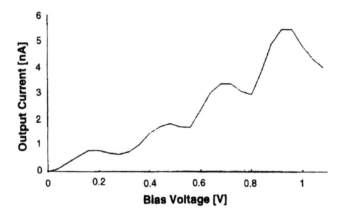

FIG. 36 *I-V* characteristics with negative resistance obtained with PbS particles.

of current on the bias voltage (Fig. 37) (Erokhin et al. 1997). The value of the voltage step depends upon the capacitance of the junction, which is related to the size of the granule. This dependence was observed experimentally, providing yet more evidence of the mono-electron nature of the phenomena. The typical voltage step is about 0.1–0.2 V, which corresponds to the capacity of 8–16×10^{-19} *F*. Such value demands dealing with particles with characteristic sizes of 2–3 nm, which corresponds to the sizes previously estimated by different methods on CdS particles.

The appearance of one or the other type of characteristics was shown to be connected to the asymmetry of the system, which is controlled by the value of the locked tunneling current, exactly as in the case of CdS particles.

When the locked current was small (0.5 nA), giving rise to a large barrier between the granule and the tip, staircase-like *V-I* characteristics were registered (Fig. 37). Instead, when the value of the locked current was higher (1 nA), characteristics with negative resistance occurred at constant-voltage steps (Fig. 36).

Similar results were also obtained when MgS and CuS particles were used for junction formation.

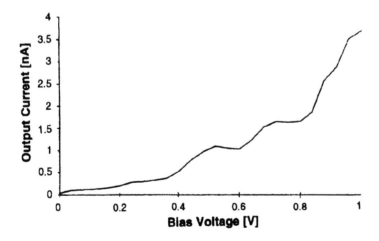

FIG. 37 *I-V* characteristics with single-electron conductivity obtained with PbS particles.

The similarity of the results obtained on systems based on nanogranules made of materials with different bulk properties allows one to conclude that the phenomena at issue are connected only with the decreased size of the granules and not with the bulk properties of the material. In fact, when dealing with such small sizes, it is probably impossible to attribute bulk properties to them, because surface states begin to play a dominant role in the electrical properties of such objects.

Several possible applications of the phenomenon are under discussion. The easiest one is to consider it an analog–digital transducer. In fact, continuous sweeping of the voltage applied to the junction results in the digital output of the current, providing, therefore, a fixed value of the current to the different voltage intervals. Moreover, it is possible to vary the unit step of the digitization, taking a granule of different sizes. The next steps in the practical realization of such a transducer will be in a synthesis of the granule between preformed sharp metal electrodes. The electrodes can be prepared using the selective etching technique of the thin and narrow metal strips deposited onto insulating substrates.

Several possible applications are proposed for systems, using three electrodes. In this case, two of them with a granule between them serve as analogs of the source and the drain in a field effect transistor. The third one, the analog of the gate electrode, serves for the application of the electric field to the granule, which varies the character of the current flow between the source and the drain. Such structures can be realized with traditional microelectronic approaches, even if the technological possibilities still do not allow forming structures capable of working at room temperature. It is more difficult to realize such structures using nanogranules. Nevertheless, the model prototype can be realized if it becomes possible to synthesize the granule between two sharp metal electrodes. In this case, the electrodes will be formed at the surface of the conductor (or semiconductor) and covered by a thin insulating layer (for example, an organic monolayer deposited by LB technique). In this case the conductive substrate will be used as a third electrode.

Apart from transistor-like devices, single-electron junctions can also be useful for sensor applications. The simplest one might be the monitoring of H_2S. Since the formation of CdS nanogranules takes place when an initial cadmium arachidate layer is exposed to this gas, we can expect the appearance of single-electron conductivity only when it is present in the atmosphere.

Other sensor applications can be considered if some sensitive biological molecules (such as antibodies or receptors) are attached to the nanogranule. If, for example, an antibody molecule is attached to it, then the granule is placed between two electrodes, and single-electron current flows between them. The step value of the coulomb staircase depends on the capacity of the junctions. When the antibody molecule binds specific antigen, the capacity value will be changed, and, therefore, the step value of the V/I characteristics will also change.

A. Ultrathin Semiconductor Layers and Superlattices

Apart from the described single-electron junctions, there is another exciting technological possibility for use of these nanoparticles. It turned out that the particles could be aggregated into very thin polycrystalline layers (Facci et al. 1994a). We will now describe some aspects of this phenomenon.

The procedure started, as in the previous case, with the deposition of the LB film of fatty acid salts with bivalent metals. Then the film was exposed to the atmosphere of H_2S

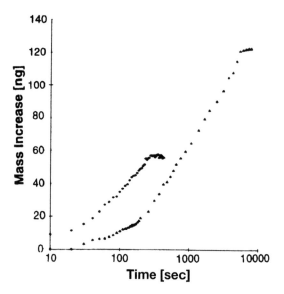

FIG. 38 Gravimmetric measurements of the nanoparticle formation reaction.

for the formation of the particles. It is worth mentioning that the completion of the reaction was controlled by quartz balance measurements (Facci et al. 1993).

The quartz balance is a tool for detecting the increase (or decrease) of the film mass deposited onto the surface of a quartz resonator, connected to the driving circuit, and registering the shift in a frequency. The dependence is expressed by the Sauerbray equation (Sauerbray 1964):

$$\frac{\Delta f}{f_0} = -\frac{\Delta m}{A\rho l}$$

where Δf is a frequency shift, f_0 is an initial frequency of the oscillations, Δm is difference in mass, and A, ρ, and l are the area of the resonator covered by the film, the quartz density, and the resonator thickness, respectively.

Using this equation, the flow of the reaction was followed by registering the frequency shift as a function of reaction time. Typical dependencies are shown in Figure 38 (Facci et al. 1994). As one can see, the curves tend to come to saturation (saturation time depends on film thickness). On the other hand, the plateau level in all cases corresponds well to the value resulting from simple calculation of the number of cadmium atoms available for the reaction in each sample. Numerical data of such estimates are presented in Table 8. The available amount of cadmium atoms was estimated in this case, taking into ac-

TABLE 8 Nanogravimetric Assay Results

Number of bilayers	Reacted H$_2$S (theory)	Reacted H$_2$S (exptl.)	Fatty acid density (theory)	Fatty acid density (exptl.)	Residual film (theory)	Residual film (exptl.)
20	3.86	3.4 ± 0.2	101.16	97 ± 5	14.3	12.4 ± 0.6
15	2.89	2.5 ± 0.1	75.87	72 ± 4	10.7	9.3 ± 0.5
10	1.93	1.6 ± 0.1	50.58	47.8 ± 2	7.1	6.5 ± 0.3

count that one atom coordinates two fatty acid head groups and that in the pH range used in this work (6.5–6.8), the amount of salt in the film must be about 70% (the remaining 30% is pure fatty acid) (Hasmonay et al. 1980).

The next step consists of the selective removal of the fatty acid from the film. It turned out that washing the samples with chloroform and drying them with nitrogen flow results in the removal of a mass amount corresponding to the amount of fatty acid present on the resonator surface (Table 8) (Facci et al. 1994a). This procedure, however, did not completely remove the film from the quartz surface, leaving a residual mass. Taking into account both the amount of reacted H_2S and the mass of the residual film (nonsoluble in chloroform, contrary to hydrocarbon chains), it could be argued that the residual film is made of CdS structures formed during the reaction.

X-ray measurements in a small-angle region had also demonstrated the formation of the semiconductor layers. X-ray patterns of initial fatty acid salt LB films contain both Bragg reflections and Kiessig fringes. After the reaction their angular position was changed, indicating the increase of the spacing and the total thickness. After washing with chloroform, both Bragg reflections and Kiessig fringes disappear (Facci et al. 1994a).

The formation of the layers was also checked by optical absorption measurements and ellipsometry. After the reaction the absorption spectrum was blue-shifted with respect to the bulk spectrum of CdS, indicating the formation of very small particles. It was also possible to estimate their sizes using Rama Krishna and Friesner theory (Rama Krishna and Friesner 1991). After washing with chloroform, the blue shift became smaller (but still remained), indicating the aggregation of the particles in the layer (Facci et al. 1994a).

The results of the ellipsometric study are presented in Table 9. As is clear from the table, the resultant average thickness of the semiconductor layer, obtained from one bilayer precursor, is about 0.8 nm. This value can be considered the thickness resolution of this technique. It is worth mentioning that among the available techniques, only molecular beam epitaxy allows one to reach such resolution. However, the proposed technique is much simpler and does not require complicated or expensive equipment.

STM imaging was performed in order to achieve a more direct understanding of the structure of the resultant semiconductor layers. For these reasons, the films were deposited onto atomically flat surfaces (freshly cleaved graphite). The image of the rather large area is shown in Figure 39. The rough surface of the sample corresponds to the aggregates of nanoparticles forming the layer. The sizes of individual particles are of the order of 3–5 nm, which corresponds well with the results of other studies. There are also rather flat areas in such films where it is possible to see atomic resolution. Figure 40 shows such an image of a CdS layer (Facci et al. 1994a). The lattice of the CdS with hexagonal symmetry is clearly visible. Disturbance of the lattice at boundaries is connected to the finite (small) sizes of the granules composing the layer. Figure 41 shows the image of the layer formed by the aggregation of PbS particles (Erokhin et al. 1997a). The symmetry of the image and lattice parameters corresponds to that of bulk PbS.

Given such technological possibilities, it was logical to try to apply them to the formation of a complicated heterostructure—semiconductor superlattices.

TABLE 9 Ellipsometric Assay Results

	Before H_2S	After H_2S	After chloroform
Film thickness per monolayer	2.7 ± 0.1	3.3 ± 0.1	0.4 ± 0.1

FIG. 39 STM image of a large area of CdS aggregated layer.

Superlattices were prepared in three different ways. The essential steps of each are as follows.

a. Deposition of the fatty acid salt LB film (with metal I); reaction for formation of nanoparticles; aggregation; deposition of fatty acid salt LB film (with metal II); reaction for formation of nanoparticles; aggregation (a schematic of the process is shown in Fig. 42). Then all the steps were repeated several two or three times.
b. Deposition of the fatty acid salt LB film (with metal I); reaction for formation of nanoparticles; deposition of fatty acid salt LB film (with metal II); reaction for formation of nanoparticles. Then all the steps were repeated several (two or three) times. Aggregation was performed when all sample layers were deposited and exposed to the reaction.
c. Deposition of the fatty acid salt LB film (with metal I); deposition of fatty acid salt LB film (with metal II). Then all the steps were repeated several (two or three) times. Particle-forming reaction and aggregation were performed when all sample layers were deposited.

First of all, only the samples, prepared according to the procedure (a) were suitable for scanning electron microscope (SEM) measurements. All other samples were very un-

FIG. 40 STM image of aggregated layers with atomic resolution.

FIG. 41   STM image of aggregated PbS layers.

stable under the electron beam. These results assume that in the case of sample preparation according to the procedure (b) or (c) the removal of the fatty acid matrix was not complete. The residual fatty acid molecules resulted in the decrease in sample stability.

SEM images of the superlattice of the type PbS–CdS–PbS are presented in the Figure 43 (Erokhin et al. 1998). There are rather uniform parts of the superlattice, while there are also disturbed regions, with some defects formed during particle formation or aggregation. The thickness of the individual layers, estimated from Figure 43a, is about 50 nm, giving an average thickness of the aggregated film corresponding to the bilayer of initial LB film of about 0.5 nm, which is consistent with the data already obtained by ellipsometry measurements. In the case of Figure 43b, the average thickness is slightly larger, which can be due to the fact that this region is disturbed.

A SEM image of the CdS–MgS–CdS–MgS–CdS superlattice is presented in Figure 44. It is clearly possible to distinguish these layers in the image. The average thickness of each layer in the superlattice is about 60 nm, which is again consistent with the ellipsometry measurements of average thickness corresponding to the bilayer of the initial film.

The proposed technique seems to be rather promising for the formation of electronic devices of extremely small sizes. In fact, its resolution is about 0.5–0.8 nm, which is comparable to that of molecular beam epitaxy. However, molecular beam epitaxy is a complicated and expensive technique. All the processes are carried out at 10^{-10} vacuum and repair extrapure materials. In the proposed technique, the layers are synthesized at normal conditions and, therefore, it is much less expansive. The presented results had demonstrated the possibility of the formation of superlattices with this technique. The next step will be the fabrication of devices based on these superlattices. To begin with, two types of devices will be focused on. The first will be a resonant tunneling diode. In this case the quantum well will be surrounded by two quantum barriers. In the case of symmetrical structure, the resonant

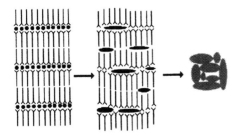

FIG. 42   Schematic of the self-aggregation process for semiconductor nanoparticles.

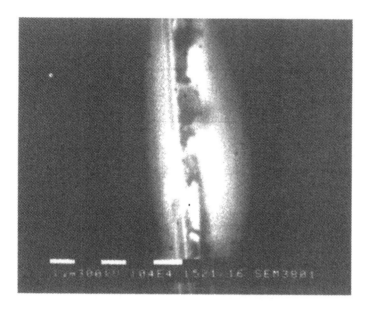

FIG. 43   SEM image of PbS–CdS–PbS superlattice.

level will appear in a quantum well, which will result in the resonant current maximum in *V/I* characteristics when the bias voltage corresponds to the level energy. The other possible device is a semiconductor laser. The laser will be similar to that fabricated with molecular beam epitaxy (Capasso et al. 2000). It does not require the recombination of the carriers, but is based on the transitions of electrons through resonant levels within quantum wells in a semiconductor superlattice. Realization of such a device will demand the formation of a complicated superlattice with a different composition and thickness of the layers in it.

Summarizing, it is possible to conclude that the technique of forming ultrasmall semiconductor particles turned out to be a powerful tool for building up single-electron junctions, even working at room temperature, as well as thin semiconductor layers and superlattices with structural features, reachable in the past only via molecular beam epitaxy.

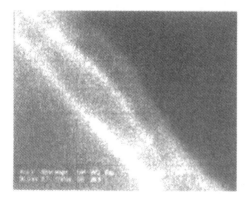

FIG. 44   SEM image of CdS–MgS–CdS–MgS–CdS superlattice.

X. DNA-BASED MULTIQUARTZ SENSOR

The commonly used hybridization technique (Kafatos et al. 1979, Martin 1985) for sequence-specific detection of DNA is sensitive to ~10 pg of DNA. However, several factors render hybridization impractical for routine testing for DNA contaminants, since it is labor intensive, time consuming, and strongly semiquantitative and usually requires a radioisotope. In addition, the specificity of the method means that some contaminating DNA may be missed.

Several other approaches for detecting nucleic acids are reported in the literature, based, for example, on the light-addressable potentiometric sensor (LAPS) (Kung et al. 1990) or on acoustic wave devices (Su et al. 1996).

It was indeed shown recently that it is possible to deposit DNA-aliphatic amine complexes onto solid substrate via the Langmuir–Blodgett (LB) technique (Erokhin et al. 1992). In this method DNA was attached to a preliminarily formed monolayer of octadecylamine (or hexadecylamine). The technique is based on the fact that at neutral pH, DNA is charged negatively while amine groups are positive (Frommer et al. 1970). Analysis of the film suggests that DNA in such a film is in a single-stranded form (Sukhorukov et al. 1993). The schematic model of the film deposited by the technique is shown in Figure 45. A single-stranded DNA layer is sandwiched between two aliphatic amine monolayers. Thus, the technique can be useful for our objectives, for it allows depision of single-stranded DNA on practically any substrate and does not demand a large quantity of DNA, since only one monolayer will be deposited. Nevertheless, there is a question of whether DNA in such a structure will hybridize. In fact, the film contains a single-stranded DNA monolayer between two amine monolayers, and it is questionable whether the upper amine monolayer will prevent hybridization with complementary DNA strands.

The aim of the study was to check this possibility. A quartz nanobalance was chosen as a sensitive tool (Sauerbrey 1964). The device allows monitoring of the mass attached to the surface of quartz oscillators (Facci et al. 1993). The method is simple, cheap, and sensitive (as it should be for practical applications) and allows one to make parallel measurements in different media, also permitting a differential scheme of measurements.

Deposited DNA was the plasmid pUC19b5 (3650 bp) extracted and purified from *E. coli* by Birnboim and Doli (Maniatis et al. 1982). As a probe for the samples, the same plasmid, transformed in to the single-stranded form by boiling during 2 min, was used (100% of continuous homology with the target). As a negative control, the same plasmid was de-

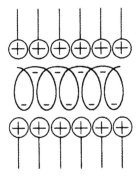

FIG. 45   Schematic representation of DNA-containing LB films.

posited and DNA of phage was used as a probe (less than 10% of the random homology) (Pharmacia Biotech, Molecular and Cell Biology Catalogue 1995–96).

The films were formed and transferred onto solid substrate by MDT trough (MDT, Russia) (Erokhin et al. 1994). A small Teflon trough (30-ml volume) was used for the film formation.

DNA solution (25 mg/mL) was used as a subphase. Octadecylamine monolayer was spread at the subphase surface and compressed to the surface-pressure value of 15 mN/m. The adsorption of DNA at the amine head groups was performed for 1 hour. Then the layer was compressed to 30 mN/m and the film was transferred onto solid substrates.

The films were deposited onto solid substrates by a horizontal lift technique. One layer was deposited for the gravimetric and fluorescence measurements. Twenty layers were deposited for x-ray study.

X-ray measurements were carried out on an automatic small-angle x-ray diffractometer with linear-position-sensitive detector (Mogilevski et al. 1984, Erokhin et al. 1995b). CuK (= 1.54 Å) radiation was used. The sample was rotated with respect to the initial beam, while the intensity was registered in all channels of the detector.

For the hybridization, samples with deposited DNA were placed in plastic envelopes containing 2 mL of the hybridization buffer (10 mM Tris-HCl, pH 7.6, 1 mM EDTA, 0.5% SDS). Twenty mL of the boiled probe was added to the same envelope. The envelopes were then sealed and placed into the water thermostat at 60°C, with stirring overnight. After the hybridization, the samples were strongly washed with distilled water for 10 min, dried, and measured. "Cold" hybridization was performed at room temperature.

After hybridization and washing, the samples were stained with DAPI (4′,6-di-amidino-2-phenylindole), which apparently associates in the minor groove of double-stranded DNA (Kapuscinski 1990). DAPI from Sigma was used. Binding of DAPI to double-stranded DNA occurs with about a 20-fold fluorescence enhancement, which usually does not occur with single-stranded DNA (Haugland 1992).

Fluorescence microscopy measurements were performed with a Zeiss Axioplan microscope (Zeiss Co. Germany) equipped with a mercury lamp and a 40× objective. Images were acquired by CCD camera CH250 (Photometrix Co., Germany) cooled at −40°C by a liquid cooling unit CH260 (Photometrix Co., Germany).

Images were acquired with 5-s exposure time after having fixed the focus. The total intensity of fluorescence was estimated by calculations the integral of the gray-level histogram.

Calibration of the quartz oscillators was performed by consequent deposition of the cadmium arachidate layers (Facci et al. 1993). The dependence of the frequency shift on the number of deposited bilayers is shown in Figure 46. The analysis of the curve reveals a sensitivity of 1.482 pg/mm^{2}Hz for our transducer.

A special setup has been developed for future applications; it allows the measurement of up to eight oscillators in parallel or up to four in differential configuration.

Two eight-channel digital multiplexers have been used to address the two ends of each oscillator. The selection operation consists of connecting the two ends of each oscillator to the driving circuit via the two multiplexers, which are driven by three digital lines allowing the selection of the eight channels. In order to check the geometry of the elementary cell containing DNA in LB film, x-ray diffraction measurements were performed. Angular position of Bragg reflections corresponds to the spacing of 5.8 nm. Taking into account that the amine used in the experiment is octadecylamine (bilayer thickness is about 4.9 nm), it appears that DNA layers incorporated into the film have the thickness of about 0.9 nm. These x-ray data, along with independent IR spectroscopy (Sukhorukov et al.

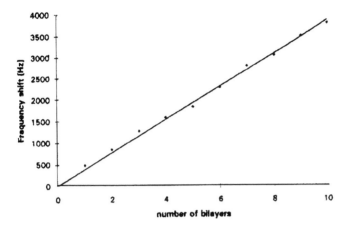

FIG. 46 Calibration curve for a gravimetric nanobalance.

1994), suggest that DNA is very likely in a single-stranded form in the film. The x-ray data confirm once again the film structure model shown in Figure 45, where a single-stranded DNA layer is sandwiched between two octadecylamine layers.

Such layer structure does not allow us to say *a priori* that hybridization of DNA will be possible, for it is protected by the octadecylamine layer. In order to control for this possibility, fluorescence measurements were performed. The first indication that hybridization was successful is that after the process, the sample surface became wettable, while before it and after "cold" hybridization it was not wettable at all. The results of the fluorescence measurements are summarized in Table 10. The results of the specific hybridization are three times more with respect to unspecific hybridization and one order of magnitude more with respect to "cold" hybridization. Thus, it appears that during a normal hybridization (100% homology) some structural changes and redistribution of the layer takes place. As a result, DNA becomes available for the specific reaction. Such a model also explains why the fluorescence level after unspecific hybridization (10% homology) is higher with respect to "cold" hybridization. Because the molecules have some mobility when the film is "warmed," some DNA from the film could be hybridized on itself, while during "cold" hybridization this is impossible.

The results of the gravimetric measurements of the hybridization are presented in Table 10. As in the case of fluorescence, normal hybridization results in a frequency shift that is much higher with respect to the unspecific and "cold" hybridization. Nonspecific hy-

TABLE 10 Results of Fluorescent and Gravimetric Measurements of DNA Hybridization Against a Reference Containing only LB-Immobilized Single-Stranded DNA/Amine Sandwich

Type of experiment	Integral intensity of fluorescence with respect to control (a.u.)	Frequency shift with respect to control [Hz]
100% homology hybridization at 60°C	31.2 ± 9.3 (100%)	688.5 ± 10.8 (100%) (1.02 ± 0.016 ng/mm²Hz)
Cold 100% homology hybridization at 22°C	3.00 ± 0.18 (9.6%)	26.3 ± 17.5 (3.8%) (0.038 ± 0.026 ng/mm²Hz)
Random 10% homology hybridization at 60°C	10.1 ± 4.8 (32.4%)	297.5 ± 70.0 (43.2%) (0.44 ± 0.104 ng/mm²Hz)

bridization (10% homology) give a frequency shift higher then does "cold" hybridization, which is also consistent with the data of fluorescence. It is interesting to note that the error (standard deviation of the measurements of 10 different samples) in the case of unspecific hybridization (difference in frequency shift of different samples) is much higher with respect to specific hybridization (both in "cold" and "warm" conditions). This fact indicates that in a case of unspecific hybridization, some random processes take place, as suggested earlier.

DNA-based biosensors are considered among the most promising elements in analytical biotechnology (Downs et al. 1987, Fawcett et al. 1988, Yevdokimov 2000).

Construction of the DNA chip (Nicolini 1996) requires a single-helix DNA immobilized at the surface of the device and sensitive to the binding of the complementary strand.

The study has shown the possibility of using a single-stranded DNA incorporated into an LB film as a sensitive layer of DNA-based biosensor. A nanogravimetric balance turned out to be a suitable candidate for a transducer of such a sensor, for it provides 0.3-ng resolution. The present study represents the first positive step toward a multiple-DNA-probe sensor with high sensitivity based on LB films and nanogravimetric balance, for it establishes its feasibility. The future development of this research will be in different directions, namely, the simultaneous detection in samples of several genes utilizing the corresponding DNA probes, the automation of genome sequencing and of course the optimization of the overall mechanics and hardware of the apparatus, until it also becomes competitive with the existing hazardous radioactive labeling techniques in terms of time, cost, and efficiency.

The fluorescence data were obtained by summarizing the intensity through the total image area. The results are the average of two different samples prepared in the same way. The gravimetric data are the average of measurements on five samples.

XI. LIGHT-EMITTING DIODES AND BATTERIES BASED ON CONDUCTING POLYMERS

The advances in synthetic methodologies for the preparation of thin films and fibers of PPVs qualify them for consideration in various applications. The overall methodology can be roughly divided into three categories: precursor approach, side-chain derivatization, and in situ polymerization. We include here only the practical and commonly used preparative methods from each category; some less common ones can be found elsewhere. The parameters considered for the synthesis of electroluminence polymers are followed as: (1) precursor polymers; (2) solubilizing groups; (3) polymers with conjugated and nonconjugated segments in the main chain; (4) nonconjugated polymers with conjugated segments in the side chain. The poly(2,5-dihexyl phenylene vinylene), poly(2-methoxy phenylene vinylene) (M-PPV), poly(2-methoxy-5-bromo-phenylene vinylene), poly(2-methoxy-5-cyano-phenylene vinylene), poly(2,3-diphenylene vinylene) (DP-PPV), and poly(1,4-napthalene vinylene) have been synthesized via the precursor approach.

A. Light-Emitting Diodes

Light-emitting diodes (LEDs) based on the electroluminescent conjugated polymers (Fig. 47) have attracted significant attention, both in academic research and industrial development, and are now on the edge of commercialization (Burroughs et al. 1990, Barth and Bässler 1997). Polymer LEDs require properties such as shown in Table 11. Recently, ef-

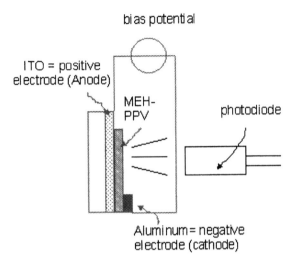

bias potential

ITO = positive
electrode (Anode)

MEH-PPV

photodiode

Aluminum = negative
electrode (cathode)

FIG. 47 Schematic for producing polymer LEDs.

forts have been made to design and synthesize electroluminescent polymers, tailor their properties, investigate the device physics, and engineer light-emitting diodes. This is an interdisciplinary field. To explain the unique phenomenon of conducting polymers, such as insulator-to-metal transition upon doping, new concepts, such as solitons, charge solitons, polarons, and bipolarons, have been introduced. Among various conducting polymers, aromatic conducting polymers such as polythiophene, poly(p-phenylene vinylene), poly(*p*-phenylene), poly(phenylene sulphide), and their derivatives have been used as the emission

TABLE 11 Physical Properties of Interest for PPV Compounds

Polymer	Abbreviation	Properties	Notes
Poly(phenylene vinylene)	PPV	Bandgap between π and $\pi^* = 2.5$–2.6 eV Intrinsic electrically conductivity $= 10^{-3}$ S/cm Emission range $= 1.8$ eV to 2.5 eV (below the bandgap) Ionization potential ≈ 5.1 eV Electron affinity ≈ 2.6 eV	Dopant compounds: iodine, ferric chloride, alkali metals (K, Ca, Mg), or acids
Poly(2-methoxy, 5-(2′-ethylhexyloxy)-1,4-phenylene vinylene)	MEH-PPV	Emission peak $= 605$ nm	*p*-type doping by sulfuric acid (H_2SO_4) *n*-type doping by sodium (electron donor) Iodine (I_2) $=$ electron acceptor $=>$ oxidizing agent

layer in polymer EL devices. A low voltage is applied over a thin film of the polymer. Subsequently the electrons and holes are injected from the electrodes; when a hole and an electron collide in the polymer, a local excited state can be formed that emits light (electroluminescence), as shown in Figure 47.

However, since the discovery of electroluminescence in PPV in 1989–1990, doped conjugated polymers in their conductive forms are no longer of prime interest. Today, the neutral or pristine conjugated polymers for semiconductor device applications, such as photovoltaic cells, field effect transistors, light-emitting diodes (LEDs), and Schottky diodes, have become the major focal points of interest. The PPVs and thiophenes family of polymers serves as a prototypical conjugated polymeric class for application as well as for fundamental understanding of the electronic processes in such conjugated polymers (Blom et al. 1997).

The basic requirements for the choice of electroluminescent polymers are: (1) The polymer should have good film-forming properties (smooth surfaces, no pinholes, minimum thickness of 50 nm). (2) The polymer film should have good thermomechanical stability. (3) The polymer films should be transparent. (4) The polymer should be amorphous. (5) The polymer should exhibit excellent heat, light, and environmental stability. (6) Special requirements for light-emitting polymers are light emission in the visible region and color tunability. Other polymers serving as charge-carrier or charge-barrier layers should not be used simultaneously with the light-emitting component, should not emit visible light, but should reversibly form radical cations (hole-transporting materials) or radical anions (electron-transporting materials), and should show charge transport. The reversibility of radical cation or radical anion formation is usually tested by cyclovoltammetry. Charge transport is usually investigated by time-of-flight measurements. Their work functions should closely match the related electrode material, which is related directly to the chemical structure of the electroluminescent polymer. (7) To control the preceding parameters, Langmuir–Blodgett and layer-by-layer adsorption techniques for the preparation of PPV conjugated polymer films have recently been developed. These demonstrate the fabrication of optically transparent PPV-containing multilayers of precursor PPV, which is converted to PPVs by thermal elimination.

B. Batteries

The batteries called "rocking-chair" systems are one of the most promising electrochemical energy storage systems, and they have a tremendous role in technical applications. Mounting concern regarding the environmental impact of throwaway technologies has caused a discernible shift away from primary batteries and toward rechargeable systems. The secondary batteries ("rechargeable systems") have the advantage of being able to operate for many charge cycles without significant loss of performance. With technologies emerging today, an even higher demand for rechargeable batteries with high specific energy and power is expected. The market for secondary batteries is growing very fast, thanks to the development of new applications encompassing such fields as games, laptop computers, cellular phones, consumer electronics, portable computers, and electric vehicles. Among the factors leading to the successful development of high-specific-energy secondary batteries is the fundamental need for high cell voltage and low-equivalent-weight electrode materials. Electrode materials must also fulfill the basic electrochemical requirements of sufficient electronic and ionic conductivity, high reversibility of the oxidation/reduction reaction, as well as excellent thermal and chemical stability within the temperature

TABLE 12 Electrochemical Parameters of 15-Monolayer Polyaniline LS Films

Material	Oxidation potential (mV)	Reduction potential (mV)	Response time (ms)
PANI	0.78, 320	130, 710	180
POT	662, 531.8, 304	627.4, 499.3, 165, 23.54	240
POAS	707, 506, 282	680, 410, 144, −5.91	230
PEOA	563.4, 374.4, 100.5	401.1, 262	280

range for a particular application. Importantly, the electrode materials must be reasonably inexpensive, widely available, nontoxic, and easy to process. Thus, a smaller, lighter, cheaper, nontoxic secondary battery is sought for the next generation of electrochemical energy storage systems. The low equivalent weights of lithium and conducting polymers make them attractive as secondary battery electrode components for improving weight ratios. Lithium and conducting polymers also provide greater energy per volume than do the traditional battery standards, nickel and cadmium.

Electronically conducting polymers represent a promising class of materials for the development of electrochemical energy storage devices, common examples of which include polyacetylene, polyaniline, polycarbazole, polypyrrole, and polythiophene. Thus, a conducting polymer is actually an electronic as well as an ionic conductor. Conventional battery electrode materials often have a distinct redox potential, more or less independent of their state of charge or discharge, while conducting polymers have, as it were, a "floating" redox potential, i.e., one that strongly depends on their state of charge or discharge (Novak et al. 1997). Thus, for conventional electrode materials the end of discharge or of charging is indicated by the fact that their potential ceases to be relatively constant (Ciric-Marjanovic and Mentus 1998). Intermediate states of charge or discharge are difficult to recognize by measurements of potential, and only deep discharge produces a distinct potential change. For conducting polymers, the potential is an indication for the state of charge or discharge (which is an advantage), even though this implies that the discharge characteristic is sloping (which is generally considered a disadvantage).

Polymers can be utilized as positive and/or negative electrodes in rechargeable cells. Figure 48 shows schematically how polymers can be used in cell assemblies as electroactive materials. In this figure, m stands for a metal, m^+ for cations, a^- for anions, p for a neutral polymer, p^- for a polymer in its reduced state, and p^+ for a polymer in its oxidized state. Electrochemically active polymers can be prepared both by electrochemical and by chemical methods. Electrochemical methods are often preferred, because they offer the advantage of a precise control of potential and state of charge of the resulting polymer. In fact, solution-cast films are also prepared for obtaining the thicker films for battery application (Fig. 49).

XII. ORGANIC PHOTOVOLTAIC CELLS

Photovoltaic (PV) solar cells, which convert incident solar radiation directly into electrical energy, today represent the most common power source for Earth-orbiting spacecraft, such as the International Space Station, where a "photovoltaic engineering testbed" (PET) is actually assembled on the express pallet. The solid-state photovoltaics, based on gallium arsenide, indium phosphide, or silicon, prove capable, even if to different extents and with

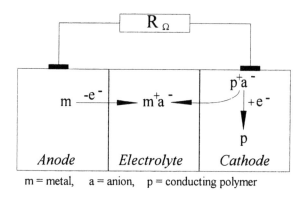

FIG. 48 Schematic of battery operation and the structures of polyanilines used as cathode electrode.

different performances, of operating in a reliable fashion at less than the 10-KW low-power range typical of the missions orbiting the Earth; the electrical power generated over many orbital cycles supports both the electrical loads and the recharge of batteries.

Sunlight is practically an inexhaustible energy source, and increasing energy demand makes it a primary source of renewable energy. Sunlight possesses a very high energy po-

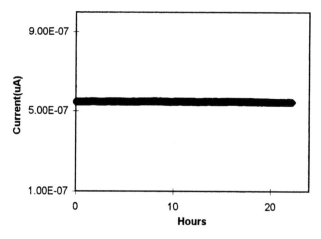

FIG. 49 Discharging effect of 30 monolayers of PANI in LiClO$_4$/PC electrolytes with Al as another electrode connected to 1-MΩ resistance that shows the measured current.

tential and is an ecologically pure and easily accessible energy source. The electrical power obtained from solar energy conversion is widely used in spacecraft power supply systems (the latest very important example is the International Space Station—ISS within the framework of ASI and ESA) and in terrestrial applications to supply autonomous customers with electrical power (portable equipment, houses, automatic meteostations, etc.). The irregular incidence of sunlight on the Earth (daily and seasonal variations) represents one of its disadvantages, together with its low energy density. For these reasons, there is the need to cover large areas with expensive semiconducting solar cells, increasing costs. Thus the electrical energy obtained in such a way is more expensive than that from conventional methods. Although, reduction of pollutant emission is a key factor in the preservation of the environment and subsequently the quality of the life itself, the increase in costs retards the development of a large-scale solar power industry. Actually, solar cells are based on inorganic semiconducting materials, namely, amorphous silicon (efficiency about 12%), multicrystalline silicon (18%), and CdTe (16%), and yield an average energy cost of about $5 per watt.

Given the actual scenario, one can state that the emerging field of nanotechnology represents new effort to exploit new materials as well as new technologies in the development of efficient and low-cost solar cells. In fact, the technological capabilities to manipulate matter under controlled conditions in order to assemble complex supramolecular structures within the range of 100 nm could lead to innovative devices (nano-devices) based on unconventional photovoltaic materials, namely, conducting polymers, fullerenes, biopolymers (photosensitive proteins), and related composites.

Among such techniques, the most promising seems to be the Langmuir–Blodgett one. As far as organic materials for photovoltaic applications, such as conducting polymers and phthalocyanines (Table 13), we note that actual research is focused on understanding the physicochemical phenomena that underline the applicability of such new materials. Several research groups around the world are trying to develop new photovoltaic cells based on unconventional materials, particularly sunlight-converting solar cells based on conducting polymers and on composites. Photoexcitation, charge injection, and/or doping induce local electronic excitations necessary for charge transport. Among them, only the doping process (intercalation) is able to induce a permanent transition to a conductive state.

As far as the photoexcitation process is concerned, we note that excited states produced by photon absorption, namely, excitons, have high relatively binding energies and do not dissociate to give electrons and holes. Therefore, the bulk process of exciton ionization is not a promising method for the development of an organic PV device. The correct energetics, which allows charge separation, can be provided either by the interfaces between molecular semiconductors or with electrodes. One promising structure is a layered heterostructure formed by an ITO glass covered with thin films of MEH-PPV and fullerene (C60). A second metallic electrode (Al, Ca, or Mg) is evaporated. The photogenerated excitons, which diffuse to the interface between the two semiconductors, ionize; therefore the electrons and the holes can be collected the actual power conversion efficiency is about 1%). Note that in organic semiconductors the carrier mobilities are generally low and the absorption depth is usually greater than the diffusion range of the photogenerated excitons. Polymers are characterized by low mobility, i.e., the speed at which charges, electrons, and holes (electrons gaps) move through the material when a voltage is applied. Polymer synthesis and functionalization allow chemists to manage the physicochemical properties of polymers. Therefore, problems can arise with high levels of illumination, and only a fraction of the excitons generated are able to reach or find the

TABLE 13 Photovoltaic Parameters of Various Tested Materials

Cell	Efficiency	Notes
Inorganic-based cells Single-junction thin-film polycrystalline cell	14%	Hydrogenated amorphous silicon (a-Si:H); cadmium Telluride (CdTe); copper indium diselenide (CuInSe$_2$)
Single-junction single crystal	30%	
Multijunction cells	> 30%	
Multijunction cells with two components		
Ga/As/CuInSe$_2$	21.3%	Year 1977/1988
GaAs/Si	31%	1988
AlGaAs/GaAs	24–28%	1988/1989
a-Si:H/CuInSe$_2$	15.6%	1988
a-Si:H/a-Si:Ge:H	13.6%	1989
GaInP2/GaAs	25%	1989
n-CdS/p-CdTe	15.8%	1993—heterojunction
Organic-based cells		
Cr/chlorophyll-a/Hg junction	0.016% (monochrom. eff.)	λ = 745 nm
Merocynine dye–absorber layer	0.7%	V_{oc} = 1.2 V J_{sc} = 1.8 mA/cm2
ITO/copper phthalocyanine (CuPC p-type)/perylene tetracarboxylic derivative (n-type)/Ag	1%	
MEH-PPV (poly(2-methoxy-5-(2′-ethyl-hexyloxy)-1,4-phenylene vinylene	< 0.05%	Pure MEH-PPV
C60/alkoxy-PPV	Energy conversion >1% 1.2% under monochrom. wave FF = 0.35	λ = 488 nm
C60/MEH-PPV (poly(2-methoxy-5-(2′-ethyl-hexyloxy)-1,4-phenylene vinylene)	2.9%	Blends of composite materials C60—acceptor MEH-PPV—donor (conducting polymer)
Photosensitivie proteins	—	Photosynthetic reaction centers, bacteriorhodopsin Under research for exploitation

interface at which the ionization can take place or occur. Usually, the efficiency of such a process can be improved by increasing the degree of purity and of crystallinity as well. This way, the ranges of exciton diffusion are greater than in the previous case. Current research focuses on the development of two different main structures. The first is based on the fabrication of a layered heterostructure, whereas thin films of two different types of organic semiconductors (p-type and n-type) are overlapped. The second approach (Fig.

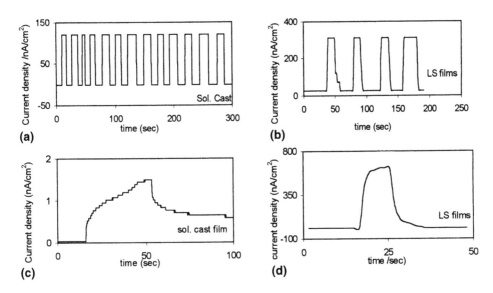

FIG. 50 Photocurrent vs. time at -0.6 V for 50-layer MEH-PPV LS films (different time scales).

50) is based on the fabrication of an interpenetrating network structure obtained by depositing thin films of composite material (namely, a starting mixture of MEH-PPV and C60).

Thus, the main aims of actual research activities are to utilize organic materials with photovoltaic properties and to set up new technologies in order to fabricate complex molecular architectures for solar-energy conversion. Special attention is being devoted to the synthesis and functionalization of organic materials (conducting polymer synthesis, molecular functionalization, nanoparticle synthesis), material processing and characterization (thin-film technologies, electrooptical study), hardware setup (electrodes, microelectrodes, Ohmic contacts, antireflection coating, thermal analysis), prototype device development (cell assembly and engineering), prototype device testing (optical setup, solar simulation, efficiency evaluation, reliability, stability), and socioeconomic impact analysis (competitiveness, costs).

The combination of different inorganic and organic materials to form heterostructures or composite multilayer structures appears as the emergent and promising strategy to fabricate organic-based diodes and cells for photovoltaic applications for space missions. In fact, photoexcited electron transfer from donor and acceptor molecular semiconductors is the basic process to obtain efficient charge generation following photoabsorption. Efficient charge separation and transport requires a suitable interface between the donor and the acceptor material, as well as the good connectivity of electrodes that our thin-film technologies appear to warrant. Our work focused on the influence of the nature of the polyelectrolytes, applied potential, type of the film, and the presence of dissolved oxygen as well as electron scavengers adsorbed on the semiconductor surface. In the case of nanocomposite materials (i.e., MEH-PPV conjugated polymer–coated TO_2 nanoparticle with fullerene composite/dye/metal contact), comparative study of the photoelectrochemical behavior of solution-cast and thin films allows us to identify the optimal unconventional material for the different demands of a particular mission and to optimize the processing of the chosen material (conducting polymers versus biological or nanoparticles or their proper

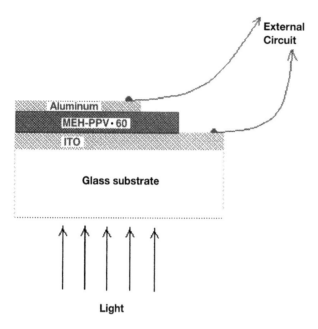

FIG. 51 Schematic of a photovoltaic cell.

combination) for the optimal fabrication of our unconventional photovoltaic cells as compared with conventional ones (Table 13).

The use of interpenetrating donor–acceptor heterojunctions, such as PPVs/C60 composites, polymer/CdS composites, and interpenetrating polymer networks, substantially improves photoconductivity, and thus the quantum efficiency, of polymer-based photovoltaics. In these devices, an exciton is photogenerated in the active material, diffuses toward the donor–acceptor interface, and dissociates via charge transfer across the interface. The internal electric field set up by the difference between the electrode energy levels, along with the donor–acceptor morphology, controls the quantum efficiency of the PV cell (Fig. 51).

XIII. HYDROGEN STORAGE IN CARBON NANOTUBE

The main impediment to the use of hydrogen as a transportation fuel is the lack of a suitable storage system. Compressed-gas storage is bulky and requires the use of high-strength containers. Liquid storage of hydrogen requires temperatures of 20 K and efficient insulation. Solid-state storage offers the advantage of safer and more efficient handling of hydrogen, but promises at most 7% hydrogen by weight and more typically 2%. Various materials, such as palladium (Pd), palladium alloy, palladium–ruthenium alloys, nanocrystalline FeTi, mechanically alloyed amorphous $Ni1_{1-x}Zr_x$ alloys, carbon nanofibers, and carbon nanotubes, are employed for the storage of hydrogen. There have been reports that certain carbon graphite nanofibers are able to absorb and retain 67 wt% hydrogen gas at ambient temperature and moderate pressure, i.e., up to 23 standard liters (2 grams of hydrogen per gram of carbon at 50–150 bars). The lowest hydrogen adsorption reported for any graphite fiber microstructure was shown to be 11 wt%. Approximately 90% of the adsorbed hydrogen can be desorbed at ambient temperature by reducing the pressure, while the balance is desorbed upon heating. Such claims are especially noteworthy, given that up to this point the typical best value of hydrogen adsorption in carbon materials has been 4%, or 0.5 H/C. A large number of research institutes and various companies are involved in the storage of hydrogen and production of full cells based on hydrogen: the Electric Power Research Institute, the American Gas Association, the Gas Research Institute, International Fuel Cells, Energy Partners, Ballard Power Systems, the Energy Research Corporation, MC Power, Westinghouse Electric Corp., Daimler-Benz, BMW, Volkswagen, Volvo, Renault, Peugeot, Siemens, Toyota, Honda, Toshiba, Mitsubishi, Fuji, and Sanyo. Fuel cell–powered cars are being researched and tested. Hydrogen is excellent for storage and would make certain sources more feasible. This would open doors to many alternative resources and begin to shift our use away from fossil fuels. Still, electrolysis and cryogenic cooling are both very expensive. Hydrogen storage is economically viable only when it is sent over very long distances, where piping hydrogen would be more efficient then sending electricity, or when a storage system is necessary, as in the case of solar or wind power. The use of Pd has revealed the restriction in storage capability due to the change in structures upon a few cycles of the adsorption–desorption process. The Pd becomes disordered after a few cycles of sorption. The Pd-Ru structure remained almost unaltered after cycling, but the disadvantage could be that the efficiency of adsorption decreases during alloy formation. Several graphite nanostructures were prepared using Fe-Cu catalysts of different compositions, in order to generate a range of fiber sizes and morphologies. The hydrogen desorption measured from these materials was found to be less than the 0.01 H/C atom, compared to the other forms of carbons. The hydrogen exposed in the metal alloy $Ni_{4-n}Zr_n$ has shown that hydrogen resides in $Ni_{4-n}Zr_n$ ($n = 4, 3, 2$) tetrahedral interstitial sites, with a maximum hydrogen ratio of 1.9. Carbon adsorption techniques rely on the affinity of carbon and hydrogen atoms. Hydrogen is pumped into a container with a substrate of fine carbon particles, where molecular forces hold it. This method is about as efficient as metal hydride technology but is much improved at low temperatures, where the distinction between liquid hydrogen and chemical bonding needs to be considered. One of the most exciting advances recently has been the announcement of carbon nanofiber and carbon nanotube technologies. There is also the claim that up to 10 wt% was achieved for hydrogen storage in single wall nanotubes. Owing to the potential importance of new materials with high hydrogen storage capacity for the worldwide energy economy, transportation systems, and interplanetary propulsion systems, carbon nanotubes can play an important role in hydrogen storage.

Iijima (1991) has focused much attention on both fundamental and applied research on carbon nanotubes since the discovery of multiwall carbon nanotubes (MWNTs) in 1991. In particular, recent progress in research on the properties of single-wall carbon nanotubes (SWNTs), such as their atomic structure and electronic properties, hydrogen storage properties, mechanical properties, and property enhancement through nanotube modification, has been outstanding, due mainly to the availability of sufficient quantities of SWNTs that can be obtained using the pulsed laser vaporization method and the electric arc technique. It has been both predicted theoretically and demonstrated experimentally that SWNTs have many interesting properties. Pores of molecular dimensions can adsorb large quantities of gases, owing to the enhanced density of the adsorbed material inside the pores, a consequence of the attractive potential of the pore wall. Dillon et al. (1997) have shown that a gas can condense to high density inside narrow SWNTs. Simonyan et al. (1999) described the adsorption of molecular hydrogen gas onto charged single-wall nanotubes by grand canonical Monte Carlo computer simulation. The present availability of various fullerene structures points up a large gap in the intermediate size range between small, highly tangled ropes of nanotubes that are currently available in short lengths. Recently, laser vaporization and electric arc methods have best for even for obtaining a continuous process for SWNT production on a commercial scale. Therefore, from an applications standpoint, emphasis is given to the production of high-purity, high-yield, low-cost, large-scale, and easily handled SWNTs for the storage of hydrogen. Recently, a novel method for synthesising SWNTs reported the catalytic hydrocarbon decomposition method, in which benzene is catalytically decomposed at 1100–12,008°C, yielding SWNTs that are similar, on a nanometer scale, to those obtained by laser vaporization and electric arc techniques. This growth method allows lower growth temperatures, permits semicontinuous or continuous preparation, and produces a large quantity of SWNTs at relatively high purity and low cost. However, subsequent experiments showed that the ends of the tubes remained open during the growth process, with highly reactive dangling bonds located around the tube ends.

A. Single-Wall Carbon Nanotube (SWNT)

Nanotubes, the tubular cousins of the spherical carbon molecules dubbed fullerenes, are stronger than steel, lightweight, and able to withstand repeated bending, buckling, and twisting and can conduct electricity as well as copper or semiconductors such as silicon. They also transport heat better than any other known material (Service 1998). With this list of qualities, the current possible uses for nanotubes include superstrong cables, wires for nanosized electronic devices in futuristic computers, charge-storage devices in batteries, and tiny electron guns for flat-screen televisions and hydrogen storage (Heer et al. 1995). The key to this potential lies in the nanotube's unique structure, which in turn depends on the unique properties of its building material and the defects that can form in the network of carbon bonds.

Single-wall nanotubes are produced using catalytic metal particles in carbon arc vaporization, catalytic decomposition of organic vapors, plasma-enhanced chemical vapor deposition, and laser vaporization techniques (Iijima 1991, Dillon et al. 1997). Typical dimensions of SWNTs are 1–2 nm in diameter and many microns in length. SWNTs can be self-organized into ropes that consist of hundreds of aligned SWNTs on a two-dimensional triangular lattice, with an intertube spacing of van der Waals gap of approximately 3.2 Å. The van der Waals gap is defined as the distance between the walls of the nearest-neighbor tubes in the bundle, which is measured from the carbon centers. Nanotubes must have open ends to allow adsorption inside the tubes; but as produced, tubes are capped with hemispherical fullerene domes containing six pentagons, required to produce closure. The tube ends may be opened by oxidation of the caps, which are more susceptible to oxidation because of the strained nature of the five-membered rings.

SWNTs have many potential advantages for hydrogen adsorption over currently available adsorbents. They have large theoretical surface areas that are on the order of those for high-surface-area activated carbons. Crystallized arrays of SWNTs have a very narrow pore-size distribution that has virtually all their surface area in the micropore region. In contrast, surface area in activated carbons is broadly distributed between macropores, mesopores, and micropores. The pore sizes in an array of tubes could be controlled by tuning the diameter of the SWNTs making up the array. Theoretical calculations by Ye et al. (1999) predicted that carbon nanotubes have very strong capillary forces for encapsulating both polar and nonpolar fluids. The filling of multiwall nanotubes with liquid lead has been experimentally observed. Wet-chemistry techniques have recently been used to open

SWNTs and to form single crystals of ruthenium metal inside the nanotubes. Dillon and coworkers (1997) used temperature-programmed desorption to study hydrogen adsorption on carbon soots containing small amounts of SWNTs. They reported high hydrogen uptake under conditions that did not induce adsorption on activated carbon or carbon soots that do not contain nanotubes. Maddox and Gubbins (1997) have modeled adsorption of argon and nitrogen in single- and double-wall nanotubes. They found that argon and nitrogen are strongly adsorbed in an SWNT of diameter 1.02 nm. Layering transitions and hysteresis were observed for double-wall nanotubes with a diameter of 4.78 nm. Model calculations for the adsorption of hydrogen in an SWNT at low-coverage zero-pressure limit were performed by Stan and Cole (1998). The quantum motion of hydrogen molecules was treated by the effective-potential method.

Ye et al. (1999) showed that hydrogen adsorption on crystalline ropes of carbon single-wall nanotubes was found to exceed 8 wt%, which is the highest capacity of any carbon material yet shown. Hydrogen was first adsorbed on the outer surfaces of the crystalline ropes. At pressures higher than about 40 bar at 80 K, however, a phase transition occurred where there was a separation of the individual SWNTs, and hydrogen was physisorbed on their exposed surfaces. The pressure of this phase transition provided a tube–tube cohesive energy for much of the material of 5 me V/C atom. This small cohesive energy is affected strongly by the quality of crystalline order in the ropes. Ye et al. (1999) also noted that the first-order phase transition in the SWNT high-capacity phase overcomes an engineering challenge in using conventional carbons for hydrogen storage. With SWNT materials, hydrogen will adsorb and desorb over a narrower range of pressure, as shown by Ye et al. in 1999, so storage systems can be designed to operate without wide pressure excursions.

Darkrim and Levesque (1998) computed hydrogen adsorption by grand canonical Monte Carlo simulations. In the simulations, interactions between hydrogen molecules resulted from a Lennard–Jones potential located at the center of mass of the molecules and a quadrupolar interaction. The hydrogen quadrupole was described by three charges: two charges q located on the protons, distance of 0.0741 nm, and one charge $-2q$ located at the center of mass ($q = 0.615 \times 10^{-26}$ esu). The cross interaction between the hydrogen molecules and the carbon atoms of the nanotube wall was a Lennard–Jones potential obtained by using Berthelot rules, which presume that carbon atoms interact by a fictitious Lennard–Jones potential. By considering Monte Carlo simulation, and under a current thermodynamic state ~ 10 MPa and 293 K), SWNTs seemed to be good adsorbents for hydrogen. This was due mainly to the favorable potential inside and outside the SWNT, which attracted gas molecules on each side of the nanotube with a minimal loss of volume. The decrease in adsorption as the SWNT diameter increases corresponds to the fact that a large part of the volume inside or outside the nanotube is out of range of the attractive forces of the solid–gas interaction. For the largest diameters, the central part of the nanotube filled with gas at the bulk density. Darkrim and Levesque (1998) stressed that the results depended on the choice of intermolecular potentials between the hydrogen molecules and the carbon atoms.

In the near future, the possible synthesis of nanotubes with solid–gas potential will be more favorable to adsorption. The effect of hydrogen overpressure on the stability of adsorbed H_2 needs to be verified in the near future. The high-purity nanotube produced by laser vaporization, catalytic decomposition, or other techniques should be investigated. It is noteworthy that the synthesis of the SWNT with defined diameters and distances between the walls is difficult to perform at present, but future synthesis routes will allow more hydrogen adsorption in the SWNT. Some theoretical calculations, such as Monte Carlo simulation, were performed for the adsorption of hydrogen with carbon nanotubes, but the real mechanisms of adsorption and desorption are still unknown. Control of these parameters, coupled with improvements in production, purification, and alignment of SWNTs, may lead to a new technology for hydrogen storage.

XIV. FUTURE CHALLENGES

The data presented here point to the successful engineering of nanotechnology based on supramolecular layer engineering of potential industrial relevance. In fact, as emphasized, the filmation process was able to induce high thermal stability and associated high lifetime and recycling, which represent a prerequisite for several processes of industrial interest. Therefore, although work is still in progress to further optimize the parameters and to evaluate in more detail, case by case, the temporal stability of thin layers within required cost effectiveness and the reproducibility within a highly competitive industrial context, this methodology clearly represents a promising general-purpose tool for the design of new industrial products and processes.

ACKNOWLEDGMENTS

This work was supported by the Italian Ministry of University and Scientific-Technological Research (Cap. 2102) through an annual allocation granted to the Fondazione EL.B.A., by research grants of the National Research Council Project on Nanotechnology to PNB-PST Elba, Fondazione Elba, and DISTBIMO University of Genova, by a research grant from the European Union to the Fondazione Elba with the framework of "Copernicus" (IC15CT960810), and by PNB-PST Elba.

REFERENCES

Agbor, MC, Petty, AP Monkman, and H Harris (1993a), Synth. Metals, 55–57, 3789.
Agbor, NE, Petty, MC, Monkman, AP, and Harris H (1993b), Synth. Metals, 55–53, 3789.
Albertus, PHJ, Lutje, SJH, Hubert, DHW, Feiters, MC, and Nolte, RJM (1998), Chem. Eur. J., 4, 871–880.
Armes, SP, and Miller, JF (1986), Synth. Met., 13, 193.
Antolini, F, Paddeu, S, and Nicolini, C (1995), Langmuir, 11, 2719.
Averin, DV, and Likharev, KK (1986a), J. Low Temp. Phys., 62, 345.
Averin, DA, and Likharev, KK (1986b), Sov. Phys. JEPT, 63, 427.
Averin, DA, Korotkov, AN, and Likharev, KK (1991), Phys. Rev. B, 44, 6199–6211.
Avrameas, S (1978), J. Immunol., 8(7), 7.
Bard, AJ, and Faulkner, LR (1980), Electrochemical, Method, Fundamental and Application, Wiley, New York, p 142.
Barner, JB, and Ruggiero, ST (1987), Phys. Rev. Lett., 59, 807.
Barth, S, and Bässler, H (1997), Physical Review Letters, 79, 4445.
Batten, JH, and R Duran, S (1998), Macromolecules, 31, 3148.
Beenakker, CW (1991), Phys. Rev. B, 44, 1646.
Beenakker, CW, Vam Houten, H, and Staring, AAM (1991), Phys. Rev. B, 44, 1657.
Ben-Jacob, E (1985), Phys. Lett., 108A, 289.
Birge, RR (1990), Ann. Rev. Phys. Chem., 41, 683.
Birge, RR (1992), IEEE Computer, 25, 56.

Bleier, H, Finter, J, Hilti, B, Hofherr, W, Mayer, CW, Minder, E, Hediger, H, and Anssermet, JP (1993), Synth. Metals, 55–57, 3605.

Blom, PWM, de Jong, MJM, and Van Munster, MG (1997), Phys. Rev. B, 55, 656.

Boussaad, S, Dziri, L, Arechabaleta, R, Tao, NJ, and Leblanc, RM (1998), Langmuir, 14, 6215.

Bousse, L, Mostarshed, Hafeman, SD, Sartore, M, Adami, M, and Nicolini, C (1994), J. Appl. Phys., 8, 4000.

Bräuchle, C, Hampp, N, and Oesterhelt, D (1991), Advanced Materials, 3(9), 420.

Briggs, J, and Zuk, R (1990), Analytical Biochemistry, 187, 220.

Burroughs, JH, Bradley, DD, and Brown, AR (1990), Nature, 347, 539.

Capasso, F, Paiella, R, Gmachl, C, Sivco, DL, Baillargeon, JN, Hutchinson, AL, Cho, AY, and Liu HC (2000), Science, 290, 1739–1742.

Carrara, S, Erokhin, V, Facci, P, and Nicolini, C (1996), in J Fendler, I Décáni (eds.), Nanoparticles in Solids and Solutions, Kluwer, Netherlands, p. 497.

Chandler, HM (1982), J. Immunol. Meth., 53, 187.

Chase, JW (1986), Ann. Rev. Biochem., 55, 103.

Chen, Z, and Birge, RR (1993), Trends Biotechnol, 11, 292.

Cheung, JH, Fou, AC, and Rubner, MF (1994), Thin Solid Films, 244, 985.

Cheung, JH, Stockton, WB, and Rubner, MF (1997), Macromolecules, 30(9), 2712.

Ciric-Marjanovic, G, and Mentus, S (1998), J. Appl. Ectrochem., 28, 103.

Crommie, MF, Lutz, CP, and Eigler, DM (1993), Science, 262, 218.

Correia dos Santos, MM, Paes de Sousa, PM, Simoes Goncalves, M, Lopes, LH, Moura, and I, Moura JJG (1999), J. Electroanalyt. Chem., 464, 76.

Darkrim F, and Levesque D (1998), J. Chem. Phys., 109(12), 4981.

Decher, G (1996), Supramolecular Chemistry, 9, 507.

Decher, G (1997), Science, 277, 1232.

Decher, G, Hong, JD, and Schmitt (1992), Thin Sold Films, 210/211, 831.

Dhawan, SK, Kumar, D, Ram, MK, Chandra, S, and Trivedi, DC (1997a), Sensors Actuator B: Chemical, 40(2–3), 99.

Dillon, AC, Jones, KM, Bekkedahl TA, Kiang, CH, Bethune, DS, and Heben, MJ (1997), Nature, 386 (27), 377.

Dorogi, M, Gomez, J, Osifchin, R, Andres, RP, and Reifenberger, R (1995), Phys. Rev. B, 52, 9071.

Downs, MEA, Kobayashi, S, and Karube, J (1987), Analyt. Lett., 20, 1897.

Dubois, JGA, Gerritsen, JW, Shafranjuk, SE, Boon, EJG, Schmid, G, and van Kempen, H (1996), Europhys. Lett. 33, 279.

Echigo, Y, Asami, K, Takahashi, H, Inouue, K, Kabata, T, Kimura, O, and Ohsawa, T (1993), Synth. Metals, 55–57, 3611.

Epstein, AJ, and MacDiarmid, AG (1989) in Kuzmany, H, Mehring, M, and Roth, S, eds. Electronic Properties of Conjugated Polymer, Springer Verlag, Berlin, 1989, p 384.

Epstein AJ, and MacDiarmid AG (1991), Chem. Macromol Sym., 51, 217.

Erokhin, V, Feigin, L, Ivakin, G, Klechkovskaya, V, Lvov, Yu, and Stiopina, N (1991), Makromol. Chem. Macromol. Symp. 46, 359.

Erokhin, V, Popov, B, Samori, B, and Yakovlev, A (1992), Mol. Cryst. Liq. Cryst., 215, 213.

Erokhin, V, Vakula, S, and Nicolini, C (1994), Thin Solid Films, 238, 88.

Erokhin, V, Facci, P, Carrara, S, and Nicolini, C (1995a), J. Phys. D: Appl. Phys. 28, 2534.

Erokhin, V, Facci, P, and Nicolini, C (1995b), Biosensors Bioelectronics, 10, 25.

Erokhin, V, Facci, P, Kononenko, A, Radicchi, G, and Nicolini, C (1996), Thin Solid Films, 284–285, 805–808.

Erokhin, V, Facci, P, Carrara, S, and Nicolini, C (1997), Biosensors and Bioelectronics, 7, 601–606.

Erokhin, V, Facci, P, Gobbi, L, Dante, S, Rustichelli, F, and Nicolini, C (1998), Thin Solid Films, 327–329, 503.

Facci, P, Erokhin, V, and Nicolini, C (1993a), Thin Solid Films, 230, 86.

Facci, P, Erokhin, V, Tronin, A, and Nicolini, C (1994a), J. Phys. Chem., 98, 13323.

Facci, P, Erokhin, V, and Nicolini, C (1994b), Thin Solid Films, 243 403–406.

Facci, P, Erokhin, V, Carrara, S, and Nicolini, C (1996), Proc. Natl. Acad. Sci, USA, 93, 10556.

Facci, P, Erokhin, V, Paddeu, S, and Nicolini, C (1998), Langmuir, 14, 193.

Fawcett, NC, Evans, J, Chien, L, and Flowers, N (1988), Analyt. Lett., 21, 1099.

Fedorov, EA, Panov, VI, Lukashev, EP, Kononenko, AA, Savinov, SV, Chernavskii, DS, Kislov, V (1994), Biol. Mem., 4, 189.

Feiters, MC, and Roeland JM Nolte, Chem. Eur. J. 1998, 4, No. 5, 871.

Ferreira, M, and Rubner, MF (1995), Macromolecules, 28, 7107.

Fou, AC, and Rubner, MF (1995), Macromolecules, 28, 7115.

Fou, AC, Onitsuka, O, Ferreira, M, Rubner, MF, and Hsieh, BR, (1996), J. Appl. Phys., 79, 7501.

Fromherz, P (1971), Biochim. Biophys. Acta, 225, 382.

Frommer, MA, Miller, IR, and Khaïat, A (1970), Adv. Experiment. Med. Biol., 7, 119.

Fukuzawa, K (1994), Appl. Optics, 33, 7489.

Fukuzawa, K, Yanagisawa, L, and Kuwano, H (1996), Sensors Actuators B, 30, 121.

Fulton, TA, and Dolan, GJ (1987), Phys. Rev. Lett., 59, 109.

Furuno, T, Takimoto, K, Kauyama, T, Ikegami, A, and Sasabe, H (1988), Thin Solid Films, 160, 145

Genies, EM, Hany, P, and Santier, C (1989), J. Appl. Electrochem., 18, 751.

Giaever, I, and Zeller, HR (1968), Phys. Rev. Lett., 20, 1504.

Glazmann, LI, and Shechter, RI (1989), J. Phys. Condens. Matter., 1, 5811.

Gritsenko, OV, and Lazarev, PI (1989), in FT Hong, ed., Molecular Electronics, Plenum Press, New York, p. 277.

Groshev, A, Ivanov, T, and Valtchinov, V (1991), Phys. Rev. Lett., 66, 1082.

Guinea, F, and García, N (1990), Phys. Rev. Lett., 65, 281.

Guryev, Oleg, Dubrovsky, Timothy, Chernogolov, Alexey, Dubrovskaya, Svetlana, Usanov, Sergei, and Nicolini, Claudio (1997), Langmuir, 13, 299.

Habig, WH, Pabst, MJ, and Jakoby, WB (1974), J. Biol. Chem., 249, 7130.

Hafemann, D, Parce, W, and McConnel, H (1988), Science, 240, 1182.

Hampp, N (1993), Nature, 366, 12.

Hasmonay, H, Vincent, M, and Dupeyrat, M (1980), Thin Solid Films, 68, 21.

Haugland, RP (1992), Handbook of Fluorescent Probes and Research Chemicals, Molecular Probs, p 222.

He, S, and Das Sarma, S (1993), Phys. Rev. B, 48, 4629.

Heer, WAH, Chatelain, A, and Ugarte, D (1995), Science, 270, 1179.

Hwang, S, Korenbrot, J, and Stoeckenius, W (1977), J. Membr. Biol., 36, 137.

Iijima, S (1991), Nature, 354, 56.

Ikonen, M, Sharonov, A, Tkachenko, N, and Lemmetyinen (1993), Adv. Mater. Optics Electron., 2, 211.

Jozefowicz, ME, Eipstein, AJ, Pouget, JP, Masters, JG, Ray, A, and MacDiarmid, AG (1991), Macromolecules, 24, 5863.

Kafatos, FC, Jones, CW, and Efstratiadis, A (1979), Nucleic Acid Res., 7, 1541.

Kapuscinski, J (1990), J. Histochem. Cytochem., 38 (9), 1323.

Kazuyuki, N, Hideo, A, Miyake, J, Nakamura, C, and Hara, M (1998), Mol. Electron. Biocomputing, 5, 773–775.

Keller, SW, Kim, HN, and Mallouk, TE (1994), J. Am. Chem. Soc., 116, 8817.

Kiselyova, OI, Guryev, OL, Krivosheev, AV, Usanov, SA, and Yaminsky, IV (1999), Langmuir, 15, 1353.

Kouwenhoven, LP, van der Vaart, NC, Johnson, AT, Kool, W, et al. (1991), Z. Phys. B, Condensed Mater., 85, 367.

Kozarac, Z, Dhathathreyan, A, and Möbius, D (1988), FEBS Lett., 229, 372.

Kung, VT, Panfili, PR, Sheldon, EL, King, RS, Nagainis, PA, Gomez, B, and Ross, DA Jr (1990).

Kuz'min, LS, and Likharev, KK (1987), JEPT Lett., 45, 495.

Laemmli, UK, Nature, 1970, 227, 680.

Lafarge, P, Pothier, H, Williams, ER, Esteve, D, et al. (1991), Langmuir, 85, 327.

Lafarge, P, Joyez, P, Esteve, D, Urbina, C, and Devoret, MH (1993), Nature, 365, 422.

Langer, JJ (1990), Synth. Metals, 36, 35.

Langmuir, I, and Schaefer, VJ (1938), J. Am. Chem. Soc., 60, 1351.

Likharev, KK, and Zorin, AB (1985), 59, 347.

Lvov, Yu, Erokhin, V, and Zaitsev, S (1991), Biol. Mem., 4(9) 1477.

Maccioni, E Radicchi, G Erokhin, V Paddeu S, Facci P, and Nicolini C (1996), Thin Solid Film, 284–285, 898.

Maddox, MW, and Gubbins, MW (1998), Surf. Sci., 395, 280–291.

Malmquist, M, and Olofsson, G (1989), US Patent 4,833,093.

Maniatis, T Fritsch, EE, and Sambrook J (1982), Molecular Cloning, Cold Spring Harbor Laboratory.

Martin, FH (1985), in Hopps, HE, and Petricciani, JC, eds. Abnormal Cells: New Products & Risk Tissue Culture Assn., Gaithersburg, MD, pp 90–93.

Martin, SJ Bradley, DDC Lane PA, and Mellor H (1999), Phys.Rev B, 59 15133.

Maxia, L Radicchi, G Pepe, IM, and Nicolini C (1995), Biophys. J. 69, 1440.

Mecke, W Zaba, B, and Möhwald, H (1987), Biochim. Biophys. Acta, 903, 166.

Mello, SV Mattoso, LHC, Faria, RM, and Oliveira, Jr., ON (1995), Synth. Metals, 71, 2039.

Méthot, M, Boucher, F Salesse, C Subirade, M, and Pézolet, M (1996), Thin Solid Films, 284–285, 627.

Michel, H Behr, J Harrenga, A, and Kannt, A (1998), Annu. Rev. Biophys. Biomol. Struct., 27, 329.

Misra, SCK Ram, MK Pandey, SS Malhotra, BD, and Chandra, S (1992), Appl. Phys. Lett., 61, 1219.

Miyasaka, T Koyama, K, and Itoh, I (1991), Science, 255, 342.

Mogilevski, L Yu, Dembo, AT Svergun, DI, and Feigin LA (1984), Crystallography, 29, 587.

Mullen, K Ben-Jacob, E Jaklevic, RC, and Shuss Z (1988), B37, 98.

Nejoh, H (1991) Nature, 353, 640.

Nicolini, C (1995), From Neural Networks and Protein Engineering to Bioelectronics, EL.B.A. Forum Series Vol. 1, Plenum Press, New York.

Nicolini, C (1996a), Thin Solid Film, 284–285, 1.

Nicolini, C (1996b), Molecular Manufacturing, EL.B.A. Forum Series Vol. 2, Plenum Press, New York.

Nicolini, C (1996c), Molecular Bioelectronics, World Scientific, Singapore.

Nicolini, C (1996d), Ann. NY Acad. Sci., 799, 297.

Nicolini, C (1997), Trends Biotechnol. 15, 395.

Nicolini, C (1998a), Biophysics of Electron transfer and molecular bioelectronics, EL.B.A. Forum Series Vol. 3, Plenum Press, New York

Nicolini, C (1998b), Ann NY Acad. Sci., 864, 435.

Nicolini, C Erokhin, V Antolini, F Catasti, P, and Facci P (1993), Biochim. Biophys. Acta, 1158 273.

Nicolini, C Sartore, M Zunino, M, and Adami M (1995), Rev. Sci. Instrument., 66, 4341.

Novak, P Muller, K Santhanam KSV, and Haas, O (1997), Chem. Rev. 207(97), 207.

Oesterhelt, D Brauchle, C, and Hampp N (1991), Quart. Rev. Bioph., 24(4), 425.

Okamura, MY Stainer, A, and Feher, G (1974), Biochemistry, 13 1394.

Ortiz de Montellano, PR (1995), In Ortiz de Montellano, ed., Cytochrome P450—Structure, Mechanism and Biochemistry, 2nd ed., Plenum Press, New York, p 201.

Pasqualli, M Pistoia, G, and Rosati R (1993), Synth. Metals, 58, 1.

Paul, EM Ricco, AJ, and Wrighton, MS (1985), J. Phy. Chem., 89, 1441.

Pepe, M, and Nicolini C (1996), J. Phtochem. Photobiol. B., 33, 191–200.

Pepe, IM Ram, MK Paddeu, S, and Nicolini, C (1998), Thin Solid Films, 327–329, 118.

Phillips, MC Hauser, H Leslie, RB, and Olandi, D (1975), Biochim. Biophys. Acta, 406, 402.

Pickett, CB, and Lu, AYH (1989), Annu. Rev. Biochem., 58, 743.

Pharmacia Biotech, Molecular and Cell Biology Catalogue (1995/1996).

Porter, TL, Thompson, D, and Bradley, M (1996), Thin Solid Films, 288, 268.

Pouget, JP, Josefowicz, ME, Epstein, AJ, Tang X, and MacDiarmid, AG (1991), Macromolecules, 24, 779.

Pouget, JP Zhao, SL Wang, ZH Oblakowski, Z Epstein, AJ Manohar, SK Wiesinger, JM, MacDiarmid, AG, and Hsu, CH (1993), Synth. Metals, 55–57, 341.

Prigodin, VN Efetov, KB, and Iida, S (1993), Phys. Rev. Lett., 71, 1230.

Qin, LC Zhou, D Krauss, AR, and Gruen, DD (1998), Appl. Phys. Lett., 72(26), 3437.

Ram, MK Sundaresan, NS, and Malhotra, BD (1993), J. Phys. Chem. 97, 11580.

Ram, MK Mehrotra, R Pandey, SS, and Malhotra, BD (1994a), J. Phys. Condens. Mater., 6, 8913.

Ram, MK Sundaresan, NS, and Malhotra BD (1994b), J. Mater. Sci. Lett., 13, 1490.

Ram, MK Carrara, S Paddeu, S, and Nicolini, C (1997a), Thin Solid Films, 302, 89.

Ram, MK Paddeu, S Carrara, S Maccioni, E, and Nicolini, C (1997b), Langmuir, 13(10), 2760.

Ram, MK Adami, M Sartore, M Paddeu, S, and Nicolini, C (1997c), Presented to Foresight Conference of Nanotechnology.

Ram, MK, Maccioni, E, and Nicolini, C (1997d), Thin Solid Films, 303, 27.

Ram, MK, Joshi, M, Mehrotra, R, Dhawan, SK, and Chandra, S (1997e), Thin Solid Films, (304), 65.

Ram, MK, Maccioni, E, and Nicolini, C (1997f), Thin Solid Films, 303, 27.

Ram, MK, Carrara, S, Paddeu, S, Maccioni, E, and Nicolini, C (1997g), Thin Solid Films, 302, 89.

Ram, MK, Mascetti, G, Paddeu, S, Maccioni, E, and Nicolini, C (1997h), Synth. Metals, 89, 63.

Ram, MK, Adami, M, Sartore, M, Sartore, M, Paddeu, S, and Nicolini, C (1999a), Synth. Metals, 100, 249.

Ram, MK, Salerno, M, Adami, M, Faraci, P, and Nicolini, C (1999b), Langmuir, 15(4), 1252.

Ram, MK, Sarkar, N, Bertoncello, P, Sarkar, A, Narizzano, R, and Nicolini, C Communicated Synth. Metal (2000).

Rama Krishna, MV, and Friesner, RA (1991), J. Chem. Phys., 95, 8309.

Ramanathan, K, Ram, MK, Malhotra, BD, Murthy, ASN (1995), J. Mater. Sci Engg. C3, 159.

Reed, MA, Randall, JN, Aggaewal, RJ, Matyi, RJ, et al. (1988), Phys. Rev. Lett., 60, 535.

Riul, A, Jr, Mattoso, LHC, Mello, SV, Telles, GD, and Oliveira, ON (1995), Synth. Metals, 71, 2060.

Roberts, G (1990), Langmuir–Blodgett Films, Plenum Press, New York

Röder, H., Hahn, E, Brune, H, Bucher, JP, Kern, K (1993), Nature, 366, 141.

Rubner, MF, and Skotheim, TA (1991), in Bredas, JL, and Silbey, R, eds., Conjugated Polymers, Kluwer, Amsterdam, pp. 363–403.

Rudzinski, WE, Lazano, L, and Walker, M (1990), 137, 3132.

Ruggiero, ST, and Barner, JB (1991), Cond. Maters, 85, 333.

Sauerbrey, GZ (1964), Z. Phys., 178, 457.

Sambrook, J, Fritsch, E, and Maniatis, T (1989), Molecular Cloning, Cold Spring Harbor Laboratory Press.

Scherr, EM, MacDiarmid, AG, Manohar, SK, Masters, JG, Sun, Y, Tang, S, Druy, MA, Glatwski, PJ, Cajipe VB, Fischer, JE, Cromack, KR, Jozefowicz, ME, Ginder, JM, MacCall, RP, and Epstein, AJ (1989), Synth. Met., 25, 649.

Schlereth, Daniela D (1999), J. Electroanalytical Chem, 464, 198.

Schmitt, J, Decher, G, Dressik, WJ, Brandow, SL, Geer, RE, Sashidhar, R, and Calvert, JM (1997), Adv. Mater, 9, 61.

Service, Robert F (1998), Science, 281, 941.

Shaik, Sason, Filatov, Michael, Schröder, Detlef, and Schwarz, Helmut (1998), Chem. Eur J., 4(2), 193.

Shen, Yi, Safinya, CR, Liang, KS, Ruppert, AF, and Rothshild, KJ (1993), Nature, 336, 48.

Shibata, A, Kohara, J, Ueno, S, Uchida, I, Mashimo, T, and Yamashita, T (1994), Thin Solid Films, 244, 736.

Shönenberger, C, van Houten, H, and Donkersloot, HC (1992a), Europhys. Lett, 20, 249.

Shönenberger, C, van Houten, H, Donkersloot, HC, van der Putten, HC, van der Putten, AMT, and Fokkink, LGJ (1992b), Physica Scripta, T45, 289–291.

Simonyan, VV, Diep, P, Johnson, JK, J. Chem. Phys., 1999, 111(21), 9778.

Smotkin, ES, Lee, C, Bard, AJ, Campion, A, Fox, MA, Mallouk, TE, Webber, SE, and White, JM (1988), Chem. Phys. Lett., 152, 265.

Stan, G, and Cole, MW (1997), J. Chem. Phys., 22, 9659–9667.

Stockton, WB, and Rubner, MF (1997), Macromolecules, 30(9), 2717.

Stone, AD, Jalabert, RA, and Alhassid, Y (1992), in Proceedings of the 14th Taniguchi Symposium, Springer-Verlag, Berlin, 39.

Storrs, M, Merhl, DJ, and Walkup, JF (1996), Applied Optics, 35, 4632–4636.

Su, H, Chong, S, and Thompson, M (1996), Langmuir, 12, 2247.

Sugiyama, Y., Inoue T, Ikematsu, M, Iseki M, and Sekiguchi, T (1997), Biochem, Biophys. Acta Biomembranes, 1326, 138.

Sukhorukov, G, Lobyshev, V, and Erokhin, V (1992), Mol. Mat., 1, 91.

Sukhorukov, G, Erokhin, V, and Tronin, A (1993), Biotechnologiya (Russian), 5, 103.

Sukhorukov, BI, Montrel, MM, Sukhorukov, GB, and Shabarchina, LI (1994) Biophysics, 39, 273–282.

Sumetskie, M (1993), Phys. Rev. B, 48, 4586.

Tian, Y, Wu, C, Kotov, N, and Fendler, JH (1994), Adv. Mater., 12, 959.

Tiede, D (1985), Biochim. Biophys. Acta, 811, 357.

Tronin, A, Dubrovsky, T, De Nitti, C, Gussoni, A, Erokhin, V, and Nicolini, C (1994), Thin Solid Films, 238, 127.

Tronin, A, Dubrovsky, T, and Nicolini, C (1995), Langmuir, 11, 385.

Turko, IV, Krivosheev, AV, and Chaschin, VL (1992), Biol. Mem, 9, 529.

Ulman, A (1991), An Introduction to Ultrathin Organic Films: From Langmuir–Blodgett to Self-Assembly, Academic Press: Boston.

Van Bentum, PJM, van Kempen, H, van de Leemput, LEC, and Teunissen, PAA (1988a), Phys. Rev. Lett., 60, 369.

Van Bentum, PJM, Smokers, RTM, and van Kempen, H (1988b), Phys. Rev. Lett., 60, 2543.

Wang, JP, Li, JR, Tao, PD, Li, XC, and Jiang, L (1994), Adv. Mater. Optics Electron., 4, 219–224.

Wang, ZH, Ray, A, MacDiarmid, AG, and Epstein, AJ (1991), Phys. Rev. B, 43, 4373.

Weiner, JS, Hess, HF, Robinson, RB, Hayes, TR, et al. (1991), Appl. Phys. Lett., 58, 2402.

Wilkins, R, Ben-Jacob, E, and Jaklevic, RC (1989), Phys. Rev. Lett., 63, 801.

Yang, LS, Shang, ZQ, Liu, YD (1991), J. Power Source, 34, 141.

Yang J, Meldrum, FC, and Fendler, JH (1995), J. Phys. Chem., 99, 5500.

Yasuda, Y, Sugino, H, and Toyotama, H (1993), ITE Technical Report, 17, 19.

Ye, Y, Ahn, C, Witham, C, Fultz, B, Liu, J, Rinzler, AG, Colbert, D, Smith, KA, and Smalley, RE (1999), Appl. Phys. Lett, 74(16), 2307.

Yevdokimov, YM (2000) 28, 77–81.

Zasadzinski, JA, Viswanathan, R, Madsen, L, Garnaes, JJ, and Schwartz, DK (1994). Science, 263, 1726.

Zeisel, D., and Hampp, N (1996), in Claudio Nicolini, ed., Molecular Manufacturing, EL.B.A. Forum Series, Vol. 2., p. 175.

Zeller, HR, and Giaever, I (1969), Phys. Rev., 181, 789.

5

Mono- and Multilayers of Spherical Polymer Particles Prepared by Langmuir–Blodgett and Self-Assembly Techniques

BERND TIEKE, KARL-ULRICH FULDA, and ACHIM KAMPES
University of Cologne, Cologne, Germany

I. INTRODUCTION

The tendency of monodisperse particles to form densely packed, ordered structures is a general phenomenon occuring in chemistry and biology, including various minerals, metal colloids, dendrimers, latex spheres, proteins, viruses, bacteria, and even gas bubbles [1–12]. Over recent decades, self-organization and controlled immobilization of mesoscale particles has attracted much attention in fundamental research as well as applied science. Fundamental research is mainly concerned with theoretical and experimental studies of particle aggregation and deposition at interfaces [13–27] and the use of microspheres as models for the study of crystallization [28], phase transitions [29–32], and fracture mechanics [33]. Recently, the physical properties and potential applications of ordered arrays of particles have become increasingly interesting. Two-dimensional (2D) periodic structures exhibit interesting optical properties [34–37] and might be suitable as optical elements, e.g., antireflective coatings [38,39], microlenses in imaging [40], physical masks for evaporation or reactive ion etching to fabricate regular arrays of micro- and nanostructures [41–47], and biocatalytic coatings [48]. Three-dimensional (3D) structures can be useful as templates to generate porous structures [49–51], precursors to produce high-strength ceramics [52,53], diffractive elements in fabricating sensors [54,55] or optical components such as gratings [56], filters [57,58], switches [59], and photonic crystals [60–64] operating in the visible region.

However, the application potential can be fully exploited only if suitable methods to control the structure and to prepare ordered particle arrays over macroscopic dimensions are available. Various methods have been tried to organize mesoscale particles in two or three dimensions. Among them are methods already being used for the organization of molecules at interfaces, while other methods were developed especially for the organization of particles:

1. Formation of Langmuir films of spherical particles at the air–liquid interface [21,22,31,43,65–84]
2. Transfer of Langmuir films of particles from the air–water interface onto solid supports, e.g., by the Langmuir–Blodgett (LB) technique [71–83]

3. Physisorption at interfaces, for example, electrostatic adsorption of charged particles at oppositely charged substrates [14,15,20,23–25,85–117]
4. Chemisorption of spherical particles at interfaces [118–120]
5. Electrophoretic deposition from colloidal solution [121–125]
6. Sedimentation from colloidal solution using ultracentrifuges [126–129]
7. Solvent evaporation from dispersion [130–138]
8. A so-called "dynamic thin laminar flow method" reported by Picard [139–141]
9. A "micromolding" technique reported by Xia and coworkers [142–145]
10. Other methods [146–149]

The number of publications on the organization of particles into 2D and 3D ordered structures has grown enormously in recent years. To keep this review concise, it is therefore restricted to spherical polymer particles, for example, latex particles, and their organization according to methods 1 to 4. Latex particles have several advantages over other particles: They are easily prepared in large quantity with reproducible quality, the size can be varied over almost the whole mesoscale range, very narrow size distributions can be obtained, the surface properties can be tailored, and various functional properties can be easily introduced. Methods 1 to 4 are advantageous over other methods because they were used for years to prepare organized molecular films, and thus an extensive knowledge on their feasibility already exists.

II. LANGMUIR FILMS OF SPHERICAL POLYMER PARTICLES

A. Film Preparation

Since the early studies of Pockels [150,151] and Langmuir [152,153] it has been known that the spreading of surfactant solutions at the air–water interface results in the formation of monomolecular layers. The monolayers—also called *Langmuir films*—are characterized by measuring their surface pressure–area (π-A) isotherms using a so-called Langmuir-type film balance. The preparation method can also be applied to spherical polymer particles and their organization in monolayers at the air–water interface. The important steps in the formation of a particle monolayer are outlined in Figure 1. A colloidal solution (particle dispersion, "latex") in ethanol (or methanol) is spread at the air–water interface, the alcohol is dissolved in the subphase, and a particle film is formed at the air–water interface. Upon subsequent film compression the surface pressure π is increased and the surface area A available for an individual particle, which is a measure of the packing density of the particle film, is decreased. Eventually, at a high value of π, a collapse of the monolayer is observed.

B. Historical Overview

The first study utilizing this method was reported by Schuller in 1966 [65]. Schuller used polystyrene latex beads that were spread on a salt-containing aqueous subphase in order to keep the particles at the interface. π-A plots of the floating particles were determined, which showed several phase regions with reproducible transition points. The author determined the particle diameters from the A-value, at which a steep rise in the isotherm occurred. Moreover, Schuller also spread millimeter-sized Styropor® particles and found isotherms of similar shape [66]. By taking pictures at different surface pressure, he was able to correlate the shape with different states of order in the monolayer. Shortly after that,

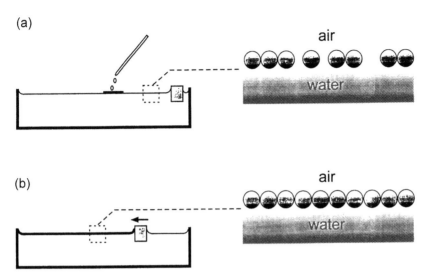

FIG. 1 Schematic showing the preparation of Langmuir films of latex particles at the air–water interface: (a) Spreading of the latex and formation of an expanded monolayer; (b) formation of the compressed monolayer.

Sheppard and Tcheurekdjian [67,68] used emulsifier-stabilized polystyrene particles 50–150 nm in diameter and spread these particles with the aid of organic liquids. π-A curves were measured using a Wilhelmy film balance. The curves showed a collapse pressure of only 17 mNm^{-1}, which was significantly lower than the one reported by Schuller, exceeding 50 mNm^{-1}. Equations were developed to calculate an average diameter of the spread particles from the projected area of an equal mass of monodisperse hard spheres of the same density and from the experimental limiting area A_0* obtained from the isotherm. The calculated diameters were in good agreement with those obtained by electron microscopic techniques. The monolayers were transferred onto solid supports using the LB technique† and viewed in an electron microscope. Tightly packed particle films one layer in thickness were found; small uncovered regions were also evident.

In subsequent years, only a few studies on Langmuir particle films appeared. Two-dimensional compression of latex particles at the heptane–water interface was studied [69], and in 1979 Garvey et al. [70] spread polystyrene latex particles stabilized with PVP at the air/2 molar aqueous sodium chloride interface. Due to the PVP coating, the particle films exhibited a large isotherm, with a slow rise in surface pressure beginning at very large A-values, which were poorly reproducible in subsequent compression/expansion cycles. Nevertheless, electron microscopy indicated a nearly close-packed film in the compressed state. Goodwin et al. [35] used 3.89-micrometer-sized polystyrene latex particles prepared by the seeded growth method reported by Chung-Li et al. [154]. Particulate monolayers were formed on a 0.5 molar aqueous NaCl solution and investigated using light scattering. Transmission of a well-collimated light beam led to diffraction patterns from which the particle size could be directly determined.

---

* For the definition of A_0 see Section II.C.

† For a description of the LB technique see Section III

In the 1980s, Kumaki [71,72] prepared monomolecular polymer particles on the water surface by spreading dilute solutions of high-molecular-weight polystyrene (Mw $> 10^6$) and studied the monolayers using a Langmuir film balance. Monolayers were also deposited on solid substrates and investigated in a transmission electron microscope (TEM). Since polystyrene does not contain any hydrophilic group, the surface pressure measured by the film balance was not due to the real decrease of the surface tension, but due to a mechanical force by compression. The isotherm showed a steep rise in surface pressure, with collapse at about 60 mNm^{-1}. At the limiting area A_0, the particles covered 56% of the water surface, but a close packing by further compression could not be reached. The macroscopic observation using the film balance was in good agreement with the TEM study, indicating that the particles were stable against compression.

In 1989, Armstrong et al. [31] spread sulfonated polystyrene microspheres at the air–water interface of a Langmuir trough. After equilibration for 1 hour or longer, the samples showed a uniform distribution of particles. Then the monolayers were compressed and re-expanded several times to study isothermal expansion melting in two dimensions. Unfortunately, no π-A curves were determined. Particles 2.88 μm in diameter showed evidence of defect-mediated melting and of an intermediate hexatic phase, while melting of the 1.01-μm-sized latex particles proceeded by a weak first-order transition. Robinson and Earnshaw [21] spread sulfonated polystyrene latex beads at the air–water interface to study colloidal aggregation in two dimensions. The authors did not compress the monolayer, but induced aggregation by adding a calcium chloride solution to the aqueous subphase, with the salt concentration being varied to change the growth conditions. Under all conditions, clusters were formed displaying statistical self-similarity. Video images at different stages of aggregation are shown in Figure 2. The measured values of the fractal dimension and cluster anisotropy were in good agreement with expectation from computer simulation. A similar study was reported by Stankiewicz et al. [22]. These authors found that the fractal dimensions increased as aggregation proceeded. Fractal dimensions were insensitive to the salt concentration of the subphase, but the aggregation rate was controlled by the salt concentration. Results for the kinetics of aggregation were in good agreement with recent scaling theory.

Yin et al. [73,74] prepared new microgel star amphiphiles and studied the compression behavior at the air–water interface. Particles were prepared in a two-step process. First, the gel core was synthesized by copolymerization of styrene and divinylbenzene in dioxane using benzoylperoxide as initiator. Microgel particles 20 nm in diameter were obtained. Second, the gel core was grafted with acrylic or methacrylic acid by free radical polymerization, resulting in amphiphilic polymer particles. These particles were spread from a dimethylformamide/chloroform (1:4) solution at the air–water interface. π-A curves indicated low compressibility above 10 mNm^{-1} and collapse pressures larger than 40 mNm^{-1}. With increase of the hydrophilic component, the molecular area of the polymer and the collapse pressure increased.

In 1994, Lenzmann et al. [43] reported thin-film micropatterning using polymeric microspheres. Polystyrene beads were spread at the air–water interface to form 2D regular arrays showing imperfections such as packing faults and fault lines between crystallites. The horizontal lifting technique (which the authors falsely denote as the Langmuir–Blodgett technique) was used to transfer the monolayer onto solid substrates. Subsequently, zinc sulphide was thermally evaporated in vacuum, and the spheres were dissolved away to leave behind a surface with peaks located where the interstitial spaces of the densely packed spheres had been.

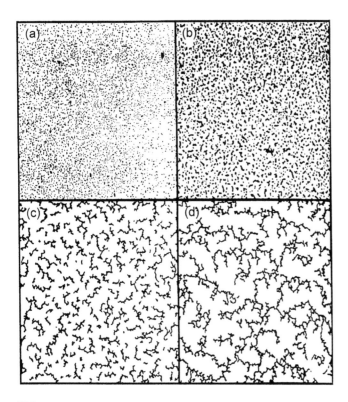

FIG. 2   Fractal growth of a layer of polystyrene particles ($D = 1.088 + 0.079$ mm) on 0.75 molar aqueous $CaCl_2$ solution after (a) 15, (b) 75, (c) 105, and (d) 135 min. (From Ref. 21, with permission from American Institute of Physics, Melville, NY.)

In the same year, Fulda and Tieke [75] reported on Langmuir films of monodisperse, 0.5-μm spherical polymer particles with hydrophobic polystyrene cores and hydrophilic shells containing polyacrylic acid or polyacrylamide. Measurement of π-A curves and scanning electron microscopy (SEM) were used to determine the structure of the monolayers. In subsequent work, Fulda et al. [76] studied a variety of particles with different hydrophilic shells for their ability to form Langmuir films. Fulda and Tieke [77] investigated the influence of subphase conditions (pH, ionic strength) on monolayer formation of cationic and anionic particles as well as the structure of films made from bidisperse mixtures of anionic latex particles.

In 1997, a Chinese research group [78] used the colloidal solution of 70-nm-sized carboxylated latex particles as a subphase and spread mixtures of cationic and other surfactants at the air–solution interface. If the pH was sufficiently low (1.5–3.0), the electrostatic interaction between the polar headgroups of the monolayer and the surface groups of the latex particles was strong enough to attract the latex to the surface. A fairly densely packed array of particles could be obtained if a 2:1 mixture of octadecylamine and stearic acid was spread at the interface. The particle films could be transferred onto solid substrates using the LB technique. The structure was studied using transmission electron microscopy.

In the following sections, the studies of Fulda and coworkers [75–77] are more extensively reviewed.

TABLE 1 Characteristic Properties of Core-Shell Latex Particles with Polystyrene Core

No.	Hydrophylic comonomer	Initiator	Particle diameter [nm]	Polydispersity	Ref.
1a	Acrylic acid	KPS	434 ± 8	1.001	77
1b	Acrylic acid	KPS	214 ± 12	1.004	77, 93
1c	Acrylic acid	KPS	440 ± 10	1.001	93, 98
2	2-Acryloxyethyl-trimethylammonium chloride	AIAP	200 ± 8	1.004	93
3	2,3-Epoxypropyl methacrylate	KPS	218 ± 12	1.009	92
4	2-Hydroxyethyl methacrylate	KPS	210 ± 24	1.019	92

C. Langmuir Films of Core-Shell Latex Particles

For the characterization of Langmuir films, Fulda and coworkers [75–77] used anionic and cationic core-shell particles prepared by emulsifier-free emulsion polymerization. These particles have several advantages over those used in early publications: First, the particles do not contain any stabilizer or emulsifier, which is eventually desorbed upon spreading and disturbs the formation of a particle monolayer at the air–water interface. Second, the preparation is a one-step process leading directly to monodisperse particles 0.2–0.5 μm in diameter. Third, the nature of the shell can be easily varied by using different hydrophilic comonomers. In Table 1, the particles and their characteristic properties are listed. Most of the studies were carried out using anionic particles with polystyrene as core material and polyacrylic acid in the shell.

The particles were spread from ethanolic dispersion. In Figure 3, a characteristic π-A curve of particles 1a is shown [75,77]. Film compression first leads to a gradual rise in

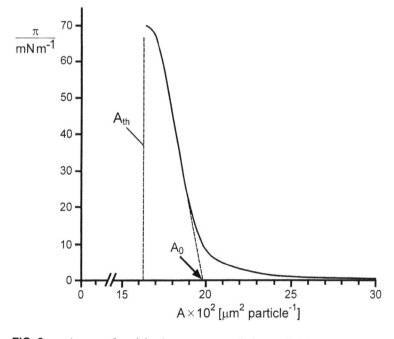

FIG. 3 π-A curve of particles 1a on aqueous subphase, pH 1.0, $T = 20°C$. (From Ref. 156.)

surface pressure π, followed by a steep rise until collapse occurs at about 65 mNm^{-1}. The curve resembles the one reported by Kumaki [72] for large polystyrene molecules, but it is very different from the one reported by Schuller [66], which already shows an increase of the π-value at an area A five to six times larger. In addition, Schuller's curve exhibits several transition steps, which are lacking in the isotherm of 1a. Probably these steps are due to the presence of surfactant molecules on the water surface, which previously desorbed from the particles. In the experiments of Fulda and coworkers, a floating barrier was used for detection of the π-value. The surface pressure thus obtained does not represent the thermodynamic surface pressure, defined as the difference between the surface tension of water without a particle layer, σ_0, and with a layer, σ_p; i.e.,

$$\pi = \sigma_0 - \sigma_p$$

but represents mainly the mechanical force from the moving barrier partially passed on to the floating barrier via the particle film. This may also explain the observation that during film compression a folding of the film already occurs at a π-value of only 30 mNm^{-1}, i.e., long before the measured collapse of 60 mNm^{-1} is reached. The actual collapse pressure is much lower and has been determined for particles 1a by Esker and Yu [155] using a Wilhelmy balance. As shown in Figure 4, the film of particles 1a collapses on aqueous subphase as soon as the π-value approaches 8 mNm^{-1}. The curve resembles the one measured by Sheppard and Tcheurekdjian [67,68].

Nevertheless, the shape of the isotherm in Figure 3 is quite similar to those of molecular films in their solid condensed state. This reflects the strong tendency of the particles to aggregate at the air–water interface. Visual inspection during spreading indicates the for-

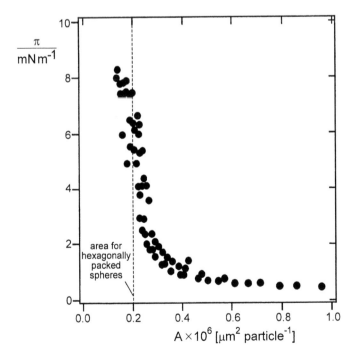

FIG. 4 π-A isotherm of carboxylated latex particles on aqueous subphase, π measured using a Wilhelmy balance. (Data taken from Ref. 155.)

220

Tieke et al.

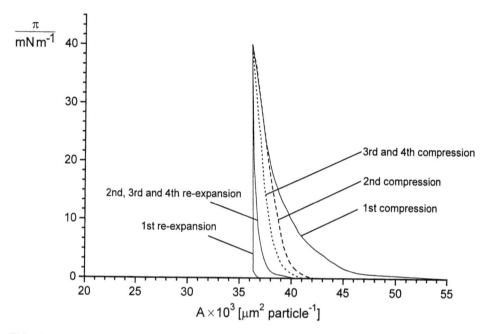

FIG. 5 Compression and re-expansion behavior of a Langmuir film of particles 2 in Table 1 on aqueous solution of KCl (conc. 1 mol L^{-1}), T = 20°C. (From Ref. 156.)

mation of large patches on the water surface 0.5–2 cm in diameter. They consist of 2D aggregated particles and are only pushed together without losing their shape upon the subsequent compression. Due to strong attractive interactions, particles even stick together if the surface pressure is released again. As shown in Figure 5, the π-value rapidly drops to zero when a compressed monolayer is re-expanded; i.e., the π-A curves are not reversible [156]. The order obtained after the first compression is frozen in and cannot be decisively changed anymore in subsequent expansion–compression cycles. Particle–particle interactions are so strong that even a rise of subphase temperature to 40°C has no decisive influence on the π-A curves.

Although the determination of the π-A curves is a macroscopic measurement, it allows one to draw some conclusions on the microscopic structure of the particle monolayer. Knowing the size of the particles, e.g., from direct observation by electron microscopy, and the number of spread particles, it is possible to calculate the theoretical area, A_{th}, a particle in hexagonal dense packing would occupy. A_{th} can be compared with the limiting area, A_0, determined from the π-A curve by extrapolation of the region of steep rise to zero surface pressure. The A_0/A_{th} ratio contains information about the structure of the particle layer at the interface. $A_0/A_{th} > 1$ indicates that the area available for an individual particle is larger than in hexagonal dense packing; i.e., free area ("holes") must exist in the monolayer. In the case of $A_0/A_{th} < 1$, no real monolayer can exist. Either the particles are aggregated into three-dimensional structures or they have partially disappeared into the subphase after spreading. As can be derived from Figure 3, the A_0/A_{th} ratio is about 1.2; i.e., one can assume true monolayer formation, with about 80% of the trough area being covered with particles [75,77].

Studying the π-A curves of various kinds of particles, we found that the A_0/A_{th} ratio is strongly influenced by the hydrophilicity of the particle shell. Nonionic particles 3 with only slightly hydrophilic oxirane groups in the shell tend to form 3D aggregates on the water surface, while particles 4 with more hydrophilic methylol groups in the shell form real monolayers [76]. Particles 1a with carboxylic acid units and particles 2 with alkylammonium groups in the shell either form real monolayers or disappear into the subphase, depending on the pH and ion concentration. This behavior is more extensively described in Section II.D.

D. Parameters Influencing Film Formation

The dependence of the monolayer formation on the pH and ion concentration of the subphase was studied in detail for particles 1a and 2 with anionic and cationic shells, respectively [77]. Let us first concentrate on particles 1a with ionizable carboxylic acid units. If the particles are spread on aqueous subphase of different pH and with no salt present, quite different A_0 values are obtained, depending on the pH (Fig. 6a). At pH < 2, the carboxylic acid units (and the sulphate groups at the chain ends originating from the persulphate initiator used for polymerization) are only slightly ionized and thus the particles are only moderately hydrophilic after spreading. At that pH, the particles stay on the surface and form a true monolayer. The A_0/A_{th} ratio is about 1.2; i.e., a surface coverage of about 80% is reached. At pH 2, the majority of the sulphate groups are already ionized (because the pK of $ROSO_3^-$ groups is 1.9), the particle surface is negatively charged, and the charged groups repel each other. Consequently, the particle shell is increased and a larger value of A_0 is obtained. At pH 3–4, a portion of the carboxylic acid units is ionized, which renders the particles very hydrophilic. Therefore, after spreading, some of the particles disappear into the subphase, and the value of A_0 is slightly decreased. Finally, for particles spread on a subphase of pH > 4, the A_0/A_{th} ratio is smaller than unity. This means that a considerable part of the particles disappears into the subphase instead of forming a monolayer. This is also apparent from the occurrence of turbid spots in the subphase where the colloidal solution was spread.

Salt addition to the subphase has a strong influence on monolayer formation, too. The effect of salt was studied by spreading particles 1a on an aqueous KCl solution of different salt concentration, with the pH of the subphase always being 5. If no salt is present at pH 5, the particles simply disappear into the subphase, as discussed earlier. However, the presence of salt causes the metal ions to penetrate the particle shell and shield the ionic groups electrostatically. Consequently, the particles become less hydrophilic and monolayer formation is improved, as indicated by the larger value of A_0. As shown in Figure 6a, a KCl concentration of 10^{-2} moles L^{-1} is sufficient to cause formation of a stable particle layer even at pH 5.

One may conclude that a particle monolayer is actually formed at each pH value of the subphase, only if care is taken that the sum concentration of protons and metal (e.g., potassium) ions in the subphase, $c_H^+ + c_K^+$, is constant. This is actually true, as shown in Figure 7, where the A_0-value of a monolayer of particles 1a is plotted against the pH of the subphase at constant sum concentration of H^+ and K^+ of 1 mol I^{-1} [77].

Monolayers of cationic particles 2 show an analogous dependence on the salt concentration of the subphase (Fig. 6b). If particles 2 are spread on a neutral subphase without any salt present, they mainly disappear into the subphase due to the large hydrophilicity of the shell. However, if KCl is added, electrostatic shielding of the alkylammonium groups by the chloride ions sets in, the hydrophilicity of the particle shell is diminished, and a stable monolayer is obtained. Different from particles 1a, the pH of the subphase has no direct

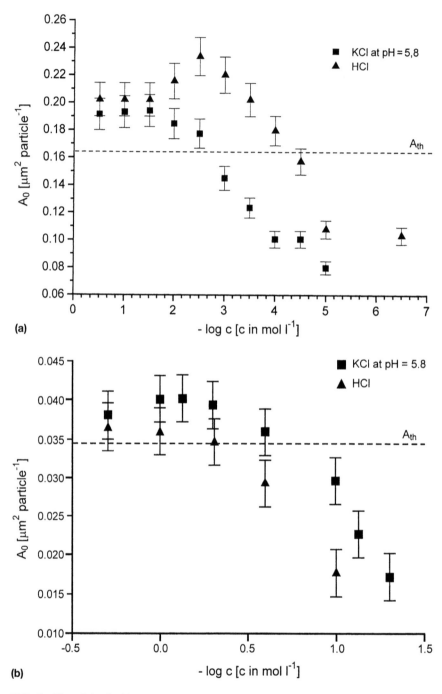

FIG. 6 Plot of the limiting area, A_0, of anionic particles (a) 1a and (b) 2 against the acid (HCl) or salt (KCl) concentration of the aqueous subphase.

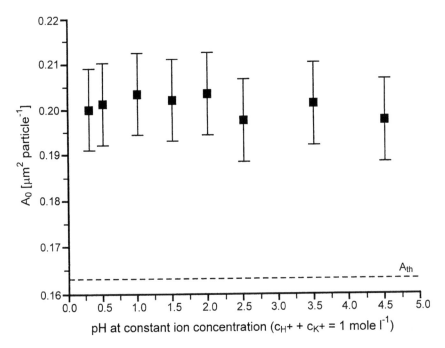

FIG. 7 Plot of the limiting area, A_0, of anionic particles 1a against the pH of the aqueous subphase, with the sum concentration of protons and potassium ions in the subphase being kept constant at 1 mol L^{-1}.

effect on the film formation of 2. The reason is that the positively charged alkylammonium groups are not able to release or take up protons. However, with the addition of an acid, anions are also introduced into the subphase, which have the same effect on monolayer formation as the salt addition discussed earlier. As a result, the limiting area A_0 of particles 2 depends on the HCl and KCl concentration of the subphase in analogy to anionic particles 1a (see Fig. 6b) [77].

The penetration of ions from the subphase into the shell of spread particles is a general phenomenon and can be used to modify and functionalize the particle surface. For example, metal ions, such as Ba^{2+} and Fe^{3+}, or cationic polyelectrolytes, such as the polycation of polyallylamine, can be adsorbed at anionic particles, while anionic water-soluble dyes, such as phthalocyanine tetrasulfonic acid and 1.4-diketo-3.6-diphenylpyrrolo[3.4-c]pyrrole-4′,4″-disulfonic acid (DPPS) [157], can be adsorbed at cationic particles. However, since only a monolayer of the dye is adsorbed, a deep coloration of the particles is not obtained unless a dye with very high absorption coefficient is used [156].

III. LANGMUIR–BLODGETT MONOLAYERS OF LATEX PARTICLES

In order to study the structure of Langmuir films of polymers spheres, most researchers deposited the films on solid substrates using the LB technique [158–162] and analyzed the structure using a microscope. A modified version of the LB method allowing the transfer of particle monolayers is outlined in Figure 8a.

(a)   (b)

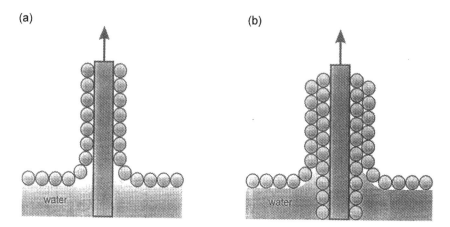

FIG. 8   Schematic showing the preparation of LB particle (a) monolayer and (b) bilayer.

First, a glass substrate is dipped into the water subphase. Then the colloidal solution is spread and the resulting particle film compressed. Finally, the substrate is vertically withdrawn from the subphase and, thereby, the particle layer transferred. Typical scanning electron micrographs of resulting films of particles 1a are shown in Figure 9 [156]. Areas of hexagonal dense packing can be seen consisting of up to about 30 particles. Between these areas small holes are apparent, which on average do not extend to twice the particle diameter. On a macroscopic scale, the particle layers appear rather homogeneous, especially because particles in a second layer are only rarely observed. The structure seen in the SEM pictures is in good agreement with the structure derived from the macroscopic π-A curves discussed earlier.

Fulda and Tieke [77] studied the effect of a bidisperse-size distribution of latex particles on the structure of the resulting LB monolayer. For this purpose, a mixed colloidal solution of particles 1a and 1b was spread at the air–water interface. Particles 1a had a diameter of 434 nm, particles 1b of 214 nm. The monolayer was compressed, transferred onto a solid substrate, and viewed in a scanning electron microscope (SEM). In Figure 10, SEM pictures of LB layers obtained from various bidisperse mixtures are shown.

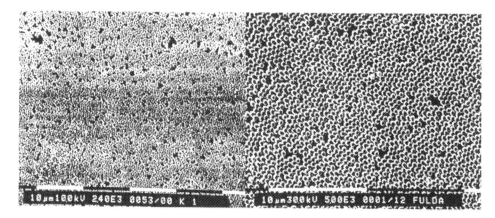

FIG. 9   Typical SEM pictures of LB monolayers of particles 1a on a glass support at different magnifications. (From Ref. 156.)

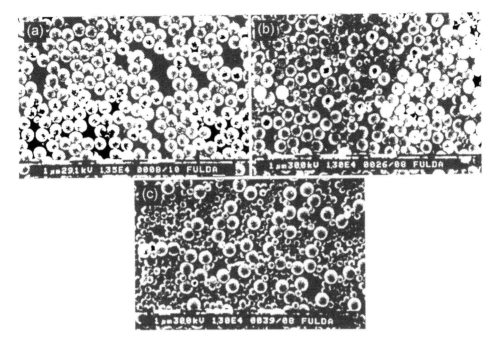

FIG. 10   SEM pictures of LB particle layers of bidisperse mixtures of 1a and 1b, with number fraction x_{1b} being (a) 0.2, (b) 0.5, and (c) 0.9. (From Ref. 156.)

If only a few small particles are present, the large ones form monolayers of their own, disturbed only very little by the small ones. The small particles are preferentially located in the cavities of the densely packed areas of the large ones (Fig. 10a). Consequently, the small particles do not contribute to the total film area and the A_0 value of the mixed monolayer is much smaller than the value calculated according to equation

$$A_{0,\text{calc}} = x_{1a} A_{0,1a} + x_{1b} A_{0,1b}$$

where $A_{0,1a}$ and $A_{0,1b}$ are the limiting areas of corresponding monolayers of particles 1a and 1b, respectively, and x_{1a}, x_{1b} are the number fractions of 1a and 1b in the mixture. If only a few large particles are present, the small ones form a layer of their own, which is disturbed only a little by the large ones. The large particles are located on top of the layer of the small ones (Fig. 10c). Now only the small particles contribute to the total film area, and thus the A_0-value of the mixed film is again smaller than $A_{0,\text{calc}}$ obtained from the equation just given. If the number fractions of small and large particles approach 0.5, the monolayers exhibit a highly disordered structure. Neither the small nor the large particles are able to form an ordered structure of their own, and instead the formation of 3D aggregates is favored (Fig. 10b). Again, the experimental A_0-value is much smaller than $A_{0,\text{calc}}$.

LB monolayers of latex particles are excellently suited for a study of the coalescence behavior of the spheres upon heating [156]. Particles with polystyrene in the core, such as 1a, exhibit a glass transition temperature T_g of $110 \pm 2°C$, which agrees with the glass transition temperature of atactic polystyrene. In Figure 11, SEM pictures are shown of LB monolayers of particles 1a after various heat treatments. While prolonged heating at temperatures below T_g does not induce any noticeable changes, annealing at T_g induces some deformation, with the adhesion to the substrate being strongly enhanced. At 120°C, particles move together and are partly fused. Band like structures are formed with a large free area in between. At 130°C, the spherical shape completely disappears, the polymer begins to flow, and the free area is decreased. Above 130°C, the viscosity of the polymer is so low that a continuous film is formed (not shown).

As discussed in Section II.D, the spreading of latex particles on a subphase containing suitable ionic compounds leads to adsorption of these ions on the particle shell so that the surface properties can be easily modified. Using this method, Fulda [156] was able to prepare catalytically active LB monolayers. Cationic particles 2 were spread on an aqueous subphase containing disodium tetrachloropalladate in 10^{-2} molar concentration, so the $PdCl_4^{2-}$ ions were immediately adsorbed at the particle shell. After LB transfer onto a solid substrate, a palladium-containing particle layer was obtained. If such a layer was subsequently immersed in a plating bath for electrodeless nickel deposition, the $PdCl_4^{2-}$ ions were rapidly reduced to Pd^0, which then was able to catalyze electrodeless nickel plating of the particles according to

$$Ni^{2+} + 2\,H_2PO_2 + 2\,H_2O \xrightarrow{Pd(0)} Ni\downarrow + H_2\uparrow + 2\,H_3PO_3$$

In Figure 12, LB monolayers of particles 2 on a glass substrate are shown after immersion in a nickel-plating bath for 15 and 40 sec. The metal coating of the particles is clearly recognizable.

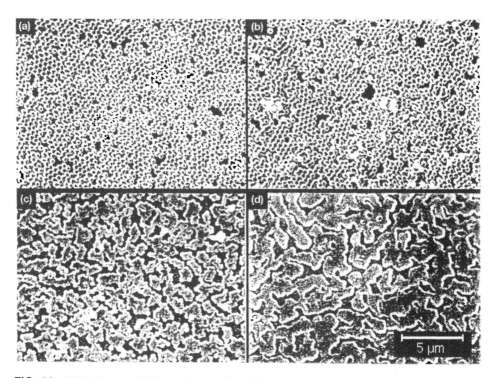

FIG. 11 SEM pictures of LB monolayer of particles 1a on a glass support, taken after heat treatment (a) below 110°C, (b) at 110°C, (c) at 120°C, and (d) at 130°C for 1h, respectively. (From Ref. 156.)

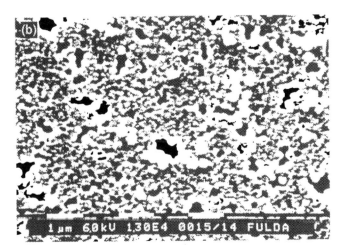

FIG. 12 SEM pictures of LB monolayers of particles 2 on a glass support, deposited from an aqueous solution containing disodium tetrachloropalladate (conc. 10^{-2} mol L^{-1}) and subsequently immersed in a plating bath for electrodeless nickel deposition for (a) 15 and (b) 40 sec. Typical SEM pictures of particle monolayers not subjected to nickel coating are shown in Figure 9. (From Ref. 156.)

IV. LANGMUIR–BLODGETT MULTILAYERS OF LATEX PARTICLES

From surfactant molecules it is known that the repeated vertical dipping of a substrate through a floating monolayer of these molecules leads to the formation of an LB multilayer on the substrate. In principle, the same procedure should also allow the preparation of multilayers of latex particles. In Figure 8b, the preparation of a particle bilayer is schematically indicated; multiple repetition should result in the formation of an LB multilayer of particles. However, if one tries to realize this concept, one immediately gets into difficulties, because the contact of the particles with the underlying substrate is very poor, and the already deposited particle layer tends to detach from the surface when the substrate is dipped into

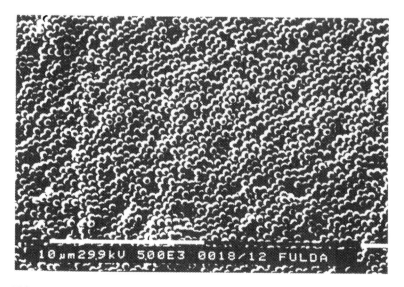

FIG. 13   SEM picture of an LB multilayer of particles 1a after transfer of five monolayers onto a glass substrate. (From Ref. 156.)

the subphase once again. As recently demonstrated [77,156], the problem can be solved by a short annealing of the first deposited layer slightly below the glass temperature of about 110°C. The particles soften and increase their contact area so that the adhesion is improved, although the spherical shape is mostly preserved (see also Fig. 11). Another method to improve adhesion is to use a highly charged substrate whose charge is opposite to that of the particle shell. Substrates of this kind can be prepared, for example, by fourfold alternating electrostatic adsorption of cationic and anionic polyelectrolyte layers on a glass plate modified with 3-aminopropylmethyl-dimethoxysilane (3-AMDS). Further details on this method are described in the Section V.

The morphology of an LB multilayer of particles 1a prepared upon fivefold dipping of the substrate is shown in the SEM picture of Figure 13. The surface is very rough and inhomogeneous. Small holes can be detected, where no particles are present, as well as regions, where more than five particles are deposited on top of each other. Small areas with dense packing are only occasionally recognizable. It appears that packing faults of individual layers accumulate in the multilayer. Any attempts to subsequently improve the regularity of the three-dimensional particle film, e.g., by annealing, failed.

V.   SELF-ASSEMBLED FILMS OF LATEX PARTICLES

In addition to the preparation of Langmuir and Langmuir–Blodgett films, the use of self-assembly techniques also plays an important role in the formation of particle films. Both physisorption, as, for example, electrostatic adsorption of charged particles from colloidal solution, and chemisorption onto a substrate have been investigated. In Section V.A, electrostatic adsorption will be reviewed; chemisorption is the subject of Section V.B.

A.   Electrostatic Adsorption

The most widely used model of adsorption is Langmuir's equation for reversible molecular adsorption [163]. However, this is inappropriate for charged latex particles, because these particles are electrostatically adsorbed and this is an irreversible process. Theoretical work, therefore, dates back to the work of Feder and Giaever [14], who argued that random sequential adsorption (RSA) validly describes the irreversible adsorption of proteins and colloidal particles on solid surfaces. In RSA, particles are brought individually to random points on the surface, where they are irreversibly placed, provided there is no overlap with any previously deposited particle. Computer simulations indicate that for monodisperse hard spheres this procedure should lead to a maximum surface coverage C_{max} of 54.7%, the so-called jamming limit or random-parking limit in two dimensions. Onoda and Liniger [19] were the first to present an analytical solution for this limit using latex particles, which were adsorbed at a substrate covered with a cationic polyelectrolyte film. The particles had a diameter of 2.95 μm and thus were large enough to study the adsorption using an optical microscope. An experimental value of the random-parking limit of 55 $\pm$ 1% was found, which agreed with the computer-generated values.

However, in subsequent studies [23–25,88–90] it was demonstrated that in reality the particle deposition is not a purely geometric effect, and the maximum surface coverage depends on several parameters, such as transport of particles to the surface, external forces, particle–surface and particle–particle interactions such as repulsive electrostatic forces [25], polydispersity of the particles [89], and ionic strength of the colloidal solution [23,88,90]. Using different kinds of particles and substrates, values of the maximum surface coverage varied by as much as a factor of 10 between the different studies.

While a good deal of recent experimental work focuses on the question of whether the observed surface coverage is in agreement with prediction by the RSA model, the target of other work [92–98] was the preparation of particle films with regular structure in two or three dimensions using electrostatic self-assembly. These studies usually refer to the work of Iler published in 1966 [85]. Iler developed a new technique by which alternate layers of positively and negatively charged colloidal particles, such as silica, alumina, and polystyrene latex, could be deposited from sols onto a smooth surface such as glass. By this means, films of controlled, uniform thickness could be built up showing interference colors. The films consisted of alternating layers of spherical particles about 100 nm in diameter and nanosized compounds such as polyvalent ions, boehmite fibrils, or water-soluble polymers. However, microscopic observations on the structure of the deposited films were not reported. In Figure 14, a modified version of Iler's method is outlined showing the exclusive adsorption of oppositely charged particles. In step A, a positively charged support (e.g., glass, mica, quartz) is dipped into a colloidal aqueous solution of anionic particles so that a monolayer is adsorbed under reversal of the surface charge. Then the substrate is withdrawn from the solution, carefully washed, and dried (step B). In step C, it is immersed in a colloidal solution of cationic particles, which are also adsorbed, forming a second layer, again under reversal of the surface charge. Finally the substrate is withdrawn, washed, and dried (step D). In order to adsorb further particle layers, steps A to D have to be repeated several times.

Fulda and coworkers [92,93], Bliznyuk and Tsukruk [94], and Serizawa and coworkers [95–97] were the first who tried to use this technique for the preparation of ordered bi- and multilayer assemblies of oppositely charged latex particles. The structure was investigated using scanning force microscopy (SFM) and SEM. In a further publication, Kampes and Tieke [98] studied the influence of the preparation conditions on the state of order of the monolayers. In the following section, the recent studies are more extensively reviewed.

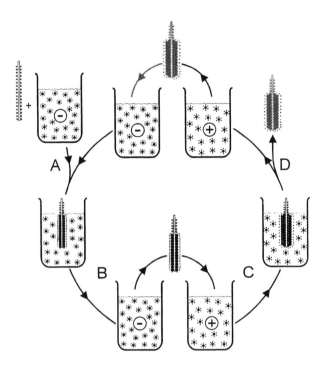

FIG. 14 Schematic showing the electrostatic self-assembly of a particle monolayer (A, B) and bi-layer (C, D). The substrate exhibits positive surface charge onto which the negatively charged parti-cles are adsorbed. Adsorption of the negatively charged particles reverts the surface charge, and pos-itively charged particles can now be adsorbed. Repetition of steps A to D allows the preparation of particle multilayers.

1. Monolayers

In order to study electrostatic adsorption of latex particles, Fulda and coworkers [93] and Kampes and Tieke [98] used smooth glass substrates pretreated with 3-AMDS and HCl, as indicated in Fig. 15. After the pretreatment, positively charged substrates were obtained, which were dipped into a colloidal solution of anionic latex particles so that these particles were adsorbed at the substrate surface. After withdrawing, washing, and drying, the coated substrates were investigated using SEM. The structure of the adsorbed layer was charac-terized by evaluating the normalized surface coverage C_s, defined as

$$C_s [\%] = (N/N_{max}) \times 100\%$$

with N being the number of particles per unit area a^2 and N_{max} being the maximum number of particles on the same area, assuming a hexagonal dense packing. N_{max} is defined as

$$N_{max} = \frac{2}{\sqrt{3}} \left(\frac{a}{d}\right)^2 - \frac{1}{2}\frac{a}{d}$$

with d being the average particle diameter.

In Figure 16, SEM pictures of substrates dipped into a colloidal solution of particles 1c for different time periods are shown [93,98]. After 10 seconds, a number of particles are adsorbed. Most of the particles are present in a nonaggregated state, indicating that single-particle adsorption is the dominating process. With dipping time, the coverage of the surface increases and the adsorbed particles form clusters of increasing size. As already pointed out by Onoda and Liniger [19], the clustering most likely results from drying before the sample is viewed in the SEM. During drying, the remaining water forms capillary bridges between the particles and provides strong enough forces to overcome electrostatic repulsion and draw the nearby particles together. After about 3 minutes, a constant C_s value of 30% is reached. The experiment was repeated for different latex concentrations, and the surface coverage was determined as a function of the dipping time [98]. As shown in the plot of Figure 17, the concentration influences only the rate of adsorption, whereas the maximum surface coverage is always about 30%.

The effect of temperature on adsorption was also studied. At low temperature (i.e., near 0°C) adsorption proceeds very slowly, while at high temperature (40°C) the surface coverage varies strongly from sample to sample, indicating side effects such as partial desorption. Final values of C_s always approached 30%, independent of the temperature. The effects of concentration and temperature on the surface coverage are clearly not compatible with a description of the adsorption behavior by Langmuir's equation.

In a further set of experiments, the pH of the latex dispersion was varied [98]. This had a strong influence on the particle adsorption, as shown in the SEM pictures of Figure 18. When the pH was lower, the surface coverage was higher. For example, at pH 3.6 a maximum C_s value of 65% was reached. The reason is that the charge densities of the substrate surface and of the particle shell are influenced by the pH. At low pH the carboxylate groups are protonated and electrostatic repulsion between the particles is reduced in favor of hydrophilic attractive interactions such as hydrogen bonding; i.e., aggregates are formed and a higher surface coverage is obtained. Low pH values also favor a higher charge density of the substrate, because all amino groups of the silane compound are protonated. This also favors the adsorption.

The SEM studies just presented indicate that particle films prepared upon electrostatic adsorption show only a short-range local ordering and a relatively low substrate coverage of 30–65%, depending on the charge density of the particle shell and the substrate surface. Bliznyuk and Tsukruk [94] came to a similar conclusion, although in their experiments the maximum surface coverage was higher. They used commercial monodisperse

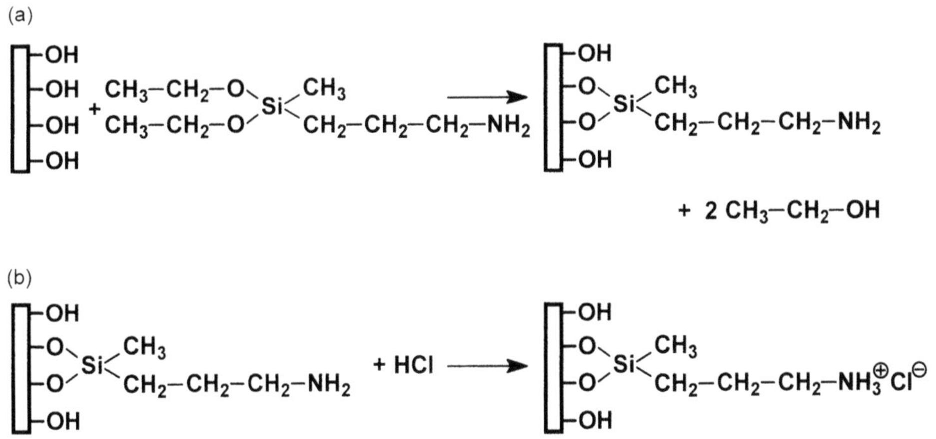

FIG. 15 Scheme of preparation of a positively charged glass substrate upon successive treatment with (a) 3-aminopropylmethyldiethoxysilane (3-AMDS) and (b) aqueous HCl.

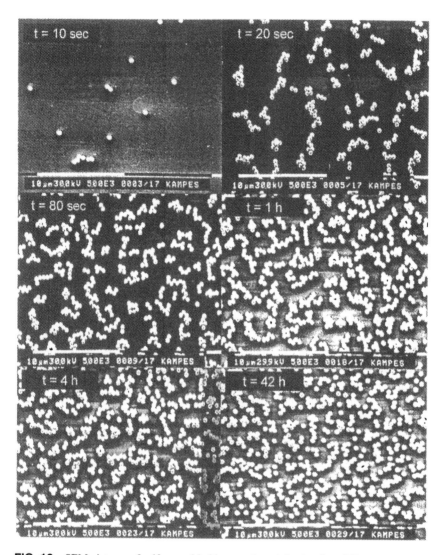

FIG. 16 SEM pictures of self-assembled layers of particles 1c after different dipping times t (concentration 3 mg/mL; T = 23.5°C; pH = 5.8). (From Ref. 98, with permission from Elsevier, Amsterdam.)

polystyrene latex with carboxyl, sulfate, or amidine surface groups 20–200 nm in diameter. With 20-nm-sized carboxylated particles, a surface coverage up to 80% could be reached on a glass substrate modified with 3-AMDS. The adsorption time had no influence on film quality for the time interval from 5 min to several hours, which is in agreement with observations made by others [98].

Serizawa and Akashi [95] analyzed the monolayer adsorption of polystyrene latex particles with cationic polyvinylamine grafted on their surface, while Serizawa et al. [96,97] used commercial anionic latex particles. Both types of particles were adsorbed on polyelectrolyte-coated substrates previously prepared by alternating adsorption of cationic and anionic polyelectrolytes such as polyallylamine hydrochloride (PAH) and polystyrene sulfonate sodium salt (PSS) according to the method described by Decher [164]. Using

SEM and a quartz crystal microbalance (QCM) as analytical tools, they found that the nanospheres adsorbed independently, in agreement with other studies [19,88], and that the time dependence of adsorption could be fitted by a Langmuir isotherm, which is in disagreement with considerations [23] and observations made by others [14,88]. The surface coverage varied between 10 and 60%, depending on the charge density of the substrate. They also studied the effect of particle size on the adsorption behavior and found the highest surface coverage, about 70%, for the largest particles, 780 nm in diameter [97].

Johnson and Lenhoff [88] used SFM for an in situ study of the localized adsorption of charged latex particles onto a mica substrate with varying adsorption times and solution ionic strengths. The initial kinetics of the adsorption process was found to be diffusion limited, and the long-term asymptotic kinetics were found to resemble those of an RSA process. The latex particles were very uniformly distributed over the substrate when the surface coverage was near saturation, and radial distribution functions indicated a pronounced short-range order at high surface coverage. A similar study was reported by Semmler et al. [90], who studied the adsorption of positively charged latex particles on a negatively charged mica surface at low ionic strength, again using SFM. Commercial polystyrene spheres with amidine head groups 40–200 nm in diameter were used. For low surface coverage, the deposition kinetics from a quiescent dispersion were found to be in good agreement with diffusion-limited adsorption. At long times, the surface coverage tended towards a maximum value of about 50%, independent of the particle concentration but decreasing with a decrease of the ionic strength. To explain this trend, a model was presented based on RSA and the effect of overlapping double layers. They also showed that the particle size polydispersity can modify the maximum surface coverage considerably. Antelmi and Spalla [99] studied the adsorption of 25-nm-sized carboxylated latex spheres at mineral surfaces of variable charge density, such as sapphire (α-Al$_2$O$_3$) and cerium oxide (CeO$_2$), by direct imaging and force measurement. The surface coverage was highest for latex dis-

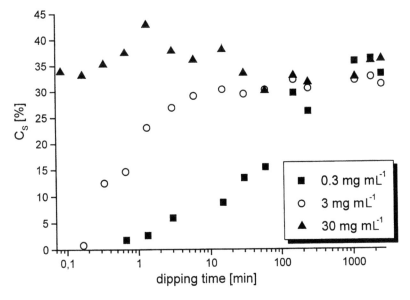

FIG. 17 Plot of normalized surface coverage C_s of self-assembled layer of particles 1c vs. dipping time of substrate for different concentrations of the latex dispersion ($T = 23.5°C$, pH = 5.8).

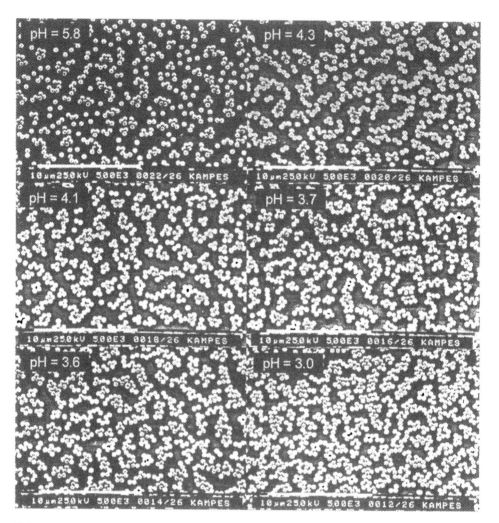

FIG. 18 SEM pictures of self-assembled layers of particles 1c prepared from latex dispersions of different pH value (substrate: glass support modified with 3-AMDS, dipping time 1 h, latex concentration 3 mg/mL; T = 23.5°C). (From Ref. 98, with permission from Elsevier, Amsterdam.)

persions at low pH, in agreement with previous studies. Adsorption to CeO_2 was consistent with classical electrostatic mechanisms, with the driving force originating from the opposite charge carried by the latex spheres and substrate over the pH range studied and no adsorption occuring above the point of zero charge (PZC) of the substrate. In contrast, significant adsorption was seen above the PZC of the sapphire substrate, where both the particles and the substrate were negatively charged. This behavior was explained by a different mechanism of adsorption originating from a chemical interaction.

2. Multilayers

Some research groups also tried to build up bi- and multilayers of latex particles by using self-assembly techniques [92–94,97]. Either the alternate adsorption technique outlined in Figure 14 was used, in which cationic and anionic particles are successively adsorbed, or a slightly modified version of successive adsorption of anionic particles and a cationic polyelectrolyte such as PAH was applied. Fulda [156] used the first approach and monitored successive adsorption of particles 1b and 2 by measuring the light transmittance of the sample after various deposition steps (Fig. 19). Generally, transmittance decreases with the number of deposition steps, indicating continuous adsorption of latex spheres. However, after the fourth, sixth, and eighth steps, respectively, a slight increase in transmittance was found, indicating particle reorganization and aggregation took place.

Additional SEM pictures taken after various deposition steps show the evolution of a completely disordered, highly amorphous three-dimensional surface coverage (Fig. 20) [93].

Since the diameters of both kinds of particles are about 200 nm, the thickness of a 10-layer assembly should be about 1.8 μm. However, the cross section of a sample subjected to 10-fold dipping indicates a thickness of only 1.2 μm (Fig. 21). Similar observations were made by others [94,97] and can be ascribed to the low substrate coverage reached in each of the dipping steps. Serizawa et al. [97] used the second approach and determined the adsorption using QCM and SEM. An SEM image after two adsorption steps indicated a very irregular, amorphous structure, in agreement with Fulda and coworkers [93].

B. Chemisorption

Adsorption of latex spheres onto surfaces does not lead to a permanent surface modification unless particles are able to form strongly adhering films upon attachment. There are reports in the literature on the covalent immobilization of latex spheres on surfaces of polyethylene, quartz, and glass. Margel et al. [118] prepared polyacrolein microspheres, which were covalently bonded to modified polyethylene surfaces carrying amine groups. Binding between spheres and substrate was accomplished by reaction of the aldehyde units

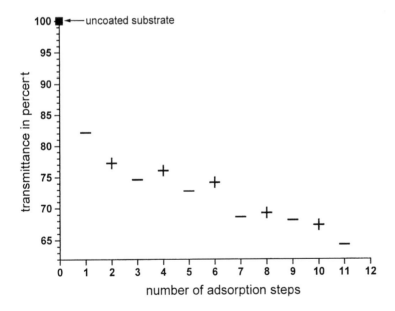

FIG. 19 Light transmittance (λ = 650 mm) of alternating multilayer of particles 1b and 2 vs. number of deposited particle layers. (+) or (−) indicate that positively or negatively charged particles, respectively, were deposited in the last adsorption step (substrate: glass modified with 3-AMDS and subsequently acidified). (From Ref. 156.)

with the amine groups of the surface to form polyvalent Schiff base bonds. The immobilization was investigated using attenuated total reflection Fourier transform–infrared spectroscopy, electron spectroscopy for chemical analysis, contact angle measurements, and SEM. The SEM images indicate single-particle adsorption and incomplete surface average, as similarly found upon electrostatic adsorption. Slomkowski and coworkers [119,120] used the same immobilization reaction to bind poly(styrene-acrolein) and poly(pyrrole-acrolein) particles to a quartz surface modified with 3-aminopropyltriethoxysilane. The SFM pictures indicate the formation of monolayers with significant interparticle attraction during deposition. A maximum surface coverage of about 60% was found. The surface-bonded latex particles were suitable for immobilization of antibodies, with retention of

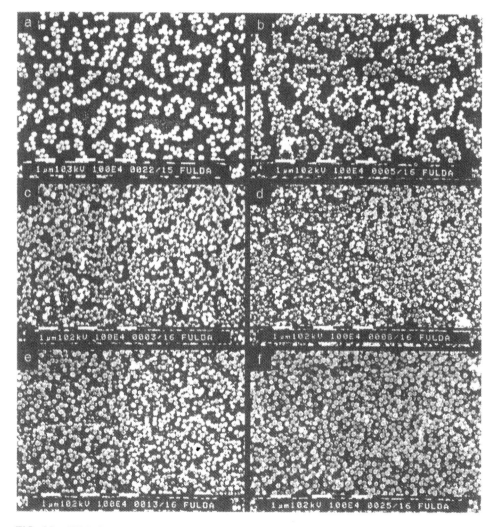

FIG. 20 SEM pictures of particle assemblies after different numbers of alternating dipping steps into dispersions of anionic particles 1b and cationic particles 2 beginning with 1b. A glass substrate modified with 3-AMDS was used and dipped with sequence (a) 1b only, (b) 1b-2, (c) 1b-2-1b, (d) (1b-2) × 2, (e) (1b-2) × 3, (f) (1b-2) × 5. (From Ref. 93, with permission from Elsevier, Amsterdam.)

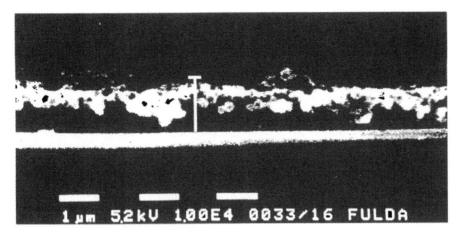

FIG. 21 Cross section of sample subjected to 10-fold alternating dipping into dispersions of 1b (5×) and 2 (5×). The bar indicates the average particle film thickness of 1.2 μm. (From Ref. 93, with permission from Elsevier, Amsterdam.)

their ability to bind antigens. The authors believe that the quartz elements covered with latex particles can be used as elements of diagnostic devices and supports for immobilization of other proteins (e.g., enzymes).

VI. CONCLUSIONS

Our review indicates that Langmuir–Blodgett and self-assembly techniques, which are suitable for preparation of organized molecular films, cannot yet be successfully applied to prepare particle films with long-range 2D and 3D order. Large problems arise from the poor control of particle–particle and particle–substrate interactions, which are responsible for aggregation and deposition at interfaces. Problems also arise from the particles themselves used by the various research groups, which differ in the surface properties and, occasionally, even contain stabilizers or emulsifiers at the surface. Problems also arise from poor knowledge on the parameters influencing the properties of particles and substrate. As a consequence, inconsistent results are obtained that render a general understanding of particle aggregation and deposition behavior at interfaces difficult.

Nevertheless, recent studies on the spreading of latex dispersions at the air–water interface at least indicate that particle films of reproducible quality can be obtained if core-shell particles prepared by emulsifier-free emulsion polymerization are used and if the properties of subphase and spreading dispersion are carefully controlled. Analyzing the spreading behavior of a variety of cationic and anionic latex particles, a strong influence of hydrophilicity and charge density of the particle shell on formation of monolayers became evident. Only particles with moderate charge density are well suited. However, good conditions for monolayer formation can be adjusted by salt addition or change of the pH of the subphase. Particles with anionic shells for example, form stable monolayers independent of the pH of the subphase if the sum concentration of protons and metal ions is kept constant. This clearly disproves the earlier opinion that a particle monolayer can be prepared only on a salt-containing subphase with higher mass density than the particles. Unfortunately, particles with moderate charge density exhibit strong attractive particle–particle interactions, such as hydrogen bonding, which dominate over electrostatic repulsion. Therefore the monolayers exhibit a frozen-in mutual arrangement of the particles, which in subsequent compression/expansion cycles cannot be broken up anymore so that a macroscopic dense packing cannot be obtained. The tendency of spread particles to rapidly adsorb ions from the subphase onto the shell can be used to modify and functionalize the particle films subsequent to their preparation and to introduce new properties such as color, biological, or catalytic activity. This is clearly an advantage of the Langmuir films.

Electrostatic self-assembly has been used to prepare 2D and 3D arrays of latex particles. Recent research on adsorption of monolayers is strongly influenced by the RSA model and the description of additional parameters influencing the adsorption. For preparation of 3D arrays, either alternate electrostatic adsorption of cationic and anionic particles or cationic polyelectrolytes and anionic particles has been applied. Since the first layer is already highly irregularly adsorbed, it does not represent a packing pattern for subsequent layers, and consequently a 3D film with only local order of the particles is obtained. In fact, completely amorphous-particle films are not easily accessible by other methods, and the highly porous latex films with large surface area might be useful for applications as separation layers, filters, or supports for catalysts. However, in view of the highly amorphous structure, any of the suggested applications as optical components are irrelevant. In fact, neither the LB technique nor the self-assembly methods presently allow one to prepare the highly ordered films of latex particles required for such applications. Quite obviously, more sophisticated techniques are needed for the preparation of organized layers with long-range order of the particles.

REFERENCES

1. A Kumar, NL Abbott, E Kim, H Biebuyck, GM Whitesides. Acc Chem Res 28:219–226, 1995.
2. E Matijevic. Langmuir 10:8–16, 1997.
3. MP Pileni. Langmuir 13:3266–3276, 1997.
4. TA Alfrey, EB Bradford, JW Vanderhoff, G Oster. J Opt Soc Am 44:603, 1954.
5. W Luck, M Klier, H Wesslau. Naturwissenschaften 50(14):485–494, 1963.
6. CD Dushkin, GS Lazarov, SN Kolsev, H Yoshimura, K Nagayama. Colloid Polym Sci 277:914–930, 1999.
7. T Okubo. Langmuir 10:1695–1702, 1994.
8. PN Pusey, W van Megen. Nature 320:340–342, 1986.
9. R Markham, KM Smith, RWG Wyckoff. Nature 159:574, 1947.
10. UB Sleytr, P Messner. J Bacteriol 170:2891–2897, 1988.
11. P Messner, D Pum, M Sára, KO Stetter, UB Sleytr. J Bacteriol 166:1046–1054 1986.
12. L Bragg, JF Nye. Proc R Soc Lond Ser A 190:474–481, 1947.
13. P Pieranski. Phys Rev Lett 45:569–572, 1980.
14. J Feder, I Giaever. J Colloid Interf Sci 78:144–154, 1980.
15. T Dabros, TGM van de Ven. Colloid Polym Sci 261:694–707, 1983.
16. P Pieranski, L Strzelecki, B Pansa. Phys Rev Lett 50:900–903, 1983.
17. AJ Hurd, DW Schaefer. Phys Rev Lett 54:1043–1046, 1985.
18. GY Onoda. Phys Rev Lett 55:226–229, 1985.
19. GY Onoda, EG Liniger. Phys Rev A 33:715–716, 1986.
20. Z Adamczyk, M Zembala, B Siwek, J Czarnecki. J Colloid Interf Sci 110:188–200, 1986.
21. DJ Robinson, JC Earnshaw. Phys Rev A 46:2045–2054, 2055–2064, 1992.
22. J Stankiewicz, MA Cabrerizo Vilchez, R Hidalgo Alvarez. Phys Rev E 47:2663–2668, 1993.
23. PR Johnson, M Elimelech. Langmuir 11:801–812, 1995.

24. M Elimelech, J Gregory, X Jia, RA Williams. Particle Deposition and Aggregation: Measuring, Modeling and Simulation. Oxford: Butterworth-Heinemann Ltd., 1995.
25. Z Adamczyk, P Warszynski. Adv Colloid Interf Sci 63:41, 1996.
26. M Trznadel, S Slomkowski. Colloid Polym Sci 274:1109–1118, 1996.
27. MR Böhmer, EA van der Zeeuw, G JM Koper. J Colloid Interf Sci 197:242–250 1998.
28. CA Murray, DG Grier. Am Sci 83:238–245, 1995.
29. DH van Winkle, CA Murray. Phys Rev A 34:562–573, 1986.
30. AE Larsen, DG Grier. Phys Rev Lett 76:3862–3865, 1996.
31. AJ Armstrong, RC Mockler, WJ O'Sullivan. J Phys Condens Matter 1:1707–1730, 1989.
32. AE Larsen, DG Grier. Nature 285:231–234, 1997.
33. AT Skjeltorp, P Meakin. Nature 335:424–426, 1988.
34. W Luck, M Klier, H Wesslau. Ber Bunsenges Phys Chem 67:75–85, 1963.
35. JW Goodwin, RH Ottewill, A Dorentich. J Phys Chem 84:1580–1586, 1980.
36. EA van de Zeeuw, GJM Koper, D Bedeaux. Progr Colloid Polym Sci 98:291–294, 1995.
37. AS Dimitrov, T Miwa, K Nagayama. Langmuir 15:5257–5264, 1999.
38. BE Yoldas, DP Partlow. Appl Opt 23:1418–1424, 1984.
39. P Hinz, H Dislich. J Non-Cryst Solids 82:411–416, 1986.
40. H Hayashi, Y Kumamoto, T Suzuki, T Hirai. J Colloid Interf. Sci 144:538–547, 1991.
41. CB Roxlo, HW Deckman, J Gland, SD Cameron, R Chianelli. Science 235:1629–1631, 1987.
42. MC Buncick, RJ Warmack, TL Ferrell. J Opt Soc Am B 4:927–933, 1987.
43. F Lenzmann, K Li, AH Kitai, HD Stöver. Chem Mater 6:156–159, 1994.
44. F Burmeister, C Schäfle, T Matthes, M Böhmisch. Langmuir 13:2983–2987, 1997.
45. J Boneberg, F Burmeister, C Schäfle, P Leiderer, D Reim, A Fery, S Herminghaus. Langmuir 13:7080–7084, 1997.
46. C Padeste, S Kossek, HW Lehmann, CR Musil, J Gobrecht, LJ Tiefenaur. Electrochem Soc 143:3890–3895, 1997.
47. F Burmeister, C Schäfle, B Keilhofer, C Bechinger, J Bonehary, P Leiderer. Adv Mater 10:495–497, 1998.
48. Z Huang, VS Thiagarajan, OK Lyngberg, LE Scriven, MC Flickinger. J Colloid Interf Sci 215:226–243, 1999.
49. OD Velev, TA Jede, RF Lobo, AM Lenhoff. Nature 389:447–448, 1997.
50. BT Holland, CF Blanford, A Stein. Science 281:538–540, 1998.
51. SH Park, Y Xia. Adv Mater 10:1045–1048, 1998.
52. MD Sacks, T Tseng. Y J Am Ceram Soc 67:526–532, 1984.
53. P Calvert. Nature 317:201, 1985.
54. JH Holtz, SA Asher. Nature 389:829–832, 1997.
55. JH Holtz, JSW Holtz, CH Munro, SA Asher. Anal Chem 70:780–791, 1998.
56. JM Weissman, HB Sunkara, AS Tse, SA Asher. Science 274:959–960, 1996.
57. PL Flaugh, SE O'Donnell, SA Asher. Appl Spectrosc 38:847–850, 1984.
58. RJ Spry, DJ Kosan. Appl Spectrosc 40:782–784, 1986.
59. SY Chang, L Liu, SA Asher. J Am Chem Soc 116:6739–6744, 1994.
60. II Tarhan, GH Watson. Phys Rev Lett 76:315–318, 1996.
61. WL Vos, M Megens, C M van Tats, J Bösecke. Phys Condens Mater 8:9503–9507, 1996.
62. H Miguez, F Meseguer, C López, A Blanco, JS Moya, J Reqena, A Milsud, V Fornés. Adv Mater 10:480–483, 1998.
63. YA Vlasov, N Yao, DJ Norris. Adv Mater 11:165–169, 1999.
64. SH Park, B Gates, Y. Xia. Adv Mater 11:462–469, 1999.
65. H Schuller. Kolloid-Z u Z Polymere 211:113–121, 1966.
66. H Schuller. Kolloid-Z u Z Polymere 216-7:380–383, 1967.
67. E Sheppard, N Tcheurekdjian. Kolloid-Z u Z Polymere 225:162–163, 1967.
68. E Sheppard, N Tcheurekdjian. J Colloid Interf Sci 28:481–486, 1968.
69. A Doroszkowski, R Lambourne. J Polymer Sci C 34:253, 1971.

70. MJ Garvey, D Mitchell, AL Smith. Colloid Polym Sci 257:70–74, 1979.
71. J Kumaki. Macromolecules 19:2258–2263, 1986.
72. J Kumaki. Macromolecules 21:749–755, 1988.
73. R Yin, X Cha, X Zhang, J Shen. Macromolecules 23:5158–5160, 1990.
74. J Shen, X Zhang, X Cha, R Yin. Makromol Chem Macromol Symp 46:157–161, 1991.
75. KU Fulda, B Tieke. Adv. Mater 6:288–290, 1994.
76. KU Fulda, D Piecha, B Tieke, H Yarmohammadipour. Progr Colloid Polym Sci 101:178–183, 1996.
77. KU Fulda, B Tieke. Supramolec Sci 4:265–273, 1997.
78. H Du, YB Bai, Zhui, LS Li, YM Chen, XY Tang, TJ Li. Langmuir 13:2538–2540, 1997.
79. NA Kotov, FC Meldrum, C Wu, JH Fendler. J Phys Chem 98:2735–2738, 1994.
80. FC Meldrum, NA Kotov, JH Fendler. J Phys Chem 98:4506–4510, 1994.
81. M Kondo, K Shinozaki, L Bergström, N Mizutani. Langmuir 11:394–397, 1995.
82. SA Iakovenko, AS Trifonov, M Giersig, A Mamedov, DK Nagesha, VV Hanin, EC Soldatov, NA Kotov. Adv Mater 11:388–392, 1999.
83. Z Horvölgyi, S Nemeth, JH Fendler. Langmuir 12:997–1004, 1996.
84. M Yanagida, T Kuri, T Kajiyama. Chem Lett 1997:911–912.
85. RJ Iler. J Colloid Interf Sci 21:569–594, 1966.
86. Z Adamczyk, B Siwek, L Szyk. J Colloid Interf Sci 174:130–141, 1995.
87. JM Meinders, HJ Busscher. Langmuir 11:327–333, 1995.
88. CA Johnson, AM Lenhoff. J Colloid Interf Sci 179:587–599, 1996.
89. Z Adamczyk, B Siwek, M Zembala, P Weronski. J Colloid Interf Sci 185:236–244, 1997.
90. M Semmler, EK Mann, J Ri˘cka, M. Borkovec. Langmuir 14:5127–5132, 1998.
91. P Carl, P Schaaf, JC Voegel, JF Stoltz, Z Adamczyk, B Senger. Langmuir 14:7267–7270, 1998.
92. KU Fulda, D Piecha, B Tieke, H Yarmohammadipour. Prog Colloid Polym Sci 101:178–183, 1996.
93. KU Fulda, A Kampes, L Krasemann, B. Tieke. Thin Solid Films 327–329:752–757, 1998.
94. VN Bliznyuk, VV Tsukruk. ACS Polymer Prepr 38(1):963–964, 1997.
95. T Serizawa, M A Kashi. Chem Lett 1997:809–810.
96. T Serizawa, H Takeshita, M Akashi. Chem Lett 1998:487–488.
97. T Serizawa, H Takeshita, M Akashi. Langmuir 14:4088–4094, 1998.
98. A Kampes, B Tieke. Mater Sci Eng C 8–9:199–208, 1999.
99. DA Antelmi, O Spalla. Langmuir 15:7478–7489, 1999.
100. M Gao, M Gao, X Zhang, Y Yang, B Yang, J Shen. Chem Commun 1994:2777–2778.
101. Y Lvov, H Haas, G Decher, H Möhwald, A Mikhailov, B Mtchedlishvily, E Morgunova, B Vainshtein. Langmuir 10:4232–4236, 1994.
102. Y Lvov, K Ariga, T Kunitake. Chem Lett 1994:2323–2326.
103. W Kong, X Zhang, M Lai Gao, H Zhou, W Li, JC Shen. Macromol Rapid Commun 15:405–409, 1994.
104. Y Lvov, K Ariga, I Ichinose, T Kunitake. Chem Commun 1995:2313–2314.
105. Y Lvov, K Ariga, I Ichinose, T Kunitake. J Am Chem Soc 117:6117–6123, 1995.
106. S Peschel, G Schmid. Angew Chem 107:1568–1569, 1995.
107. M Gao, X Zhang, B Yang, F Li, J Shen. Thin Solid Films 284–285:242–245, 1996.
108. Y Sun, E Hao, X Zhang, B Yang, M Gao, J Shen. Chem Commun 1996:2381–2382.
109. Y Lvov, K Ariga, M Onda, I Ichinose, T Kunitake. Langmuir 13:6195–6203, 1997.
110. J Schmitt, G Decher, WJ Dressick, SL Brandow, RE Geer, R Shashidhar, JM Calvert. Adv Mater 9:61–65, 1997.
111. Y Sun, E Hao, X Zhang, B Yang, J Shen, L Chi, H Fuchs. Langmuir 13:5168–5174, 1997.
112. T Yonezawa, S Onoue, T Kunitake. Adv Mater 10:414–416, 1998.
113. A Rosidian, Y Liu, RO Claus. Adv Mater 10:1087–1091, 1998.
114. F Caruso, DN Furlong, K Ariga, I Ichinose, T Kunitake. Langmuir 14:4559–4565, 1998.

115. T Yonezawa, H Matsune, T Kunitake. Chem Mater 11:33–35, 1999.
116. FG Aliev, MA Correa-Duarte, A Madedoe, JW Ostrander M Giersig, LM Liz-Marzan, NA Kotov. Adv Mater 11:1006–1010, 1999.
117. X Zhang, J Shen. Adv Mater 11:1139–1143, 1999.
118. S Margel, E Cohen, Y Dolitzky, O Sivan. J Polym Sci Pt A: Polym Chem 30:1103–1110, 1992.
119. S Slomkowski, D Kowalczyk, M Trzadel. Trends Polym Sci 3:297–304, 1995.
120. S Slomkowski, B Miksa, M Trznadel, D Kowalczyk, FW Wang. ACS Polymer Prepr 37(2):747–748, 1996.
121. M Giersig, P Mulvaney. Langmuir 9:3408–3413, 1993.
122. M Böhmer. Langmuir 12:5747–5750, 1996.
123. M Trau, DA Saville, IA Aksay. Science 272:706–709, 1996.
124. SR Yeh, M Seul, BI Shraiman. Nature 386:57–59, 1997.
125. Y Solomentsev, M Böhmer, JL Anderson. Langmuir 13:6058–6068, 1997.
126. R Mayoral, J Requena, JS Moya, C López, A Cintas, H Miguez, F Meseguer, L Vàzquez, M Holgado, A Blanco. Adv Mater 9:257–260, 1997.
127. LN Donselaar, AP Philipse, J Suurmond. Langmuir 13:6018–6025, 1997.
128. H Miguez, F Meseguer, C López, A Mifsud, JS Moya, L Vàzquez. Langmuir 13:6009–6011, 1997.
129. A van Blaaderen, R Ruel, P Wiltzins. Nature 385:321–323, 1997.
130. ND Denkov, OD Velev, PA Kralchevsky, IB Ivanov, H Yaoshimura, K Nagayama. Langmuir 8:3183–3190, 1992.
131. CD Dushkin, K Nagayama, T Miwa, PA Kralchevsky. Langmuir 9:3695–3701, 1993.
132. GS Lazarov, ND Denkov, OD Velev, PA Kralchevsky, K Nagayama. J Chem Soc Faraday Trans 90:2077–2083, 1994.
133. M Yamaki, J Higo, K Nagayama. Langmuir 11:2975–2978, 1995.
134. R Micheletto, H Fukuda, M Ohtsu. Langmuir 11:3333–3336, 1995.
135. AS Dimitrov, K Nagayama. Langmuir 12:1303–1311, 1996.
136. S Rakers, LF Chi, H Fuchs. Langmuir 13:7121–7124, 1997.
137. KP Velikov, F Durst, OD Velev. Langmuir 14:1148–1155, 1998.
138. CD Dushkin, GS Lazarov, SN Kotsev, H Yoshimura, K Nagayama. Colloid Polym Sci 277:914–929, 1999.
139. G Picard, I Nevernov, D Allicata, L Pazdernik. Langmuir 13:264–276, 1997.
140. G Picard. Langmuir 13:3226–3234, 1997.
141. G Picard. Langmuir 14:3710–3715, 1998.
142. E Kim, Y Xia, GM Whitesides. Adv Mater 8:245–247, 1996.
143. SH Park, D Qin, Y Xia. Adv Mater 10:1028–1032, 1998.
144. SH Park, B Gates, Y Xia. Adv Mater 11:462–466, 466–469, 1999.
145. SH Park, Y Xia. Langmuir 15:266–273, 1999.
146. MM Burns, JM Fournier, JA Golovchenko. Science 249:749–754, 1990.
147. CA Mirkin, RL Letzinger, RC Mucic, JJ Storkoff. Nature 382:607–609, 1996.
148. AP Alivisatos, KP Johnsson, X Peng, TE Wilson, CS Loweth, MP Bruchez, PG Schultz. Nature 382:609–611, 1996.
149. S Neser, C Bechinger, P Leiderer, T Palberg. Phys Rev Lett 79:2348–2351, 1997.
150. A Pockels. Nature 43:437–439, 1891.
151. A Pockels. Nature 50:223–224, 1894.
152. I Langmuir. J Am Chem Soc 39:1848–1906, 1917.
153. I Langmuir. Trans Faraday Soc 15:62–74, 1920.
154. Y Chung-Li, JW Goodwin, RH Ottewill. Prog Colloid Polym Sci 60:163–175, 1976.
155. A Esker, H Yu. University of Madison, Wisconsin, 1995, unpublished result.
156. KU Fulda. PhD dissertation, University of Cologne, 1998.
157. F Saremi, G Lange, B Tieke. Adv Mater 8:923–926, 1996.

158. KB Blodgett. J Am Chem Soc 57:1007–1022, 1935.
159. GL Gaines. Insoluble Monolayers at Liquid–Gas Interfaces. New York: Interscience, 1966.
160. GG Roberts. Langmuir Blodgett Films. New York: Plenum Press, 1990.
161. A Ulman. Ultrathin Organic Films. Boston: Academic Press, 1991.
162. RH Tredgold. Order in Thin Organic Films. New York: Cambridge University Press, 1994.
163. I Langmuir. J Am Chem Soc 40:1361–1403, 1918.
164. G Decher. Science 277:1232–1237, 1997.

6

Studies of Wetting and Capillary Phenomena at Nanometer Scale with Scanning Polarization Force Microscopy

LEI XU and MIQUEL SALMERON Lawrence Berkeley National Laboratory, Berkeley, California

I. INTRODUCTION

Wetting phenomena are related to the contact between liquids and solids. Capillary phenomena are related to the curvature of surfaces. Both kinds of phenomena originate from the existence of a surface energy or surface tension. This is the energy needed to create a unit area of surface or interface. Wetting and capillary phenomena are extremely common in nature and have been studied for many years. They are important in our daily lives, in science, and in technology. Many industrial operations require the spreading of a liquid on a solid, for example, coating and painting, plant protection, gluing, oil recovery from porous rocks, lubrication, and liquid crystal displays. The properties of the interfaces determine to a large extent the way they interact with the environment. Surface forces arise from the loss of bonding from the missing half of the substrate and can be felt at distances determined by the range of the atomic or molecular interaction potentials. These distances can be several hundred angstroms. It is clear then that surface tension is not the same for very thin liquid films and their bulk counterparts.

It is well known that when liquid droplets form on a flat substrate they adopt spherical cap shapes (neglecting gravity effects) with a contact angle θ. This angle depends solely on the interfacial energies as described by the Young's equation:

$$\gamma_{LV} \cos\theta = \gamma_{SV} - \gamma_{SL} \tag{1}$$

where L, V, and S denote liquid, vapor, and solid, respectively. Usually γ_{LV} is written without indices, simply as γ. The wetting properties of a surface by a given liquid are described by means of the so-called spreading coefficient, defined as:

$$S = \gamma_{SV} - \gamma_{SL} - \gamma \tag{2}$$

For a nonvolatile liquid, where transport through the gas phase is negligible, one could substitute the first term in the equation, γ_{SV}, with γ_{SO}, which corresponds to the interface energy between the dry solid and vacuum. If $S > 0$, the liquid will wet the surface; if $S < 0$, a finite contact angle will exist, determined by Eq. (1).

If the atomic or molecular interaction potentials are known, the surface energy for very thin films can be calculated by assuming additivity and integrating over the finite thickness of the film below the surface plus that of the semi-infinite substrate underneath.

The value so obtained will differ from that of the bulk liquid, γ, by a thickness-dependent term $P(e)$. For a nonvolatile liquid film, the free energy F (per unit area) will then be expressed as [1]:

$$F(e) = \gamma_{SL} + \gamma + P(e) \tag{3}$$

A quantity called disjoining pressure Π was introduced by Derjaguin [2,3] as the natural canonical conjugate of the film thickness, e:

$$\Pi(e) = \frac{-\partial F}{\partial e} = \frac{-\partial P}{\partial e}$$

or, equivalently,

$$P(e) = \int_e^\infty \Pi(x)\,dx \tag{4}$$

The disjoining pressure characterizes the wetting properties at short ranges. S and Π are related by:

$$S = \int_0^\infty \Pi(x)\,dx = P(e \to 0) \tag{5}$$

In a real liquid, the nature of the disjoining pressure can be a complicated function of the distance, due to the simultaneous contribution of several types of forces [1,4–6]. For two bodies with flat surfaces separated by a distance z, the van der Waals interaction, which varies as $-z^{-6}$ (or $-z^{-7}$, if one considers retardation) for single atoms and molecules, gives rise to a power law of the form:

$$P(z) = \begin{cases} \dfrac{-A}{12\pi z^2} & \text{nonretarded} \\[2mm] \dfrac{-B}{3z^3} & \text{retarded} \end{cases} \tag{6}$$

where A and B are constants.

The constant A is called the Hamaker constant. As mentioned already, these laws can be deduced assuming additivity and integrating over all the atoms and molecules of the interacting bodies and the interposed medium. A more exact deduction can be made using the dielectric model, as in the Lifshitz theory [3]. Although the van der Waals interaction is always attractive between individual atoms in a vacuum, in a medium it can give rise to either attractive or repulsive forces. This is reflected in the sign of A.

Although van der Waals forces are present in every system, they dominate the disjoining pressure in only a few simple cases, such as interactions of nonpolar and inert atoms and molecules. It is common for surfaces to be charged, particularly when exposed to water or a liquid with a high dielectric constant, due to the dissociation of surface ionic groups or adsorption of ions from solution. In these cases, repulsive double-layer forces originating from electrostatic and entropic interactions may dominate the disjoining pressure. These forces decay exponentially [5,6]:

$$\Pi(z) = C \exp(-2\kappa_D z) \qquad (\kappa_D z > 1) \tag{7}$$

where κ_D^{-1} is the screening length of ions in the solution or Debye length.

There are other forces that come into play when the thickness of the liquid film is in the nanometer range and the size of the molecule is no longer negligible. Short-range oscillatory forces arise then because the liquid molecules feel the presence of the walls of the substrate and are forced to form a layered structure near the interface. These forces are also called structural or solvation forces [6].

Water is a special liquid that forms unique bonds involving protons between the oxygen atoms of neighboring molecules, the so-called hydrogen bond. The solvation forces are then due not simply to molecular size effects, but also and most importantly to the directional nature of the bond. They can be attractive or hydrophobic (hydration forces between two hydrophobic surfaces) and repulsive or hydrophilic (between two hydrophilic surfaces). These forces arise from the disruption or modification of the hydrogen-bonding network of water by the surfaces. These forces are also found to decay exponentially with distance [6].

Other forces can arise as a result of elastic strain on the growing film, which can be due to a surface-induced ordering in the first few layers that reverts to the bulk liquid structure at larger distances. This elastic energy is stored in intermolecular distances and orientations that are stretched or compressed from the bulk values by the influence of the substrate at short distances [7]. Similar phenomena are well known to occur in the growth of epitaxial layers in metals and semiconductors.

If S and the disjoining pressure are known, the general wetting behavior of a nonvolatile liquid on a surface may be quantitatively described [8]. Brochard-Wyart et al. have summarized the various cases when van der Waals interactions dominate the long-range forces (1). Generally, if the long-range disjoining pressure P decays monotonously, depending on the sign of S, three cases of wetting are found:

1. Complete wetting if $S > 0$, $P > 0$. In this case, droplets spread to a uniform film or pancake.
2. Pseudo-partial wetting if $S > 0$ and $P < 0$. A macroscopic droplet with a finite contact angle can form on top of a flat film, while a microscopic droplet will spread to form a pancake or a diluted gas.
3. Partial wetting if $S < 0$. The droplets make a finite contact angle.

The shape of a droplet or of the front end of a film can be determined from the surface energies and interaction forces between the interfaces. These also determine the equilibrium thickness of a liquid film that completely wets a surface. The calculation is done by minimization of the free energy of the total system. In a two-dimensional case the free energy of a cylindrical droplet can be expressed as [5]:

$$F(z) = F_0 + \int dx \left[-S + \frac{\gamma}{2} \left(\frac{dz}{dx} \right)^2 + P(z) + g(z) \right] \qquad (8)$$

where the second term comes from the expansion of the length element, assuming that dz/dx is small, and $g(z)$ describes the gravitational effects. Supersaturation effects can be introduced by adding a term of the form $\Delta\mu z$, where $\Delta\mu$ is the difference in chemical potentials of the liquid and vapor phases. For nonvolatile liquids and for a fixed volume of the droplets, minimization of $F(z)$ leads to an important first integral (neglecting gravity effects):

$$\frac{\gamma}{2} \left(\frac{dz}{dx} \right)^2 = P(z) - \lambda z - S \qquad (9)$$

where λ is a Lagrange multiplier. As an example, for a small contact angle, if the disjoining pressure is due to the nonretarded van der Waals forces, solving Eq. (9) gives a hyperbolic profile of the droplets near the contact line [1,5].

In many cases the potential $P(z)$ is small compared with the surface energy of the liquid and the droplet shape is very close to a spherical cap. If the height e and the radius of curvature R at the top of the droplets can be measured, an effective contact angle can be defined through the expression:

$$\theta^2 = \frac{2e}{R} \tag{10}$$

From Eq. (9), the following relation between the effective contact angle and the disjoining pressure under constant volume can be obtained:

$$\theta^2 = \theta_0^2 + \frac{2}{\gamma} \cdot [P(e) + e \cdot \Pi(e)] \tag{11}$$

where θ_0 is the macroscopic contact angle. It is clear that the effective contact angle is influenced by the disjoining pressure. It can be shown that a similar expression is valid when the droplets are very close to shallow spherical caps [9].

Capillary phenomena are due to the curvature of liquid surfaces. To maintain a curved surface, a force is needed. In Eq. (8), this force is related to the second term in the integral. The so-called Laplace pressure due to the force to maintain the curved surface can be expressed as, $P_L = 2\gamma/r$, where r is the curvature radius. The combination of the capillary effects and disjoining pressure can make a liquid film climb a wall.

II. EXPERIMENTAL STUDIES OF WETTING PHENOMENA

Many experimental methods have been developed to study wetting phenomena. With an optical microscope the contact angle of droplets larger than a few micrometers can be measured and the spreading of liquids can be followed. However, to study the effects of the disjoining pressure, one needs films and droplets of submicrometer dimensions and, therefore, instruments with nanometer resolution. Several techniques possessing such resolution in the direction perpendicular to the surface exist today and are used to study the effects of the disjoining pressure. We will describe them briefly in the following.

The surface forces apparatus (SFA) can measure the interaction forces between two surfaces through a liquid [10,11]. The SFA consists of two curved, molecularly smooth mica surfaces made from sheets with a thickness of a few micrometers. These sheets are glued to quartz cylindrical lenses (~10-mm radius of curvature) and mounted with their axes perpendicular to each other. The distance is measured by a Fabry–Perot optical technique using multiple beam interference fringes. The distance resolution is 1–2 Å and the force sensitivity is about 10 nN. With the SFA many fundamental interactions between surfaces in aqueous solutions and nonaqueous liquids have been identified and quantified. These include the van der Waals and electrostatic double-layer forces, oscillatory forces, repulsive hydration forces, attractive hydrophobic forces, steric interactions involving polymeric systems, and capillary and adhesion forces. Although cleaved mica is the most commonly used substrate material in the SFA, it can also be coated with thin films of materials with different chemical and physical properties [12].

Infrared (IR) spectroscopy and ellipsometry are used to measure the thickness of thin films with angstrom resolution. Ellipsometry has proven to be very useful in studies of dynamic processes of spreading and of the shape of pancakes [13–16]. Gravitational forces and capillary forces can be used as control parameters for wetting liquids to measure disjoining pressure [3,17]. For volatile liquids, the vapor pressure can be used to measure the disjoining pressure through adsorption isotherm techniques [18]. Steplike adsorption isotherms [19] and layered spreading [16] have been observed. The thickness of these layers is comparable to the molecular diameter, indicating that the disjoining pressure is oscillatory [16].

In recent years, high-resolution x-ray diffraction has become a powerful method for studying layered structures, films, interfaces, and surfaces. X-ray reflectivity involves the measurement of the angular dependence of the intensity of the x-ray beam reflected by planar interfaces. If there are multiple interfaces, interference between the reflected x-rays at the interfaces produces a series of minima and maxima, which allow determination of the thickness of the film. More detailed information about the film can be obtained by fitting the reflectivity curve to a model of the electron density profile. Usually, x-ray reflectivity scans are performed with a synchrotron light source. As with ellipsometry, x-ray reflectivity provides good vertical resolution [14,20] but poor lateral resolution, which is limited by the size of the probing beam, usually several tens of micrometers.

Understanding wetting at smaller lateral scales is important in many applications, such as in corrosion, which occurs at localized regions of the surface. As a result, there is a need for techniques that allow the study of liquid droplets or films with nanometer resolution. A less fundamental but very practical advantage of high lateral resolution is that it opens the way for studies of materials that cannot be prepared as very flat samples over large areas. This need has recently been fulfilled thanks to the explosive development of scanning probe microscopy (SPM) techniques, in particular, the atomic forces microscope (AFM). The AFM has been used to study oscillatory hydration forces between tips and $CaCO_3$ and $BaSO_4$ substrates in water [21]. Tapping-mode AFM has also been used in the imaging of liquid droplets and films [22–25]. In these AFM studies, however, the tip enters in contact with the surface, which makes it difficult to avoid a strong perturbation of the liquid by the imaging tip.

So while imaging of solids or rigid substrates has been revolutionized by SPM, the study of liquids had to avoid contact. Scanning interferometric apertureless microscopy (SIAM), developed by Wickramasinghe et al. at IBM, has been used to image liquid droplets on a transparent substrate [26]. SIAM uses a vibrating scanning probe tip in close proximity (noncontact) to a sample surface. The variations of the scattered electric field caused by the tip and the local optical properties of the sample are measured by interferometry. In their original paper the authors demonstrated the imaging of small oil droplets with a lateral resolution of about 10 Å. However, the requirement that the sample be transparent limits the application of the method.

Other noncontact AFM methods have also been used to study the structure of water films and droplets [27,28]. Each has its own merits and will not be discussed in detail here. Often, however, many noncontact methods involve an oscillation of the lever in or out of mechanical resonance, which brings the tip too close to the liquid surface to ensure a truly nonperturbative imaging, at least for low-viscosity liquids. A simple technique developed in 1994 in the authors' laboratory not only solves most of these problems but in addition provides new information on surface properties. It has been named scanning polarization force microscopy (SPFM) [29–31]. SPFM not only provides the topographic structure, but allows also the study of local dielectric properties and even molecular orientation of the liquid. The remainder of this paper is devoted to reviewing the use of SPFM for wetting studies.

III. SCANNING POLARIZATION FORCE MICROSCOPY (SPFM)

A. Principle of Operation

SPFM uses the same technology as the atomic force microscope, which is based on sensing the force between sharp tips and the surface, the use of piezoelectric scanners, and feedback control electronics. In the AFM a sharp tip (with typical radius between 100 and 1000 Å) is used to "feel" by contact the topography of the surface. The tip is mounted on a flexible cantilever with a spring constant on the order of 1 N/m. The images are obtained by adjusting the separation between the sample and the cantilever such that its bending stays constant while the tip is scanned over the surface. The cantilever bending is usually sensed by the angular deflection experienced by a beam of light reflected off its backside. Due to the load applied in the standard contact operation mode, elastic deformations produce a finite tip–surface contact region. The diameter of this region determines the resolution achievable in this AFM mode. One must take into account that due to attractive, capillary, and adhesive forces, it is difficult to apply arbitrarily small forces, with the goal of achieving high resolution. The force is typically a few tens of nanonewtons, which for typical tips and sample materials produces nanometer-size contact diameters.

In contact mode, liquids are virtually impossible to image, because the mechanical contact of the tip deforms the surface. It is also possible for the liquid to wet the tip and form a capillary neck around it.

These problems have been overcome in the SPFM mode. In this mode a bias voltage (dc or ac) is applied to the conductive cantilever (Si_3N_4 levers and other insulating levers can be made conductive by coating them with a thin metal film). The opposite charges generated at the tip and surface cause their mutual attraction, and this bends the lever toward the surface. Because electrostatic forces are long range, they provide a means of imaging at a distance that is large enough to minimize perturbation of the liquid surface. The concept of SPFM is schematically shown in Figure 1 [32]. The term *polarization force* stresses the physical origin of the interactions that give rise to contrast in the images and that can be exploited by performing dielectric spectroscopy. Typically, tip–sample operating distances are a few hundred angstroms when the applied bias is on the order of a few volts. This produces an attractive force in the nanonewton (nN) range. The advantage of the long-range noncontact SPFM operation comes at the price of diminished lateral resolution compared to the contact mode. This resolution is on the same order as the tip–surface separation, i.e., a few hundred angstroms. This is still far better than the typical optical resolution of several micrometers. The vertical resolution of the SPFM, on the other hand, is as good as that of other SPM probes, i.e., in the angstroms region, as determined by the noise of the apparatus.

B. Dependence of Electrostatic Force on Distance

From simple electrostatic arguments it is clear that the force can be written in the following form [33]:

$$F = -\frac{1}{2}\cdot\frac{\partial C}{\partial z}\cdot(V_{tip} - \phi)^2 \approx -4\pi\varepsilon_0\cdot\frac{\varepsilon - 1}{\varepsilon + 1}\cdot f\!\left(\frac{R}{z}\right)\cdot(V_{tip} - \phi)^2 \qquad (12)$$

where C is the system capacitance, z is the tip–sample distance, ε is the local dielectric constant, and R is the tip radius. V_{tip} is the applied bias and ϕ is the local contact potential difference so that $V_{tip} - \phi$ represents the total voltage drop. During imaging, the attractive force (lever deflection) is kept constant by the feedback control electronics, and a constant force

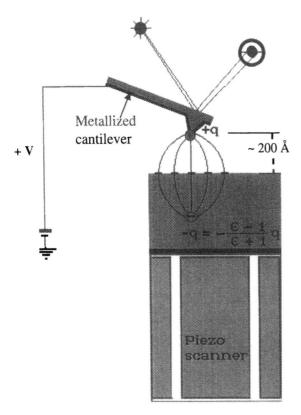

+ V

FIG. 1 Schematic representation of the operation of the scanning polarization force microscope (SPFM). An electrically biased AFM tip is attracted toward the surface of any dielectric material. The polarization force depends on the local dielectric properties of the substrate. SPFM images are typically acquired with the tip scanning at a height of 100–300 Å. (From Ref. 32.)

image is obtained. In dc mode images, dielectric constant and surface potential contributions are mixed with topographic information. The function $f(R/z)$ depends on the geometry of the tip and the sample. For a point charge on a flat surface, f is of the form $1/z^2$, but for a real tip the dependence is more complicated. Although an analytical expression for the function $f(R/z)$ cannot be obtained even for simple tip shapes [34,35], the following approximation is useful in many cases. The tip and lever can be approximated by a sphere of radius R (the tip apex) separated by a distance z from the surface, and a flat parallel condenser plate at a distance $z + D$, representing the lever. This is shown schematically in Figure 2.

These two elements are treated as two capacitors in parallel. The capacity of a sphere relative to the sample can be calculated exactly by the method of images. Even then, however, a complicated expression is obtained that must be calculated numerically. Fortunately, when the tip–surface distance is sufficiently smaller than the tip radius, an approximate expression for the capacitance can be found by integrating the contributions of flat infinitesimal rings of spherical surface centered at the apex [36]. The result is:

$$C(z) \approx 2\pi\varepsilon_0 R \times \left[\ln\left(\frac{R}{z}\right) + a\right] \tag{13}$$

where $a \approx 2$.

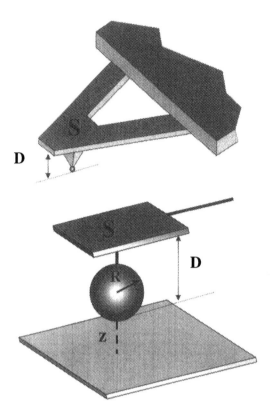

FIG. 2 Model for the calculation of electrostatic forces. The tip–lever system (top) is approximated by a sphere of radius equal to the apex tip radius R and (bottom) a flat plate of area S equal to that of the support lever.

The capacitance of the supporting lever is estimated to be that of a flat plate with area S at a distance $z + D$ from the surface. Its contribution is $C_{lever} \approx \varepsilon_0 S/(z + D)$, and the force due to the lever is $F_{lever} \propto -\partial C_{lever}/\partial z \approx C_{lever}/(z + D)$. An estimation of the orders of magnitude involved is useful. For $R = 1000$ Å and for $z = R/10$ (i.e., for $z = 100$ Å), we get $C_{tip} \approx 20$ aF, while using $S = 25 \times 50$ μm^2 (the moving part of the lever) and $D = 4$ μm for the lever we get $C_{lever} = 2.7$ fF. So it is clear that at typical operating distances, the capacitance is dominated by the lever (by a factor of ~100) and not by the tip. An actual measurement of the capacitance would yield an even higher value (pF) because it would include the nonmoving part of the lever and its support chip. The force, however, is comparable, with $\partial C_{lever}/\partial z \approx \partial C_{tip}/\partial z$ at $z = R/10$. Moreover, while $\partial C_{lever}/\partial z$ stays constant during imaging (because $z \ll D$), $\partial C_{tip}/\partial z$ contains all the topographic information. So we conclude that at typical imaging distances (~100 Å) and for typical tips used in AFM, the capacitance is due to the whole lever, while the varying part of the force is due to the tip apex. This is of course why high-resolution imaging is possible at all in SPFM.

The sphere–plane capacitor model gives a useful approximate expression for the function $f(R/z)$. Equation [13] shows that in the region $0 < z/R < 1$, which is typical in SPFM imaging, f can be approximated by a $1/z$ dependence. The planar lever adds a nearly constant term. Thus for the range $0 < z < R$, we have the following approximate function for the electrostatic force:

$$F \approx \left(\frac{A \cdot R}{z} + B \right) \cdot (V - \phi)^2 \; \text{Newton} \cdot \text{volt}^{-2} \tag{14}$$

The constants A and B are both on the order of 10^{-11}. This simple relation is indeed found experimentally, as shown in the Figure 3. For larger values of z/R, the functional form of f deviates from $1/z$.

C. Effect of Dielectric Constant on Topographic Heights

As we have seen, the constant-force images depend on the local dielectric constant. We will now discuss the effect of the dielectric constant by calculating the relation between true and measured heights of a flat parallel film on a surface of different dielectric constant.

Let's consider the simple case of a point charge at a distance z over a film of thickness L and dielectric constant ε_1 on a substrate of dielectric constant ε_2 [29]. In this case the force can be calculated exactly by the multiple-image method. The result is:

$$F = \frac{1}{4\pi\varepsilon_0} \frac{\gamma_1 q^2}{(2z)^2} (1 + \xi) \tag{15}$$

with

$$\xi = \frac{(\gamma_1^2 - 1)}{\gamma_1} \sum_{i \geq 0} \frac{\gamma_1^i \gamma_2^{i+1}}{[1 + (i + 1)L/z]^2}$$

where $\gamma_1 = (\varepsilon_1 - 1)/(\varepsilon_1 + 1)$ and $\gamma_2 = (\varepsilon_1 - \varepsilon_2)/(\varepsilon_1 + \varepsilon_2)$. The measured or apparent height of a film L_{app} can be calculated from the different heights of the tip over the surface

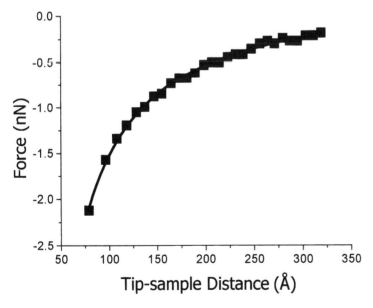

FIG. 3 Dependence of the electrostatic force on tip–surface distance. The experimental data (square points) can be fit reasonably well with a $A/z + B$ function (solid curve) predicted by the model in the previous figure for tip–sample distances smaller than the tip radius.

with and without the dielectric film (for the same force). The preceding equations yield the following results: At very close range, when $z \ll L$, we get the obvious result that the apparent L_{app} and the real height L are very similar. This, however, is not the interesting case for liquids, since for films a few tens of angstroms thick, such a close distance would result in a strong perturbation of the film by the tip. In addition, for most typical cantilevers, with spring constants of ~1 N/m, the tip would be within the instability range and jump to contact. SPFM images are typically taken at heights of 100–300 Å, i.e., $L/z < 10$. In these conditions L_{app} varies with ε_1, as shown in Figure 4. For metallic substrates ($\varepsilon_2 = \infty$), L_{app}/L increases with ε_1 in a manner that follows closely the ratio $(\varepsilon_1 - 1)/\varepsilon_1$. This approximation can be obtained from Eq. (15) by expanding the sum in powers of L/z and retaining only the linear terms. It predicts, for example, that for saturated alkane-chain films with $\varepsilon_1 \approx 2$ (e.g., thiols on gold), the measured height should be a factor of 2 smaller than the real height. For a nonmetallic substrate the dependence is also shown. In this case the ratio L_{app}/L can be larger than 1. The general rule is that $L_{app}/L < 1$ if $\varepsilon_1 < \varepsilon_2$, and $L_{app}/L > 1$ if $\varepsilon_1 > \varepsilon_2$.

A perturbative approach to Eq. (12) has recently been developed by Gomez-Monivas et al. [37]. For a dielectric film on top of a flat metallic surface, these authors find that the electrostatic force is a convolution of the instrumental resolution with an effective profile Z_{eff}:

$$Z_{eff}(x,y) = \int_0^{Z(x,y)} \frac{\varepsilon(x,y,z) - 1}{\varepsilon(x,y,z)} \, dz \qquad (16)$$

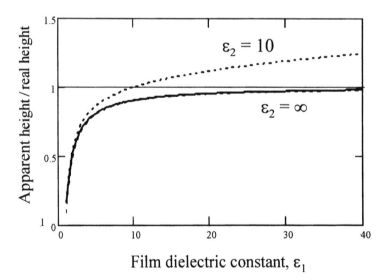

FIG. 4   Calculated ratio between the apparent L_{app} (measured by SPFM) and real heights L of a flat film on a substrate of dielectric constant ε_2, as a function of film dielectric constant, ε_1. The ratio is calculated with the multiple-image method, assuming a scanning height of a point charge tip of $z > 10L$. Calculations for two different values of the substrate dielectric constant ε_2 are shown. The measured height can be larger or smaller than the real height, depending on whether $\varepsilon_1 > \varepsilon_2$ or $\varepsilon_1 < \varepsilon_2$. For a metallic or high-dielectric-constant substrate, the curve is very closely approximated by an $(\varepsilon_1 - 1)/\varepsilon_1$ dependence (dotted curve, almost on top of the solid one).

where $\varepsilon(x,y,z)$ is the dielectric constant of the film as a function of position and $Z(x,y)$ is the surface topography. This formula gives the same $(\varepsilon - 1)/\varepsilon$ approximation as earlier for a homogenous film.

More subtle effects of the dielectric constant and the applied bias can be found in the case of semiconductors and low-dimensionality systems, such as quantum wires and dots. For example, band bending due to the applied electric field can give rise to accumulation and depletion layers that change locally the electrostatic force. This "force spectroscopy" character has been shown by Gekhtman et al. in the case of Bi wires [38].

The fact that the dielectric constant depends on the frequency gives SPFM an interesting spectroscopic character. Local dielectric spectroscopy, i.e., the study of $\varepsilon(\omega)$, can be performed by varying the frequency of the applied bias. Application of this capability in the RF range has been pursued by Xiang et al. in the study of metal and superconductor films [39,40] and dielectric materials [41]. In these applications a metallic tip in contact with the surface was used.

Rotational spectroscopy of polar molecules is a promising potential of the technique that is worth pursuing. Rotational modes, however, occur at extremely high frequency, in the gigahertz range for free molecules. This puts severe restrictions on the shape and shielding of the tips and wiring. Waveguide-type designs for the microfabricated tips are necessary. Its feasibility was demonstrated by van der Weide et al. [42]. For the present status of tips available today, only experiments in the low-frequency range of the spectrum can be performed routinely. We will present examples of the exploitation of this capability in studies of ionic mobility in the sections that follow.

D. Separation of Topography and Contact Potential

As we have shown, the polarization force depends not only on the topography [through the $f(R/z)$ term] and dielectric constant ε, but also on the local contact potential ϕ. As we shall see now, ac bias modulation and lock-in detection allow these contributions to be separated.

Experiments using ac bias modulation for the purpose of separating topography and contact potential were first carried out by Schönenberger et al. [43] and later by Yokoyama et al. [44]. When the cantilever is driven by a voltage of frequency ω, the force detected by the lever can be expressed as:

$$
\begin{aligned}
F &= B \cdot f\left(\frac{R}{z}\right)(V_{\text{tip}} \sin(\omega t) - \phi)^2 \\
&= B \cdot f\left(\frac{R}{z}\right)\left[-\frac{1}{2} V_{\text{tip}}^2 \cos(2\omega t) - 2V_{\text{tip}}\phi \sin(\omega t) + \frac{1}{2} V_{\text{tip}}^2 + \phi^2\right] \qquad (17) \\
&= F(2\omega) + F(1\omega) + F(0\omega)
\end{aligned}
$$

where

$$
B = -4\pi\varepsilon_o \cdot \frac{\varepsilon - 1}{\varepsilon + 1}
$$

The amplitudes of the signals with frequencies 2ω and 1ω can easily be measured using lock-in amplifiers. The 2ω signal is used as control signal in the feedback electronics so that its value is kept constant. The feedback output is then the topographic image [modulated by $\varepsilon(x,y)$]. Since the term $B \cdot f(R/z)$ in Eq. (17) is kept constant by the feedback, the 1ω signal obtained simultaneously with another lock-in amplifier provides a map of surface charge or surface potential distribution. Alternatively one can add a dc offset to null the $F(1\omega)$ signal, as in the Kelvin method for work function measurements. Names such as scanning Kelvin probe microscopy [45] and scanning surface potential microscopy [46] have been used for the technique. As we shall see, the surface potential can provide important information on the charge distribution or dipole orientation.

IV. WETTING STUDIES WITH SPFM

In spite of the short time since its implementation, SPFM has already produced a number of interesting results in the area of liquid surfaces, which will be discussed in the following sections. First we show the use of SPFM to measure droplet shapes and contact angles, which extends a popular and venerable physical chemistry method of surface characterization into the realm of nanoscale dimensions. Deviations of the contact angle of small droplets relative to the macroscopic values are expected. As we have seen, this contains information on the surface forces or the disjoining pressure. Then we present two examples where the SPFM resolution was sufficient to observe molecular-scale structures near the edge of liquid drops. In the third subsection we focus on the most important liquid on Earth, water. First, we present studies involving the structure of water and ice films formed on mica. Second, we show results on the adsorption of water on ionic substrates leading to solvation and deliquescence of the substrates. We conclude with an example of a corrosion reaction of sulfuric acid on aluminum substrates.

A. Disjoining Pressure Effects on the Contact Angle of Small Droplets

Ideally one would like to visualize the molecular-scale details at the edge of a droplet to obtain direct information about the molecular nature of wetting. This is not always possible, particularly when these details have dimensions below ~300 Å, the resolution limit of SPFM. However, the height and curvature of a droplet can usually be measured accurately. These parameters can then be used to obtain an effective contact angle, as defined in Eq. (10). We present here a few examples of this type of study.

Droplets of various liquids were prepared in several ways. For example, a macroscopic drop was first deposited on the substrate and then absorbed from an edge using filter paper. In other cases a macroscopic drop was blown away with a jet of N_2 or air. These processes leave a surface that appears dry to the naked eye but still contains many tiny droplets that can be observed with SPFM. If the droplets are of aqueous solutions, the water vapor pressure in the chamber, with which they readily equilibrate, determines their size. For liquids with low vapor pressure, films and droplets can be formed by condensation from a warmed reservoir.

In order to determine the disjoining pressure, Eqs. (10) and (11) are used, and $P(z)$ is modeled by the expression:

$$P(z) = -\frac{A}{12\pi z^2} + B \exp\left(-\frac{z}{C}\right) \tag{18}$$

The first term is related to the van der Waals interaction, with A being the Hamaker constant. The second term includes other forces that decay exponentially with distance. As discussed, these may include double-layer, solvation, and hydration forces. In our data analysis, B and C were used as fitting variables; the Hamaker constant A was calculated using Lifshitz theory [6].

1. Aqueous KOH Solutions

Figure 5 shows a typical SPFM image of KOH solution (we will use the notation aq.KOH from now on) droplets on highly oriented pyrolytic graphite (HOPG) [47]. The images were obtained in ambient conditions of room temperature and relative humidity around 35–40%. Atomic steps on the HOPG can be observed in the images. The lateral dimensions of the droplets range from hundreds to thousands of angstroms, and the heights from tens of angstroms to ~100 Å. Although droplets can also be found in the flat terraces, they prefer to adsorb on the step edges.

Although it is known that trapping of liquid droplets occurs at surface defects [48], it is still amazing that atomic steps of only one atom's height can anchor drops as large as a few thousand angstroms containing millions of molecules. The droplet distribution in this experiment is also interesting: Along a given step, their size and separation are nearly uniform. Also, large droplets tend to be located near steps limiting large terraces. This is likely the result of hydrodynamic instabilities during removal of the liquid. As the liquid film is being displaced by the air jet (or removed by osmotic transfer to the filter paper), it first breaks down into disconnected fragments on the surface in the form of strips attached to the steps. Further liquid removal thins out the strips to the point where they become unstable and break into droplets.

Figure 6 shows droplets of KOH solution on mica produced by similar methods. In both cases the drop profiles are very close to a spherical cap. In Figure 7 we have plotted the effective contact angle as a function of droplet height. The deviation from the macroscopic contact angle with decreasing droplet volume can clearly be seen.

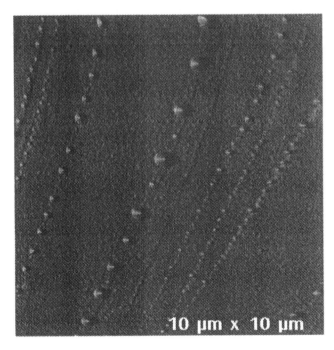

FIG. 5 SPFM image of droplets of an aqueous KOH solution deposited on an HOPG surface. Notice that droplets attach preferentially to steps of the graphite surface. Droplets tend to be the same size and regularly separated on a given step. Larger droplets are found on steps delimiting large terraces.

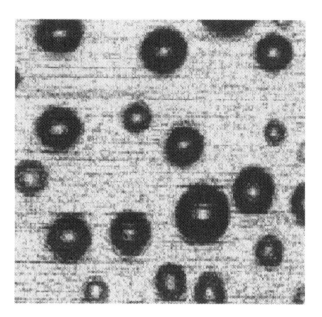

FIG. 6   1.5-μm × 1.5-μm SPFM image of droplets of aqueous KOH solution on the cleavage surface of mica.

For the aq.KOH–graphite system, the van der Waals interaction should be repulsive, because Lifshitz theory predicts a negative Hamaker constant A, which we calculated to be approximately -7.7×10^{-20} J. Using this value, the fit gives:

$$P(z) = \frac{7.7 \times 10^{-20}}{12\pi z^2} - 0.024 \exp\left(\frac{-z}{40}\, \text{Å}\right) J/m^2$$

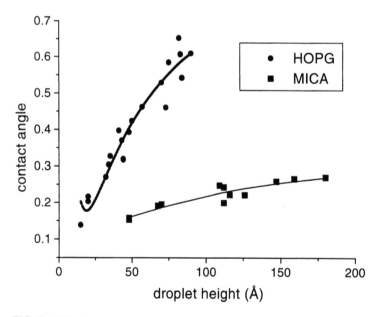

FIG. 7   Effective contact angle of the aqueous KOH droplets on HOPG and mica as a function of droplet height. Solid lines correspond to fits obtained using the disjoining pressure given by Eq. (18).

The exponential in the second term could be explained by the hydrophobic interaction between the two surfaces, which is attractive and exponentially dependent on the distance. We see that the hydrophobic force dominates the disjoining pressure in the aq.KOH–graphite system; therefore the effective contact angle increases with drop height. The macroscopic contact angle obtained with this fitting agrees with the one measured with standard optical methods. We have used 72 mJ/m^2 for the surface tension of the KOH solution in the fitting. By comparison, the value of 24 mJ/m^2 indicates that the hydrophobic interaction is indeed quite strong.

For aq.KOH on mica, the dependence of the effective contact angle on droplet height is much weaker than that for the aq.KOH–graphite system. The estimated Hamaker constant of the van der Waals interaction for this system is -1.9×10^{-20} J (repulsive), and the fitting gives:

$$P(z) = \frac{1.9 \times 10^{-20}}{12\pi z^2} - 0.003 \exp\left(\frac{-z}{54}\, \text{Å}\right) \text{J/m}^2$$

In this case, the hydrophobic interaction is very weak compared to that of aq.KOH–graphite system. In spite of this, it still dominates the disjoining pressure.

2. Sulfuric Acid

Figure 8 shows H$_2$SO$_4$ droplets on mica at very low humidity ($<5\%$). The contact line around the drops is smooth and circular, revealing that no pinning has occurred [49]. Although the area between drops is flat, it does not correspond to clean mica but to a liquid film covering it of a few monolayers thickness. This is deduced from the hysteresis in the force versus distance experiments, where the tip is brought into contact with the surface and then pulled off.

The observation of liquid drops on top of a film of the same substance is often referred to as *pseudo-partial* [1] or *autophobic* wetting. This behavior is not surprising in this

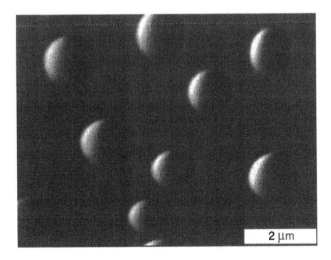

Sulfuric acid on mica

FIG. 8 SPFM image showing H$_2$SO$_4$ droplets deposited on mica at low humidity ($<5\%$). The region between drops is covered by a thin acid film (see text).

case due to the large surface energy of cleaved mica. Values of 300–600 mJ/m^2 have been reported for the surface energy of freshly cleaved mica [50], compared with 53 mJ/m^2 for the surface tension of sulfuric acid.

When plotted as a function of drop size (Fig. 9), the contact angle was found to decrease with increasing drop height. A different analysis of these data was performed in the original paper. In that case the maximum slope near the drop edge was used, as well as a direct inversion of the droplet shape. The data could be fit to an empirical $1/z^2$ function. In the present analysis we use the method of the effective contact angles defined earlier, together with Eq. (18). For the Hamaker constant A, we calculated a value of approximately -2×10^{-20} J. However, the best fit to Eq. (18) is for a pure exponential decay of the form:

$$P(z) = -0.0014 \exp\left(\frac{-z}{250} \, \text{Å}\right) \text{J/m}^2$$

This potential is weaker and of longer range than that for the aq.KOH/mica system. The reason for the negligible contribution of the van der Waals term is not understood at present.

3. Glycerol

Glycerol is well known for its strong hydrogen bonding in the liquid phase and its high viscosity. To study its wetting properties, we deposited glycerol on mica by condensation from its vapor in standard laboratory air [9].

On freshly cleaved mica, condensation begins with the formation of droplets (Figure 10a). Their shapes are not spherical, but resemble flat ellipsoids. As condensation continues the droplets grow vertically and then laterally. However, when their heights reach ~100 Å, they spread and become pancakes with flat tops. From that point, the pancakes decrease in height while expanding laterally. Just before the film completely covers the surface, the relative contrast or height is 30 Å. Interestingly, the edges of the patches tend to have a

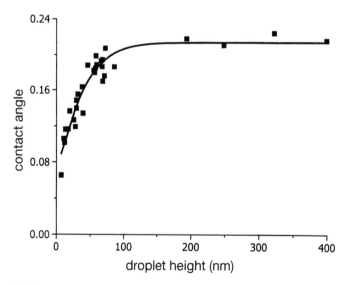

FIG. 9 Effective contact angle of H_2SO_4 on mica as a function of droplet height. Solid line shows fit with disjoining pressure given by Eq. (18).

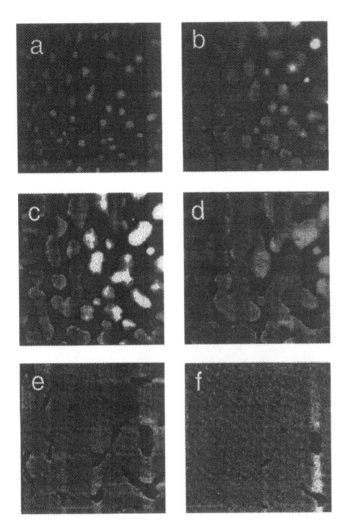

FIG. 10 17-μm × 17-μm SPFM images showing the evolution of the structures formed by condensation of glycerol from its vapor on a surface of freshly cleaved mica. Droplets with roughly flat tops grow initially until their heights reach ~100 Å. They then expand laterally and coalesce, forming flat films. The height of these films decreases slowly down to ~30 Å just before closure of the film in (f). Experiments performed under normal ambient conditions of 50% RH and 21°C. (From Ref. 9.)

polygonal shape, which might indicate that glycerol has an ordered structure in epitaxial relationship with the mica substrate.

A very different behavior is observed when condensation occurs on a mica surface that has been exposed to air for a few hours (we will refer to it as "contaminated mica"). In this case glycerol forms droplets in the shape of spherical caps, indicating that it does not completely wet the surface. This behavior is similar to that of water, which we present in detail later. The contact angle of water on mica surfaces increased from 0° on the freshly cleaved surface to a small value between 2 and 3° on the contaminated mica [51].

On the contaminated and slightly hydrophobic surface, the spherical droplets grow continuously with time, as shown in the sequence of images in Figure 11. This behavior is in sharp contrast to the case of freshly cleaved mica, in which flat pancakes form. The shape of the droplets is very close to a spherical cap, as shown by the linear correlation coefficient of 0.991 found from a fit to a sphere. When two droplets grow close enough to touch each other, they coalesce, as can be seen in several images in Figure 11.

The effective contact angle values (θ) measured as a function of drop height are shown in Figure 12. A rapid decrease of θ is observed when the height falls below 200 Å. This indicates that the potential $P(z)$ is negative. An exponential dependence of the disjoining pressure with distance gives a very good fit, as shown in the inset of Figure 12, which shows a semilog plot of $P(z) - zP'(z)$ vs. z. As in the case of sulfuric acid, we find that the van der Waals term in Eq. (18) is negligible. The fit gives:

$$P(z) = -6.4 \times 10^{-5} \exp\left(\frac{-z}{50 \text{ Å}}\right) \text{ J/m}^2$$

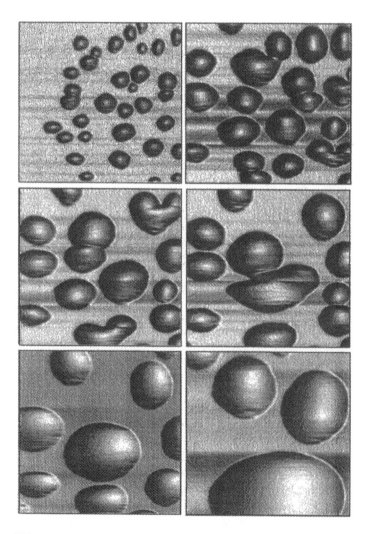

FIG. 11 15-μm $\times$ 15-μm SPFM images showing the condensation of glycerol on contaminated mica. Droplets with a nearly spherical cap shape are formed and increase indefinitely in size as condensation continues. When two drops establish contact, they coalesce into a larger drop. (From Ref. 9.)

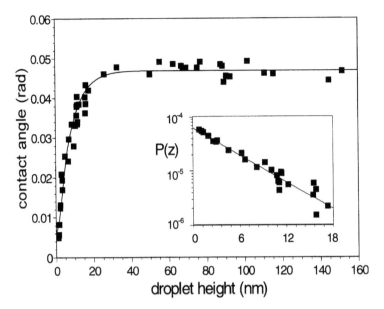

FIG. 12 Plot of the effective contact angle θ of the glycerol drops on contaminated mica vs. drop height e. A rapid increase occurs up to 200 Å (corresponding to a drop with a base diameter of ~2.2 μm), where it reaches a constant value θ_0. Inset: Plot of $(\theta_0^2 - \theta^2)/(1 + e/\delta)$ vs. e. This plot shows that the potential due to the long-range forces depends exponentially on the distance. (From Ref. 9.)

The exponential dependence is indicative of hydrophobic attractive forces between the glycerol–air interface and the glycerol–mica interface. This potential gives a negative disjoining pressure of ~1 atm at z close to 0. The strength of the force appears to be 100 times lower than that typical for water between hydrophobic surfaces [6].

B. Layering

In this section we discuss two examples in which the molecular-scale structure of the film is manifested in the form of features that can be resolved with SPFM. These are steps and terraces that appear at the edge of droplets and films, when the molecules form uniform layers during spreading. Layering is due to the substrate hard wall that forces the layers of molecules next to it to adopt a periodic structure with the dimensions of the molecular diameter. This has been observed many times with the surface force apparatus [6]. It has also been observed with ellipsometry when the terraces are many tens of micrometers wide [52,53]. With SPFM molecular terraces can be observed at much higher resolution, so we can study at a much finer scale the phenomena of wetting and dewetting. In addition to the topographic images, SPFM can provide simultaneous images of surface contact potential, which provides additional useful information on the structure of molecules, particularly when these are polar. The two examples presented next are liquid crystals and perfluoropolyether lubricants. Later we will see more applications of the contact potential mapping in studies of the most important polar molecule, water.

1. Liquid Crystals

The wetting and spreading properties of liquid crystals (LC) on solid substrates are of interest due to their use in display devices. In addition, these interesting liquids provide a good testing ground for fundamental theories of wetting. Films of 8CB (4′-*n*-octyl-4-

cyanobiphenyl) have been studied by optical microscopy [54], x-ray reflectivity [20,55], ellipsometry [7,20], second-harmonic generation [56], scanning tunneling microscopy (STM) [57], atomic force microscopy (AFM) [58], and surface forces apparatus (SFA) [59].

Bulk 8CB exists in three phases. With increasing temperature, there is a transition at 21.5°C from crystalline to the smectic-A phase; at 33.5°C, there is a second-order transition to a nematic phase; and at 40.5°C, there is a weak first-order transition to an isotropic liquid [60]. In bulk, the 8CB molecules are paired into dimers (1.4 times longer than a single molecule) with antiparallel dipole moments. In the smectic-A phase, the dimers form layers spaced by 31.7 Å, with the dimers aligned perpendicular to the layers [61]. Ellipsometry and x-ray-reflectivity have been used to study the spreading and structure of macroscopic (in x, y) films of 8CB on silicon wafers [20,25]. It was found that a prewetting film is formed during spreading in the smectic and nematic phases. The prewetting film is composed of a monolayer of single 8CB molecules with a thickness of 8 Å connected to a 41-Å-thick film. This film consists of an ~33-Å-thick bilayer on top of the 8-Å monolayer. Both monolayer and trilayer extend over millimeter distances.

For our studies of the wetting properties of 8CB, a small amount of liquid was deposited directly on the substrate [62]. For studies of the nematic and isotropic phases, 8CB was condensed from its vapor while keeping the Si substrate at the appropriate temperature (the vapor was produced by heating an 8CB reservoir to 80°C in front of the substrate). When studying the spreading of submicrometer-size deposits of 8CB, we observed a behavior similar to that found at macroscopic scales [20]: 8CB forms droplets and pancakes on the surface in the smectic phase, wets the surface in the nematic phase, and dewets it in the isotropic phase. However, we found that rapid cooling to room temperature, combined with the high viscosity of 8CB, for days freezes the liquid in the phase corresponding to the substrate temperature during deposition.

Molecular multilayer spreading was observed with SPFM. This is shown in the top image of Figure 13, which was obtained near the edge of a large spreading drop. It shows a layered structure with 32-Å steps (all z values have an associated error bar of ± 1 Å), the thickness of a smectic bilayer. The cross section in the middle of Figure 13, in the direction of the arrow shown in the top image, shows the perfect layering of the smectic bilayers. Layering near the edge of a spreading droplet has never been observed before with other techniques. In a sequence of time-lapsed images, the layers can be seen advancing at an average speed of 20–30 Å/s at room temperature. The layers advance on top of the prewetting trilayer, which extends over millimeter distances in front of the droplet. Although the stable state of 8CB on silicon is in the form of a pancake, we observed metastable nanometer-scale droplets surrounded by layered structures. The bottom image of Figure 13 shows an example of this. The droplet is not spherical. A higher-resolution image shows that the lower part of the droplet is also layered with steps 32 Å high.

The vapor deposition method at variable substrate temperature provides additional insight into the structure and wetting properties of 8CB in its various phases. If the substrate temperature is between 41 and 33°C, flat islands 32 Å thick are formed if only a small amount of 8CB is condensed on the surface. The size of these flat islands increases with deposition time while their height remains constant until a uniform layer is formed. If more 8CB is deposited, droplets form on top of the film. This is shown in the image of Figure 14.

On samples prepared with a substrate temperature above 41°C, only spherical droplets were observed. Although no change could be observed in the droplets for several days, we noticed that by applying a strong attractive electrostatic force, the tip could induce

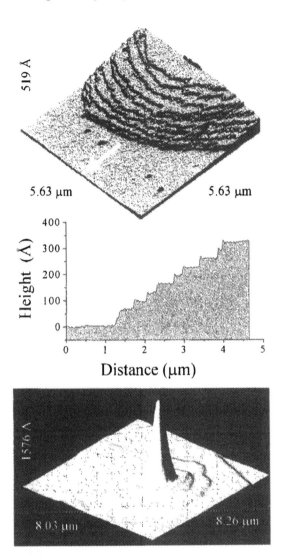

FIG. 13 *Top*: SPFM image of the spreading front of a smectic drop of 8CB liquid crystal on a Si wafer, showing a layered structure. Each layer is 32 Å thick. The layers advance in the direction of the arrow at the rate of 20–30 Å/s at room temperature. *Middle*: Profile of the droplet front showing the steps. *Bottom*: Drop and surrounding smectic layers. Vertical scale is greatly exaggerated. (From Ref. 62.)

the droplets to spread and form pancakes. This indicates that the droplets are indeed in a metastable state.

To determine whether the 8CB droplets condensed above 41°C (trapped in the isotropic phase) sit on a trilayer or on bare silicon, we used the AFM tip to mechanically spread the droplets and thus accelerate their conversion to a stable configuration. The SPFM images shown in Figure 15 were obtained after such tip-induced spreading. A layered structure with ~32-Å-high steps typical of the smectic phase is obtained. The first, or bottom, layer is 41 Å thick, while the layers above it are all 32 Å thick. This indicates that the bottom layer of the film is a trilayer and that the remaining substrate is dry silicon, i.e.,

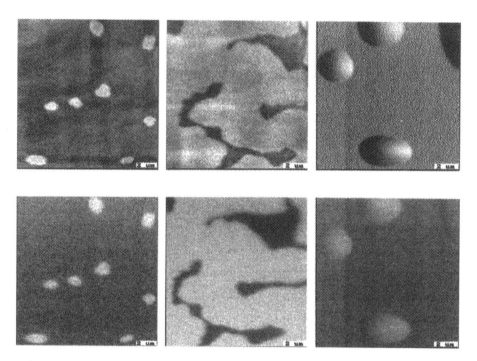

FIG. 14 *Top*: SPFM topographic images showing various stages of the vapor-deposited 8CB film on a Si substrate at a temperature between 33 and 41°C. The evaporation time increases from seconds to minutes from the left to the right images. Flat 32-Å-thick islands grow laterally until they cover the substrate completely. Further deposition (right image) produces droplets on top of the flat film. *Bottom*: Simultaneously acquired contact potential images. The gray scale is proportional to the local surface potential. On the islands, the potential is 40 mV higher than that of the outside region. On the drops, the potential is 10 mV higher than that of the surrounding area. (From Ref. 62.)

not covered with either a monolayer or trilayer of 8CB. We can thus conclude that the frozen isotropic droplets sit on bare silicon. By contrast, the flat layer of Figure 14, formed by evaporation between 33 and 41°C, is 32 Å high, indicating that it consists of a bilayer on top of the monolayer. The monolayer must have formed very rapidly during the first stages of exposure to the vapor.

The simultaneously acquired contact potential images provide important structural information on the liquid film. The droplets formed at temperatures over 41°C have a surface potential 50 mV (± 3) higher than the surrounding background (not shown). The 32-Å film formed between 41 and 33°C has a potential about 40 mV higher than the surrounding monolayer, and the droplets formed above that film have a potential 10 mV higher than the bilayer film (Fig. 14). In the structure formed by tip-induced spreading, the bottom trilayer has a surface potential 40 mV higher than the surrounding substrate, while the bilayer above it is 7 mV more positive. The third bilayer adds 2 mV, and the fourth adds ~1 mV. Layers above the fourth make negligible contributions. The total potential increase at the top of pancakes or multilayers is ~50 mV higher than that of the bare silicon substrate, which is similar to that of the frozen isotropic droplets. This indicates that the isotropic droplets have similarly oriented bottom layers and that the remaining 8CB molecules above them do not contribute to the droplet potential.

The information obtained from the simultaneously acquired topographic and surface potential images allows us to propose a model for the structure of 8CB in various wetting situations. We have seen in Figure 14 that the potential of the first bilayer (32 Å thick) is 40 mV higher than the monolayer it sits on. The same contact potential difference is measured between the trilayer (41 Å thick) and the bare silicon (Fig. 15). These two observations indicate that the monolayer bound to silicon does not contribute appreciably to the surface potential. This in turn indicates that the molecular dipole moment is parallel to the surface. Since the dipole originates in the charge distribution between the nitrogen and benzene rings, only this part needs to be parallel to the surface. This is consistent with the 8-Å thickness of the monolayer observed by ellipsometry. The 8 Å is contributed by the alkane group at a ~30° with the surface, giving a height of 4 Å, plus the 4–5 Å of the flat-lying biphenyl group.

Since the 8CB dimers in the bulk have no net dipole moment, we conclude that the 40 mV of the first bilayer on top of the monolayer originates from a distortion of the 8CB pair that misaligns their dipoles so as to give a net value. One possible mechanism for that distortion is the van der Waals interaction between the alkane tail chains. This tends to make the chains in the monolayer (partially sticking up) parallel with the alkane chains of the next bilayer. This would affect one of the 8CB monomers in the dimer. To create a 40-mV potential change, the dipoles in the dimer must misalign on the order of ~3°, if we use the known dipole moment of 8CB and the surface density given by the molecular dimensions. In subsequent bilayers, the distortion decays rapidly, as manifested by their decreasing contribution to the surface potential. Another possibility is that the flat-lying 8CB molecules of the isolated monolayer tilt up by a few degrees when a bilayer is on top. We can not distinguish between the two at present.

2. Hard-Disc Lubricants

Magnetic hard discs are used in computers to store information in the form of magnetic bits on a ferromagnetic alloy film. To protect the magnetic medium, a hard carbon coating about

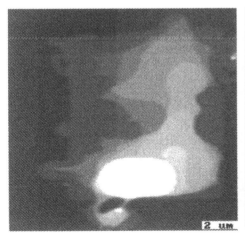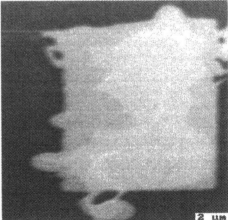

FIG. 15 *Left*: SPFM image obtained after spreading an 8CB droplet on silicon by contact with the AFM tip scanning over a square area. Several layers can be observed. The first one is 41 Å thick, and the following ones are 32 Å thick. The bright spot near the bottom is the remaining unspread droplet. *Right*: Simultaneously acquired contact potential image. The potential in the first layer is 40 mV higher than that of the substrate, and that of the second layer is 7 mV higher than the first. The third and fourth layer potentials are 2 and 1 mV higher, respectively. Above that, the increase is negligible. (From Ref. 62.)

100 Å thick is deposited on top. This is then covered with a molecularly thin film of lubricant to minimize wear during start–stop contacts and to passivate the disc surface against contamination and corrosion. High-molecular-weight perfluoropolyalkylether (PFPE) polymers are widely used for this purpose. In order to improve surface bonding, the PFPEs are modified with specific functional end groups. All these molecules have similar backbone structures, namely:

$$X—CF_2—[(O—CF_2—CF_2)_n—(O—CF_2)_m]—O—CF_2—X$$

where $m/n \sim 1.1$. They differ from one another in the length of the chain and in the end group, X. For the PFPEs studied here, ZDOL, ZDOL-TX, and Tetraol, X = CH_2OH, — $CH_2OCH_2CH_2OCH_2CH_2OH$, and —$CH_2OCH(CH_2OH)_2$, respectively.

A substantial amount of research has been carried out on these lubricants to understand the processes of wetting (spreading and diffusion) [63–68] and its opposite, dewetting [28] and to determine the conformation of the molecules on the disc surface. [66,67,69] Using ellipsometry it has been found that lubricants with functional end groups (e.g., Zdol) show a layering structure during spreading, in contrast to the smooth profiles of lubricant films without functional end groups (e.g., Z03 where X = CF_3) [63–68]. Unfortunately, the limited lateral resolution of the optical methods prevented investigation of the nanometer-scale structure of the spreading film.

We performed SPFM studies on hard disc coupons covered with lubricant films of different types and thicknesses [70]. No evidence of dewetting was observed with films of submonolayer average thickness (17 Å for Zdol and 14 Å for Zdol-TX). Dewetting was observed, however, when the film thickness was above one monolayer. The structure of these films during dewetting differs depending on the type of lubricant.

On discs coated with Zdol films 62 Å thick or more, droplets with lateral dimensions of tens of micrometers were observed in some areas of the surface, as shown in Figure 16. Apart from that, however, most of the surface was uniform unless disturbed by contact with the SPFM tip. After such contact, a droplet was usually formed due to capillary effects. Once formed, the droplet slowly grows in volume, indicating that the film was in a metastable state. However, no detectable layering in the surrounding region was observed.

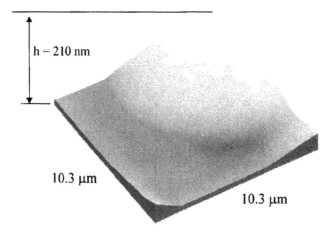

h ≈ 210 nm

10.3 μm

10.3 μm

FIG. 16 SPFM image of a droplet formed as a result of dewetting of Zdol on an amorphous hard carbon substrate film. No layering around the drop was observed. (From Ref. 70.)

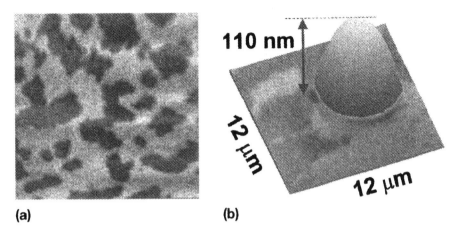

(a)   **(b)**

FIG. 17 SPFM images showing the structures formed during dewetting of a Zdol-TX film with nominal thickness of 50 Å. (a) 15-μm × 15-μm image of a region with a partially covering mobile layer. (b) Three-dimensional image acquired in a different region showing a droplet growing at the expense of lubricant material from the partially covering layer. (From Ref. 70.)

In the case of Zdol-TX, however, layered dewetting was observed for films thicker than one monolayer. Typical images are shown in Figure 17. On a nominally 86-Å-thick Zdol-TX film, a 50-Å-thick layer was observed partially covering the surface (Fig. 17a). These films were often found to be connected to big droplets, as shown in Figure 17b. A similar behavior was observed for Z-Tetraol. The example in Figure 18 corresponds to a 56-Å-thick film. The thickness of the layer partially covering the surface is approximately 25 Å, close to that expected for a monolayer of Zdol-TX. In both cases, the films can easily be influenced by the tip, not only in contact as in the example of Figure 18b, but also by the attractive force during SPFM imaging (see later discussion), indicating their weakly

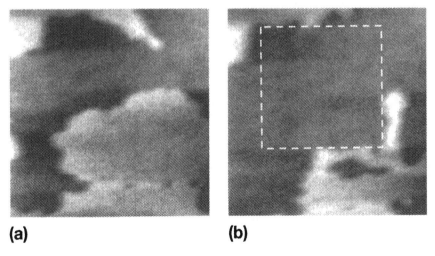

(a)   **(b)**

FIG. 18 (a) 10-μm × 10-μm SPFM image from a nominally 56-Å-thick film of Z-Tetraol. A 25-Å-thick layer partially covers the surface. (b) SPFM image acquired after contact scanning in the region marked with a box. (From Ref. 70.)

bound nature. The mobile lubricant layers correspond to the second layer above the bonded first layer.

Usually no contrast in the first layer was observed, due to its continuous nature. In some cases, however, the carbon overcoat was not completely covered by the lubricant. In these cases three contrast levels could be seen in the images, as in the example shown in Figure 19, corresponding to a Zdol-TX film with a nominal thickness of 86 Å. The lowest contrast level (the darkest regions in the images, marked 0) corresponds to the exposed carbon coating. The next region is ~25 Å higher and corresponds to the first, strongly bound lubricant layer (marked 1 in Figure 19a), and the third region is ~75 Å higher (marked 2) and corresponds to the mobile lubricant layer with a thickness of 50 Å. Only layers 1 and 2 are visible in Figure 17. The mobility of the second-layer molecules due to the attractive forces exerted by the tip can be seen in these images.

Spontaneous dewetting takes place by the formation of a droplet connected to the second layer. As a function of time, the droplets increase in volume while the area covered by the second layer decreases. In cases where the bare substrate was partly exposed, the diffusion and aggregation of second-layer molecules into droplets preserved the exposed regions of the substrate, as shown in Figure 20.

The layering of the lubricant films was explained by the interaction among the lubricant molecules and between the molecules and the substrate. The fact that the thickness of

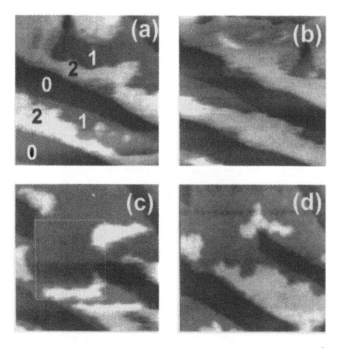

FIG. 19   13.5-μm × 13.5-μm SPFM images of a nominally 86-Å-thick Zdol-TX film on a carbon-coated disc. An uncovered (possibly contaminated) area of the substrate and two layers of lubricant are observed, which are labeled 0, 1, and 2, respectively. Only layers 1 and 2 are seen in Figure 17. Layer 2 is weakly bound on top of layer 1 and can be moved by the attractive force of the biased tip, as shown in (a) and (b). Layer 1 is bound strongly to the carbon film substrate. The images in (c) and (d) were obtained after contact scanning in the region marked by the square. Notice the change in the structure of layer 2 from (c) to (d), due to the electrostatic force from the tip. (From Ref. 70.)

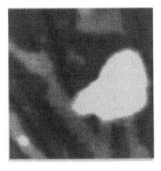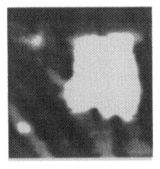

FIG. 20 Dewetting of Zdol-TX on a nonuniformly covered surface. As in the previous figure, bare substrate regions are exposed (darker areas). Only material in layer 2 dewets to form a droplet (bright area). Diffusion of molecules from layer 2 occurs on top of layer 1 and does not fill the uncovered, possibly contaminated, substrate regions. (From Ref. 70.)

the mobile layer is larger than the cross-sectional diameter of the molecules (~6–7 Å) indicates that the molecules are not lying completely flat. The reason for the difference in the thickness of the mobile second layers for Zdol-TX and Z-Tetraol is not yet understood, although it must be related to the differences in the functional end group for the two molecules. For both Zdol-TX and Z-Tetraol, the molecules in the first layer are bound to the carbon film. Zdol-TX molecules have the tendency to form "bilayers," which gives a thicker second layer of ~50 Å, while Z-Tetraol molecules appear to form only a single molecular layer.

In contrast to the case of the 8CB liquid crystal, no contact potential differences between first and second layers were observed with these lubricants. This indicates that there is no special orientation of the dipole active end groups, or perhaps that the end groups form hydrogen-bonded pairs with neighboring molecules so as to give no net dipole moment.

C. Structure of Water on Moist Surfaces

Water is the most abundant liquid on Earth. Its interaction with other materials, in particular with their surfaces, is of paramount importance. The formation of water films on the surface of materials is a usual occurrence in moist environments. These films can modify the surface properties and induce chemical reactions. Since the diameter of the water molecule is about 2.5 Å, and since the lateral resolution of SPFM is at best 100 Å, flat, molecularly smooth substrates are convenient. One such substrate is the cleavage surface of muscovite mica, an alumino-silicate mineral commonly found in soils with formula $KAl_2(Si_3AlO_{10})(OH)_2$. It has a layered structure with entire formula units stacked in layers of 10-Å thickness. The K^+ ions are located in planes between layers and provide a weak ionic bond between them. These are the easy cleavage planes of mica. They expose O atoms arranged in hexagonal patterns with 5.2-Å periodicity, formed by the bases of corner-sharing SiO_4 and $(AlO_4)^-$ tetrahedra. The cleavage surface of mica is known to be hydrophilic, and water readily spreads on a freshly prepared surface.

In humid environments, the ellipsometry results of Beaglehole and Christenson [71] indicate that a film is formed with an average thickness of ~2 Å when the relative humidity (RH) is ~50%. As discussed earlier, however, the lateral resolution of ellipsometry is limited by the size of the light spot on the sample, typically several tens of micrometers. In addition, thickness values are obtained from the measured changes in polarization angle by assuming values for the optical refraction index. Since this index is unknown for sub-monolayer or even monolayer amounts of water, reported thickness values should be taken with caution.

When the surface of mica was imaged by SPFM, no contrast could be observed as the humidity changed. This could be due to: (a) the dimensions of any water structures present being below the lateral resolution limit of SPFM, or (b) high molecular mobility at room temperature such that any structure formed below saturation is time-averaged (image acquisition time of several minutes). We found, however, that contrast could be created by perturbing the surface by a brief contact with the imaging tip. The contact induces capillary condensation of water around the tip. The capillary condensation can easily be seen in the humidity-dependent pull-off force that is necessary to remove the tip. This force is determined largely by the Laplace pressure, which arises from the negatively curved part of the meniscus. The force measured in one such experiment is plotted in Figure 21. As can be seen, the pull-off force is small below 20% RH but increases rapidly thereafter. After reaching a maximum at about 30%, the pull-off force decreases slowly. This decrease was explained as being due to the increasing influence of the positive part of the meniscus when its size increases.

Although most of the water in the meniscus evaporates once the tip has been retracted, residual structures can be observed in a radius of several tens of micrometers (depending on humidity and contact time) around the original contact point. For the tip radius and loads used in these experiments, the contact radius is approximately 10 Å. The residual structures are in the form of flat islands and sometimes droplets. In our first experiments the perturbation created by a brief tip contact was not fully appreciated. Accidental tip contacts during approach of the tip to the surface do often occur. In such cases the tip is subsequently moved to an adjacent area, several micrometers away, to study the "unperturbed" surface. However, as stated already, the perturbed areas can extend over tens of micrometers away from the contact point. Droplets can be observed when the relative humidity is

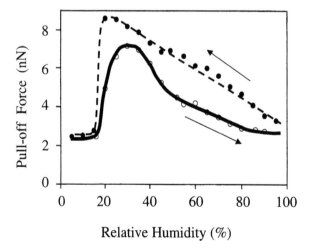

FIG. 21 Pull-off force between a hydrophilic Si_3N_4 tip and mica as a function of the relative humidity (RH) at room temperature. The spring constant of the lever is 0.1 N/m. Estimated tip radius ~200 Å. Hysteresis is observed between increasing (open circles) and decreasing (closed circles) humidity. (From Ref. 51.)

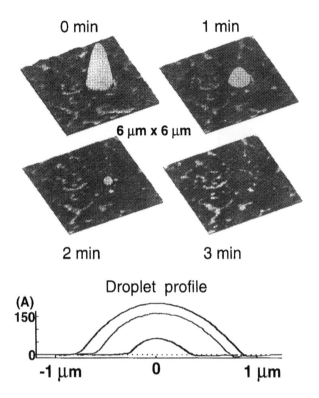

FIG. 22   SPFM images (6 μm × 6 μm) showing the evaporation of a droplet of water left on the surface of mica after breaking the capillary neck formed around the AFM tip during contact. The mica surface is slightly hydrophobic after having been exposed to air for a couple of hours. The droplet evaporates in a few minutes at 58% humidity. Profiles of the droplets are shown at the bottom. The contact angle is around 2.6°. Two-dimensional water structures can also be seen in the area surrounding the droplet. (From Ref. 51.)

high (above 50%) and when the mica has been exposed for some time to the environment, so it is presumably contaminated. The contact angle of these droplets is between 2° and 3°. This is reminiscent of similar observations in the case of glycerol, also a strong hydrogen-bonded liquid. The droplets are unstable and evaporate in a few minutes at 58% RH, as shown in the example of Figure 22. Flat islands can also be seen around the droplet in these images. Their apparent height is around 2 Å. They form contiguous structures on freshly cleaved mica. On aged mica (exposed to air for several hours), the islands have irregular shapes or sometimes form beautiful patterns of interconnected ribbons, as shown in Figure 23. Like the droplets, the islands disappear as well, but on a much longer time scale, lasting from minutes to hours when RH > 30%. Figure 24 shows an example of the evolution of the size and shape of the islands produced after a contact of several seconds. These observations indicate that the water structures are metastable. In our papers we referred to the metastable islands as phase-II, while phase-I designates the surrounding unperturbed water film.

The phase-II islands have several interesting properties. First, they have a positive surface potential relative to the surrounding unperturbed water film. The potential is highest immediately after formation and decays with time to zero. Another interesting property is the shape of the islands. Their boundaries are often polygonal, bending in angles of 120°,

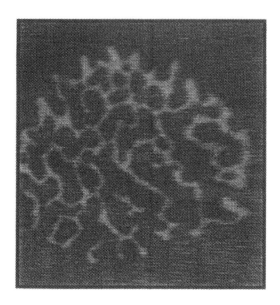

10 μm x 10 μm

FIG. 23 SPFM image of a network of interconnected water channels formed after 5 seconds of tip contact at 40% RH, with a mica surface contaminated as a result of exposure to the ambient air for about 2 hours. Notice that many angles between segments are close to 120°. The area covered by the water structures increases with contact time. (From Ref. 51.)

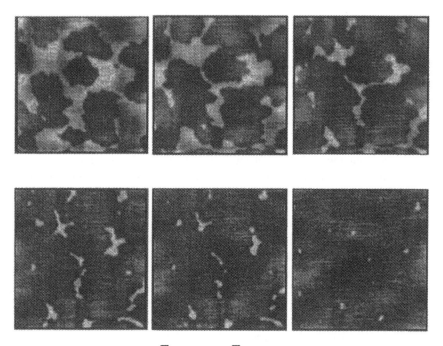

5 μm x 5 μm

FIG. 24 Evolution of the shape and size of the water structures formed after a tip contact of 5 seconds on freshly cleaved mica. The SPFM images were acquired, from top left to bottom right, at 5, 30, 50, 70, 130, and 150 minutes after formation. (From Ref. 51.)

as shown in Figure 25. Similar angles are observed in the bends of the narrow ribbons in Figure 23. By combining SPFM images with lattice-resolved images of the mica substrate obtained in contact mode immediately after, it was shown that the directions of the boundaries between phase-I and phase-II are closely related to the lattice orientation of the mica surface, as shown in Figure 25. A histogram of angular orientations is shown at the bottom of Figure 25. This observation suggests that the water films might have a solid or icelike structure in epitaxial relationship with the mica.

The topographic contrast of the islands disappears above 80% RH [72], and no islands can be observed after tip contact at this and higher humidity. These observations can be explained by assuming that the metastable islands formed by capillary condensation correspond to a second or higher layer of water above the first one (phase-I). As we shall see

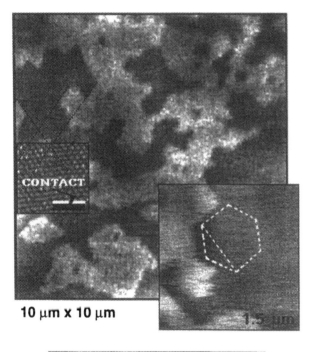

10 μm x 10 μm

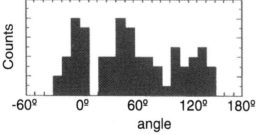

FIG. 25 SPFM images of water-film structures on mica. The boundaries tend to have polygonal shapes, as shown in the smaller image (a hexagon is drawn for visual reference). The directions are strongly correlated with the mica lattice. The smaller inset shows a contact AFM image obtained after the SPFM images. A histogram of the angles of the water-film boundaries with the mica lattice is shown at the bottom. (From Refs. 31 and 32.)

next, at around 80% RH the first monolayer reaches saturation and second and successive layers of water start to condense. In these conditions the extra water formed by capillary condensation is no longer visible above the background.

Stimulated by these observations, Odelius et al. [73] performed molecular dynamic (MD) simulations of water adsorption at the surface of muscovite mica. They found that at monolayer coverage, water forms a fully connected two-dimensional hydrogen-bonded network in epitaxy with the mica lattice, which is stable at room temperature. A model of the calculated structure is shown in Figure 26. The icelike monolayer (actually a warped molecular bilayer) corresponds to what we have called phase-I. The model is in line with the observed hexagonal shape of the boundaries between phase-I and phase-II. Another result of the MD simulations is that no free OH bonds stick out of the surface and that on average the dipole moment of the water molecules points downward toward the surface, giving a ferroelectric character to the water bilayer.

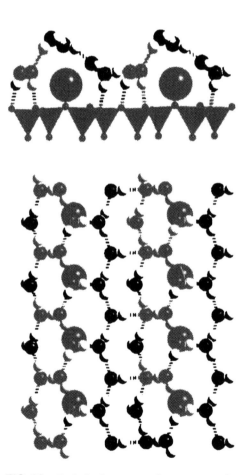

FIG. 26   Optimized structure of a water monolayer on mica obtained from molecular dynamic simulations by Odelius et al. The water molecules and the first layer of silica tetrahedra of the mica substrate are shown in a side view in the top. K ions are the large dark balls. The bottom drawing shows a top view of the water. Oxygen atoms are dark, hydrogen atoms light. Notice the ordered icelike structure and the absence of free OH groups. All the H atoms in the water are involved in a hydrogen bond to another water molecule or to the mica substrate. (From Ref. 73.)

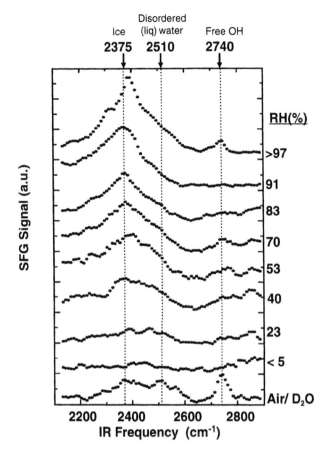

FIG. 27 Sum frequency generation spectra in *ssp* polarization of a deuterated water (D₂O) film on mica as a function of the relative humidity (RH) at room temperature (296 K). Above 40% RH, the spectrum is very similar to that of ice. The free OD stretching mode can be seen above 91% RH. The bottom trace corresponds to the spectrum of the bulk waver/vapor interface. (From Ref. 72.)

Confirmation of these results and models came from experiments performed with sum frequency generation (SFG) vibrational spectroscopy [72]. Briefly, SFG is a second-order nonlinear optical process in which a tunable infrared laser beam is mixed with a beam of visible light to generate a sum frequency output. In the dipole approximation, the process is allowed only in a medium lacking inversion symmetry, such as a surface or an interface. As a result, this technique is very useful in surface-structure studies. SFG spectra of water on mica as a function of humidity at room temperature are shown in Figure 27, along with a spectrum of a bulk liquid water surface. Deuterated water was used in these experiments to separate the spectrum from the contribution of modes due to OH groups in the mica lattice. The spectra show that: (a) very little signal is obtained below 20% RH; (b) between 20 and 90% the spectrum is very similar to that of ice; (c) little or no free OD stretch mode (i.e., non-hydrogen-bonded) intensity at 2740 cm^{-1} is observed below 91% RH; and (d) the non-hydrogen-bonded OD stretch mode appears suddenly and strongly above 91% RH. It is possible that the humidity values in these SFG experiments are higher by 5% due to laser heating (estimated to be around 1 K). The small intensity at 2740 cm^{-1}

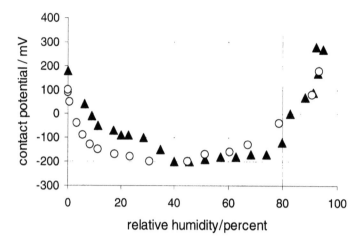

FIG. 28 Changes in contact potential of mica relative to a hydrophobic tip as a function of relative humidity. The tip–sample distance during measurements was maintained at 400 Å. At room temperature the potential first decreases by about 400 mV. At ~30% RH it reaches a plateau and stays approximately constant until about 80% RH. At higher humidity the potential increases again, eventually becoming more positive than the initial dry mica surface. The changes in surface potential can be explained by the orientation of the water dipoles described in the previous two figures.

for RH values between 20 and 80% could be due to noise, small hydrocarbon contamination, or residual free OD stretch at boundary edges of incomplete phase-I domains. However, this signal disappears at saturation, when RH ≈ 90%. These findings are fully consistent with the MD simulation. The presence of free OH groups with the H pointing out into the vacuum after completion of the icelike bilayer explains why the surface potential of the metastable islands of phase-II is positive.

Recent experiments on the surface potential of mica as a function of humidity performed by Bluhm et al. [74] provide additional support for these models. In these experiments, the tip of the SPFM was maintained at a fixed distance from the sample by keeping constant the 2ω amplitude by feedback. The 1ω amplitude was then used to measure the surface potential [see Eq. (17)]. The results are shown in Figure 28. There is a rapid decrease in potential from the dry mica value to a value 0.4 volts lower, which is reached at RH ~ 30%. After this, the potential decreases only slightly or remains constant in a plateau extending up to 80% RH. This observation is in line with the previous results and models, where the first water bilayer (phase-I) has no free OH pointing out to the vacuum and should therefore have a negative surface potential relative to dry mica. The observed decrease of 0.4 volts is smaller than that expected on the basis of the MD simulation model, which is 2.5 volts. This, however, is not surprising, since the calculated value is for a perfectly ordered bilayer. In addition, we expect some K^+ solvation. A few percentage of K ions becoming solvated would make the surface more positive by a few tenths of a volt, partially offsetting the decrease caused by the oriented dipoles of the icelike bilayer.

As can be seen, above 80% RH the surface potential increases and becomes positive. This is again in line with the appearance at high humidity of free OH groups with the H pointing out into the vacuum. In an interesting new finding, if the mica substrate is cooled below 0°C, the surface potential above 80% RH changes in the opposite direction and becomes negative. These observations seem to indicate that below 0°C, the film grows dipole-oriented for several layers, continuing the ferroelectric ordering of the first ice bilayer [74]. Additional SFG experiments at low temperature are needed to firmly corroborate this model.

SPFM has also been applied to studies of the growth of ice films on mica. These recent studies reveal that above a few bilayers of solid ice, a liquid film is present at temperatures as low as $-20°C$. We will not review these new results here, and the reader is directed to the original publications [75].

D. Ion Solvation, Mobility, and Exchange

The frequency-dependent spectroscopic capabilities of SPFM are ideally suited for studies of ion solvation and mobility on surfaces. This is because the characteristic time of processes involving ionic motion in liquids ranges from seconds (or more) to fractions of a millisecond. Ions at the surface of materials are natural nucleation sites for adsorbed water. Solvation increases ionic mobility, and this is reflected in their response to the electric field around the tip of the SPFM. The schematic drawing in Figure 29 illustrates the situation in which positive ions accumulate under a negatively biased tip. If the polarity is reversed, the positive ions will diffuse away while negative ions will accumulate under the tip. Mass transport of ions takes place over distances of a few tip radii or a few times the tip–surface distance.

Surface ions are thus expected to substantially contribute to the polarization force at low frequencies. Also, one expects different ions to have different solvation properties and mobility. These phenomena can be explored by SPFM. They are important in surface reactions, ionic exchange processes between surface and bulk ions, rock weathering, ion sequestration, and other environmental problems.

We have performed ionic mobility studies on mica and in alkali halide surfaces. Here we shall describe some results obtained on mica with different surface ions. Alkali halides will be discussed in detail in the next section.

Muscovite mica naturally contains K^+ ions on its cleavage surface. A fraction of these ions become solvated and mobile in humid environments. Experiments were per-

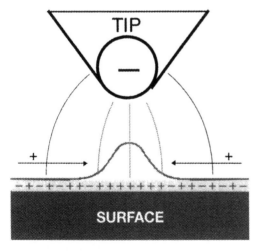

FIG. 29 Schematic diagram of a negatively charged tip attracting positively charged ions under it. Diffusion of surface ions is enhanced by solvation in humid environments.

formed to determine the response time of the polarization force as a function of humidity. To that effect a square bias voltage such as the one in Figure 30 was applied to the tip. The response time of the force to the sudden changes of bias voltage can easily be measured. The K ions of the mica surface were then exchanged by other ions by washing the mica in appropriate solutions. For example, in acids or in pure water, K^+ is replaced by H^+. In Mg- and Ca-sulfate solutions, Mg^+ and Ca^+ ions replace K^+, as verified by x-ray photoelectron spectroscopy [76]. The response time for these different ions as a function of humidity is shown in the graph in Figure 31. It is clear that H^+ (or H_3O^+) ions are much more mobile than K^+, while these ions are more mobile than Mg^+ and Ca^+.

E. Water Adsorption on Alkali Halides

Adsorption of water on salt crystals plays a key role in many atmospheric and environmental processes. Alkali halides in particular play an important role in the first stages of drop growth in clouds. To understand the atomistic details of the wetting and dissolution processes that take place in these crystals, we applied SPFM to the study of the adsorption of water vapor on single crystal surfaces and the role of surface defects, such as steps.

The (100)-oriented crystals of various alkali halides were prepared by cleavage and placed in a humidity-controlled chamber housing the SPFM microscope. Two types of ex-

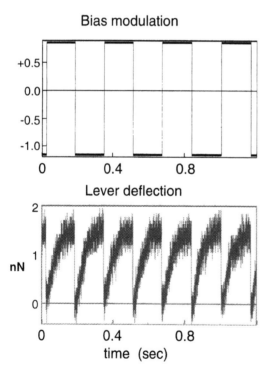

FIG. 30 *Top*: Square voltage waveform applied to the tip. *Bottom*: Corresponding changes in tip deflection (converted to force after multiplying by the lever spring constant). There is a net attractive force for both the positive and the negative cycles, but it takes time to reach the final force value. Note that the square-wave voltage is not symmetrical around zero. An offset is applied to compensate for the contact potential difference between the tip and the surface. This offset is dependent on humidity and is equal to the potential difference between the tip and the sample. (From Ref. 78.)

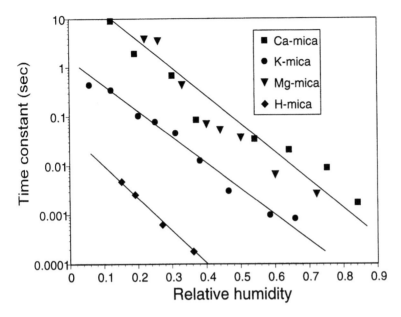

FIG. 31 Semilog plot of the force–response time measured over mica and ion-exchanged mica surfaces as a function of relative humidity. The time constant characterizes the mobility of different surface ions at different humidity values. (From Ref. 76.)

periments were performed as a function of relative humidity: (a) contact potential changes, $\Delta\phi$, and ionic mobility or, more precisely, response time, τ; and (b) topographic and contact potential mapping. The first type was performed with the tip at a fixed position, at a height of about 1000 Å over the surface. A square-wave bias plus an offset dc potential was applied to the tip. The response time was determined from the exponential change of the lever deflection (i.e., electrostatic force) to the abrupt change in bias, similar to the experiments on mica described in the previous section. The contact potential was determined from the offset voltage that was necessary to obtain a symmetrical force response, i.e., the same force in the positive and negative cycles. In this manner, both $\Delta\phi$ and τ were measured for each humidity value. The results, shown in Figure 32 for the case of KF, indicate that there is a characteristic relative humidity (RH = A) that separates two different water adsorption regimes. The values of A for several salts studied are listed in Table 1.

In the low-humidity regime (RH < A), $\Delta\phi$ increases monotonically. When RH = A, the K salts show a decrease in $\Delta\phi$, followed by a constant plateau for RH > A. For NaCl, $\Delta\phi$ continues to increase but at a lower rate. The response time τ also shows a different behavior below and above A, which is manifested by a change in the slope of $\log(\tau)$.

Images acquired in 2ω and 1ω ac-SPFM modes reveal the following changes: Below point A, there is little topographic change, as seen in both contact AFM and 2ω SPFM images. However, above A, rapid step motion occurs. The straight step edges that are present on the dry surface after preparation become curved. Small structures and sharp corners disappear. At even higher humidity values (RH = B), the deliquescence point is reached and the surface becomes featureless and smooth, due to the formation of a liquid film. A series of images illustrating this is shown in Figure 33 for NaCl [77].

The contact potential image (1ω) shows a striking contrast difference below and above point A. At low humidity there is a strong enhancement of $\Delta\phi$ at the step edges; i.e.,

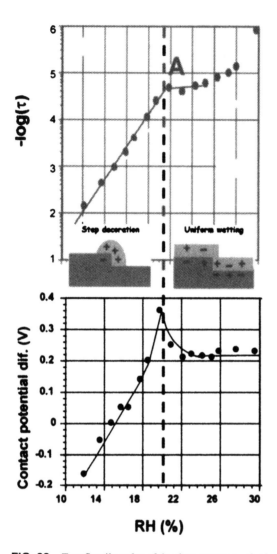

FIG. 32 *Top*: Semilog plot of the time constant τ for ionic motion as a function of RH for KF. *Bottom*: Simultaneously measured contact potential. At a critical humidity *A*, there is a break or a change in slope in these two surface properties. Below *A*, water solvates preferentially cations at the step edges. Above *A*, the rates of dissolution (solvation) of anions and cations are similar and water uni-

TABLE 1 Relative Humidity Values at Which Changes in the Slope of Surface Potential (Point *A*) Occur

	Salt			
	NaCl	KF	KCl	KBr
Point *A*	46%	20%	68%	52%
Point *B*	73%	25%	87%	86%

At point *B*, the steps disappear due to surface dissolution. This is the deliquescence point.
Source: Ref. 78.

Relative humidity (%)

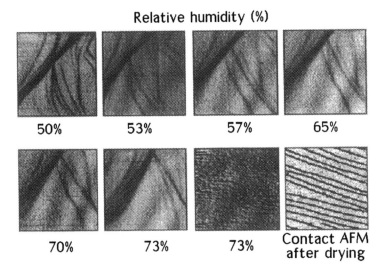

50%   53%   57%   65%

70%   73%   73%   Contact AFM
                                   after drying

FIG. 33   4-μm × 4-μm SPFM images of NaCl acquired as the humidity increases from 50 to 73%. A large multiatomic step (darkest shade) about 50 Å high runs approximately along one diagonal. Many smaller steps join this large step. The smaller steps evolve quickly, many of them disappearing as the humidity increases. At the highest humidity (73%), the features become unstable and the image changes abruptly to one without any features. This corresponds to the deliquescence point, and the surface is now covered by a liquid film. A contact AFM image (2-μm × 2-μm) acquired after lowering the humidity to under 20% is shown at the bottom right. Only monoatomic steps are observed. (From Ref. 77.)

the steps are more positive than the surrounding terraces. The enhancement reaches a maximum somewhere around RH $\approx A/2$ and decays to zero at point A. Examples of such images are shown in Figure 34.

Based on these results, a model of the water adsorption on alkali halide surfaces was proposed [78], which we summarize here. Point A corresponds to the saturation of the first water monolayer, a result that agrees with the infrared spectroscopy results of Ewing and Peters [79]. At lower humidity (RH < A), water solvates cations preferentially at the step edges. The mobile cations remain electrostatically bound in the vicinity of the negatively charged steps, giving rise to a decorative ribbon and making the local surface potential positive. The average surface potential also becomes more positive. At point A both cations and anions dissolve at similar rates, removing the fixed negative charge accumulated at the steps and triggering rapid step motion. In order for step motion to take place, mass transport of both ionic species (cations and anions) must occur. Since water dissolves mostly positive ions before point A is reached, no motion or only slow motion of the steps is expected. Anion solvation after point A produces a decrease in surface potential and also a change in the response time. Rapid displacement of step edges can now take place, since both cations and anions are dissolving, thus making mass transport possible.

Finally, upon reaching the deliquescence point, the surface step structure collapses and a thick liquid film is formed.

F. Corrosion

We conclude this review with an example of the application of SPFM to corrosion studies. Atmospheric chemical corrosion constitutes a severe threat to the structural integrity of

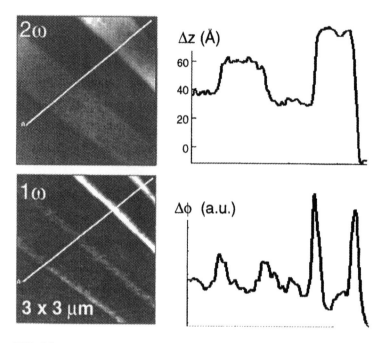

FIG. 34 3-μm × 3-μm SPFM image of a KCl surface at a relative humidity below point *A* (68%). *Top*: Topographic image (2ω). *Bottom*: 1ω-amplitude image acquired simultaneously. The contrast in this image is due exclusively to local contact potential. Cursor profiles are shown on the right side. Notice the large enhancement at the step edges, where the potential becomes positive relative to that of the surrounding terraces. (From Ref. 78.)

many metals and is responsible for considerable economic loss [80–84]. For example, sulfuric acid droplets from acid rain or dew deposition are the principal cause of the corrosion of aluminum. Since atmospheric corrosion reactions occur at the solid–liquid interface and involve submicrometer-thick liquid films and droplets, studying these reactions is difficult. SPFM, in combination with infrared reflection absorption spectroscopy (IRAS), has been used to follow in detail the physical and chemical changes occurring during the corrosion of aluminum by sulfuric acid droplets in humid environments [85]. In situ IRAS studies revealed that hydrated aluminum sulfate is the principal product of the corrosion reaction. The rate of corrosion, measured by the appearance of hydrated aluminum and sulfate ions, was found to be strongly dependent on the relative humidity (RH). At 30% RH the corrosion rate remained low and constant over time, whereas at 80% RH and above, more than a fivefold increase in the rate was observed.

SPFM experiments were performed on sulfuric acid deposited on the surface of aluminum films on silicon. A macroscopic droplet was first deposited and then rapidly dispersed using a jet of gas. This produced submicrometer-sized droplets. The initial concentration of the sulfuric acid ranged from 20 to 98 wt.%. However, the acid droplets equilibrate rapidly with the ambient water vapor. For example, at room temperature and RH = 30%, the concentration of sulfuric acid is 55 wt%; at 90% RH, it is 20 wt%. The increase in droplet volume as they equilibrate with the ambient humidity is shown in Figure 35.

After 30–60 minutes at RH > 90%, the droplets undergo a dramatic morphological change. After reaching maximum swelling, the droplets suddenly spread, as shown in Figure 36. Separate contact-angle measurements showed that dilute sulfuric acid droplets have

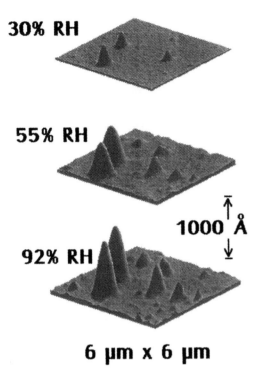

FIG. 35 6-μm $\times$ 6-μm SPFM images of sulfuric acid droplets on an aluminum substrate. Swelling of the droplets is observed as a result of increasing the relative humidity (RH). (From Ref. 85.)

larger contact angles with the Al surface than do concentrated droplets, so we can rule out changes in wettability as a source of the spreading. In addition, in experiments performed on gold, which does not react with sulfuric acid, the droplets only swelled and did not spread over a 2-hour period at 90% RH. The spreading coincides with the onset of the corrosion reactions, as indicated by the rapid growth of the SO_4^{-2} and $Al(H_2O)_x^{+3}$ infrared absorption bands.

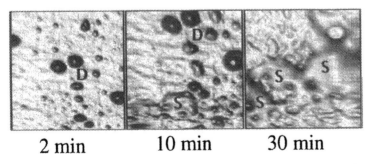

FIG. 36 Morphological changes of the sulfuric acid droplets as a function of time at about 94%. The three droplets (marked *D*) are the same as those shown in Figure 35. After a period of time of about 10 or more minutes, the droplets spread and coalesce (*S*). Spreading coincides with the corrosion reaction of the aluminum substrate. Image size is 10 μm $\times$ 10 μm. (From Ref. 85.)

90% RH   **79% RH**   **32% RH**

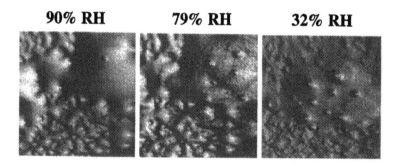

FIG. 37 10-μm × 10-μm SPFM images showing the phase separation between the hydrated aluminum sulfate and the excess unreacted sulfuric acid as the humidity is lowered. The aluminum sulfate salt precipitates, forming a solid lentillike deposit. At the lowest RH (32%), segregated acid droplets are surrounded by flat (~20 Å-thick) pancakes of liquid. (From Ref. 85.)

When the humidity is subsequently decreased to 20%, several morphologically distinct phases are detected. As shown in Figure 37a, a uniform salt-acid solution drop is seen at high humidity (RH > 90%). As the RH decreased to 79%, small droplets start to segregate (Fig. 37b). The number and size of these droplets increased as the RH decreased further (Fig. 37c). These droplets remained liquid even at RH < 20%, strongly suggesting that they consist of unconsumed sulfuric acid phase-separated from the aluminum sulfate solution during the drying process. This phase separation is indeed supported by the IRAS results.

The solid precipitate could be easily detected by contact AFM. An image taken in this mode at 20% RH is shown in the bottom image of Figure 38. The image reveals holes that penetrate the sulfate salt layer down to the underlying substrate. The holes are filled with the segregated acid droplets, as evidenced by the SPFM image shown in the top image of Figure 38, taken just before the contact AFM image. The images reveal that, in addition to the droplets filling the holes, there is a flat liquid pancake ~20 Å high surrounding the

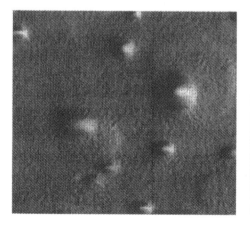 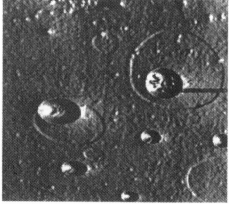

FIG. 38 *Left*: Enlarged SPFM image of the area showing the liquid droplets after drying of the reaction solution (previous figure). *Right*: Contact image acquired immediately after. The craters in this image are filled with the liquid acid droplets. The flat liquid pancake surrounding the droplets fills a shallow depression (~10 Å) in the sulfate precipitate. (From Ref. 85.)

droplets. These pancakes are contained in shallow depressions (~10 Å deep) surrounding the pits.

The following model of the corrosion process can be proposed based on the wealth of data provided by the combined application of SPFM, contact AFM, and IRAS: At low RH, the principal corrosion product, hydrated aluminum sulfate, is solid. It acts as a diffusion barrier between the acid and the aluminum substrate and prevents further corrosion. The phase separation observed between the acid and the salt at low RH strongly suggests that the salt inhibits further corrosion once it precipitates. At high RH, on the other hand, aluminum sulfate forms a liquid solution. Sulfuric acid mixes with this solution and reaches the underlying substrate, where further reaction can occur. The fluid sulfate solution also wets the surface better and thus spreads the sulfuric acid. The two processes assist each other, and the corrosion proceeds rapidly once the critical RH of 80–90% is reached.

V. CONCLUSION

An AFM-based technique, scanning polarization force microscopy, has been developed that is capable of imaging liquid films and droplets with a resolution of a few hundred angstroms in the xy plane and 1 angstrom in z. This technique operates on the principle of electrical polarization of the substrate and the ensuing electrostatic forces. It can be operated in the dc and ac modes to provide a wealth of information. Topographic mapping and surface potential mapping can be performed simultaneously. In addition, the frequency of the ac bias can be varied to distinguish between different surface processes that have characteristic response times. At present, with frequencies in the range of up to several kilohertz, SPFM provides a very useful tool to study ionic mobility, particularly on surfaces that can adsorb water from moist environments. In that respect SPFM should be a powerful tool in environmental research, in the weathering of rocks and other materials that are present in the form of microparticulates in the atmosphere. SPFM has demonstrated its usefulness in studies of disjoining pressure, by measuring the shape of nanometer-size droplets on surfaces.

Further developments promise an even brighter future for the technique. We are currently working on extending the frequency to the megahertz range, which might open the way to spectroscopic and dielectric studies of polar molecules. As we have discussed, several successful attempts at implementing high-frequency modulation have already been carried out in several laboratories.

ACKNOWLEDGMENTS

This work was supported by the Director, Office of Science, Office of Basic Energy Sciences, Materials Sciences Division, of the U.S. Department of Energy under contract number DE-AC03-76SF00098.

REFERENCES

1. F Brochard-Wyart, JM di Meglio, D Quere, PG de Gennes. Langmuir 7:335–338, 1991.
2. BV Derjaguin. Kolloid Zh 17:205–216, 1955.
3. BV Derjaguin, NV Churaev, VM Muller, JA Kitchener. Surface Forces. New York: Consultants Bureau, 1987.
4. JF Joanny, PG de Gennes. CR Acad Sci II, Mec Phys Chim Sci Univers Sci Terre 299:605–608, 1984.

5. PG de Gennes. Rev Mod Phys (USA) 57:827–863, 1985.
6. J Israelachvili. Intermolecular and Surface Forces. New York: Academic Press, 1985.
7. MP Valignat, S Villette, J Li, R Barberi, R Bartolino, E Dubois-Violette, AM Cazabat. Phys Rev Lett 77:1994–1997, 1996.
8. LD Landau, LP Pitaevski, EM Lifshitz. Statistical physics. 3rd ed. New York: Pergamon Press, 1980.
9. L Xu, M Salmeron. J Phys Chem B 102:7210, 1998.
10. JN Israelachvili, GE Adams. Nature 262:774–776, 1976.
11. JN Israelachvili, RK Tandon, LR White. Nature 277:120–121, 1979.
12. JL Parker, HK Christenson, BW Ninham. J Phys Chem 92:4155–4159, 1988.
13. JF Joanny. Journal de Mécanique Théorique et Appliquée (spec. issue):249–271, 1986.
14. J Daillant, JJ Benattar, L Bosio, L Leger. Europhys Lett 6:431–436, 1988.
15. F Heslot, AM Cazabat, P Levinson. Phys Rev Lett 62:1286–1289, 1989.
16. S Villette, MP Valignat, AM Cazabat, FA Schabert, A Kalachev. Physica A 236:123–129, 1997.
17. AM Cazabat. Experimental Aspects of Wetting. In: J Charvolin, JF Joanny, J Zinn-Justin, eds. Liquid at Interfaces. North-Holland: Elsevier Science, 1990, pp 371–414.
18. M Schick. Introduction to Wetting Phenomena. In: J Charvolin, JF Joanny, J Zinn-Justin, eds. Liquid at Interfaces. North-Holland: Elsevier Science, 1990, pp 415–498.
19. AM Cazabat, N Fraysse, F Heslot, P Carles. J Phys Chem 94:7581–7585, 1990.
20. S Bardon, R Ober, MP Valignat, F Vandenbrouck, AM Cazabat, J Daillant. Phys Rev E 59:6808–6818, 1999.
21. JP Cleveland, TE Schaffer, PK Hansma. Phys Rev B 52:R8692–8695, 1995.
22. T Pompe, A Fery, S Herminghaus. Langmuir 14:2585–2588, 1998.
23. T Pompe, A Fery, S Herminghaus. J Adhes Sci Technol 13:1155–1164, 1999.
24. S Sheiko, E Lermann, M Möller. Langmuir 12:4015–4024, 1996.
25. S Bardon, MP Valignat, AM Cazabat, W Stocker, JP Rabe. Langmuir 14:2916–2924, 1998.
26. F Zenhausern, Y Martin, HK Wickramasinghe. Science 269:1083–1085, 1995.
27. M Luna, J Colchero, AM Baro. J Phys Chem B 103:9576–9581, 1999.
28. HI Kim, CM Mate, KA Hannibal, SS Perry. Phys Rev Lett 82:3496–3499, 1999.
29. J Hu, XD Xiao, M Salmeron. Appl Phys Lett 67:476–478, 1995.
30. J Hu, XD Xiao, DF Ogletree, M Salmeron. Surf Sci 344:221–236, 1995.
31. J Hu, XD Xiao, DF Ogletree, M Salmeron. Science 268:267–269, 1995.
32. M Salmeron, L Xu, J Hu, Q Dai. MRS Bull 22:36–41, 1997.
33. Y Martin, DW Abraham, HK Wickramasinghe. Appl Phys Lett 52:1103–1105, 1988.
34. HW Gao, AM Baro, JJ Saenz. J Vac Sci Technol B 9:1323–1328, 1991.
35. S Belaidi, F Lebon, P Girard, G Leveque, S Pagano. Appl Phys A 66:S239–S243, 1998.
36. P Frantz, N Agrait, M Salmeron. Langmuir 12:3289–3294, 1996.
37. S Gomez-Monivas, JJ Saenz, R Carminati, JJ Greffet. Appl Phys Lett (submitted).
38. D Gekhtman, ZB Zhang, D Adderton, MS Dresselhaus, G Dresselhaus. Phys Rev Lett 82:3887–3890, 1999.
39. T Wei, XD Xiang, WG Wallace-Freedman, PG Schultz. Appl Phys Lett 68:3506–3508, 1996.
40. I Takeuchi, T Wei, F Duewer, YK Yoo, XD Xiang, V Talyansky, SP Pai, GJ Chen, T Venkate-san. Appl Phys Lett 71:2026–2028, 1997.
41. C Gao, XD Xiang. Rev Sci Instrum 69:3846–3851, 1998.
42. DW van der Weide. Appl Phys Lett 70:677–679, 1997.
43. C Schönenberger, SF Alvarado. Phys Rev Lett 65:3162–3164, 1990.
44. H Yokoyama, MJ Jeffery, T Inoue. Jap J Appl Phys Part 2 (Letters) 32:L1845–1848, 1993.
45. AK Henning, T Hochwitz, J Slinkman, J Never, S Hoffmann, P Kaszuba, C Daghlian. J Appl Phys 77:1888–1896, 1995.
46. M Fujihira. Annu Rev Mater Sci 29:353–380, 1999.
47. J Hu, RW Carpick, M Salmeron, XD Xiao. J Vac Sci Technol B 14:1341–1343, 1996.

48. L Leger, JF Joanny. Rep Prog Phys 55:431–486, 1992.
49. F Rieutord, M Salmeron. J Phys Chem B 102:3941–3944, 1998.
50. SS Sheiko, AM Muzafarov, RG Winkler, EV Getmanova, G Eckert, P Reineker. Langmuir 13:4172–4181, 1997.
51. L Xu, A Lio, J Hu, DF Ogletree, M Salmeron. J Phys Chem B 102:540–548, 1998.
52. S Villette, MP Valignat, AM Cazabat. J Mol Liq 71:129–135, 1997.
53. MP Valignat, N Fraysse, AM Cazabat, F Heslot, P Levinson. Thin Solid Films 234:475–477, 1993.
54. MF Grandjean. CR Acad Sci 166:165–167, 1917.
55. PS Pershan. J Phys Colloq 50:C7-1, 1989.
56. MB Feller, W Chen, YR Shen. Phys Rev A 43:6778–6792, 1991.
57. F Stevens, DL Patrick, VJ Cee, TJ Purcell, TP Beebe Jr. Langmuir 14:2396–2401, 1998.
58. J Fang, CM Knobler, H Yokoyama. Physica A 244:91–98, 1997.
59. A Artsyukhovich, LD Broekman, M Salmeron. Langmuir 15:2217–2223, 1999.
60. O Mondain-Monval, HJ Coles, T Claverie, JR Lalanne, JP Marcerou, J Philip. J Chem Phys 101:6301–6317, 1994.
61. SHJ Idziak, CR Safinya, RS Hill, KE Kraiser, M Ruths, HE Warriner, S Steinberg, KS Liang, JN Israelachvili. Science 264:1915–1918, 1994.
62. L Xu, M Salmeron, S Bardon. Phys Rev Lett 84:1519–1522, 2000.
63. VJ Novotny. J Chem Phys 92:3189–3196, 1990.
64. TM O'Connor, YR Back, MS John, BG Min, DY Yoon, TE Karis. J Appl Phys 79:5788–5790, 1996.
65. X Ma, CL Bauer, MS John, J Gui, B Marchon. Phys Rev E 60:5795–5801, 1999.
66. X Ma, J Gui, B Marchon, MS John, CL Bauer, GC Rauch. IEEE Trans Magn 35:2454–2456, 1999.
67. X Ma, J Gui, L Smoliar, K Grannen, B Marchon, MS Jhon, CL Bauer. J Chem Phys 110:3129–3137, 1999.
68. X Ma, J Gui, L Smoliar, K Grannen, B Marchon, CL Bauer, MS Jhon. Phys Rev E 59:722–727, 1999.
69. VJ Novotny, I Hussla, JM Turlet, MR Philpot. J Chem Phys 90:5861–5868, 1989.
70. L Xu, DF Ogletree, M Salmeron, H Tang, J Gui, B Marchon. J Chem Phys 112:2952–2957, 2000.
71. D Beaglehole, HK Christenson. J Phys Chem 96:3395–3403, 1992.
72. PB Miranda, L Xu, YR Shen, M Salmeron. Phys Rev Lett 81:5876–5879, 1998.
73. M Odelius, M Bernasconi, M Parrinello. Phys Rev Lett 78:2855–2858, 1997.
74. H Bluhm, T Inoue, M Salmeron. Surf Sci 462:L599–L602, 2000.
75. H Bluhm, M Salmeron. J Chem Phys 111:6947–6954, 1999.
76. L Xu, M Salmeron. Langmuir 14:5841–5844, 1998.
77. Q Dai, J Hu, M Salmeron. J Phys Chem B 101:1994–1998, 1997.
78. M Luna, F Rieutord, NA Melman, Q Dai, M Salmeron. J Phys Chem A 102:6793–6800, 1998.
79. GE Ewing, SJ Peters. Surf Rev Lett 4:757–770, 1997.
80. TE Graedel, RP Frankenthal. J Electrochem Soc 137:2385–2394, 1990.
81. TE Graedel. J Electrochem Soc 136:204–212, 1989.
82. TE Graedel. J Electrochem Soc 136:193–203, 1989.
83. RT Foley, TH Nguyen. J Electrochem Soc 129:464–467, 1982.
84. TH Nguyen, RT Foley. J Electrochem Soc 129:27–32, 1982.
85. Q Dai, J Hu, A Freedman, GN Robinson, M Salmeron. J Phys Chem 100:215–220, 1996.

7

Nanometric Solid Deformation of Soft Materials in Capillary Phenomena

MARTIN E. R. SHANAHAN   National Centre for Scientific
Research/School of Mines Paris, Evry, France

ALAIN CARRÉ   Corning S.A., Avon, France

I. INTRODUCTION

The spreading of liquids on solid surfaces is of considerable interest and importance in many fields of activity. The dynamic aspects of spreading are particularly relevant in several practical applications in industry, such as coating, adhesive bonding, printing, and composite manufacturing. When inks or paints are applied, they must wet their substrates before solidifying. Similarly, good spreading is required to ensure interfacial contact between two phases when applying a polymeric adhesive or in the manufacturing of composite materials.

Capillary phenomena are also essential in tribology and in many biological systems, such as blood circulation and eye irrigation, involving the formation and persistence of the lachrymal film.

The equilibrium at the solid (S), liquid (L), and vapor (V) triple line is described by Young's equation [1] in the form

$$\gamma_{SV} = \gamma_{SL} + \gamma \cos \theta_0 \tag{1}$$

where γ_{SV}, γ_{SL}, and γ represent interfacial tensions for the solid/vapor, solid/liquid, and liquid/vapor interfaces, respectively, and θ_0 is the equilibrium contact angle between the tangent planes to the S/L and L/V boundaries at the three-phase line, or triple line. One simple way to derive Young's equation is to consider the local force balance parallel to the solid surface, although more rigorous approaches exist based on surface thermodynamics [2–4].

When a liquid spreads spontaneously on a (flat) rigid solid, in either the wetting or dewetting mode, a dynamic equilibrium is set up in which the nonequilibrated Young force, corresponding to excess capillary energy, causes triple-line motion. Simultaneously, viscous dissipation in the liquid, chiefly near the wetting front, acts as an energy sink and moderates spreading speed [5]. Although many commonly encountered solids are, to all intents and purposes, infinitely rigid as far as capillary forces are concerned, soft solids such as certain classes of polymers and gels can be deformed significantly. Deformations involved are generally in the mesoscopic range (of the order of γ/μ, where γ is the surface tension of the liquid and μ is the shear modulus of the solid [6]), yet they can have marked macroscopic effects. A "wetting ridge" is formed at the triple line, as the wetting front advances or recedes, this ridge accompanies the motion leading to energy dissipation due to the hysteretic nature of the soft substrate and the strain cycle undergone [7]. Following a number of experimental studies [8–14] it is now well established that the wetting ridge, typically of the order of tens of nanometers in height, can be so important in its consequences that viscous dissipation within the liquid becomes virtually negligible.

The process of "viscoelastic braking" just described has certain parallels with the dynamic adhesion of elastomers. When, for example, a rubber strip is peeled from a rigid substrate, the effective, or apparent, work of adhesion, W, is usually much greater than the intrinsic, or reversible, energy of adhesion, W_0, given by the Dupré equation [15]:

$$W_0 = \gamma_1 + \gamma_2 - \gamma_{12} \tag{2}$$

where γ_1, γ_2 represent the surface free energies of the solids in contact and γ_{12} is their interfacial free energy.

As has been known for many years now, the extra energy is associated with the viscoelastic dissipation occurring in the deformed rubber in the vicinity of the fracture front [16–19]. It turns out that the effective energy of adhesion can often, to a good degree of approximation, be expressed as W_0 multiplied by a factor involving viscoelastic properties of the polymer (dependence on fracture rate and temperature). The degree of cross-linking of the elastomer is also relevant [20]; even in the case of quasi-static adhesion as measured by the JKR test [21], it has been shown that apparent adhesion can increase with average inter-cross-link molecular weight, M_c (at least in the fracture mode, if not in the formation mode) [22].

In this chapter, we will review the consequences of solid deformation in the kinetics of the spreading of a liquid on a soft material, in both wetting and dewetting modes. The influence of solid deformation induced by the liquid surface tension will be shown in the case of a liquid drop placed on a soft elastomeric substrate and in the case of an unstable liquid layer dewetting on a soft rubber. The impact of solid deformation on the kinetics of the wetting or dewetting of a liquid will be analyzed theoretically and illustrated by a few concrete examples. The consequences of solid deformation in capillary flow will be also analyzed.

II. BASIC THEORY

A. Capillarity and Elasticity at the Triple Line

We shall first consider the local behavior near a solid/liquid/vapor triple line (NB: the vapor could be replaced by a second liquid, immiscible with the first, but we shall refer to "vapor" in order to avoid possible confusion). Figure 1 corresponds to a schematic representation of the wetting front. Although horizontally (i.e., parallel to the undeformed solid surface), components of interfacial tensions balance (at least at equilibrium), vertically, $\gamma \sin \theta$ is counteracted by a stress field within the solid, leading to the formation of what we term a *wetting ridge*. Angle θ may be either the static equilibrium angle, θ_0, or a dynamic angle during spreading, $\theta(t)$, where t is time. The following argument also applies to the latter case, since spreading phenomena will occur at a rate far below the speed of sound in the various phases and local force equilibrium is therefore ensured.

More complete analyses have been effected [6,23], but we shall present here a simple scaling procedure in order to isolate the essential nature of the wetting ridge. Consider a zone of typical linear dimension ω in the vicinity of the triple line, which corresponds to the solid disturbed by the capillary force $\gamma \sin \theta$. Within this region, a stress of order of

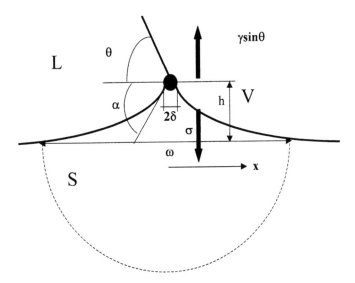

FIG. 1 Schematic representation of disturbed triple-line region on a soft solid.

magnitude σ results. For local, vertical equilibrium to ensue, we have:

$$\omega\sigma \approx \gamma \sin\theta \tag{3}$$

Adopting h as the height of the wetting ridge, we have local strain, ε, given in order of magnitude by:

$$\varepsilon \approx \frac{h}{\omega} \tag{4}$$

Taking the solid to the Hookean, we have $\sigma = \varepsilon Y$, with Y being Young's modulus. Using Eqs. (3) and (4), we obtain as a scaling law:

$$h \approx \frac{\gamma \sin\theta}{Y} \tag{5}$$

This simple expression really states that the perturbation to the solid is of dimension determined by the ratio of liquid surface tension to solid Young's modulus (i.e., capillarity versus elasticity) modulated by the angle of application of γ. Equation (5) demonstrates why the vertical component of surface tension is often neglected. Typically, a "hard" solid such as a metal presents a value of Y of the order of 100 GPa, and an organic liquid, a γ of the order of 3×10^{-2} Nm^{-1}. With a contact angle of 90°, $h \approx 10^{-2}$ Å: clearly a value without any physical meaning since we are well below the scale of applicability of continuum mechanics. However, many soft (and also dissipative—this is important in the following) solids, such as gels and elastomers, present Young's moduli of the order of 1 MPa (or less). With similar liquids, we may expect a value of h of about 30 nm or more—still small but nonnegligible. As we shall see, this mesoscopic effect may lead to macroscopic consequences.

The foregoing suggests a scale for h but not the shape of the wetting ridge. More detailed analysis [6] leads to the form:

$$h(x) = \frac{2(1 - \nu^2)\gamma \sin\theta}{\pi Y} \ln\left|\frac{L}{x}\right| \tag{6}$$

where ν is Poisson's ratio for the solid, x is (horizontal) distance from the triple line, and L is a cutoff distance that can be related to a datum depth within the solid (this constant is a usual consequence of 2-dimensional stress analysis). Due to the logarithmic dependence of $h(x)$ on x, there is a cutoff at $|x| \approx \delta$. For $|x| < \delta$, Eq. (6) no longer applies, and the behavior of the solid within this region is no longer linearly elastic [6]. Note from Figure 1 that the "true" contact angle is now somewhat greater than θ. However, in the context of wetting it may be shown that the conventional definition of this quantity, i.e., the angle subtended between the L/V and the *undisturbed* solid surface, remains the angle to be employed in subsequent analysis (unless we are dealing with an extremely small sessile drop (radius of the order of microns or less) and the solid is exceedingly soft [24]).

In the case of elastomeric solids, which we shall treat later, $\nu = 0.5$, thus Eq. (6) simplifies to:

$$h(x) = \frac{\gamma \sin \theta}{2\pi\mu} \ln \left| \frac{L}{x} \right| \qquad \text{for } |x| > \delta \qquad (7)$$

where μ is the shear modulus of the elastomer ($Y = 3\mu$).

B. Evidence of the "Wetting Ridge"

Direct evidence for the existence of the wetting ridge, whose size can be estimated from Eq. (5) and whose form corresponds to Eq. (7), has been obtained [8]. The solid substrate used was a silicone rubber of low Young's modulus, $Y = 1.9$ MPa (RTV 615, General Electric Co.) and the liquid, tricresyl phosphate (TCP, Aldrich Chemicals). Scanning white-light interferometric microscopy was used to investigate topographical modifications near the triple line elastomer/TCP/air. Figure 2 shows the surface profiles obtained with the undisturbed, smooth elastomeric surface (arithmetic mean roughness of about 2 nm) and

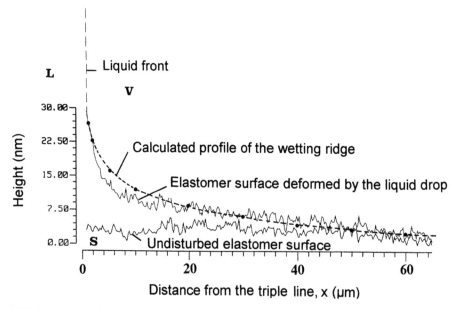

FIG. 2   Results of white-light interferometric microscopy observation near the triple line of a drop of tricresyl phosphate (TCP) on a soft silicone rubber (RTV 615, General Electric Co.).

that perturbed by the proximity of a wetting triple line. By estimating the distance L, from Eq. (7), as being about 60 μm, it was found that the ridge height, h_0, was approximately 30 nm. As can be seen in Figure 2, the theoretical profile of the wetting ridge, given by Eq. (17), is in good agreement with experimental observation.

Considering Figure 1, it may now be argued that the equilibrium contact angle is no longer θ_0, since the angle subtended between the tangents of the L/V and S/L interfaces at the triple line is now somewhat greater, $\theta_0 + \alpha$. However, a semivariational analysis has been effected that leads to the conclusion that Young's equation (with θ_0 defined in the usual way), with respect to the undisturbed solid surface, should still be valid in nearly all circumstances [24].

C. Dynamic Considerations

Unless we are considering an exceedingly small sessile drop on a very soft solid, as stated earlier, the conventional contact angle, θ_0, as defined by the Young equation, is still the pertinent value to be considered at equilibrium and the existence of a wetting ridge is of rather academic interest. However, in the context of spreading (wetting or dewetting), the solid deformation can have a very marked effect on *dynamic* behavior. As the triple line moves toward equilibrium, a wetting ridge accompanies it on a soft solid. The advancing wetting front raises the solid immediately ahead of its path and, after passing, releases the strained solid. As a result, a surface layer of solid [of thickness of the order of L as given by Eq. (7)—typically of the order of 50 μm in the case of a soft elastomer] undergoes a strain cycle. If the solid were to be purely elastic, all strain energy would be restored after passage of the triple line, and the wetting ridge would not contribute to the overall dynamic behavior. However, soft elastomers (and other soft solids) are, in general, viscoelastic and therefore lossy. As a result, the strain cycle results in energy loss by dissipation, and this must be allowed for in the dynamic energy balance of spreading.

A rigorous approach to evaluating this dissipated energy has previously been undertaken [7] and later improved to allow for the behavior of practical elastomers [9–12]. However, we shall present here a simplified argument in order to arrive at the essential scaling law and will subsequently quote the more exact version.

Consider the work, E_1, effected when the vertical component of liquid surface tension, $\gamma \sin \theta$, "lifts" the local solid to a height h—that of the wetting ridge. This work, per unit length of triple line, is simply $h\gamma \sin \theta$. Using Eq. (5), we obtain:

$$E_1 \approx \frac{\gamma^2 \sin^2 \theta}{Y} \tag{8}$$

In fact, the horizontal component of liquid surface tension, $\gamma \cos \theta$, is not without effect, since the triple-line region now "protrudes" from the bulk solid. This component leads to a stretching of the surface layer of solid on the vapor-phase side of the triple line for $\theta < \pi/2$ or on the liquid side for $\theta > \pi/2$. By analogy with Eq. (8), we have a second contribution to the work effected, E_2:

$$E_2 \approx \frac{\gamma^2 \cos^2 \theta}{Y} \tag{9}$$

and thus the overall work supplied by the liquid surface tension and transferred to the solid (per unit length of triple line) is given by:

$$E \approx E_1 + E_2 \approx \frac{\gamma^2}{Y} \tag{10}$$

As the wetting front advances at speed U, the solid undergoes a strain cycle at a variety of frequencies, f, the local frequency depending on the distance of the element of solid from the contact line at the moment under consideration. The solid the furthest from the contact line, yet still perturbed by the presence of the three-phase line, is at a distance of ca. ω and thus "feels" a strain cycle at frequency U/ω. At the other extreme, near the lower cutoff at $|x| = \delta$, the frequency is ca. U/δ. The latter frequency will be dominant, since it is in the direct vicinity of the three-phase line that the solid is strained the most. As a consequence, and using Eq. (10), we can define the rate at which work is being done as:

$$\dot{E} \approx \frac{\gamma^2 U}{Y\delta} \tag{11}$$

A viscoelastic solid may be characterized by a complex tensile modulus, $Y^*(f)$, a function of strain rate frequency, f,

$$Y^*(f) = Y'(f) + iY''(f) \tag{12}$$

where Y' and Y'' are, respectively, storage and loss components of the modulus and $i^2 = -1$. The fraction of strain energy dissipated in a cycle is given by the loss tangent, $\tan \Delta(f)$ $[= Y''(f)/Y'(f)]$, or, alternatively, $\tan \Delta(U)$, since $f \approx U/\delta$ in the present context. We may thus write an expression giving the quantity of energy dissipated per unit time and per unit length of triple line, $\dot{D}$:

$$\dot{D} \approx \dot{E} \tan \Delta(f) \approx \frac{\gamma^2 U Y''(U)}{Y Y'(U)\delta} \tag{13}$$

where $Y = |Y^*(f)|$.

Equation (13) gives the basic scaling law for dissipation. But if a more rigorous treatment is applied, we obtain [7]:

$$\dot{E} \approx \frac{\gamma^2 U}{2\pi\mu\delta} \tag{14}$$

Given that for a rubber compound, $Y = 3\mu$, comparison of Eqs. (11) and (14) shows that the simple scaling argument is in agreement with the more exact expression to within a factor of about 2.

However, use of the loss tangent in Eq. (13) is an oversimplification. Instead, we shall make use of an empirical dissipation law obtained in the study of rate-dependent adhesion of elastomers (19). Maugis and Barquins [19], in order to explain the dynamic adhesion of elastomers in a variety of test configurations (peel, flat punch, etc), proposed the following relation:

$$G - W_0 = W_0 \Phi(a_T v) \tag{15}$$

where G is the strain energy release rate, W_0 is the Dupré energy of adhesion [15], $\Phi(a_T v)$ is a dimensionless function of separation rate v, and a_T is the Williams, Landel, and Ferry time–temperature shift factor [25]. It was shown that $\Phi(a_T v)$ varied as v^n, where the constant n was of the order of 0.5–0.6. Given that similar elastomeric materials will be considered in wetting and dewetting, we replace Eq. (13) by Eq. (16):

$$\dot{D} \approx \frac{\gamma^2 U}{2\pi\mu\delta} \left(\frac{U}{U_0}\right)^n \tag{16}$$

where U_0 is constant, which may be taken as a characteristic speed.

D. Triple-Line Motion

The preceding description applies to energy dissipation during triple-line motion, irrespective of the overall context of the process. In practical terms, the three-phase line can either advance, as in the case of wetting, or recede, as in dewetting. Both wetting and dewetting are of considerable importance. The former is essential in, for example, the manufacture of composite materials when fibers must be adequately (completely) covered by a polymeric resin. Dewetting is necessary in the application of windscreen wipers on a road vehicle! In the case of spin casting or coating, the aim is to obtain a homogeneous, thin layer of a liquid, before solidification, on a given substrate. If the equilibrium contact angle of the liquid, θ_0, is nonzero and the applied film is too thin, spontaneous dry patches may form and ruin the uniformity of the coating. This is an undesirable case of dewetting. In both wetting and dewetting, it will be shown how deformation of a soft substrate may markedly slow down movement of the triple line.

E. Wetting of a Sessile Drop

We can consider the spreading of a sessile drop on a soft, lossy substrate rather like the advance of a "negative" crack and thus use fracture mechanics concepts, as was the case in the derivation of Eq. (15) for the separation of an elastomer from a rigid solid. The term *negative* is used since the spreading of a drop leads to the *creation* of solid/liquid interface rather than separation.

Equilibrium at the triple line satisfies Young's equation [Eq. (1)], but if the contact angle as a function of time, $\theta(t)$, is greater than θ_0, then there is a net spreading force acting per unit length of triple line and given by [5]:

$$F = \gamma_{SV} - \gamma_{SL} - \gamma \cos \theta(t) = \gamma[\cos \theta_0 - \cos \theta(t)] \tag{17}$$

By analogy with Eq. (15), this may be expressed as:

$$W_0 - G = \gamma[\cos \theta_0 - \cos \theta(t)] \tag{18}$$

where

$$W_0 = \gamma_{SV} + \gamma - \gamma_{SL} = \gamma(1 + \cos \theta_0) \tag{19}$$

and

$$G = \gamma[1 + \cos \theta(t)] \tag{20}$$

Thus the "strain energy release rate" is effectively an "instantaneous" value of Dupré's energy of adhesion, with $\theta = \theta(t)$ instead of the equilibrium value. The sign reversal in the left-hand side of Eq. (18) when compared to Eq. (15) is due simply to the fact that we have a "closing crack" with a spreading liquid.

The equivalent of the right-hand side of Eq. (15), the dissipation, is composed of two additive terms. One is due to viscous shear within the liquid, and the other is due to viscoelastic losses in the wetting ridge. We may thus write:

$$W_0 - G \approx \frac{3\eta \ell U}{\theta(t)} + \frac{\gamma^2}{2\pi\mu\delta} \left(\frac{U}{U_0} \right)^n \tag{21}$$

The first term on the right-hand side of Eq. (21), due to viscous dissipation, involves liquid viscosity, η, and a logarithmic ratio of drop contact radius and a molecular cutoff, ℓ. This viscous term [5] is, however, small compared to the viscoelastic dissipation, and we shall henceforth neglect it. Rearranging Eq. (21) without the viscous contribution gives:

$$U \approx U_0 \left[\frac{2\pi\mu\delta}{\gamma^2} (W_0 - G) \right]^{1/n} = U_0 \left[\frac{2\pi\mu\delta}{\gamma} (\cos\theta_0 - \cos\theta(t)) \right]^{1/n} \qquad (22)$$

Practically, contact angles are measured as a function of time t after deposition of liquid drops. The spreading speed, U, is given by dr/dt, with r being the drop radius at time t. In the following, the dynamic contact angle will sometimes be expressed as a function of the spreading speed, U [$\theta(t) \equiv \theta(U)$].

III. WETTING DYNAMICS

A. Evidence of Viscosity-Independent Spreading

As illustrated in Figure 3, contact angle $\theta(t)$ as a function of time t for three liquids of widely varying viscosity η but similar surface tension γ was measured on an elastomer. Drops of 2 μL were placed on a silicone rubber (a two-component silicone, RTV 630, General Electric Co.) at 20°C. The contact angles were measured with a Ramé–Hart contact-angle goniometer equipped with a video camera connected to a video recorder and printer. Contact angles $\theta(t)$ were measured as a function of time after deposition, t. It was possible to analyze up to 24 frames/s. Each contact-angle value is the average of five measurements, the standard deviation is of the order of 0.5 degree or less.

The shear modulus of the elastomer, μ, was 1.5 MPa. The liquids used were N-methylpyrrolidone, NMP ($\eta = 2$ cP, $\gamma = 41.2$ mN·m^{-1}), NMP containing 17.5 wt% of polyimide, PI (Whitford Plastics Ltd) ($\eta = 150$ cP, $\gamma = 41.5$ mN·m^{-1}), and NMP containing 28 wt% of PI ($\eta = 1250$ cP, $\gamma = 42.8$ mN·m^{-1}). The rubber samples were placed in a glass box saturated in vapor of the relevant liquid to eliminate any interference due to evaporation of NMP from the drops and to prevent water vapor pickup from the atmosphere.

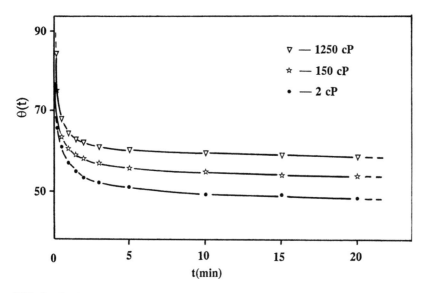

FIG. 3 Evolution with time, t, of contact angle, of NMP, with NMP containing 17.5 wt% of polyimide and NMP containing 28 wt% of polyimide.

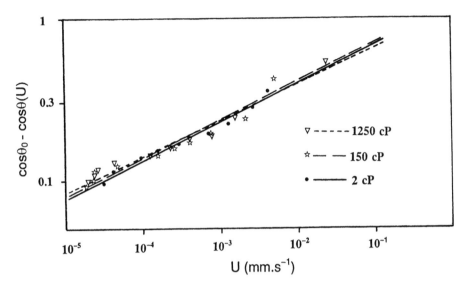

FIG. 4   Variation of $\cos \theta_0 - \cos \theta(U)$ as a function of the spreading speed U, both on logarithmic scales, corresponding to viscoelastically controlled spreading for the three liquids of Figure 3.

Although the three liquids have very different viscosities, it is remarkable to observe that the general variation of contact angles after deposition follows essentially similar dependence with time t. The contact angles reach relatively stable values after 20 minutes, whatever their viscosity. However, equilibrium contact angles, θ_0, were measured after 2 h, when the systems had apparently reached equilibrium. A viscosity-dependent regime would have allowed a more quickly attained stable value for the least viscous fluid ($\eta = 2$ cP) and a longer time to equilibrium for the highest value of viscosity ($\eta = 1250$ cP). Therefore, the spreading dynamics of the three liquids is not governed by viscosity. According to Eq. (21), replacing W_0 and G by the Young–Dupré equation ($W_0 = \gamma(1 + \cos \theta_0)$ and $G = \gamma(1 + \cos \theta(U))$ and neglecting viscous dissipation we obtain:

$$\cos \theta_0 - \cos \theta(U) = \frac{\gamma}{2\pi\mu\delta} \left(\frac{U}{U_0} \right)^n \tag{23}$$

Equation (23) may be written as:

$$\log[\cos \theta_0 - \cos \theta(U)] = n \log U + \log\left(\frac{\gamma}{2\pi\mu\delta U_0^n} \right) \tag{24}$$

Since the second member on the right-hand side is essentially constant for the solid and liquids considered, we should expect a plot of $\log[\cos \theta_0 - \cos \theta(U)]$ versus $\log U$ to give straight lines of gradient n that are directly superposable, irrespective of liquid viscosity.

Figure 4 gives the results corresponding to Figure 3 plotted in this manner. As can be seen, the relationship between $\log[\cos \theta_0 - \cos \theta(U)]$ and $\log U$ is satisfactorily linear within a range covering approximately three decades of speed. In addition, the results obtained for the three viscosities of 2, 150, and 1250 cP lie on the same line, within the limits of experimental scatter, thus showing that the motion is governed essentially by the viscoelastic properties of the (same) substrate and not by liquid viscosity. The values of gradient n are, respectively, 0.24, 0.24, and 0.22 for the viscosities 2, 150, and 1250 cP, giving an average value of 0.23 ± 0.01. These values would seem to be rather low, but it should be pointed out that the elastomer was not preswollen for the experiments reported here, although the experiments were effected in atmospheres saturated with the vapor of the relevant liquid to eliminate potential evaporation effects. Typical value of n found in adhesion experiments are in the range of 0.5–0.6 [19]. It has been hypothesized that swelling of the substrate by the liquid influences the behavior of the solid material and modifies the value of n, as will be demonstrated and interpreted in the next subsection.

B. Swelling Effect

To investigate the influence of swelling of the substrate by the contacting liquid, the contact angle θ of sessile drops of tricresylphosphate, TCP (drop volume 2 μL, viscosity η = 70 cP, surface tension = 40.9 mN · m^{-1}), has been measured as a function of time after deposition, t, on flat, smooth, horizontal surfaces of soft and rigid solids at 20°C. The method of measurement of contact angle is the same as in Section III.A.

Although TCP is a liquid of low volatility at 20°C, solid samples were placed in a closed glass box to avoid any interference from the room atmosphere.

In the case of relatively high-modulus solids (Teflon PFA, Du Pont de Nemours, μ ≈ 250 MPa, and fused silica, Quartz et Silice, μ ≈ 30 GPa), equilibrium contact angles are attained after only 15 seconds or less, as shown in Figures 5 and 6. For these two rigid materials, the kinetics of spreading are very similar, although the contact angle on silica is small compared with that on Teflon PFA. This confirms that viscous dissipation is relatively short lived, even for small equilibrium contact angles.

Figure 5 also gives the variation of the contact angle, $\theta(t)$, on an elastomer (two-component silicone rubber, RTV 630, General Electric Co.) having a low shear modulus, μ, equal to 1.5 MPa. In the first set of experiments, in order to avoid any interference produced by moderate swelling of the elastomer by TCP *during* spreading, the rubber samples were immersed in the liquid for several days prior to the wetting experiments until equilibrium

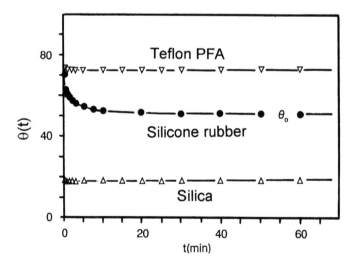

FIG. 5 Evolution of the contact angle, $\theta(t)$, of TCP on preswollen silicone rubber (RTV 630, General Electric Co.), Teflon PFA, and fused silica.

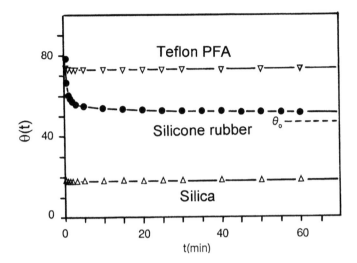

FIG. 6 Evolution of the contact angle, $\theta(t)$, of TCP on nonpreswollen silicone rubber (RTV 630, General Electric Co.), Teflon PFA, and fused silica.

swelling was obtained (noted by gravimetry). The samples were then carefully dried superficially before depositing drops of the same liquid on the surface.

Much slower spreading occurs with the rubbery material, with approximately 30 minutes being necessary to achieve equilibrium. This is attributed to local deformation of the substrate leading to viscoelastic braking of the spreading of TCP. The hypothesis is corroborated by line a of Figure 7. This line has been obtained by plotting the difference be-

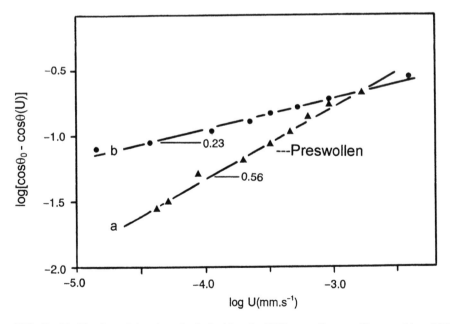

FIG. 7 Verification of the viscoelastic braking for TCP spreading on silicone rubber (RTV 630, General Electric Co.). Line *a* with preswelling, line *b* without preswelling by TCP.

tween the cosines of equilibrium contact angle, θ_0, and the instantaneous contact angle, $\theta(U)$, as a function of the liquid spreading speed, U ($U = dr/dt$, where r is the drop contact radius). Verification of Eq. (24) with a gradient $n = 0.56$ suggests that the spreading dynamics of TCP on the silicone elastomer is effectively governed by the formation of a wetting ridge at the liquid-drop periphery, with a phenomenon of viscoelastic dissipation in the wetting ridge. The dissipative properties of the elastomer have also been determined using the rolling cylinder adhesion test [12]; the value of n was found to be 0.55, which is in good agreement with the value deduced from wetting kinetic measurements.

Figure 6 presents the variation of the contact angle, $\theta(t)$, on the same rubber, but this time the elastomer had not been preswollen by TCP. In comparison with the spreading kinetics on rigid solids (Teflon PFA, fused silica), we observe the same qualitative behavior as before, i.e., a relatively slower variation of the contact angle $\theta(t)$ on the elastomer. However, after the main variation of the contact angle occurring within the first 30 minutes, we observe a very slow, yet continuous, decrease of $\theta(t)$ over a period of several hours before a stable value for θ_0 is obtained.

Application of Eq. (24) leads to the line b of Figure 7. This result suggests that the spreading dynamics of TCP on the silicone rubber is again governed by the formation of a wetting ridge. However, the value of the gradient, $n = 0.23$, is rather low in comparison, on one hand, with values typically found in dynamic adhesion studies [12,19,26] and, on the other hand, with the formerly discussed experiments using the preswollen rubber. It is suspected that moderate swelling of the rubber by TCP, which is of the order of 1 wt% at equilibrium, may influence the spreading kinetics. Swelling may be responsible for the low value of n. Further evidence for interference due to swelling is the slow and very long decay of $\theta(t)$ for several hours before a stable equilibrium value, θ_0, is reached, with this period of time being reduced to 30 minutes when the rubber is preswollen.

To separate purely viscoelastic effects from swelling effects in the set of experiments reported in Figure 6, a simple model of diffusion of TCP in the rubber substrate has been developed.

Rubber swelling modifies the liquid/solid work of adhesion, W_0, because in addition to the initial liquid/solid interactions, liquid diffusion into the solid produces supplementary liquid/liquid interactions, liquid molecules having passed through the liquid/solid interface. Therefore, to the initial work of adhesion in the absence of swelling, W_0, an additional term corresponding to a fraction of the cohesion energy of the liquid, 2γ, should be added. If t is the time of diffusion, the work of adhesion at t, $W_0(t)$, can then be expressed as

$$W_0(t) = W_0 + 2\lambda(t)\gamma \tag{25}$$

with $\lambda(t) \ll 1$. In Eq. (25) neither changes of the volume of the liquid and solid phases nor possible variations of the solid/liquid interface area are taken into account, given the low degree of swelling of the rubber by TCP (≤ 1 wt%) and the relatively minor impact of swelling on contact angle at equilibrium.

To evaluate the time-dependent function, $\lambda(t)$, a simple model of diffusion is proposed. Starting from Langmuir adsorption theory, we consider that liquid molecules having diffused into the elastomer are localized on discrete sites (which might be free volume domains). In these conditions, we can deduce the rate of occupation of these sites by TCP with time. Only the filling of the first layer of the sites situated below the liquid/solid interface at a distance of the order of the length of intermolecular interaction, i.e., a few nanometers, needs to be considered to estimate $\lambda(t)$.

As in a monolayer adsorption process, we consider that the rate of filling of sites by TCP molecules follows first-order kinetics. If N_0 represents the total number of free sites per unit area at time $t = 0$, and $N(t)$ is the number of sites available at time t, then $dN(t)/dt = -kN(t)$, where k is the rate constant of the adsorption process. Therefore, $N(t)$ decreases as $N_0 \exp(-kt)$, and the number of sites occupied by TCP molecules at t becomes $[N_0 - N(t)]$, a quantity that determines directly the parameter $\lambda(t)$ in Eq. (25). So $W_0(t)$ can be written as

$$W_0(t) = W_0 + aN_0(1 - e^{-kt})\gamma \tag{26}$$

where a is a constant.

Using the Young–Dupré equation, $W = \gamma(1 + \cos\theta)$, and introducing the boundary condition of $\theta = \theta_0$ for $t = 0$ (no swelling) and $\theta_{0\infty}$ (contact angle at equilibrium with swelling), we obtain:

$$\cos\theta_0(t) = \cos\theta_0 + (\cos\theta_{0\infty} - \cos\theta_0)(1 - e^{-kt}) \tag{27}$$

Considering that $e^{-kt} \approx 1 - kt$ (for relatively low values of t) and that $\cos\theta \approx 1 - \theta^2/2$ (for $\theta \leq 1$ rd) allows us to deduce an approximate expression for Eq. (27):

$$\theta_0(t) \approx \theta_0 - k\frac{\theta_0^2 - \theta_{0\infty}^2}{2\theta_0}t \tag{28}$$

This last equation explains the linear decay of $\theta(t)$ observed between 30 and 60 minutes (Fig. 8). Now taking the experimental values for $\theta_0(t = 0)$ and $\theta_{0\infty}$ ($t \to \infty$) as 53° and 47° allows us to estimate the rate constant of the diffusion process of TCP, k, which is found to be of the order of 9×10^{-5} s^{-1}. It can be observed that the variation of θ_0 from 53° to 47° due to swelling is minor. This validates the use of the simplifying hypotheses leading to Eq. (28).

The accuracy of constant k has been evaluated by comparing the experimental values of $\theta(t)$ with the values deduced from Eq. (28) for contact times greater than 30 minutes shown in Table 1. Good agreement between both series of values justifies the simple model of diffusion of TCP in silicone rubber.

Considering that the evolution of the contact angle after 30 minutes results only from rubber swelling, it is possible to separate purely viscoelastic braking from the swelling effect by taking $\theta^*(t) = \theta(t) + \alpha t$ as being the theoretical contact angle in the absence of swelling (curve c, Fig. 8).

The use of $\theta^*(t)$ in Eq. (24) is shown in Figure 9, where the swelling free linear fit (line c) has a gradient n of 0.59 and is situated parallel and near to the preswollen data (line a). This result indicates that the low gradient value of $n = 0.23$ is related to the slight swelling of the silicone rubber by TCP. The swelling produces a slow and small decay of the equilibrium contact angle with time.

Wetting experiments performed using a silicone rubber and TCP as a liquid have shown the existence of two phenomena. Using the rubber in both its unswollen and preswollen states, it has been demonstrated that the major effect observed is a slowing down of spreading after deposition of the liquid drop, as compared with behavior on rigid solids. This effect is related mainly to viscoelastic dissipation occurring in the wetting ridge in the vicinity of the triple line. There is also a modification of the spreading kinetics due to swelling of the elastomer by the liquid and subsequent modification of interfacial interactions. This secondary phenomenon increases the time necessary to attain the equilibrium value θ_0. As a result, the value of the factor n of Eq. (24) is apparently reduced. Variation

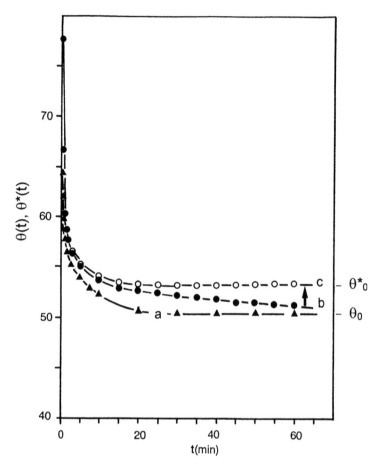

FIG. 8 Evolution of the contact angle, $\theta(U)$, of TCP on silicone rubber (RTV 630, General Electric Co.), (a) preswollen by TCP, (b) silicone rubber, and (c) on silicone rubber after allowing for swelling effects.

of interfacial interactions with time due to swelling has a direct impact on the Dupré work of adhesion at equilibrium. However, swelling may affect also the damping properties of the elastomer. The small horizontal shift between lines *a* and *c* of Figure 9 may be due to a reduction of the loss modulus, Y'', produced by the swelling of the rubber. This slight shift is also observed with the cylinder adhesion test performed on swollen and nonswollen rubber tracks.

TABLE 1 Comparison Between Values of $\theta(t)_{exp}$ and Theoretical Values Deduced from Eq. (28) for $t > 30$ min ($k = 9 \times 10^{-5}$ s^{-1} $\theta_0 = 53°$, and $\theta_{0\infty} = 47°$)

	t/s						
	1800	2100	3600	7200	10800	14400	∞
$\theta(t)_{exp}$	52.2	52.0	51.3	50.7	50.2	49.7	47.0
$\theta_0(t)_{th}$	52.2	52.1	51.5	50.3	49.4	48.7	47.0

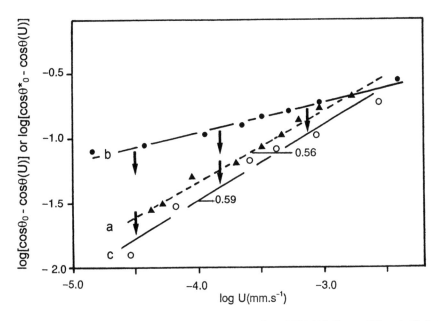

FIG. 9   Variation of the gradient n with rubber swelling (RTV 630, General Electric Co.). After correction of swelling effect on θ_0, line b becomes c. Line a refers to preswollen rubber.

It has been shown that the kinetics of the spreading of a liquid on a rubber is largely dependent on viscoelastic dissipation in the wetting ridge of the substrate near the triple line. This behavior may, in practice, be slightly altered by moderate swelling that modifies the solid/liquid interactions for long contact times.

IV.   DEWETTING DYNAMICS: HARD VERSUS SOFT SUBSTRATES

A.   Hard Substrate

Spreading dynamics plays a key role in numerous applications. However, controlling the dewetting of liquids may be potentially more important for some industrial uses.

Let us consider a circular "puddle" of liquid, L, on solid, S, in the presence of liquid vapor, V. The puddle is of radius R_0 and small initial thickness e_0. We assume that holes nucleate spontaneously in the puddle and grow with radius $r(t)$ as time t passes because the equilibrium contact angle, θ_0, is nonzero. The liquid is unstable as a wetting film. Equilibrium thickness of a film, e_c, is given by [27,28]

$$e_c = 2K^{-1} \sin \frac{\theta_0}{2} \tag{29}$$

where K^{-1} is the Laplace or capillary length [$K^{-1} = (\gamma/\rho g)^{1/2}$, in which ρ is liquid density and g is gravitational acceleration]. When the film thickness e_0 is less than e_c, it is unstable, and holes may nucleate spontaneously due to thermally induced fluctuations. We will consider the range of $r(t)$ much smaller than R_0.

From simple considerations of the local force balance near the triple line, or of free energy changes, the global driving force for dewetting per unit length of triple line, F_m, is given by

$$F_m = \gamma + \gamma_{SL} - \gamma_{SV} = \gamma(1 - \cos\theta_0) \tag{30}$$

where γ_{SV} and γ_{SL} are the solid/vapor and solid/liquid interfacial tensions, respectively.

At the early stage of dewetting, when $r(t) \ll R_0$, rims form around dry dewetting zones and the braking force results mainly from shearing of the liquid contained in the wedge of the liquid rim surrounding a dry zone (see Fig. 10). The profile of the rim is not very different from the profile of a spreading drop, at least at the early stages of the dewetting process when the width of the rim stays below K^{-1}, neglecting gravitation.

In these conditions, the two edges of the rim contribute to the dissipation [27,28], and the global braking force, F_v, satisfies

$$F_v \approx \frac{6\eta\ell U}{\theta(U)} \tag{31}$$

assuming that the dynamic contact angles are equal at the two edges of the rim.

Equating the driving and braking forces, we obtain:

$$U \approx \frac{\gamma\theta(U)(1 - \cos\theta_0)}{6\eta\ell} \tag{32}$$

Therefore, the dewetting speed appears to be constant and, since $U = dr(t)/dt$, where $r(t)$ is the hole, or dry-zone, radius, Eq. (32) leads to:

$$r(t) \approx \theta(U)\left[\frac{\gamma(1 - \cos\theta_0)}{6\eta\ell}\right]t \tag{33}$$

However, on rigid substrates, the growth of dry zones is accompanied by a rim of excess liquid with width λ (Fig. 10). As the dewetting proceeds, λ increases. For short times and $\lambda < K^{-1}$, the growth of dry patches is controlled only by surface tension forces and the dewetting speed is constant. A constant dewetting speed of 8 mm·s^{-1} has been measured when a liquid film of tricresyl phosphate (TCP) dewets on Teflon PFA, a hard fluoropolymer of low surface free energy ($\mu = 250$ MPa, $\gamma = 20$ mJ·m^{-2}).

B. Soft, Viscoelastic Substrate

For a deformable yet purely elastic solid, the energy input corresponding to Eq. (14) would be restored after passage of the triple line, but solids that are sufficiently soft for the wetting ridge to be significant are, in general, lossy, and thus a fraction of the strain energy is

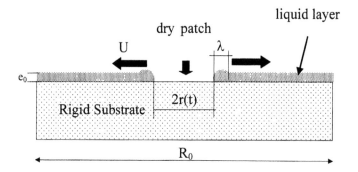

FIG. 10 Formation of a dry patch (zone) in an unstable liquid film on a rigid substrate.

dissipated during spreading. The dissipation function previously adopted with success is of the form $(U/U_0)^n$, where U_0 and n are constants [8–12].

Dewetting on soft materials in the following part specifically concerns us. In dewetting, the contact angle stays reasonably constant (and close to the equilibrium value, θ_0). Energy dissipation due to liquid viscosity can then be expressed as $k\eta U^2$, where k is a constant and η is the liquid viscosity [5]. As the triple line recedes, the capillary energy supplied (per unit time and per unit length of the triple line) is given by:

$$\dot{F} = U(\gamma + \gamma_{SL} - \gamma_S) = U\gamma(1 - \cos\theta_0) \tag{34}$$

where the second equality follows from Young's equation. Equating source and sinks, we obtain:

$$U\gamma(1 - \cos\theta_0) = k\eta U^2 + \frac{\gamma^2 U}{2\pi\mu\delta}\left(\frac{U}{U_0}\right)^n \tag{35}$$

which is the fundamental relation describing dewetting on a viscoelastic substrate. The viscous contribution to energy dissipation is minor and will henceforth be neglected. Since $U = dr/dt$, where r is the hole, or dry-patch, radius and t is time, Eq. (35) leads to:

$$r(t) \approx U_0\left[\frac{4\pi\mu\delta\sin^2\left(\frac{\theta_0}{2}\right)}{\gamma}\right]^{1/n} t \tag{36}$$

The technique used to study dewetting dynamics on materials consists of making a flat, smooth elastomer surface. A liquid "puddle" is deposited within a 50-mm-diameter ring of 0.1-mm-thick plasticized adhesive paper adhering to the substrate. The adhesive paper acts as a spacer. A microscope slide is drawn over the liquid to obtain a liquid film of ca. 0.1-mm thickness. At this thickness, the liquid film is unstable, being much less than the equilibrium value, e_c, of ca. 1.5 mm calculated from Eq. (29). Nucleation of dry patches occurs spontaneously for the most part, presumably due to surface defects or thermal agitation energy.

The occurrence and growth of dewetted holes, or dry patches, is followed using a video camera rigged up to a low-power microscope. Using a video recorder, it is possible to analyze up to 24 frames/second.

For the experiments reported here, the liquid used was tricresyl phosphate (TCP) (Aldrich Chemical Co.), which has the advantage of low volatility ($\gamma = 40.9$ mN·m^{-1}, $\rho = 1.14 \times 10^3$ kg·m^{-3} and viscosity, $\eta = 0.07$ Pa·s at ambient temperature).

Figure 11 shows top views of a dry zone growing as function of time, t, in a film of TCP on a silicone rubber (RTV 615, General Electric Co.) whose Young's modulus is 2.1 MPa. One interesting feature of the dewetting phenomenon on a soft material is that there is no rim visible at the periphery of the growing dry patch, with the relatively low speed of dewetting allowing the rapid equilibrium of Laplace's pressure in the liquid film.

In order to study effects of cross-linking density, the polymer used was a two-component transparent silicone elastomer (RTV 615, General Electric Co.). By varying the ratio of the two components, resin and cross-linking agent, constituting the final rubber, it was possible to alter the degree of cross-linking, or average inter-cross-link molecular weight, M_c (measured in gm·mole^{-1}). The cross-linking density was modified by varying the curing agent content from 2 to 20 parts (by weight) for 100 parts of resin. Although no experiments were undertaken to measure M_c directly, elastic moduli Y were assessed for the elastomers using hardness measurements (in all cases but for the softest rubber, where

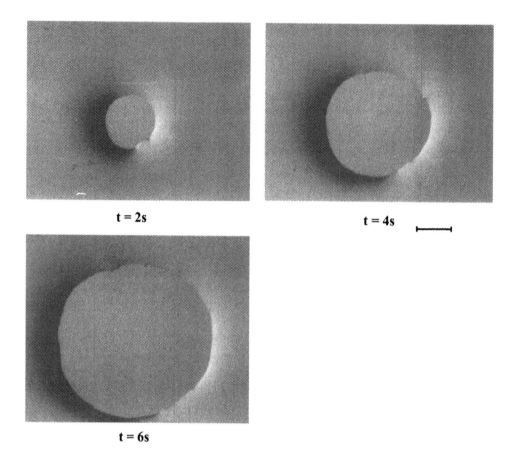

FIG. 11 Dry patch growing in a film of TCP on silicone rubber (RTV 615, General Electric Co., Y = 2.1 MPa). The origin of time, t, is the nucleation of the patch (the scale represents 1 mm). Note the absence of rim around the dry zone.

hardness could not be readily measured—estimation in this case was made using a simple tensile test). Bearing in mind that the relationship between Young's modulus Y and shear modulus μ for an elastomer is simply $Y = 3\mu$, M_c could be estimated from the classic equation of rubber elasticity [29]:

$$M_c = \frac{\tilde{\rho}RT}{\mu} \tag{37}$$

where $\tilde{\rho}$ is polymer density, R and T are, respectively, the gas constant and the absolute temperature ($\tilde{\rho} = 1.02 \times 10^3$ kg $\cdot$ m^{-3} for all cross-link densities studied).

The equilibrium contact angle, θ_0, of TCP on the silicone rubber was found to be 47 $\pm$ 3°, with a small apparent dependence on M_c, which may possibly be linked to a slight degree of swelling of the elastomer by the liquid, being more accentuated in the case of higher M_c [30]. The variability being nevertheless small, it will be neglected in the following. Data concerning the elastomer in its different states of crosslinking are presented in Table 2.

Some comparative dewetting experiments were conducted on a fluoropolymer, Teflon PFA (du Pont de Nemours and Co.) representing a relatively rigid substrate (μ = 250 MPa) with similar surface characteristics (surface free energy $\gamma_S = 20$ mJ $\cdot$ m^{-2}).

TABLE 2 Data Concerning the Silicone Elastomer in Its Different States of Cross-linking

Y (MPa)	μ (MPa)	M_c (gm·mole^{-1})	$\theta_0(°)$
0.01	0.003	750 000	44
0.65	0.217	11 500	45
1.8	0.6	4 150	47
2.1	0.7	3 550	50

In Figure 12 we present results of hole formation in the TCP layer on the silicone elastomer for different values of M_c, together with a summary of those on Teflon PFA for purposes of comparison. The radius of the hole, $r(t)$, is given as a function of time, t, from hole initiation. Results are the averages of five experiments for the silicone elastomer. It is clear that, in all cases, the relationship is linear, in accordance with Eq. (36); and what is more, the gradient increases with the elastic modulus of the polymer, Y (or μ). The Teflon PFA has the highest dewetting speed of 8 mm·s^{-1}, and the silicone rubber of Young's modulus of 0.01 MPa has the lowest, at 0.048 ± 0.007 mm·s^{-1}.

We therefore have qualitative evidence for the dependence of the dewetting speed on the elastic properties of the substrate. Dependence of wetting on the elastic modulus was previously suggested in the case of thin substrates [31]. It may be conjectured that cross-linking affects the surface properties of the elastomer and, therefore, wettability. However,

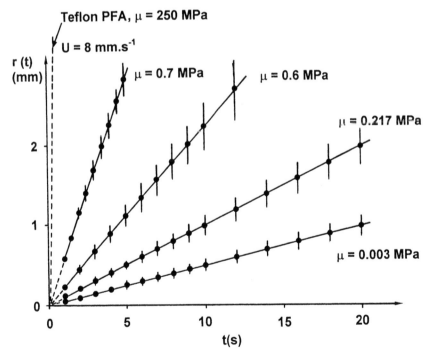

FIG. 12 Radius, $r(t)$, of dry patches growing in an unstable TCP film on silicone rubber (RTV 615, General Electric Co.), of different shear modulus, μ. Teflon PFA is a hard fluoropolymer of similar surface free energy ($\gamma = 20$ mJ·m^{-2}).

this effect is small, as shown by the equilibrium contact angles of $47 \pm 3°$ observed with TCP and reported earlier. The large differences in observed dewetting speeds are, however, far too great to be attributable to any small changes in wettability.

The viscoelastic character must also be taken into account, and we shall attempt to elucidate the behavior quantitatively or at least semiquantitatively. We shall start by rewriting Eq. (36) in the form

$$U \approx U_0 \left[\frac{4\pi\mu\delta \sin^2\left(\frac{\theta}{2}\right)}{\gamma} \right]^{1/n} \tag{38}$$

The values of γ and θ_0 are, respectively, totally and virtually independent of degree of cross-linking. The shear modulus, μ, is directly related to M_c via Eq. (37). As for U_0, we must make some assumptions, since the dissipative factor of Eq. (16) is empirical rather than theoretical. Although not entirely rigorous, we shall assimilate the dissipation factor approximately with the viscoelastic loss factor of the substrate, $\tan \Delta = Y''/Y'$, where Y' and Y'' are, respectively, storage and loss moduli. It is well know that $\tan \Delta(f)$ depends on frequency, f, and also possibly on amplitude, among other things; therefore a full appraisal of energy losses within the wetting ridge would be exceedingly complex. We shall thus limit ourselves to a simple treatment. Since U is a variable of the system and not of the solid alone, we shall tentatively associate only U_0 and n, the constants characterizing the loss behavior, with $\tan \Delta$, and write:

$$\tan \Delta \sim U_0^{-n} \tag{39}$$

It has previously been found experimentally [22], albeit for a different elastomer (a synthetic polyisoprene), that to a good degree of approximation there is a linear relationship between $\tan \Delta$ and average inter-cross-link molecular weight, M_c, such that $\tan \Delta \propto M_c$. We therefore write:

$$U_0 \sim (\tan \Delta)^{-1/n} \sim M_c^{-1/n} \tag{40}$$

With the preceding, we may now write Eq. (38) in the form

$$U \approx \alpha \left[\frac{4\pi\delta \sin^2\left(\frac{\theta}{2}\right)}{\bar{\rho}RT\gamma} \right]^{1/n} \mu^{2/n} \tag{41}$$

where α is constant.

Although the value of n is not accessible experimentally from dewetting experiments, earlier work involving wetting suggested a value of ca. 0.6. We assume that n remains unaffected by the degree of cross-linking: this would seem to be a reasonable supposition given that a range of elastomers previously studied gave similar values of n in the range of 0.5–0.6 [8,9,11,12,14]. Taking $n = 0.6$, the exponent of μ in Eq. (41) is equal to 10/3. Figure 13 shows dewetting speed, U, as a function of $\mu^{10/3}$, and Table 3 gives relevant data.

The relationship can be seen to be acceptably linear. Linear regression analysis gives the relation:

$$U = 1.51 \times \mu^{10/3} + 4.79 \times 10^{-2} \tag{42}$$

where U is in mm·s^{-1} and μ is in MPa, with a regression coefficient of $r = 0.95$. From the foregoing, simple theory, the line would be expected to go through the origin. The corresponding one-parameter least squares analysis leads to $U = 1.7\mu^{10/3}$.

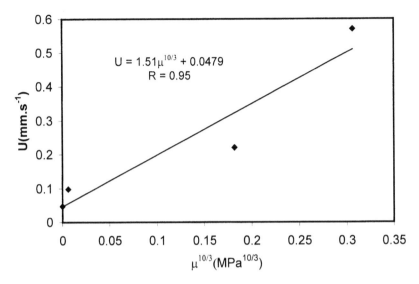

FIG. 13 Linear regression between the dewetting speed, U, and $\mu^{10/3}$, where μ is the shear modulus of the silicone rubber (RTV 615, General Electric Co.).

The preceding semiquantitative analysis seems reasonably satisfactory, but we have nevertheless neglected any possible variations in δ with degree of cross-linking. Notwithstanding any potential error introduced by this neglect, difficult to counteract with the data presently available, let us estimate δ for the hardest elastomer, viz. that of Young's modulus equal to 2.1 MPa (the elastomer in this state of cross-linking has been previously studied, and we therefore have more information about it). From the preceding one-parameter least squares analysis and Eq. (38) we have:

$$U \approx 1.7\mu^{10/3} \approx U_0 \left[\frac{4\pi\delta \sin^2\left(\dfrac{\theta}{2}\right)}{\gamma} \right]^{5/3} \mu^{5/3} \tag{43}$$

The values of θ_0 and γ are known and for the elastomer of Young's modulus of 2.1 MPa, $U_0 = 8 \times 10^{-1}$ mm·s^{-1} [12]. We can then evaluate δ at ca. 20 nm. This value is perhaps a little high but of the same order of magnitude as earlier estimated [6]. Thus, despite some necessary approximations and simplifying hypotheses, we arrive at a semiquantitative explanation of the relationship between dewetting and therefore, presumably, wetting speed and the molecular structure of the elastomeric substrate.

TABLE 3 Modulus Data and Spreading Speed, U, for the Silicone Elastomer

μ (MPa)	$\mu^{10/3}$ (MPa)$^{10/3}$	U (mm·s^{-1})
0.003	5.14×10^{-9}	$(4.8 \pm 0.7) \times 10^{-2}$
0.217	6.2×10^{-3}	$(9.8 \pm 1.2) \times 10^{-2}$
0.600	0.182	0.22 ± 0.03
0.700	0.306	0.57 ± 0.004

V. CONSEQUENCES OF SOLID DEFORMATION IN CAPILLARY FLOW

In the wetting and dewetting kinetics studies described earlier, the solid substrate was a flat and smooth surface. However, the solid deformation due to the action of the vertical component of the liquid surface tension may be expected to act in any geometry. For example, viscoelastic braking is involved in the sliding of a liquid drop on a tilted rubber track [32].

The wetting ridge may also form in a capillary tube if the material has a sufficiently low Young's modulus (a few MPa).

In a small-diameter capillary tube, wetting forces produce a distortion of the free liquid surface, which takes a curvature. Across a curved liquid surface, a difference of pressure exists and the variation of pressure across the surface is the Laplace or capillary pressure, given by:

$$\Delta P = \frac{2\gamma \cos\theta_0}{r} \tag{44}$$

where r is the capillary tube radius, γ is the liquid surface tension, and θ_0 is the equilibrium contact angle of the liquid on the solid.

Laplace's pressure produces the capillary rise inside a small tube. We propose to examine now the consequences of the solid deformation in capillary flow.

In order to simplify the analysis, we will consider the capillary flow of a liquid L in a horizontal small tube (diameter much smaller than the capillary length), in order to avoid complications due to gravity effects (Fig. 14).

The flow rate inside a capillary may be obtained from Poiseuille's law, which relates the liquid flow rate, dV/dt (V = volume, t = time), to the difference of pressure ΔP "pushing" the liquid inside the tube and to the dimensions of the tube, in particular, its radius, r.

According to Poiseuille's law, the flow rate is given by:

$$\frac{dV}{dt} = \frac{\pi r^4 \, \Delta P}{8\eta H} = \pi r^2 \frac{dH}{dt} \tag{45}$$

where η is the viscosity of the liquid and H is the length of the liquid column inside the capillary.

In the simple case where the tube is horizontal and supposing that $P_i = P_e$ (Fig. 14), Eqs. (44) and (45) lead to the speed of displacement of the liquid inside the tube:

$$\frac{dH}{dt} = \frac{\gamma r \cos\theta_0}{4\eta H} \tag{46}$$

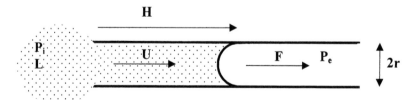

FIG. 14 Capillary flow in a small horizontal tube.

Equation (46) takes into consideration only the viscous drag due to Poiseuille flow inside the tube.

In the case where the material used to make the capillary tube is a soft rubber, Eq. (46) does not apply any more, due to the viscoelastic braking induced by the displacement of the wetting ridge. The viscoelastic braking force, f_v, per unit of length of the triple line depends on the flow speed U according to:

$$f_v = \frac{\gamma^2}{2\pi\mu\delta} \left(\frac{U}{U_0}\right)^n \tag{47}$$

where the parameters μ, δ, and U_0 have the same meaning as in Eq. (23).

On the wetting perimeter of the soft capillary, this braking force becomes:

$$F_v = \frac{\gamma^2}{2\pi\mu\delta} \left(\frac{U}{U_0}\right)^n \times 2\pi r = \frac{\gamma^2 r}{\mu\delta} \left(\frac{U}{U_0}\right)^n \tag{48}$$

The net force "pushing" the liquid inside the capillary reduces to:

$$F = 2\pi r\gamma \cos\theta_0 - \frac{\gamma^2 r}{\mu\delta} \left(\frac{U}{U_0}\right)^n \tag{49}$$

The effective capillary pressure is now given by:

$$\Delta P = \frac{2\gamma \cos\theta_0}{r} - \frac{\gamma^2}{\pi\mu\delta r} \left(\frac{U}{U_0}\right)^n \tag{50}$$

and as a result, the flow rate corresponds to:

$$U = \frac{dH}{dt} = \frac{r^2}{8\eta H} \left[\frac{2\gamma \cos\theta_0}{r} - \frac{\gamma^2}{\pi\mu\delta r} \left(\frac{U}{U_0}\right)^n\right] \tag{51}$$

Let us denote as U_1 the flow rate when only viscosity intervenes according to the Poiseuille analysis and as U_2 the flow rate in the presence of the additional viscoelastic braking. U_2 and U_1 are related according to:

$$U_2 = U_1 - \frac{\gamma^2 r}{8\pi\eta\mu\delta H} \left(\frac{U_2}{U_0}\right)^n \tag{52}$$

Equation (52) allows us to estimate the impact of viscoelastic braking on the capillary flow rate. As an example, we will consider that the liquid is tricresyl phosphate (TCP, $\gamma = 50$ mN·m^{-1}, $\eta = 0.07$ Pa·s). The viscoelastic material is assumed to have elastic and viscoelastic properties similar to RTV 615 (General Electric, silicone rubber), i.e., a shear modulus of 0.7 MPa ($Y = 2.1$ MPa), a cutoff length of 20 nm, and a characteristic speed, U_0, of 0.8 mm·s^{-1} [30]. TCP has a contact angle at equilibrium of 47° on this rubber.

A simple resolution of Eq. (52) is feasible if the factor n is taken as 0.5. Then, the relationship between U_2 and U_1 is a second-degree equation:

$$x^2 + \frac{\gamma^2 r}{8\pi\eta\mu\delta H U_0^{\frac{1}{2}}} x - U_1 = 0 \tag{53}$$

where $x = (U_2)^{1/2}$.

The flow rates U_1 and U_2 are given as a function of the length H in Figure 15 for a capillary tube having a radius of 0.1 mm. The impact of the viscoelastic braking is particularly strong for small values of the penetration depth (length H).

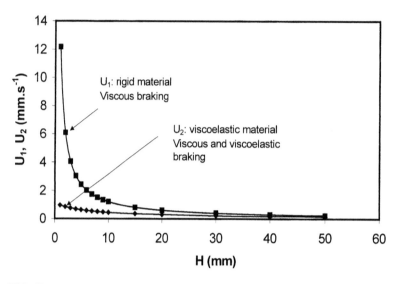

FIG. 15   Flow rate of TCP in a horizontal capillary tube ($r = 0.1$ mm) in the case of purely viscous braking (U_1, rigid material) and in the case of viscoelastic braking (U_2, soft rubber).

VI.   CONCLUSIONS

Wetting and dewetting on elastomeric substrates are controlled not only by surface properties of the liquid and solid phases and liquid viscosity, but also by bulk properties of the polymer. The component of liquid surface tension acting perpendicularly to the (undisturbed) solid surface leads to local nanometric deformation and the creation of a wetting ridge of height of order of magnitude γ/μ, where γ is liquid surface tension and μ is solid shear modulus. Existence of the wetting ridge has been demonstrated by using white-light interferometric microscopy. In the wetting or the dewetting mode, as triple-line motion occurs, this ridge must accompany the wetting/dewetting front, leading to a strain cycle of the solid surface, which causes (viscoelastic) energy dissipation. This dissipation can outweigh viscous losses and thus control spreading speed, as verified in several examples.

The degree of cross-linking of an elastomer, or inter-cross-link molecular weight, M_c, has an influence both on the elastic moduli and the dissipation properties of the material, with both factors entering into the viscoelastic braking term in spreading. By varying the ratio of resin to cross-linking agent in a silicone rubber, it has been shown how spreading speed decreases with increasing M_c. A simple theory has been developed in which a linear relationship between spreading speed, U, and $\mu^{10/3}$, where μ is the shear modulus, is predicted for the elastomer in question.

The theory of viscoelastic braking in liquid spreading exposes the various possibilities that may exist for controlling wetting or dewetting speeds by changing solid rather than liquid properties. Applications may exist in the fields of contact lenses, printing, and vehicle tire adhesion.

The consequences of the wetting ridge in the capillary penetration of a liquid into a small-diameter tube have been evaluated. Viscoelastic braking reduces the liquid flow rate when viscoelastic dissipation outweighs the viscous drag resulting from Poiseuille flow.

REFERENCES

1. T Young. Phil Trans Roy Soc London 95:65, 1805.
2. RE Johnson. J Phys Chem 63:1655, 1959.
3. RE Collins, CE Cooke. Trans Faraday Soc 55:1602, 1959.
4. MER Shanahan. In: KW Allen, ed. Adhesion 6. London: Applied Science, 1982, p 75.
5. PG de Gennes. Rev Mod Phys 57:827, 1985.
6. MER Shanahan, PG de Gennes. In: KW Allen, ed. Adhesion 11. London: Elsevier Applied Science, 1987, p 71.
7. MER Shanahan. J Phys D Appl Phys 21:981, 1988.
8. A Carré, JC Gastel, MER Shanahan. Nature 379:432, 1996.
9. A Carré, MER Shanahan. C R Acad Sci Paris 317(II):1153, 1993.
10. MER Shanahan, A Carré. Langmuir 10:1697, 1994.
11. A Carré, MER Shanahan. Langmuir 11:24, 1995.
12. MER Shanahan, A Carré. Langmuir 11:1396, 1995.
13. A Carré, MER Shanahan. Langmuir 11:3572, 1995.
14. MER Shanahan, A Carré. J Adhesion 57:179, 1996.
15. A Dupré. Théorie Mécanique de la Chaleur. Paris: Gauthier-Villars, 1869, p 369.
16. AN Gent, RP Petrich. Proc Roy Soc London A310:433, 1969.
17. AN Gent, J Schultz. J Adhesion 3:281, 1972.
18. EH Andrews, AJ Kinloch. Proc Roy Soc London A332:385, 1973.
19. D Maugis, M Barquins. J Phys D Appl Phys 11:1989, 1978.
20. A Carré, J Schultz. J Adhesion 17:135, 1984.
21. KL Johnson, K Kendall, AD Roberts. Proc Roy Soc London A324:301, 1971.
22. MER Shanahan, F Michel. Int J Adhesion Adhesives 11:170, 1991.
23. GR Lester. J Colloid Sci 16:315, 1961.
24. MER Shanahan. J Phys D Appl Phys 20:945, 1987.
25. JD Ferry. Viscoelastic Properties of Polymers. 2nd ed. New York: Wiley, 1970, p 314.
26. N Zaghzi, A Carré, MER Shanahan, E Papirer, J Schultz. J Polym Sci Polym Phys 25:2383, 1987.
27. F Brochard, C Redon, F Rondelez. C R Acad Sci Paris 306:1143, 1988.
28. C Redon, F Brochard, F Rondelez. F Phys Rev Lett 66:175, 1991.
29. LRG Treloar. The Physics of Rubber Elasticity. Oxford: Clarendon, 1949.
30. A Carré, MER Shanahan. J Colloid Interface Sci 191:141, 1997.
31. MER Shanahan. Rev Phys Appl 23:1031, 1988.
32. MER Shanahan, A Carré. C R Acad Sci Paris 1(IV):263, 2000.

8

Two-Dimensional and Three-Dimensional Superlattices: Syntheses and Collective Physical Properties

MARIE-PAULE PILENI   Université Pierre et Marie Curie, LM2N, Paris, France

I. INTRODUCTION

Self-assembled nanocrystals and their specific properties represent an area between macroscopic and microscopic physics that has progressed rapidly over the last five years, with the activity in this area expanding exponentially during this time [1–36]. The first self-assembled 2D and 3D superlattices were observed with Ag_2S and CdSe [1–4], and a large number of groups have now succeeded in forming various self-organized nanocrystals of silver [5–16], gold [15–27], cobalt [28,29], and cobalt oxide [30,31]. Except for CdSe [2] and cobalt [28,29] nanocrystals, most structures are formed from particles coated with dodecanethiol. By changing the experimental conditions in the deposition technique, various states of self-organization have been observed: At low particle concentration in an initial bulk suspension of oil, very large rings, which separate areas covered with monolayers from the bare surface, were obtained with silver [14,15], gold [17], and CdS [32] nanoparticles. These patterns have been explained in terms of wetting properties. Large "wires" of silver nanocrystals have been observed [9], and simultaneously it was seen that self-organization changes with the length of the alkyl chains [33] used to coat the particles. It has been demonstrated that the 3D superlattices, termed *aggregates* in the present paper, are usually self-organized in a crystalline phase in an FCC structure [1–6,10], and in some cases the structure may be hexagonal [8,34]. More recently it has been demonstrated that the physical properties of silver [6,35] and cobalt [28,29] nanocrystals organized in 2D and/or 3D superlattices differ from those of isolated nanoparticles. Collective properties are observed and with silver nanocrystals, it has been shown that these properties depend on the structure of the self-organization (hexagonal or square network) [36].

   In this review, we describe collective physical properties due to self-organization of nanocrystals in 2D and 3D superlattices.

II. SYNTHESIS OF NANOCRYSTALS [37,38]

Nanocrystals are fabricated by using reverse micelles. Functionalized surfactants are employed.

A.   Synthesis of Silver Nanocrystals [5,6]

Two reverse micelles [39] are mixed with the water content fixed at $w = [H_2O]/[AOT] = 40$. The first one is made with 30% Ag(AOT), 70% Na(AOT), and the second one with N_2H_4 with 100% of Na(AOT). The overall concentration of hydrazine is 7×10^{-2} M. After adding dodecanthiol to the solution (1 μL/1 mL), addition of ethanol to the micellar solution induces flocculation of the silver dodecanthiol–coated particles. Then the solution is filtered and the precipitate is easily redispersed in hexane. After extraction, the size distribution is reduced from 40% to 30%. However, the polydispersity remains rather large. To reduce this, size-selected precipitation is used. Size-selected precipitation, SSP, is a well-known technique for separating mixtures of copolymers and homopolymers during the synthesis of sequenced copolymers. It has been used for extraction of nanosized crystals elsewhere [40]. This method is based on the mixture of two miscible solvents differing by their ability to dissolve the surfactant alkyl chains. The silver-coated particles are highly soluble in hexane and poorly soluble in pyridine. Thus, pyridine is progressively added to hexane solution containing the silver-coated particles. At a given volume of added pyridine (roughly 50%), the solution becomes cloudy and a precipitate appears. This corresponds to agglomeration of the largest particles as a result of their greater van der Waals interactions. The solution is centrifuged and an agglomerated fraction rich in large particle collected, leaving the smallest particles in the supernatant. The agglomeration of the largest particles is reversible, and the precipitate, redispersed in hexane, forms a homogeneous clear solution. This procedure induces a decrease in the size distribution. By repeating the same procedure, very small particles are obtained with a very low polydispersity (15%).

B.   Synthesis of Cobalt Nanocrystals [28,29]

A reverse micellar solution of 0.25 M Na(AOT) and 2×10^{-2} M Co(AOT)$_2$ is mixed with 0.25 M Na(AOT) micelles containing 2×10^{-2} M sodium tetrahydroboride, NaBH$_4$, as reducing agent. The two micellar solutions keep the same water content, i.e., the same diameter ($w = 10$). The synthesis is carried out in the presence of air. Immediately after mixing, the micellar solution remains optically clear, and its color turns from pink to black, indicating the formation of colloidal particles. They are extracted from reverse micelles, under anaerobic conditions, by covalent attachment of either trioctyl phosphine [41] or lauric acid [42]. They are then redispersed either in pyridine or in hexane, respectively. A size selection as described earlier is made. This chemical surface treatment highly improves the stability of cobalt exposed to air. Thus cobalt nanocrystals are stored without aggregation or oxidation for at least one week.

III.   FABRICATION OF 2D SUPERLATTICES

A.   Monolayers of Silver Nanocrystals

At low particle concentration (2.5×10^{-5} M), deposition of a solution drop on a HOPG grid induces formation of a 2D network. The TEM pattern (Fig. 1A) clearly shows nanoparticles having 5-nm average diameter and organized in a hexagonal network with an average distance between particles of 1.8 nm. A total interdigitation of the dodecane alkyl chains takes place. Inset Figure 1A shows a similar deposit of silver nanocrystals on amorphous graphite. Better self-organization is obtained by using HOPG as the substrate. By replacing HOPG by Au(111), the topographic STM image (Fig. 1B) shows similar self-or-

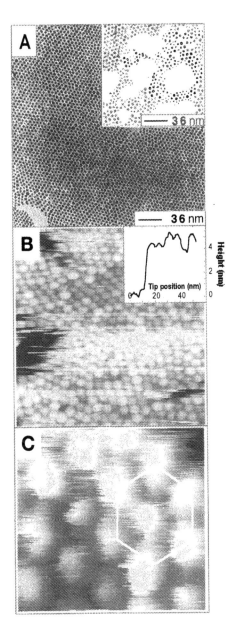

FIG. 1 Patterns of 5-nm silver nanocrystal observed by deposition on cleaved graphite (A), on amorphous graphite (inset A), on Au(111) substrate (B and C) (scan size, 136 × 136 nm, V_t = 2.5V, I_t = 0.8nA). Inset B: Cross section of the SRM image shown in B.

ganization with particles arranged in a compact hexagon. High-resolution STM (Fig. 1C) confirms the presence of well-ordered nanocrystals with a hexagonal network pattern. The cross-sectional analysis (inset Fig. 1B) shows a close-packed monolayer with about 5-nm thickness. This value is very close to the average diameter of the particles determined by TEM. From this, it is concluded that the dodecane alkyl chains do not sit on the Au substrate in their zigzag configurations, for otherwise the average height along the vertical

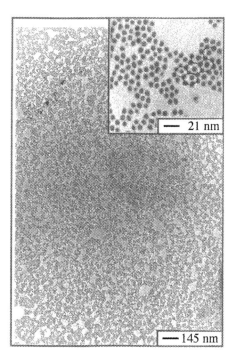

FIG. 2 Cobal nanocrystals deposited on a carbon grid.

z-axis observed by STM would be around 6 nm (5 nm + 1.8 nm). The average substrate–particle distance is less than 1.8 nm, and the dodecane alkyl chains are oriented mainly along the substrate. These data are in good agreement with those obtained by AFM in tapping mode with silver sulfide nanocrystals [3].

B. Self-Organization of Cobalt Nanocrystals in 2D Superlattices

Under deposition of cobalt nanocrystals, self-assemblies of particles are observed and the nanocrystals are organized in a hexagonal network (Fig. 2). However, it can be seen that the grid is not totally covered. We do not have a simple explanation for such behavior. In fact, the size distribution, which is one of the major parameters in controlling monolayer formation, is similar to that observed with the other nanocrystals, such as silver and silver sulfide. One of the reasons could be that the nanocrystals have magnetic properties, but there is at present no evidence for such an assumption.

IV. "SUPRA-CRYSTALS" IN FCC STRUCTURE MADE OF NANOCRYSTALS

The formation of a 3D lattice does not need any external forces. It is due to van der Waals attraction forces and to repulsive hard-sphere interactions. These forces are isotropic, and the particle arrangement is achieved by increasing the density of the "pseudo-crystal," which tends to have a close-packed structure. This imposes the arrangement in a hexagonal network of the monolayer. The growth in 3D could follow either an HC or FCC structure. The fourfold structure excludes the possibility of having an HC structure. Of course, for very large "supra"-crystals, as described later, stacking faults cannot be excluded. It is important to note than the nanocrystals described next are obtained spontaneously only by decreasing their size distribution.

A. Silver Nanocrystals [5,6]

With silver particles, similar behavior is observed with formation of rather large aggregates (Fig. 3A) made of nanocrystals. Figure 3B shows that the nanocrystals are arranged either in a hexagonal or a cubic arrangement. The transition from one structure to another is abrupt, and there is a strong analogy with "atomic" polycrystals with a small grain called *nanocrystals*. Each domain or grain has a different orientation. This clearly shows that the stacking of nanoparticles is periodic and not random. The "pseudo-hexagonal" structure corresponds to the stacking of a {110} plane of the FCC structure. In the same pattern, a fourfold symmetry that is again characteristic of the stacking of {011} planes of the cubic

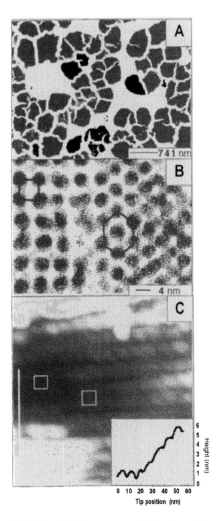

FIG. 3   FCC "supra-crystals" made of 5-nm silver nanocrystals.

structure and cannot be found in the hexagonal structure. This is because there is no direction in a perfect hexagonal compact structure for which the projected positions of the particles could take this configuration. This is confirmed by TEM experimental measurements made at various tilt angles. By tilting the sample, it is always possible to find an orientation for which the stacking appears to be periodic. Hence by tilting a sample having a pseudo-hexagonal structure, a fourfold symmetry is obtained. From these results, it can be concluded that rather small (a few micrometers) aggregates of silver particles are formed by stacking monolayers in a face-centered cubic arrangement. The cell parameter can be deduced from either the fourfold symmetry or "hexagons." In the first case, the cell parameter is to 9 ± 1 nm. In a perfect crystal, the lattice of a hexagon corresponding to a $\{1,1,0\}$ plane is constant. The lattice varies with $a = b = 9$ nm and $c = 6.6$ nm. This change in one of the lattices can be attributed to a distortion of the FCC structure. Recently [43], simulations of a 3D superlattice made of gold particles coated with dodecanethiol reveal the formation of a tetragonally distorted FCC structure at room temperature. Hence, the appearance of a distorted structure can be observed and a good agreement between the experimental data and the simulations is obtained.

As observed with graphite, silver nanocrystals can be arranged in a 3D superlattice on Au(111) substrate. The STM topography markedly differs. Figure 3C shows particles organized with a fourfold symmetry. This confirms the data presented earlier and is characteristic of the stacking (011) plane of cubic structures. The cross section (inset Fig. 3C) shows a good correlation between the height along the vertical z-axis and the stacking of monolayers. Several cross sections between two layers of particles arranged in 3D superlattices have been measured. The distance between two layers is always found to be between 2 nm and 2.8 nm. This value is lower than that of coated particles (6.8 nm = 5 + 1.8), indicating that the particles of the second layer sit in the center of the triangle formed

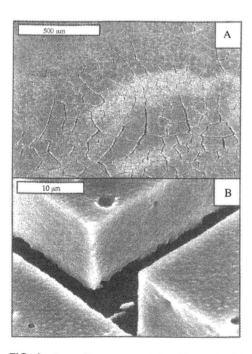

FIG. 4 5-nm silver nanocrystals self-organized in very large crystalline structure.

by three particles of the first one. The height along the vertical z-axis, corresponding to the cross section shown in inset Figure 3C, indicates that the alkylchains interdigitate in a zigzag configuration. Observation of almost these layers and the fourfold symmetry are characteristic of an FCC structure. Hence, silver nanocrystals on Au(111) substrate self-assemble in an FCC structure.

By using the same procedure, but with slow evaporation of the solvent, well-defined FCC "supra-crystals" made of nanoparticles are observed (Fig. 4) [44].

V. COLLECTIVE PROPERTIES OF NANOCRYSTALS SELF-ORGANIZED IN 2D AND 3D SUPERLATTICES

A. Collective Optical Properties

Concerning optical properties of colloidal silver nanocrystals, the absorption spectra of metal colloidal dispersions exhibit broad bands in the UV-visible range due to the excitation of plasma resonances or interband transitions. The UV-visible absorption spectra of a fairly dilute dispersion of colloidal particles can be calculated from the "Mie" theory [45–50]. The experimental [6,51] and simulated [50,52] absorption spectra show a decrease in the plasmon band intensity and an increase in the bandwidth with decreasing particle size. The optical absorption spectra of silver-coated nanocrystals dispersed in hexane and having an average diameter of 4 nm and 5 nm, respectively, is given in Figure 5. A well-defined plasmon peak centered at 2.9 eV is observed. The simulated spectra (dashed line) indicate that the resonance peak is close to a Lorentzian. The misfit observed at high energy is due to the interband transitions ($4d$–$5sp$) [53]. Such optical response has been well described by the quasi-static approximation of the Mie theory [54,55], where the optical extinction cross section is given mainly by the dipole absorption and shows a narrow plasmon resonance influenced by the contribution of the interband transitions. As predicted, an increase in the bandwidth with decreasing particle size is observed, and the peak position is centered at 2.9 eV (Fig. 5).

When the silver nanocrystals are organized in a 2D superlattice, the plasmon peak is shifted toward an energy lower than that obtained in solution (Fig. 6). The covered support is washed with hexane, and the nanoparticles are dispersed again in the solvent. The absorption spectrum of the latter solution is similar to that used to cover the support (free particles in hexane). This clearly indicates that the shift in the absorption spectrum of nanosized silver particles is due to their self-organization on the support. The bandwidth of the plasmon peak (1.3 eV) obtained after deposition is larger than that in solution (0.9 eV). This can be attributed to a change in the dielectric constant of the composite medium. Similar behavior is observed for various nanocrystal sizes (from 3 to 8 nm).

Reflective spectra under polarized light perpendicular (s) or parallel (p) to the plane of incidence at various incident angles, θ, are recorded. Under s-polarization, the electric field vector is always directed along the major axis of the spheroidal particles and the longitudinal surface plasmon is investigated [parallel to the substrate]. This tends to be insensitive to the plasmon mode oriented perpendicular to the substrate [56] and does not provide any information on the optical surface anisotropy. Conversely, under p-polarization, the electric field has two components: One is along the minor axis growing with increasing θ angle. The other is along the major axis and decreases with increasing θ angle. Hence, the (p) polarization provides information on the optical film anisotropy [83].

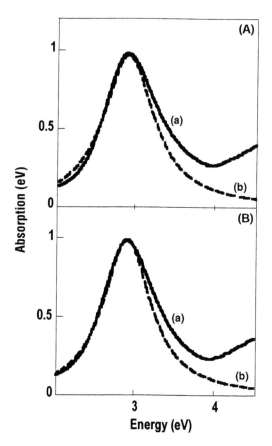

FIG. 5 Absorption spectra of coated silver particles dispersed in hexane and considered as isolated: Average particle size 4 nm (A) and 5 nm (B). The dashed lines (b) are the Lorentzian simulations of the absorption spectra (a).

Under *s*-polarization light, the optical spectra of 5-nm nanoparticles (Fig. 7A), recorded at various incident angles θ do not change with increasing θ. They are characterized by a maximum centered at 2.9 cV, which is similar to that observed for isolated particles (Fig. 5B). However, the plasmon resonance peak remains asymmetrical, as observed under nonpolarized light (Fig. 6).

Under *p*-polarization, the optical spectra markedly change with the incident angle θ (Fig. 7B): At low θ values, one peak is observed. On increasing θ, a new plasmon resonance peak appears at high energy. At θ = 60°, the two peaks are well defined: One is centered at 2.8 eV [close to that observed for isolated particles (2.9 eV)], whereas the other one is centered at 3.8 eV. Hence, when particles are organized in a hexagonal superlattice with 1.8 nm as average distance between particles, a new resonance peak appears at high energy, whereas the one (2.8 eV) close to that of isolated spherical particles (2.9 eV) still remains. These data are highly reproducible and do not depend on coverage defects.

To determine if the resonance peak is due to self-organization, optical spectra of disorganized nanocrystals (see TEM pattern inset Fig. 8) are recorded under *s*- and *p*-polarization. Under *s*-polarization, the optical spectrum obtained at θ = 60° (Fig. 8A) shows one resonance peak at 2.7 eV. This is attributed to the surface plasmon parallel to the substrate.

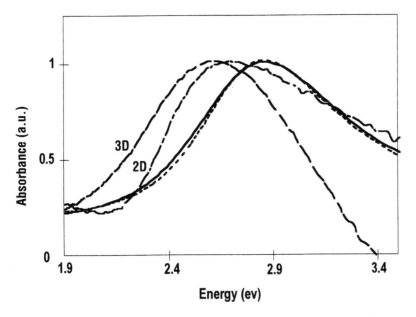

FIG. 6 Absorption spectra of 5-nm nanocrystals either dispersed in hexane (--- and ---) or deposited in 2D (— – —) and 3D (– –) superlattices.

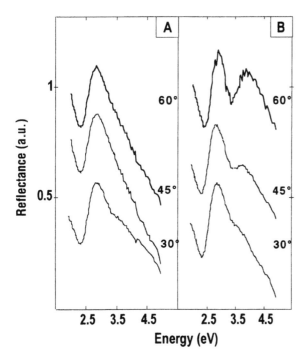

FIG. 7 Absorption spectra under polarization *s* (A) and *p* (B) of 5-nm nanocrystals self-organized in a hexagonal network on HOPG.

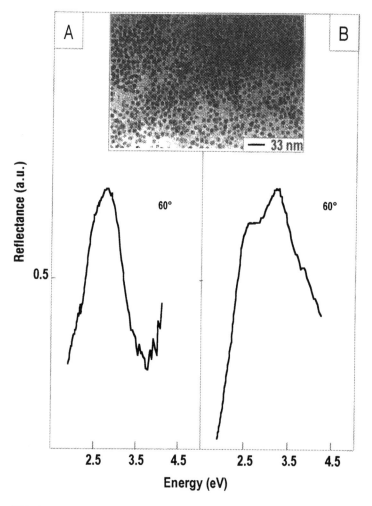

FIG. 8   Absorption spectra under polarization *s* (A) and *p* (B) of 5 nm dispersed on HOPG.

Conversely, under *p*-polarization (Fig. 8B), a splitting of the optical spectra with increasing θ is observed: One of the peaks is shifted toward high energy compared to that of isolated spherical particles, whereas the second one is at lower energy.

Inset Figure 8 shows that the sizes of most of the particles remain similar. However, it can be observed that, in part of the TEM pattern, some particles form chains, which are either close together or coalesced. The presence of the two plasmon modes (Fig. 8B) indicates an optical anisotropy due to the nonspherical shape of the particles, as shown on the TEM grid (inset Fig. 8).

From comparison of the optical properties of particles deposited on the same substrate and differing by their organization (Figs. 7 and 8) it can be concluded that the appearance of the resonance peak at 3.8 eV is due to the self-organization of the particles in a hexagonal network. This can be interpreted in terms of mutual dipolar interactions between particles. The local electric field results from dipolar interactions induced by particles at a given distance from each other. Near the nanocrystals, the field consists of the applied field plus a contribution due to all other dipoles and their images. The calculation for finite-size clusters gives, at a qualitative level, a correct explanation for the difference between the *s*- and the *p*-polarization spectra: Under *s*-polarization one resonance peak is observed, whereas under *p*-polarization an additional peak toward high energy appears [35].

B. Collective Optical Properties of Particles Self-Assembled in Multilayers

The UV-visible spectrum (Fig. 6) of the aggregates described earlier shows a 0.25-eV shift toward lower energy of the plasmon peak with a slight decrease in the bandwidth (0.8 eV) compared to that observed in solution (0.9 eV). As observed earlier with monolayers, by washing the support, the particles are redispersed in hexane and the absorption spectrum remains similar to that of the colloidal solution used to make the self-assemblies.

According to simulated spectra, the increase in the dielectric constant induces a shift to lower energy and an increase in the bandwidth of the plasmon peak. For particles organized in multilayers, each silver nanoparticle is surrounded by 12 other clusters, whereas in a monolayer it has six neighbors. Thus in 3D superlattices, the dielectric constant surrounding each silver particle increases, inducing a larger plasmon peak shifted toward the lower energy. These data confirm the effect of the medium dielectric constant of the particle when organized in 2D and 3D, and show a decrease in the plasmon peak bandwidth that could be due to an increase in the mean free path of silver-particle conduction electrons through a 2-nm barrier. This is rather surprising, because the average distance between silver nanocrystals is 2 nm, and we would not expect a tunneling electron effect through such a large barrier. This will be confirmed in the next section.

C. Electron Transport Properties of Nanocrystals Either Isolated or Self-Assembled in 2D and 3D Superlattices

When a single particle is deposited on the Au substrate, the STS measurement shows a double tunneling junction (Fig. 9A). On increasing the applied voltage, small capacitances of the junctions are charged up and the detected current is close to zero. Above a certain threshold voltage, the electron turns through the system and the current increases with the applied voltage. The plot of dI/dV versus V (inset Fig. 9A) clearly shows that the derivative reaches zero at zero bias. This nonlinear $I(V)$ spectrum and the zero value of dI/dV at zero voltage are characteristic of the well-known Coulomb blockade effect. The $I(V)$ curve shows a gap of 2 V at zero current (Fig. 9A). This indicates that the ligands are sufficiently good electrical insulators to act as tunnel barriers between particles and the underlining substrate. As shown in Figure 9A, no Coulomb staircase is obtained. In the present experimental conditions, we are operating at constant-current mode, which imposes a tip–particle distance.

The tip–particle distance, using ($V_{bias} = -1$ V, $I_{tunnel} = 1$ nA) as tunnel parameters does not correspond to that needed to observe the Coulomb staircase. The particle–substrate distance is fixed by the coating with dodecanethiol. Hence the two tunnel junctions are characterized by fixed parameters. Similar Coulomb blockade behavior has been observed [58,59].

When silver particles are self-organized in a 2D superlattice on Au(111) substrate (Fig. 1B), the recorded $I(V)$ curve is that given in Figure 9B. For large biases, the detected current is reduced by more than one order of magnitude compared to that observed for isolated particles. This indicates an increase in the ohmic contribution to the current. The

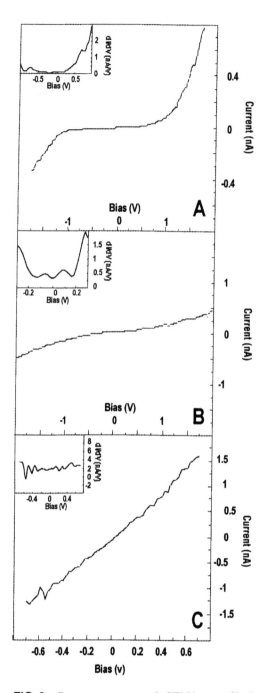

FIG. 9 Constant-current-mode STM image of isolated (A), self-organized in closed-packed hexagonal network (B) and in FCC structure (C) of silver nanoparticles deposited on Au(111) substrate (scan size: A: 17.1 × 17.1 nm, V_t = 1 V, I_t = 1 nA; C: 136 × 136 nm, V_t = 2.5 V, I_t = 0.8 nA; E: 143 × 143 nm, V_t = −2.2 V, I_t = 0.72 nA).

shape of the $I(V)$ curve changes drastically, its nonlinearity decreases, and it markedly differs from that observed with isolated particles. The Coulomb gap is very low (0.45 V) compared to that obtained with isolated particles (2 V). This indicates that the tunneling-current contribution to the junction decreases and additional conducting paths through the array appear. This means that electron-tunneling transport occurs either via the particle to the substrate or to the neighboring particles. This is highly supported by differentiation of the $I(V)$ curve. Inset Figure 9B shows metallic conduction behavior with nonzero current at zero bias. In fact, the dI/dV value is far from zero, as observed for a Coulomb blockade. From the $I(V)$ and (dI/dV) (V) plots (Fig. 9B and its inset), it can be concluded that when particles are arranged in 2D superlattices, the tunneling current exhibits both metallic and Coulomb contributions simultaneously. This indicates that lateral tunneling between adjacent nanocrystals is very important and significantly contributes to the electron transport process. These data are in good agreement with those published. Rimbert et al.[60] measured, at 40 mK, $I(V)$ curves when Al islands linked by $Al/Al_2O_3/Al$ are either isolated or organized in 2D. A change in the shape of the $I(V)$ curves and in the Coulomb gap was observed. This has been attributed to an increase in the number of connections from one Al island to another. Ohgi et al. [61] studied the electronic transport properties of 5-nm gold nanocrystals coated with dihexane-thiol (this corresponds to a ligand barrier of 1.4 nm). They found that the Coulomb gap decreases with increasing coverage of particles on the substrate. An inductive metallic-like response is observed when propane-thiol–coated silver particles are deposited on a Langmuir trough and subjected to compression to decrease the interparticle spacing down to 6 Å [62]. The major difference between these papers [58,60,61] and the present data is that the average distances between islands and/or particles were either not controlled or rather short. In the present case, the average distance between particles is large (between 1.8 nm and 2 nm), and the nanocrystals are self-organized in a closed-packed hexagonal network. This means that, even for rather large distances between particles, lateral electron transport takes place through ligand shells acting as an insulating barrier.

When particles are arranged in an FCC structure, as shown in Figure 3, the $I(V)$ curve shows a linear ohmic behavior (Fig. 9C). The detected current, above the site point, markedly increases compared to data obtained with a monolayer made of nanocrystals (Fig. 9C). Of course, the $dI/dV(V)$ curve is flat (inset Fig. 9C). This shows a metallic character without Coulomb blockade or staircases. There is an ohmic connection through multilayers of nanoparticles. This effect cannot be attributed to coalescence of nanocrystals on the gold substrate, for the following reasons:

1. The particles remain spherical, as observed by TEM. This has been extensively described in our previous paper [5,6,35]
2. Let us come back to the sample preparation: A drop of solution containing silver nanoparticles dispersed in hexane is deposited on the substrate. The nanocrystals can be removed by washing the substrate and collected in hexane. The absorption spectrum of silver particles recorded before and after deposition remains the same. This indicates that coalescence does not take place. Similar behavior was observed by using HOPG as a substrate [6,35].

Thus, it is concluded that the FCC structure induces an increase in the tunneling rate; i.e., the resistance decreases between particles. The tunneling between adjacent particles is a major contribution to the conduction. This inhibits the Coulomb blockade in the tunneling $I(V)$ measurements, and thus the 3D superlattices yield an increased tunneling current.

These results could be explained as an increase in the dipole–dipole interactions along the z-axis, which could favor the electron tunneling from the tip to the substrate via several layers of particles arranged in a FCC structure. Furthermore, the Fermi level of nanocrystals subjected to a given bias is perturbed.

The electron transport properties described earlier markedly differ when the particles are organized on the substrate. When particles are isolated on the substrate, the well-known Coulomb blockade behavior is observed. When particles are arranged in a close-packed hexagonal network, the electron tunneling transport between two adjacent particles competes with that of particle–substrate. This is enhanced when the number of layers made of particles increases and they form a FCC structure. Then ohmic behavior dominates, with the number of neighbor particles increasing. In the FCC structure, a direct electron tunneling process from the tip to the substrate occurs via an electrical percolation process. Hence a "micro-crystal" made of nanoparticles acts as a metal.

D. Collective Magnetic Properties of Cobalt Nanocrystals

Magnetization of cobalt nanocrystals dispersed in toluene shows a super-paramagnetic behavior at room temperature. At 3 K, the nanoparticles are ferromagnetic. The magnetization curves of nanoparticles dispersed in toluene (Fig. 10A) and deposited on a substrate (Fig. 10B) are recorded. When 0.01% volume fraction of nanocrystals is dispersed in toluene, the saturation magnetization is not reached at $2T$. It is estimated to be 120 emu/g from the extrapolation of the plot of M/H versus H. As usually observed for nanoparticles in solution, it is lower than that of the bulk phase for cobalt (162 emu/g). The ratio of remanence to saturation magnetization, M_r/M_s, is 0.45 and the hysteresis field is 0.11 tesla. Magnetization curves are recorded for particles deposited on an HOPG substrate and subjected either to a parallel or to a perpendicular applied field. Figure 10B clearly shows a change in the shape of the hysteresis loop compared to the isolated particles (Fig. 10A). Saturation magnetization is reached at 0.75 tesla and is equal to 120 emu/g, and the coercive field decreases to 0.06 tesla. These changes in magnetization properties cannot be attributed to coalescence of the aggregates. This is clearly shown from the TEM pattern over a long distance. This is attributed to an increase in the interactions between nanocrystals deposited on a substrate. The origin of these interactions is dipolar. The exchange coupling can be excluded because of the overly large distance between the edge of two neighboring particles (2 nm), and there is no coalescence, as is clearly shown from the TEM pattern. The magnetic dipolar interactions could be due either to the self-organization with collective flip of magnetization between adjacent particles or to a high local volume fraction of particles randomly dispersed. To distinguish between these two possibilities, hysteresis loops have been recorded under various directions of the applied field toward the substrate (Fig. 10B).

1. When the substrate is parallel to the applied field, the remanence-to-saturation-magnetization ratio is 0.60. The hysteresis loop is squarer than that obtained with the particles dispersed in solution.
2. When the substrate is perpendicular to the applied field, the hysteresis loop is smoother. The remanence-to-saturation-magnetization ratio, M_r/M_s, decreases to 0.4.

This clearly shows that, for a given saturation magnetization, the remanence magnetization, M_r, markedly varies with the orientation of the applied field. This change is at-

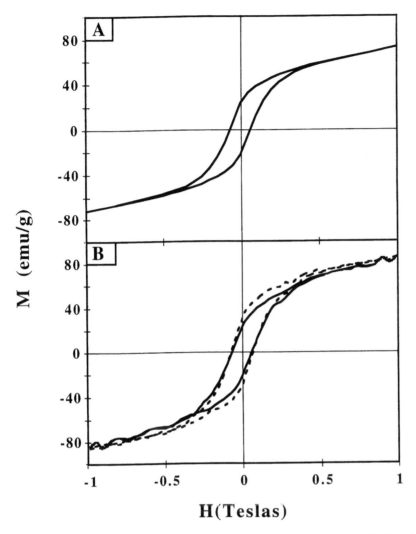

FIG. 10 Hysteresis magnetization loops obtained at $T = 3$ K. (A) Diluted liquid solution of cobalt nanoparticles in hexane. (B) Cobalt nanoparticles deposited onto freshly cleaved graphite (HOPG) and dried under argon to prevent oxidation. Substrate parallel (---) and perpendicular (—) to the field.

tributed to magnetic dipole–dipole interactions induced by the high vicinity of the particles on the substrate. We confirmed these data by numeric simulation recently reported, considering each particle as spherical of uniaxial symmetry, with the easy axes randomly distributed [63]. The magnetization curve has been calculated for two-dimensional lattices depending on the orientation of the field. From this model it is concluded than the difference of the hysteresis loop as measured with an applied field parallel or normal to the sample surface is due mainly to the dipolar interactions. This demonstrates that collective magnetic properties observed when particles are arranged in 2D superlattices are due to an increase of the dipole–dipole interactions when particles are fixed at a given distance between each other.

ACKNOWLEDGMENTS

I thank Drs. A. Courty, L. Motte, C. Petit, V. Russier, and A. Taleb for their fruitful contributions to this work.

REFERENCES

1. Motte, L.; Billoudet, F.; Pileni, M. P. J. Phys. Chem. 1995, 99, 16425.
2. Murray, C. B.; Kagan, C. R.; Bawendi, M. G. Science 1995, 270, 1335.
3. Motte, L.; Billoudet, F.; Lacaze, E.; Pileni, M. P. Adv. Mater. 1996, 8, 1018.
4. Motte, L.; Billoudet, F.; Lacaze, E.; Douin, J.; Pileni, M. P. J. Phys. Chem. B 1997, 101, 138.
5. Taleb, A.; Petit, C.; Pileni, M. P.; Chem. Mater. 1997, 9, 950.
6. Taleb, A.; Petit, C.; Pileni, M. P.; J. Phys. Chem. B 1998, 102, 2215.
7. Harfenist, S. A.; Wang, Z. L.; Alvarez, M. M.; Vezmar, I.; Whetten, R. L. J. Phys. Chem. 1996, 100, 13904.
8. Harfenist, S. A.; Wang, Z. L.; Whetten, R. L.; Vezmar, I.; Alvarez, M. M.; Adv. Mater. 1997, 9, 817.
9. Chung, S. W.; Markovich, G.; Heath, J. R.; J. Phys. Chem. B 1998, 102, 6685.
10. Korgel, B. A.; Fullam, S.; Connely, S.; Fitzmaurice D.; J. Phys. Chem. B 1998, 102, 8379.
11. Korgel, B. A.; Fitzmaurice, D. Adv. Mater. 1998, 10, 661.
12. Wang, Z. L.; Harfenist, S. A.; Whetten, R. L.; Bentley, J.; Evans; N. D. J. Phys. Chem. B 1998, 102, 3068.
13. Wang, Z. L.; Harfenist, S. A.; Vezmar I.; Whetten, R. L.; Bentley, J.; Evans, N. D.; Alexander, K. B. Adv. Mater. 1998, 10, 808.
14. Ohara, P. C.; Heath, J. R.; Gelbart, W. M. Angew. Chem. Int. Engl. 1997, 36, 1078.
15. Vossmeyer, T.; Chung, S.; Gelbart, W. M.; Heath, J. R. Adv. Mater. 1998, 10, 351.
16. Vijaya Sarathy, K.; Raina, G.; Yadav, R. T.; Kullkarni, G. U.; Rao, C. N. R. J. Phys. Chem. B 1997, 101, 9876.
17. Ohara, P. C.; Leff, D. V.; Heath, J. R.; Gelbart, W. M. Phys. Rev. Lett. 1995, 75, 3466.
18. Murthy, S.; Wang, Z. L.; Whetten, R. L. Philos. Mag. Lett. 1997, 75, 321.
19. Hostetler, M. J.; Stokes, J. J.; Murray, R. W. Langmuir 1996, 12, 3604.
20. Badia, A.; Cuccia, L.; Demers, L.; Morin, F.; Lennox, R. B. J. Am. Chem. Soc. 1997, 119, 2682.
21. Whetten, R. L.; Khoury, J. T.; Alvarez, M. M.; Murthy, S.; Vezmar, I.; Wang, Z. L.; Cleveland, C. C.; Luedtke, W. D.; Landman, U. Adv. Mater. 1996, 8, 428.
22. Brust, M.; Bethell, D.; Schiffin, D. J.; Kiely, C. J. Adv. Mater. 1995, 9, 797.
23. Fink, J.; Kiely, C. J.; Bethell, D.; Schiffrin, D. J. Chem. Mater. 1998, 10, 922.
24. Kiely, C. J.; Fink, J.; Brust, M.; Bethell, D.; Schiffrin, D. J. Nature, 1998, 396, 444.
25. Brown, L. O.; Hutchison, J. E. J. Am. Chem. Soc. 1999, 121, 882.
26. Lin, X. M.; Sorensen, C. M.; Klabunde, K. J. Chem. Mater. 1999, 11, 198.
27. Schaaff, T. G.; Hafigullin, M. N.; Khoury, J. T.; Vezmar, I.; Whetten, R. L.; Cullen, W. G.; First, P. N.; Gutierrez-Wing, C.; Ascensio, J.; Jose-Yacaman, M. J. J. Phys. Chem. B 1997, 101, 7885.
28. Petit, C.; Taleb, A.; Pileni, M. P. J. Phys. Chem. B 1999, 103, 1805.
29. Petit, C.; Taleb, A.; Pileni, M. P. Adv. Mater. 1998, 10, 259.
30. Yin, J. S.; Wang, Z. L. J. Phys. Chem. B 1997, 101, 8979.
31. Yin, J. S.; Wang, Z. L. Phys. Rev. Lett. 1997, 79, 2570.
32. Maenosono, S.; Dushkin, C. D.; Saita, S.; Yamaguchi, Y. Langmuir 1999, 15, 957.
33. Motte, L.; Pileni, M. P. J. Phys. Chem. B 1998, 102, 4104.
34. Luedtke, W. D.; Landman, U. J. Phys. Chem. 1996, 100, 13323.
35. Taleb, A.; Russier, V.; Courty, A.; Pileni, M. P. Phys. Rev. B, 1999, 59, 13350.
36. Russier, V.; Pileni, M. P., Surf. Sci., 1999, 425, 313.
37. Pileni, M. P.; J. Phys. Chem. 1993, 97, 6961.

38. Pileni, M. P. Langmuir, 1997, 13, 3266.
39. Pileni, M. P., ed. Reactivity in Reverse Micelles. Elsevier, Amsterdam, 1989.
40. Wilson, W. L.; Szajowski, P. F.; Brus, L. E. Science 1993, 262, 1242.
41. Petit, C.; Taleb, A.; Pileni, M. P. Adv. Mater. 1998, 10, 259.
42. Russier, V.; Petit, C.; Legrand, J.; Pileni, M. P.; Phys. Rev. B 2000, 62, 3910.
43. Luedtke, W. D.; Landman, U. J. Phys. Chem. 1996, 32, 13324.
44. Courty, A.; Feimon, C. and Pileni, M. P. Adv. Mat., 2001, 13, 254.
45. Bohren, C. F., Huffman, D. R., Eds. Absorption and Scattering of Light by Small Particles. Wiley, New York, 1983.
46. Charlé, K. P.; Frank, F.; Schulze, W. Ber. Bunsenges. Phys. Chem. 1984, 88, 354.
47. Mie, G.; Ann. Phys. 1908, 25, 377.
48. Greighton, J. A., Eaton, D. G. J. Chem. Soc. Faraday Trans. 2 1991, 87, 3881.
49. Hövel, H.; Fritz, S.; Hilger, A.; Kreibig, U.; Vollmer, M. Phys. Rev. B 1993, 48, 18178.
50. Persson, B. N. J. Surf. Sci. 1993, 281, 153.
51. Petit, C.; Pileni, M. P. J. Phys. Chem. 1993, 97, 12974.
52. Alvarez, M. A.; Khoury, J. T.; Shaaf, T. G.; Shafigullin, M. N.; Vezmar, I., Whetten, R. L. J. Phys. Chem. B 1997, 101, 3706.
53. Yamaguchi, T.; Ogawa, M.; Takahshi, H.; Saito N., Anno, E. Surf. Sci. 1983, 129, 232.
54. Bohren, C. F.; Huffman, D. R., eds. Absorption and Scattering of light by Small Particles. Wiley, New York, 1983.
55. Ruppin, R. Surf. Sci. 1983, 127, 108.
56. Kennerly, S. W.; Little, J. W.; Warmack, R. J.; Ferrell, T. L. Phys. Rev. B 1984, 29, 2926.
57. Bobbert, P. A.; Lieger, J. V. Physica A 1987 147, 115.
58. Petit, C.; T. Cren, T.; Roditchev, D.; Sacks, W.; Klein, J.; Pileni, M. P. Adv. Mat. 1999, 11, 1108.
59. Simon, U. Adv. Mat. 1998, 10, 1487.
60. Rimbert, A. J.; Ho, T. R.; Clarke, J.; Phys. Rev. Lett. 1995, 74, 4714.
61. Ohgi, T.; Sheng, H. Y.; Nejoh, H. Appl. Surf. Sci. 1998, 130, 919.
62. Markovich, G.; Collier, C. P.; Heath, J. R. Phys. Rev. Lett. 1998, 80, 3807.
63. Russier, V., Petit, C.; Legrand, J., Pileni, M. P. Appl. Surf. Sci. 2000, 164, 260.

9

Molecular Nanotechnology and Nanobiotechnology with Two-Dimensional Protein Crystals (S-Layers)

UWE B. SLEYTR, MARGIT SÁRA, DIETMAR PUM, and BERNHARD SCHUSTER   Universität für Bodenkultur Wien, Vienna, Austria

I. INTRODUCTION

One of the key challenges in material sciences is the technological utilization of self-assembly systems, wherein molecules spontaneously associate under equilibrium conditions into reproducible aggregates and supramolecular structures joined by noncovalent bonds. Although molecular self-assembly is the governing principle in the morphogenesis of biological systems, few molecular species have yet been exploited for the controlled assembly of defined nanostructures. However, in recent years, self-assembly of molecules into monomolecular arrays has grown into a scientific and engineering discipline that crosses the boundaries of several established fields in the nanosciences.

In this chapter we describe the basic principles involved in the controlled production and modification of two-dimensional protein crystals. These are synthesized in nature as the outermost cell surface layer (S-layer) of prokaryotic organisms and have been successfully applied as basic building blocks in a biomolecular construction kit. Most importantly, the constituent subunits of the S-layer lattices have the capability to recrystallize into isoporous closed monolayers in suspension, at liquid–surface interfaces, on lipid films, on liposomes, and on solid supports (e.g., silicon wafers, metals, and polymers). The self-assembled monomolecular lattices have been utilized for the immobilization of functional biomolecules in an ordered fashion and for their controlled confinement in defined areas of nanometer dimension. Thus, S-layers fulfill key requirements for the development of new supramolecular materials and enable the design of a broad spectrum of "nanoscale" devices, as required in molecular nanotechnology, nanobiotechnology, and biomimetics [1–3].

II. GENERAL PRINCIPLES

A. Occurrence and Ultrastructure of S-Layers

Cell envelopes of prokaryotic organisms (archaea and bacteria) are characterized by the presence of two distinct components: the cytoplasmic membrane, which constitutes the inner layer, and an outer supramolecular layered cell wall (for reviews see Ref. 4), which presumably evolved by selection in response to specific environmental and ecological pressures. One of the most remarkable features of prokaryotic cell envelopes is the presence of monomolecular arrays of protein and glycoprotein subunits termed *S-layers* [5–7]. Their identification on selected organisms was originally considered to represent a rather unique cell wall component [5,8]. Nevertheless, today S-layers are recognized as one of the most commonly observed prokaryotic cell surface structures. They have until now been identified in hundreds of different species belonging to all major phylogenetic groups of bacteria and represent an almost universal feature of archaea (for compilation, see Refs 7 and 9). Because S-layers are ubiquitous and, where present, one of the most abundant of cellular proteins, it is presumed they have a vital function. So far, a great variety of functions have been identified [1,7,10,11]. S-layers function as protective coats, as structure involved in cell adhesion and surface recognition, as molecular sieves, as molecular and ion traps, as a scaffolding for enzymes, and as virulence factors. Moreover, in archaea, which generally possess S-layers as exclusive cell wall component outside the cytoplasmic membrane [12], the monomolecular arrays even determine cell shape and can direct cell division [13]. Most important, S-layers can contribute to virulence when present on pathogenic bacteria.

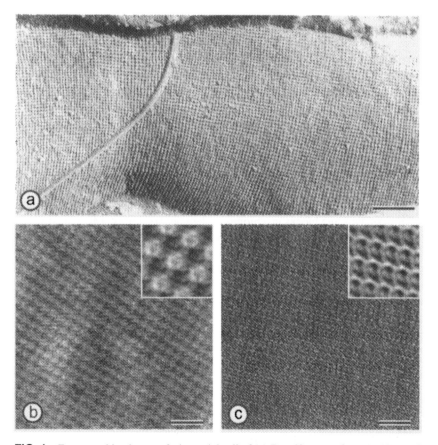

FIG. 1 Freeze-etching image of a bacterial cell of (a) *Desulfotomaculum nigrificans* (bar, 100 nm). Atomic force micrographs of the S-layer proteins of (b) *Bacillus sphaericus* CCM 2177 and (c) *Bacillus stearothermophilus* PV72/p2 recrystallized in monolayers on silicon wafers. Bars, 50 nm. The insets in (b) and (c) show the corresponding computer-image reconstructions.

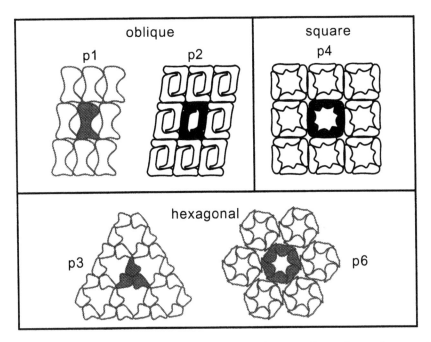

FIG. 2 Schematic drawing of different S-layer lattice types detected on prokaryotes. The regular arrays exhibit either oblique (p1, p2), square (p4), or hexagonal lattice symmetry (p3, p6). The morphological units are composed of one, two, three, four, or six identical subunits. (Modified from Ref. 59.)

It is difficult to detect S-layered cells without using electron microscopy. The most useful preparation techniques are freeze-etching of intact cells (Fig. 1a) or negative staining of cell wall or envelope preparations [5,7,14]. Analyses of freeze-etched preparations of a great variety of archaea and bacteria have shown that the crystalline arrays completely cover the cell surface at all stages of cell growth and division [5,15,16]. High-resolution studies on the mass distribution of S-layer lattice are generally performed on negatively stained preparations or unstained, thin, frozen foils. Both two- and three-dimensional analyses, including computer-image enhancement, have revealed structural information down to approximately 1 nm [7,17–20]. More recently, high-resolution images of the surface topography of S-layers could be obtained using atomic force microscopy in a wet cell (Fig. 1b, c) [2,3,21–25]. The topographical images obtained strongly resembled the three-dimensional reconstructions of S-layer lattices derived from electron microscopical studies. S-layer lattices show oblique (p1, p2), square (p4), or hexagonal (p3, p6) symmetry with center-to-center spacings of the morphological units of approximately 2.5–35 nm (Fig. 2). Depending on the lattice type, the morphological units constituting the crystalline array consist of one, two, three, four, or six identical protein or glycoprotein subunits. However, some S-layer lattices apparently are composed of two subunit proteins. Most monomolecular arrays are generally 5–10 nm thick, with an inner surface more corrugated than the outer surface. In S-layers of archaea, pillar-like domains on the inner corrugated surface frequently are observed that are associated or even integrated in the cytoplasmic membrane [20,26]. High-resolution microscopical and scanning force microscopy studies have demonstrated that S-layers are highly porous monomolecular arrays, with pores occupying up to 70% of the surface area (Fig. 3).

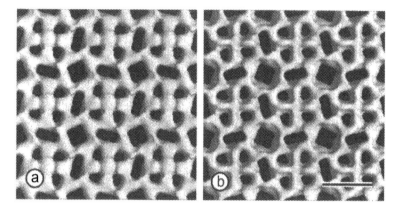

FIG. 3 Three-dimensional model of the protein mass distribution of the S-layer of *Bacillus stearothermophilus* NRS 2004/3a [(a) outer, (b) inner face]. The square S-layer is about 8 nm thick and exhibits a center-to-center spacing of the morphological units of 13.5 nm. The protein meshwork composed of a single protein species shows one square-shaped, two elongated, and four small pores per morphological unit. (Modified from Ref. 7.)

Since S-layers are protein lattices composed of identical subunits, they exhibit pores of identical size and morphology. In many S-layers, two or even more distinct classes of pores (usually in the 2- to 8-nm range) could be observed. In many species of bacteria, individual strains exhibit great diversity with respect to lattice symmetries and center-to-center spacings of the morphological units. In some organisms two or more superimposed S-layer lattices composed of different S-layer proteins have been identified [7,9]. These structural data, including chemical studies and sequence comparison of protein sequences [11], indicate that S-layers are nonconservative structures and are of limited taxonomical value. More recently it was demonstrated that even individual strains are capable of synthesizing different S-layer (glyco)proteins (for review, see Ref. 11).

B.   Isolation, Molecular Biology, and Chemical Characterization of S-Layers

Since S-layers are present in gram-positive and gram-negative archaea and bacteria, they can be associated with quite different supramolecular cell envelope structures [10,11,17]. In gram-positive bacteria and archaea, the S-layer lattice is linked to the wall matrix, which is composed mainly of peptidoglycan and secondary cell wall polymers or of pseudomurein, respectively. In gram-negative bacteria, attachment involves components of the outer membrane (e.g., lipopolysaccharides). In archaea lacking a rigid wall layer, the S-layer, as the only wall component, is closely associated with the plasma membrane. Due to the diversity in the supramolecular structures of prokaryotic cell envelopes, S-layers differ considerably in their susceptibility to disruption into constituent subunits [7,9,27–29]. Most commonly in gram-positive organisms a complete disintegration of S-layers into monomers can be obtained by treatment of purified cell wall fragments with hydrogen-bond-breaking agents (e.g., urea or guanidine hydrochloride). In particular, S-layers from gram-negative bacteria may also disintegrate upon application of metal-chelating agents (such as EDTA, EGTA, cation substitution: e.g., Na^+ or Li^+ replacing Ca^{2+}) or detergents (at pH < 4.0). In certain cases, even washing cells with deionized water can lead to dissociation of the S-layer lattice

[30–32]. The various extraction and disintegration experiments revealed that the intersubunit bonds in the S-layer lattices are stronger than those binding the crystalline arrays to the supporting envelope layer [5,28]. This property is seen as a major requirement for continuous recrystallization of the lattice during cell growth and division. Some archaeal S-layers were shown to be highly resistent to common denaturating agents, indicating the possible presence of covalent intersubunit bonds [12,17]. With many solubilized S-layers it has been demonstrated that isolated subunits reassemble into lattices identical to those observed on intact cells upon removal of the disrupting agents (see also Section II.C).

Chemical analysis and comparison of the primary structure of a variety of S-layer proteins (Table 1) has revealed a similar overall composition [2,7,9,11]. Most S-layers are composed of weakly acidic proteins or glycoproteins, with isoelectric points between pH 3 and 6. For *Methanothermus fervidus*, however, a pI value for the S-layer protein of 8.4 [31] and for Lactobacilli in the range of 9–10 have been determined [33]. S-layer proteins have a high amount of glutamic and aspartic acid (together approximately 15 mol%), 40–60% hydrophobic amino acids, 10% lysine, but only a few or no sulfur-containing amino acids. Circular dichroism measurements and/or secondary structure predictions after amino acid sequence analysis show that typically about 40% of the amino acid residues are organized as β-sheet and 20% as α-helix structure. Most α-helical segments are arranged in the N-terminal part of the S-layer proteins. For the remaining part of the sequences, mainly short β-strands connected by loops and turns have been predicted [11].

Furthermore, hydrophilic and hydrophobic amino acids do not form extended clusters. Posttranslation modifications of S-layer proteins include removal of the signal peptide [11,33,34], phosphorylation [35], and glycosilation [36–38]. In bacterial S-layer glycoproteins, the carbohydrate moieties strongly resemble the O-antigen polysaccharides of gram-negative bacteria [39]. They are polymers of linear or branched repeating sequences of two

TABLE 1 Properties of S-Layers

The S-layer lattices can have oblique (p1, p2), square (p4), or hexagonal (p3, p6) symmetry.
The center-to-center spacing of the morphological unit can be 3–35 nm.
The lattices are generally 5–20 nm thick (in archaea, up to approximately 70 nm).
The outer surface is generally less corrugated than the inner surface.
The S-layer lattices exhibit pores of identical size and morphology.
In many S-layers, two or even more distinct classes of pores are present.
The pore sizes range from approximately 2 to 8 nm.
The pores can occupy 30–70% of the surface area.
The relative molecular mass of constituent subunits is in the range of 40,000–200,000.
These are weakly acidic proteins (pI ~4–6), except for *Methanothermus fervidus* (pI = 8.4) and lactobacilli (pI > 9.5).
Large amounts of glutamic and aspartic acid (about 15 mol%) are present.
There is a high lysine content (about 10 mol%).
There are large amounts of hydrophobic amino acids (about 40–60 mol%).
Hydrophilic and hydrophobic amino acids do not form extended clusters.
In most S-layer proteins, about 20% of the amino acids are organized as α-helices and about 40% occur as β-sheets.
Aperiodic foldings and β-turn content may vary between 5 and 45%.
Posttranslational modifications of S-layer proteins include cleavage of N- or C-terminal fragments, glycosylation, and phosphorylation of amino acid residues.

Source: Refs. 2, 7, 10, 11, 18, 28.

to six monosaccharide units and include a wide range of hexoses, deoxy- and amino sugars, uronic acids, or even sulfate and phosphate residues. The chain length may vary from a few sugars to approximately 150 monosaccharide residues. The glycan chains can be attached to the protein moiety by N- or O-glycosidic linkages. Among them, novel linkage types, such as β-glucose → tyrosine, β-galactose → tyrosine, or β-N-galactosamine → threonine/serine have been identified (for recent review see Refs. 2 and 37). In comparison to bacterial glycan structures, glycoproteins in archaea consist almost exclusively of short carbohydrate chains of heterosaccharides with up to 10 sugar residues and no repeating units [40,41]. In halobacteria, up to three different glycan species with up to 25 glycosylation sites per S-layer monomer can be present [41]. For a most recent compilation of glycan structures of selected strains of bacteria and archaea see Ref. 2. Although the amino acid composition of S-layer proteins shows no significant differences, sequencing of S-layer genes from organisms of all phylogenetic branches (Table 2) revealed that sequence identities are rare [2,11,33,42]. Nevertheless, it is quite obvious that common structural principles must exist in S-layer proteins (e.g., the ability to form intersubunit bonds and to self-assemble into 2D arrays, the formation of hydrophilic pores with low unspecific adsorption, and the interaction with the underlying cell envelope layers). Sequencing of S-layer genes from strains belonging to the same species, such as *Lactobacillus acidophilus* [33], *Bacillus stearothermophilus* [43–46], or *Campylobacter fetus* [47], revealed that evolutionary relationship plays an important role in the sequence identity of functionally homologous domains. The identities observed at the amino-termini of S-layer proteins of gram-positive and gram-negative bacteria are a consequence of the binding specificities for components of the supporting wall layers (secondary wall polymers and lipopolysaccharides) [1,11]. For example, the S-layer proteins SbsA and SbsC from *B. stearothermophilus* PV72/p6 and ATCC 12980 are bound via their N-terminal region to an identical type of cell wall polymer that is covalently linked to the peptidoglycan backbone [48,49]. The N-terminal parts of both S-layer proteins shows an identity of 85%, and more than 70% of the N-terminal 240 amino acids are organized as short α-helices. Using sequence comparison, S-layer homologous domains (SLH) [50] have been identified at the amino-terminal region of several S-layer proteins [2,11]. Triplicated forms of SLH-domains have also been observed on the C-terminal part of cell-associated exoenzymes [51] and other exoproteins [51,52]. Although only preliminary information on the structural interactions of SLH domains is available, there is strong evidence that these regions are involved in anchoring the proteins permanently or transiently to the cell surface [50,53,54]. Most recently the complete structure of the secondary cell wall polymer functioning as a binding site for the SLH motifs of the S-layer protein from *Bacillus sphaericus* CCM 2177 was elucidated by nuclear magnetic resonance analysis [55].

C. Assembly and Morphogenesis of S-Layers

Since S-layers possess a high degree of structural regularity and are composed of a single proteinaceous subunit species, they represent very appealing models for studying the morphogenesis of a supramolecular structure during cell growth. Studies on the morphogenesis of S-layer lattices have focused on the question how the S-layer proteins assemble on supporting supramolecular structures into two-dimensional arrays during cell growth.

1. Self-Assembly in Vivo

Electron micrographs of freeze-etched preparations clearly demonstrated that S-layers completely cover the cell surface during all stages of cell growth and division [5,56,57].

TABLE 2 Survey of S-Layer Proteins Whose Amino Acid Sequences Are Known

Species	Strain	Gene/ Protein	No. of amino acids, including N-terminal leader peptide/N-terminal leader peptide	Lattice type[a]	GenBank accession no.
Aeromonas hydrophila	TF7	ahs	467/19	S	L37348
Aeromonas salmonicida	A450	vapA	502/21	S	M64655
Bacillus anthracis	Sterne derivative substrain 9131	sap	814/29	O[c]	Z36946
		eag	862/29	O	X99724
Brevibacillus brevis (Bacillus brevis)	47	owp	1,004/24	—	M14238
		MWP	1,053/23	H	M19115
Brevibacillus brevis (Bacillus brevis)	HPD31	HWP	1,087/23	H	D90050
Bacillus licheniformis	HM105	olpA	874/29	—	U38842
Bacillus sphaericus	P1	sequence 8	1,252/30	S	A45814
Bacillus sphaericus	2362	gene 125	1,176/30	S	M28361
		gene 80	745 (silent)	—	—
Bacillus sphaericus	CCM 2171	sbpA	1,268/30	S	AF211170
Bacillus stearothermophilus	PV72/p6	sbsA	1,228/30	H	X71092
	PV72/p2	sbsB	920/31	O	X98095
Bacillus stearothermophilus	ATCC 12980	sbsC	1,099/30	O	AF055578
Bacillus stearothermophilus	12980	sbsD	904/30	O	AF228338
Bacillus thuringiensis	CTC	ctc	842/29	—	AJ012290
Bacillus thuringiensis, ssp. galleriae	NRRL 4045	slpA	821/29	—	AJ249446
Campylobacter fetus, ssp. fetus	—	sapA	933/none	H, S[b]	J05577
Campylobacter fetus, ssp. fetus	23B	sapAl	920/none	H, S[b]	L15800
Campylobacter fetus, ssp. fetus	82-40LP3	sapA2	1,109/none	H, S[b]	S76860
Campylobacter fetus, spp. fetus	84–91	sapB	936/none	—	U25133
	CIP 5396T	sapB2	1,112/none	—	AF048699
Campylobacter rectus	314	crs	1,361/none		AF010143
Caulobacter crescentus	CB15	rsaA	1,026/none	H	M84760

TABLE 2   (*Continued*)

Species	Strain	Gene/ Protein	No. of amino acids including N-terminal leader peptide/N-terminal leader peptide	Lattice type[a]	GenBank accession no.
Clostridium thermocellum	NCIMB 10682	*slpA*	1,036/26	O	U79117
Corynebacterium glutamicum	ATCC 17965	*csp2*	510/30	H	X69103
Deinococcus radiodurans	—	HPI gene	1,036/31	H	M17895
Halobacterium halobium	—	*csg*	852/34	H	J027/67
Haloferax volcanii	—	—	828/34	H	M62816
Lactobacillus acidophilus	ATCC 4356	*slpA*	444/24	O	X89375
		slpB	456 (silent)	—	X89376
Lactobacillus brevis	ATCC 8287	—	465/30	O	Z14250
Lactobacillus crispatus	JCM 5810	*cbsA*	440/30	O	AF001313
Lactobacillus helveticus	CNRZ 892	*slpH1*	440/30	O	X91199
Lactobacillus helveticus	CNRZ 1269	*slpH2*	440/30	O	X92752
Methanococcus voltae	—	*sla*	565/12	H	M59200
Methanosarcina mazei	S-6	*slgB*	652/31	H	X77929
Methanothermus fervidus	DSM 2088	*slgA*	593/22	H	X58297
Methanothermus sociabilis	DSM 3496	*slgA*	593/22	H	X58296
Rickettsia prowazekii	Brein 1	*spaP*	1,612/32	—	M37647
Rickettsia rickettsii	R	*p120*	1,645/32	—	X16353
Rickettsia typhii	Wilmington	*slpT*	1,645/32	—	L04661
Serratia marcescens	Isolate 8000	*slaA*	1,004/none	—	AB007125
Staphylothermus marinus	F1	—	1,524/putative	—	US7967
Thermoanaerobacter kivui (Acetogenium kivui)	DSM 2030	*slp*	762/26	H	M31069
Thermus thermophilus	HB8	*slpA*	917/27	H, S	X57333

[a] H, hexagonal; S, square; O, oblique.

[b] In *Campylobacter fetus* subsp. *fetus,* the lattice type was found to be dependent on the molecular weights of the S-layer subunits (H. 97,000; S. 127,000 and 149,000).

[c] Presumably.

Source: Refs. 2 and 11.

Freeze-etching preparations of rod-shaped cells also generally show a characteristic orientation of the lattice with respect to the longitudinal axis of the cell (Fig. 1a). This characteristic structural feature was considered as strong evidence that S-layers are "dynamic closed surface crystals" with the intrinsic capability to continuously assume a structure of low free energy during cell growth [5,58]. In this context it is interesting to note that a closed S-layer lattice on an average-sized, rod-shaped prokaryotic cell consists of approximately 5×10^5 monomers [9]. This implies that at high growth rates (20–30 min), 400–500 copies of a protein or glycoprotein subunit have to be synthesized by a cell per second, translocated to the cell surface, and incorporated into the S-layer lattice. With S-layer from various bacteria it was demonstrated that distinct surface properties of the subunits (charge distribution, hydrophobicity, specific interactions with components of the supporting envelope layer) are essential for the proper orientation of the S-layer subunits during local insertion in the course of lattice growth [11,59]. Valuable information about the mechanisms involved in the development and maintenance of crystalline arrays of macromolecules on a growing cell surface also came from reconstitution experiments with isolated S-layers on cell surfaces from which they had been removed (homologous reattachment) or on those of other organisms (heterologous reattachment) [57,60,61]. Results of homologous and heterologous recrystallization experiments of S-layers clearly demonstrated that the formation of the regular lattices resides entirely in the subunits themselves and is not affected by the matrix of the supporting envelope layers.

Labeling experiments with colloidal gold/antibody and fluorescent antibody marker methods have been performed for elucidating the accurate incorporation sites of constituent subunits in "closed surface crystals" during cell growth [57,59]. As predicted by Harris and coworkers [62–64], dislocations and disclinations could serve as sites for incorporation of new subunits in closed lattices that grow by "intussusception" [59]. Since both types of lattice faults can be observed on high-resolution freeze-etching images of intact cells, it can be assumed that the rate of growth of S-layer lattices by the mechanism of nonconservative climb of dislocations depends on the number of dislocations present and the rate of incorporation of new subunits at these sites.

2. Self-Assembly in Vitro

Valuable information about the morphogenesis of S-layer lattices on intact cells was derived from self-assembly studies of isolated S-layer subunits in vivo [57,59]. Isolated S-layer subunits from a great variety of bacteria have been demonstrated to reassemble into lattices identical to those observed on intact cells upon removal of the disrupting agent used for their isolation. Depending on the intrinsic properties of the S-layer protein or glycoprotein and the recrystallization conditions (e.g., ionic strength, ion composition, temperature, pH value, protein concentration, concentration of associated polymers), isolated S-layer subunits may recrystallize into flat sheets, open-ended cylinders, or closed vesicles (Figs. 4 and 5). So far the most detailed studies on the kinetics of the in vitro self-assembly processes in suspension, the shape of self-assembly products, the charge distribution, and the topographical properties of the outer and inner surface were performed with S-layers from different *Bacillaceae* [57–59,65]. Studies on the kinetics of the self-assembly process of S-layer proteins from *B. stearothermophilus* NRS 1536/3c involving light-scattering and cross-linking experiments revealed that a rapid initial phase and a slow consecutive process of higher than second order exists [66]. In the rapid initial phase, oligomeric precursors composed of 12–16 subunits ($M_n < 10^6$) were formed that in a second stage fused and recrystallized. This recrystallization process led to sheetlike self-assembly products (Fig. 4)

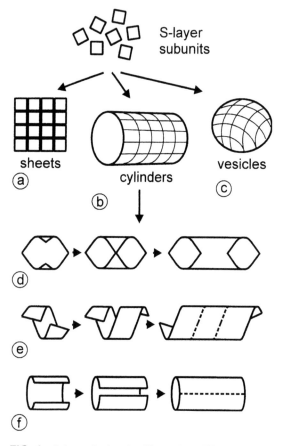

FIG. 4 Schematic drawing illustrating different self-assembly routes of S-layer subunits leading to the formation of (a) flat sheets, (b) and (d) to (f), cylinders, and (c) spheres. (Modified from Ref. 59.)

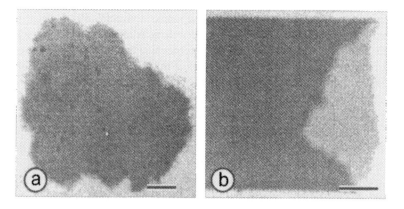

FIG. 5 Electron micrographs of negatively stained preparations of S-layer self-assembly products. (a) flat sheet (bar, 100 nm), (b) open-ended cylinder (bar, 200 nm).

exhibiting the same square lattice as observed on intact cells [66]. Many S-layers from Bacillaceae assemble in suspension into double-layer sheets or cylinders. In such self-assembly products the individual monolayers may be linked to each other either with their inner (more corrugated) or their outer (smoother, charge-neutral) surface [67]. Recently it was demonstrated that secondary cell wall polymers associated with the isolated S-layer proteins from different *B. stearothermophilus* strains can have a significant influence on the self-assembly process [54,68]. For example, the S-layer protein SbsB, which possesses three SLH-motifs comprising the secondary cell wall polymer-binding domain reassembles into flat mono- and double-layer sheets with 1- to 3-μm size. Interestingly, upon addition of the purified high-molecular-mass secondary cell wall polymer (which functions as an anchoring structure for the S-layer protein in the bacterial cell wall), the in vitro self-assembly was inhibited and the isolated S-layer protein was kept in a water-soluble state. This observation indicated that the polymer chains not only have the potential of anchoring the S-layer lattice to the cell wall properly but can also act as spacers between the individual S-layer subunits. Such functional specificities appear essential for preventing self-assembly of S-layer subunits during translocation from sites of synthesis to sites of lattice growth.

Most important for many applications of S-layer lattices in molecular nanotechnology, biotechnology, and biomimetics was the observation that S-layer proteins are capable of reassembling into large coherent monolayers on solid supports (e.g., silicon wafers, polymers, metals) at the air/water interface and on Langmuir lipid films (Fig. 6) (see Sections V and VIII).

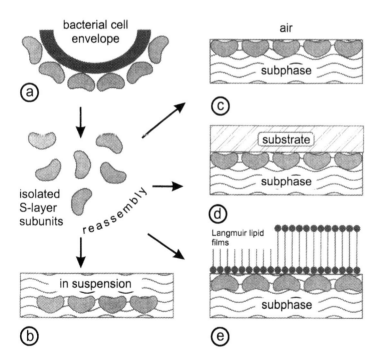

FIG. 6 (a) Schematic illustration of the recrystallization of isolated S-layer subunits into crystalline arrays. The self-assembly process can occur (b) in suspension, (c) at the air/water interface, (d) on solid supports, and (e) on Langmuir lipid films. (Modified from Ref. 59.)

III. S-LAYERS FOR THE PRODUCTION OF ULTRAFILTRATION MEMBRANES

A. Molecular Sieving Properties of S-Layer Lattices

In order to determine the size of the largest type of pore passing through S-layer lattices of *Bacillus stearothermophilus* strains, permeability studies were performed according to the space technique [69]. For this purpose, native and glutaraldehyde-treated S-layer containers were prepared that resembled the shape of whole bacterial cells [70]. Such S-layer containers were obtained by extracting the plasma membrane of whole cells by mild detergent treatment, by removing most of the cytoplasma content by extensive washing, and by degradation of nucleic acids with DNAse and RNAse. Native S-layer containers consisted of the outer S-layer, the peptidoglycan-containing layer, and the inner S-layer. The last resulted from an S-layer protein pool accumulated in the peptidoglycan-containing layer that was released during the preparation procedure and could assemble on the inner face of the rigid cell wall layer [71]. For introducing covalent bonds between the S-layer subunits, S-layer containers were treated with glutaraldehyde, and the peptidoglycan was subsequently degraded with lysozyme. This led to glutaraldehyde-treated S-layer containers that consisted merely of an outer and an inner S-layer with a 10- to 15-nm-wide distance between them. Both native and cross-linked S-layer containers were suspended in solutions of proteins with increasing molecular masses, which were selected as test molecules for the permeability studies. After common incubation, only proteins smaller than the lattice pores were detected inside the S-layer containers (Fig. 7a). Independent of the presence of the peptidoglycan-containing layer, an identical and sharp molecular mass cutoff was obtained for the square and hexagonal S-layer lattices of *Bacillus stearothermophilus* strains NRS 1536/3c and PV72/p6. This was expressed by the fact that carbonic anhydrase (CA; M_r 30,000; molecular size 4.1 by 4.1 by 4.7 nm) could pass through the pores in these crystalline arrays, whereas ovalbumin (OVA; M_r 43,000; molecular size 4.5 nm) was rejected

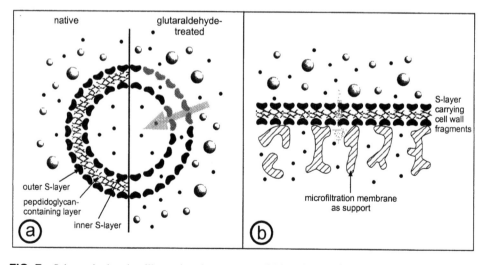

FIG. 7 Schematic drawing illustrating the structure of (a) native or glutaraldehyde-treated S-layer containers as prepared for permeability studies according to the space technique and of (b) S-layer ultrafiltration membranes.

by at least 90% [70]. As derived from the results of the permeability studies, the molecular mass cutoff (defined as 90% rejection for a distinct molecule species) was in the range of 40,000, indicating that the size of the largest type of pore was 4–5 nm. The presence of the peptidoglycan-containing layer had no influence on the molecular mass cutoff, which confirmed that the S-layer lattices exclusively determine the rejection properties of S-layer containers. The sharp molecular mass cutoffs and the exclusion limits further revealed that S-layers are isoporous molecular sieves that work in the range of ultrafiltration membranes produced by synthetic polymers. In contrast to S-layer lattices, conventional ultrafiltration membranes possess an amorphous structure and a pore size distribution with pores differing in size by as much as one order of magnitude [72].

B. Production and Rejection Characteristics of S-Layer Ultrafiltration Membranes (SUMs)

To produce the first type of biomimetic ultrafiltration membrane (Fig. 7b) and to exploit the precise molecular sieving properties of S-layer lattices for technological applications, S-layer-carrying cell wall fragments from *B. stearothermophilus* strains were deposited on a microfiltration membrane out of a suspension in a pressure-dependent procedure [73]. The concentration of the S-layer-carrying cell wall fragments was adjusted in a way that a coherent layer was generated on the surface of the microporous support. Although a broad screening was performed involving numerous cross-linking agents with different bridge lengths, glutaraldehyde was found to be the most efficient one. To increase the chemical stability of cross-linked S-layer lattices, Schiff bases formed by the reaction of glutaraldehyde with ε-amine groups from lysine were reduced with sodium borohydride [74]. SUMs produced of S-layer-carrying cell wall fragments from *B. stearothermophilus* strains exhibited identical exclusion limits as determined for native and glutaraldehyde-treated S-layer containers by applying the space technique [70,73]. The sharp molecular mass cutoff of this biomimetic ultrafiltration membrane confirmed that S-layer lattices work as isoporous molecular sieves. In comparison to the S-layer lattices from *B. stearothermophilus* strains, the square S-layer lattice from *B. sphaericus* CCM 2120 (Table 3) and the oblique S-layer lattice from *B. coagulans* E38-66 exhibited cutoff levels shifted slightly to the lower molecular mass range [75,76]. These S-layer lattices allowed free passage for myo-

TABLE 3 Rejection Characteristics of SUMs Prepared of S-Layer-Carrying Cell Wall Fragments from *B. sphaericus* CCM 2120

Protein	M_r	Molecular size (in nm)	pI	% R	pH value of the protein solutions
Ferritin	440,000	12	4.3	100	7.2
Bovine serum albumin (BSA)	67,000	$4.0 \times 4.0 \times 14.0$	4.7	100	7.2
Ovalbumin (OVA)	43,000	4.5	4.6	95	4.6
Carbonic anhydrase (CA)	30,000	$4.1 \times 4.1 \times 4.7$	5.3	80	5.3
Myoglobin (MYO)	17,000	$4.4 \times 4.4 \times 2.5$	6.8	0	6.8

The rejection coefficient (R) was calculated according to the following equation: $R = \ln (C_r/C_o)/\ln (V_o/V_r)$. C_r or V_r represent the protein concentration in the retentate or the volume of the retentate; C_o is the concentration of the protein in the solution before filtration; V_o is the initial volume of the feed. The pH value of each protein solution was immediately measured after dissolving the proteins in distilled water.

globin (MYO; M_r 17,000; molecular size 4.4 by 4.4 by 2.5 nm) and rejected CA (M_r 30,000; molecular size 4.1 by 4.1 by 4.7 nm) to either 80% or 90%.

C. Importance of the Native Charge Distribution for the Integrity of S-Layer Lattices

Acetylation and succinylation of free amine groups (primarily ε-amine groups from lysine) of the S-layer proteins from various Bacillaceae converted them into either neutral or negatively charged groups, which led to partial or complete disintegration of these crystalline arrays [67,77]. In contrast, modification of amine groups with the monofunctional imidoesters ethylacetimidate or 2-iminothiolane did not interfere with the integrity of S-layer lattices, which can be explained by the fact that generated amidines had a pK value of about 12 [78], which was higher than the pK values of the ε-amine groups from lysine (~10). Consequently, modification of S-layer lattices with monofunctional imidoesters had no influence on the native charge distribution in these crystalline arrays. By modification of the S-layer lattices with 2-iminothiolane, sulphhydryl groups were introduced that could be exploited for further modification reactions or for covalent binding of sulphhydryl groups containing enzymes such as β-galactosidase [79]. After disintegration of S-layer lattices modified with imidoesters with guanidine hydrochloride and removing the disrupting agent by dialysis, regularly structured self-assembly products with a shape identical to those obtained with the native S-layer protein were formed [67,77]. Although both imidoesters are monofunctional, cross-linking reactions are known to occur under mild alkaline conditions (pH 8–9) if free amine groups are located at a distance of 0.3–0.4 nm [78]. Since S-layer lattices from Bacillaceae disintegrate into their constituent subunits at pH values over 9, modification reactions had to be carried out at pH 8.5, which actually led to partial cross-linking of these crystalline arrays. Modification of free carboxylic acid groups with 1-ethyl-3,3′(dimethylaminopropyl)carbodiimde (EDC) and glycine methyl ester (GME) resulted in the loss of the regular lattice structure. Depending on the S-layer used, either partial solubilization of the modified S-layer protein (of up to 50%) or complete cross-linking was observed. The GME-modified S-layer protein, which was not cross-linked and could therefore be extracted with guanidine hydrochloride, had lost the ability to reassemble into regularly structured lattices. Complete cross-linking of the S-layer lattice with EDC as a zero-length cross-linker indicated that amine and carboxylic acid groups from adjacent S-layer subunits are located closely enough for direct electrostatic interactions. Complete cross-linking of S-layer lattices from Bacillaceae was achieved with homobifunctional imidoesters and N-hydroxy succinimide esters of different cross-linking spans [67,76,80].

D. Surface Properties of Native S-Layer Lattices from Bacillaceae

S-layer lattices from *B. stearothermophilus*, *B. coagulans*, and *B. sphaericus* strains are, like most other S-layers, highly anisoptropic structures. The outer S-layer surface is rather smooth, whereas the inner S-layer surface is much more corrugated [55]. The anisotropic nature was also confirmed by sequencing the genes encoding the S-layer proteins from *B. stearothermophilus* strains and *B. sphaericus* CCM 2177 and by investigating the structure–function relationship of selected S-layer proteins [11,46]. Affinity studies with proteolytic cleavage fragments and recombinant truncated S-layer proteins demonstrated that the N-terminal part either comprises typical SLH-motifs [50] or does not recognize distinct types of secondary cell wall polymers as anchoring structures in the rigid cell wall layer

[46,48,54,55,68]. In the case of the S-layer proteins SbsA and SbsC from *B. stearother-mophilus* PV72/p6 and ATCC 12980, the N-terminus possesses a high number of lysine and arginine residues, which endows this part with a positive net charge (pI of the mature SbsC 5.73; pI for the segment comprising amino acids 31 to 270 is 9.13). This positive net charge is in clear contrast to that of the remaining part of the SbsC sequence, which comprises amino acid 271 to 1,099 and has a pI of 4.88. Secondary structure prediction further revealed that up to 70% of the amino acids of the N-terminal part are organized as short α-helices [46]. The remaining part of the sequence consists of β-sheets and loops. The positively charged N-terminal part possesses a high tyrosine content, is located on the inner S-layer surface, and interacts with the net negatively charged secondary cell wall polymer via direct electrostatic interactions and hydrogen bonds. The net charge of the outer S-layer surface was finally evaluated by electron microscopical examination of whole cells or S-layer-carrying cell wall fragments after labeling with polycationic ferritin (PCF; pI ~ 11). This spherical topographical marker has a molecular size of 12 nm and is therefore completely excluded from the pores in S-layer lattices from Bacillaceae. In contrast to most bacterial cell surfaces, which are net negatively charged, the outer S-layer surfaces could not be labeled with PCF, indicating that there is not an excess of net negatively charged groups [67]. Labeling of S-layers with cytochrome c (M_r 12,000; molecular size 2.5 by 2.5 by 3.7 nm), which is also positively charged (pI 10.8) but in contrast to ferritin small enough to pass through the pores, revealed the absence of free carboxylic acid groups inside the pores [75]. Finally, the oblique S-layer lattice from *B. coagulans* E38-66 was labeled with polycationic carbonic anhydrase (PCA; pI 9.5; obtained by modification of free carboxylic acid groups with EDC and *N,N'*-dimethylpropylamine) [77]. This protein was selected because, due its molecular size (4.1 by 4.1 by 4.7 nm), it is just within the dimension of the pores but, according to the permeability studies, too large to pass through. The amount of PCA adsorbed to the outer S-layer surface was 270 μg/mg S-layer protein, which corresponded to 1 PCA molecule per S-layer subunit, with a molecular mass of 97,000. These findings indicated that PCA was adsorbed by carboxylic acid groups exposed in indentations of the corrugated S-layer protein network that were not accessible for binding of the large PCF molecules [77].

E. Surface Properties of Glutaraldehyde-Treated S-Layer Lattices

In comparison to the outer surface of native S-layer lattices from Bacillaceae, those cross-linked with glutaraldehyde could be labeled with PCF. Moreover, after chemical modification with glutaraldehyde, the smaller positively charged cytochrome c was adsorbed inside the pores [75]. Since, on the other hand, chemical modification of EDC-activated carboxylic acid groups with GME resulted in loss of PCF and cytochrome c binding, it became evident that adsorption of the positively charged marker molecules was due to the presence of free carboxylic acid groups that were neutralized by free amine groups in the native state of the S-layer lattices. As demonstrated with SUMs prepared from S-layer-carrying cell wall fragments from *B. coagulans* E38-66, adsorption of a monolayer of densely packed PCF molecules on the outer S-layer surface caused flux losses for particle-free distilled water of 10% of the initial flux. Flux losses were calculated according to the following equation: $R_f = (1 - J_a/J_0)$ (R_f = relative flux loss; usually given $\times$ 100 in %; J_a is the water flux after filtration of the respective protein solution; J_0 is the water flux before filtration of the protein solution = initial water flux) [81]. On the other hand, adsorption of cytochrome c inside the pores caused flux losses of up to 80%, which can only be explained by severe pore plugging.

As described earlier, SUMs are produced by depositing S-layer-carrying cell wall fragments on microporous supports, cross-linking of the S-layer protein with glutaraldehyde, and reducing Schiff bases with sodium borohydride [74]. Thus, in SUMs the outer S-layer surface is exposed to the ambient environment, whereas the inner S-layer surface is blocked due to binding to the peptidoglycan-containing layer (Fig. 7b) [73]. Since mainly ε-amine groups from lysine are involved in the cross-linking reaction with glutaraldehyde, carboxylic acid groups from acidic amino acids of the S-layer protein remain available for further modification reactions to change the surface properties. However, the major goal of the chemical modification studies was to produce SUMs with tailor-made surface properties for specific demands in downstream processing [76]. To correlate the number of functional groups introduced per nm^2 S-layer lattice (outer S-layer surface, including the pore areas) leading to either charged, neutral hydrophilic, or neutral hydrophobic S-layer surfaces with the adsorption properties of SUMs, chemical modification reactions were performed in parallel on suspended S-layer-carrying cell wall fragments for exact quantification of bound nucleophiles and on SUMs for protein adsorption studies. In the case of S-layer-carrying cell wall fragments, the S-layer lattice was cross-linked with the cleavable homobifunctional amine group–specific imidoester 3,3′dithiobis(succinimidyl)propionate (DSP) instead of glutaraldehyde [76]. For determining the number of free carboxylic acid groups available for chemical modification in the DSP-cross-linked S-layer lattice, carboxylic acid groups were activated with EDC and modified with different nucleophiles. Subsequently, the disulphide bonds in DSP were cleaved with dithiothreitol, and the modified S-layer subunits were extracted with guanidine hydrochloride and subjected to amino acid analysis. Introduced amine groups were quantified via the increase in leucine by using butyloxycarbonyl-1-leucine N-hydroxysuccinimidester (BOC-Leuse) as an amine group–specific modifying agent. For producing either neutral hydrophilic, neutral hydrophobic, or positively or negatively charged S-layer surfaces, the following nucleophiles were covalently linked to the EDC-activated carboxylic acid groups of the S-layer protein from *B. sphaericus* CCM 2120: glycine methyl ester (GME), glucosamine (GAM), hexadecylamine (HDA), ethylenediamine (EDA), and glutamic dimethyl ester (GDM) [76,82]. For increasing the number of free carboxylic acid groups available for further modification reactions, ester groups from GDM covalently bound to the S-layer protein were liberated by mild alkaline hydrolysis (pH 12.5), which also caused deamination of some asparagine and glutamine residues of the S-layer protein and led to a higher number of free carboxylic acid groups than introduced by chemical modification. The results of the modification studies, summarized in Table 4, show that the EDC-activated carboxylic acid groups could be rather quantitatively modified.

SUMs prepared of S-layer-carrying cell wall fragments from *B. sphaericus* CCM 2120 were treated in the same way as suspended S-layer-carrying cell wall fragments. But in contrast to the latter, SUMs were used for filtration experiments and protein adsorption studies. As described earlier, protein adsorption was expressed in terms of flux losses after filtration of defined volumes of selected protein solutions [76,82]. SUMs produced from S-layer-carrying cell wall fragments from *B. sphaericus* CCM 2120 rejected OVA and BSA to 95 and 100%, respectively; CA was retained to 80%, whereas MYO was small enough to pass through the pores in this square S-layer lattice (Table 3).

Experiments with SUMs demonstrated that the net negatively charged BSA and OVA showed the lowest adsorption on negatively charged SUMs, but flux losses increased with the negative charge density [(Table 4; standard < GME/H < GDM/H < 2(GDM/H)].

TABLE 4 Chemical Modification of SUMs Produced from S-Layer-Carrying Cell Wall
Fragments from *B. sphaericus* CCM 2120

Modification reactions	Residues introduced/nm^2 S-layer surface (SUM)	BSA	OVA	CA	MYO
None (standard)	1.6 ~ COO$^-$	5	0	70	55
GME/H	1.8	25	10	75	45
GDM/H	3.5 ~ COO$^-$	30	10	85	70
2 (GDM/H)	7.0 ~ COO$^-$	40	15	90	80
EDA	1.6 ~ NH$_3^+$	40	70	15	50
GDM/H/EDA	3.5 ~ NH$_3^+$	55	80	25	65
GME	1.6	40	40	15	25
GDM/H/GAM	3.5	35	30	20	25
HDA	1.6	55	55	30	55
2 (GDM/H)/HDA	7.0	70	65	60	70

The influence of attaching different nucleophiles to the EDC-activaed carboxylic acid groups from the S-layer protein on the adsorption of selected test proteins was evaluated via relative flux losses $(1 - R_f$, given $\times$ 100 in %) of SUMs after filtration of the respective protein solution (BSA, OVA, CA, MYO).

GME glycine methyl ester; GDM glutamic dimethyl ester; EDA ethylendiamine; GAM glucosamine; HDA hexadecylamine; BSA bovine serum albumin; OVA ovalbumin; CA carbonic anhydrase; MYO myoglobin; H: alkaline hydrolysis for converting ester groups from GME or GDM into free carboxylic acid groups that were subsequently activated with EDC for further modifications.

On neutral SUMs, flux losses correlated with the hydrophobicity of introduced nucleophiles (GDM/H/GAM < GME < HDA < 2(GDM/H)/HDA). As expected, BSA and OVA were strongly adsorbed on positively charged SUMs, and protein adsorption increased with the positive charge density (EDA < GDM/H/EDA). A similar observation that electrostatic interactions between the protein molecules in solution and the SUM surface determine the extent of protein adsorption was made for the net positively charged CA. This protein was strongly adsorbed on negatively charged SUMs, and flux losses correlated with the negative charge density [(standard < GME/H < GDM/H < 2(GDM/H)]. As expected, adsorption was significantly reduced on positively charged SUMs. However, flux losses increased with the positive charge density (EDA < GDM/H/EDA). MYO was close to its isoelectric point and interacted with all types of SUMs, but the lowest adsorption for this protein was observed on neutral and neutral hydrophilic SUMs (GDM/H/GAM ~ GME < HDA < 2(GDM/H)/HDA) (Table 4). However, contact angle measurements revealed that SUMs are basically hydrophobic, which could only be slightly influenced by modification of free carboxylic acid groups with different nucleophiles. The hydrophobic structure of SUMs may be derived from both the presence of a high number of hydrophobic amino acids in S-layer proteins and the incorporation of glutaraldehyde in the course of cross-linking the S-layer lattice [76,82]. Adsorption experiments were generally performed in distilled water. Electrostatic interactions between SUMs and dissolved proteins were completely interrupted in buffered solutions, such as phosphate buffered saline [82].

To conclude, a strong correlation was found to exist between the net charge of the proteins in solution, the net charge of the SUM surface, and the extent of protein adsorption, which was expressed in terms of flux losses measured after filtration of the different protein solutions. Moreover, in the case of charge-neutral SUMs, flux losses increased with the hydrophobicity of the nucleophiles bound to the S-layer lattice. All proteins caused higher flux losses on SUMs modified with HDA than on those modified with GME or

GAM (Table 4). Another interesting correlation was found to exist between the size of the protein molecules that were rejected by the S-layer lattice and the maximum flux losses determined. For example, flux losses of 60% were measured after filtration of BSA solutions using positively charged SUMs, while flux losses of only about 10% were determined after filtration of ferritin solutions. Both proteins have a similar isoelectric point (Table 3), but, in comparison to the spherical ferritin, BSA is significantly smaller and a long molecule (4 by 4 by 14 nm). These findings led to the conclusion that despite the fact that BSA was rejected by the pores in the S-layer lattice from *B. sphaericus* CCM 2120, the long molecules could penetrate the pore openings to some extent, thereby leading to similarily high flux losses as caused by cytochrome c, which was small enough for passing through the pores [77].

IV. S-LAYERS AS MATRIX FOR THE IMMOBILIZATION OF FUNCTIONAL MACROMOLECULES

A. General Introduction

For immobilization of biologically active macromolecules, mostly carboxylic acid groups from the S-layer protein were activated with EDC, which could subsequently react with free amine groups from enzymes, antibodies, or ligands such as streptavidin and protein A. From the amount, the molecular mass, and the size of macromolecules that could be bound to the respective S-layer lattice, the molecular mass of the S-layer subunits, and the area occupied by one morphological unit in the S-layer lattice, it was derived that most macromolecules formed a dense monolayer on the outer surface of the S-layer lattices [1,2,83,84]. With ferritin this could actually be visualized by electron microscopical investigation of negatively stained, freeze-etched, or freez-dried preparations of self-assembly products or S-layer-carrying cell wall fragments (Fig. 8) [85]. Most frequently, the ferritin monolayer

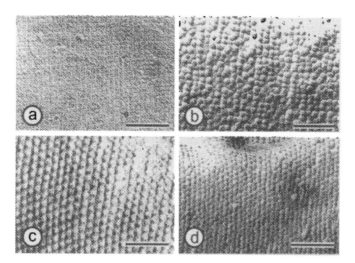

FIG. 8 Electron micrographs of freeze-etched preparations of whole cells from (a, b) *Bacillus sphaericus* CCM 2120 exhibiting a square S-layer lattice or from (c, d) *Thermoanaerobacter thermohydrosulfuricus* L111-69 carrying a hexagonally ordered S-layer lattice. (a, c) Native S-layer lattices; (b, d) S-layer lattices after covalent binding of ferritin to carbodiimide-activated carboxylic acid groups of the S-layer protein. Bars, 100 nm.

reflected the symmetry of the underlying S-layer lattice. If enzymes were immobilized, the introduction of spacer molecules such as 4-amino butyric acid, 6-amino caproic acid, or GDM was advantageous regarding the retained enzymic activity [86]. In the case of S-layer lattices composed of glycoprotein subunits, the surface-located carbohydrate chains were also exploited for covalent binding of functional macromolecules. Either vicinal hydroxyl groups were activated with cyanogen bromide or they were cleaved with periodate. Although the hydroxyl groups were arranged in high density, the binding capacity and the retained enzymic activity were significantly lower than what was obtained by linking the enzymes to the EDC-activated carboxylic acid groups from the S-layer protein. For determining the immobilization capacity of S-layer lattices, S-layer-carrying cell wall fragments from *Thermoanaerobacter thermohydrosulfuricus* L111-69 and *B. sphaericus* CCM 2120 were used. The S-layer lattice from *Th. thermohydrosulfuricus* L111-69 shows hexagonal lattice symmetry, with a center-to-center spacing of the morphological units of 14.2 nm. The molecular mass of the glycosylated S-layer subunits is 120,000 [87]. The center-to-center spacing of the morphological units of the square S-layer lattice from *B. sphaericus* 2120 is 12.5 nm. One morphological unit cell consists of four identical subunits, with a molecular mass of 127,000 [76].

B. Immobilization of Enzymes

As model systems, the enzymes invertase (M_r 270,000), glucose oxidase (M_r 150,000), naringinase (α-L-rhamnosidase; M_r 110,000), and β-glucosidase (M_r 66,000) were linked to the hexagonally ordered S-layer lattice from *Th. thermohydrosulfuricus* L111-69 [85,86]. In the case of invertase, 1,000 μg/mg S-layer protein were immobilized, which corresponded to an average of 2.7 enzyme molecules per morphological unit, resembling a dense monolayer. The retained enzymic activity was in the range of 70%. The amount of glucose oxidase that could be linked to the EDC-activated carboxylic acid groups was 550 μg/mg S-layer protein, showing that two to three enzyme molecules were bound per hexameric unit cell. In the case of this enzyme, the retained activity was 35% and could be increased to 60% when the enzyme was immobilized via 4-amino butyric acid or 6-amino caproic acid. Immobilization via spacer molecules was particularly advantageous for enzymes that were small enough to entrapped inside the pores or to penetrate the pore openings to some extent. For example, β-glucosidase, with a molecular mass of 66,000, retained only 16% activity after direct coupling to the EDC-activated carboxylic acid groups of the S-layer protein [86]. A tenfold increase in the activity to 160% was achieved when this enzyme was immobilized via spacers. However, the amount of immobilized enzyme was the same in both cases (240 μg/mg S-layer protein). The K_M value for free β-glucosidase and the immobilized enzyme was 3.1 mM. From the significant increase in activity it may be derived that immobilization via spacers prevented multipoint attachment and increased the distance between the enzyme molecules and the crystalline matrix. Similar observation was made for peroxidase, which has a molecular mass of 40,000 and was small enough to be entrapped inside the pores [86]. Naringinase (α-L-rhamnosidase) was directly linked either to the EDC-activated carboxylic acid groups of the S-layer protein, to 6-amino caproic acid as spacer, or, alternatively, to diazotized *p*-phenylendiamine. The last was introduced into the S-layer lattice by reaction of EDC-activated carboxylic acid groups with this nucleophile. Depending on the immobilization method, either 40 or 280 μg naringinase was bound per milligram of S-layer protein, which corresponded to 0.3 and 2 enzyme molecules per hexameric unit cell, respectively. The retained activity was either 60 or 80% [86]. For immobilization of β-galactosidase, a sulphhydryl group containing enzyme, the S-layer lattice was first modified with 2-iminothiolane, which introduced 56 sulphhydryl groups per S-layer subunit. After activation of sulphhydryl groups with 2,2′-dipyridyl disulphide, 550 μg β-galactosidase could be bound per milligram of S-layer protein, which on average corresponded to 3.6 enzyme molecules per hexameric unit cell. The immobilized enzyme had retained 25% of its initial activity [79].

C.   Immobilization of Ligands and Immunoglobulins

Protein A (M_r 42,000) which was originally isolated from the cell wall of *Staphylococcus aureus*, recognizes the Fc part of most mammalian antibodies and is a ligand that leads to an oriented binding of immunoglobulins [88]. In order to functionalize the S-layer lattice either protein A was directly linked to the EDC-activated carboxylic acid groups of the S-layer protein from *Th. thermohydrosulfuricus* L111-69 and *B. sphaericus* CCM 2120 or it was immobilized via spacer molecules [87,89]. The immobilization capacity of both S-layer lattices was either 550 μg or 700 μg protein A/mg S-layer protein. Derived from the area occupied by the morphological units and from the molecular mass of the constituent S-layer subunits, this binding capacity corresponded to 320 or 410 ng protein A per cm^2 S-layer surface. Assuming a Stoke's radius of 5 nm for protein A [90], the theoretical saturation capacity for a planar surface lies in the range of 90 ng/cm^2, which demonstrated that the binding capacity of the S-layer lattices was three to four times higher than the values calculated with the Stoke's radius for a planar surface. The observed difference can be explained by the shape of protein A and by the distribution of free amine groups in this ligand. On the one hand, protein A is an extremely long molecule; on the other hand, lysine residues are concentrated at one end, the so-called X region. Considering the extremely high density of amine groups at one end of this molecule and the high binding density on the S-layer lattices, it became evident that the protein A molecules were immobilized in an oriented manner, with their long axis perpendicular to the S-layer lattice. This provided evidence that a crystalline immobilization matrix with uniformly aligned and oriented functional groups favors the immobilization of macromolecules in a well-defined manner [87,89].

Another approach for preparing S-layer affinity matrices can be seen in the immobilization of streptavidin (M_r 66,000), which is a tetramer, and each subunit has one binding site for biotin. Since well-established protocols are available for biotinylation of any type of substance and the biotin–streptavidin bonds are among the strongest noncovalent bonds known in nature, this system is particularly attractive for functionalizing solid supports or lipid layers. Since covalent binding of this ligand to S-layer lattices via EDC-activated carboxylic acid groups did not lead to a monolayer of densely packed molecules, the S-layer protein from *B. sphaericus* CCM 2120 was first modified with EDC/EDA, and introduced amine groups were subsequently exploited for binding of sulfo *N*-hydroxysuccinimide biotin. After incubation with streptavidin, such modified S-layer lattice could bind 800 ng per cm^2, which corresponded to a monolayer of densely packed molecules of this ligand [89].

Both S-layer affinity matrices, obtained by binding protein A or either streptavidin, were exploited for immobilization of native or biotinylated human IgG [89]. In comparative studies, human IgG was directly linked to the EDC-activated carboxylic acid groups from the S-layer protein. Depending on the orientation and the state of the IgG molecules (compact or expanded Fab regions), the theoretical binding capacity of a planar surface lies between 240 ng/cm^2 in lying position and 650 ng/cm^2 in upright position [91]. The binding capacity of the S-layer protein A affinity matrix for human IgG was 700 ng per cm^2, which corresponded to the theoretical value given for a monolayer of uniformly oriented

IgG molecules, with the Fc region bound to protein A and the Fab regions in the compact state (Fig. 9b). When human IgG was linked to the EDC-activated carboxylic acid groups, the binding capacity was 375 ng/cm^2 S-layer lattice, indicating the formation of a monolayer of randomly oriented covalently bound IgG molecules (Fig. 9a). On S-layer lattices modified with streptavidin, 150 ng biotinylated IgG/cm^2 S-layer surface could be immobilized. Depending on the orientation and the state of the IgG molecules, this binding capacity represented 23–63% of the theoretical saturation capacity of a planar surface (Fig. 9c) [89]. To evaluate the binding capacity of human IgG immobilized to the S-layer lattice from *B. sphaericus* CCM 2120 by the three different methods, alkaline phosphatase–labeled anti-human IgG was applied in different concentrations and bound anti-human IgG was finally determined by using *p*-nitrophenyl phosphate as a substrate for alkaline phosphatase. As shown in Figure 9d, the highest values were obtained for the S-layer protein A affinity matrix, which was in accord with the highest binding density of human IgG on such a matrix. On the other hand, the directly bound IgG and biotinylated IgG that was immobilized on the S-layer streptavidin matrix led to comparable absorbance values over the whole concentration range.

D. Affinity Microparticles

Affinity microparticles (AMPs) were obtained by cross-linking the S-layer lattice on S-layer-carrying cell wall fragments with glutaraldehyde, reducing Schiff bases with sodium borohydride, and immobilizing protein A as an IgG-specific ligand [92]. Thus, AMPs rep-

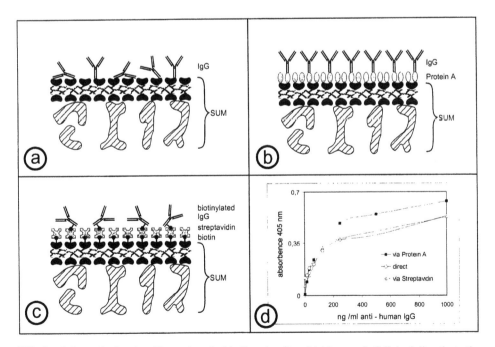

FIG. 9 Schematic drawing illustrating the binding density of (a) human IgG linked directly to the carbodiimide-activated carboxylic acid groups of the S-layer protein from *Bacillus sphaericus* CCM 2120, (b) human IgG immobilized via the Fc-specific ligand protein A, (c) biotinylated human IgG bound to the S-layer lattice modified with streptavidin. (d) Results demonstrating the biological activity (binding capacity) of the immobilized human IgG for anti-human IgG.

resent 1-μm-large cup-shaped structures with an identically oriented outer and inner S-layer lattice on both surfaces of the peptidoglycan-containing layer available for immobilization of the ligand. Both the shape of AMPs and the presence of two identically oriented S-layer lattices endow these particles with an extremely high surface-to-volume ratio. In previous studies, AMPs were used as escort particles in affinity cross-flow filtration [87,92]. Affinity cross-flow filtration is a technique applied in downstream processing that combines a highly specific adsorption step on microparticles or microspheres with a membrane separation step [93]. For isolation of IgG from serum or hybridoma cell culture supernatants, AMPs were permanently circulated in a cross-flow apparatus incorporating an ultrafiltration membrane (nominal molecular mass cutoff of 300,000), which completely rejected AMPs. During the first step, the IgG-containing solution was pumped into the cross-flow apparatus, and IgG could bind to protein A on AMPs. After extensive washing to remove contaminants showing no affinity to protein A, the protein A–IgG complex was cleaved by decreasing the pH value to 3.5. Eluted IgG was collected in the filtrate, and the solution was immediately neutralized. In contrast to many particles proved for their applicability as escort particles in affinity cross-flow filtration, AMPs showed a high stability toward shear forces and revealed no leakage of the covalently bound ligand. Due to these features AMPs were used for the isolation of human IgG from human IgG–human serum albumin mixtures, from serum, or from hybridoma cell culture supernatants [92]. However, one drawback to using protein A as affinity ligand can be seen in the fact that it has only low affinity to (monoclonal) mouse antibodies.

E. S-Layers as Novel Matrix for Dipstick-Style Solid-Phase Immunoassays

SUMs produced from S-layer-carrying cell wall fragments from *B. sphaericus* CCM 2120 were used as a novel immobilization matrix in the development of dipstick-style solid-phase immunoassays [94]. The monoclonal antibodies were linked to the EDC-activated carboxylic acid groups of the S-layer protein. Independent of the type of antibody, 370 ng could be immobilized per cm^2 SUM surface, which corresponded to a monolayer of randomly oriented antibody molecules (Fig. 9a). The specific advantage of SUMs as immobilization matrix in comparison to other polymers can be seen in the presence of a dense monolayer of covalently bound monoclonal antibodies (catching antibodies) on the surface of the S-layer lattice, which prevents nonspecific adsorption and diffusion-limited reactions [89]. After immobilization of the monoclonal antibodies, small discs (1–3 mm in diameter) were punched out and sandwiched between Teflon foils, leaving the SUM surface exposed for further chemical reactions. Three different types of SUM-based dipsticks have been developed: (1) for diagnosis of type I allergies (determination of IgE in whole blood or serum against the major birch pollen allergen); (2) for quantification of tissue-type plasminogen activator (t-PA) in patients' whole blood or plasma for monitoring t-PA levels in the course of thrombolytic therapy after myocardal infarcts; (3) for determination of interleukins in whole blood or serum to clarify whether intensive care patients suffer under traumatic or septic shock. In the case of the IgE dipstick, recombinant birch pollen allergen (rBet v 1) was bound to the monoclonal antibody BIP 1. After incubation in whole blood or serum, IgE was finally quantified via an anti-IgE–alkaline phosphatase conjugate using a substrate that formed an IgE-concentration-dependent colored precipitate on the SUM surface [94]. Either the intensity was evaluated with a reflectometer, which allowed exact quantification, or semiquantitative (visual) evaluation was performed with a color card. With the SUM-based IgE dipstick it was possible to differentiate between RAST classes 1 to 6, which corresponds to IgE concentrations of 0.35 to over 100 PrU. Direct immobilization of rBet v 1 (M_r 17,000) to the EDC-activated carboxylic acid groups from the S-layer protein led to a significantly reduced sensitivity, which can be explained by the fact this allergen is small enough to be entrapped inside the pores and is therefore not accessible to the large IgE molecules in the first reaction step [87,94]. The relevant concentration range for the t-PA dipstick was 1,000 to 2,000 ng/mL plasma. The detection limit for the interleukin dipstick was 100 pg/mL serum. The advantages of S-layers as immobilization matrix for producing solid-phase immunoassays may be summarized as follows: (1) Since S-layers are crystalline arrays of identical protein or glycoprotein subunits, they have repetitive surface properties down to the subnanometer scale; (2) S-layer proteins have a high density of functional groups that are located on the outermost surface of these crystalline arrays; (3) because of the molecular size of the catching antibody, immobilization can occur only on the outermost surface, thereby preventing diffusion-limited reactions during further incubation and binding steps; (4) since the catching antibody is covalently linked to the S-layer lattice, no leakage problems arise during the test procedure; (5) S-layers do not unspecifically adsorb plasma or serum components, which makes blocking steps as required for immunoassays with conventional matrices unnecessary; (6) stable, concentration-dependent precipitates are formed on the S-layer surface by using appropriate substrates for peroxidase- or alkaline phosphatase–conjuated antibodies in the final detection step.

F. Biosensors Based on S-Layer Technology

S-layer lattices have been used as immobilization matrices for a broad range of macromolecules in the development of amperometric and optical biosensors [95–98]. For the fabrication of single-enzyme sensors, such as the glucose sensor, glucose oxidase was covalently bound to the outer S-layer surface from SUMs (Fig. 10a). The electrical contact to the sensing layer was established by sputtering a thin layer of platinum or gold onto the enzyme layer. The whole assembly was pressed against a solid gold plate in order to increase the mechanical stability of this composite structure. The analyte reached the sensing layer through the highly porous microfiltration membrane. In the course of the enzymatic reaction, gluconic acid and hydrogenperoxide are produced under consumption of oxygen. The glucose concentration was determined by measuring the current of the electrochemical oxidation of hydrogen peroxide. For the construction of multienzyme sensors, a different building principle had to be developed, since the simultaneous immobilization of different enzyme species generally leads to an uncontrollable competition for the available activated binding sites on the S-layer lattice (Fig. 10b). For producing multienzyme biosensors, each enzyme species was bound on samples of S-layer fragments individually. Subsequently the different enzyme-loaded S-layer-carrying cell wall fragments were mixed in suspension, deposited on the microfiltration membrane, and sputter-coated with platinum or gold. This approach allowed the optimization of the immobilization for each enzyme species individually and the amount of enzymes bound. This fabrication method led to a well-structured sandwich of thin monomolecular enzyme layers, where protective layers with or without surface charge could easily be integrated. On the basis of this technique, several multienzyme sensors, such as the sucrose sensor with three enzymes species (invertase–mutarotase–glucose oxidase) and a cholesterol sensor (with cholesterol esterase and cholesterol oxidase), were developed (Table 5).

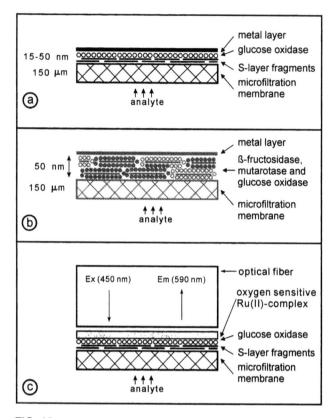

FIG. 10 Schematic drawing of the design of (a) an S-layer-based amperometric glucose sensor, (b) a sucrose sensor, and (c) an optical glucose sensor.

Later on, such S-layer-based sensing layers were also used in the development of optical biosensors (optodes), where the electrochemical transduction principle was replaced by an optical one [97] (Fig. 10c). In this approach an oxygen-sensitive fluorescent dye (ruthenium(II) complex) was immobilized on the S-layer in close proximity to the glucose oxidase–sensing layer [97]. The fluorescence of the Ru(II) complex is dynamically quenched by molecular oxygen. Thus, a decrease in the local oxygen pressure as a result of

TABLE 5 S-Layer-Based Amperometric Biosensors

Sensor type	Immobilized enzymes	Application
Monoenzyme layer		
Glucose	Glucose oxidase	Medical care
Ethanol	Alcohol oxidase	Medical care
Xanthine	Xanthine oxidase	Food technology
Multienzyme layer		
Maltose	Maltase and glucose oxidase	Biotechnology
Sucrose	β-Fructosidase and mutarotase and glucose oxidase	Biotechnology
Cholesterol	Cholesterol esterase and cholesterol oxidase	Medical care

the enzymatic reaction can be detected and transformed into a measurable signal. In an alternative approach, the design of the oxygen-sensitive S-layer optode was modified by separating the biological sensing layer from the optical sensing element. The biological sensing layer was made of an enzyme-loaded SUM that had already been optimized in the development of amperometric biosensors. The fluorescent dyes were bound in a thin foil of a polymeric matrix. This design allowed one to optimize the components individually. Later, a broad range of biologically sensitive layers based on oxidases could be used with the same optical system. S-layers as immobilization structures for biologically active molecules were also used in the development of an infrared optical biosensor [99]. In this device, an S-layer that had been recrystallized on the cylindrical part of an infrared transparent optical fiber (chalkogenide fiber) was used to bind glucose oxidase. The glucose concentration was determined from the infrared spectrum of gluconic acid.

G. S-Layers for Vaccine Development

The development of vaccines based on S-layer technologies has focused on two strategies: (1) exploiting S-layers present on pathogenic organisms, and (2) use of S-layer lattices as carrier/adjuvants for vaccination and immunotherapy [100,101].

Studies on S-layers present on the cell envelopes of a great variety of pathogenic organisms [100] revealed that these crystalline arrays can represent important virulence factors. Most detailed studies have been performed on the fish pathogenic bacteria *Aeromonas salmonicida* and *Aeromonas hydrophila* [102] and the human pathogen *Campylobacter fetus* subsp. *fetus* [103] and *Bacillus anthracis* [104]. For example, whole-cell preparations or partially purified cell products are currently used as attenuated vaccines against various fish pathogens [102,105].

Another line of vaccine development exploited the suitability of recrystallized S-layers as carrier/adjuvant for the production of conjugated vaccines [100,101,106,107]. Generally in conjugated vaccines, the antigens or haptens are bound by covalent linkages to a protein (e.g., tetanus or diphteria toxoids) present as monomers in solution or dispersed as unstructured aggregates [108,109]. Consequently, the use of regularly structured S-layer self-assembly products as immobilization matrices represents a completely new approach [110,111]. Due to the crystalline nature of S-layers, the functional groups available for binding haptens or antigens occur on each S-layer subunit in identical positions and orientations. A great variety of chemical reactions have been reported for covalent attachments of ligands, to either the protein or the glycan moiety of the S-layer (glyco)protein lattices [85,110,112]. Subcutaneous and interperitoneal administration of S-layer vaccines did not cause observable trauma or side effects. Most relevant immune responses to S-layer–hapten conjugates were also observed following oral/nasal application [106,107]. Depending on the type of S-layer preparations, the antigenic conjugates induced immune responses of a predominantly cellular or predominantly humoral nature [113–115]. Studies with a variety of carbohydrate haptens (e.g., tumor-associated oligosaccharides, blood group–specific oligosaccharides) and proteinaceous allergens (e.g., the birch pollen allergen Bet v 1) strongly indicated that significant differences with respect to T- and B-cell responses can be elicited, depending on the type of S-layer preparation used or on whether coupling has been performed to native or to chemically cross-linked self-assembly products. Secondary and tertiary immunization can be performed using the same hapten(s) coupled to different, immunologically non-cross-reactive S-layers. This would circumvent the tolerance problem frequently observed with toxoid carriers.

Although the experiments conducted so far have indicated that the present S-layer vaccine technology is suitable for a broad spectrum of applications, cloning and characterization of genes encoding S-layer proteins and the use of different prokaryotic and eukaryotic expression systems opens new possibilities for S-layer vaccine research. In a variety of S-layer proteins, domains have been identified that allow foreign epitopes insertion without hindering self-assembly into regular arrays [116,117]. A further strategy for the development of new model vaccines is based on liposomes stabilized by S-layers that represent biomimetic structures, copying the supramolecular principle of virus envelopes (see also section VIII.A) [80,118,119].

V. RECRYSTALLIZATION AT THE LIQUID–AIR AND LIQUID–SOLID INTERFACES

One of the most fascinating properties of S-layer proteins is their capability to reassemble in suspension [57], at the liquid–air interface [120,121], at solid surfaces [22], at spread lipid monolayers [120–122], and on liposomes [118,119,123]. This occurs after removal of the disrupting agent used in the dissolution procedure. In general, a complete disintegration of S-layer lattices on bacterial cells can be achieved using high concentrations of chaotropic agents (e.g., guanidine hydrochloride, urea), by lowering or raising the pH value, or by applying metal-chelating agents (e.g., EDTA, EGTA) or cation substitution. The formation of monolayers at surfaces and interfaces is a key feature of S-layers exploited in a broad spectrum of nanobiotechnological applications (for review see Refs. 2 and 3). As determined by electron and scanning force microscopy, recrystallization starts at several distant nucleation points on the surface and proceeds in-plane until neighboring crystalline areas meet (Fig. 11) [120,121]. In this way, a closed mosaic of randomly aligned monocrystalline S-layer

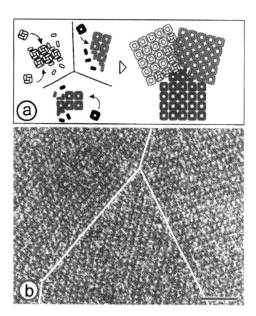

FIG. 11 Schematic drawing illustrating crystal growth of S-layers at the air–water and solid–liquid interfaces and on lipid films. Starting from nucleation sites, the crystalline domains grow until a closed monolayer is formed (a). A grain boundary in the S-layer of *Bacillus coagulans* E38/v1 is shown in the transmission electron micrograph in (b). Bar, 50 nm.

domains is formed. Size and shape of the individual domains depend on the particular S-layer species used, on the properties of the S-layer protein solution (e.g., temperature, pH value, ionic composition, and/or ionic strength), and, for recrystallization on solid supports, on the surface properties of the substrate. For example, the size of individual monocrystalline domains obtained from S-layer proteins from *B. coagulans* E38/v1 may be as large as 10–20 μm in diameter, while S-layer protein from *B. sphaericus* CCM 2177 forms much smaller crystalline domains [120,121]. In general, highly hydrophobic surfaces (contact angle greater than 80°) are better suited for the formation of large-scale closed S-protein monolayers than less hydrophobic or hydrophilic supports [22]. Silanization procedures with different compounds, such as octadecyltrichlorosilane (OTS) and hexamethyldisilazane (HMDS) can be used to produce hydrophobic surfaces on silicon or glass [22]. For practical reasons, most investigations have been carried out on silicon wafers with a native or plasma-induced oxide layer or on photoresist-coated silicon wafers. Further materials suitable for the formation of S-layer protein monolayers are noble metal such as gold, glass, cellulose, a broad range of polymers (e.g., Nuclepore membranes), graphite, highly oriented pyrolytic graphite, and mica. Due to the asymmetry of the surface properties of S-layers, the S-layer protein monolayers are most often oriented with their outer (more hydrophobic) surface face against the hydrophobic support.

Scanning force microscopy (SFM) is the only tool that allows one to image S-layer protein monolayers on solid supports at molecular resolution (Fig. 1b and c) [22–25]. In particular, SFM in contact mode under water with loading forces in the range of <500 pN leads to an image resolution in the subnanometer range (0.5–1.0 nm).

S-layers that have been recrystallized on solid surfaces are the basis for a variety of applications in nanomanufacturing and in the development of miniaturized biosensors [124]. All these applications require suitable methods to pattern recrystallized S-layer protein monolayers on solid surfaces while at the same time the functional and structural integrity of the protein layer in the untouched regions is maintained. The controlled binding of biologically functional molecules [95–98] and of metallic or semiconducting nanoparticles as well as the precipitation of monodisperse nanoparticle arrays [125,126] is a further requirement for using S-layers as molecularly precisely defined templates in the nanobiosciences. These developments are described in more detail in the following.

VI. S-LAYER AS TEMPLATES FOR THE FORMATION OF REGULARLY ARRANGED NANOPARTICLES

In recent years, metallic, or semiconducting nanoparticles have attracted much attention, since their electronic and optical properties revealed a broad range of new applications in material science [127]. This is particularly true when the nanoparticles are deposited in large numbers on a substrate in a regular arrangement. Although the top-down approach of down-scaling microlithographic procedures is successful to some extent, it is definitely not very attractive for emerging technologies, since the generating of an etching mask with feature sizes below 100 nm requires electron-beam writing, which is a slow, sequential process. In contrast, parallel processing and bottom-up approaches, particularly when based on biomolecular templating, have to be used to acquire the highest efficiency in the fabrication process [127].

Douglas and coworkers were the first one that described a bottom-up approach based on S-layers as templates for the formation of perfectly ordered arrays of nanoparticles [128]. The S-layer lattice was used primarily to generate a nanometric lithographic mask for the subsequent deposition of metals. In this approach a thin Ta-W film was deposited on S-layer fragments derived from *Sulfolobus acidocaldarius* and ion-milled, leading to regularly arranged holes in the metal film. This S-layer shows hexagonal lattice symmetry with a repeating distance of the morphological units of 22 nm. Under ion-milling, the S-layer/metal heterostructure exhibited differential metal removal and rearrangement. The remaining metal layer showed pores with the same periodicity as the underlying S-layer lattice. Later, in a basically similar approach, a nanostructured titanium oxide layer was derived from the same S-layer as used before [129]. The S-layer fragments (<1 μm in diameter) had been deposited onto smooth graphite surfaces and coated with a 3.5-nm titanium oxide layer. After oxidation in air, the metal film was ion-milled at normal incidence, yielding a thin nanoporous metal mask. Recently, a further development of this nanolithographic approach led to the generation of orderd arrays of nanometric titanium clusters. Low-energy electron-enhanced etching was used to transfer the structure of the S-layer of *S. acidocaldarius* onto a silicon surface [18,130]. After etching, the S-layer mask was removed and the patterned surface intentionally oxidized in an oxygen plasma. After deposition of a thin layer (~1.2 nm) of titanium on the oxidized surface, an ordered array of nanometric metal clusters was formed in the etched holes.

Most recently, extended nanoparticle arrays were fabricated directly, without the use of nanometric lithographic etching masks, by chemical reduction of metal salts on S-layer protein monolayers on solid supports or on S-layer self-assembly products [125,126]. Inorganic superlattices of monodisperse CdS nanoparticles with either oblique (on the S-layer of *B. coagulans* E38/v1) or square (on the S-layer of *B. sphaericus* CCM 2177) lattice symmetries were fabricated by exposing self-assembled S-layer lattices to Cd ion solutions followed by slow reaction with hydrogen sulfide. Precipitation of the inorganic phase was confined to the S-layer pores, yielding CdS superlattices with prescribed symmetries and lattice parameters. The nanoparticles were 4–5 nm in size. Electron diffraction patterns confirmed the zinc-Blende crystal structure of CdS. In a similar procedure, a square superlattice of uniform 4- to 5-nm-size gold particles with 13.1 nm repeat distance was fabricated by exposing the square S-layer lattice of *B. sphaericus* CCM 2177 with preinduced thiol groups to a tetrachloroauric(III) acid solution (Fig. 12) [126]. Transmission electron mi-

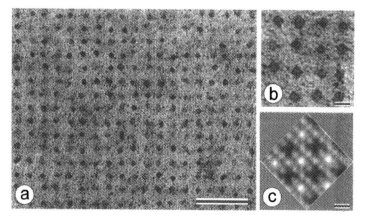

FIG. 12 Transmission electron micrograph of gold nanoparticles on the square S-layer of *Bacillus sphaericus* CCM 2177. Bar, 50 nm. The nanoparticles resemble the morphology of the square S-layer lattice, as shown in (b) a zoomed view of a subregion in (a). A corresponding computer-image reconstruction of a scanning force micrograph of the native S-layer is shown in (c). Bar for (b) and (c), 10 nm. (Modified from Ref. 59.)

croscopical studies showed that the gold nanoparticles were formed in the pore region during electron irradiation of an initially grainy gold coating covering the whole S-layer lattice. The shape of the gold nanoparticles resembled the morphology of the pore region (Fig. 12b and c). By electron diffraction and energy dispersive x-ray analysis, the crystallites were identified as gold [Au(0)]. Electron diffraction patterns showed typical diffraction rings at radii of 1.442 Å and 1.230 Å. Thus it was concluded that the gold nanoparticles were crystalline, but in the long-range order not crystallographically aligned. Later, reduction of the metal salt by the electron beam was replaced by chemical reduction with hydrogen sulfide. This modification allowed one to obtain ordered arrays of nanoparticles without the need for electron irradiation. These experiments were repeated with a broad range of different metal salts, including $PdCl_2$, $NiSO_4$, $KPtCl_6$, $Pb(NO_3)_2$, and $K_3Fe(CN)_6$ [131].

Regular arrays of platinum were achieved by chemical reduction of a platinum salt that had been deposited onto the S-layer of *Sporosarcina ureae* [132]. This S-layer exhibits square lattice symmetry with a lattice constant of 13.2 nm. Transmission electron microscopy revealed the formation of well-separated metal clusters with an average diameter of ~1.9 nm. Seven cluster sites per unit cell were observed. UV-VIS spectrometry was used to study the growth kinetics of the clusters.

These experiments have clearly demonstrated that the S-layer approach features adjustable lattice constants and control over template surface properties by chemical modifications. Current state-of-the-art methods for self-assembly of nanoparticle arrays, which generally involve bifunctional linkers, molecular recognition, or Langmuir–Blodgett techniques, do not offer the control and flexibility of the S-layer approach. The S-layer technology provides a unique possibility to self-assemble highly regular arrays of closely spaced nanoparticles over large substrate areas (at least 10 μm) and thus provide for the first time a fabrication technology for self-assembly of prototypes for molecular memory devices.

VII. WRITING WITH MOLECULES

Not only has the scanning force microscope (SFM) proven its capability to image biological structures with molecular resolution, it has also been used as a "molecular assembler" for positioning atoms on a flat surface [133,134]. Since these experiments are usually done in ultrahigh vaccum and at a temperature of 4 K in order to "freeze" the atoms on the surface in position, it is obvious that such environmental conditions are unsuitable for biological applications. Since all biomolecules require natural environments for maintaining their function, the unavoidable strong thermal fluctuations would not permit assembly of molecular devices on a chemically inert surface by SFM. Furthermore, most practical applications, such as writing with molecules, require a geometrically and physicochemically precisely defined binding matrix as patterning structure on the molecular level. The use of S-layers as immobilization matrix for biomolecules, together with their ability to form extended monolayers on solid substrates, has shown that these protein layers are perfectly suited for this application. In parallel to the classic work of Eigler and coworkers at the atomic level [133,134], it was suggested to transfer molecules with the SFM tip on S-layers (Fig. 13) [124]. For this purpose, polycationic ferritin (PCF) electrostatically bound to negatively charged domains on the S-layer could be used as model system. For moving PCF molecules, an electrically negatively charged (metal-coated) SFM tip could be used, while release would require reversing the polarity. Due to the crystalline arrangement of the negatively charged sites on the S-layer lattice, the PCF molecules can be held in a regular

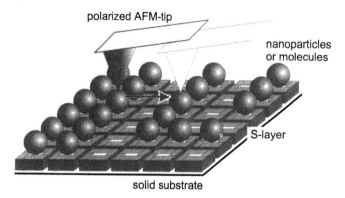

FIG. 13 Schematic drawing of writing with molecules on S-layers. The tip of a scanning force microscope is used to drag-and-drop positively charged ferritin molecules.

array resembling the periodicity of the underlying lattice. In this way it should be possible to generate geometric patterns by "writing with molecules." Removing the molecules from the molecular matrix would result in a "negative writing," while depositing them on an originally blank S-layer surface would be a "positive writing." Reading the information would only require scanning the structure with the SFM tip again or imaging it via high-resolution electron microscopy and automated image analysis. Theoretically, the storage capacity of such a device would be one terabit of binary information on 1 cm² for a square S-layer lattice with a lattice spacing of 10 nm. A major advantage of using molecules instead of atoms for this kind of application would be found in the capability of replicating the written information by using metal shadowing or decoration techniques, as known from electron microscopical preparation techniques.

VIII. S-LAYER AS SUPPORTING STRUCTURE FOR FUNCTIONAL LIPID MEMBRANES

Since it is well known that a great variety of biological processes are membrane mediated, increasing interest has been focused on the meso- and macroscopic reconstitution of biological membranes to utilize their specific features, e.g., as sensing components. In particular, functional transmembrane proteins (e.g., ion channels, carriers, pore-forming proteins, proton pumps) have a broad potential for bioanalytical, biotechnological, and biomimetic applications. On the other hand, investigations are impeded primarily by a low stability of artificial planar and spherical lipid membranes [135,136]. Consequently, there is a strong demand to develop systems that reinforce such fragile structures without interfering with their functional aspects, such as the structural and dynamic properties of lipid membranes.

In this context it is interesting to note that archaea, which possess S-layers as exclusive cell wall components outside the cytoplasmic membrane (Fig. 14), exist under extreme environmental conditions (e.g., high temperatures, hydrostatic pressure, and salt concentrations, low pH values). Thus, it is obvious one should study the effect of proteinaceous S-layer lattices on the fluidity, integrity, structure, and stability of lipid membranes. This section focuses on the generation and characterization of composite structures that mimic the supramolecular assembly of archaeal cell envelope structures composed of a cytoplasmic membrane and a closely associated S-layer. In this biomimetic structure, either a tetraether

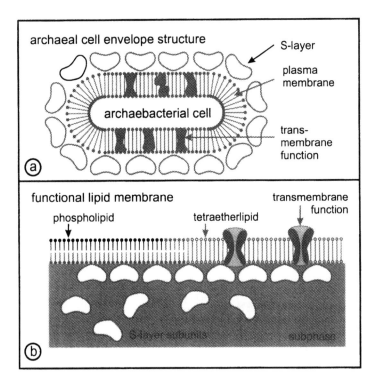

FIG. 14 Schematic illustration of an archaeal cell envelope structure (a) composed of the cytoplasmic membrane with associated and integral membrane proteins and an S-layer lattice, integrated into the cytoplasmic membrane. (b) Using this supramolecular construction principle, biomimetic membranes can be generated. The cytoplasmic membrane is replaced by a phospholipid or tetraether lipid monolayer, and bacterial S-layer proteins are crystallized to form a coherent lattice on the lipid film. Subsequently, integral model membrane proteins can be reconstituted in the composite S-layer-supported lipid membrane. (Modified from Ref. 124.)

lipid monolayer or a phospholipid mono- or bilayer replaces the cytoplasmic membrane (Fig. 14). The isolated bacterial S-layer protein subunits are crystallized on one or both sides of the lipid film [2,101,124]. Finally, functionalization of the S-layer-supported lipid membranes is obtained by incorporation of membrane-active peptides and by reconstitution of transmembrane proteins.

A. Crystallization of S-Layers on Lipid Membranes and Liposomes

In the first step, lipid model membranes have been generated (Fig. 15) on the air/liquid interface, on a glass micropipette (see Section VIII.A.1), and on an aperture that separates two cells filled with subphase (see Section VIII.A.2). Further, amphiphilic lipid molecules have been self-assembled in an aqueous medium surrounding unilamellar vesicles (see Section VIII.A.3). Subsequently, the S-layer protein of *B. coagulans* E38/v1, *B. stearothermophilus* PV72/p2, or *B. sphaericus* CCM 2177 have been injected into the aqueous subphase (Fig. 15). As on solid supports, crystal growth of S-layer lattices on planar or vesicular lipid films is initiated simultaneously at many randomly distributed nucleation

assembly of lipid molecules to form membranes | injection of S-layer subunits into the subphase | applied biophysical methods to the various S-layer/lipid structures

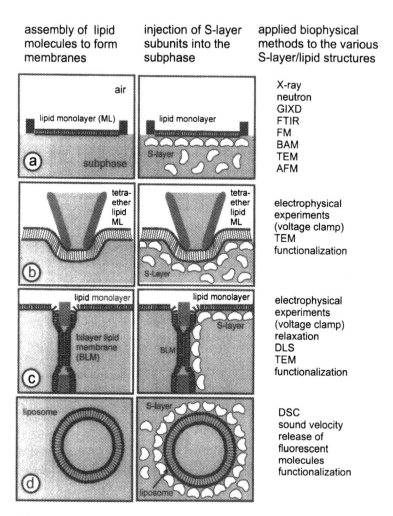

X-ray
neutron
GIXD
FTIR
FM
BAM
TEM
AFM

electrophysical
experiments
(voltage clamp)
TEM
functionalization

electrophysical
experiments
(voltage clamp)
relaxation
DLS
TEM
functionalization

DSC
sound velocity
release of
fluorescent
molecules
functionalization

FIG. 15 Schematic illustrations of lipid membranes and films generated by different procedures involving crystallization of S-layer protein, which has been injected into the subphase. A coherent S-layer lattice can be generated on (a) phospholipid monolayers floating on the air/subphase interface; (b) tetraether lipid monolayers generated in a Langmuir trough clamped by a glass micropipette (tip-dip technique); (c) phospholipid bilayers spanning a Teflon aperture that separates two compartments filled with subphase; and (d) spontaneously self-assembled spherical phospholipid bilayer membranes (liposomes). On the right side, biophysical methods applied for characterization of the supramolecular structures are listed. (Abbreviations: X-ray and neutron: x-ray and neutron reflectivity; GIXD: grazing incidence x-ray diffraction; FTIR: Fourier transform Infrared spectroscopy; FM: fluorescence microscopy; BAM: Brewster-angle microscopy; TEM: transmission electron microscopy; AFM: atomic force microscopy; DLS: dynamic light scattering; DSC: differential scanning microcalorimetry.)

points and proceeds in-plane until the crystalline domains meet (Table 6). Finally, a closed, coherent mosaic of crystalline areas with mean diameters of 1 to several tens of micrometers has been formed on the lipid/subphase interface [121].

The S-layer lattices of *B. coagulans* E38/v1 and *B. stearothermophilus* PV72/p2 are composed of subunits with a molecular mass of 97,000, show oblique lattice symmetry

TABLE 6 Specific Features of S-Layer Proteins Governing Interactions with Lipid Membranes

S-layer proteins adsorb preferentially at lipid films in the liquid-expanded phase[M] [138]

Crystalization is observed only at the liquid-condensed phase[M] [138]

Dynamic crystal growth is initiated at several distant nucleation points, the individual monocrystalline areas grow in all directions until the front edges of neighboring crystals meet[M] [138]

Electrostatic interactions are the dominant binding forces[M,L] [118,122]

"Primary-" and "secondary binding sites" with different specificity exist on the S-layer protein[M] [122]

The S-layer protein interacts with the head groups of the lipid molecules[M] [139]

Size and charge of the lipid head groups and subphase conditions affect the crystallization[M] [122]

The orientation of the S-layer lattice is determined by the subphase conditions[M] [122]

Coating positively charged liposomes resulted in inversion of the ζ-potential[L] [119]

S-layer protein was crystallized on: [M]lipid monolayers; and [L]liposomes.

(base angle 80°) with a center-to-center spacing of the morphological units of 9.4 nm and 7.4 nm, and a thickness of 5 nm. Scanning force microscopical studies revealed ~4-nm-size pores in the S-layer lattices. The S-layer protein SbsB from *B. stearothermophilus* PV72/p2 has been sequenced and cloned [45]. According to the sequence data, the pI of the mature SbsB is 5.0, whereas the S-layer protein of *B. coagulans* E38/v1 revealed an apparent pI value of 4.3, as determined by isoelectrical focusing [122]. The S-layer protein of *B. sphaericus* CCM 2177, with a molecular mass of 127.000, exhibits square lattice symmetry with a lattice constant of 13.1 nm and a thickness of ~8 nm. The pI value of this S-layer protein has been determined to be 4.2 [137]. The largest pores within the protein lattice exhibit a diameter of 4.5 nm, as determined by scanning force microscopy.

In order to obtain more information about the interaction between the S-layer protein and lipid membranes and to study the influence on the fluidity of the lipid membrane, different biophysical methods have been performed, including dual-label fluorescence microscopy (FM), Fourier transform Infrared (FTIR) spectroscopy [138], Brewster-angle microscopy (BAM) [124], x-ray and neutron reflectivity and grazing incidence x-ray diffraction (GIXD) [139,140], fluorescence recovery after photobleaching (FRAP) studies [137], transmission electron microscopy (TEM) [120,121,141,142], atomic force microscopy (AFM) [21], sound velocity and density measurements [143], dynamic light scattering (DLS) [144], differential scanning microcalorimetric studies (DSC) [123], and relaxation and electrophysical experiments [141,142,145] (Fig. 15).

1. Langmuir Lipid Films

Langmuir lipid films, the simplest model membranes, are very suitable to study the recrystallization process of S-layer subunits, because the charge and the phase state of the surface layer can be controlled reproducibly. Moreover, the impact of the S-layer lattice on the structure of the lipid layer can be investigated in detail (Fig. 15a). In all crystallization experiments on Langmuir films, the first step was to generate a monomolecular film of amphiphilic molecules on an air/subphase interface (Fig. 16). Experiments were done with commercially available phospholipid molecules, since a lot of knowledge on the physical parameters (e.g., area per lipid molecule, phase transition pressure) has accumulated over recent decades [146–148]. The physical conditions of these so-called Langmuir films could be controlled by the temperature, the surface pressure, and the composition of the subphase (e.g., pH value, ionic strength, valence of ions) [149,150]. A certain amount of lipid was carefully placed on the air/subphase interface between two barriers (Fig. 16a). The surface

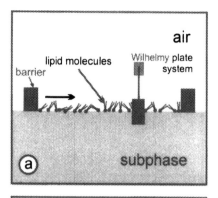

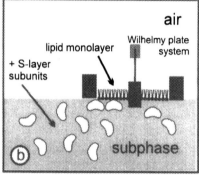

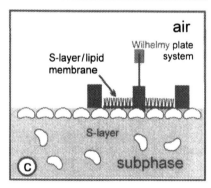

FIG. 16 Formation of a Langmuir lipid monolayer at the air/subphase interface and the subsequent crystallization of S-layer protein. (a) Amphiphilic lipid molecules are placed on the air/subphase interface between two barriers. Upon compression between the barriers, increase in surface pressure can be determined by a Wilhelmy plate system. (b) Depending on the final area, a liquid-expanded or liquid-condensed lipid monolayer is formed. (c) S-layer subunits injected in the subphase crystallized into a coherent S-layer lattice beneath the spread lipid monolayer and the adjacent air/subphase interface.

pressure, measured by a Wilhelmy plate system [151], increased if one barrier was moved toward the other one as the area per lipid molecule decreased (Fig. 16b). At constant temperature, the surface-pressure-versus-area diagram (isotherm) showed breaks that correspond to phase transitions from gaseous to liquid-expanded to the liquid-condensed phase upon compression of the Langmuir film. In the S-layer crystallization experiments, pure distilled water adjusted to the desired pH value between 4 and 9, sometimes with low amounts of sodium ions or divalent calcium ions, was used as subphase. Langmuir films in the liquid-expanded (isotropic) and liquid-condensed (anisotropic) phase were used as crystallization matrix for S-layer proteins.

Generally, the recrystallization of S-layer protein into coherent monolayer on phospholipid films was demonstrated to depend on (1) the phase state of the lipid film, (2) the nature of the lipid head group (size, polarity, and charge), and (3) the ionic content and pH of the subphase [122,138] (Table 6).

It has been shown by FM that the phase state of the lipid exerted a marked influence on S-layer protein crystallization [138]. When the 1,2-dimyristoyl-*sn*-glycero-3-phosphoethanolamine (DMPE) surface monolayer was in the phase-separated state between liquid-expanded and ordered, liquid-condensed phase, the S-layer protein of *B. coagulans* E38/v1 was preferentially adsorbed at the boundary line between the two coexisting phases. The adsorption was dominated by hydrophobic and van der Waals interactions. The two-dimensional crystallization proceeded predominately underneath the liquid-condensed phase. Crystal growth was much slower under the liquid-expanded monolayer, and the entire interface was overgrown only after prolonged protein incubation.

Transmission electron microscopy studies revealed that the nature of the lipid head group also affected the formation of the lattice of the S-layer protein from *B. coagulans* E38/v1 at the lipid film [122]. Coherent S-layer monolayers could be obtained on lipid films composed of lipids with zwitterionic head groups in the presence of calcium ions and on cationic lipids if the lipid films were in the liquid-condensed phase. In contrast, the S-layer protein crystallized poorly under most lipids with negatively charged head groups and under lipids with unsaturated chains. The crystallization process could be facilitated by adding a small portion of positively charged surfactants such as hexadecylamine (HDA) [118,145] or lipid analogs [144] to zwitterionic lipid membranes. These results demonstrated that electrostatic interaction is the primary coupling force between certain domains on the S-layer protein and certain lipid head groups (Table 6).

To investigate the effect of the attached S-layer lattice from *B. coagulans* E38/v1 on the hydrophobic part of the lipid film, FTIR spectroscopy has been performed [138]. As indicated by characteristic frequency shifts of the methylene stretch vibrations on DMPE and 1,2-dipalmitoyl-*sn*-glycero-3-phosphoethanolamine (DPPE), protein crystallization affected the order of the alkane chains and drove the fluid lipid into a state of higher order (Table 7). This effect was pronounced for Langmuir films adjusted to a low surface pressure (5 mN/m), whereas injection of S-layer protein underneath a condensed monolayer (~28 mN/m) resulted in only a slight increase in the segmental alkane chain order [138]. A more detailed study on the coupling of an S-layer lattice from *B. coagulans* E38/v1 and *B. sphaericus* CCM 2177 to lipid monolayers generated from DPPE using x-ray reflectivity and GIXD demonstrated for the lipid chains before and after protein crystallization only a small rearrangement in the hydrophobic part [139]. In quantitative terms, the molecular density of a lipid molecule in the surface monolayer decreased by less than 1%. This small area reduction was accompanied by a slight decrease of the chain tilt angle from ϑ ~9.5° to ~6°. Moreover, these results provided definite evidence that the S-layer protein did not interpenetrate the hydrophobic section of the lipid monolayer, a conjecture that was drawn from indirect evidence obtained in earlier fluorescence microscopic work [138].

In another set of experiments, the integrated electron densities in the region of the DPPE lipid monolayer measured by GIXD before and after S-layer crystallization (*B. coagulans* E38/v1 and *B. sphaericus* CCM 2177) have been compared. The electron densi-

TABLE 7 Impact of Crystallized S-Layer Lattices on Lipid Membranes

A significant change of the lipid head group interactions is observed[L] [123]

The lipid head groups are tilted toward the surface normal of the membrane[B,M] [139,144]

No interpenetrating of S-layer protein in the membrane hydrophobic region occurs[M,B] [138,139, 141,142]

No impact on the hydrophobic thickness of the lipid membrane occurs[B] [141,142]

The fluid lipid film is driven into a state of higher order, especially at low surface pressure[M] [138]

S-layer coating of liposomes leads to an ordering effect of the lipid molecules[L] [123]

The liquid-ordered gel-like state of liposomes is stabilized by S-layer coating[L] [123]

The attachment of S-layers results in a decreased membrane tension[B] [144]

A significant increase of the previously negligible surface viscosity is observed[B] [144,145]

Highest mobility of lipid probe molecules occurs in S-layer-supported bilayers (compared to silane- or dextran-supported bilayers)[B] [137]

Increase in conformational freedom of the hydrophobic core at temperatures below 29°C occurs[L] [143]

S-layer cover prevents the formation of inhomogenities in the bilayer[B] [137]

S-layer protein was crystallized on: [M]lipid monolayers; [B]lipid bilayer membranes; and [L]liposome.

ties were essentially identical in the region of the hydrophobic chains. By contrast, the electron density after S-layer crystallization was slightly, but significantly, larger in the head group region of the lipid monolayer [139,149]. This is very likely due to partial insertion of protein, presumably amino acid side chains, into the lipid head group region, at least to the phosphate moieties and probably further beyond. Further analysis suggested a replacement of ~1.2 H_2O molecules per lipid by peptide material upon S-layer protein crystallization. Consequently, the lipid head groups, which were oriented essentially parallel to the interface in the unperturbed DPPE monolayer [152], had to reorient during protein crystallization. To accommodate the interpenetrating amino side chains, the orientation of the lipid head groups was tilted toward the surface normal in at least some of the lipids in the monolayer that are directly associated with the peptide. This was likely to facilitate an enhancement of electrostatic interactions between the amine functions on the lipid and the anionic amino acid side chain groups [122], which are presumably buried in pockets on the surface of the S-layer protein lattice [77,84]. The increase in electron density within the head groups upon peptide insertion, consistent with an interpretation that almost one amino acid side chain per three to four lipids inserts into the monolayer, left two possibilities open. It might well be that amino acid side chains interpenetrated the DPPE monolayer rather homogeneously, such that each side chain was in contact with a number of lipid head groups and all head groups were affected. Alternatively, peptides might cluster within the lipid's head group region, such that some head groups were strongly affected while some were not [139]. The effect of the S-layer-induced head group motion on the dynamic surface roughness of lipid membranes is also discussed in Section VIII.A.2.

The anisotropy in the physicochemical surface properties and differences in the surface topography of S-layer lattices allowed the determination of the orientation of the monolayers with respect to different surfaces and interfaces. Since in S-layers used for crystallization studies the outer surface is more hydrophobic than the inner one, the protein lattices were generally oriented with their outer face against the air–water interface [120,121]. Crystallization studies with the S-layer protein from *B. coagulans* E38/v1 at different lipid monolayers [122] revealed that the S-layer lattice is attached to lipid monolayers with its outer face if the pH value of the subphase was close to the isoelectric point of ~4.3 for the S-layer subunits. This orientation was also found at clean water surfaces [120] and at various hydrophobic solid supports [22]. Thus, the second important force besides an electrostatic interaction determining the orientation of S-layer subunits with respect to the interface must be a hydrophobic interaction, as has also been suggested for other systems [153]. However, because this binding of the S-layer protein toward the lipid interface occurred more unspecifically, the contact points located on the outer face of the protein have been called "secondary binding sites." In contrast, the S-layer was orientated with its inner surface toward the lipid monolayer whenever the pH value of the buffer was in the range of 5.7–9 (with or without calcium ions) [54,122]. An exception to this was observed, with one cationic lipid film at a pH value of 5.7 as the S-layer lattice was oriented with its outer surface toward the lipid monolayer. Since S-layer lattices at dipolar lipid interfaces were quite strongly coupled, it was assumed that cationic sites located on the lipid head groups or associated with them interact with two to three anionic contact points, the so-called "primary binding sites" at the inner surface of the S-layer protein. This number of binding sites is thought to be more specific, in that binding is subject to steric constrains [122].

Langmuir films have been generated not only from phospholipids but also from tetraether lipids (Fig. 14b). Tetraether glycerophospho- and glycolipids are typical for archaea, where they may constitute the only polar lipids of the cell envelope [154,155]. Tetraether lipids are membrane-spanning lipids, a single monolayer has almost the same thickness as a phospholipid bilayer.

Electrophysical parameters like conductance and capacitance of plain and S-layer-supported glycerol dialkyl nonitol tetraether lipid (GDNT) monolayer have been measured by voltage clamp methods in order to investigate the S-layer/lipid interaction and to prove the stability of the composite membrane [141]. In brief, a small patch of the lipid film was clamped on a glass micropipette in which an electrode was placed. A second electrode was placed at the opposite side of the lipid film, in the subphase. Voltage functions were applied across the lipid film and the corresponding transmembrane current was measured (Fig. 17). Voltage clamp measurements indicated the formation of a tight GDNT monolayer on the tip of the glass micropipette. Even a decrease in conductance was observed upon the crystallization of the S-layer protein from *B. coagulans* E38/v1 [141]. Thus, in accordance with FTIR studies [138], the S-layer apparently did not penetrate or rupture the lipid monolayer.

2. Planar Bilayer Lipid Membranes

The use of lipid bilayers as a relevant model of biological membranes has provided important information on the structure and function of cell membranes. To utilize the function of cell membrane components for practical applications, a stabilization of lipid bilayers is imperative, because free-standing bilayer lipid membranes (BLMs) typically survive for minutes to hours and are very sensitive to vibration and mechanical shocks [156,157]. The following concept introduces S-layer proteins as supporting structures for BLMs (Fig. 15c) with largely retained physical features (e.g., thickness of the bilayer, fluidity). Electrophysical and spectroscopical studies have been performed to assess the application potential of S-layer-supported lipid membranes. The S-layer protein used in all studies on planar BLMs was isolated from *B. coagulans* E38/v1.

Folded membranes were generated out of two phospholipid monolayers at the air/subphase interface [158]. A thin Teflon septum separated the two half-cells of the setup which was also made of Teflon. This septum, with an orifice 50–200 μm in diameter, was

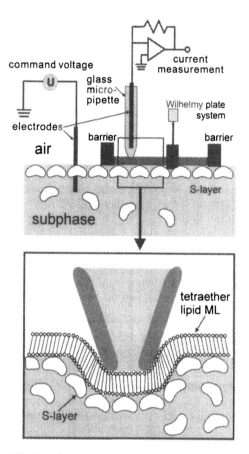

FIG. 17 Schematic illustration of the setup for a tip-dip experiment. First glycerol dialkyl nonitol tetraether lipid (GDNT) monolayers are compressed to the desired surface pressure (measured by a Wilhelmy plate system). Subsequently a small patch of the monolayer is clamped by a glass micropipette and the S-layer protein is recrystallized. The lower picture shows the S-layer/GDNT membrane on the tip of the glass micropipette in more detail. The basic circuit for measurement of the electric features of the membrane and the current mediated by a hypothetical ion carrier is shown in the upper part of the schematic drawing.

pretreated with a small portion of long-chain alkane. Raising the level of the subphase and, thus, the floating lipid monolayer within the two half-cells to above the orifice led to bilayer formation [159]. The BLM, composed of 1,2-diphytanoyl-*sn*-glycero-3-phosphocholine (DPhPC) and small amounts of HDA, could be characterized by measuring the conductance, the capacitance (Table 8), and the breakdown voltage and by data obtained with reconstitution experiments. Comparative voltage clamp studies on plain and S-layer-supported BLMs revealed no significant difference in the capacitance [142]. As also previously observed with clamped GDNT films [141], the conductance of the composite membrane decreased slightly upon crystallization of the S-layer proteins (Table 8). Thus, the attached S-layer lattice did not interpenetrate or rupture the folded BLM.

Painted BLMs were made of a solution of the desired phospholipid in a long-chain alkane (*n*-decane or hexadecane) or squalene [160]. Again, a Teflon cuvette was separated into two compartments by a septum with an orifice 0.8–3 mm in diameter. The orifice was pretreated with the phospholipid, and the compartments were filled with subphase to above the orifice. Subsequently, a small drop of the phospholipid dissolved in *n*-decane was put on a Teflon loop or a small brush and was stroked up the orifice [159]. The front side of one compartment contained a glass window, allowing optical observation of the membrane and its annulus. The thinning of the membranes could be observed through a microscope.

The effect of an attached S-layer lattice on the boundary potential, the capacitance, and the conductance of the DPhPC/HDA membrane was found to be negligible (Table 8), meaning that the BLM was not forced by the attached S-layer lattice to considerable structural rearrangements. A voltage pulse in the microsecond range was applied across the BLM and the current relaxation was measured. This voltage pulse acted as an external force and may cause the formation of a water-filled pore in the lipid membrane. The kinetics of the pore widening and the subsequent rupture process of the membrane could be recorded. Plain BLMs showed a fast linear increase of the pore in time, indicating an inertia-limited defect growth. The attachment of an S-layer caused a slow exponential increase of the induced pore during rupture, indicating a viscosity-determined widening of the pore [145]. This allowed the determination of the two-dimensional viscosity of the composite S-layer/lipid membrane. S-layer-supported BLMs have been found to be significantly more viscous compared to BLMs with absorbed proteins like actin [161] or polyelectrolytes like high-molecular-mass poly-L-lysine [162]. The slow opening velocity and thus the calculated high viscosity might reflect a high number of contact sites (e.g., repetitive domains of the associated S-layer lattice) per unit membrane, as has been suggested for polymers with different density of hydrophobic anchors [163].

To investigate the dynamic surface roughness of the bilayer membranes upon crystallization of the S-layer protein from *B. coagulans* E38/v1, DLS was applied [144]. Beside the lipid head group motions, which has also been discussed in Section VIII.A.1, the out-of-plane vibrational motion of the center of mass of lipid and the collective undulation of the bilayer define the dynamic surface roughness of the membrane [164]. It turned out, that for plain BLMs, which were composed of zwitterionic 1,2-dielaidoyl-*sn*-glycero-3-phosphocholine (DEPC) and positively charged amphiphile dioctadecyl-dimethylammonium bromide (DODAB), the collective motions were dominated by the membrane tension rather than by the membrane curvature energy. The symmetric crystallization of S-layer protein from *B. coagulans* E38/v1 at both sides of the BLM caused a considerable reduction of the membrane tension (Table 7). The membrane bending energy increased by three

TABLE 8   Electrophysical Parameters of Plain, S-Layer-Supported, and SUM-Supported Bilayer Lipid Membranes

	$A \times 10^4$ (cm^2)	G_m (pS)	G_s (μS/cm^2)	C_m (pF)	C_s (μF/cm^2)
Plain folded bilayer [142]	1.32	115 ± 7.2	0.87 ± 0.05	112 ± 7.2	0.84 ± 0.05
Folded bilayer + S-layer [142]	1.32	72 ± 12.0	0.54 ± 0.09	110 ± 6.0	0.83 ± 0.05
Plain painted bilayer [145]	50.3	131	0.03	2010	0.40
Painted bilayer + S-layer [145]	50.3	106	0.02	2075	0.41
Bilayer on SUM [172]	8.6	65 ± 8.0	0.08 ± 0.01	523 ± 85	0.61 ± 0.1

A = approximate area of the bilayer lipid membrane; G_m = membrane conductance; G_s = specific membrane conductance; C_m = membrane capacitance; C_s = specific membrane capacitance.
S-layer is crystalline bacterial cell surface layer of *Bacillus coagulans* E38-66/v1; SUM is S-layer ultrafiltration membrane (*Bacillus sphaericus* CCM 2120).

orders of magnitude over that reported for erythrocyte or vesicular membranes; this is of the same order as that of a shell composed of a 5-nm-thick polyethylene layer [164]. The partial penetration of the protein into the head group region caused a dehydration in this region, which in turn might force the lipid chains into a state of higher molecular order [138]. This higher order might provide a major contribution to the observed lowering of tension, i.e., a facilitation of the transverse shear motions of the lipids. A striking experimental result was the observed S-layer protein–induced reduction of the lateral tension, whereas the opposite behavior was observed for the binding of a streptavidin layer to a zwitterionic BLM via biotinylated lipids, where an increase by a factor of 3 was observed [165]. This indicated significant differences in the interaction mechanisms between the lipid molecules in membrane with the S-layer protein compared with a layer of streptavidin.

A second mechanism that is likely to contribute to the lowering of the surface tension was a possible protein-induced lateral segregation of the cationic lipid molecules. Cationic DODAB may become enriched at sites where the inner face of the attached S-layer protein exhibited an excess of opposite charges. This will result in pillar-like clusters surrounded by a large excess of fluid, highly mobile zwitterionic DEPCs. The presence of the rigid cationic pillars might contribute to a synchronization of the transversal shear motions in the two opposite DODAB-depleted DEPC monolayers and thus give rise to higher amplitudes (and consequently lower tension) in comparison to the BLM prior to the protein crystallization. In addition, the previous increased dynamic surface roughness induced by the addition of DODAB became smoother as the cationic amphiphiles separated into phases that interact with certain domains on the S-layer. Thus, the P-N dipole of the DEPC-molecules within the phases depleted of DODAB returned to their initial orientation, perpendicular to the membrane normal [144]. In line with these results on planar lipid bilayers, phase separations were also proposed in S-layer-coated liposomes composed of 1,2-dipalmitoyl-*sn*-glycero-3-phosphocholine (DPPC), HDA, and cholesterol (see Section VIII.A.3), for lipid domains with different degrees of mobility have been observed [123].

3. Liposomes

Due to the intrinsic amphiphilic character, lipid molecules can spontaneously self-assemble in an aqueous environment to form spherical structures, the so-called liposomes (Fig. 15d). In the case of a single bilayer encapsulating the aqueous core, one speaks of unilamellar vesicles [166]. Due to their structure, chemical composition, and uniform colloidal size, liposomes exhibit special membrane and surface characteristics. These include bilayer phase behavior, elastic properties and permeability, charge density, and attachment of special ligands [167]. In addition to these physicochemical properties, liposomes exhibit many biological characteristics, including (specific) interactions with biological membranes and various cells [168]. These properties have led to several industrial applications with liposomes as drug delivery vehicles in medicine, adjuvants in vaccination, signal enhancers/carriers in medical diagnostics as well as in analytical biochemistry, and solubilizers for various ingredients [167].

In order to enhance the stability of liposomes and to provide a biocompatible outermost surface structure for controlled immobilization (see Section IV), isolated monomeric and oligomeric S-layer protein from *B. coagulans* E38/v1 [118,123,143], *B. sphaericus* CCM 2177, and the SbsB from *B. stearothermophilus* PV72/p2 [119] have been crystallized into the respective lattice type on positively charged liposomes composed of DPPC, HDA, and cholesterol. Such S-layer-coated liposomes are spherical biomimetic structures (Fig. 18) that resemble archaeal cells (Fig. 14) or virus envelopes. The crystallization of S-

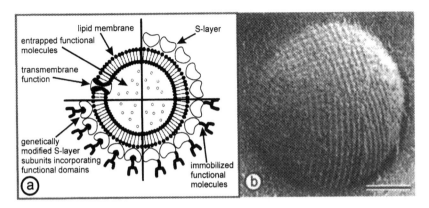

FIG. 18 Schematic drawing of a liposome with entrapped functional molecules, coated with an S-layer lattice, that can be used as immobilization matrix for functional molecules. Alternatively, liposomes can be coated with genetically modified S-layer subunits incorporating functional domains. (Modified from Ref. 59.) (b) Electron micrograph of a freeze-etched preparation of an S-layer-coated liposome (bar, 100 nm).

layer subunits from *B. coagulans* E38/v1 did not affect the morphology of the liposomes [118,123]. As shown by freeze-drying, the S-layer protein from *B. stearothermophilus* PV72/p2 completely covered the liposome surface and bound with its outer surface to such liposomes, leaving the N-terminal region exposed to the ambient environment. Moreover, zeta-potential measurements strongly indicated that the SbsB subunits had bound with their outer charge-neutral surface to the positively charged liposomes [119].

The thermotropic phase behavior of plain DPPC/HDA/cholesterol and S-layer-coated (*B. coagulans* E38/v1) liposomes was characterized by DSC [123]. Both preparations revealed a broad transition around 50°C due to the chain-melting from a liquid-ordered gel-like phase to a liquid-ordered fluid phase, as described for DPPC/cholesterol mixtures [169]. The slightly higher phase transition temperature for the S-layer coated liposomes was explained by an increased intermolecular order (Table 7). This explanation was also supported by sound-velocity studies on plain and S-layer-coated liposomes [143]. The increase in molecular order is also in accord with studies on Langmuir monolayers [138–140]. Based on the DSC data, it was proposed that different lipid domains of phospholipids exhibited different degrees of mobility, particularly when the S-layer protein was cross-linked with positively charged lipid molecules. In addition, the lower enthalpy of the lipid alkyl chain melting found for S-layer-coated liposomes was consistent with a change in the head group interaction [123], as has also been suggested from x-ray data on Langmuir films [139].

B. SUM-Supported Lipid Membranes

S-layer ultrafiltration membranes (SUMs) are isoporous structures with very sharp molecular exclusion limits (see Section III.B). SUMs were manufactured by depositing S-layer-carrying cell wall fragments of *B. sphaericus* CCM 2120 on commercial microfiltration membranes with a pore size up to 1 μm in a pressure-dependent process [73]. Mechanical and chemical resistance of these composite structures could be improved by introducing inter- and intramolecular covalent linkages between the individual S-layer subunits. The uniformity of reactive groups on the surface and within the pores of the S-layer provided the possibility for chemical modifications to obtain differently charged, hydrophilic, or hydrophobic SUMs with tailored molecular sieving and antifouling characteristics [76,82].

In general, lipid membranes generated on a porous support combine the advantages of possessing an essentially unlimited ionic reservoir on each side of the bilayer lipid membrane and of easy manual handling. However, surface properties like roughness or great differences in pore size significantly impaired the stability of the attached BLM. In this section the strategy to use an S-layer as stabilizing and biocompatible layer between the BLM and the porous support is described. This composite structure has been characterized by voltage clamp and reconstitution experiments and compared with BLMs generated on microfiltration membranes.

The SUM was covered by a polymer film with an orifice of approximately 0.3 mm in diameter on each side, and subsequently a folded BLM was generated from a DPhPC/1,2-dipalmitoyl-*sn*-glycero-3-phosphatidic acid (DPPA) monolayer on the side facing the SUM (Fig. 19). Interestingly, no pretreating of the orifice with any alkane or lipid was required, as is imperative for all other BLM techniques. Thus, an accumulation of such compounds could be excluded, and the physicochemical properties of the membrane and

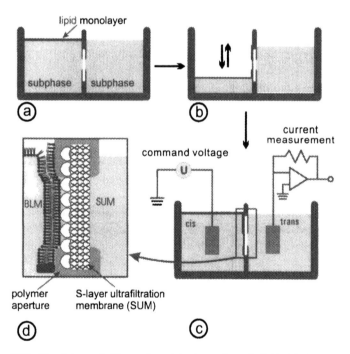

FIG. 19 Schematic illustration of the formation of a folded BLM on an S-layer ultrafiltration membrane (SUM). (a) First, both compartments are filled with subphase. Subsequently, a lipid monolayer is generated on the air/subphase interface of the compartment facing the S-layer lattice. (b) By lowering the level of the subphase, a first lipid monolayer is attached to the S-layer lattice exposed on the SUM. In the following, a bilayer is formed by raising to the initial level. The basic circuit for measuring the electrical features of the membrane and the current flowing through a hypothetical pore is shown in (c). The BLM generated on the S-layer-faced side of the SUM is shown in (d). The actual thickness of the microfiltration membrane is much larger than that of the BLM.

consequently their functioning will not be impaired, as has been reported for other lipid membranes that contained traces of alkane or alcohol [170,171]. The electrophysical intrinsic features are summarized in Table 8. The composite SUM-supported bilayers were tight structures with breakdown voltages well above 500 mV during their whole lifetime of about 8 hours [172]. For a comparison, DPhPC/DPPA membranes on a plain nylon microfiltration support revealed a lifetime of about 3 hours. If voltage ramps were applied, the BLM on the microfiltration membrane ruptured at a magnitude of about 210 mV. The specific capacitance as well as reconstitution experiments revealed that the lipid membrane on the SUM consisted of two layers, for the pore-forming protein α-hemolysin could be assembled whereas no pore formation was observed with the BLM an the microfiltration membrane [172].

C. Solid-Supported Lipid Membranes

Solid-supported membranes were developed in order to overcome the fragility of free-standing BLMs, but also to enable biofunctionalization of inorganic solids (e.g., semiconductors, gold-covered surfaces) for use at sensing devices [173,174]. There are several ways to generate solid-supported lipid membranes. Membranes can be covalently coupled to or separated from solids by ultrathin layers of water or soft polymer cushions (Fig. 20a). Another approach is to use the strong chemisorption of thiolipids on gold surfaces to generate the first lipid monolayer [175,176]. Subsequently, a second phospholipid layer could be generated by the Langmuir–Schaefer technique or by vesicle fusion (Fig. 20) [153,177–179]. However, the various types of solid-supported lipid membranes often show considerable drawbacks, because there is a limited ionic reservoir at one side of the membrane, some membranes appear to be leaky (noninsulating), and extramembrane domains of large-membrane proteins might become denatured by the inorganic support. To overcome the problem of stability and to maintain the structural and dynamic properties the lipid membrane, S-layer proteins have been studied to elucidate their potential as separating ultrathin layer (Fig. 20c).

Composite structures composed of a lipid bilayer or a GDNT monolayer attached to a solid support and an S-layer cover at the opposite side have been generated as follows [21]. After compressing the DPPE monolayer on a Langmuir trough into the liquid-condensed phase, a thiolipid-coated solid support (gold-coated glass cover slip) was placed horizontally onto the monolayer. Subsequently the S-layer protein from *B. sphaericus* CCM 2177 was injected into the subphase and assembled into a closed crystalline monolayer at the DPPE film, as demonstrated by AFM. In a second model system, a GDNT monolayer was generated on a Langmuir trough, and a glass cover slip was placed horizontally onto the monolayer. Again, the S-layer protein from *B. sphaericus* CCM 2177 was injected beneath the GDNT monolayer to form a coherent S-layer lattice. Both assemblies were removed from the liquid–air interface by the Langmuir–Schaefer technique. These composite lipid/S-layer structures were stable enough to allow lifting from the air/water interface and rinsing with deionized water without any disintegrating effect [21]. Prolonged storage of the whole assembly was possible only when an S-layer was used as stabilizing structure.

Supported lipid bilayers on planar silicon substrates have been formed using S-layer protein from *B. coagulans* E38/v1 and from *B. sphaericus* CCM 2177 as support onto which 1,2-dimyristoyl-*sn*-glycero-3-phosphocholine (DMPC) (pure or mixtures with 30% cholesterol) or DPPC bilayers were deposited by the Langmuir–Blodgett-technique (Fig.

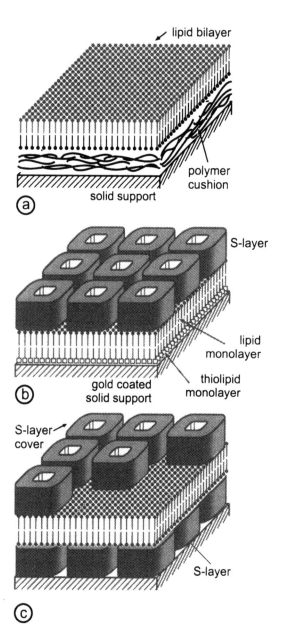

FIG. 20 Schematic drawing illustrating the concept of S-layer-stabilized solid-supported lipid membranes. In (a), a soft polymer cushion (actual thickness is much larger than that of the lipid bilayer) is present as a separating, biocompatible layer. In (b), the generation of the lipid bilayer makes use of the strong chemisorption of thiolipids to gold. The second leaflet and the S-layer that had recrystallized at the lipid film before in a Langmuir trough was transferred onto the thiolipid-coated solid support. (c) As an alternative, an S-layer is located between the solid support and the lipid layer. Optionally, the external leaflet of the lipid bilayer can be stabilized by the attachment of an S-layer cover. (Modified from Ref. 3.)

20c) [150,180]. Lateral diffusion of fluorescence lipid probes in these layers have been investigated by FRAP [137]. In comparison with hybrid lipid bilayers (lipid monolayer on alkylsilanes; Fig. 20b) and lipid bilayers on dextran (Fig. 20a), the mobility of lipids was highest in the S-layer-supported bilayers (Table 7). The type of S-layer protein used as support caused no significant difference in mobility of the probe molecules. In S-layer-supported DPPE bilayers, clear cracks could be seen below the transition temperature of DPPE, whereas above this temperature inhomogeneous round structures were formed [137]. In another set of experiments, the supported bilayers have been covered by S-layer protein (Fig. 20 b, c). The crystallization of S-layer protein was visualized in large scale by TEM and AFM. The S-layer cover induced an enhanced mobility of the probe in the lipid layer, as demonstrated by FRAP. Furthermore it was noticed that the S-layer lattice cover could prevent the formation of cracks and other inhomogenities in the bilayers (Table 7) [137].

D. Functionalization of S-Layer-Supported Lipid Membranes

As model system, the function of the ion-carrier valinomycin and the pore-forming protein α-hemolysin, reconstituted in planar S-layer-supported lipid membranes, has been investigated by voltage clamp methods. Both functional molecules exhibited the same ion selectivity and channel conductance as reconstituted in corresponding free-standing lipid membranes. But most important, functionalized S-layer/lipid structures revealed the advantage of an enhanced long-term stability [141,142]. Incorporation experiments using valinomycin as ion carrier revealed that the increase in conductance was less pronounced for the S-layer-supported than for the plain GDNT monolayer, indicating differences in the local accessibility and/or in the fluidity of the lipid membrane [141].

In reconstitution experiments, the self-assembly of the pore-forming protein α-hemolysin of *Staphylococcus aureus* (αHL) [181–183] was examined in plain and S-layer-supported lipid bilayers. Staphylococcal αHL formed lytic pores when added to the lipid-exposed side of the DPhPC bilayer with or without an attached S-layer from *B coagulans* E38/v1. The assembly of αHL pores was slower at S-layer-supported compared to unsupported folded membranes. No assembly could be detected upon adding αHL monomers to the S-layer face of the composite membrane. Therefore, the intrinsic molecular sieving properties of the S-layer lattice did not allow passage of αHL monomers through the S-layer pores to the lipid bilayer [142].

Reconstitution experiments revealed that the lipid membrane composed of DPhPC/DPPA on the SUM consisted of two layers, because αHL could be assembled into functional pores [172]. The opening and closing behavior of even single αHL pores could be measured, although the traces were noisier compared to folded membranes. The specific conductance for single reconstituted αHL pores was found to be slightly higher when incorporated in folded BLMs than in a SUM. In accordance with studies on αHL reconstituted in folded DPhPC bilayers [184–186], both reconstitution assays revealed a higher specific conductance of the αHL pore when a positive potential was applied compared to the same but negative one. The present results indicated that the S-layers of the SUM represent a water-containing and biocompatible layer for the closely attached lipid bilayer and also provide a natural environment for protein domains protruding from the lipid bilayer.

For functionalization of S-layer-coated liposomes, the S-layer lattice was stabilized by cross-linking the S-layer subunits with either periodate-oxidized raffinose or bis(sulfo-succinimidyl) suberate (BS^3). Both are amine group–specific reagents that were large and hydrophilic enough to be completely rejected by the liposomal membrane. Moreover, both reagents could initiate covalent bonds between part of the hexadecylamine molecules and the S-layer subunits of *B. stearothermophilus* PV72/p2. For introducing biotin residues into the S-layer lattice cross-linked with periodate-oxidized raffinose or BS[3], *p*-diazobenzoyl biocytin was freshly prepared from *p*-aminobenzoyl biocytin, which preferably reacts

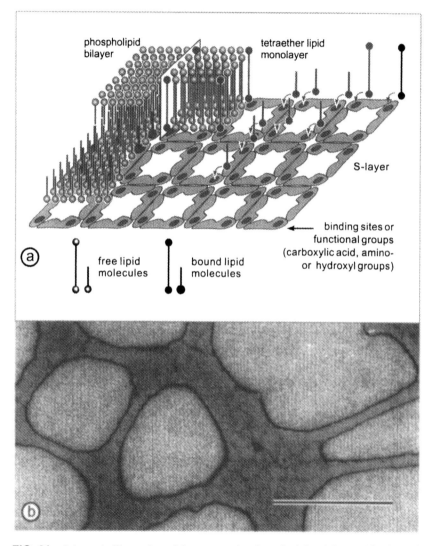

FIG. 21 Schematic illustration of the supramolecular principle of the *semifluid membrane* model, composed of an S-layer-supported phospholipid bilayer or tetraether lipid monolayer (a). The proportion of the lipid molecules that can be linked covalently to the porous S-layer lattice or that interact by noncovalent forces with domains of the S-layer protein subunits significantly modulates the lateral and transverse diffusion of the free lipid molecules and consequently the fluidity of the membrane. (b) Electron micrograph of a negatively stained preparation. Isolated S-layer subunits from *Bacillus coagulans* E38/v1 have been crystallized on a glycerol dialkyl nonitol tetraether lipid (GDNT) monolayer. The composite structure completely covers the holes of the polymer film on the electron microscope grid. Bar, 400 nm. (Modified from Ref. 124.)

with the phenolic residues from tyrosine or with the imidazole ring from histidine. By applying this method, two biotin residues accessible for further streptavidin binding were introduced per S-layer subunit [80]. Both avidin and streptavidin formed a dense monolayer on the surface of such modified S-layer-coated liposomes, which was also confirmed by labeling with an avidin–ferritin conjugate. Immobilized streptavidin was finally exploited for binding of biotinylated IgG. In general, S-layer-coated liposomes completely covered with streptavidin as affinity layer represent versatile targeting and delivery systems [80].

E. Stability of S-Layer-Supported Lipid Membranes

It is now evident that S-layer lattices interact via distinct protein domains with some lipid head groups, leading to major rearrangements within this region (Table 6). In such composite structures the hydrophobic alkyl chains are driven to a state of higher molecular order, although this effect was not pronounced. The higher area per free lipid molecule, i.e., the lipids that do not interact with the S-layer, resulted in a higher mobility, as determined by in-plane lateral diffusion, out-of-plane vibrational motion, and collective undulation (Table 7). Thus, the term *semifluid membrane* (Fig. 21) has been introduced for composite S-layer-supported lipid films [121].

The stability of planar and spherical S-layer-supported membranes has been studied by the application of electrical, mechanical, and thermal challenges and by the release of entrapped carboxyfluoresceine (CF) (Table 9). A significantly lower tendency of Langmuir lipid films to rupture during handling procedures was observed for S-layer-supported than for plain ones [21,121]. Moreover, in contrast to plain BLMs, applying pressure from the S-layer face of composite BLMs, a significantly higher pressure was required to induce bulging of the membrane, as determined by monitoring the change in capacitance [145]. The release of CF entrapped in S-layer-coated liposomes was measured by fluorescence spectroscopy and compared to their uncovered counterparts. S-layer-coated liposomes clearly showed enhanced stability properties against thermal and mechanical challenges and against shear forces, for considerably less release of CF was observed [119]. In reconstitution experiments, the S-layer-supported lipid membrane revealed a decreased tendency to rupture in the presence of a high amount of αHL [142] and valinomycin [141], as determined by measuring the conductance of the lipid membrane. In addition, SUM-supported

TABLE 9 Stability of Composite S-Layer/Lipid Membranes

Observed behavior	Lipid structure and shape
Composite aperture-spanning membranes resist electron beam under low-dose conditions	Phospholipid monolayer [120, 121]
A significant increase in the mechanical stability of solid—supported lipid membranes is observed	Planar phospholipid bilayer [137], tetraether lipid monolayer [21]
A higher hydrostatic pressure has to be applied to induce irreversible breakdown of bilayer membranes	Planar phospholipid bilayer [145]
Liposomes reveal an enhanced stability against mechanical and thermal challenges	Spherical unilamellar liposomes [119]
An enhanced long-term stability is observed in reconstitution experiments	Planar phospholipid bilayer [141,142]

bilayer showed an enhanced stability, for the tendency to rupture in the presence of a high concentration of αHL again was significantly reduced.

These results have demonstrated that the biomimetic approach of copying the supramolecular principle of archaeal cell envelopes opens new possibilities for exploiting functional lipid membranes at meso- and macroscopic scales. Moreover, this technology has the potential to initiate a broad spectrum of developments in such areas as sensor technology, diagnostics, biotechnology, and electronic or optical devices.

IX. PATTERNING OF S-LAYERS RECRYSTALLIZED ON SOLID SUPPORTS

Standard optical lithography using deep ultraviolet (DUV) radiation has proven to be a powerful tool to pattern S-layer protein monolayers on silicon substrates for a broad range of applications [187,188]. For example, the specific binding of biologically active molecules on precisely defined target areas (e.g., electrodes) on a substrate is possible when the binding matrix, namely, the S-layer, is available only there. In fact, this lithographic approach requires the complete removal of the protein layer at certain areas while maintaining the structural and functional integrity of the remaining S-layer lattice. For patterning an S-layer that had been recrystallized as a monolayer on a silicon wafer, the lattice is brought in direct contact with a microlithographic mask and exposed to deep ultraviolet radiation of a pulsed argon fluoride (ArF) excimer laser (wavelength of emitted light 193 nm; pulse duration ~8 ns) (Fig. 22a) [187,188]. The mask is preferably made of a thin layer of chromium on synthetic quartz and may have feature sizes down to 200 nm. Scanning force microscopy (SFM) was used to demonstrate that the S-layer had been removed specifically from the silicon surface in the exposed regions but retained its structure in the unexposed areas (Fig. 22c). And SFM was also used to determine the step height (~8 nm) between exposed and unexposed S-layer areas. It was found that the step height was in perfect agreement with results obtained from the thickness of incomplete S-layer protein monolayers on silicon substrates. At the beginning of these studies the patterned S-layers showed interference fringes at the edges in the unexposed regions. Although the narrow spacing of the interference fringes demonstrated that the S-layers themselves were not the limiting factor in this surface imaging process, it became clear that the crucial step in this procedure is careful drying of the protein layer. While excess water prevents the direct contact of the mask with the S-layer and thus leads to interference fringes upon exposure, a certain amount of residual water is necessary for maintaining the structural integrity of the protein lattice. In particular, drying the S-layer in a stream of dry nitrogen at room temperature has proven to eliminate this unwanted effect. The patterning process was performed in several shots of 100–200 mJ/cm^2 each (pulse frequency 1 Hz). Subsequently, the remaining unexposed S-layer areas could be used either to bind enhancing ligands [189] or to enable electroless metallization [190] in order to form a layer that allows a final patterning process of the silicon by reactive ion etching. Since S-layers are only 5–10 nm thick and consequently much thinner than conventional resists, a considerable improvement in edge resolution in the fabrication of submicron structures can be expected. As an alternative to their application as resist material for microlithographic patterning, the unexposed S-layer areas may also be used for selectively binding intact cells (e.g., neurons), lipid layers, or biologically active molecules, as required for the development of biosensors [191–196].

Finally, it is interesting to note that under exposure to krypton fluoride (KrF) excimer laser radiation supplied in several shots of ~350 mJ/cm^2, the S-layer is not ablated but car-

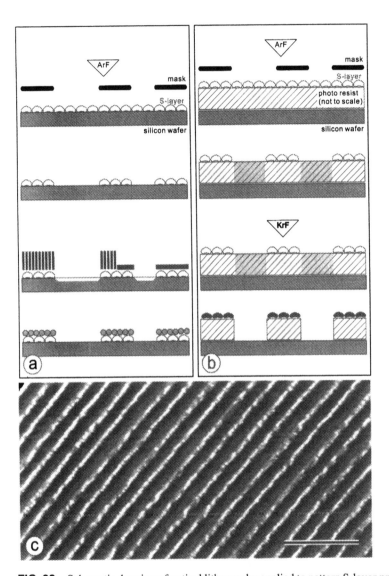

FIG. 22 Schematic drawing of optical lithography applied to pattern S-layer protein monolayers on silicon substrates. (a) A microlithographic mask is brought in direct contact with the S-layer. Upon irradtion with an ArF excimer laser, the S-layer is removed in the exposed regions but remains unaffected in the unexposed areas. Unexposed S-layer areas can be used either to bind enhancing ligands or to enable electroless metallization. In both cases a layer is formed that allows a patterning process by reactive ion etching. Alternatively, unexposed S-layers may also be used for selectively binding biologically active molecules that would be necessary for the fabrication of miniaturized biosensors or biocompatible surfaces. (b) In the two-layer resist approach, the S-layer is removed by ArF excimer laser radiation in the open regions of the mask. In a second step, KrF laser radiation is used to pattern the polymeric resist, with the S-layer as high-resolution lithographic mask. Due to the lower sensitivity of the S-layer toward KrF radiation, the polymeric resist is patterned before the protecting S-layer is burnt down. (c) Scanning force micrograph of the patterned S-layer of *Bacillus sphaericus* CCM 2177. Bar, 2 μm. (Modified from Ref. 59.)

bonized in the exposed areas [187]. This was demonstrated by SFM, which showed a dense layer of carbonized material between the unexposed S-layer regions. Since the protein layer was not ablated but burnt down by KrF radiation, the heat dissipation into the unexposed material was noticeable, as determined from the smooth edge profile of the pattern. The different sensitivity of S-layer protein toward ArF and KrF radiation was used for high-resolution patterning of polymeric resists (Fig. 22b). S-layers that had been formed on top of a spin-coated polymeric resist (on a silicon wafer) were first patterned by ArF radiation and subsequently served as a mask for a blank exposure of the resist by a single shot of KrF irradiation. This two-step process was possible because, in contrast to the polymeric resist, the S-layer is not sensitive to a single KrF pulse. The thinness of the S-layer causes very steep side walls in the developed polymeric resist, as demonstrated by high-resolution scanning electron microscopy.

Preliminary experiments with electron-beam writing and ion-beam projection lithography have demonstrated that the S-layer may also be patterned by these techniques in the sub-100-nm range (unpublished results). The combination of ion-beam projection lithography and S-layers as resist might become important in the near future, since ion beams allow the transfer of smaller features into S-layer lattices compared to optical lithography.

A completely different approach in obtaining S-layer-coated and uncoated areas on solid supports is the use of microcontact printing (μCP) for modifying the surface properties prior to S-layer protein recrystallization (Fig. 23). Microcontact printing makes use

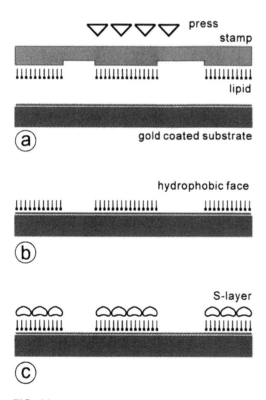

FIG. 23 Schematic drawing of using microcontact printing for obtaining hydrophobic areas on a gold-coated substrate. After pattern transfer (a and b), incubation with an S-layer protein solution (c) leads to the formation of a protein monolayer on the hydrophobic areas only.

of the strong chemisorption of thiolipids on gold surfaces [197]. The exposed end of the lipid molecules is usually chemically modified by the introduction of functional groups, such as carboxyl, hydroxyl, or methyl groups. Terminal methyl groups provide a hydrophobic and hydroxyl groups a hydrophilic surface characteristic. And µCP is a stamping technique in which the lipid molecules are used as ink. In this way it is possible to generate a pattern of well-defined hydrophilic and hydrophobic areas onto a gold-coated substrate. This approach is particularly interesting for the fabrication of S-layer arrays, since several S-layer protein species have already been shown to recrystallize only on hydrophobic substrates [22,198,199]. Although the fluidity of the ink usually limits the feature size of stamped areas to the micrometer range, µCP seems to be a perfect method in combination with S-layer technology to fabricate micrometer-size bioanalytical sensors.

X. CONCLUSIONS AND PERSPECTIVES

It is now evident that S-layers represent the most common cell surface component of prokaryotic organisms. Since their construction principle is based on a single constituent protein (or glycoprotein) subunit with the intrinsic ability to assemble into closed isoporous lattices, on cell surfaces S-layers can be considered the simplest protein membranes developed during biological evolution. Because S-layer lattices possess repetitive physicochemical properties down to the subnanometer scale, they represent structures that exist at the ultimate resolution limit for the molecular functionalization of surfaces and interfaces. As studies on the structure, chemistry, genetics, morphogenesis, and function of S-layer lattices have progressed, their broad application potential has been recognized. Today, most applications developed for using S-layers depend on the in vitro self-assembly capabilities of isolated S-layer subunits in suspension on the surface of solids (e.g., metals, polymers, and silicon wafers), Langmuir-lipid films, and liposomes. Since the functional groups on S-layer lattices are aligned in well-defined positions and orientations, a broad spectrum of very precise chemical modifications can be applied. These unique features can also be exploited for a defined binding of different-size functional molecules. In particular, the possibility of immobilizing or growing other materials on top of recrystallized S-layer lattices with an accurately spatially controlled architecture opens up many new possibilities in supramolecular engineering and nanofabrication.

Moreover, an important line of development is presently directed toward the genetic manipulation of S-layer proteins or glycoproteins. The possibility of modifying and changing the natural properties of S-layer proteins by genetic engineering techniques, such as directed mutagenesis, opens new horizons for the specific tuning of their structure and functional features. Generating truncated S-layer proteins incorporating specific functional domains of other proteins while maintaining the self-assembly capability should, among other things, lead to new ultrafiltration membranes, affinity structures, enzyme membranes, ion-selective binding matrices, microcarriers, biosensors, diagnostics, biocompatible surfaces, and vaccines.

Another important area of future development concerns copying the supramolecular principle of cell envelopes of archaea, which have evolved in the most extreme and hostile ecosystems. This biomimetic approach is expected to lead to new technologies for stabilizing functional lipid membranes and their use at the mesoscopic and macroscopic scales [200]. Along the same line, liposomes coated with S-layer lattices resemble archaeal cell envelopes or virus envelopes. Since liposomes have a broad application potential, particularly as carrier for drug targeting and drug delivery or as vehicles for gene therapy, this possibility for modulating the surface properties appears to be very important.

Finally, feasibility studies have clearly demonstrated that S-layer technologies have a great potential for nanopatterning of surfaces, biological templating, and the formation of arrays of metal clusters, as required in nonlinear optics and molecular electronics.

ACKNOWLEDGMENTS

Part of this work conducted by our research group was supported by the Austrian Science Fund (Projects P-14419, P-14689, and P-12938), the Austrian Federal Ministry of Education, Science, and Culture, the Federal Ministry of Transport, Innovation and Technology, and the Federal Ministry for Economy and Labour.

REFERENCES

1. UB Sleytr, M Sára. Trends Biotechnol 15:20–26, 1997.
2. UB Sleytr, P Messner, D Pum, M Sára. Angew Chemie Int Ed 38:1034–1054, 1999.
3. D Pum, UB Sleytr. Trends Biotechnol 17:8–12, 1999.
4. TJ Beverdige, LL Graham. Microbiol Rev 55:684–705, 1991.
5. UB Sleytr. Int Rev Cytol 53:1–26, 1978.
6. UB Sleytr, P Messner, D Pum, M Sára. Crystalline Bacterial Cell Surface Layers. Berlin: Springer Verlag, 1988.
7. UB Sleytr, P Messner, D Pum, M Sára, eds. Crystalline Bacterial Cell Surface Proteins. Austin, TX: Landes/Academic, 1996.
8. RGE Murray. In: TJ Beveridge, SF Koval, eds. Advances in Bacterial Paracrystalline Surface Layers. New York: Plenum Press, 1993. pp 3–9.
9. P Messner, UB Sleytr. Adv Microb Physiol 33:213–275, 1992.
10. UB Sleytr, TJ Beveridge. Trends Microbiol 7:253–260, 1999.
11. M Sára, UB Sleytr. J Bacteriol 182:859–868, 2000.
12. H König. Can J Microbiol 34:395–406, 1988.
13. D Pum, P Messner, UB Sleytr. J Bacteriol 173:6865–6873, 1991.
14. UB Sleytr, P Messner, D Pum. Methods Microbiol 20:29–60, 1988.
15. UB Sleytr, P Messner. Annu Rev Microbiol 37:311–339, 1983.
16. UB Sleytr, AM Glauert. In: JR Harris, ed. Electron Microscopy of Proteins. Vol 3. London: Academic Press, 1982, pp 41–76.
17. TJ Beveridge. Curr Opin Struct Biol 4:204–212, 1994.
18. UB Sleytr. FEMS Microbiol Rev 20:5–12, 1997.
19. S Hovmöller. In: TJ Beveridge, SF Koval, ed. Advances in Bacterial Paracrystalline Surface Layers. New York: Plenum Press, 1993, pp 13–21.
20. W Baumeister, G Lembcke. J Bioenerg Biomembr 24:567–575, 1992.
21. B Wetzer, D Pum, UB Sleytr. J Struct Biol 119:123–128, 1997.
22. D Pum, UB Sleytr. Supramolec Sci 2:193–197, 1995.
23. DJ Müller, W Baumeister, A Engel. J Bacteriol 178:3025–3030, 1996.
24. DJ Müller, D Fotiadis, A Engel. FEBS Lett 430:105–111, 1996.
25. A Engel, CA Schoenenberger, DJ Müller. Curr Opin Struct Biol 7:279–284, 1997.
26. J Mayr, A Lupas, J Kellermann, C Eckerskorn, W Baumeister, J Peters. Curr Biol 6:739–749, 1996.
27. SF Koval, RGE Murray. Can J Biochem Cell Biol 62:1181–1189, 1984.
28. UB Sleytr, P Messner. Annu Rev Microbiol 37:311–339, 1983.
29. P Messner, UB Sleytr. In: IC Hancock, I Poxton, eds. Bacterial Cell Surface Techniques. Chichester, UK: Wileys, 1988, pp 97–104.

30. Z Pei, RT Ellison III, RV Lewis, MJ Blaser. J Biol Chem 263:6414–6420, 1988.
31. G Bröckl, M Behr, S Fabry, R Hensel, H Kaudewitz, E Biendl, H König. Eur J Biochem 199:147–152, 1991.
32. P Kosma, T Wugeditsch, R Christian, S Zayni, P Messner. Glycobiology 5:791–796, 1995.
33. HJ Boot, PH Pouwels. Mol Microbiol 21:1117–1123, 1996.
34. UB Sleytr, P Messner, D Pum, M Sára. Mol Microbiol 10:911–916, 1993.
35. SR Thomas, TJ Trust. J Mol Microbiol 21:1117–1123, 1995.
36. P Messner. Glycoconj J 14:3–11, 1997.
37. P Messner, C Schäffer. In: RJ Doyle, ed. Glycomicrobiology. New York: Plenum Press, 2000, pp. 93–125.
38. P Messner, UB Sleytr. Glycobiol 1:545–555, 1991.
39. C Schäffer, T Wugeditsch, C Neuninger, P Messner. Microbial Drug Resistance 2:17–23, 1996.
40. Sumper, FT Wieland. In: J Montreuil, JFG Vliegenthart, H Schachter, eds. Glycoproteins. Amsterdam: Elsevier, 1995, pp 455–473.
41. M Sumper. In: TJ Beveridge, SF Koval, eds. Advances in Paracrystalline Bacterial Surface Layers. New York: Plenum Press, 1993, pp 109–117.
42. B Kuen, W Lubitz. In: UB Sleytr, P Messner, D Pum, M Sára, eds. Crystalline Bacterial Cell Surface Proteins. Austin, TX: Landes/Academic, 1996, pp 77–111.
43. B Kuen, UB Sleytr, W Lubitz. Gene 145:115–120, 1994.
44. B Kuen, M Sára, W Lubitz. Mol Microbiol 19:495–503, 1995.
45. B Kuen, A Koch, E Asenbauer, M Sára, W Lubitz. J Bacteriol 179:1664–1670, 1997.
46. M Jarosch, EM Egelseer, D Mattanovich, UB Sleytr, M Sára. Microbiology 146:273–281, 2000.
47. J Dworkin, MJ Blaser. Mol Microbiol 26:433–440, 1997.
48. EM Egelseer, K Leitner, M Jarosch, C Hotzy, S Zayni, UB Sleytr, M Sára. J Bacteriol 180:1488–1495, 1998.
49. M Sára, B Kuen, HF Mayer, F Mandl, KC Schuster, UB Sleytr. J Bacteriol 178:2108–2117, 1996.
50. A Lupas, H Engelhardt, J Peters, U Santarius, S Volker, W Baumeister. J Bacteriol 176:1224–1233, 1994.
51. E Leibovitz, M Lemaire, I Miras, S Salamitou, P Beguin, H Ohayon, P Gounon, M Matuschek, K Sahm, H Bahl. FEMS Microbiol Rev 20:127–133, 1997.
52. M Lemaire, H Ohayon, P Gounon, T Fujino, P Beguin. J Bacteriol 177:2451–2459, 1995.
53. S Mesnage, E Tosi-Couture, A Fouet. Mol Microbiol 31:927–936, 1999.
54. W Ries, C Hotzy, I Schocher, UB Sleytr. J Bacteriol 179:3892–3898, 1997.
55. N Ilk, P Kosma, M Puchberger, EM Egelseer, HF Mayer, UB Sleytr, M Sára. J Bacteriol 181:7643–7646, 1999.
56. UB Sleytr, AM Glauert. J Ultrastruct Res 50:103–116, 1975.
57. UB Sleytr, P Messner. In: H Plattner, ed. Electron Microscopy of Subcellular Dynamics. Boca Raton, FL: CRC Press, 1989, pp 13–31.
58. UB Sleytr, R Plohberger. In: W Baumeister, W Vogell, eds. Microscopy at Molecular Dimensions. Berlin: Springer-Verlag, 1980, pp 36–47.
59. UB Sleytr, M Sára, D Pum. In: A Ciferri, ed. Supramolecular Polymerization. New York: Marcel Dekker, 2000, pp. 177–213.
60. UB Sleytr. Nature 257:400–402, 1975.
61. UB Sleytr. In: O Kiermayer, ed. Cell Biology Monographs. Vol. 8. New York: Springer-Verlag, 1981, pp 1–26.
62. WF Harris, LE Scriven. Nature 228:827–828, 1970.
63. WF Harris. Sci Am 237:130–145, 1977.
64. FRN Nabarro, WF Harris. Nature 232:423–425, 1971.
65. P Messner, D Pum, UB Sleytr. J Ultrastruct Mol Struct Res 97:73–88, 1986.
66. R Jaenicke, R Welsch, M Sára, UB Sleytr. Biol Chem Hoppe-Seyler 366:663–670, 1985.

67. M Sára, UB Sleytr. J Bacteriol 169:2804–2809, 1987.
68. M Sára, C Dekitsch, HF Mayer, E Egelseer, UB Sleytr. J Bacteriol 180:4146–4153, 1998.
69. R Scherrer, P Gerhardt. J Bacteriol 107:718–735, 1971.
70. M Sára, UB Sleytr. J Bacteriol 169:4092–4098, 1987.
71. A Breitwieser, K Gruber, UB Sleytr. J Bacteriol 174:8008–8015, 1992.
72. W Blatt. In: P Meares, ed. Membrane Separation Processes. 1976, pp 81–120.
73. M Sára, UB Sleytr. J Membrane Sci 33:27–49, 1987.
74. UB Sleytr, M Sára. US Patent 4,752,395, 1988.
75. M Sára, D Pum, UB Sleytr. J Bacteriol 174:3487–3493, 1992.
76. S Weigert, M Sára. J Membrane Sci 106:147–159, 1995.
77. M Sára, UB Sleytr. J Bacteriol 175:2248–2254, 1993.
78. JK Inman, RN Perham, GC Du Bois, E Apella. Methods Enzymol 91:559–580, 1983.
79. M Sára, UB Sleytr. Appl Microbiol Biotechnol 38:147–151, 1992.
80. C Mader, S Küpcü, UB Sleytr, M Sára. Biochim Biophys Acta 1463:142–150, 2000.
81. B Matthiasson. J Membrane Sci 16:23–36, 1983.
82. S Weigert, M Sára. J Membrane Sci 121:185–196, 1996.
83. M Sára, UB Sleytr. Micron 27:141–156, 1996.
84. M Sára, UB Sleytr. Prog Biophys Mol Biol 65:83–111, 1996.
85. M Sára, UB Sleytr. Appl Microbiol Biotechnol 30:184–189, 1989.
86. S Küpcü, C Mader, M Sára. Biotechnol Appl Biochem 21:275–286, 1995.
87. C Weiner, M Sára, UB Sleytr. Biotechnol Bioeng 43:321–330, 1994.
88. JJ Langone. Adv Immunol 32:157–252, 1982.
89. A Breitwieser, S Küpcü, S Howorka, S Weigert, C Langer, K Hoffmann-Sommergruber, O Scheiner, UB Sleytr, M Sára. Biotechniques 21:918–925, 1996.
90. J Sjöquist, B Meloun, H Hjelm. Eur J Biochem 29:572–578, 1972.
91. M Kim, K Saito, S Furusaki, T Sugo, I Ishigaki. J Chrom 586:27–33, 1991.
92. C Weiner, M Sára, G Dasgupta, UB Sleytr. Biotechnol Bioeng 4:55–65, 1994.
93. DC Herak, EW Merrill. Biotechnol Prog 6:33–40, 1990.
94. A Breitwieser, C Mader, I Schocher, K Hoffmann-Sommergruber, O Scheiner, W Aberer, UB Sleytr, M Sára. Allergy 53:786–793, 1998.
95. A Neubauer, D Pum, UB Sleytr. Anal Lett 26:1347–1360, 1993.
96. A Neubauer, C Hödl, D Pum, UB Sleytr. Anal Lett 27:849–865, 1994.
97. A Neubauer, D Pum, UB Sleytr, I Klimant, OS Wolfbeis. Biosens Bioelectron 11:315–323, 1996.
98. A Neubauer A Neubauer, S Pentzien, S Reetz, W Kautek, D Pum, UB Sleytr. Sensors Actuators 40:231–236, 1997.
99. K Taga, R Kellner, U Kainz, UB Sleytr. Anal Chem 66:35–39, 1993.
100. P Messner, FM Unger, UB Sleytr. In: UB Sleytr, P Messner, D Pum, M Sára, eds. Crystalline Bacterial Cell Surface Proteins. Austin TX: Landes/Academic, 1996, pp 161–173.
101. UB Sleytr, H Bayley, M Sára, A Breitwieser, S Küpcü, C Mader, S Weigert, FM Unger, P Messner, B Jahn-Schmid, B Schuster, D Pum, K Douglas, NA Clark, JT Moore, TA Winningham, S Levy, I Frithsen, J Pankovc, P Beale, HP Gillis, DA Choutov, KP Martin. FEMS Microbiol Rev 20:151–175, 1997.
102. WW Kay, TJ Trust. Experientia 47:412–414, 1991.
103. MJ Blaser, EC Gotschlich. J Biol Chem 265:14529–14535, 1990.
104. S Mesnage, M Weber-Levy, M Haustant, M Mock, A Fouet. Infect Immun 67:4847–4850, 1999.
105. JC Thornton, RA Garduno, SG Newman, WW Kay. Microbiol Pathog 11:85–99, 1991.
106. AJ Malcolm, P Messner, UB Sleytr, RH Smith, FM. In: UB Sleytr, P Messner, D Pum, M Sára, eds. Immobilized Macromolecules: Application Potentials. London: Springer-Verlag, 1993, pp 195–207.

107. AJ Malcolm, MW Best, RJ Szarka, Z Mosleh, FM Unger, P Messner, UB Sleytr. In: TJ Beveridge, SF Koval, eds. Advances in Bacterial Paracrystalline Surface Layers. New York: Plenum Press, 1993, pp 219–233.
108. R Schneerson, JB Robbins, SC Szu, Y Yang. In: R Bell, G Torrigiani, eds. Towards Better Carbohydrate Vaccines. Chichester, UK: Wiley, 1987, pp 307–332.
109. M Sing, D O'Hagan. Nature Biotechnol 17:1075–1081, 1999.
110. UB Sleytr, W Mundt, P Messner, RH Smith, FM Unger. US Patent 5,043,158, 1991.
111. UB Sleytr, W Mundt, P Messner. Eur Patent 03064 73 B1, 1989.
112. P Messner, MA Mazid, FM Unger, UB Sleytr. Carbohydr Res 233:175–184, 1992.
113. B Jahn-Schmid, P Messner, FM Unger, UB Sleytr, O Scheiner, D Kraft. J Biotechnol 44:225–231, 1996.
114. B. Jahn-Schmid, M Graninger, M Glozik, S Küpcü, C Ebner, FM Unger, UB Sleytr, P Messner. Immunotechnology 2:103–113, 1996.
115. B Jahn-Schmid, U Siemann, A Zenker, B Bohle, P Messner, FM Unger, UB Sleytr, O Scheiner, D Kraft, C Ebner. Int Immunol 9:1867–1874, 1997.
116. J Smit, WH Bingle. US Patent 5,500,353, 1996.
117. W Lubitz, UB Sleytr. PCT/EP 97/00432, 1997.
118. S Küpcü, M Sára, UB Sleytr. Biochim Biophys Acta 1235:263–269, 1995.
119. C Mader, S Küpcü, M Sára, UB Sleytr. Biochim Biophys Acta 1418:106–116, 1999.
120. D Pum, M Weinhandl, C Hödl, UB Sleytr. J Bacteriol 175:2762–2766, 1993.
121. D Pum, UB Sleytr. Thin Solid Films 244:882–886, 1994.
122. B Wetzer, A Pfandler, E Györvary, D Pum, M Lösche, UB Sleytr. Langmuir 14:6899–6906, 1998.
123. S Küpcü, K Lohner, C Mader, UB Sleytr. Molec Membrane Biol 15:69–74, 1998.
124. D Pum, UB Sleytr In: UB Sleytr, P Messner, D Pum, M Sára, eds. Crystalline Bacterial Cell Surface Proteins. Austin, TX: Landes/Academic, 1996, pp 175–209.
125. W Shenton, D Pum, UB Sleytr, S Mann. Nature 389:585–587, 1997.
126. S Dieluweit, D Pum, UB Sleytr. Supramolec Sci 5:15–19, 1998.
127. A TenWolde. Nanotechnology: Towards a Molecular Construction Kit. The Hague: STT, 1998.
128. K Douglas, NA Clark. Appl Phys Lett 48:676–678, 1986.
129. K Douglas, G Devaud, NA Clark. Science 257:642–644, 1992.
130. TA Winningham, HP Gillis, DA Choutov, KP Marzin, IT Moore, K Douglas. Surf Sci 406:221–228, 1998.
131. S Dieluweit. Zwei-Dimensionale Proteinkristalle (S-Schichten) als Matrix zur Biomineralisation und biomimetische Membranen. PhD Thesis, Universität für Bodenkultur, Vianna, 1999.
132. M Mertig, R Kirsch, W Pompe, H Engelhardt. Eur Phys J D 9:45–48, 1999.
133. DM Eigler, EK Schweizer. Nature 344:524–526, 1990.
134. DM Eigler, CP Lutz, WE Rudge. Nature 352:600–603, 1991.
135. HF Knapp, W Wiegräbe, M Heim, R Eschrich, R Guckenberger. Biophys J 69:708–715, 1995.
136. X Lu, A Leitmannova-Ottova, TH Tien. Bioelectrochem Bioenerg 39:285–289, 1996.
137. E Györvary, B Wetzer, UB Sleytr, A Sinner, A. Offenhäusser, W Knoll. Langmuir 15:1337–1347, 1999.
138. A Diederich, C Sponer, D Pum, UB Sleytr, M Lösche. Coll Surf B: Biointerfaces 6:335–346, 1996.
139. M Weygand, B Wetzer, D Pum, UB Sleytr, N Cuvillier, K Kjaer, PB Howes, M Lösche. Biophys J 76:458–468, 1999.
140. M Weygand, M Schalke, PB Howes, K Kjaer, J Friedmann, B Wetzer, D Pum, UB Sleytr, M Lösche. J Mater Chem 10:141–148, 1999.
141. B Schuster, D Pum, UB Sleytr. Biochim Biophys Acta 1369:51–60, 1998.

142. B Schuster, D Pum, O Braha, H Bayley, UB Sleytr. Biochim Biophys Acta 1370:280–288, 1998.
143. T Hianik, S Küpcü, UB Sleytr, P Rybár, R Krivánek, U Kaatze. Coll Surf A: Physicochem Eng Aspects 147:331–339, 1999.
144. R Him, B Schuster, UB Sleytr, TM Bayerl. Biophys J 77:2066–2074, 1999.
145. B Schuster, UB Sleytr, A Diederich, G Bähr, M Winterhalter. Eur Biophys J 28:583–590, 1999.
146. M Lösche, H Möhwald. J Colloid Interface Sci 131:56–67, 1989.
147. D Möbius, H Möhwald. Adv Mater 1:19–24, 1991.
148. H Möhwald. In: R Lipowsky, E Sackmann, eds. Structure and Dynamics of Membranes. Amsterdam: Elsevier, 1995, pp 161–212.
149. D Marsh. Biochim Biophys Acta 1286:183–223, 1996.
150. JA Zasadzinski, R Viswanathan, L Madson, J Garnaes, KD Schwartz. Science 263:1726–1733, 1994.
151. MC Petty, WA Barlow. In: G Roberts, ed. Langmuir–Blodgett films. Austin, TX: Academic Press, 1990, pp 93–132.
152. DA Pink, M Belaya, V Levadny, B Quinn. Langmuir 13:1701–1711, 1997.
153. A Ulman. An Introduction to Ultrathin Organic Films. From Langmuir–Blodgett to Self-Assembly. San Diego, CA: Academic Press, 1991.
154. Y Koga, M Nishihara, H Morii, M Akagawa-Matsushita. Microbiol Rev 57:164–182, 1993.
155. F Paltauf. Chem Phys Lipids 74:101–139, 1994.
156. B Raguse, V Braach-Maksvytis, BA Cornell, LG King, PDJ Osman, RJ Pace, L Wieczorek. Langmuir 14:648–659, 1998.
157. TH Tien, A Ottova-Leitmannova. Membrane Biophysics as Viewed from Experimental Bilayer Lipid Membranes: Planar Lipid Bilayers and Spherical Liposomes. New York: Elsevier Science, 2000.
158. M Montal, P Mueller. Proc Natl Acad Sci USA 69:3561–3566, 1972.
159. W Hanke, WR Schlue. In: DB Sattelle, ed. Biological Techniques Series. London: Academic Press, 1993, pp 60–78.
160. R Fettiplace, LGM Gordon, SB Hladky, J Requena, H Zingsheim, DA Haydon. In: ED Korn, ed. Methods of Membrane Biology. New York: Plenum Press, 1975, Vol. 4, pp 1–75.
161. M Lindemann, M Steinmetz M Winterhalter. Prog Colloid Polymer Sci 105:209–213, 1997.
162. A Diederich, G Bähr, M Winterhalter. Langmuir 14:4597–4605, 1998.
163. A Diederich, M Strobel, W Maier, M Winterhalter. J Phys Chem B 103:1402–1407, 1999.
164. E Sackmann. In R Lipowsky, E Sackmann, eds. Structure and Dynamics of Membranes. Amsterdam: Elsevier, 1995, pp 213–304.
165. R Hirn, R Benz, TM Bayerl. Phys Rev E 59:5987–5994, 1999.
166. D Papahadjopoulos. Liposomes and their use in biology and medicine. Ann NY Acad Sci 308:1–412, 1978.
167. DD Lasic. In: R Lipowsky, E Sackmann, eds. Structure and Dynamics of Membranes. Amsterdam: Elsevier, 1995, pp 491–519.
168. DD Lasic. Liposomes. Am Sci 80:20–31, 1992.
169. TPW McMullen, RN McElhaney. Biochim Biophys Acta 1234:90–98, 1995.
170. K Lohner. Chem Phys Lipids 57:341–362, 1991.
171. FJ Weber, JAM de Bont. Biochim Biophy. Acta 1286:225–245, 1996.
172. B Schuster, D Pum, M Sára, O Braha, H Bayley, UB Sleytr. Langmuir 17:499–503, 2001.
173. E Sackmann. Science 271:43–48, 1996.
174. E Sackmann, M Tanaka. Trends Biotechnol 18:58–64, 2000.
175. AL Plant. Langmuir 9:2764–2767, 1993.
176. AL Plant, M Gueguetechkeri, W Yap. Biophys J 67:1126–1133, 1994.
177. I Langmuir, VJ Schaefer. J Am Chem Soc 59:1406–1417, 1937.

178. R Naumann, EK Schmidt, A Joncyk, K Fenderl, B Kadenbach, T Liebermann, A Offen-
      häusser, W Knoll. Biosens Bioelectron 14:651–662, 1999.
179. EK Schmidt, T Liebermann, M Kreiter, A Joncyk, R Naumann, A Offenhäusser, E Neumann,
      A Kukol, A Maelicke, W Knoll. Biosens Bioenerg 13:585–591, 1998.
180. KJ Blodgett. J Am Chem Soc 57:1007–1022, 1935.
181. S Bhakdi, R Füssle, J Tranum-Jensen. Proc Natl Acad Sci USA. 78:5475–5479, 1981.
182. S Bhakdi, U Weller, I Walev, E Martin, D Jonas, M Palmer. Med Microbiol Immunol
      182:167–175, 1993.
183. S Bhakdi, J Tranum-Jensen. Microbiol Rev 55:733–751, 1991.
184. G Menestrina. J Membr Biol 90:177–190, 1986.
185. YE Korchev, CL Bashford, GM Alder, JJ Kasianowicz, CA Pasternak, J Membr Biol
      147:233–239, 1995.
186. YE Korchev, GM Alder, A Bakhramov, CL Bashford, BS Joomun, EV Sviderskaya. J Membr
      Biol 143:143–151, 1995.
187. D Pum, G Stangl, C Sponer, W Fallmann, UB Sleytr. Colloids and Surfaces B: Biointerfaces
      8:157–162, 1996.
188. D Pum, G Stangl, C Sponer, K Riedling, P Hudek, W Fallmann, UB Sleytr. Microelectron Eng
      35:297–300, 1997.
189. GN Taylor, RS Hutton, SM Stein, HE Katz, ML Schilling, TM Putvinski. Microelectr Eng
      23:259–263, 1994.
190. JM Calvert. J Vac Sci Technol B 11:2155–2163, 1993.
191. P Fromherz, A Offenhäusser, T Vetter, J Weis. Science 252:1290–1293, 1991.
192. P Fromherz, H Schaden. Eur J Neurosci 6:1500–1504, 1994.
193. A Offenhäuser, C Sprössler, M Matsuzawa, W Knoll. Biosensors Bioelectronics 12:819–826,
      1997.
194. C Sprössler, D Richter, M Denyer, A Offenhäusser. Biosensors Bioelectronics 13:613–618,
      1998.
195. C Sprössler, M Denyer, S Britland, W Knoll, A Offenhäusser. Phys Rev E 60:2171–2176,
      1999.
196. WH Baumann, M Lehmann, A Schwinde, R Ehret, M Brischwein, B Wolf. Sensors Actuators
      B55:77–89, 1999.
197. GM Whitesides, JP Mathias, CT Seto. Science 254:1312–1319, 1991.
198. UB Sleytr, D Pum, W Fallmann, G Stangl, IMS. Austrian Patent 373/97, 1997.
199. UB Sleytr, D Pum, W Fallmann, G Stangl, H Löschner. PCT/AT98/00050, 1998.
200. B Schuster, UB Sleytr. Rev Mol Biotechnol 74:233–254, 2000.

10

DNA as a Material for Nanobiotechnology

CHRISTOF M. NIEMEYER   University of Bremen, Bremen, Germany

I. INTRODUCTION

The essence of chemical science finds its full expression in the words of that epitome of the artist-scientist, Leonardo da Vinci: "Where Nature finishes producing its own species, man begins, using natural things and with the help of this nature, to create an infinity of species." Nobel laureate Jean-Marie Lehn uses this quotation to look on the future and perspectives of supramolecular chemistry [1]. Starting from the pioneering work of J. F. Pedersen, D. J. Cram, and J.-M. Lehn, the research on supramolecular aggregates, "supermolecules," held together by weak, noncovalent interactions such as hydrogen bonding, electrostatic forces, and van der Waals forces, has been developed over the past 30 years to become today's well-established discipline that numerous research groups are devoted to [1–3]. One distinguishing characteristic of supramolecular chemistry is the interdisciplinary background of the investigators engaged in this field, ranging from biology to biochemistry, organic, inorganic, and physical chemistry to physics, mathematics, and engineering sciences. This enormous expanse of scientific background is necessary for an initial understanding of the principles of nature, how its fascinating, complex, and functional supramolecular devices are formed from small molecular building blocks by means of self-assembly processes. Clearly, the ultimate goal of synthetic supramolecular chemistry goes beyond the pure desire to learn from nature to utilize natural principles in order to generate entirely novel devices and materials of enhanced performance. Such biomimetics are potentially useful for sensory applications, catalysis, membrane transport, medicinal or engineering sciences, to name a few.

The investigation of supramolecular systems and self-organization suggests that to utilize natural processes, we must first explore and understand nanometer-scale systems, and in a second step use the concepts and principles applied in such systems to fabricate novel synthetic assemblages. For this purpose, the development of a "nanochemistry" is crucial [4]. Another strong motivation to elaborate nanometer-scale systems is based on commercial requirements to produce microelectronics and micromechanical devices of increasingly minimized dimensions. Current technologies hardly allow the generation of microcircuits smaller than 100 nanometers, but, in the words of Nobel physicist Richard Feynman [5], "There is [still] plenty of room at the bottom."

With the advent of today's "nanotechnology" research, devoted to the generation of nanometer-scale structural and functional elements, it is apparent that a further reduction of the available microsystems by "engineering-down," using, for instance, photolithographic methods, is becoming increasingly uneconomical. Thus, new "engineering-up" strategies are currently being explored for the assembly of small molecular building blocks for creating larger devices, preferably via self-organization [6]. The individual modular components employed need to be supplied with programmed recognition capabilities resulting from their specific constitution, configuration, conformation, and dynamic properties. Cells, as nature's ultimate molecular machines, use ensembles of proteins, nucleic acids, and other macromolecules that have been tailored by billions of years of evolution to perform highly complicated tasks. Thus, early researchers have suggested the use of biological macromolecules as components in nanostructured systems [7–9]. For a number of reasons, DNA is particularly suitable as a construction material in nanotechnology [10–13]. This chapter is dedicated to the utilization of nucleic acids, in particular DNA, as a material to fabricate nanostructured as well as mesoscopic micrometer-scale supramolecular architecture. Applications of the synthetic DNA-based systems will be described. This novel area of research is considered a segment of the broad field of "molecular biotechnology." Obviously, it is closely associated with the various, more conventional technical aspects, such as molecular cloning, synthetic and recombinant DNA, and protein expression, which are, however, already extensively covered in numerous prominent review articles and textbooks [14]. This chapter will instead focus on the use of DNA as a construction material for the fabrication of synthetic supramolecular assemblies, and thus it shall contribute to the establishment of a novel discipline that might be descriptively termed *biomolecular nanotechnology* or, perhaps, *nanobiotechnology*.

II. PROPERTIES OF DNA AND NATIVE NUCLEIC ACID–BASED NANOSTRUCTURES

To understand the extraordinary potential for DNA to be utilized as a material in construction processes, the general properties of this biomolecule will first be discussed. In addition, examples of naturally occurring nucleic acid–based nanostructures will be described that are of great importance both for cellular processes and conventional applications in molecular biotechnology.

The DNA molecule is a linear oligomer (polymer) of nucleotides, each containing a phosporylated sugar moiety attached to a base, adenine (A), guanine (G), cytosine (C), or thymine (T). Two antiparallel sugar–phosphate backbones wrap around each other to form a linear double helix held together by specific Watson–Crick hydrogen bonds, A–T and G–C. Despite its simplicity, this interaction leads to the greatest specificity known, and the stringent hybridization of two complementary DNA molecules allows any unique DNA sequence of about 20 nucleotides to be specifically detected in a target with the complexity of a mammalian genome containing approximately 3×10^9 base pairs. Thus, the simple A–T and G–C interaction allows the convenient programming of DNA receptor moieties, highly specific for the complementary nucleic acid. The power of DNA as a molecular tool is enhanced by our ability to synthesize virtually any DNA sequence by automated methods [15] and to amplify any DNA sequence from microscopic to macroscopic quantities by the polymerase chain reaction (PCR) [16]. Another very attractive feature of DNA is the great mechanical rigidity of short double helices [17] so that they behave effectively like a rigid rod spacer between two tethered functional molecular components on both ends. Moreover, DNA displays a comparably high physicochemical stability. This is impressively illustrated by the recent sequencing analysis of DNA samples extracted from bones of a Neanderthal skeleton [18], indicating intactness of the biomolecules even after a "storage period" of more than 50,000 years. Finally, nature provides a complete toolbox of highly specific biomolecular reagents, ligases, nucleases, and other DNA-modifying enzymes that allow for processing of the DNA material with atomic precision and accuracy on the angstrom level (Fig. 1). No other (polymeric) material currently known offers these advantages for constructions in the range from about 5 nm up to several micrometers.

The constitution of nucleic acid–based complexes is of fundamental importance for living systems. To elucidate some basic principles and purposes, in the following the native supramolecular ensembles of proteins and nucleic acids involved in biological processing of information will be briefly described. One should be aware that these systems also play an essential role in molecular biotechnology, in particular, for the genetic engineering of recombinant proteins and transgenic organisms.

To use DNA for the storage of genetic information, evolution was forced to develop an efficient condensation strategy for achieving the dense packing of the long polymeric molecule. Therefore, the genomic DNA within the cellular nuclei in eukaryotes is organized in three levels to form the chromatin structure. The *nucleosome* is the first level of supramolecular organization. The nucleosome particle consists of a protein core of eight hi-

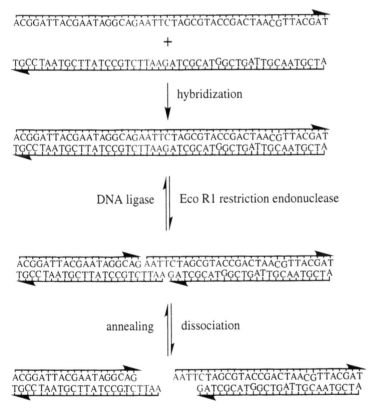

FIG. 1   Two single-stranded DNA oligomers with complementary nucleotide sequences hybridize to form a double helix. The double-stranded DNA (dsDNA) can be restricted using sequence-specific endonucleases, for instance, Eco R1, cleaving the phosphate backbone at a distinct recognition site (highlighted). The resulting two dsDNA fragments, each containing a 5′-cohesive (sticky) end, dissociate and reanneal via specific Watson–Crick base pairing. The nicked double helix can be covalently closed using DNA ligase. For simplification, linear double helical DNA stretches are represented by lines. The arrowheads indicate the 3′-ends of the DNA strands.

stone molecules forming a discoid structure of 7-nm thickness and 11 nm in diameter. The large number of basic amino acids within the histone proteins provide attractive interaction sites for the negatively charged phosphate backbone of the DNA. Thus, the histone octamer serves as a carrier to bind 165 base pairs (bp) of DNA, wrapped around the protein core in 1.8 turns of a left-handed DNA superhelix (Fig. 2a). Due to this arrangement, the contour length of the DNA is condensed about tenfold; i.e., the regular length of 56 nm is compacted to 5.6 nm. One additional histone molecule locks the nucleosome, leading to a particle of about 11-nm diameter. These particles are lined up like pearls on a DNA chain with linking DNA fragments of about 19-nm length (Fig. 2b). In a second level of organization, the chained nucleosome beads form 30-nm-thick filamentous structures; and in a third level, the filaments are organized to form radial loops. As a consequence of this supramolecular packing, the overall length of the DNA can be condensed as much as 8000-fold in the chromosome.

The amplification of genetic information, i.e., the replication of parent DNA molecules, is achieved by means of an entire set of enzymes. The major players are a DNA-gyrase for unwinding of the double helix, proteins to separate the two antiparallel DNA strands at the replicational junction, single-stranded binding proteins (SSB) that prevent the

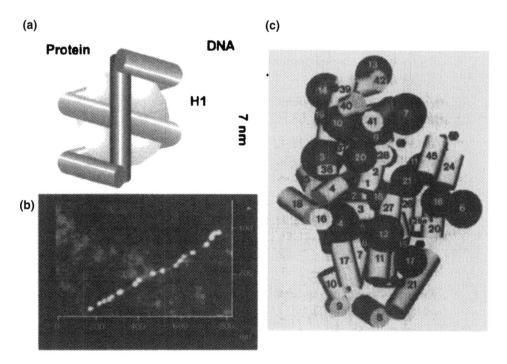

FIG. 2 Native supramolecular DNA nanostructures. (a) Schematic drawing of the nucleosome core particle. (b) An AFM image of native chromatin fibers, illustrating that nucleosome particles and linker DNA are organized like pearls on a chain. (The image is reprinted with kind permission from Ref. 205; copyright 1993 American Chemical Society.) (c) Schematic representation of the 30S subunit of the *E. coli* ribosome, indicating the location of the 21 proteins (dark spheres) and the double-helical portions of the RNA backbone (white cylinders). The model was generated from protein–RNA crosslinking studies. Reprinted with kind permission from Ref. 206; copyright 1988 Oxford University Press.

single strands from reannealing, RNA- and DNA-polymerases to incorporate the monomeric nucleotide building blocks, RNA-nuclease for the removal of primer molecules, and DNA-ligase for covalently linking Okazaki fragments. As an example of the high cooperativity of this process, DNA-polymerase III (Pol-III) may be considered, which is itself a supramolecular multienzyme complex consisting of at least seven proteins [19]. Since Pol-III is not capable of unwinding the DNA double helices, other proteins need to cooperate. The rep-protein unwinds and separates the DNA strands while moving along the leading strand. It is followed by several molecules of SSB, preventing the reformation of base pairs.

During other phases of the cell cycle, the processing of genetic information is achieved by various supramolecular protein aggregates. For the first step, the sense-strand of the DNA is transcribed to the corresponding messenger RNA (mRNA) by means of the RNA-polymerase enzyme. As an example, the *E. coli* RNA-polymerase holoenzyme, consisting of five protein subunits, is one of the largest soluble enzymes known. It has a molecular weight of 480 kDa. Due to its size of about 15×25 nm, it can be directly observed by electron or scanning force microscopy [20–22]. During mRNA synthesis, the cleavage of the double-stranded DNA template is necessary for transcribing the sense-strand. As a consequence, the transcription process leads to a supertwisting of the DNA template, and thus the enzyme needs to summon up enormous forces, which can range up to 25 piconewtons, as determined recently using an optical tweezer technique [23]. These measurements indicated that RNA-polymerase is stronger than the mechanoenzymes myosin or kinesin.

Finally, to produce the structural and functional devices of the cell, polypeptides are synthesized by ribosomal translation of the mRNA. The supramolecular complex of the *E. coli* ribosome consists of 52 protein and three RNA molecules. The power of programmed molecular recognition is impressively demonstrated by the fact that all of the individual 55 ribosomal building blocks spontaneously assemble to form the functional supramolecular complex by means of noncovalent interactions. The ribosome contains two subunits, the 30S subunit, with a molecular weight of about 930 kDa, and the 1590-kDa 50S subunit, forming particles of about 25-nm diameter. The resolution of the well-defined three-dimensional structure of the ribosome and the exact topographical constitution of its components are still under active investigation. Nevertheless, the localization of the multiple enzymatic domains, e.g., the peptidyl transferase, are well known, and thus the fundamental functions of the entire supramolecular machine is understood [24].

The examples of native DNA–protein supramolecules just described demonstrate that nature uses self-organized assemblages of functional macromolecules to accomplish the process of self-replication. The exploitation of the natural principles is crucial, both for the fundamental understanding of the cellular processes and for the utilization of distinct complexes in molecular biotechnology, for instance, the use of cellular equipment for the in vitro translation of mRNA or the synthesis of recombinant proteins with an extended genetic code [25]. Various components of the cellular DNA processing machinery are at least in part characterized down to the molecular level by the aid of x-ray crystallography [26]. These studies are particularly meticulous, since they allow for detailed insight into how the macromolecular components interrogate each other. It is fascinating to discover how the enormous specificity of binding results from the collection of low-specific, noncovalent contacts between distinct amino acids of the protein and the nucleotide bases and the phosphate backbone of the DNA by hydrogen bonding, electrostatic, and hydrophobic interactions.

III. SYNTHETIC DNA NANOSTRUCTURES

Now being aware of the unique properties of DNA and some central principles and meanings of nucleic acid–based nanostructures, approaches toward the utilization of DNA as a material for synthetic nanoconstructions will be discussed. This section begins by describing semisynthetic conjugates of DNA and proteins and their use as tools in the fabrication of oligofunctional supramolecular bioconjugates. Following that, applications of DNA hybridization with respect to material sciences will be discussed. Finally, the state of the art in the construction of synthetic 1D-, 2D-, and 3D DNA nanostructures will be summarized.

A. Semisynthetic DNA–Protein Conjugates

The concept of using DNA as a framework for the precise spatial arrangement of molecular components was initially suggested by Seeman [8]. For example, three-dimensional (3D) DNA networks (Fig. 3) might be useful as matrices for the immobilization of DNA-recognition proteins. This strategy should improve the crystallization properties of both cognate proteins and DNA scaffolds, and thus it should significantly simplify experimental crystallization protocols, which currently limit x-ray crystal structure determination of biological compounds. As suggested by Seeman and Robinson [27], 3D DNA networks might also be useful for the construction of electronic memory devices by selective positioning of molecular wires and switches, i.e., conducting organic polymers and redox-active cations. Due to the molecular dimensions of such biochips, largely improved memory densities of about 4 million Gbytes/cm^3 might be attained.

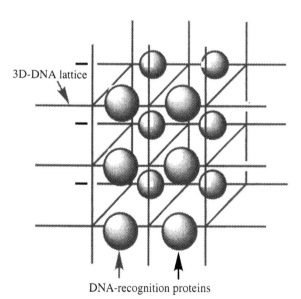

3D-DNA lattice

DNA-recognition proteins

FIG. 3   Three-dimensional (3D) DNA networks as crystallization matrices for the selective binding of DNA-recognition proteins [8]. The latter, represented by shaded spheres, bind specifically to recognition sites of the six-connected, cubic DNA lattice. For simplification, linear double-helical DNA stretches are represented by lines.

1. Oligonucleotide-Directed Assembly Using DNA–Streptavidin Conjugates

The initial experimental demonstration that molecular compounds can be selectively positioned along a nucleic acid scaffold was reported by Niemeyer et al. [28]. In this publication an approach was described allowing the arrangement of enzymes and antibodies along nucleic acids to generate novel supramolecular bioconjugates and nanometer-scale protein arrays. For this purpose, conjugates 2 of single-stranded DNA oligonucleotides and the protein streptavidin (STV) have been synthesized (Fig. 4). The tetrameric protein STV binds the small water-soluble molecule biotin (vitamin H) rapidly and with extraordinary specificity [29]. Since biotinylated materials are often commercially available or can be prepared with a variety of mild biotinylation procedures, biotin–STV conjugates form the basis of many diagnostic and analytical tests [30,31]. Another great advantage of STV is the extreme chemical and thermal stability of the protein. The covalent attachment of an oligonucleotide moiety to STV provides a specific recognition domain for a complementary nucleic acid sequence in addition to the four native biotin-binding sites. These bispecific binding capabilities allow the DNA–STV conjugates 2 to serve as versatile molecular adaptors in a variety of applications, such as the fabrication of nanostructured protein arrays. Such supramolecular aggregates are accessible by positioning several adapter molecules along a single-stranded RNA or DNA carrier molecule containing a set of complementary sequences (Figs. 4, 5).

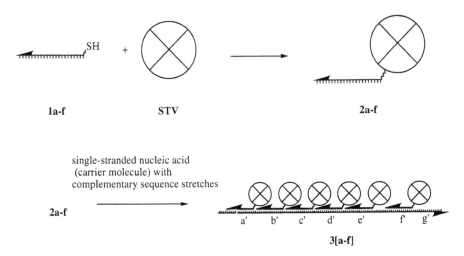

FIG. 4   Schematic drawing of the synthesis of DNA–streptavidin conjugates and "oligonucleotide-directed assembly of macromolecules" [28]. Stoichiometrically controlled, nanostructured supramolecular aggregates are generated from DNA–STV conjugates 2, obtained by covalent coupling of 5′-thiol-modified oligonucleotides 1 and STV. The 3′ end of the oligonucleotide is indicated by an arrowhead, the spacer chains between DNA and protein are represented by wavy lines. Conjugates 2 with distinct nucleotide sequences, e.g., **a** through **f**, self-assemble in the presence of a single-stranded nucleic acid carrier molecule, containing complementary sequence stretches, to form supramolecular aggregates 3. This strategy was used for the spatially controlled positioning of proteins [28,34] and later applied to the fabrication of "nanocrystal molecules" [60,74] from gold clusters, as indicated in Fig. 11.

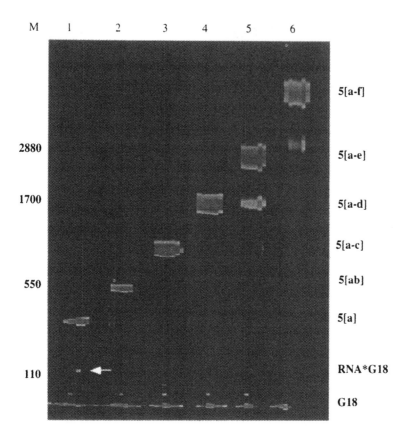

FIG. 5 Supramolecular self-assembly of DNA-STV hybrids **2** studied by fluorometric gel-shift analysis. The image shows a nondenaturing gel-electrophoretic separation of supramolecular aggregates **3** obtained from the successive addition of DNA–STV conjugates **2a–f** to an RNA carrier molecule. The aggregates are detected by means of a fluorescently labeled oligonucleotide probe. The weak band of uncomplexed RNA in lane 1 is indicated by an arrow. An increasing number of protein components bound to the carrier leads to an increasing signal strength (grayscale presentation of the originally blue fluorescein bands). Pentameric and hexameric DNA-protein aggregates display enhanced dissociation, likely due to the absence of double-helix-stabilizing cations during the course of electrophoresis. Lane M depicts the electrophoretic mobility of a DNA molecular weight marker (GeneScan-2500 Rox, length in base pairs). (From Ref. 34, with kind permission.)

Almost any type of biotinylated compound can be arranged by means of the DNA-STV adapter molecules **2**. Examples so far range from proteins such as antibodies and enzymes [28,32], low-molecular-weight compounds such as ionic groups, fluorophores and peptides [33], to inorganic metal nanoclusters [34]. As a particularly attractive feature, the tetravalency of the protein allows the DNA–STV conjugates to be functionalized with two different types of molecules [33]. This strategy is schematically drawn in Fig. 6. The modification of the DNA–STV hybrids with two different biotinylated components is achieved by initial coupling with a macromolecular functional compound, such as an enzyme, and then the remaining biotin-binding sites are saturated with low-molecular-weight modulators. The coupling of suitable modulators, such as positively charged biotinylated peptides, allows fine-tuning of the bioconjugate's nucleic acid hybridization properties [33]. More-

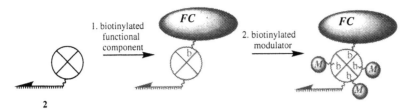

2

FIG. 6 Successive coupling of two different biotinylated compounds with the DNA–STV conjugates **2** [33]. In a first step, a macromolecular functional component (*FC*, represented by the shaded ellipse), such as a biotinylated enzyme or oligonucleotide, is coupled. In a second step, a biotinylated low-molecular-weight modulator (*M*, represented by the shaded sphere) is coupled to the remaining free biotin-binding sites. The modulator is used to modify the conjugate's hybridization properties or to supplement its functionality.

over, extra versatility is obtainable via the employment of modulators containing additional functional moieties, such as fluorophors or cofactor groups, chemically linked with the biotin. This supplements the functionality of the streptavidin bioconjugates, for instance, to exhibit high sensitivity as a detection tool in biosensor applications or to enhance the catalytic turnover of enzymatic transformations [33].

Chemical and structural features of the nucleic acid constituents employed in DNA-directed self-assembly play an essential role, since they determine the aggregation efficiency of the individual components [28,32,34,35]. For example, solid-phase hybridization studies have shown that the attachment of the voluminous STV to an oligonucleotide leads to up to a five fold decrease in hybridization kinetics. However, the same study revealed that the individual oligonucleotides attached to the streptavidin molecules are responsible for the large, nearly 100fold variations in hybridization efficiency, depending on the specific nucleotide sequences [35]. This is remarkable, since the sequences of the oligomers had been chosen for their similar melting temperature (T_m) and, thus, a comparable thermodynamic stability of the corresponding duplexes (ΔG). This indicated that the sequence-specific hybridization efficiency is highly dependent on the presence of secondary structures, such as the formation of intramolecular hairpin loops or intermolecular homodimers [35]. Structural influences are particularly important for nucleic acid hybridization reactions proceeding with kinetic control, for instance, in the assembly of temperature-sensitive biotinylated enzymes or antibodies during the reversible and site-selective immobilization using DNA microarrays [32].

The influences of nucleic acid secondary structure are even more complex in the supramolecular assembly of several components using a suitable single-stranded nucleic acid carrier backbone (Fig. 4). At moderate temperatures, the carrier strand forms an intramolecular secondary structure, displaying a thermodynamic stability comparable with the double helices formed by intermolecular hybridization. Typically, the hybridization of DNA–STV conjugates with an RNA carrier reaches thermodynamic equilibrium in one hour at 37°C [34], leading to the formation of a mixture containing the uncomplexed carrier and also the protein conjugate-bound carrier. This equilibrium can be shifted by means of helper oligonucleotides, which bind to uncomplexed sequence stretches of the carrier, thus disrupting its secondary structure. This leads to a completion of the previously incomplete supramolecular aggregation of several compounds [34,36]. Moreover, the chemical nature of the carrier strand significantly affects the assembly yields. The comparison of RNA and DNA carriers revealed that DNA molecules are superior templates for the supramolecular assembly, due to the lower stability of the intramolecular folding [36]. To overcome the problem of secondary structure influences, synthetic DNA analogs should be very useful in supramolecular synthesis. For instance, peptide nucleic acids [37] will, despite their currently high costs, soon be considered as carrier molecules of choice due to their high specificity of binding, great (bio)chemical and physical robustness, and low tendency to form secondary structures.

2. Oligofunctional DNA–Protein Nanostructures

A major initial motivation for the development of oligonucleotide-directed assembly of macromolecules was the fabrication of stoichiometrically and spatially defined aggregates of various enzymes [28]. The synthesis of an oligofunctional bioconjugate **4** of several enzymes, schematically depicted in Figure 7a, is of great interest as a target for studying multienzyme complexes (MECs). In biological systems, MECs appear both as supramolecular assemblies and as covalent conjugates of several catalytic proteins. As an example, the pyruvat-dehydrogenase MEC is comprised of 48 polypeptide chains. Such MECs offer mecha-

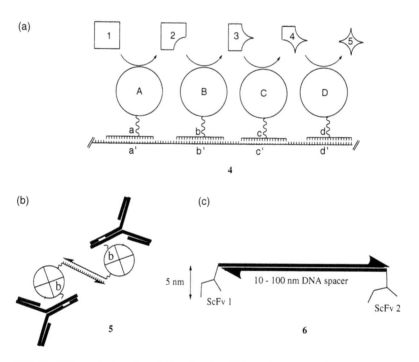

FIG. 7 Schematic drawing of oligofunctional bioconjugates. (a) Synthetic multienzyme aggregate **4**, generated from oligonucleotide-tagged enzymes A–D by means of nucleic acid hybridization. The enzymes catalyze a sequential four-step transformation of substrate 1 to the product 5. (b) Bispecific antibody aggregate **5**, consisting of two complementary DNA–STV conjugates **2**, each previously coupled with an individual biotinylated immunoglobulin. The functionality of this aggregate was shown by specific immunosorption [28]. (c) Bispecific antibody aggregate **6** consisting of two single-chain antibody fragments (ScFv) and a DNA spacer. The antibody aggregates have potential applications as specific reagents in immunoaffinity-based diagnostic assays or as molecular tools for the study of surface topologies.

nistic advantages during the multistep catalytic transformation of a substrate. Reactions limited by the rate of diffusional transport are accelerated by the immediate proximity of the catalytic centers. Furthermore, the "substrate channeling" of intermediate products avoids the occurrence of side reactions. Earlier model systems for fundamental research on substrate channeling processes within MECs had been restricted to heterobifunctional conjugates, accessible by means of chemical cross-linking or genetic engineering [38–40].

Another highly attractive application of nucleic acid–directed assembly of proteins is the fabrication of oligovalent antibodies (Fig. 7b, c). Due to the presence of multiple binding domains, bispecific antibodies are known to have a remarkably higher affinity for the target structure than regular monospecific antibodies [41–43]. However, to ensure sterically undisturbed binding of the two domains, the use of conventional bispecific antibodies requires that both target epitopes be in close proximity. In contrast, the supramolecular antibody aggregates, schematically drawn in Fig. 7b,c, contain a DNA molecule that is both the structural backbone and a variable rigid spacer. Thus, it is possible to control the spatial arrangement of the binding sites within a wide range from about 10 to several hundred nanometers. This should allow novel applications, because DNA-based antibodies might be used for specific recognition, even when the target epitopes are not nearby and/or reveal only weak antibody–antigen interactions. Moreover, one can construct specific supramolecules containing several binding sites spatially arranged in a fashion complementary to the target's topography. This further enhances the specificity of binding and thus causes improved signal-to-noise ratios in analytical applications.

Early experimental realizations of oligospecific DNA antibodies 5 based on DNA–STV conjugates (Fig. 7b) have shown the feasibility of the concept [28]. However, further refinements and, in particular, detailed investigations of the binding characteristics require that the aggregates be constructed with stringent regioselective control over the positioning of the protein binding sites on the nucleic acid backbone. Since both the biotinylation of immunoglobulins and the oligonucleotide coupling with the STV is nonregioselective, this system is probably not suitable for model studies. As a promising alternative, the incorporation of single-chain antibody fragments (ScFv) [44–46] in supramolecular aggregates 6 may be considered (Fig. 7c). These proteins can be generated from antibody-coding gene fragments derived from hybridoma cell lines or mRNA libraries [47] by means of recombinant methods. Moreover, the use of suitable cloning vectors allows site-specific incorporation of reactive groups, such as cysteine or histidine residues, which can subsequently be utilized for selective chemical coupling. An additional advantage of ScFv antibody fragments results from their small dimensions of only about 4 nm. This ensures that the distance between the binding sites within the supramolecular complex 6 is predominantly determined by the structure of the DNA backbone.

One great virtue of using DNA as a linker between two protein molecules is that a wide variety of enzymes can be subsequently used to probe or change the nature of the linker. As an example, enzymatic restriction and ligation may serve as ways of determining the mode of attachment and the mobility or stability of attached targets. Thus, DNA–antibody complexes might be applied as "molecular rulers," for instance, useful for exploring the lateral mobility of membrane proteins or the topology of biological and artificial objects. Because DNA has a vast amplification potential using PCR techniques, these experiments should be feasible even at the level of single cells and even in rather complex biological preparations. This allows the detection of nucleic acids at sensitivity levels far below those available for the detection of proteins by conventional immunological methods, such as the antibody-based enzyme-linked immunosorbent assay (ELISA).

The detection of a few hundred protein molecules can be attained by a combination of the ELISA with the amplification power of PCR. This method, termed immuno-PCR [48] (IPCR), is based on the coupling of specific antibodies with a DNA reporter fragment to be amplified by PCR. Recently, novel oligovalent DNA–STV conjugates **7** have been developed as reagents for IPCR (Fig. 8) [49]. The self-assembly of bis-biotinylated dsDNA and STV reproducibly generates populations of individual oligomeric complexes, as studied by nondenaturing gel electrophoresis and atomic force microscopy (AFM). Most strikingly, the oligomers dominantly contain bivalent STV molecules bridging two adjacent DNA fragments to form linear nanostructures. Trivalent STV branch points occur with a lower frequency and the presence of tetravalent STV is scarce in the supramolecular networks (Fig. 9a). Therefore, the oligomeric conjugates have a large residual biotin-binding capacity, and they can be further functionalized, for instance, by the coupling of biotinylated immunoglobulins. Both the pure and the antibody-modified DNA–STV oligomers reveal a superior performance as reagents in IPCR [49]. The employment of oligovalent reagents leads to about a 100-fold sensitivity enhancement compared with the conventional IPCR procedure [50].

The self-assembled oligomeric DNA–STV networks are suitable reagents for IPCR, since a high amount of DNA fragments are intrinsically linked with a similar amount of target-binding sites, thus allowing for affinity enhancements by increased avidity [51]. To generate supramolecular complexes of even higher versatility, an efficient attachment of large numbers of molecular devices can be attained from the incorporation of covalent DNA–STV conjugates [28] as distinct building blocks in self-assembled oligomeric networks [52]. This extends the capabilities of the DNA–STV aggregates to serve as a molecular framework for the attachment and selective positioning of biotinylated functional moieties, such as proteins [32], inorganic metal nanoclusters [34], and low-molecular-weight peptides and fluorophores [33]. Moreover, the dsDNA-STV oligomers **7** are suitable model systems for basic studies on self-assembled nanoparticle networks and might even be used for the fabrication of ion-switchable nanoarchitecture [53]. Since the oligomers are formed by statistical self-assembly, distinct supramolecular species might be isolated on a prepar-

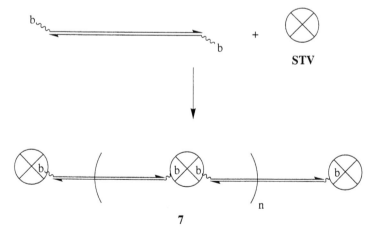

FIG. 8 Synthesis of oligomeric DNA–STV conjugates **7** from 5′,5′-bis-biotinylated DNA and STV [49]. Note that the schematic structure of **7** is simplified, since a portion of the STV molecules function as tri- and tetravalent linker molecules between adjacent DNA fragments.

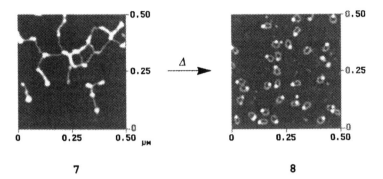

FIG. 9   AFM images of oligomeric DNA–STV conjugates **7** (left). The supramolecular network of **7** can be disrupted by thermal treatment, leading to the formation of DNA-STV nanocycles **8** (right). (From Ref. 54 with kind permission.)

ative scale and further functionalized. The greatest challenge, however, is to gain control over the self-assembly process. The first progress in this regard has recently been attained. The oligomeric networks **7** can effectively be transformed into supramolecular DNA-STV nanocircles **8** by simple thermal treatment (Fig. 9) [54]. The synthesis of cyclic DNA structures has previously been reported with respect to their utilization as constituents for nanostructured systems [55–58]. These perspectives, however, are considerably expanded, since the endogeneous protein molecule within the supramolecular DNA-STV nanocircles can be conveniently used for the attachment of functional molecular devices.

B. DNA-Based Nanocluster Assembly

Two key publications describing the assembly of DNA-derivatized gold colloids [59,60] have largely extended the common interest in the concept of DNA hybridization-based self-assembly of molecular compounds. The work of Mirkin et al. [59] and Alivisatos et al. [60] was motivated by the requirements of a material science research involving the generation of well-defined arrangements of nanocrystal metal clusters (quantum dots). Such systems are being investigated for new material properties, and potential applications are anticipated, for instance, in the field of laser technology [61–64]. Current research in this field is devoted to generating superlattices or quantum dot molecules in which crystallites from different materials are spatially assembled at will. The conventional strategies applied are often based on crystallization, monolayer-based self-assembly, or synthetic chemical methods (see Chapter 8).

1. Macroscopic, Repetitive DNA–Nanocluster Composites

The group of C. A. Mirkin collaborating with R. L. Letsinger uses DNA hybridization to generate repetitive nanocluster materials (Fig. 10). In the initial publication, two noncomplementary oligonucleotides were coupled in separate reactions with 13-nm gold particles via thiol adsorption [59]. A DNA duplex molecule containing a double-stranded region and two cohesive single-stranded ends, which are complementary to the particle-bound DNA, was used as a linker. The addition of the linker duplex to a mixture of the two oligonucleotide-modified clusters led to the aggregation and slow precipitation of a macroscopic DNA-colloid material. The reversibility of this process was demonstrated by the temperature-dependent changes of the UV/VIS spectrophotometric properties [59]. Since the clus-

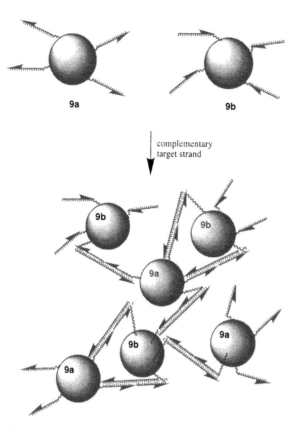

FIG. 10 Assembly of gold nanocrystals by means of DNA hybridization [65]. Two batches of gold clusters are derivatized with noncomplementary oligonucleotides, either via 5'- (**9a**) or 3'- (**9b**) thiol groups. The nanoparticles are mixed with single-stranded nucleic acid targets. In the presence of complementary target strands, the DNA-modified nanoclusters form three-dimensionally linked networks by which a change of the optical property of the nanocluster network occurs. Note that within the network only a heterodimeric "A–B" linkage is present. This allows the generation of binary composites, in which two types of particles, e.g., of different size [69] or different materials [70], are interconnected. Applications of similar systems may include sensors for nucleic acid diagnostics [65,66]. The 3' ends of the oligonucleotide compounds are indicated by an arrowhead. The wavy lines represent the spacer chains between DNA and the gold particle, represented by shaded spheres.

ters contain multiple DNA molecules, the supramolecular aggregates developing were well ordered and three-dimensionally linked, as judged from transmission electron microscopy (TEM) studies. Images of two-dimensional, single-layer aggregates reveal close-packed assemblies of the colloids with uniform particle separations of about 6 nm, corresponding to the length of the DNA linker duplex [59].

One approach to applying the nanocluster assembly concerns the generation of a sensor made from a web of DNA and gold particles [65,66]. This macroscopic-scale device changes color upon detection of particular DNA target strands: Two sets of DNA-coated cluster probes are combined with single-stranded target molecules containing oligonucleotide sequences that are either fully, partially, or not complementary to the nanoparticle-bound DNA. In the presence of the complementary target to be detected, the two probes are connected and consequently form a three-dimensionally linked, oligomeric DNA-gold particle hybrid web (Fig. 10). The formation of this network changes the particle distance and, thus, the electronic and optical properties of the metal clusters. As a result, a color change occurs. In principle, this strategy might be used for simple and cheap sensors in biomedical diagnostics, e.g., for the detection of nucleic acids from pathogenic organisms. Although detailed studies of the optical phenomena are still in progress [67], several analytical applications have already been reported [68].

More recent work from this group concerns the generation of binary networks. For example, gold clusters of either 40-nm or 5-nm diameter were initially modified with individual 12-mer oligonucleotides containing a thiol group at either the 5' or 3' end. Subsequently, the two cluster preparations were assembled using a complementary 24-mer oligonucleotide [69]. Due to the specificity of Watson–Crick base pairing, only heterodimeric "A–B" composites with alternating particle sizes are formed (Fig. 10). In the case of an excess of one particle, satellite-like aggregate structures can be generated, as determined by TEM. Using a similar approach, oligonucleotide functionalized CdSe/ZnS quantum dots have recently been incorporated into a binary nanoparticle network also containing gold nanoclusters [70]. As investigated by TEM, these hybrid metal/semiconductor assemblies exhibit an A–B structure. Moreover, fluorescence or electronic absorption spectroscopy studies reveal initial indications of cooperative optical and electronic phenomena within the network materials [67,71].

2. Individual Supramolecular Assemblies of Nanoclusters

The approach of Mirkin and coworkers is leading to novel hybrid materials with promising electronic and optical properties, potentially useful for sensor and also for technical applications with respect to material sciences [72,73]. Nevertheless, possible limitations may result from the lack of stoichiometric control during the assembly process. To control the architecture of materials, spatially defined arrangements of molecular devices are required. For example, to organize metal and semiconductor nanocrystals into ultrasmall electronic devices, one may consider a linear aggregate of several individual components, a structural analog to the supramolecular protein aggregate 3 in Fig. 4. As initially demonstrated for proteins [28] and more recently for gold nanoclusters [60,74], the rational construction of stoichiometrically defined nanoscale assemblies can be achieved from building blocks, each containing a single nucleic acid moiety.

Following this strategy, Alivisatos et al. [60] synthesized well-defined monoadducts from commercially available 1.4-nm gold clusters containing a single reactive maleimido group. By coupling with thiolated 18-mer oligonucleotides, the nanocrystals were supplied with an individual "codon" sequence. Upon addition of a single-stranded DNA template molecule containing complementary codons in any order whatsoever, a self-assembly of nanocrystal molecules occurred (Fig. 11). Purified DNA–nanocluster conjugates were assembled to generate the head-to-head and head-to-tail homodimeric target molecules **10a** and **10b**. The TEM characterization of the aggregates revealed that approximately 70% of the complexes show the dimeric structure expected. The center-to-center distances observed in the two isomers are about 3–10 nm and 2–6 nm in **10a** and **10b**, respectively, and are consistent with model calculations. Also, the trimeric molecules (**10c**) show the structures expected, as judged from TEM images. In recent work of this group, a variety of supramolecular nanocrystal molecules has been synthesized by means of DNA-directed assembly [74]. The preparations, containing up to three nanoclusters of different size organized in multiple fashions, were purified by electrophoresis subsequent to the self-assembly and were characterized by TEM. These studies indicated that the nanocrystal molecules

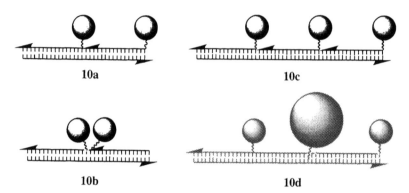

FIG. 11 Assembly of nanocrystal molecules using DNA hybridization [60]. Conjugates from gold particles (represented as shaded spheres) and oligonucleotide codons are organized to supramolecular assemblies by the addition of a template strand. The use of 3'- or 5'-derivatized oligonucleotides allows the fabrication of head-to-head (**10a**) or head-to-tail (**10b**) homodimers. A template containing the complementary sequence in triplicate affects the formation of the trimer **10c**. Aggregate **10d** reveals a limited flexibility, due to the unnicked double-helical backbone [74].

reveal a high flexibility when the DNA backbone is nicked (e.g., **10c**). In the case of **10d**, however, the unnicked double helix significantly lowers the flexibility. UV/VIS absorbance measurements indicated changes in the spectral properties of the nanoparticles as a consequence of the supramolecular organization.

Using two independent biomolecular recognition systems, the hybridization of complementary DNA and RNA, and strong binding of biotin by STV, the DNA–STV conjugates **2** (Fig. 4) have been used to organize gold nanoclusters (Fig. 12) [34]. In this work, 1.4-nm gold clusters were derivatized with a biotin substituent and subsequently coupled with the conjugates **2**. The metal–protein hybrids self-assemble in the presence of a complementary single-stranded nucleic acid carrier molecule, thereby generating novel biometallic nanostructures, such as **11**. Since the DNA–STV conjugates can be used like a molecular construction kit, functional protein components can easily be incorporated into the biometallic nanostructures. The proof of feasibility was achieved by the synthesis of construct **12** containing an immunoglobulin molecule (Fig. 12). The functionality of this supramolecule was demonstrated by specific immunosorption to a surface-immobilized complementary antigen and subsequent TEM analysis. Such experiments impressively demonstrate the applicability of DNA hybridization for the nanoconstruction of novel hybrid systems that may eventually serve as interface structures between electronic and biological systems.

C. Nanostructured Molecular Scaffolds from DNA

The use of DNA hybridization just described opens up a novel, uncomplicated, yet powerful strategy for supramolecular synthesis: Many different devices are connected to a distinct sequence codon and are subsequently organized on a suitable template strand. The utilization of appropriate nucleic acid scaffolds should even allow the fabrication of highly complex supramolecular structures by means of a modular construction kit. For approximately 20 years, the work of Seeman and coworkers [8,27] have been engaged in the rational construction of 1D, 2D, and 3D DNA frameworks. They use branched DNA

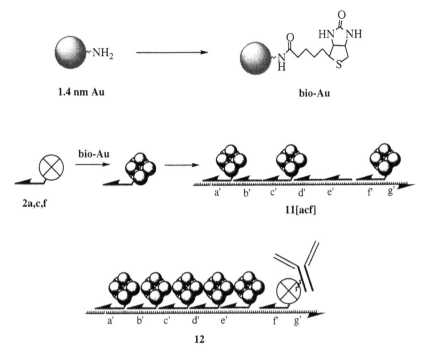

FIG. 12 Fabrication of biometallic aggregates by means of DNA–STV adapters **2**. Monoamino-modified 1.4-nm gold clusters are derivatized with a biotin moiety, and the biotinylated clusters (bio-Au) are coupled with DNA–STV adducts **2**. The resulting hybrids, gold-labeled **2a, 2c,** and **2f,** are assembled with the RNA carrier to form the supramolecular aggregate **11[acf]**. In this synthesis, helper oligonucleotides **1b, 1d,** and **1e** had been employed to obtain higher aggregation yields [34,36]. An antibody-containing, functional biometallic construct **12** was fabricated from gold-labeled **2a–2e** and a conjugate of **2f** and a biotinylated immunoglobulin, previously coupled in separate reactions [34].

molecules containing three, four, and more double-helical arms (termed "DNA junctions"; see **13, 14** in Figs. 13, 14). These modules are similar to the structure of the replicational junction or the Holliday junctions of genetic recombination. They are easily accessible on a preparative scale by briefly heating stoichiometric mixtures of the oligonucleotide components and subsequent stepwise cooling. In the following, the use of DNA junctions and related building blocks for the fabrication of pure DNA nanostructures will be discussed.

1. Synthesis of Periodic DNA Materials

The synthesis of repetitive structures from DNA [75–79] was initially attempted using three-arm junctions [75]. Enzymatic connection of the cohesive ends of **13** (Fig. 13) using DNA ligase yielded a mixture of linear and cyclic oligomers, starting with the trimer. Since double-helical arms with a length of 5–7 nm in **13** behave like rigid rods, the fraction of cyclized products allows conclusions about the variations in the valence angle between the ligation arms of the junction molecule. The following experiments using four- [76], five-, and six-arm junctions [77] confirmed the high flexibility of the modules. Thus, these "simple" motifs are not suitable for the assembly of large repetitive constructs. Therefore, Seeman et al. started to explore a different class of DNA modules, termed "double-crossover" (DX) molecules. This class of motifs contains two junctions connected by two double-

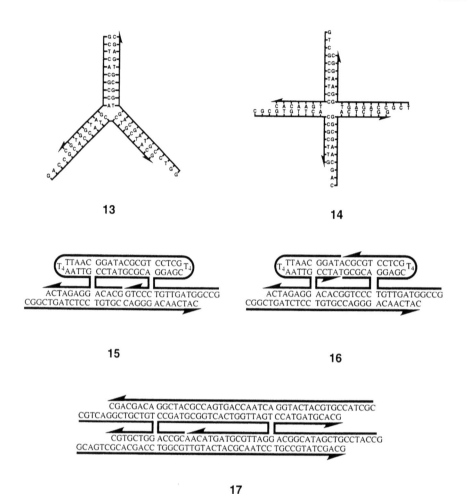

FIG. 13 Synthetic DNA motifs for the construction of DNA framework: Three- [75] (**13**) and four-arm (**14**) DNA junction [8]; DNA double-crossover (DX) molecules **15, 16** were used for initial studies of enzymatic oligomerization [79]. The DX motif **17**, containing four cohesive ends of individual nucleotide sequence, was used for the construction of two-dimensional DNA crystals [80].

helical DNA arms, and is differentiated by the relative orientation of their helix axes as well as by the number of double-helical half-turns between the two crossovers (see, for instance, **15** in Fig. 13). The ligation of DX model compound **15** (termed "DAE molecule"), whose second domain is closed with dT_4 loops, led almost exclusively to the production of linear oligomers [79]. This fact was attributed to the reduced flexibility of the ligation arms and was not observed for control compound **16**, in which the nicked backbone of the second helical domain permits higher flexibility. Due to the rigidity of the DAE molecules, comparably long oligomers containing up to 17 monomers are formed during ligation. In contrast, the more flexible junctions (such as **13**) hardly formed oligomers larger than heptamers.

As proposed earlier, the utilization of structurally related DX motifs, e.g., triangle-shaped modules containing an additional ligation arm, should allow the construction of periodic lattices from DNA [78]. Recently, a set of publications demonstrated the feasibility of using DX motifs for DNA nanoconstructions [80–82]. DX-based triangle modules were

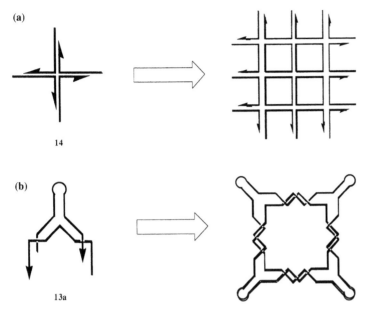

FIG. 14 Construction of periodic framework and geometric objects from DNA. (a) Construction of two-dimensional DNA lattices from tetravalent four-arm DNA junctions (**14**) [8]. (b) Synthesis of a macrocyclic molecule from bivalent three-arm DNA junctions (**13a**) containing two cohesive ends [83]. For simplification, linear double-helical stretches are represented by parallel lines.

enzymatically oligomerized to yield extended 1D structures, as determined by AFM [81]. Also, a 2D periodic lattice structure was generated using tetravalent DX modules (see **17** in Fig. 13). The reversible annealing yielded "two-dimensional DNA crystals" of up to 2 × 8 μm and a uniform thickness of about 1-2 nm [80]. These structures were also characterized by means of AFM (Fig. 15b). This technique even allowed proof that the surface features of the two-dimensional DNA crystals can be modified by enzymatic treatments [82].

2. Synthesis of Individual Objects from DNA

Individual supramolecules of DNA with a topology of geometric objects [83–86] and molecular knots [87,88] can be constructed from modules whose valence arms are supplied with individual cohesive ends. For example, the synthesis of a DNA molecule containing four vertices proceeds readily from three-arm junctions (**13a** in Fig. 14) with approximately 15% yields [83]. However, the construction of a more complex object with the connectivity of a cube containing eight vertices [84] was only 1% efficient. The DNA cube (Fig. 15c) contains edges of about 7-nm length. Each of the six "faces" is formed by a cyclized oligonucleotide, connected by two turns of a double helix with the four neighboring strands. To increase product yields and to allow for automated synthesis of DNA objects, a solid-support methodology was developed [85], permitting the synthesis of entire segments of DNA constructs on Teflon surfaces. The power of this method was demonstrated by the synthesis of a polyhedron containing 24 vertices, proceeding in 1% yield [86]. The "truncated octahedron" is a complex 14-catenan with a molecular weight of about 790 kDa, containing six square and eight hexagonal faces. Due to the lack of suitable physical characterization methods, all DNA constructs described so far have been analyzed by biochemical methods. For this purpose, the oligonucleotides used for construction were de-

(a)

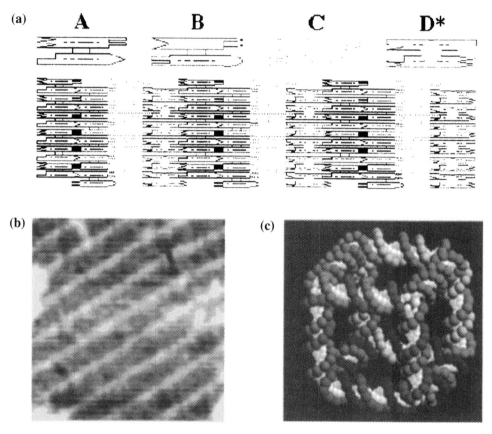

FIG. 15 Advances in the construction of molecular architecture from DNA. (a) Schematic representation of a set of DX molecules designed to form a 2D lattice. Compound D contains a DNA hairpin that acts as a topographic marker detectable by AFM. (b) AFM image of the 2D DNA crystal obtained from the four DX molecules shown in (a). The stripes indicate that the D compounds have a regular distance of about 65 nm, corresponding to the 4 × 16 nm dimensions of the four DX molecules within the lattice [80]. (c) Representation of a DNA cube, formed from six different cyclic strands. Each edge of the cube is formed by two turns of a double-helical DNA molecule [84]. (Images courtesy of N. C. Seeman. For further images and details, see: http://seemanlab4.chem.nyu.edu.)

signed with an individual restriction site. Therefore, each double-helical edge can be identified by specific endonuclease cleavage. Digestion of breakdown products by 3'-exonuclease and electrophoretic analysis allows verification of the integrity and connectivity of intermediate and final products. The actual three-dimensional shape of the molecules is currently unknown.

Recent advances of the Seeman group led to the construction of a nanomechanical device from DNA [89]. In this molecular apparatus, the ion-dependent transition of B-DNA into the Z-conformation is used to alter the distance between two DNA DX domains attached to the switchable double helix. Atomic displacements of about 2–6 nm were attained. Ionic switching of nanoparticles by means of DNA supercoiling has also been reported [53]. Additional advances regarding the use of DNA is nanomechanical devices have been reported by Fritz et al., who showed that an array of cantilevers can be used to translate DNA hybridization into a nanomechanical response [90]. Moreover, Yurke et al. reported on a molecular machine both made up of DNA and fueled with DNA. In this device, a DNA molecule functions as a tweezer that can be opened and closed by the addition of single-stranded DNA [91].

The preceding examples demonstrate the significant progress achieved in the fabrication of 2D and 3D artificial DNA structures. However, the use of such framework to realize dense protein and biochip assemblies, suggested earlier by Seeman, has not yet been accomplished. In this context, it should be noted that Smith and coworkers have recently reported on the organization of several proteins along a 1D double-stranded DNA fragment [92]. Their approach is based on the specific binding of DNA (cytosin-5)-methyltransferases to distinct recognition sequences within double-helical DNA (see Fig. 20). Covalent adducts of the enzyme and dsDNA are formed if the synthetic DNA base analog 5-fluorocytosine (^{F}C) is present in the recognition site. Using two representative methyltransferases, M.$HhaI$ and M.$MspI$, the sequence-specific covalent attachment of the enzymes at their target sites, G^{F}CGC and ^{F}CCGG, respectively, was demonstrated. By means of recombinant techniques, the methyltransferases can be modified with additional binding domains, such as peptide antigens. Thus, the authors note that their concept should be useful for generating DNA–protein conjugates applicable as chromatin models or other macromolecular devices.

IV. MESOSCOPIC DNA STRUCTURES AND MICROTECHNOLOGY

Although the theory of producing synthetic DNA nanostructures is well ahead of experimental confirmation [93], the work of the Seeman group described earlier impressively demonstrates the great potential of this approach. In the following, the use of DNA in the fabrication of somewhat larger, mesoscopic structures and aggregates will be described. Moreover, recent developments concerning the integration of DNA-based methods and materials in microtechnology will be discussed.

A. DNA-Templated Synthesis

The electrostatic and topographic properties of the DNA molecule can be utilized to synthesize nano-, meso-, and microscopic aggregates. Pioneering work in this area was carried out by Coffer and coworkers [94–97]. They used the negatively charged phosphate backbone of the DNA double helix to accumulate Cd^{2+} ions, which were subsequently treated with Na_2S to form CdS nanoparticles. In the initial study, solutions of calf thymus DNA and Cd^{2+} ions were mixed, and after that molar amounts of Na_2S were used to initiate the CdS nanoparticle formation. High-resolution TEM (HRTEM) analysis revealed that the particles generated had an average diameter of about 5.6 nm [94]. Since the actual role of the DNA was vague from these studies, later experiments were carried out to elucidate potential influences of the base sequences of the DNA template employed [95]. In fact, it was found that in particular the adenine content of the DNA affects the size of the nanoparticles formed. These findings provide indirect evidence for the template mechanism suggested. In further studies, attempts were made to synthesize surface-bound mesoscopic nanoparticle aggregates [96,97]. For this purpose, the circular plasmid DNA pUCLeu4, forming a circle of about 375-nm diameter containing 3455 base pairs, was used as a template. The DNA was mixed with Cd^{2+} ions in solution; subsequently, the Cd^{2+}-loaded DNA was adsorbed to an amino-modified glass surface. The formation of the CdS nanostructures was then carried out by H$_2$S treatment [96]. HRTEM analysis revealed that CdS particles of an about 5-nm diameter had developed that were assembled close to the circular DNA backbone. Measurements of its circumference indicated the intactness of some of the DNA molecules; however, other aggregates with various shapes were also observed.

The use of DNA as a template to fabricate mesoscale structures was also demonstrated in a recent work of Torimoto and coworkers. They used preformed, positively charged 3-nm CdS nanoparticles with a thiocholine-modified surface to be assembled into chains by using the electrostatic interaction between positively charged nanoparticle surfaces and the phosphate groups of DNA. As determined by TEM analysis, the CdS nanoparticles were arranged in a quasi-one-dimensional dense packing. This revealed interparticle distances of about 3.5 nm, which is almost equal to the height of one helical turn of the DNA double strand [98].

Similar to the concept of Coffer et al., Tour and coworkers have used DNA to congregate macromolecular ammonium ions [99]. In this approach, C$_{60}$ fullerene molecules were modified with an N,N-dimethylpyrrolidinium iodide substituent to yield 18 (Fig. 16). The fullerene derivative was chosen as a complexing agent for the electrostatic binding to the phosphate backbone of DNA, since fullerenes can be directly imaged by TEM without the need for heavy-metal shadowing or other staining techniques. 18 was mixed with plasmid DNA, and TEM analysis of the resulting DNA–fullerene hybrid materials revealed that the complexation had significantly altered the structure of the DNA. In the case of circular plasmid DNA templates, it was found that the diameter of the hybrid complexes were condensed by about five-fold, while the thickness of the double helix ranged between 15 and 30 nm, instead of 2 nm for native DNA. This condensation is likely a consequence of extensive hydrophobic interactions between the fullerene moieties attached to the DNA. However, it is very well known that pure polyamines, such as spermin and spermidin, already induce a significant packing of DNA structures, and the dynamics of DNA condensates at the solid–liquid interface has recently been studied by AFM [100].

Blessing et al. have applied the spermin-induced packing of DNA to generate calibrated nanometric particles [101]. They synthesized the polymerizable cation 19 (Fig. 16) from natural cysteine and spermine precursor molecules. Since spermine binds to the minor groove of B-DNA, the latter serves as a matrix during air-induced thiol/disulfide oligomerization of 19. Determination of the oxidation rates indicated that disulfide formation is completed within 2 hours in the presence of DNA, while the sulfides in pure 19 are stable for days. As a consequence of DNA-templated oligomerization, a physical collapse accompanied by a chemical stabilization of the 19/DNA adduct, occurs. This leads to the formation of 50 ± 15 nm particles, as determined by laser light scattering. The nanometric monomolecular DNA particles are potentially useful for gene delivery.

An early example of using DNA as a template for the aggregation of organic molecules was reported by Gibbs and colleagues [102]. They demonstrated that porphyrins containing cationic side chains form long-range structures on a DNA template, revealing the helical sense of the nucleic acid, as determined by circular dichroism measurements. Later studies used various spectroscopic methods to gain insights into the kinetics and thermodynamics of the supramolecular DNA–porphyrin assemblies [103,104]. The use of DNA as a nanotemplate for the spontaneous assembly of dye aggregates was recently reported by Seifert and coworkers [105]. The symmetric cationic cyanine dye 20 (Fig. 16), consisting of two benzothiazole groups linked via a pentamethine bridge, dimerizes in the presence of dsDNA containing alternating A/T residues. The dimerization induces a shift

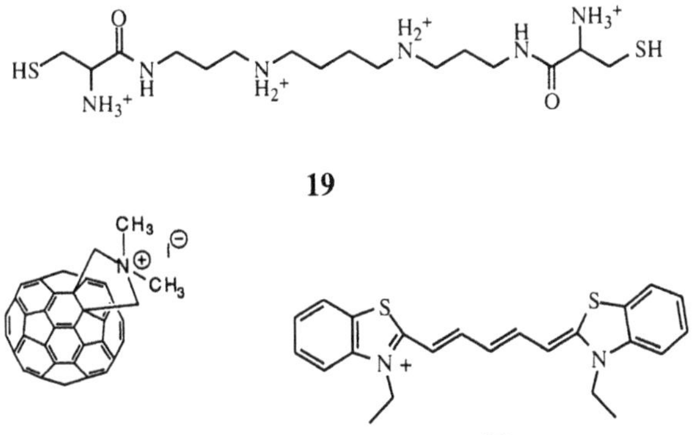

19

18   **20**

FIG. 16 Organic ligands used in DNA-templated assembly reactions. C_{60}-*N,N*-Dimethylpyrrolidinium iodide **18** was assembled on a dsDNA template to yield DNA-fullerene hybrid materials [99]. The polymerizable cation **19**, synthesized from cysteine and spermine precursors, binds to the minor groove of plasmid DNA. The resulting adducts physically collapse during air-induced thiol/disulfide oligomerization of **19**, leading to the formation of 50-nm particles [101]. Cyanine dye **20** dimerizes in the presence of dsDNA and the dimers bind cooperatively to the minor groove of the double helix, leading to the formation of extended helical cyanine dye aggregates [105].

of the absorption maximum from 647 to 590 nm and a quenching of the fluorescence. As indicated from variation of the template sequences and viscometric analysis, the dimers bind to the minor groove of the double helix. Strikingly, the dimer binding is highly cooperative. This means that the binding of one dimer greatly facilitates the binding of a second dimer, leading to the formation of extended helical cyanine dye aggregates in the case of longer DNA templates. Thus, the DNA structure precisely controls the spatial dimensions of the supramolecular aggregate.

The examples just presented give initial impressions of how DNA can be utilized as a template in the synthesis of nanometric and mesoscopic aggregates. However, the studies emphasize the importance of fundamental research on the interaction between DNA and the various binders, such as metal and organic cations. Of particular importance are the consequences of binding events on the structure and topology of the nucleic acid components involved.

B. DNA as a Material in Microelectronics

In addition to its use as a nanotemplate, DNA might also be used to fabricate micrometer-scale elements, potentially useful in microelectronics. A descriptive example of this approach was published by Braun and coworkers [106]. They used a dsDNA molecule, λ-DNA 16 μm in length, containing two cohesive ends to bridge the distance between two microelectrodes. For this, gold electrodes separated by a gap of about 12–16 μm were prepared on a glass support by means of standard photolithography. Subsequently, the two electrodes were modified with an individual capture oligonucleotide, each complementary to one of the cohesive ends of the λ-DNA, and the 16-μm dsDNA fragment was allowed to hybridize. To confirm successful interconnection of the electrodes, the λ-DNA was fluoresently labeled, and the hybridization process was monitored by fluorescence microscopy. Next, the sodium ions bound to the phosphate backbone of the λ-DNA were exchanged with Ag^+ ions, and the latter were chemically reduced by hydroquinone. The small silver aggregates formed along the λ-DNA backbone were then used as catalysts for further reductive deposition of silver, eventually leading to the formation of a silver nanowire. This micrometer-size element with a typical width of 100 nm had a granular morphology, as judged from AFM images. Two-terminal electrical measurements of the Ag nanowire revealed nonlinear, history-dependent *I–V* curves, possibly a result of polarization or corrosion of the individual 30- to 50-nm Ag grains comprising the wire.

construction of 2-D DNA networks

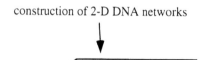

DNA-network as scaffold for the deposition of inorganic materials:

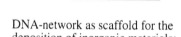

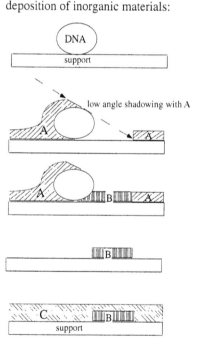

FIG. 17 Utilization of two-dimensional DNA networks as scaffolds for the production of electrical circuits [107]. Two-dimensional DNA networks are generated on suitable solid supports. Subsequently, the network is used as a scaffold for the production of replicas made of inorganic materials. For this purpose, the DNA is shadowed with substance A under a low angle of incidence, leading to an uncovered track along the DNA double helix. A layer of a conducting material **B**, such as gallium arsenide or indium phosphite, can be deposited by metallo-organic CVD. After selective removal of substance **A**, the remaining wires of **B** can be embedded in a second conductor **C**. Following this strategy, e.g., field effect transistor elements, should be attainable. (Adapted from Ref. 107.)

Artificial DNA structures might be used in chip construction in order to realize circuit sizes below 100 nm. A concept of Di Mauro and Hollenberg [107] is based on the construction of DNA networks on suitable solid supports using, for instance, oligonucleotides as initiation points for enzymatic DNA synthesis or hybridization. These networks are then used as scaffolds for the deposition of conducting materials such as gallium arsenide and indium phosphite via CVD procedures (chemical vapor deposition). Shadowing techniques, well established for the preparation of biological samples for electron microscopy studies, should allow the preparation of basic microelectronic elements (Fig. 17). As noted by the authors, the calculable advantage of "DNA technology" would be due to the enormous precision of DNA biosynthesis and hybridization in the process of matrix generation. Thus, the accuracy of photolithographic techniques should be improved by about two orders of magnitude. Furthermore, since the size of the DNA double helix is only 2 nm in diameter, the lower size limit of the circuit structures should also be substantially reduced.

A DNA-based fabrication technique, currently under development at UC San Diego and the U.S. company Nanogen/Nanotronics, should allow the carrying out of the controlled organization of complex molecular structures within defined perimeters of silicon or semiconductor structures produced by classical microfabrication techniques [108]. Esener and coworkers use microelectronic template arrays with microlocations, spots of about 50 μm to 80 μm functionalized with capture oligonucleotides, to organize DNA-tagged devices within selected microlocations on the array (Fig. 18). Both, the site-selective attachment of the capture oligonucleotides to the solid support as well as the hybridization of the DNA-tagged devices, such as DNA derivatized 20–200 nm microspheres, can be achieved or assisted by directed electrophoretic transport. This technique, electric field directed nucleic acid hybridization, was developed by Nanogen [109]. Based on the negative charges

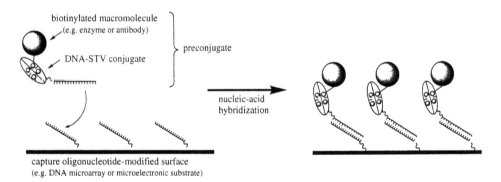

FIG. 18 Schematic representation of DNA-directed immobilization (DDI) [28,32]. In this example, covalent DNA–STV conjugates are coupled with a biotinylated enzyme or antibody by mixing of the two compounds. Note that basically any oligonucleotide-tagged compound, such as biomolecules, metal nanoclusters, or latex microspheres, can be used instead of the protein–DNA–STV preconjugate shown. Complementary capture oligonucleotides, immobilized on solid supports, are used as positioning elements on the surface, which can comprise a single capture species or multiple spots, each containing an individual capture sequence (DNA microarray). The preconjugate is then allowed to bind to its complement by formation of specific Watson–Crick base pairs. Note that due to the enormous specificity of DNA base pairing, many compounds can be site-specifically immobilized simultaneously in a single step. In a related approach, the capture DNA-modified surface was patterned using photolithography [110].

of the DNA backbone, it allows a rapid DNA transport, site-selective concentration, and accelerated hybridization reactions to be carried out on active microelectronic arrays. Moreover, capture oligonucleotide-functionalized surfaces can be patterned with a high-energy UV write ($\gamma = 255$ nm) using a conventional photolithography masking procedure prior to the DNA-directed immobilization. DNA molecules exposed to the UV light loose their ability to hybridize with the complementary DNA and thus are no longer capable of binding the devices to be immobilized [110].

C. Nucleic Acid–Functionalized Microstructured Surfaces

The foregoing examples give initial impressions of how DNA can be used to functionalize solid supports with respect to applications, for instance, in the field of microelectronics. In the following, the use of laterally structured substrates is addressed. In particular, the generation and applications of microstructured supports functionalized with nucleic acid molecules will be discussed.

1. DNA Microarrays

Microstructured arrays of capture oligonucleotides are currently of tremendous interest with respect to bioanalytical applications [111–113]. The fabrication of the arrays can be achieved by two different approaches (Fig. 19). Up to 10,000 DNA fragments, prepared by enzymatic or chemical syntheses, can be covalently immobilized on an activated glass support by means of automated dispensing or plotting devices [114]. Moreover, highly structured lateral oligonucleotide libraries on glass supports are accessible by initially modifying the surface with photolabile protection groups [115]. Illumination through a microstructured photomask leads to the deprotection of selected areas, to which the first phosphoramidite building block is covalently attached. Since the coupled nucleotides also contain photolabile protection groups, the iterative repetition of the process generates new patterns, leading to two-dimensionally structured oligonucleotide arrays. As an example, an array made up of 256 octanucleotides on a surface area of 1.3×1.3 cm was synthesized by 16 reaction cycles in 4 hours [116]. Currently available fabrication routines allow the production of arrays containing more than 300,000 DNA oligomers of length of up to 25 nucleotides, and the next array generations already envisioned will accommodate several millions of oligomer probes.

Much effort in microarray technology is devoted to technical improvements of reliable and highly sensitive detection methods, for instance, to enable the analysis of even trace amounts of target nucleic acids, possibly without the necessity of a previous amplification step. In microarray analyses, fluorophor labels have practically replaced radiolabels, commonly used for highly sensitive detection and quantification purposes. Fluorescence scanners currently available allow the quantitation of sub-attomol amounts of fluorophors with a dynamic range of more than three decades. However, the use of labels in the detection of nucleic acid hybridization generally has drawbacks. Homogeneously labeled sample materials are required, and stringent washing steps are necessary to remove unbound materials subsequent to hybridization. Instead of this endpoint determination, a real-time hybridization analysis would be advantegous. Optical [117] or mass-sensitive methods [118], often based on layers of metallic transducers, are currently beeing investigated for their suitability as label-free detection principles. Although not yet sufficiently sensitive, the use of impedance spectroscopy [119] is very attractive, since it would open a way to realize visions of DNA chips with integrated electronics for signal analysis. Matrix-assisted laser desorption ionization time of flight mass spectrometry (MALDI-TOF MS) is also a

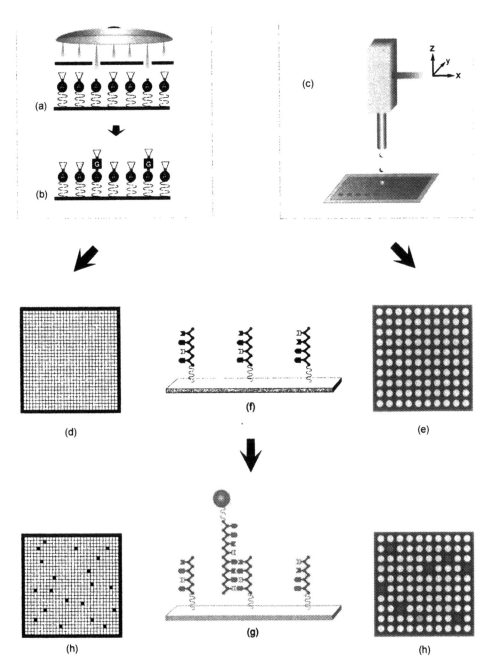

FIG. 19 The making and reading of DNA microarrays. High-density oligonucleotide arrays (d) are attainable using Affymetrix photolithographic on-chip synthesis (a, b). The deposition of DNA fragments, previously synthesized by chemical or enzymatic methods, using automated plotting or dispensing devices, allows the production of microarrays with a lower density of hybridization spots (c, e). However, this technique allows the immobilization of long pieces of DNA, for instance, obtained from clone libraries or by PCR. Following the hybridization of fluorophor-labeled cRNA or cDNA samples (f, g), the resulting hybridization patterns (h) are analyzed using a fluorescence scanner. (From Ref. 112 with kind permission.)

powerful method for the "readout" of microarrays [120]. Moreover, further sensitivity enhancement in microarray analyses might result from current progress in the area of single-molecule detection and also from chip-based amplification methods [121].

Despite the impressive technical advances in microarray development, there are still many basic questions within this new field of research, for instance, concerning the topology of DNA-functionalized surfaces or the details of the molecular interactions proceeding at the interphase. For instance, Southern and coworkers have investigated how the shape, length, and base composition of the double helices during the process of formation affect the efficiency of hybridization. They have also studied the effects of surface density and of oligonucleotide and spacer lengths on the hybridization yields [113]. Recently, DNA monolayers formed through self-assembly of thiol-modified oligonucleotides on gold substrates have been characterized. For this purpose, conventional hybridization assays using radiolabeled oligonucleotides as well as XPS (x-ray photoelectron spectroscopy) [122], neutron reflectivity [123], electrochemical methods such as cyclovoltammetry [124–127], and AFM [126,128] have been used. These studies reveal insights into the density of surface coverages and the orientation of the DNA fragments attached. For example, the single-stranded oligonucleotides within the monolayers are oriented horizontally, suggesting multiple contacts with the surface [122,123]. In contrast, terminally attached double-helical fragments are oriented almost perpendicular to the surface [123,126]. Interestingly, single-stranded oligomers can also be "put upright" by treating the DNA-functionalized surface with C_6-thioalkanoles [122,123].

2. Applications of Microstructured Biofunctionalized Surfaces

The initial motivation for the development of DNA microarrays resulted from efforts during early states of the Human Genome Project to develop powerful alternative methodologies for the sequencing of nucleic acids. For this, "sequencing by hybridization" (SBH) was considered a promising approach [129]. The SBH technique allows the de-novo determination of an unknown nucleic acid sequence by means of hybridization analysis using a comprehensive DNA array containing all possible 65,536 ($= 4^8$) octanucleotides. However, during the experimental exploration, SBH turned out to be associated with severe unexpected practical problems. Nevertheless, the microarrays were found to be suitable for immediate applications in the fields of gene expression analysis and the determination of nucleic acid sequences actually known with respect to mutational screenings. In these applications, the particular strength of microarray analysis results from the highly redundant measurement of many parallel hybridization events, thus leading to an extraordinary level of assay validation. The probe redundancy does not mean the attachment of identical probe molecules on several sites of the chip, but rather the presence of multiple probes of different sequence with specificity for the same nucleic acid target.

The ultimate goal of microarray-based expression analysis is to acquire a comprehension of the entire cellular process, in order to exploit and to standardize the multidimenisional relations between genotype and phenotype. However, an increasingly important parameter, which has not yet been substantially taken into account, is the role of cellular translation. This means that mRNA expression data need to be correlated with the assortment of proteins actually present in the cell. One approach is based on the use of microarrays containing double-stranded DNA probes for the analysis of DNA–protein interaction and, thus, the detection and identification of DNA-binding proteins by means of fluorescence [130] or mass spectrometry analysis [131]. Moreover, substantial efforts are currently under way to develop protein, antibody, or even cell arrays, applicable to the correlation of genome and proteome research [132]. As an example, solid-phase-bound peptide arrays fabricated by combinatorial techniques, such as "spot synthesis" [133], are used for the screening of antibodies or other binding proteins [134]. The immobilization of immunoglobulins to glass surfaces has been achieved by photolithographical techniques [135]; more recently, in a number of publications, robotic deposition of the multiple proteins was used to manufacture protein microarrays [136–140]. One application of the protein biochips is in proteome research, for instance, to directly and specifically assay G protein activation by receptors [141]. Another application is for high-throughput gene expression and antibody screening [137]. The application of microarrays containing multiple spots of specific proteins, in particular antibodies, as miniaturized multianalyte immunosensors is also highly attractive [142]. Miniaturization of ligand-binding assays not only reduces costs by decreasing reagent consumption, but can simultaneously exceed the sensitivity of macroscopic techniques [143]. Thus, it is not surprising that an increasing number of research groups are currently exploiting the novel platform of microchip systems with respect to clinical diagnostics [138–140,144].

The immobilization of proteins on surfaces in a spatially defined way is obstructed by the general instability of most biomolecules, which often prevents a stepwise, successive immobilization of multiple delicate proteins. Thus, a single regioselective process under chemically mild conditions for the attachment of multiple compounds is required to circumvent this problem. The concept of DNA-directed immobilization (DDI) provides an optimal solution to the process of highly parallel immobilization (Fig. 18) [28,32]. Using DDI, the process of lateral surface patterning is carried out on the level of nucleic acids. Because of the exceptionally high chemical and physical stability of DNA, microarrays of different oligonucleotides can be prepared by successive attachment strategies and by spatially separated photolithographic processes (Fig. 19). Moreover, DNA-functionalized solid supports, once fabricated, can be stored almost indefinitely and can also be recovered by alkaline or heat treatment subsequent to hybridization. Thus, DNA microarrays can ideally be utilized as an immobilization matrix for the parallel, site-selective immobilization of many different DNA-tagged proteins and other functional compounds.

As an early demonstration of the feasibility of DNA-directed immobilization, several immunoglobulins were coupled with an oligonucleotide using the DNA–STV conjugates 2, and a mixture of the DNA-tagged immunoglobulins was allowed to site-selectively hybridize to an array of complementary capture oligonucleotides [28]. The successive self-sorting during the immobilization was established by means of specific immunosorption of target molecules. Similar experiments, carried out later with oligonucleotide-tagged enzymes, indicated that the DDI not only proceeds with extraordinary site selectivity due to the unique specificity of Watson–Crick base pairing, but also has additional advantages [32]. Quantitative measurements indicated that DDI proceeds with a higher immobilization efficiency than conventional immobilization techniques, such as the binding of the biotinylated proteins to streptavidin-coated surfaces or direct physisorption. These findings can be attributed to the reversible formation of the rigid, double-stranded DNA spacer between the surface and the proteins. Moreover, as indicated from surface-plasmon resonance (SPR) measurements [145], DDI allows a reversible functionalization of sensor surfaces with reproducible amounts of proteins [32]. Series of more than 150 cycles of hybridization/regeneration have been carried out with various DNA-tagged proteins, demonstrating that the immobilized protein can be completely removed by alkaline denaturation of the DNA double helix. These results show that DDI should be advantageous not only for recovery and reconfiguration of expensive sensor chips, but also for the generation of miniaturized multiplex sensor elements. In addition, since DDI is not limited in terms of the number and nature of compounds to be arrayed, its use to fabricate highly functionalized micro- and nanostructured surface architecture can be anticipated.

The latter conclusion is currently being realized: Mirkin and coworkers have reported on the DNA-directed immobilization of gold nanoparticles to form supramolecular surface architecture [146]. Similarly, Möller et al. [147], as well as our group [148], have used specific nucleic acid hybridization to immobilize gold nanoparticles on solid supports. A recent publication of the Mirkin group clearly indicates that this approach can be utilized for the highly sensitive scanometric detection of nucleic acids in DNA-microarray analyses. The scanometric detection is based on the gold-particle-promoted reduction of silver ions and allows for an about 100-fold increased sensitivity compared to fluorescence detection [149]. The use of colloidal gold nanoparticles also allows for the signal enhancement in the DNA hybridization detection using quartz crystal microbalance [150] and surface plasmon resonance [151]. The latter report suggests that due to about a 1000-fold improvement in sensitivity, the detection limit of surface plasmon resonance [152] now begins to approach that of traditional fluorescence-based DNA hybridization detection.

A highly interesting route to the generation of covalent nucleic acid–protein conjugates, which has great potential for impact on the fabrication of protein biochips, was reported in 1997 by Nemoto et al. [153] and also the group of Szostak [154]. They demonstrated a principle that is based on the in vitro translation of mRNA, covalently modified with a puromycin group at its 3′ end. The peptidyl-acceptor antibiotic puromycin covalently couples the mRNA with the polypeptide chain grown at the ribosome particle. This in situ conjugation of the informative (mRNA) with the functional (polypeptide) moiety can be applied not only to the selective isolation of mRNAs from large combinational libraries using in vitro selection. Moreover, as pursued by the U.S. company Phylos [155], the puromycin conjugation strategy might provide a powerful method for the production of protein microarrays.

Besides the lateral functionalization of microstructured solid supports, a nanostructuring of the surfaces can also be achieved using the DNA-directed immobilization approach. For this, surface-bound nucleic acids of the appropriate length are used to generate nanostructured complexes from various molecular building blocks, for instance, enzymes or other functional proteins (Fig. 4). These supramolecules should prove particularly useful as synthetic multienzyme complexes for enzyme process technology. Further prospects result from the direct integration of functional nucleic acid components into micro- and nanostructured surface architecture, for instance, by using catalytic DNA enzymes [156], switchable and allosteric ribozymes [157,158], or nucleic acids capable of specifically binding to target molecules, termed *aptamers* [159]. As an example, surface-bound aptamers with specificity for the binding of L-adenosine have been used as highly selective recognition elements in biosensors [160].

3. Electron Transfer and DNA Computing

The utilization of DNA surpasses the options of fabricating structural, mechanical, optical, or catalytic elements, discussed earlier. As a simple yet creative example, the use of DNA to conceal messages was recently reported [161]. In "genomic steganography," the secret-message DNA strand contains an encoded message flanked by PCR primer sequences. The secret-message molecule was hidden in a microspot of nonsense messages, i.e., denatured human DNA. Microdots containing 100 copies of the secret message had been attached to full stops in a printed letter and posted through regular mail. Subsequent PCR amplification and sequencing led to successful decoding of the message. In addition, applications in molecular biotechnology are currently under investigation in which nucleic acids are used to generate devices capable of performing electrical and numerical operations.

The initial observation that a charge transport can occur along the DNA double helix [162] propagated tremendous research activities [163]. Various groups are engaged to learn how the transport of charges, electrons or holes, operates over long and short distances within the DNA. The development of model systems is currently a crucial point to allow for deeper insights into the electron transfer kinetics. The transfer rates currently determined are varying from micro- to picoseconds, depending on the model system investigated. The goals that motivate such studies are intriguing. First of all, the electron migration through the π-stacked DNA base pairs is of fundamental relevance, since this mechanism might be involved in long-range oxidative damage of GG sequences and thus, in turn, might also contribute to cellular events, such as DNA repair processes. Furthermore, the DNA-mediated charge transport might also be utilized as a diagnostic tool. Since the electron transfer capability is strongly dependent on an intact DNA double helix, the hybridization of complementary single-stranded molecules should be detectable by means of direct electrochemical measurement [164–167].

Barton and coworkers have shown that proteins can in fact modulate the DNA electron transfer [168]. Methyltransferases are enzymes that recognize distinct DNA sequences, e.g., 5'-G*CGC-3', and effect methylation by extruding the target base cytosine (*C) completely out of the DNA duplex while the remainder of the double helix is left intact. The methyltransferase *Hha* I-DNA complex is a well-characterized example, revealing that the structure of the DNA is significantly but locally distorted [169,170]. In a recent study, Rajski et al. used DNA duplex **20** containing the M.*Hha* I binding site between two oxidizable 5'-GG-3' sites [168] (Fig. 20). The duplex contains a complementary strand, selectively 5'-modified with a Rh^{3+} intercalator that can function as a photooxidant. Upon

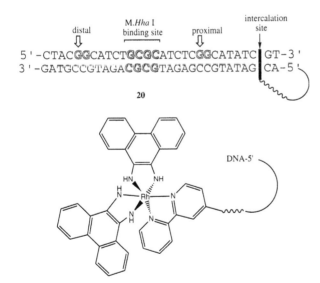

FIG. 20 DNA duplex **20** used for studies regarding the protein-modulated DNA electron transfer. Methyltransferase M.*Hha*I is capable of binding to the shadowed recognition site between two oxidizable 5'-GG-3' sites (outlined letters). The complementary strand of the duplex contains the Rh^{3+} intercalator, [Rh(phi)$_2$bpy']$^{3+}$, at its 5' end, which can function as a photooxidant. (Adapted from Ref. 168.)

irradiation at 365 nm, photooxidation and subsequent cleavage occurs at preferential sites. As determined by electrophoretic analysis of the cleavage products, a significant decrease in distal 5′-GG-3′ damage occurred in an enzyme-dependent fashion. In contrast, the efficiency of photo-oxidative damage at the proximal 5′-GG-3′ site was unchanged, even at high M.*Hha* I concentrations, indicating that long-range oxidative damage to DNA is highly sensitive to protein-induced DNA distortions [168].

The preceding results give rise to the development of electrochemical sensors for the study of protein–DNA interactions. A recent work of Corn and colleagues might supplement this elaboration [171]. They presented a procedure for creating DNA arrays on metal transducer surfaces, using a multistep chemical modification procedure. On gold surfaces, photolithography is used to generate an array of spots that are first surrounded by a hydrophobic background that allows mechanical deposition of aqueous DNA solutions. Subsequently, the hydrophobic background is replaced by one that resists the non-specific adsorption of proteins during in situ SPR imaging. The utility of such surface arrays was demonstrated by surface plasmon resonance imaging of the specific adsorption of single-stranded binding protein (SSB) to an oligonucleotide array created by this procedure [171].

The gold surface-bound DNA arrays might also be used in DNA computing. This field was initiated by Adleman [172], who proposed that nucleic acid hybridization could be used to solve instances of difficult mathematical problems, known as NP-complete problems. Corn and coworkers have adapted these ideas to combinatorial mixtures of DNA molecules attached to surfaces, and they proposed to perform logical manipulations of large sets of data by the hybridization and enzymatic manipulation of the attached oligonucleotides [173–176]. For instance, they demonstrated a word design strategy for DNA computing on surfaces that is based on 16-mer oligonucleotide "DNA words" attached to chemically modified gold substrates. By linking these words together into word strands, the longer DNA molecules required to make large combinatorial sets of oligonucleotides can be created [173]. To solve computational problems, they developed a set of "operations" to manipulate the surface-bound DNA words [175]: "Mark" taggs subsets of the DNA words by the hybridization of complementary words; "Unmark" untaggs the marked molecules by enzymatic digestion of the DNA double helices. More complex operations can be performed using DNA ligase enzymatic action [176], previously applied in the heterogeneous ligation of surface-bound oligonucleotides related to SBH applications [177]. In the context of DNA computing, the "Surface Word Append" operation selectively appends additional DNA words onto immobilized word strands. The "Two-Word Mark and Destroy" operation selectively removes singly marked two-word DNA strands from the surface in the presence of doubly marked two-word strands. These ligation-based operations, which may also be used in conjunction with PCR amplification for multiword "readout" [175], are essential for manipulation of large combinatorial sets of linked DNA word strands required for DNA computing [176]. The feasibility of DNA computing on surfaces has recently been demonstrated by solving an NP-complete problem [178]. In a commentary, DNA computing is envisaged as a step toward the development of organic computing devices implanted within a living body that can integrate signals from several sources and compute a response in terms of an molecular delivery device for a drug or signal [179]. In a recent publication, Seeman and coworkers showed that a one-dimensional algorithmic self-assembly of DNA triple-crossover molecules [180] can be used to execute a logical operation on a string of binary bits [181].

V. CONCLUSIONS

This chapter describes the use of DNA as a material in nanosciences. Due to its unique recognition capabilities, physicochemical stability, mechanical rigidity, and high-precision processibility, DNA is a highly promising construction material for the fabrication of nanostructured DNA scaffolds as well as for the selective positioning of conducting proteins, metal and semiconductor nanoclusters, and other molecular devices. Additional advantages of the material DNA result from its processibility in aqueous solutions, possibilities of convenient biological decomposition, and the reversibility of the Watson–Crick base pairing with an optional fixation of the structures using selective cross-linking methods [182]. The prospect of high economic profits in nanotechnology [62] justifies the recently required financial expenditure accompanied with the use of synthetic oligonucleotides. Although current examples are typically carried out with nano- to femtomole amounts of DNA, advances in the (bio)synthesis of DNA, for instance, using isothermal "rolling circle amplification" [183,184], may soon allow for preparative work on a larger scale. In addition, other rapid developments are currently in progress: synthetic DNA derivatives [185,186] and specific DNA ligands, such as polyamides [187,188], polysaccharides [189], and synthetic DNA-binding peptides [190], are being developed.

   X-ray crystallography, the ultimate tool in characterization of biological macromolecules, is often not applicable due to limited quantities and purity or, worse, due to inappropriate crystallization properties of the particular components under investigation. Thus, the choice and optimization of appropriate analytical methods is of particular importance for the further development of DNA nanotechnology. Fortunately, the various microscopy methods, high-resolution electron microscopy [191], and, in particular, scanning probe microscopy methods, such as AFM, are constantly being improved. The last, for instance, has already been extensively applied in the characterization of biomolecules [192]. Using AFM, nucleic acid components have been studied with respect to their enzymatic processing [21], mobility on solid supports [193], crystallization at interphases [194], and mechanical manipulations [195,196]. In addition, the use of AFM has recently proven advantageous in the characterization of synthetic supramolecular DNA–protein conjugates [49,53,54] and DNA nanostructures [80–82]. Better resolution of the AFM might be attained by using carbon nanotubes as nanoprobes in scanning force microscopy [197]. Once attached to the AFM cantilever, nanotubes not only reveal ideal structural and mechanical properties to serve as a molecular tool for imaging nanostructures, but also can be functionalized by covalent methods [198] or by the adsorption of biomolecular components, such as oligonucleotides [199] or proteins [200].

   The refinement of other analytical methods, such as electrophoresis [34,36], the various techniques of optical spectroscopy [103–105], and nuclear magnetic resonance [201], is supplemented by the recent advances in real-time affinity measurements [152,202], contributing to the understanding of biomolecular reactivity. Taken together, the improvement of analytical methods will eventually allow a comprehensive characterization of the structure, topology, and properties of the nucleic acid–based supramolecular components under consideration for distinctive applications in nanobiotechnology.

   The nanostructured molecular arrangements from DNA developed by Seeman may find applications as biological encapsulation and drug-delivery systems, as artificial multienzymes, or as scaffolds for the self-assembling nanoscale fabrication of technical elements. Moreover, DNA–protein conjugates may be anticipated as versatile building blocks in the fabrication of multifunctional supramolecular devices and also as highly functionalized, micro- and nanostructured surface architecture. It is safe to predict that these developments will profit from the current progress of surface nanostructuring, for instance, by means of the "dip-pen" nanolithography using the AFM tip to write alkanethiols with less than 30-nanometer linewidth resolution on gold substrates [203,204]. As a particular strength of using nucleic acid technology for construction purposes, this approach is not limited to biomolecular components, and thus nucleic acid–based hybrid composites will allow entirely novel applications in material sciences, enzyme process technology, biomedical analysis, and other fields of supramolecular chemistry and molecular biotechnology.

ACKNOWLEDGMENTS

I wish to thank my coworkers, whose names are listed in the bibliography, for their excellent and motivated contributions. Michael Adler and Steven Lenhert are gratefully acknowledged for critical reading of the manuscript and invaluable discussion. I thank Professor Dietmar Blohm for his generous support. This work was financially supported by Deutsche Forschungsgemeinschaft, Fonds der Chemischen Industrie, and Tönjes-Vagt Stiftung Bremen.

REFERENCES

1. J-M Lehn. Supramolecular Chemistry: Concepts and Perspectives. Weinheim, Germany: VCH, 1995.
2. F Vögtle. Supramolecular Chemistry. Chichester, England: Wiley, 1991.
3. JL Atwood, JED Davies, DD MacNicol, F Vögtle. Comprehensive Supramolecular Chemistry. Oxford: Elsevier, 1996.
4. D Philp, JF Stoddart. Angew. Chem. Int. Ed. Engl. 35:1154–1196, 1996.
5. RP Feynman. In: HD Gilbert, ed. Miniaturization. New York: Reinhold, 1961, pp 282–296.
6. GM Whitesides, JP Mathias, CT Seto. Science 254:1312–1319, 1991.
7. KE Drexler. Proc. Natl. Acad. Sci. USA 78:5275–5278, 1981.
8. NC Seeman. J. Theor. Biol. 99:237–247, 1982.
9. KE Drexler. Nanosystems: Molecular Machinery, Manufacturing, and Computation. New York: Wiley, 1992.
10. NC Seeman. Angew. Chem. Int. Ed. 37:3220–3238, 1998.
11. NC Seeman. Trends Biotechnol. 17:437–443, 1999.
12. CM Niemeyer. Angew. Chem. Int. Ed. 36:585–587, 1997.
13. CM Niemeyer. Curr. Opin. Chem. Biol. 4:609–618, 2000.
14. D Voet, JG Voet. Biochemistry. New York: Wiley, 1994.
15. MH Caruthers. Science 230:281–285, 1985.
16. KB Mullis. Angew. Chem. Int. Ed. 33:1209–1213, 1994.
17. PJ Hagerman. Annu. Rev. Biophys. Chem. 17:265–286, 1988.
18. M Krings, A Stone, RW Schmitz, H Krainitzki, M Stoneking, S Pääbo. Cell 90:19–30, 1997.
19. CS McHenry. Ann. Rev. Biochem. 57:519–550, 1988.
20. HG Hansma, M Bezanilla, F Zenhausern, M Adrian, RL Sinsheimer. Nucleic Acids Res. 21:505–512, 1993.
21. S Kasas, NH Thomson, BL Smith, HG Hansma, XS Zhu, M Guthold, C Bustamante, ET Kool, M Kashlev, PK Hansma. Biochemistry 36:461–468, 1997.
22. NH Thomson, BL Smith, N Almqvist, L Schmitt, M Kashlev, ET Kool, PK Hansma. Biophys. J. 76:1024–1033, 1999.
23. MD Wang, MJ Schnitzer, H Yin, R Landick, J Gelles, SM Block. Science 282:902–907, 1998.
24. The structure of the ribosome's large subunit has recently been resolved at 2.4-Å resolution by means of x-ray crystallography: N Ban, P Nissen, J Hansen, PB Moore, TA Steitz. Science

289:905–920, 2000. The atomic structure of this subunit and its complexes with substrate analogs revealed the enzymatic activity of the rRNA backbone. Thus, the ribosome is in fact a ribozyme: P Nissen, J Hansen, N Ban, PB Moore, TA Steitz. Science 289:920–930, 2000. Atomic structure of the ribosome's small 30S subunit, resolved at 5 Å: WM Clemons Jr, JL May, BT Wimberly, JP McCutcheon, MS Capel, V Ramakrishnan. Nature 400:833–840, 1999. The 8-Å crystal structure of the 70S ribosome reveals a double-helical RNA bridge between the 50S and the 30S subunit: GM Culver, JH Cate, GZ Yusupova, MM Yusupov, HF Noller. Science 285:2133–2136, 1999.

25. D Mendel, VW Cornish, PG Schultz. Annu. Rev. Biophys. Biomol. Struct. 24:435–462, 1995.
26. Examples: (a) nucleosome: K Luger, AW Mader, RK Richmond, DF Sargent, TJ Richmond. Nature 389:251–260, 1997; (b) DNA polymerases: CA Brautigam, TA Steitz. Curr. Opin. Struct. Biol. 8:54–63, 1998; (c) single-stranded binding protein: Y Shamoo, AM Friedman, MR Parsons, WH Konigsberg, TA Steitz. Nature 376:362–366, 1995; (d) restriction endonucleases: RA Kovall, BW Matthews. Curr. Opin. Chem. Biol. 3:578–583, 1999; (e) DNA ligase: S Shuman. Structure 4:653–656, 1996; (f) DNA helicases: MC Hall, SW Matson. Mol. Microbiol. 34:867–877, 1999; (g) zinc-finger proteins: Y Choo, JW Schwabe. Nat. Struct. Biol. 5:253–255, 1998.
27. BH Robinson, NC Seeman. Protein Eng. 1:295–300, 1987.
28. CM Niemeyer, T Sano, CL Smith, CR Cantor. Nucl. Acids Res. 22:5530–5539, 1994.
29. PC Weber, DH Ohlendorf, JJ Wendoloski, FR Salemme. Science 243:85–88, 1989.
30. M Wilchek, EA Bayer. Methods Enzymol. 184:14–45, 1990.
31. M Wilchek, EA Bayer. Methods Enzymol. 184:51–67, 1990.
32. CM Niemeyer, L Boldt, B Ceyhan, D Blohm. Anal. Biochem. 268:54–63, 1999.
33. CM Niemeyer, B Ceyhan, D Blohm. Bioconjug. Chem. 10:708–719, 1999.
34. CM Niemeyer, W Bürger, J Peplies. Angew. Chem. Int. Ed. 37:2265–2268, 1998.
35. CM Niemeyer, W Bürger, RMJ Hoedemakers. Bioconjugate Chem. 9:168–175, 1998.
36. CM Niemeyer, L Boldt, B Ceyhan, D Blohm. J. Biomol. Struct. Dynamics 17:527–538, 1999.
37. PE Nielsen. Acc. Chem. Res. 32:624–630, 1999.
38. L Bulow, K Mosbach. Trends Biotech. 9:226–231, 1991.
39. C Lindbladh, M Persson, L Bulow, K Mosbach. Eur. J. Biochem. 204:241–247, 1992.
40. C Lindbladh, RD Brodeur, G Lilius, L Bulow, K Mosbach, PA Srere. Biochemistry 33:11684–11691, 1994.
41. AG Cook, PJ Wood. J. Immunol. Methods 171:227–237, 1994.
42. MF Fanger. Bispecific Antibodies. New York: Springer, 1995.
43. D Neri, M Momo, T Prospero, G Winter. J. Mol. Biol. 246:367–373, 1995.
44. F Breitling, S Dübel, T Seehaus, I Klewinghaus, MDH Little. Gene 104:147–153, 1991.
45. G Winter, C Milstein. Nature 349:293–299, 1991.
46. MDH Little, F Breitling, S Dübel, P Fuchs, M Braunagel. J. Biotechnol. 41:187–195, 1995.
47. A Nissim, HR Hoogenboom, IM Tomlinson, G Flynn, C Midgley, D Lane, G Winter. EMBO J. 13:692–698, 1995.
48. T Sano, CL Smith, CR Cantor. Science 258:120–122, 1992.
49. CM Niemeyer, M Adler, B Pignataro, S Lenhert, S Gao, LF Chi, H Fuchs, D Blohm. Nucleic Acids Res. 27:4553–4561, 1999.
50. CM Niemeyer, M Adler, D Blohm. Anal. Biochem. 246:140–145, 1997.
51. M Mammen, S-K Choi, GM Whitesides. Angew. Chem. 110:2908–2953, 1998; Angew. Chem. Int. Ed. Engl. 37:2754–2794, 1998.
52. CM Niemeyer, M Adler, S Gao, LF Chi. Bioconjug. Chem. In press, 2001.
53. CM Niemeyer, M Adler, S Lenhert, S Gao, H Fuchs, LF Chi. Chem. Bio. Chem. 2:260–264, 2001.
54. CM Niemeyer, M Adler, S Gao, LF Chi. Angew. Chem. Int. Ed. 39:3055–3059, 2000.
55. J Shi, DE Bergstrom. Angew. Chem. Int. Ed. 36:111–113, 1997.

56. H Kuhn, VV Demidov, MD Frank-Kaminetskii. Angew. Chem. 111:1544–1547, 1999; Angew. Chem. Int. Ed. 38:1446–1449, 1999.

57. MS Shchepinov, KU Mir, JK Elder, MD Frank-Kamenetskii, EM Southern. Nucleic Acids Res. 27:3035–3041, 1999.

58. M Scheffler, A Dorenbeck, S Jordan, M Wüstefeld, G v. Kiedrowski. Angew. Chem. Int. Ed. 38:3312–3315, 1999.

59. CA Mirkin, RL Letsinger, RC Mucic, JJ Storhoff. Nature 382:607–609, 1996.

60. AP Alivisatos, KP Johnsson, X Peng, TE Wilson, CJ Loweth, MP Bruchez Jr, PG Schultz. Nature 382:609–611, 1996.

61. H Weller. Angew. Chem. Int. Ed. 35:1079–1081, 1996.

62. D Bethell, DJ Schiffrin. Nature 382:581, 1996.

63. AP Alivisatos. Science 271:933–937, 1996.

64. RF Service. Science 271:920–922, 1996.

65. R Elghanian, JJ Storhoff, RC Mucic, RL Letsinger, CA Mirkin. Science 277:1078–1081, 1997.

66. JJ Storhoff, R Elghanian, RC Mucic, CA Mirkin, RL Letsinger. J. Am. Chem. Soc. 120:1959–1964, 1998.

67. JJ Storhoff, AA Lazarides, RC Mucic, CA Mirkin, RL Letsinger, GC Schatz. J. Am. Chem. Soc. 122:4640–4650, 2000.

68. RA Reynolds, CA Mirkin, RL Letsinger. J. Am. Chem. Soc. 122:3795–3796, 2000.

69. RC Mucic, JJ Storhoff, CA Mirkin, RL Letsinger. J. Am. Chem. Soc. 120:12674–12675, 1998.

70. GP Mitchell, CA Mirkin, RL Letsinger. J. Am. Chem. Soc. 121:8122–8123, 1999.

71. S-J Park, AA Lazarides, CA Mirkin, PW Brazis, CR Kannewurf, RL Letsinger. Angew. Chem. Int. Ed. 39:3845–3848, 2000.

72. RF Service. Science 277:1036–1037, 1997.

73. JJ Storhoff, CA Mirkin. Chem. Rev. 99:1849–1862, 1999.

74. CJ Loweth, WB Caldwell, X Peng, AP Alivisatos, PG Schultz. Angew. Chem. Int. Ed. 38:1808–1812, 1999.

75. R-I Ma, NR Kallenbach, RD Sheardy, ML Petrillo, NC Seeman. Nucl. Acids Res. 14:9745–9753, 1986.

76. ML Petrillo, CJ Newton, RP Cunningham, R-I Ma, NR Kallenbach, NC Seeman. Biopolymers 27:1337–1352, 1988.

77. Y Wang, JE Mueller, B Kemper, NC Seeman. Biochemistry 30:5667–5674, 1991.

78. J Qi, X Li, X Yang, NC Seeman. J. Am. Chem. Soc. 118:6121–6130, 1996.

79. X Li, Y Xiaoping, Q Jing, NC Seeman. J. Am. Chem. Soc. 118:6131–6140, 1996.

80. E Winfree, F Liu, LA Wenzler, NC Seeman. Nature 394:539–544, 1998.

81. X Yang, LA Wenzler, J Qi, X Li, NC Seeman. J. Am. Chem. Soc. 120:9779–9786, 1998.

82. F Liu, R Sha, NC Seeman. J. Am. Chem. Soc. 121:917–922, 1999.

83. J-H Chen, NR Kallenbach, NC Seeman. J. Am. Chem. Soc. 111:6402–6407, 1989.

84. J Chen, NC Seeman. Nature 350:631–633, 1991.

85. Y Zhang, NC Seeman. J. Am. Chem. Soc. 114:2656–2663, 1992.

86. Y Zhang, NC Seeman. J. Am. Chem. Soc. 116:1661–1669, 1994.

87. SM Du, H Wang, Y-C Tse-Dinh, NC Seeman. Biochemistry 34:673–682, 1995.

88. J Chen, NC Seeman. Nature 386:137–138, 1997.

89. C Mao, W Sun, Z Shen, NC Seeman. Nature 397:144–146, 1999.

90. J Fritz, MK Baller, HP Lang, H Rothuizen, P Vettiger, E Meyer, H Guntherodt, C Gerber, JK Gimzewski. Science 288:316–318, 2000.

91. B Yurke, AJ Turberfield, AP Mills, Jr., FC Simmel, JL Neumann. Nature 406:605–608, 2000.

92. SS Smith, LM Niu, DJ Baker, JA Wendel, SE Kane, DS Joy. Proc. Nat. Acad. Sci. USA 94:2162–2167, 1997.

93. NC Seeman, J Chen, SM Du, JE Mueller, Y Zhang, T-J Fu, Y Wang, H Wang, W Zhang. New J. Chem. 17:739–755, 1993.

94. JL Coffer, SR Bigham, RF Pinizzotto, H Yang. Nanotechnology 3:69–76, 1992.
95. SR Bigham, JL Coffer. Colloids Surfaces A 95:211–219, 1995.
96. JL Coffer, SR Bigham, X Li, RF Pinizzotto, YG Rho, RM Pirtle, IL Pirtle. Appl. Phys. Lett. 69:3851–3853, 1996.
97. JL Coffer. J. Cluster Sci. 8:159–179, 1997.
98. T Torimoto, M Yamashita, S Kuwabata, T Sakata, H Mori, H Yoneyama. J. Phys. Chem. B 103:8799–8803, 1999.
99. AM Cassel, WA Scrivens, JM Tour. Angew. Chem. Int. Ed. 110:1528–1531, 1998.
100. MY Ono, EM Spain. J. Am. Chem. Soc. 121:7330–7334, 1999.
101. T Blessing, J-S Remy, J-P Behr. J. Am. Chem. Soc. 120:8519–8520, 1998.
102. EJ Gibbs, I Tinoco, J, MF Maestre, PA Ellinas, RF Pasternack. Biochem. Biophys. Res. Commun. 157:350–358, 1988.
103. RF Pasternack, JI Goldsmith, S Szep, EJ Gibbs. Biophys. J. 75: 1024–1031, 1998.
104. RF Pasternack, EJ Gibbs, PJ Collings, JC dePaula, LC Turzo, A Terracina. J. Am. Chem. Soc. 120:5873–5878, 1998.
105. JL Seifert, RE Connor, SA Kushon, M Wang, BA Armitage. J. Am. Chem. Soc. 121:2987–2995, 1999.
106. E Braun, Y Eichen, U Sivan, G Ben-Yoseph. Nature 391:775–778, 1998.
107. E Di Mauro, CP Hollenberg. Adv. Mater. 5:384–386, 1993.
108. SC Esener, D Hartmann, MJ Heller, JM Cable. DNA-assisted micro-assembly: a heterogeneous integration technology for optoelectronics. Proc. SPIE Critical Reviews of Optical Science and Technology, Heterogeneous Integration, San Jose, CA, 1998, pp 65–68.
109. CF Edman, DE Raymond, DJ Wu, E Tu, RG Sosnowski, WF Butler, M Nerenberg, MJ Heller. Nucleic Acids Res. 25:4907–4914, 1997.
110. SC Esener. 1999. The approach of this group is described in: S Bains. Science 279:2043–2044, 1998. Additional information can be found in the Internet at: http://soliton.ucsd.edu/~hartmann/DNA.html
111. JD Hoheisel. TIBTECH 15:465–469, 1997.
112. CM Niemeyer, D Blohm. Angew. Chem. Int. Ed. 38:2865–2869, 1999.
113. EM Southern, K Mir, M Shchepinov. Nat. Genet. 21:5–9, 1999; see also the series of review articles published in this supplement of Nat. Genet.
114. M Schena, D Shalon, RW Davis, PO Brown. Science 270:467–470, 1995.
115. Recent publications concerning: (a) the chemistry of photolithographical DNA synthesis: GH McGall, AD Barone, M Diggelmann, SPA Fodor, E Gentalen, N Ngo. J. Amer. Chem. Soc. 119:5081–5090, 1997; MC Pirrung, L Fallon, GH MacGall. J. Org. Chem. 63:241–246, 1998; M Beier, JD Hoheisel. Nucleosides Nucleotides. In press, 1999; (b) the surface immobilization of presynthesized DNA fragments and the synthesis of linkers: M Beier, JD Hoheisel. Nucleic Acids Res. 27:1970–1977, 1999.
116. AC Pease, D Solas, EJ Sullivan, MT Cronin, CP Holmes, SPA Fodor. Proc. Natl. Acad. Sci. USA 91:5022–5026, 1994.
117. Surface plasmon resonance: (a) SJ Wood. Microchem. J. 47:330–337, 1992; (b) analysis of enzymatic DNA manipulations: P Nilsson, B Persson, M Uhlen, PA Nygren. Anal. Biochem. 224:400–408, 1995; (c) elipsometry: DE Gray, SC Case-Green, TS Fell, PJ Dobson, EM Southern. Langmuir 13:2833–2842, 1997; (d) optical grating coupler: FF Bier, FW Scheller. Biosens. Bioelectron. 11:669–674, 1996; (e) resonance mirror: HJ Watts, D Yeung, H Parkes. Anal. Chem. 67:4283–4289, 1995; (f) reflectometric interference spectroscopy; M Sauer, A Brecht, K Charisse, M Maier, M Gerster, I Stemmler, G Gauglitz, E Bayer. Anal. Chem. 71:2850–2857, 1999.
118. Nucleic acid hybridization can be detected by means of the piezoelectric QCM (= quartz crystal microbalance): (a) Y Okahata, Y Matsunobu, K Ijiro, M Mukae, A Murakami, K Makino. J. Am. Chem. Soc. 114:8299–8300, 1992; (b) S Yamaguchi, T Shimomura. Anal. Chem. 65:1925–1927, 1993; (c) K Ito, K Hashimoto, Y Ishimori. Anal. Chim. Acta 327:29–35, 1996;

(d) Hybridization using QCM-bound DNA multilayers: F Caruso, E Rodda, DN Furlong, K Niikura, Y Okahata. Anal. Chem. 69:2043–2049, 1997; (e) detection of single mismatches by means of QCM-bound PNA probes: J Wang, PE Nielsen, M Jiang, X Cai, JR Fernandes, DH Grant, M Ozsoz, A Beglieter, M Mowat. Anal. Chem. 69:5200–5202, 1997; (f) hybridization with QCM-bound DNA dendrimers: J Wang, M Jiang, TW Nilsen, RC Getts. J. Am. Chem. Soc. 120:8281–8282, 1998; (g) amplified QCM assay of DNA using oligonucleotide-functionalized liposomes or biotinylated liposomes: F Patolsky, A Lichtenstein, I Willner. J. Am. Chem. Soc. 122:418–419, 2000; (h) detection of the enzymatic activity of a DNA polymerase at surfaces of a QCM: K Niikura, M Hisao, Y Okahata. J. Am. Chem. Soc. 120:8537–8538, 1998; for electrochemical analysis of hybridization, see: J Wang. Chem. Eur. J. 5:1681–1685, 1999.

119. R Hintsche, M Paeschke, A Uhlig, R Seitz. In: FW Scheller, F Schaubert, J Fredrowitz, eds. Frontiers in Biosensors: Fundamental Aspects. City of Public. Birkhaeuser Verlag, 1997, pp 267–283; Sensing and amplification of oligonucleotide–DNA interactions by means of impedance spectroscopy in the presence of the [Fe(CN)6]3-/4- redox couple: A Bardea, F Patolsky, A Dagan, I Willner. Chem. Commun. 21–22, 1999; F Patolsky, E Katz, A Bardea, I Willner. Langmuir 15:3703–3706, 1999.

120. (a) Review: E Nordhoff, F Kirpekar, P Roepstorff. Mass Spectrom. Rev. 15:67–138, 1996; (b) enzymatic elongation of surface-immobilized oligomers with dideoxy-terminators allows us to detect single nucleotide point mutations: GS Higgins, DP Little, H Köster. Biotechniques 23:710–714, 1997; (c) sequencing of DNA by means of MALDI-MS: DJ Fu, K Tang, A Braun, D Reuter, B Darnhofer-Demar, DP Little, MJ O'Donnell, CR Cantor, H Köster. Nat. Biotechnol. 16:381–384, 1998; (d) nucleic acids of more than 500 kDa can be detected using an infrared laser: S Berkenkamp, F Kirpekar, F Hillenkamp. Science 281:260–262, 1998.

121. (a) Single-molecule detection using fluorescence [review: S Weiss. Science 283:1676–1683, 1999] has recently been applied to the analysis of DNA hybridization [W Trabesinger, GJ Schutz, HJ Gruber, H Schindler, T Schmidt. Anal. Chem. 71:279–283, 1999] and RNA transcription [AM Femino, FS Fay, K Fogarty, RH Singer. Science 280:585–590, 1998]; (b) surface-immobilized DNA can be amplified by isothermal "rolling circle amplification": PM Lizardi, X Huang, Z Zhu, P Bray-Ward, DC Thomas, DC Ward. Nat. Genet. 19:225–232, 1998.

122. TM Herne, MJ Tarlov. J. Am. Chem. Soc. 119:8916–8920, 1997.

123. R Levicky, TM Herne, MJ Tarlov, S Satija. J. Am. Chem. Soc. 120:9798–9792, 1998.

124. AB Steel, TM Herne, MJ Tarlov. Anal. Chem. 70:4670–4677, 1998.

125. SO Kelley, JK Barton, NM Jackson, MG Hill. Bioconjugate Chem. 8:31, 1997.

126. SO Kelley, JK Barton, NM Jackson, L McPherson, A Potter, EM Spain, MJ Allen, MG Hill. Langmuir 14:6781, 1998.

127. SO Kelley, NM Jackson, MG Hill, JK Barton. Angew. Chem. 111:991–996, 1999. Angew. Chem. Int. Ed. Engl. 38:941–945, 1999.

128. LT Mazzola, SP Fodor. Biophys. J. 68:1653–1660, 1995.

129. (a) W Bains, G Smith. J. Theor. Biol. 135:303–307, 1988; (b) R Drmanac, I Labat, I Brukner, R Crkvenjakov. Genomics 4:114–128, 1989; (c) K Khrapko, LY, A Khorlyn, V Shick, V Florentiev, AD Mirzabekov. FEBS Lett. 256:118–122, 1989; (d) CR Cantor, A Mirzabekov, E Southern. Genomics 13:1378–1383, 1992.

130. ML Bulyk, E Gentalen, DJ Lockhart, GM Church. Nat. Biotechnol. 17:573–577, 1999.

131. E Nordhoff, AM Krogsdam, HF Jorgensen, BH Kallipolitis, BF Clark, P Roepstorff, K Kristiansen. Nat. Biotechnol. 17:884–888, 1999.

132. Current reviews on proteome research: (a) A Persidis. Nat. Biotechnol. 16:393–394, 1998; (b) A Dove. Nat. Biotechnol. 17:233–236, 1999; (c) F Lottspeich. Angew. Chem. Int. Ed. 38:2476–2492, 1999.

133. R Frank. Tetrahedron 48:9217–9232, 1992.

134. A Kramer, J Schneider-Mergener. Methods Mol. Biol. 87:25–39, 1998.

135. LF Rozsnyai, DR Benson, SPA Fodor, PG Schultz. Angew. Chem. Int. Ed. Engl. 31:759, 1992.
136. D Guschin, G Yershov, A Zaslavsky, A Gemmell, V Shick, D Proudnikov, P Arenkov, A Mirzabekov. Anal. Biochem. 250:203–211, 1997.
137. A Lueking, M Horn, H Eickhoff, K Bussow, H Lehrach, G Walter. Anal. Biochem. 270:103–111, 1999.
138. LG Mendoza, P McQuary, A Mongan, R Gangadharan, S Brignac, M Eggers. Biotechniques 27:778–780, 1999.
139. CA Rowe, LM Tender, MJ Feldstein, JP Golden, SB Scruggs, BD MacCraith, JJ Cras, FS Ligler. Anal. Chem. 71:3846–3852, 1999.
140. CA Rowe, SB Scruggs, MJ Feldstein, JP Golden, FS Ligler. Anal. Chem. 71:433–439, 1999.
141. C Bieri, OP Ernst, S Heyse, KP Hofmann, H Vogel. Nat. Biotechnol. 17:1105–1108, 1999.
142. RP Ekins, FW Chu. Clin. Chem. 37:1955–1967, 1991.
143. JW Silzel, B Cercek, C Dodson, T Tsay, RJ Obremski. Clin. Chem. 44:2036–2043, 1998.
144. NH Chiem, DJ Harrison. Clin. Chem. 44:591–598, 1998.
145. M Malmquist. Nature 361:186–187, 1993.
146. TA Taton, RC Mucic, CA Mirkin, RL Letsinger. J. Am. Chem. Soc. 122:6305–6306, 2000.
147. R Möller, A Csaki, JM Köhler, W Fritzsche. Nucleic Acids Res. 28:E91, 2000.
148. CM Niemeyer, B Ceyhan, S Gao, LF Chi, S Peschel, U Simon. Colloid. Polym. Sci. 279:68–72, 2001.
149. TA Taton, CA Mirkin, RL Letsinger. Science 289:1757–1760, 2000.
150. F Patolsky, KT Ranjit, A Lichtenstein, T Willner. Chem. Commun. 1025–1026, 2000.
151. L He, MD Musick, SR Nicewarner, FG Salinas, SJ Benkovic, MJ Natan, CD Keating. J. Am. Chem. Soc. 122:9071–9077, 2000.
152. JM Brockman, BP Nelson, RM Corn. Annu. Rev. Phys. Chem. 51:41–63, 2000.
153. N Nemoto, E Miyamoto-Sato, Y Husimi, H Yanagawa. FEBS Lett. 414:405–408, 1997.
154. RW Roberts, JW Szostak. Proc. Natl. Acad. Sci. USA 94:12297–12302, 1997.
155. http://www.phylos.com
156. RR Breaker, GF Joyce. Chem. Biol. 1:223–229, 1994.
157. MP Robertson, AD Ellington. Nat. Biotechnol. 17:62–66, 1999.
158. GA Soukup, RR Breaker. Proc. Natl. Acad. Sci. USA 96:3584–3589, 1999.
159. M Famulok, G Mayer, M Blind. Acc. Chem. Res. 33:591–599, 2000.
160. F Kleinjung, S Klussmann, VA Erdmann, FW Scheller, FF Bier. Anal. Chem. 70:328–331, 1998.
161. CT Clelland, V Risca, C Bancroft. Nature 399:533–534, 1999.
162. CJ Murphy, MR Arkin, Y Jenkins, ND Ghatlia, SH Bossmann, NJ Turro, JK Barton. Science 262:1025–1029, 1993.
163. Review articles: (a) RE Holmlin, PJ Dandliker, JK Barton. Angew. Chem. Int. Ed. 36:2715–2730, 1997; (b) SO Kelley, JK Barton. Met. Ions Biol. Syst. 36:211–249, 1999; (c) SO Kelley, JK Barton. In: A Sigel, eds. Metal Ions in Biological Systems. Vol. 36. New York: Marcel Dekker, 1999, pp 211–249; (d) ME Nunez, JK Barton. Curr. Opin. Chem. Biol. 4:199–206, 2000.
164. TJ Meade, JF Kayyem, SE Fraser. US Patent 5,591,578, 1995.
165. SO Kelley, EM Boon, JK Barton, NM Jackson, MG Hill. Nucleic Acids Res. 27:4830–4837, 1999.
166. D Porath, A Bezryadin, S de Vries, C Dekker. Nature 403:635–638, 2000.
167. EM Boon, DM Ceres, TG Drummond, MG Hill, JK Barton. Nat. Biotechnol. 18:1096–1100, 2000.
168. SR Rajski, S Kumar, RJ Roberts, JK Barton. J. Am. Chem. Soc. 121:5615–5615, 1999.
169. X Cheng, S Kumar, J Posfai, JW Pflugrath, RJ Roberts. Cell 74:299–307, 1993.
170. M O'Gara, S Klimasauskas, RJ Roberts, X Cheng. J. Mol. Biol. 261:634–645, 1996.
171. JM Brockman, AG Frutos, RM Corn. J. Am. Chem. Soc. 120:8044–8051, 1999.

172. LM Adleman. Science 266:1021–1024, 1994.
173. AG Frutos, Q Liu, AJ Thiel, AM Sanner, AE Condon, LM Smith, RM Corn. Nucleic Acids Res. 25:4748–4757, 1997.
174. Q Liu, AG Frutos, AJ Thiel, RM Corn, LM Smith. J. Comput. Biol. 5:269–278, 1998.
175. LM Smith, RM Corn, AE Condon, MG Lagally, AG Frutos, Q Liu, AJ Thiel. J. Comput. Biol. 5:255–267, 1998.
176. AG Frutos, LM Smith, RM Corn. J. Am. Chem. Soc. 120:10277–10282, 1999.
177. N Broude, T Sano, CL Smith, CR Cantor. Proc. Natl. Acad. Sci. USA 91:3072–3076, 1994.
178. Q Liu, L Wang, AG Frutos, AE Condon, RM Corn, LM Smith. Nature 403:175–179, 2000.
179. M Ogihara, A Ray. Nature 403:143–144, 2000.
180. TH LaBean, H Yan, J Kopatsch, F Liu, E Winfree, JH Reif, NC Seeman. J. Am. Chem. Soc. 122:1848–1860, 2000.
181. C Mao, TH LaBean, JH Reif, NC Seeman. Nature 407:493–496, 2000.
182. J Woo, ST Sigurdsson, PB Hopkins. J. Am. Chem. Soc. 115:3407–3415, 1993.
183. SL Daubendiek, K Ryan, ETJ Kool. J. Am. Chem. Soc. 117:7818–7819, 1995.
184. M Frieden, E Pedroso, ET Kool. Angew. Chemie 38:3654–3657, 1999.
185. PE Nielsen. Annu. Rev. Biophys. Biomol. Struct. 24:167–183, 1995.
186. M Egli. Angew. Chem. Int. Ed. 35:1894–1909, 1996.
187. ME Parks, EE Baird, PB Dervan. J. Am. Chem. Soc. 118:6153–6159, 1996.
188. JW Trauger, EE Baird, PB Dervan. Nature 382:559–561, 1996.
189. KC Nicolaou, K Ajito, H Komatsu, BM Smith, T Li, MG Egan, L Gomez-Paloma. Angew. Chem. Int. Ed. Engl. 34:576–578, 1995; Angew. Chem. Int. Ed. Engl. 34:576–578, 1995.
190. M Pellegrini, RH Ebright. J. Am. Chem. Soc. 118:5831–5835, 1996.
191. E Delain, A Fourcade, J-C Poulin, A Barbin, D Coulaud, EL Cam, E Paris. Microsc. Microanal. Microstruct. 3:457–470, 1992.
192. HG Hansma, KJ Kim, DE Laney, RA Garcia, M Argaman, MJ Allen, SM Parsons. J. Struct. Biol. 119:99–108, 1997.
193. M Argaman, R Golan, NH Thomson, HG Hansma. Nucl. Acid. Res. 25:4379–4384, 1997.
194. JD Ng, YG Kuznetsov, AJ Malkin, G Keith, R Giege, A McPherson. Nucleic Acid Res. 25:2582–2588, 1997.
195. W-L Shaiu, DD Larson, J Vesenka, E Henderson. Nucleic Acids Res. 21:99–103, 1993.
196. GU Lee, LA Chrisey, RJ Colton. Science 266:771–773, 1994.
197. H Dai, JH Hafner, AG Rinzler, DT Colbert, RE Smalley. Nature 384:147–151, 1996.
198. SS Wong, E Joselevich, AT Woolley, CL Cheung, CM Lieber. Nature 394:52–55, 1998.
199. SC Tsang, Z Guo, YK Chen, MLH Green, HAO Hill, TW Hambley, PJ Sadler. Angew. Chem. Int. Ed. 36:2198–2200, 1997.
200. F Balvoine, P Schultz, C Richard, V Mallouh, TW Ebbesen, C Mioskowski. Angew. Chem. Int. Ed. 38:1912–1915, 1999.
201. Y Yang, M Kochoyan, P Burgstaller, E Westhof, D Faulhammer, M Famulok. Science 274:1343, 1996.
202. RL Rich, DG Myszka. Curr. Opin. Biotechnol. 11:54–61, 2000.
203. RD Piner, J Zhu, F Xu, S Hong, CA Mirkin. Science 283:661–663, 1999.
204. S Hong, J Zhu, CA Mirkin. Science 286:523–525, 1999.
205. MJ Allen, XF Dong, TE Oneill, P Yau, SC Kowalczykowski, J Gatewood, R Balhorn, EM Bradbury. Biochemistry 32:8390–8396, 1993.
206. D Schüler, R Brimacombe. EMBO J. 7:1509–1513, 1988.

11
Self-Assembled DNA/Polymer Complexes

VLADIMIR S. TRUBETSKOY   Mirus Corporation, Madison, Wisconsin

JON A. WOLFF   University of Wisconsin–Madison, Madison, Wisconsin

I. INTRODUCTION

Studies of DNA-based nanoassemblies are a rapidly growing area of materials science. The reason behind this rapid growth is mostly a purely practical one: creating new materials that exploit unique polymeric properties of the polyanion. DNA is a polymer that serves as a carrier of genetic information in all living organisms. Since the molecular machinery behind transcription, translation, and replication of DNA has been clarified in great detail, the emphasis has shifted to the exploitation of DNA's unique properties as an information carrier and a polymeric molecule in engineered artificial structures. Several areas of materials science would benefit from the development of the corresponding DNA-based technology. For example, DNA strands immobilized on a chip were proposed for design of nanocircuits [1]. A similar approach was proposed for construction of electronic memory devices [2]. Finally, DNA-based materials play important role in the design of gene delivery vehicles for gene therapy [3], an application for which the size and the surface properties of a DNA assembly are of crucial importance.

As we invent new technologies manipulating genetic information, these technologies require new materials with extended properties. Evolving technologies such as genomics and gene therapy require the expansion of the repertoire for the familiar DNA molecule. For example, DNA dendrimers demonstrated increased melting stability at elevated temperatures, which suggests their use in DNA chip technology [4]. Similarly, the preparation of negatively charged nanosize particles of condensed DNA might find unexpected applications as nonviral carriers in gene therapy [5]. Morphology and preparation methodology impose properties on these supramolecular assemblies.

Progress in biology requires corresponding progress in materials science with respect to older biological molecules. Properties of these molecules or supramolecular structures based on these molecules should be stretched to meet the demands of new technologies. One route in this direction involves pursuing new developments in supramolecular assemblies. *Molecular assembly* can be defined as the spontaneous association of molecules into thermodynamically or kinetically stable, structurally well-defined aggregates joined by noncovalent bonds. In fact, native double-stranded DNA can be classified as a supramolecular assembly, since its two strands are connected via hydrogen bonding. DNA helices can interact with other molecules via different types of noncovalent bonds, such as hydrogen and electrostatic. All of these cases will be covered in this review.

II. STRUCTURE OF DNA IN SOLUTION

The DNA structure involves two polyanionic phosphodiester strands linked together by hydrogen bonding of base pairs. The strands can be separated by a denaturation process (melting). The melting temperature increases with an increase in guanine (G)–cytosine (C) content, since this base pair possess three hydrogen bonds as compared to just two for the adenine (A)–thymine (T) pair.

Polymorphism of double helices was studied mainly by x-ray difraction of its humidified crystals (fibers). Several characteristic structures were observed, depending on solution conditions. Native DNA can be found in A, B, and C conformations with 11, 10, and 9.33 nucleotides per turn, respectively [6]. Synthetic sequences can form more exotic structures, such as left-handed Z-DNA. The A and B types are most common with interphosphate distances of 5.9 Å and 7 Å, respectively. Structural differences between these two types are determined by the conformational isomerization of the deoxyribose ring. The diameter of the double helix for the A type is wider, with a 3.5-Å hollow well situated along the helix axis. This hollow cylinder does not exist in B DNA, and its helix axis goes through the base pairs for B DNA. The diameter for B DNA is smaller as compared to the A form (approximately 2 nm). Interphosphate distance (charge density) along the phosphodiester chain is approximately 0.7 nm. However, DNA contains two negative charges per monomer unit (base pair). Transition from B to A form usually takes place with an increase of solution ionic strength and dehydration. In solution, the B form adopts a slightly relaxed conformation, yielding 10.3–10.6 nucleotides per turn (as opposed to precisely 10 nucleotides in DNA crystals), with bases located perpendicular to the helix axis. The bases are tilted slightly toward the axis in the A form. It is believed that the B form is predominant in vivo. However, this might not be true when DNA is complexed with certain ligands (for example, cationic lipids and surfactants).

Topologically, DNA in solutions can be linear, closed circular, or supercoiled. The polymer is semiflexible, which can be described by its persistence length (vaguely defined as the minimal length of a macromolecule that can be bent half a turn at the expense of a kT of elastic energy [7]). The persistence length of double stranded DNA is 50–60 nm in dilute aqueous solutions of low ionic strength [8]. This parameter decreases to approximately 35 nm in high salt concentrations. For comparison, the persistence length of single-stranded DNA is only 1.4 nm [9]. Native DNA in solution can demonstrate both random coil and rigid rodlike conformational properties in aqueous solutions, depending on conditions [10]. Despite this apparent stiffness, DNA can be bent upon electrostatic complexation with cations. As demonstrated recently, simple monovalent cations located in the minor groove mainly along AT-enriched sequences may play a significant role in this process. These places of increased cation density contribute to DNA bending and, eventually, to DNA condensation by counterions [11].

The length of natural DNA may be extremely large. The size of the largest chromosome DNA of *Drosofila* fruitfly is 6.2×10^7 base pairs, representing a single DNA molecule about 2 centimeters long [12]. The size of a human genome is 3×10^9 base pairs, which yields a total DNA length of about 1 meter. However, when solving practical problems of DNA nanochemistry, building blocks can be significantly smaller. For example, constructing complexes for nonviral gene delivery, plasmid DNA of 1–20 kilobases is used as a starting material, which yields 340–7800 nm of fully extended polymer (contour length). Unusual solution conditions, nucleotide sequences, or chemical modifications might result in DNA conformations and structures that differ significantly from the usual A and B forms as well from canonical Watson–Crick base pairing.

III. INTERACTIONS OF DNA MOLECULES

Due to its unique chemical composition and structure, DNA can interact with a plethora of chemical structures via numerous types of bonds. This property ultimately defines the ability of DNA fragments to serve as the building blocks in the complex three-dimensional self-assembled structures. Following we list four major types of polymer/DNA interactions that can lead to formation of supramolecular structures:

1. Hydrogen bonding between DNA bases and a polymer
2. Polymer binding via minor group binding
3. Interactions with neutral and anionic polymers (osmotic stress)
4. Electrostatic interactions

We will briefly discuss the properties of the structures obtained as a result of DNA/polymer interactions, with emphasis on electrostatic ones. The reason for this preference is purely practical, since it has been a major route for the preparation of various DNA structures involved in gene transfer in vitro and, more recently, in vivo.

A. DNA/Polymer Complexes Based on Hydrogen Bonding

This type of DNA/polymer complex includes DNA alone, since both DNA strands are linked via hydrogen bonding. Also included are DNA assemblies containing sequence blocks that do not participate in double-helix formation.

1. Formation of DNA Periodic Arrays Based on Branched Motifs Held Together by Sticky Ends

Branching of DNA molecules occurs naturally during the process of recombination, for example, Holliday junctions [13]. Four-way junctions occurring at double crossover recombination points can serve as starting points for more complex DNA branching structures. The invention of immobile Holliday junctions [14] with the addition of DNA sticky-end technology has allowed the synthesis of DNA cubic networks and other three-dimensional structures of branched DNA [15]. It has been suggested that such types of three-dimensional DNA structures can be used as scaffolds in the design of molecular electronic components [16]. This area of DNA nanotechnology was extensively covered in a recent review by Seeman [17].

2. Triplex and Quadruplex DNA Helices

Interest in the synthesis of poly- and oligonucleotides that can bind double-stranded DNA has been stimulated by the idea of expression control of specific genes at the transcription level [19,20]. Among structural types of sequence-specific duplex-bound agents, the most interesting are linear triplex-forming oligonucleotides that bind in the major groove [18] and peptide nucleic acids that can locally denature the double helix and form complexes with only one strand [21]. AT and CG base pairs involved in a regular Watson–Crick pairing can form triplets T/A/T and C+/G/C, with the corresponding nucleotide on the third strand based on hydrogen bonding (Hoogsteen bonding). Cytosines in the third strand should be protonated (in C+/G/C complex), so the whole complex is stable at lower pH. Some nucleotide chemical modifications are required to keep the triplex stable at neutral pH. The third strand occupies its position in DNA's major groove, and the original two strands stay in B form [22]. Nonnucleotide linkers (such as hexakis ethylene glycol) are frequently used to connect the third strand of complementary oligonucleotides to the initial double helix.

Some very unusual nucleotide bonding has been discovered in so-called triplet repeat sequences [23]. Two neighboring bases in one strand complex with a single base of the opposite strand in a complex, which is called triad DNA. This type of bonding requires three bases, but unlike triplexes the three bases are within only two DNA strands.

DNA quadruplexes represent the next level in the hierarchy of polynucleotide nanoassemblies. This area is derived from experiments with guanine-rich sequences, which are found in telomeres. It has been demonstrated that such sequences can form quadruplexes [24]. The core of such structures is a G quartet, a structure with four guanine bases locked in a four-ring cycle in which each base maintains hydrogen bonding with two adjacent bases. Cytosine-rich sequences were also shown to form four-stranded structures that are structurally different from guanine quadruplexes. In a model oligonucleotide d(TCC-CCC), two double strands of C+/C intercalate each other (I motif) [25]. Continuous repeats of guanine quartets were recently shown to bring together two intact DNA duplexes side by side (forming a "synapsis") in physiological conditions [26]. In recent developments, such systems have been refined to allow such "synapsable" duplexes to discriminate between self- and nonself-duplexes [27].

B. Assemblies Based on the Insertion into Sequences of Some Building Blocks that Do Not Participate in Double-Helix Formation

The approach to studying the sequence-specific aggregation of gold nanospheres with small oligonucleotides is aimed at designing DNA-based materials that can be used as sequence-specific sensors. The procedure was developed to attach single-stranded oligonucleotides to the nanosphere surface via mercaptoalkyl linker [28]. The suspension of nanoparticles can then be specifically aggregated, with formation of a macroprecipitate. The formation of the precipitate was observed if nucleotides possessed complementary "sticky" ends. Four bases mismatching in a 12 base-pair overlap was found to be enough to prevent such aggregation. Later this principle was further developed into a method of probing of single-stranded oligonucleotides in solution. The addition of complementary nucleotides precipitated gold particles with immobilized complementary oligonucleotides and changed the solution color from pink to purple [29].

The spatial position of gold nanocrystals linked to the single-stranded oligonucleotides can be controlled entirely by Watson–Crick base-pairing interactions [30]. Smaller (5 nm) and larger (10 nm) gold nanospheres linked to thiolated complementary oligonucleotides were allowed to hybridize in different combinations on the complementary sequences in solution. The synthetic strategies can easily be verified using electron microscopy. Up to 90% of resulted assemblies were found in the correct combinations.

Streptavidin–single-stranded DNA covalent conjugates were described as the building blocks for assembling nanostructured scaffolds [31]. The amount and type of biotinylated ligands were used to modulate the affinity of duplex formation between solid-phase-bound nucleic acid templates and DNA–streptavidin conjugates. This system has been proposed for the design of "fine-tuned" sequence detection systems.

Shchepinov et al. have recently provided another interesting example of DNA nanoassemblies based on noncomplementary inserts within oligonucleotide dendrimers [4]. DNA dendrimers with 2, 3, 6, 9, and 27 arms were synthesized, thus achieving DNA nonlinearity. Doubling and trebling inserts based on pentaethyleneglycol phosphoramidites were used as branching points. These structures demonstrated unusually high melting stabilities and were suggested for use in oligonucleotide array/DNA chip technology.

In another example of DNA-aided design of nanostructures, closed circular plasmid DNA was used as a template to form a string of cadmium sulfide (CdS) nanocrystals, which are used in the fabrication of semiconductors [32]. High-resolution electron microscopy and electron diffraction demonstrated formation of cadmium sulfide nanocrystals nucleated along the DNA backbone. In this case DNA served only as a physical template for the nucleation of crystallization. Possible applications involve fabrication of quantum wires, rings, and other semiconductor elements. A further modification of this method entails the surface of CdS nanoparticles being cationized with thiocholine to ensure electrostatic interaction with DNA [33]. The semiconductor nanocrystals were found to form quasi-one-dimensional tight packing along the DNA backbone, with an average distance of 3.5 nm between crystal centers, which corresponds to the length of 10 base pairs.

A somewhat similar approach has been used for the formation of nanosize wires stretching between gold electrodes [34]. Lambda-DNA was positioned between two electrodes, with immobilized oligonucleotides complementary to lambda-DNA' sticky ends. Silver (Ag^+) ions were deposited on the stretched DNA "bridges," followed by reduction of absorbed ions to metallic silver with hydroquinone. The resulting silver clusters formed on DNA strands were found to be 100 nm in diameter and were capable of conducting the electric current.

C. DNA Assemblies Based on Minor Group Binding

Certain compounds can specifically interact with DNA by fitting into its minor groove via a combination of van der Waals, hydrogen, and electrostatic forces. Such compounds usually possess antitumor and antimicrobial activity [35]. This is still a realm of small molecules; however, some initial attempts to include polymeric conjugates into such interactions have been reported. For example, a covalent conjugate between minor groove binder and 3' terminus of single-stranded oligonucleotide was described [36]. The minor groove binding motif included three repeating 1,2-dihydro-3*H*-pyrrolo(2,3-*e*)indole-7-carboxylate units that were linked to octatymidylate, $(dT)_8$. The conjugate upon recombination with complementary chain yielded a duplex with increased melting stability as compared with nonconjugated duplex, apparently due to the "clamping" of complementary oligos in the minor groove binding part of the conjugate.

D. Assemblies Based on the DNA Condensation Phenomenon

DNA can undergo coil–globule transition when treated with a variety of agents. In terms of polymer science, this process can be formalized as an extension of classic Flory–Huggins theory of polymer precipitation in poor solvents [37]. In terms of biological significance, this phenomenon offers a unique opportunity to compact the voluminous polymer in order to fit it into space-limited compartments of cell nuclei and viral particles. Reduction of DNA volume upon condensation in vivo usually can reach three to six orders of magnitude [7]. Cationic polypeptides and low-molecular-weight polyamines serve as condensing reagents in vivo. However, cations are not the only agents that cause DNA collapse. Any agent capable of modification of DNA segment/segment or segment/solvent interactions can potentially serve as a condensing agent [38]. Major classes of DNA condensing agents are: (1) agents that lower solvent dielectric constant, (2) osmotic stress (crowding) agents, (3) multivalent cations (both low molecular weight and polymeric). Most condensing agents that lower the dielectric permeability of water are small molecule solvents (for example, alcohols). Consequently, this type of condensation is not the subject of the present review. Alternatively, many agents of the second and the third cases are polymers, and the products of their interactions with DNA form structured nanoassemblies.

E. Structures Based on DNA Condensation Under Osmotic Stress

This type of DNA condensation can be classified only formally as a product of DNA/polymer interactions, since no binding between these two components has been observed. Polymers that cause DNA condensation serve in this case as phase separation agents and concentrate DNA in the aqueous phase in high concentration. The presence of a certain amount of salt is required to overcome phosphate repulsion.

Interaction of DNA with neutral or negatively charged polymers in aqueous solutions results in phase separation and the formation of condensed DNA phases. Structures based on these interactions might be interesting from the point of view of DNA nanoassemblies. Since osmotic stress is experimentally caused by adding a neutral polymer to a DNA solution, the resulting structures can be formally classified as products of the DNA/polymer interaction. A typical example of such interactions is DNA precipitation from salt-containing solutions upon addition of polyethyleneglycol (PEG). This process historically was termed psi-condensation (from polymer and salt) [39]. In this case neutral polymers exert osmotic stress on DNA helices, effectively "squeezing" them into a separate and highly ordered phase. X-ray scattering of condensed DNA fibers showed a long-range positional order of helices at higher osmotic pressures (hexagonal columnar liquid crystals) that gradually disappears when the osmotic pressure drops [10]. It is interesting that at high osmotic pressures (above 160 atm), DNA is forced to adopt an A conformation. The A–B transition was observed when the pressure is decreased [40]. The use of negatively charged polymers as crowding agents also has been reported [for example, with poly(glutamic acid)] [7].

F. DNA Assemblies Based on Counterion Condensation

Neutralization of DNA phosphates with certain cations leads to polymer collapse and consequent condensation (coil–globule transition in terms of polymer physicochemistry). It has been shown that in aqueous media any cation with a valence greater than 3^+ (for example, such low-molecular-weight molecules as spermidine and spermine) can cause condensation [38]. In water/alcohol mixtures, due to decreased dielectric constant, cations with charge as low as 2^+ can effectively condense DNA [41]. Electrostatic condensation is responsible for DNA condensation in vivo via its complexation with positively charged proteins and polyamines. This type of condensation is responsible for DNA packing inside cell nuclei and virions and yields remarkable polymer density. For example, DNA density in the head of T7 bacteriophage is increased by a factor of 10^4 as compared with free polyanion in solution, which results in a DNA concentration of 450 mg/mL for encapsulated polymer (Fig. 1) [42]. Upon condensation, DNA adopts a set of characteristic morphologies, some of which can be observed in nature [43]. More important for materials science, such structures can be reproducibly formed in dilute DNA solutions in vitro.

1. Mechanisms of Counterion Condensation

It has been proposed that a combination of various molecular forces is responsible for the phenomenon of condensation [38]. Probably, electrostatic forces play the most important role in the screening of phosphate repulsion by counterions, thus helping in the collapse of

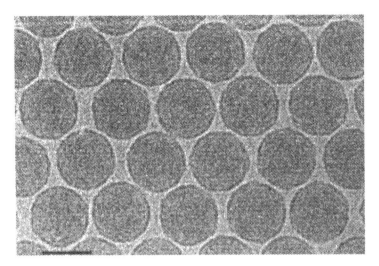

FIG. 1 Monolayer of T7 phage heads. Cryoelectron micrograph of tailless mutant of T7 bacterio-phage. Bar: 50 nm. (From Ref. 42. Copyright © 1997 Cell Press.)

the helices. During titration of T7 bacteriophage DNA with spermidine it was found that condensation occurs when 90% of the phosphate charges were neutralized by the counterion [41]. This value ideally coincides with the theoretical prediction derived from Manning's theory of counterion interactions [44]. Later this value was directly measured using electrophoresis of DNA plasmids in the presence of various cations [45]. It is important to stress that DNA collapse is a consequence of charge neutralization and not of counterion binding to DNA. Another important factor to keep in mind is that the DNA/cation complex completely condenses at the $+/-$ charge ratio of 0.9, and there is excess of net negative charge in such complexes. This charge excess, as will be demonstrated later, can be utilized for the self-assembly of multilayered polyion structures based on DNA.

Multivalent cations not only screen negatively charged phosphates but also induce short-range attraction between DNA strands via correlated fluctuations of the ion atmosphere, thus collapsing the whole polymer [46–47]. The concept of ionic cross-bridging of multivalent cations between neighboring phosphates as the cause of DNA condensation represents an approach that is very close to the theory of correlated fluctuations [48–50]. Another widely accepted point of view in this area is that mostly long-range hydration forces are responsible for DNA collapse [51]. It has been noted that DNA condensation can be explained by a combination of the preceding mechanisms [38].

An attempt to quantify DNA condensation has been undertaken by Manning [44]. The theory formally describes the interactions of a polyelectrolyte with counterions in solution as approximation of point charges beaded on a semiflexible string. This approach enabled the calculation of the counterion concentration necessary to cause the coil/globule transition. In general, such a transition was found to be dependent on the counterion valence, the dielectric susceptibility of the medium, and the polyelectrolyte linear charge density in the absence of counterions. Spontaneous collapse of the polyion chain occurs upon reaching a certain threshold charge density upon titration with counterions. However, when applied to DNA, the theory predicts that monovalent cations (such as Na^+) can reduce DNA linear charge density in water by 76%, divalent cations by 88%, and trivalent (e.g., spermidine, see earlier) by 90%. Interestingly, divalent cations condense DNA in water/methanol (50%/50%) mixtures, where they neutralize up to 91% of the phosphate charge [41]. A refinement of the existing theory was recently published by Manning and Ray [52].

2.   A New Method for Assessing DNA Condensation

There was a need to develope a fast, simple, and reliable method to monitor DNA condensation. Previously used methods for detection of DNA condensation, which included electron microscopy [53], static and dynamic light scattering [41], circular dichroism [54], fluorescent microscopy [55], and intercalation of fluorescent probes [56–57] were laborious and qualitatively imprecise. The method of measuring the changes in ethidium bromide fluorescence upon interaction with condensing agents has been especially popular. However, its quantitative results are known to be dependent on the type of condensing agent [58]. Moreover, the degree of fluorescence of the ethidium bromide/DNA complex is strongly dependent on the structures of condensing cations, thus impeding the titration of the DNA with multivalent cations.

We have developed a quantitative method for assessing condensation based on the concentration-dependent self-quenching of fluorescent moieties covalently attached to DNA [59]. For this purpose, the plasmid DNA was labeled with fluorescein and rhodamine LabelIT reagents developed by Mirus Corporation, which allow direct covalent binding of fluorophores to the DNA backbone. As discussed earlier, DNA upon condensation undergoes a coil–globule transition accompanied by a dramatic decrease in DNA dimensions that leads to a significant increase in the concentration of fluorescent groups attached to the DNA. Such crowding of fluorophores is known to lead to a substantial decrease in fluorescent intensity (quenching). Extremely precise determinations of 100% DNA condensation point were performed using fluorescein- and rhodamine-labeled DNA and a variety of multivalent cations [59]. This technique also allowed us some deeper insights into the mechanism of DNA condensation caused by different cationic agents. Using mixtures of labeled and unlabeled DNA molecules, it was possible to determine the relative number of DNA molecules that are intimately condensed together in one particle. It was found that the extent of quenching for spermine-induced condensation was related to the percentage of labeled DNA, whereas the quenching for polylysine-induced quenching was not affected by the percentage of labeled DNA. This finding suggests that several DNA molecules are condensed together by spermine, while only one or two DNA molecules are condensed together by polylysine.

3.   Structure of Cation-Condensed DNA

When dilute aqueous solutions of DNA and counterion are mixed together, DNA forms particles with remarkable toroid and/or rod morphology. This most interesting aspect of the DNA condensation phenomenon relates to nanotechnology, since the dimensions of such structures are generally below the 100-nm range. DNA toroids were first observed upon condensation with poly-L-lysine (PLL) [60] and spermidine [61]. Spermidine-condensed DNA formed perfect toroid particles with an 80-nm diameter. It was noted by many researchers that treating DNA with different condensing agents results in the formation of particles of nearly the same morphology and dimensions. The same structures were found upon gentle rupture of T7 bacteriophage heads [42], implying that the same mechanism of DNA condensation exists in biological systems. Rods also can be found among morphologies of counterion-condensed DNA with dimensions similar to toroids. Bloomfield believes that the deviation from toroidal morphology rests in the nonpolarity of the solvent or condensing agent [62]. However, recent the demonstration of toroidal particles formed in nonpolar solvents may argue against this hypothesis [63]. Other authors claim that the difference in DNA length is responsible for the prevalence of either toroids or rods. Marquet et al. have found that spermine-condensed linear DNA longer than 630 base pairs (bp) condensed mainly into toroid particles, whereas shorter DNA preferentially formed rods. A minimum DNA length of 630 bp was found to be sufficient to nucleate the growth of toroidal particles [64]. Similar conclusions were made from the study of (Co (NH3))$^{+3}$–condensed DNA [65]. Recent new work detailing mechanisms of toroid formation sheds additional light on the formation of toroids and rods. Attempts to "freeze" structures intermediate between toroids and rods were undertaken in recent work on the structure of condensed DNA using atomic force microscopy [66]. The authors claim that rod morphology is primary and toroids are formed by the opening up of the rods. Others, however, claim that toroids are formed by blunt-end circularization of rods [67,68]. This claim is substantiated by the fact that rod length and toroid circumference are usually close. Computer-aided simulation of the folding of stiff polymers demonstrated that toroid configuration is the most stable one [68]. Rods and toroid–rod intermediate structures were also found among metastable states in this work.

It has been found experimentally that not only DNA but any other polyelectrolyte collapses into compact structures upon charge neutralization. However, the stiffness of polymer chain in solution seems to dictate the morphology of the resulting particles. Spheroids were found to be the major type of structure upon charge neutralization of the polyelectrolytes other than DNA. Electron microscopic images of nanosize spheroids with an average diameter of 20 nm were reported [69] as a result of neutralization of polyanion poly(styrenesulfate) with polycation poly(diallyldimethylammonium chloride). Unlike DNA toroids, these particles exhibited a large degree of size heterogeneity.

Similar to the situation with DNA structures formed under osmotic stress, DNA strands in cation-condensed bundles were found to be hexagonally packed and to possess liquid crystalline order. For example, spermine and spermidine-condensed samples were found to contain a cholesteric phase [70]. Surprisingly, DNA condensed with $(Co(NH_3)_6)^{+3}$ failed to exhibit a liquid crystalline ordering [47].

4. Nanoparticles Containing Condensed DNA: Morphology and Formation

Toroids and rods are the most frequently observed shapes in specimens of condensed DNA. The observations were performed using electron microscopy [53,71] and, more recently, atomic force microscopy [72]. On most images, toroidal structures with a diameter of 40–60 nm are clearly visible. The diameter and the length of the rods usually coincide with the thickness and the circumference of the toroids [68]. It was shown that DNA is circumferentially wrapped in torus particles [62,73]. One particle can contain up to 60 kbp of DNA [74]. If DNA used for particle preparation is short enough, the single toroid may contain multiple DNA molecules. The kinetics of DNA condensation by spermine and spermidine was studied in detail using stopped-flow light-scattering methods [75]. It has been shown that upon mixing, at least two phases of DNA condensation are exhibited: the fast stage takes place in the millisecond range, which can be attributed to intramolecular condensation, and the slow stage (tens of seconds) is recognized as intermolecular DNA association. Apparently these processes are concentration dependent. It was concluded that pure monomolecular condensation with low-molecular-weight polyamine can be registered only at DNA concentrations less than 0.2 µg/mL [76].

Morphologies other than toroids and rods (mostly fibrous structures) can be observed using electron microscopy in alcohol/water mixtures. Under these conditions, DNA frequently loses its B-form conformation (secondary structure) and adopts the A form [77]. The formation of toroidal particles was demonstrated directly using very large DNA molecules fluorescently labeled with intercalating dyes. Melnikov et al. used T4 bacteriophage DNA labeled with 4,6-diamidino-2-phenylindole intercalator dye with a contour length of several tens of microns. This enabled the visualization of the coil–globule transition upon the addition of cations by fluorescent microscopy [55]. Brewer et al. used lambda-phage DNA attached to a 1-μm latex bead held steady by an optical trap in the stream of a solution containing different polycations. When complexed with hexa-arginine, an oligocation, the whole complex decondensed in a reasonable time by merely immersing the bead into the solution without the cation [78].

Though in general DNA/cationic lipid particles have a tendency to form structures significantly different from the usual DNA/cation complex (see later), under some conditions condensed DNA particles of toroidal morphology can also be found upon interaction with cationic lipids. Such particles, with a diameter of approximately 70 nm, were found recently by Xu et al. These particles were prepared with cationic lipid DOTAP at a high lipid/DNA ratio and separated from excess lipid by ultracentrifugation in sucrose gradient [79]. DNA inside these particles was completely condensed, as judged from ethidium bromide exclusion and nuclease protection studies.

The structures of PLL/DNA complexes were studied in detail in [80] using atomic force microscopy (AFM). The structural difference between particles obtained with an excess and a deficiency of polycation was clearly demonstrated. Completely condensed spheroids with a diameter of less than 100 nm were obtained upon condensation of DNA with PLL (molecular weight 10 kDa) at a lysine/nucleotide ratio of 2 (Fig. 2c). When PLL was added to DNA at a lysine/nucleotide ratio of 0.5, the particles demonstrated typical "daisy-shaped" structures (Fig. 2b), with condensed core and uncondensed DNA strands forming stabilizing corona. The existence of such structures was also confirmed in Ref. 72. It is interesting to note that condensed structures appeared in AFM images as spheroids and not as toroids, with some residues of uncondensed DNA (Fig. 2c, arrows). The presence of uncondensed material may be explained by interactions with negatively charged mica (used in AFM studies as a substrate), which can cause partial decondensation. In this work toroids were seen only when a conjugate of PLL with protein asialo-orosomucoid (designed for targeted gene delivery) was used for DNA condensation (Fig. 2d). Possibly, additional interactions of protein component of the polycation with mica substrate help to stabilize toroidal structures.

As has been reiterated in many studies, the morphology of condensed DNA structures is dependent only slightly, if at all, on the type of multivalent cation used for condensation. This includes the situation with several natural cationic proteins. For example, histones from different species and protamine were used to condense a number of different DNA molecules (linear, supercoiled plasmid, closed circular relaxed, size range 500–5243 bp) [81]. Results indicated that, regardless of polycation or DNA type, all complexes exhibited toroid and rod morphology of similar dimensions. The only parameter to have a major influence on the size of resulting structures was ionic strength: the thickness of the average rods and toroids was observed to increase from 16.5 nm to 29 nm when ionic strength was increased from 1.5 mM to 100 mM NaCl. The authors concluded that the morphology and dimensions of condensed DNA particles is determined mainly by the balance of electrostatic and van der Waals forces, with a minimal contribution of specific interactions.

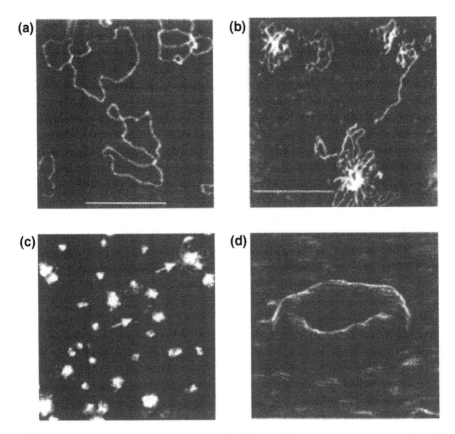

FIG. 2 Atomic force microscopy of plasmid DNA at different stages of condensation with polycations. (a) Circular plasmid DNA; (b) DNA condensed with poly-L-lysine (PLL, mol. wt. 4 kDa) at a
DNA phosphate/lysine ratio of 2:1; (c) DNA condensed with PLL (mol. wt. 10 kDa) at a phos-
phate/lysine ratio of 1:2; (d) toroid of DNA condensed with PLL-asialo-orosomucoid conjugate at a
phosphate/lysine ratio of 1:6. [(a) and (d) from Ref. 80, copyright © 1998 Oxford University Press;
(b) and (c) reprinted with permission from Ref. 66, copyright © 1999 American Chemical Society.]

Interestingly, the morphology of DNA/cation particles was found not to be dependent on the charge density of DNA. Benbasat [82] studied the condensation of bacteriophage phi W14 DNA, which contains the positively charged base α-putrescinylthymine in
its sequence and hence possesses decreased charge density. Average interphosphate spacing were increased to 2.2 Å, as compared with 1.7 Å for regular B DNA. The modified
DNA required significantly higher spermidine concentrations to achieve complete condensation, in agreement with predictions from Manning's theory [44]. However, toroid dimensions differed only slightly from the ones prepared with regular B DNA. The chirality
of the condensing agent also seems to be important. Reich et al. [83] reported that poly(L-
lysine) binds and condenses DNA preferably as compared with poly(D-lysine) and poly(L-
ornithine), a PLL homolog.

5. Stability of DNA-Based Nanoassemblies

The stability of the complexes based on DNA condensation is a legitimate issue for
nanoassemblies based on DNA. Under in vivo conditions, DNA as a constituent of a gene

delivery vector would encounter a number of conditions adverse to its structural stability, such as an extreme pH or the presence of nucleases. Under natural conditions, it has been demonstrated that complexation with polyamines stabilizes DNA against a variety of denaturing conditions, including temperature denaturation [84], shear stress [85], and nuclease degradation [86]. Evidently, specific mechanisms may play a role in each of these stabilization mechanisms. Generally, however, entropic changes derived from the release of small ions upon polyamine/DNA interaction are responsible for stabilization of the whole complex [87].

Since DNA assemblies based on condensation form mostly nanosize particles (lyophilic colloids), it is colloid stability that is important for maintaining their integrity in an aqueous environment. Researchers usually did not discriminate between DNA condensation and DNA precipitation. It was a common point of view that all condensed DNA species precipitate from aqueous solutions at reasonable DNA concentrations [38]. Ultrasensitive static light-scattering methods were able to detect monomolecular DNA condensation at concentrations less than 0.2 µg/mL. However, at DNA concentrations greater than 0.5 µg/mL, this stage was indistinguishable from total aggregation [76].

Two major types of stabilization mechanisms are described for submicron particles: (1) charge stabilization, where surface charge forms a repulsive screen that prevents the particles from flocculation, and (2) steric stabilization, where a surface repulsive screen is formed by solvent-compatible flexible polymeric chains attached to the particle's surface.

For charged systems the solution stability of dispersed colloid suspensions is described by Derjaguin–Landau–Verwey–Overbeek (DLVO) theory [88]. This theory quantitatively characterizes the balance between repulsive electrostatic and attractive van der Waals forces in maintaining colloid particles in the dispersed state. As predicted by DLVO theory, low-molecular-weight salt decreases the stability of charged colloids, which flocculate upon the addition of the aforementioned threshold concentrations of salt (critical coagulation concentration). Simple mixing of DNA and cation aqueous solutions leads to DNA condensation and particle formation. However, the stability of such particles is strictly dependent on the input charge ratio, the solution conditions, and the properties of polycation (its charge density in particular). The behavior of DNA/cation aqueous mixtures can be described by the phase diagrams similar to the ones demonstrated by Lasic for DNA/cationic lipid mixtures [7]. DNA/cation complexes usually flocculate (precipitate) in the area of electroneutrality ($-/+$ ratio $= 1$). However, the extent of such flocculation is dependent on the DNA and salt concentrations in the system. When the DNA is present in excess ($-/+$ ratio > 1), DNA is partially condensed and the particles are usually stable. The same is true in some specific conditions when cation is in excess ($-/+$ ratio < 1). However, the flocculation area (range of charge ratios) around electroneutrality strictly depends on the nature of the cation and on solution conditions. For example, it requires 1 mM spermidine to completely precipitate DNA at 1 mg/mL (in 25 mM NaCl solution), but 100 mM spermidine takes it back into solution [47]. In the case of polycations, the flocculation range can be substantially more narrow, especially at lower DNA and salt concentrations [89]. Surface-charge reversal and consequent peptization is a likely mechanism for the resolubilization of flocculated condensed DNA material. However, there is no consensus among authors on this point. Pelta et al., for example, believe that cation-driven resolubilization is due to the screening of short-range electrostatic attraction [47].

The charge on the surface of colloid particles is an important parameter, and DNA/cation self-assembled complexes are no exception. It can be measured experimentally as the ζ-potential or electrokinetic potential (the potential at the surface of shear between charged surface and electrolyte solution). The zeta-potential is considered to be an experimental approximation of Stern layer potential, especially for lyophobic colloids and at low and moderate ionic strengths [88]. Upon titration of DNA with an increasing amount of polycation and consequent DNA condensation, the ζ-potential of the particles formed reverses its charge from negative to positive [90] reflecting the fact that a certain excess of polycation is present on the surface of condensed DNA particles. This excess and, hence, the numerical value of the ζ-potential is of prime importance for the colloid stability of the DNA complexes.

6. Stoichiometry of Polyelectrolyte Complexes

Stoichiometry of polyelectrolyte complexes has been studied in great detail for a number of years, employing different systems of synthetic polyelectrolytes. In the vast majority of polyacid/polybase pairs, a 1:1 charge ratio was found in complex formation. However, significant deviations from 1:1 stoichiometry have been detected under several circumstances: (1) extreme differences in component charge density; (2) inaccessibility of countercharges to interact, due to chain branching; (3) pH sensitivity of either polyion [91]. Polymer chemists studying polyelectrolyte complexes usually believe that 1:1 stoichiometric complexes are insoluble, admitting thought that at low concentrations macroscopic phase separation might not occur and stable quasi-soluble colloidal particles can be formed [69].

We have studied stoichiometry of complex formation between DNA and PLL and found that the complexes formed in low-salt-buffer solutions are of a 1:1 charge ratio [92]. The same 1:1 stoichiometry was found experimentally for DNA complexed with other synthetic polycations of different nature in low-salt aqueous solutions [such as poly (diallyldimethylammonium chloride), poly(dimethylimino)ethylene(dimethylimino)ethylene-1,4-dimethylphenylmethyl ene dichloride, and poly(4-vinyl-N-methylpyridinium bromide)] [93].

Nonhomogenous distribution of charge inside fully condensed DNA/polycation particles has been discussed earlier, in Section III.F.1. Since it takes only 0.9 equivalents of positive charges to condense DNA in aqueous environment, there should be an excess of positive charge on the surface of completely condensed DNA/polycation particles. This fact opens the opportunity to form multilayered structures based on alternations of polyelectrolytes of opposite charge. Details of this approach of DNA/polyion self-assembly are discussed in Section III.F.9.

7. Methods of DNA Particle Assembly

It has been acknowledged in the literature that the composition of DNA/polycation self-assembled complexes does not coincide with the composition of the initial mixture [89]. Specifically, when polycation is present in excess, DNA usually binds polycation in a 1:1 charge ratio, leaving the rest of the polycation free in solution. As we have shown, this 1:1 DNA/PLL (Mw = 34 kDa) complex can be isolated from the initial input mixture by ultracentrifugation in a density gradient [92]. Interestingly, using the fluorescence quenching method, the excess of polycation has been shown to be in dynamic equilibrium with the polycation in complex with DNA [92]. This dynamic equilibrium is required for maintaining the complexes in a dispersed state: The separation from an excess of the polycation led to flocculation of the 1:1 DNA/PLL complex. Apparently, the whole process is dependent on PLL molecular weight and low-molecular-weight salt concentration. Generally, PLL of higher molecular weight can keep condensed DNA in solution at higher salt concentrations. It is necessary to note that there is some inconsistency in understanding the term *insoluble polyelecrolyte complex.* Most polymer chemists consider stable colloid 1:1 (stoichiometrical) complexes insoluble and distinguish them from truly soluble nonstoichiometric "soluble symplexes," which can maintain their solubility at high polymer concentrations (see, for example, Ref. 92). Here we would consider stoichiometric DNA complexes soluble as long as they are not macroscopically flocculated.

As has been mentioned earlier, DNA/polycation particles formed with an excess of polycation are positively charged and possess positive electrokinetic (ζ) potential. However, low-molecular-weight polycations (for example, spermine) upon complexation with DNa fail to stabilize condensed DNA globules, thus causing its precipitation [89]. As judged from ζ-potential measurements, these short cations fail to render DNA particles positively charged [90]. These experiments, however, were carried out in a concentration range below DNA resolubilization described in Ref. 47 and hence in DNA precipitation conditions (+/− charge ratio = 70). Our measurements of the ζ-potential of DNA/spermine complexes in resolubilization conditions (+/− charge ratio = 2500) demonstrated slightly positive (+8 mV) value. Upon addition of cations, the surface charge approached zero, causing flocculation. Thus, the addition of relatively long cationic molecules or small molecules in conditions of great charge excess can make condensed DNA particles positively charged, thus stabilizing them in low-salt aqueous solutions. Tang and Szoka have found that pentalysine (charge +5) can achieve positive ζ-potentials of DNA/cation complexes at reasonable +/− charge ratio (around 50) [90]. Increasing the molecular weight of PLL can decrease the polycation concentration required for positive surface charge necessary for the particle stabilization. Surface positive charge necessary to stabilize DNA/polycation particles of fully condensed DNA can conceivably be generated by constant exchange of free and DNA-complexed polycations and looping of polycations on the surface. The presence of increased salt concentrations apparently increases the probability of dissociation of significant segments of polycations from condensed DNA particles, generation of uncompensated negative charges, and cross-bridging of polycations to neighboring particles, resulting in complete flocculation. Thus, the addition of an excess of polycation to DNA aqueous solution in the absence of flocculating salt concentrations leads to the formation of a colloidally stable suspension of condensed DNA particles.

8.   Condensed DNA Particle Assembly Using Template Polymerization of Counterions

As mentioned earlier, low-molecular-weight cations with a valence under +3 fail to condense DNA in aqueous solutions under normal conditions. Among "special" conditions it is worth mentioning that divalent cations can condense DNA in water/alcohol mixtures where the dielectric constant is lower [94]. Supercoiled (but not relaxed or linearized) plasmid DNA can be condensed by Mn^{+2} cation in water [95]. However, in normal solvent conditions, cationic molecules with a charge under +3 can be polymerized in the presence of DNA and the resulting polymers can cause DNA condensation into compact structures. Such approach is known in synthetic polymer chemistry as a template polymerization [96]. During this process, monomers (which are initially weakly associated with the template) are positioned along template's backbone, thus promoting their polymerization. Weak elecrostatic association of the nascent polymer and the template becomes stronger with the polymer's chain growth. Trubetskoy et al. used two types of polymerization reactions to achieve DNA condensation: step polymerization and chain polymerization [97] (Fig. 3). Bis(2-aminoethyl)-1,3-propanediamine (AEPD), a tetramine with approximately 2.5 positive charges per molecule at pH 8, was polymerized in the presence of plasmid DNA using

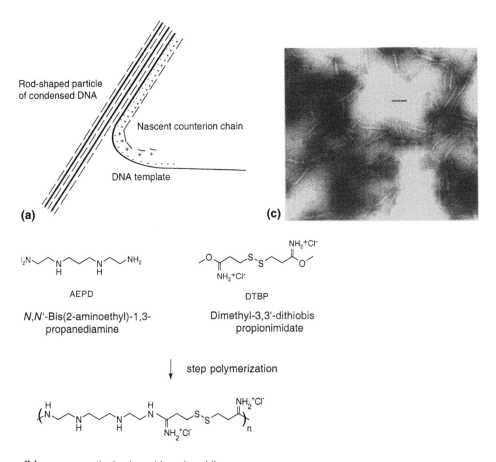

FIG. 3 Condensed DNA particle assembly using template polymerization of cations. (a) Principle of DNA condensation by counterion polymerization on DNA template; (b) chemistry of polycation formation; (c) e microscopy of condensed DNA particles, bar = 100 nm. [(c) was reprinted from Ref. 97, copyright © 1998 Oxford University Press.]

disulfide-cleavable amino-reactive cross-linkers dithiobis(succinimidyl propionate) and dimethyl-3,3′-dithiobispropionimidate. Both reactions yielded DNA/polymer complexes with significant retardation in agarose electrophoresis gels, demonstrating significant binding and DNA condensation. Treatment of the polymerized complexes with 100 mM dithiothreitol (DTT) resulted in the pDNA returning to its normal supercoiled position following electrophoresis, thus proving cleavage of the backbone of the condensing polymer.

The template-dependent polymerization process was also tested using a 14-mer peptide encoding the nuclear localizing signal (NLS) of SV40 T antigen (CGYGP-KKKRKVGGC) as a cationic "macromonomer." Cysteine residues on the termini were cross-linked with the sulfhydryl-reactive homobifunctional cross-linker 1,4-di(3′,2′-pyridyldithio-(propionamido)butane) (DPDPB). Analysis of this reaction mixture using protein gel electrophoresis in nonreducing conditions revealed a ladder of bands starting from peptide dimers and ranging up to multimers of 10–20 peptides in length. Substantially longer polymers were obtained using chain polymerization of cationic acrylates in the presence of DNA.

Our experiments indicated that template DNA remained intact and functionally active after template polymerization. The luciferase-encoded plasmid pCI Luc recovered from the reduced complexes after dialysis was able to express luciferase at levels comparable to pCILuc that had not undergone template polymerization but was also reduced and dialyzed.

DNA condensation was monitored using dimeric cyanine dye (TOTO), which is able to fluoresce only upon intercalation between stacked nucleotide bases. Upon DNA condensation the dye usually dissociates from DNA and is no longer fluorescent. When the AEPD was polymerized with DTBP in the presence of DNA, complete inhibition of TOTO intercalation was achieved when using molar ratios of AEPD to DNA base of greater than 20:1. When the DNA was added one hour after mixing of AEPD and DTBP, TOTO protection was substantially reduced, providing additional evidence for the template dependence of polymerization.

Usually the decrease in TOTO signal during template step polymerization was accompanied by increases in scattered-light intensity, indicating DNA particle flocculation. As expected, rates of DNA condensation and particle flocculation were found to be dependent on the DNA concentration and DNA/monomer ratio. The addition of 20 mM glycine (a monofunctional amine) at early times arrested polymerization and substantially attenuated the decrease in TOTO signal and turbidity. Formation of small individual particles of condensed DNA was detected using dynamic light scattering during the course of template polymerization prior to the formation of large aggregates. It was concluded that cationic chain growth on template DNA results in condensed DNA particle formation that can be modeled as occurring in three phases.

During the initial phase of polymerization, less than 90% of the DNA's negative charge was neutralized. Short segments of growing polycations forms particles with a centralized globular core and a halo of uncondensed DNA strands ("daisy"-shaped particles [72]). In the second phase, the DNA was almost completely condensed and the ζ-potential was lower than -10 mV. Electron microscopy indicated that the particles formed during this phase were regular toroids and rods, which are similar to the condensed DNA particles obtained by the addition of a charge excess of preformed polycations to DNA. Dynamic light scattering analysis indicated that the majority of the particles were relatively small (less than 150 nm) and nonaggregated. In the third phase, particle sizing indicated that the particles started to form aggregates. The ζ-potential of the particles remained near neutral, which explains the aggregation according to colloid stabilization theory. A positive ζ-potential from an excess of cations would have charge-stabilized the particles and prevented the aggregation, as occurs when a charge excess of polycation is mixed in low-salt aqueous solutions with DNA. However, template-assisted chain growth is limited by the availability of the free template. Thus when, at later stages of the template process, all of the DNA is covered with cationic polymer and condensed, extensive aggregation of DNA particles occurs as stabilizing excess of free DNA is consumed by growing chains of polymeric counterion.

Interestingly, this behavior of the reaction mixture can be prevented by employing another principle of particle stabilization: steric protection. Inclusion of pegylated comonomer (PEG-AEPD) into the reaction mixture did enable the formation of nonaggregating DNA particles. It also caused the particles to form "worm"-like structures (as judged by transmission electron microscopy) that have previously been observed with DNA complexes formed from block copolymers of PLL and PEG [98].

Blessing et al. used a bisthiol derivative of spermine and the reaction of thiol–disulfide exchange to promote chain growth. The presence of DNA accelerated the polymerization reaction, as measured by the rate of disappearance of free thiols in the reaction mixture [99]. Template polymer serving as a catalyst for polymerization is known to be one of the characteristic features of classical template polymerization reaction [100]. Polymorphic toroids with a diameter of 50 nm occasionally flattened out into 100-nm rods were detected by electron microscopy as primary morphological motifs in this case. However, these particles were also found to aggregate in physiological salt solution. In a more recent publication a similar method has been used for template-assisted dimerization of cationic detergents [101]. In this particular case, a DNA-dependent template process converted thiol-containing detergent possessing a high critical micelle concentration into a dimeric lipid-like molecule with apparently low water solubility.

9. "Caging" of Polycation-Condensed DNA Particles

Stability is the issue of prime concern for DNA nanoassemblies based on DNA condensation phenomena designed for practical applications. Unfortunately, the stability of such systems in aqueous solutions is generally low, because they can easily engage in polyion exchange reactions [102]. The process of exchange consists of two stages: (1) rapid formation of a triple complex, (2) slow substitution of one same-charge polyion with another. At equilibrium conditions, the whole process eventually results in the formation of a new binary complex and an excess of a third polyion. The presence of a low-molecular-weight salt can greatly accelerate such exchange reactions, which often result in the complete disassembly of condensed DNA particles [103]. Hence, it is desirable to obtain more colloidally stable structures, where DNA would stay in its condensed form in complex with corresponding polycation independent of environmental conditions. Here we describe some approaches aimed to stabilize condensed DNA structures in terms of colloid chemistry [92].

As we mentioned earlier, complete DNA condensation upon neutralization of only 90% of the polymer's phosphates results in the presence of uncompensated positive charges on the surface of DNA particles. If the polycation contains such reactive groups as primary amines (PLL, for example), this means that these reactive groups on the surface can be modified as a result of a chemical reaction. This situation opens practically limitless possibilities of modulating colloidal properties of DNA particles via regular chemical modifications of the complex. It is important to stress the differences of this approach with DNA particle formation via template polymerization. In the latter case, DNA/polycation complexes obtained on DNA templates never reach $+/-$ charge ratio over 1, which can be judged by the flocculation of the reaction mixture upon reaching the neutrality point and by the measurement of the ζ-potential of the resulting complexes (which remains negative), demonstrating the fact that the excess of reactive groups on the particle's surface is never formed.

In contrast, reactive positively charged groups are available for chemical modification when preformed polycations are used for condensed DNA particle formation. We have demonstrated the utility of such reactions using a traditional DNA/PLL system reacted with a cleavable cross-linking reagent, dimethyl-3,3′-dithiobispropionimidate (DTBP), which reacts with primary amino groups with the formation of amidines [92]. The salt-induced flocculation of DNA/PLL complexes described earlier was used as a test for such surface chemical modifications. Amidine formation resulting from DTBP reaction with PLL primary amines preserves positive charge on the surface of condensed DNA particles.

It has been found that DTBP cross-linking substantially increased the salt stability of the complexes. The salt stabilization is reversed upon the addition of DTT, which cleaves the bifunctional reagent, indicating that it is not due to the conversion of the amines to amidines and is dependent upon the cross-linking. Similar results were achieved with other polycations, including poly(allylamine), and histone H1.

One possible explanation of the phenomenon by which cross-linking increases salt stability is that additional polycation is recruited to the particle's surface, thus ensuring so-called electrosteric stabilization [88]. However, upon cross-linking, neither ζ-potential nor particle size was significantly altered as compared to non-cross-linked DNA/PLL complexes. For example, particle size and ζ-potential for PLL/DNA (6:1 charge ratio) complex were found to be 61.6 nm and +59 mV for non-cross-linked vs. 62.1 nm and +53 mV for cross-linked complex, respectively. Likewise, electron microscopic analysis indicated that the cross-linked particles were toroids and rods similar in size and shape to non-cross-linked particles. Moreover, direct composition measurements of cross-linked and initial DNA/PLL complexes purified by ultracentrifugation in sucrose gradient yielded the same 1:1 $-/+$ charge ratio, indicating no recruiting of additional polycation to the particle's surface.

Another possible explanation of the increased colloidal stability of cross-linked particles is the prevention of salt-induced interparticle cross-bridging by cationic polymers. Such cation bridging between DNA strands has been considered a general mechanism for DNA condensation [48,49]. Interpolymer covalent links forming a "cage" around condensed DNA particles evidently can prevent the salt-induced linkages between neighboring particles (Fig. 4). Other experiments confirmed this mechanism. Cross-linking markedly decreases the exchange of unlabeled PLL in solution with fluorescein-labeled PLL in 3:1 PLL/DNA complex, as detected by fluorescence quenching. It is known that high sodium chloride concentrations and the addition of strong polyanions with high charge density (for example, dextran sulfate) can displace DNA from its complexes with polycations [57]. Subsequently, we have found that cross-linking with DTBP prevents these agents from displacing the DNA in PLL complexes. Thus one may conclude that the cross-linked particles are more resistant to salt-induced aggregation because a polyamine molecule on one particle is less able to cross-bridge to another particle. In essence, cross-linking "freezes" the polycations on the particle, "caging" the whole complex.

Recently the use of another bifunctional reagent, glutaraldehyde, has been described for the stabilization of DNA complexes with cationic peptide CWK18 [104]. The authors of this paper, however, limited the study to the protective effects toward nuclease degradation.

10. Surface-Charge Reversal of Condensed DNA Particles

The caging approach just described could lead to more colloidally stable DNA assemblies. However, this approach does not change the particle surface charge. Caging with bifunctional reagents preserves the positive charge of the amino group, although for many practical applications negative surface charge would be more desirable. The phenomenon of surface recharging is well-known in colloid chemistry and is described in great detail for lyophobic/lyophilic systems (for example, silver halide hydrosols) [88]. The addition of polyion to a suspension of latex particles with an oppositely charged surface leads to the permanent absorption of this polyion on the surface and, upon reaching appropriate stoichiometry, changing the surface charge to the opposite one. This whole process is salt dependent, with flocculation occuring upon reaching the neutralization point.

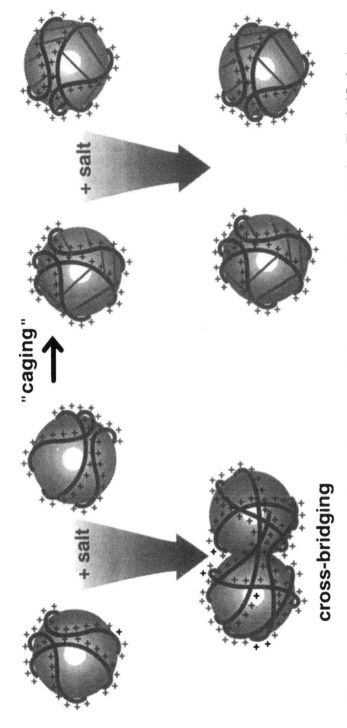

FIG. 4 "Caging" of the DNA particle condensed with PLL. (Reprinted from Ref. 92, copyright © 1999 American Chemical Society.)

It is known that synthetic polyelectrolytes of opposite charge form "fuzzy" layers on macrosurfaces [105] and surfaces of latex particles [106]. Due to the flexibility of poly-electrolyte chains, charges are not quenched completely during polyanion/polycation reaction, especially when the reaction is carried out in a low-salt environment. Hence, it is possible to achieve the layering of one polyion on top of underlaying counterion by taking advantage of some uncompensated surface charges. With appropriate washing steps it is possible to achieve multiple layering. Strict 1:1 stoichiometry of anionic and cationic groups was observed for strong polyelectrolytes, such as poly(styrene sulfonate) and poly(allylamine) [106]. Recently, macrosurfaces were repetitively coated with alternating layers of DNA and poly(allylamine) [107]. However, for weak (pH-sensitive) polymers deviation from this stoichiometry can be expected. Interlayer distance is dependent on the conditions of layer formation, such as the salt concentration and the molecular weight of the polymers.

DNA/polycation complexes in aqueous solutions behave similarly to colloidal suspensions; hence, similar surface-charge manipulations also seem possible for them. Recently, we have demonstrated that similar layering of polyelectrolytes can be achieved on the surface of DNA/polycation particles [5]. The principal DNA/polycation complex used in this study was DNA/PLL (1:3 charge ratio) formed in low-salt 25 mM HEPES buffer and recharged with increasing amounts of various polyanions. The DNA particles were characterized after the addition of a third polyion component to the DNA/polycation complex using a new DNA condensation assay [59] and static light scattering. It has been found that certain polyanions, such as poly(methacrylic acid) and poly(aspartic acid), decondensed DNA in DNA/PLL complexes. Suprisingly, polyanions of lower charge density, such as succinylated PLL (SPLL) and poly(glutamic acid), even when added in 20-fold charge excess to condensing polycation (PLL), did not decondense DNA in DNA/PLL (1:3) complexes. Further studies have found that displacement effects are salt dependent and that polyglutamate but not SPLL displaces DNA at higher sodium chloride concentrations. Measuring the ζ-potential of DNA/PLL particles during titration with SPLL revealed the change of particle surface charge at approximately the charge equivalency point. Thus, it can be concluded that the addition of low-charge-density polyanion to the cationic DNA/PLL particles results in particle surface-charge reversal while maintaining the condensed DNA core intact.

The effect of the additional layers of polyion on the particle size was determined using dynamic light scattering. In this experiment, the next polyion of opposite charge was added at charge excess to underlying complexes in water. For example, an excess of SPLL (ranging from 1.25 to 3.5 $-/+$ ratio) was added to DNA/PLL (1:1.5) complex in water. Consecutively, the same excess of PLL was added to DNA/PLL/SPLL (1:1.5:2) complex. For DNA/PLL/SPLL/PLL particles, each layer increased the particle diameter approximately 10 nm on average (Fig. 5). After addition of the third layer, polyion (PLL) DNA was still found condensed inside this quaternary complex. Atomic force microscopy was used to visualize the size and shape of the DNA/PLL/SPLL complexes (Fig. 6). It revealed the DNA/PLL/SPLL complexes to be nonaggregated spheroids 50 nm in diameter on average. In complete agreement with dynamic light scattering, the sizes of recharged particles did not differ significantly from those of the starting DNA/PLL complexes.

The stoichiometry of the recharged DNA/PLL/SPLL particles was studied using sucrose-gradient ultracentrifugation of fluorescently labeled polyion complexes in 25 mM HEPES buffer. Rhodamine-labeled DNA (Rh-DNA) and either fluorescein-labeled PLL (Fl-PLL) or SPLL (Fl-SPLL) were used to determine their relative amounts within DNA

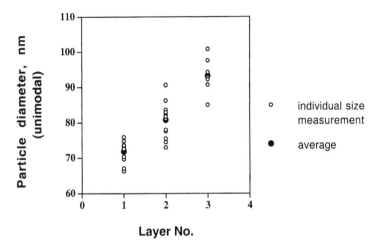

FIG. 5 Sizing of DNA/PLL (layer no. 1), DNA/PLL/SPLL (layer no. 2), and DNA/PLL/SPLL/PLL (layer no. 3) complexes in water (dynamic light scattering, unimodal mode). Core DNA/PLL particles were prepared in 1:1.2 charge ratio at a DNA concentration of 20 μg/mL in water. Each additional polyion was added in 1.5–4 charge excess. PLL and SPLL with Mw of 210 kDa were used for this experiment. (Reprinted from Ref. 5, copyright © 1999 Oxford University Press.)

complexes. Optical densities at 495 nm and 595 nm indicated the amount of rhodamine and fluorescein label, respectively. After ultracentrifugation of the Rh-DNA/Fl-PLL/SPLL (charge ratio 1:3:10) or Rh-DNA/PLL/Fl-SPLL (input charge ratio 1:3:10) mixtures, all of the Rh-DNA appeared at a centrifugation tube's bottom as precipitated complex. Recovered complexes were solubilized in 2.5 M NaCl, and their visible spectra were compared

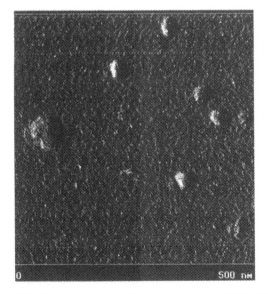

FIG. 6 Atomic force microscopic images of DNA/PLL/SPLL complexes (1:3:10 initial ratio) absorbed on mica in 25 mM HEPES, 1 mM NiCl₂, pH 7.5. (Reprinted from Ref. 5, copyright © 1999 Oxford University Press.)

to a 1:1 standard mixture of Rh-DNA and Fl-PLL or Fl-SPLL in the same high-salt solution. Precipitated triple complex was found to possess 1:1:1 charge ratio stoichiometry. Precipitation occurred, apparently due to the loss of excess SPLL, which served as charge stabilizer to the whole complex.

Thus, this study demonstrates that negatively charged particles containing condensed DNA can be formed by recharging DNA/polycation particles with certain polyanions. The relative difference of condensing polycation and recharging polyanion charge densities and bulk solution salt concentration seems to play a crucial role in the ability of DNA to remain condensed in the particle's core. The size of the DNA particles was found to increase, with each additional polyion assuming an onion-like structure for the complexes. The 1:1:1 charge stoichiometry found for the isolated DNA/PLL/SPLL complexes makes the whole DNA particle look like "fuzzy" polyelectrolyte multilayered assemblies on macrosurfaces, for which 1:1 +/− stoichiometry also was found [105].

11. DNA Condensation with Cationic Block Copolymers

Another mechanism for the stabilization of lyophilic colloid particles, steric stabilization, can be employed within DNA complexes. Such DNA particles are stabilized by repulsive layers of dense hydrophilic polymers on their surfaces that prevent them from flocculation or electrostatic interaction with any other charged surface. Several attempts to prepare particles of condensed DNA with such block- and comblike copolymers as poly(cation)-*block*-(neutral/hydrophilic) polymers have been described [98,108,109]. Most frequently, polyethyleneglycol (PEG) was used as a hydrophilic block in such conjugates. However, other neutral hydrophilic polymers (e.g., hydrophilic polyacrylamides [98]) were used, with similar results. This approach was aimed to exploit the superb steric protection properties of PEG. Despite the presence of this bulky neutral chain, the resulting PEG-PLL copolymer is still able to condense DNA, as judged by ethidium bromide exclusion from DNA upon complexation with the copolymer [98]. Such complexes were reported to be extremely soluble in aqueous solutions, even near the charge neutralization point, which is important for gene delivery applications. Flexible hydrophilic chains protruding into solution from the DNA/polycation core were suggested as a basic structure [102,110]. Structurally these complexes can be drastically different from regular DNA/polycation complexes. Atomic force microscopy of the DNA complexes with copolymer of cationic methacrylate with neutral poly-*N*-(2-hydroxypropyl) methacrylamide revealed spheroidal particles with a diameter of 60 nm. Suprisingly, the DNA complexes with PEG-PLL copolymer were found not to be spheroidal but to possess extended wormlike structures [98], possibly exemplifying hydrogen-bond formation between PEG chains and the mica substrate used for imaging.

12. Complexes of DNA with Cationic Lipids and Detergents

The recent surge of interest in structures of DNA/cationic lipid (CL) complexes has been caused by the discovery that such structures can promote gene transfer to cells in vitro and in vivo and, hence, serve as possible gene medicines to treat hereditary diseases [111,112]. Like any polyanion, DNA strongly interacts with cationic amphiphiles, with effective condensation as judged by the displacement of intercalating dyes [113]. Generally, the same colloid stability rules governing the stability of DNA/polycation complexes can be applied to DNA/CL complexes [7]. However, there seems to be a major difference in the colloidal behavior of DNA/CL complexes: These properties do not affect the complex's microstructure, which is of paramount importance for its biological activity.

The colloidal properties of DNA/CL complexes were found to be dependent mainly on the charge ratio and were independent of lipid composition [79]. However, the type of cationic lipid was found to be of primary importance for the microstructures of the complexes as well as for biological activity. The colloidal properties of polyanion/CL (surfactant) complexes were studied recently in a system that did not contain DNA [114]. However, conclusions made as a result of this study are very characteristic and can be applicable to DNA-containing systems. Poly(methacrylate-co-N-isopropylacrylamide) polyanions with various charge densities were complexed with cationic detergents tetradecyl- and dodecyltrimethylammonium bromide in aqueous medium. The complexes were found to form highly ordered nanostructures using small-angle x-ray scattering. The system's structure was found to undergo consecutive transition from cubic to hexagonal close packing of spheres (detergent micelles) with a decrease of the polyanion charge density. Hence, merely by changing polyion charge density it is possible to control the microstructure of ordered phases in the systems, based on oppositely charged polyion and surfactant.

Early models of DNA/CL complexes suggested either DNA encapsulated into the interior of elongated cationic liposome formed by the fusion of smaller vesicles (56) or fibers of DNA helices coated with CL [115]. Later, many researchers began to note characteristic highly regular striated structures of DNA/CL complexes on transmission electronic micrographs [116–119]. Detailed studies of these structures yielded a model with DNA helices sandwiched between CL bilayers [118]. Using the mixture of cationic lipid dioleoyl trimethylammoniumpropane (DOTAP) and neutral phospholipid dioleoylphosphatidyl ethanolamine (DOPE) (1:1 molar), the authors demonstrated that highly regular arrays of DNA helices are tightly packed between CL bilayers, with an interlayer distance of 6.5 nm

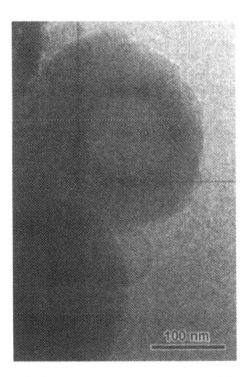

FIG. 7   Electron micrograph of cationic lipid/DNA complexes (example of striated lamellae). (Reprinted from Ref. 117, copyright © 1999 Biophysical Society.)

(Fig. 7). That space is enough to accommodate just one monolayer of B DNA. Small-angle x-ray scattering also detected long-range order in the position of the axes of the DNA helices.

Recently, the scope of these studies has been extended to include DNA/CL mixtures purified from the excess of components and to connect the structural changes with corresponding gene transfer activity. Xu et al. studied the structure of DOTAP/DNA mixtures with excesses of DNA and CL (79). Depending on the excess of negatively charged DNA or positively charged cationic lipid, the authors succeeded in purifying two distinct types of DNA/CL complexes with negative and positive ζ-potential. For negatively charged complexes, cryoelectron micrographs revealed striated structures with a spacing of 5.9 nm, which the authors attributed to ordered DNA helices layered between lipid bilayers, as previously described [118]. However, the highest gene transfer activity was observed for purified positively charged complexes, which lacked striated morphology in the lipid phase but were rich in toroidal structures with dimensions similar to DNA/polycation complexes. The authors hypothesized on the role of high-curvature structures in the biological activity of DNA/CL complexes. It has been demonstrated that cationic lipids organized into bilayers can promote DNA ordering.

There have been some attempts to study DNA/CL complexes by cryoelectron microscopy. This technique allowed the timing of different stages in the process of interaction of DNA with cationic components. Templeton et al. [120] have developed an improved procedure of extruding DOTAP/cholesterol (1:1, molar) mixture allowing generation of so-called "vase structures"—completely invaginated vesicles with an excess of surface area. Upon the addition of plasmid DNA, the inversion of part of the vesicles was observed, with complete engulfment of the DNA. The thickness of newly formed complex of DNA sandwiched between two bilayers was found to be 10.5 nm, which is in a good agreement with previous observations. Other support of formation of well-ordered two-dimensional structures of DNA absorbed on cationic membranes came from studies of Huebner et al. [117], who studied the complex formation of cationic lipid 3b(N-(N,N'-dimethylaminoethane)-carbamoyl] cholesterol (DC-Chol) with linearized plasmid DNA. According to their measurements the two bilayers/DNA sandwich thickness was 13.5 (04 nm). They found similar DNA-induced invagination phenomena, though their liposomes were regular spheres. Interestingly, since these regular liposomes did not exhibit an excess of surface area, upon interaction with DNA some bilayer ruptures were observed with characteristic unclosed bilayers.

One of the major recent advances in the area of self-assembled DNA-based nanostructures was the demonstration of the fact that a hydrophilic polymer such as DNA can be solubilized in nonpolar organic solvents by complexing with cationic detergents and lipids [121–123]. Such complexation results in the formation of water-insoluble complexes that can be dispersed in aqueous medium to form particles with diameters of less than 100 nm. In organic phase, DNA retains a double-helix architecture. Interestingly, the secondary structure can be reversibly changed from B to C conformation merely by adding water to the chloroform/ethanol (4:1 v/v) mixture. Upon casting from organic solution, DNA/lipid phase can be fabricated into film where DNA strands are positioned in hexagonal packing with a 4.1-nm interhelix distance. It seems that applying mechanical force to such films can increase the ordering of the DNA. X-ray diffraction data showed that DNA fibers can be aligned parallel by stretching the film.

Complexes of DNA with cationic detergents were found to behave similarly to the DNA/CL: They also are soluble in nonpolar organic solvents and form condensed DNA structures, with one notable exception. The morphology of some of the DNA/detergent complexes formed in organic solvents was found to be remarkably similar to what was observed in aqueous solutions. For example, the formation of toroidal structures (100–200 nm in diameter) was detected, using atomic force microscopy, upon complexation of DNA with dodecyldimethylammonium chloride in chloroform and chloroform/acetone (2:1) mixture [64]. The authors explain this phenomenon by noting that double-stranded DNA is a prestressed elastic polyanion, which invariably collapses upon neutralization of its charges in any solvent conditions. It also has been observed that the longer detergent's alkyl chain favors the dissolution of DNA/detergent complex in organic solvents: DNA complexes with didodecyldimethylammonium are more soluble in toluene than corresponding complexes with a cetyl derivative [123].

Peculiar DNA architecture was demonstrated in 25% aqueous ethanol when DNA was complexed with series of cationic detergents in the presence of poly(glutamic acid) [124]. Electron microscopy and x-ray scattering demonstrated that DNA can pack cetyltrimethylammonium bromide molecules into rodlike micelles, which form a hexagonal lattice. Interestingly, circular dichroism spectroscopy revealed that in these complexes DNA adopts left-handed conformation.

G. Basic Proteins Compact DNA in Vivo

DNA condensation in nature provides a means to compact enormous amounts of genetic information encoded in this giant molecule. As has been mentioned earlier, the total length of DNA in higher mammals might reach 1 meter. A variety of cationic proteins and polyamines serve as condensing agents in cells' nuclei and virions' heads (Fig. 1). For example, inside the sperm nucleus DNA is condensed with protamine—arginine-rich highly basic protein [125]—into compact toroids 60 nm in diameter. The absence of supercoiling and the presence of a strong condensing agent (protamine) allows compacting of the whole mammalian genome into the very small nuclei of a sperm cell, which is thought to be one of the most compact DNA structures found in nature [126]. A variety of small specialized proteins (histones) help to condense DNA in the nuclei of somatic cells. The organization of DNA in such nuclei represents an elaborate multistage nanoassembly process in which DNA is initially wrapped around individual core histones. Particles containing histone octamers with complexed DNA are called *nucleosomes* and represent structures with a diameter of 11 nm. The nucleosomes are further packed into fibers 30 nm in diameter (chromatin), which are further condensed in chromosomes. Various divalent metal cations present in vivo are believed to help DNA condensation [11]. It is believed that the decondensation of DNA that is necessary for gene expression is regulated by phosphorylation and acetylation of histones.

IV. CONCLUSIONS

DNA nanoassemblies represent a new perspective for materials science. First, one can exploit the polymer properties of DNA as fabrication material to design new supramolecular polymer assemblies. Second, significant progress is achieved in applications where DNA is used not only as construction material but also as a functional carrier of genetic information (e.g., in nonviral gene transfer applications). Various interactions of DNA double strands can be employed to assemble such structures. Interstrand hydrogen bonding and stable branching junctions were used to fabricate three-dimensional DNA structures, with possible application in the preparation of nano-wires and electronic switches. DNA counterion condensation results in the self-assembly of DNA strands into structures with distinct morphologies regardless of the particular condensing agent. This approach was employed to prepare novel multilayered and colloidally stable compact DNA particles with possible applications as gene transfer vectors. Condensed DNA particles can be formed with pre-formed polycations as well as with small monomeric cations, which can be polymerized along the DNA backbone using a template polymerization process. Further progress in this area will lead to dramatic applications in the fabrication of DNA-based nanoassemblies and a variety of therapeutic applications, such as gene therapy.

ABBREVIATIONS

PEG, poly(ethyleneglycol)
PLL, poly(L-lysine)
SPLL, succinylated poly(L-lysine)
DOTAP, dioleoyl 1,2-diacyl-3-trimethylammonium-propane
AFM, atomic force microscopy
AEPD, bis(2-aminoethyl)-1,3-propanediamine
NLS, nuclear localization sequence
DPDPB, 1,4-di[3',2'-pyridyldithio-(propionamido)butane]
DTBP, dimethyl-3,3'-dithiobispropionimidate
HEPES, 4-(2-hydroxyethyl)-1-piperazineethansulfonic acid
Rh, rhodamine
Fl, fluorescein
CL, cationic lipid
DC-chol, 3b(N-(N,N'-dimethylaminoethane)carbamoyl)cholsterol
DTT, dithiotreithol

REFERENCES

1. CM Niemeyer. Appl. Phys. A 68:119–124, 1999.
2. BH Robinson, NC Seeman. Protein Eng. 1:295–300, 1987.
3. JA Wolff, VS Trubetskoy. Nat. Biotechnol. 16:421–422, 1998.
4. MS Shchepinov, KU Mir, JK Elder, MD Frank-Kamenetski, EM Southern. Nucleic Acids Res. 27:3035–3041, 1999.
5. VS Trubetskoy, A Loomis, JE Hagstrom, VG Budker, JA Wolff. Nucleic Acids Res. 27:3090–3095, 1999.
6. W Saenger. Principles of Nucleic Acid Structure. New York: Springer-Verlag, 1984.
7. DD Lasic. Liposomes in Gene Delivery. Boca Raton, FL: CRC Press, 1997.
8. HG Elias. Macromolecules. New York: Plenum Press, 1984.
9. VA Bloomfield, DM Crothers, I Tinoco. Physical Chemistry of Nucleic Acids. New York: Harper & Row, 1974.
10. R Podgornik, HH Strey, DC Rau, VA Parsegian. Biophys. Chem. 57:111–121, 1995.
11. L McFail-Isom, CC Sines, LD Williams. Curr. Opin. Struct. Biol. 9:298–304, 1999.
12. L. Stryer. Biochemistry. New York: W.H. Freeman and Co., 1988, p. 82.
13. R Holliday. Genet. Res. 5:282–304, 1964.
14. NR Kallenbach, RI Ma, NC Seeman. Nature 305:829–831, 1983.
15. J Chan, NC Seeman. Nature 350:631–633, 1991.
16. BH Robinson, NC Seeman. Prot. Eng. 1:295–300, 1987.
17. NC Seeman. Annu. Rev. Biophys. Biomol. Struct. 27:225–248, 1998.

18. C Helene. Anticancer Drug Design 6:569–584, 1991.
19. K Ryan, ET Kool. Chem. Biol. 5:59–67, 1998.
20. HE Moser, PB Dervan. Science 254:645–650, 1987.
21. PE Nielsen, M Egholm, RH Berg, O Buchargdt. Science 254:1497–1500, 1991.
22. E Wang, J Feigon. In: S Neidle, ed. Structures of Nucleic Acid Triplexes. New York: Oxford University Press, 1999, pp 355–388.
23. VV Kuryavyi, TM Jovin. Nature Genet. 9:339–341, 1995.
24. JR Williamson, MK Raghuraman, TR Cech. Cell 59:871–880, 1989.
25. K Gehring, JL Keroy, M Gueron. Nature 363:561–565, 1993.
26. EA Venczel, D Sen. J. Mol. Biol. 257:219–224, 1996.
27. RP Fahlman, D Sen. J. Am. Chem. Soc. 121:11079–11083, 1999.
28. CA Mirkin, RL Letsinger, RC Mucic, JJ Storhoff. Nature 382:607–611, 1996.
29. R Elghanian, JJ Storhoff, RC Mucic, RL Letsinger, CA Mirkin. Science 277:1078–1081, 1997.
30. CJ Loweth, WB Caldwell, X Peng, AP Alivisatos, PG Schultz. Angew. Chem. Int. Ed. 38:1808–1812, 1999.
31. CM Niemeyer, B Ceyhan, D Blohm. Bioconjugate Chem. 10:708–719, 1999.
32. JL Coffer, SR Brigham, X Li, RF Pinizzotto, YG Rho, RM Pirtle, IL Pirtle. Appl. Phys. Lett. 69:3851–3853, 1996.
33. T Torimoto, M Yamashita, S Kuwabata, T Sakata, H Mori, H Yoneyama. J. Phys. Chem. B 103:8799–8803, 1999.
34. E Braun, Y Eichen, U Sivan, G Ben-Yoseph. Nature 391:775–778, 1998.
35. U Pindur, G Fisher. Curr. Med. Chem. 3:379–406, 1996.
36. EA Lukhtanov, IV Kutyavin, RB Meyer. Bioconjugate Chem. 7:564–567, 1996.
37. CB Post, BH Zimm. Biopolymers 21:2123–2137, 1981.
38. VA Bloomfield. Curr. Opion Struct. Biol. 6:334–341, 1996.
39. LS Lerman. Proc. Natl. Acad. Sci. USA 68:1886–1890, 1971.
40. SM Lindsay, SA Lee, JW Powell, T Weidlich, C de Marco, GD Lewen, NJ Tao, A Rupprecht. Biopolymers 27:1015–1021, 1988.
41. RW Wilson, VA Bloomfield. Biochemistry 18:2192–2196, 1979.
42. ME Cerritelli, N Cheng, AH Rosenberg, CE McPherson, FP Booy, AC Steven. Cell 91:271–280, 1997.
43. KE Richards, RC Williams, R Calendar. J. Mol. Biol. 78:255–259, 1973.
44. GS Manning. Quart. Rev. Biophys. 11:179–246, 1978.
45. C Ma, VA Bloomfield. Biopolymers 35:211–216, 1995.
46. R Marquet, C Houssier. J. Biomol. Struct. Dyn. 9:159–167, 1991.
47. J Pelta, F Livolant, JL Sikorav. J. Biol. Chem. 271:5656–5662, 1996.
48. M Olivera de la Cruz, M Belloni, M Delsanti, JP Dalbiez, O Spalla, M Drifford. J. Chem. Phys. 103:5781–5791, 1995.
49. E Raspaud, M Olivera de la Cruz, JL Sikorav, F Livolant. Biophys. J. 74:381–393, 1998.
50. JA Schellman, N Parthasarathy. J. Mol. Biol. 175:313–329, 1984.
51. DC Rau, VA Parsegian. Biophys. J. 61:246–259, 1992.
52. GS Manning, J Ray. J. Biomolec. Struct. Dynam. 16:461–475, 1998.
53. DK Chattoraj, LC Gosule, JA Schellman. J. Mol. Biol. 121:327–337, 1978.
54. LC Gosule, JA Schellman. J. Mol. Biol. 121:311–326, 1978.
55. SM Mel'nikov, VG Sergeev, K Yoshikava. J. Am. Chem. Soc. 117:2401–2408, 1995.
56. H Gershon, R Girlando, SB Guttman, A Minsky. Biochemistry 32:7143–7151, 1993.
57. Y Xu, FC Szoka Jr. Biochemistry 35:5616–5623, 1996.
58. KD Stewart. Biochem. Biophys. Res. Comm. 152:1441–1446, 1988.
59. VS Trubetskoy, PM Slattum, JE Hagstrom, JA Wolff, VG Budker. Anal. Biochem. 267:309–313, 1999.
60. M Haynes, RA Garrett, WB Gratzer. Biochemistry 9:4410–4416, 1970.

61. LC Gosule, JA Schellman. Nature 259:333–335, 1976.
62. VA Bloomfield. Biopolymers 44:269–282, 1998.
63. VG Sergeev, OA Pyshkina, AV Lezov, AB Mel'nikov, EI Ryumtsev, AB Zezin, VA Kabanov. Langmiur 15:4434–4440, 1999.
64. R Marquet, A. Wyart, C Houssier. Biochim. Biophys. Acta 909:165–172, 1987.
65. PG Arscott, AZ Li, VA Bloomfield. Biopolymers 30:619–630, 1990.
66. R Golan, LI Pietrasanta, W Hsieh, HG Hansma. Biochemistry 38:14069–14070, 1999.
67. TH Eickbush, EN Moudrianakis. Cell 13:295–306, 1978.
68. H Noguchi, S Saito, S Kidoaki, K Yoshikawa. Chem. Phys. Lett. 261:527–533, 1996.
69. H Dautzenberg, J Hartmann, S Grunewald, F Brnd. Ber. Bunsenges. Phys. Chem. 100:1024–1032, 1996.
70. JL Sikorav, J Pelta, F Livolant. Biophys. J. 67:1387–1392, 1994.
71. UK Laemmli. Proc. Natl. Acad. Sci USA 72:4288–4292, 1975.
72. DD Dunlap, A Maggi, MR Soria, L Monaco. Nucleic Acids Res. 25:3095–3101, 1997.
73. YY Vengerov, TE Semenov, SA Streltsov, VL Makarov, AA Khorlin, GV Gursky. FEBS Lett. 180:81–84, 1985.
74. NV Hud, MJ Allen, KH Downing, J Lee, R Balhorn. Biochem. Biophys. Res. Commun. 193:1347–1354, 1993.
75. D Porschke. Biochemistry 23:4821–4828, 1984.
76. J Widom, RL Baldwin. Biopolymers 22:1559–1620, 1983.
77. PG Arscott, C Ma, JR Wenner, VA Bloomfield. Biopolymers 36:345–364, 1995.
78. LR Brewer, M Corzett, R. Balhorn. Science 286: 120–123, 1999.
79. Y Xu, SW Hui, P. Frederik, FC Szoka Jr. Biophys. J. 77:341–353, 1999.
80. HG Hansma, R Golan, W Hsieh, CP Lollo, P Mullen-Ley, D Kwoh. Nucleic Acids Res. 26:2481–2487, 1998.
81. M Garcia-Ramirez, JA Subirana. Biopolymers 34:285–292, 1994.
82. Benbasat, JA. Biochemistry 23:3609–3619, 1984.
83. Z Reich, Y Ittah, S Weinberger, A Minsky. J. Biol. Chem. 265:5590–5594, 1990.
84. PM Vertino, RJ Bergeron, PF Cammanaugh Jr, C Porter. Biopolymers 26:691–703, 1987.
85. DT Hung, LJ Marton, DF Deen, RH Shafer. Science 221:368–370, 1983.
86. U Bachrach, G Eilon. Biochim. Biophys. Acta 179:494–496, 1969.
87. D Esposito, P Del Vecchio, G Barone. J. Am. Chem. Soc. 119:2606–2613, 1997.
88. Shaw, DJ. Introduction to Colloid and Surface Chemistry. Oxford: Butterworth-Heinemann, 1991.
89. AV Kabanov, FC Szoka Jr, LW Seymour. In: PF AV Kabanov, LW Seymour, eds. Interpolyelectrolyte Complexes for Gene Delivery: Polymer Aspects of Transfection Activity. Chichester, UK: Wiley, 1998, pp 197–218.
90. MX Tang, FC Szoka Jr. Gene Therapy 4:823–832, 1997.
91. B Philipp, H Dautzenberg, KJ Linow, J Kotz, W Davydoff. Prog. Polym. Sci. 14:91–172, 1989.
92. VS Trubetskoy, A Loomis, PM Slattum, JE Hagstrom, VG Budker, JA Wolff. Bioconjugate Chem. 10:624–628, 1999.
93. T Schindler, E Nordmeier. Macromol. Chem. Phys. 198:1943–1972, 1997.
94. H Votavova, D Kucerova, J Felsberg, J Sponar. J. Biomol. Struct. Dynam. 4:477–489, 1986.
95. C Ma, VA Bloomfield. Biophys. J. 87:1678–1681, 1994.
96. HT van de Grampel, YY Tan, G Challa. Macromolecules 23:5209–5216, 1990.
97. VS Trubetskoy, VG Budker, LJ Hanson, PM Slattum, JA Wolff, LE Hagstrom. Nucleic Acids Res. 26:4178–4185, 1998.
98. MA Wolfert, EH Schacht, V Toncheva, K Ulbrich, O Nazarova, LW Seymour. Human Gene Ther. 7:2123–2133, 1996.
99. T Blessing, JS Remy, JP Behr. J. Am. Chem. Soc. 120:8519–8520, 1998.
100. G Odian. Principles of Polymerization. New York: Wiley, 1991.

101. M Ouyang, JS Remy, FC Szoka Jr. Bioconjugate Chem. 11:104–112, 2000.
102. AV Kabanov, VA Kabanov. Adv. Drug Delivery Rev. 30:49–60, 1998.
103. VA Izumrudov, SI Kargov, MV Zhiryakova, AB Zezin, VA Kabanov. Biopolymers 35:523–531, 1995.
104. RC Adam, KG Rice. J. Pharm. Sci. 739–746, 1999.
105. G Decher. Science 277:1232–1237, 1997.
106. GB Sukhorukov, E Donath, S Davis, H Lichtenfeld, F Caruso, VI Popov, H Mohvald. Polym. Adv. Technol. 9:759–767, 1998.
107. J Lang, M Liu. J. Phys. Chem. B 103:11393–11397, 1999.
108. LW Seymour, K Kataoka, AV Kabanov. In: PF AV Kabanov, LW Seymour, eds. Cationic Block Copolymers as Self-Assembling Vectors for Gene Delivery. Chichester, UK: Wiley, 1998, pp 219–240.
109. S Katayose, K Kataoka. Bioconjugate Chem. 8:702–707, 1997.
110. LW Seymour, K Kataoka, AV Kabanov. In: PF AV Kabanov, LW Seymour, eds. Interpolyelectrolyte Complexes for Gene Delivery. Chichester, UK: Wiley, 1998, pp 197–218.
111. PL Felgner, TR Gadek, M Holm, R Roman, HW Chan, M Wenz, JP Northrop, GM Ringold, M Danielsen. Proc. Natl. Acad. Sci. USA 84:7413–7417, 1987.
112. EB Nabel, G Plautz, GJ Nabel. Science 249:1285–1288, 1990.
113. SJ Eastman, C Siegel, J Tousignant, AE Smith, SH Cheng, RK Scheule. Biochim. Biophys. Acta 1325:41–62, 1997.
114. S Zhou, C Berger, F Yeh, B Chu. Macromolecules 31:8157–8163, 1998.
115. B Sternberg, F Sorgi, L Huang. FEBS Lett. 356:361–366, 1994.
116. I Solodin, CS Brown, MS Bruno, CY Chow, EH Jang, RJ Debs, TD Heath. Biochemistry 34:13537–13544, 1995.
117. S Huebner, BJ Battersby, R Grimm, G Cevc. Biophys. J. 76:3158–3166, 1999.
118. JO Radler, I Koltover, T Salditt, CR Safinya. Science 275:810–814, 1997.
119. DD Lasic, H Strey, MCA Stuart, R Podgornic, PM Frederik. J. Am. Chem Soc. 119:832–833, 1997.
120. N Smyth Templeton, DD Lasic, PM Frederik, HH Strey, DD Roberts, GN Pavlakis. Nat. Biotechnol. 15:647–652, 1997.
121. K Ijiro, Y Okahata. J. Chem. Soc. Chem. Commun. 1339–1341, 1992.
122. K Tanaka, Y Okahata. J. Am. Chem. Soc. 118:10679–10683, 1996.
123. SM Mel'nikov, B Lindman. Langmiur 15:1923–1928, 1999.
124. R. Ghirlando, EJ Wachtel, T Arad, A Minsky. Biochemistry 31:7110–7119, 1992.
125. R Balhorn. J. Cell. Biol. 93:298–305, 1982.
126. WS Ward. Biol. Reprod. 48:1193–1201, 1993.

12

Supramolecular Assemblies Made of Biological Macromolecules

NIR DOTAN   Glycominds Ltd., Maccabim, Israel

NOA COHEN, ORI KALID, and AMIHAY FREEMAN   Tel Aviv University, Tel Aviv, Israel

I. INTRODUCTION

The potential inherent in the design and fabrication of complex nanostructures has attracted wide attention in recent years, resulting in numerous publications on the construction of one-, two-, and three-dimensional arrays via the self-assembly of synthetic organic molecules [1]. The analogous use of biological molecules, such as DNA and proteins, as "building blocks" for the in vitro construction of nanostructures is less developed, in some contradiction to the potential inherent in the following advantages offered by these biological macromolecules:

"Built-in" functional specific recognition and binding, e.g., antigen binding by antibodies or hybridization of nucleic acids;

Homogeneous molecular population for a wide range of macromolecules presenting variety of molecular weights and shapes;

Detailed structural data and modeling tools are available;

Composite nanostructures comprising biological structural elements and bioactive elements e.g., enzymes, antibodies, receptors, may be designed and fabricated;

Molecular biology techniques provide means for the design and production of new macromolecules exhibiting envisaged structural functionality as well as predesigned modification of natural structural elements.

It must be mentioned, however, that while considering the use of biological macromolecules as "building blocks," the question of their relative instability under the intended working conditions may turn out to be a limiting factor.

Most published work on the design and fabrication of nanostructures from biological macromolecules relate to DNA and proteins; the use of other biopolymers, such as cyclodextrins [2], was far less developed. Because the use of DNA is intensively covered in Chapter 10 of this volume (also see Ref. 3), as well as recently described by Seeman [4,5] this chapter will focus on proteins as a potential tool for the construction of nanostructures. Hence this chapter is focused on literature that may provide a basis for the identification of guidelines, methodologies, and examples having potential for further development of new protein-based composite nanostructures integrating structural and bioactive components.

II. SUPRAMOLECULAR ASSEMBLIES MADE OF NATURAL STRUCTURAL PROTEINS

Self-assembling systems made of proteins provide many biological systems with essential structural elements, including viral envelopes, bacterial S-layers, microtubules, collagens, and keratins (see Ref. 6 for a review). Detailed studies on in vitro self-assembly optimization were carried out for several systems as a part of an ongoing effort to elucidate the in vivo mechanism. Data reported from such studies may provide an optional basis for the design and in vitro fabrication of nanostructures made of natural proteins (Fig. 1).

Reconstruction of bacterial S-layers has been extensively investigated by Sleytr et al. (see Chapter 9 and Refs. 7–9) and will not be discussed here.

Microtubules are highly dynamic, hollow cylindrical supramolecular assemblies, 25 nm in diameter and reaching lengths of several microns, made of tubulin. Tubulin consists of two nonidentical 50,000 molecular weight polypeptides, designed as α and β. When tubulin molecules assemble into microtubules, protofilaments about 4–5 nm in diameter and made of tubulin polypeptides are aligned in rows obtained via the association of the β part of one dimer with the α part of the other. In vitro, the self-assembly of microtubules is a two-step process: nucleation and elongation, where microtubule-organizing centers nucleate microtubules and orient them [10–12].

Collagens are made of rigid triple-helical structural elements. Three protein chains are wound around each other as a "superhelix," 1.5 nm in diameter and 28 nm in length, to generate this element [6]. These elements may be packed into cylindrical fibrils in an entropy-driven crystallization process. In vitro enzymatic generation of supersaturated monomer solution by cleavage of carefully isolated procollagen led to fibril formation with a lag phase and a propagation phase. The diameter of the fibrils thus obtained was temperature dependent, generating at 37°C fibrils with the same diameter and flexibility as formed in the in vivo process. Lower temperatures resulted in the formation of larger fibrils. Collagen fibrils may be 0.18 μ in diameter and 60–150 μ long [13].

Keratins are made of filaments, approximately 10 nm in diameter and hundreds of nanometers in length, via assembly of rod-shaped, coiled-coil proteins. Filament formation is initiated by the creation of a dimer comprising monomeric units 44–54 nm in length. Such dimers may form three types of lateral interactions leading to filament formation from equimolar amounts of acidic and basic dimers. In vitro assembly involves the correct alignment of two, three, or four dimers into a nucleus for further, rapid filament assembly [6].

Myosin is a fibrous contractile protein, assembled in vivo into muscular, thick filaments. Myosin is a relatively large protein (molecular weight 520,000), made of two "heavy" chains (220,000) and four "light" chains (17,000–22,000). The heavy chains interact to form two distinctive domains: a pair of globular heads (15 nm long and 9 nm wide) and an α-helical coiled-coil rod domain (150 nm long and 2 nm wide). This rod is a two-stranded coiled-coil motif made by two α-helical chains interwined around each other (Ref. 14 and references cited therein). Myosin may be enzymatically cleaved into two well-defined fragments: heavy meromyosin (140,000), soluble in low-ionic-strength buffers, and light meromyosin (80,000), insoluble in low-ionic-strength buffers.

Myosin may be extracted via high-ionic-strength buffers and purified. Synthetic thick filaments of myosin spontaneously assemble upon lowering the ionic strength of its solution, exhibiting the morphological characteristics of native thick filaments. This process initiates with myosin monomers assembled into parallel dimers. The dimers assemble into antiparallel tetramers, the tetramers into octamers, and the octamers into minifilaments comprising 16 molecules [14]. In low-salt conditions, these minifilaments are assembled into synthetic filaments 1.6 μ long and 15 nm in diameter [14]. A 29-residue region of the myosin "rod" has been shown to be necessary for filament formation [15].

The observation that a repeating, relatively short sequence of amino acids directs the spontaneous self-assembly of a large protein is shared by other structural proteins from mammalian systems [6] as well as of plant origin [16]. Hence, such sequences may inspire the construction of nanostructures made of polypeptides and small proteins, as discussed later.

It should be mentioned here that the self-assembly of native proteins may also be effective in the construction of inorganic–organic multicomponent structures, as demonstrated by the biominerlization processes [17].

III. SUPRAMOLECULAR ASSEMBLIES MADE OF NATURAL BINDING PROTEINS

Binding proteins such as antibodies, receptors, and lectins may spontaneously aggregate in presence of bi- or multivalent specific ligands. This process, however, often results in assemblies that lack well-defined morphological and structural elements.

In vitro specific cross-linking of lectins by natural branched oligosaccharides or their synthetic analogs resulted in several cases of ordered crystalline structures. These lectins included plant-derived lectins, such as wheat germ agglutinin [18] and soybean agglutinin [19], as well as galectin derived from heart muscle [20]. Distinctive protein lattices with morphologies depending on the lectin and the bi- or oligosaccharides employed were obtained and could be characterized by electron microscopy and x-ray crystallography [20,22,25].

IV. SUPRAMOLECULAR ASSEMBLIES MADE OF POLYPEPTIDES

Polypeptides are small proteins comprising less than 100 amino acids [23]. Their capability to present defined three-dimensional structural motifs affecting self-assembly will strongly depend on their amino acid sequence [6,24]. Several natural or nature-inspired sequences were shown to be very effective in affecting dimerization and self-assembly, including the "leucine zipper" motif [25]. Such dimerizations are directed predominantly by accumulating complementary hydrophobic and electrostatic interactions, effected via two different mechanisms. The first is based on a regional effect: Complex formation is first stabilized by hydrophobic interactions, followed by structural elongation via electrostatic interactions located at the edges of the polypeptide chain [26]. In the second, repeating sequences of alternating and mixed hydrophobic and electrostatic groups affected salt-dependent self-assembly [27]. Such amphiphilic polypeptide segments, approximately 15 amino acids long, separated approximately 50 amino acids in proteins sequences from their "antisense" homologs ("antisense homology boxes"), were found to have a natural role in protein-folding mechanisms [28].

The "leucine zipper" motif was recently successfully employed for the construction of native-like triple-stranded coiled-coil in solution [29].

Due to their small size, many short polypeptides do not exhibit in solution a well-defined and stable three-dimensional structure, thus limiting potentially useful self-assembly-affecting motifs. Chemical modifications leading to the addition of unnatural side-chain interactions, such as the incorporation of transition-metal complexes or crown ethers complexes, were shown to provide solutions to this problem [24,30]. In parallel, the introduction of appropriate ligands via chemical modification of amino terminal groups of polypeptides led in a number of cases to the self-assembly of three polypeptides into an artificial triple helix via a metal-ion complexation mechanism. This and similar approaches were demonstrated to be useful for the generation of relatively small, cagelike structures hosting metallic ions [24,31].

Salt-dependent self-assembly of relatively short ionic polypeptides (8–32 amino acid sequences) into three-dimensional structures has recently been described [32].

Methodologies for the de novo design and synthesis of polypeptides were recently developed. The preparation of periodic polypeptides, polypeptides containing artificial amino acids, polypeptides exhibiting rodlike structures, and hybrids of natural and artificial polypeptide segments was recently described [33].

Cyclic artificial polypeptides comprised of alternating D and L amino acids were recently synthesized and self-assembled into nanotubes exhibiting lengths of 1000–1500 Å, formed at an air–water interface [34,35].

Examples representing the very wide range of self-assembled protein structures obtained as just described are presented in Figure 1. These examples demonstrate that the size and shape of self-assembled nanostructures made of proteins primarily depend on the molecular mechanism effecting self-assembly and are not merely an amplified reflection of the shape and size of the starting "building block."

Building Blocks   *In-Vitro* Self-assembled Structure

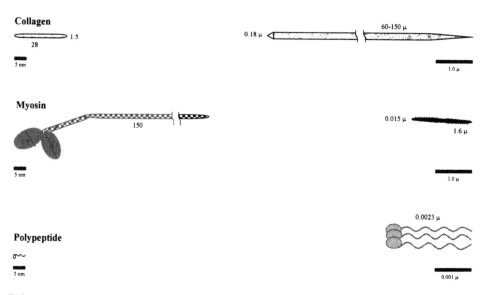

FIG. 1   Examples of shape, geometry and size of self-assembled protein made structures demonstrating that the size and shape of in vitro obtained nanostructures depend on the mechanism of self-assembly rather than on the size and shape of the "building block." Note that the scale of the "building blocks" (left-hand side) is in nanometers, while the scale of the structures obtained (right-hand side) is in microns.

V. ORIENTATION AND PATTERNING OF PROTEINS ON SOLID SURFACES

The feasibility of two-dimensional oriented immobilization of self-assembled proteins on solid supports was demonstrated for several systems. Antibody-mediated orientation of bacteriorhodopsin was successfully applied for the construction of highly oriented purple membrane. Bispecific antibodies exhibiting two different binding sites, one to a specific side of bacteriorhodopsin and the other to a phospholipid hapten deposited on a metal electrode, were employed. Step-by-step growth of oriented layers was thus made possible by first treating the electrode with the phospholipid hapten, oriented antibody immobilization, and finally oriented bacteriorhodopsin immobilization. Furthermore, by using two kinds of bi-specific antibodies, two versions of bacteriorhodopsin orientation were made possible: one "facing out" and the other "facing in" [36].

Two-dimensional protein layer orientation could be also effected by metal-ion coordination: Monolayer of iminodiacetate-Cu(II) lipid was successfully employed as substrate for oriented immobilization of proteins naturally displaying histidine residues on their surface [37]. Affinity-resin-displaying Ni(II) complexes could also be successfully employed for oriented protein immobilization [38].

The high affinity of avidin or streptavidin for biotin provided an effective tool for protein oriented immobilization. Biotin-containing dioctadecylamine molecules, forming hollow cylinders with a diameter of 27 nm and a few microns in length, served as a template for the oriented self-assembly of streptavidin into helical arrays located at the tube surface [39]. Biotinylated lipid vesicles also served as a template for the positioning of ferritin–avidin conjugates [40]. A step-by-step assembly of a protein multilayer array was demonstrated on the surface of liposomes displaying biotinylated lipids. First, streptavidin was specifically bound, followed by saturation of its remaining binding sites with bifunctional monobiotinylated ligand. A specific group, e.g., sugar moiety recognized by the lectin concanavalin A, could thus affect the orientation of concanavalin A as a second protein layer. Extension of this procedure to three-layer assembly was also readily demonstrated [41].

Patterning of enzyme monolayers on a solid surface was carried out by photoactivation of immobilized monolayer of "caged"-biotin derivatives in selected areas. Specific oriented binding of enzyme–avidin conjugates could be readily made to the photoactivated zones [42]. Oriented immobilization of G-protein-coupled receptors on a solid surface was also made possible on a biotinylated surface by first immobilizing streptavidin, followed by the immobilization of biotinylated G-protein-coupled receptor [43].

More complexed structures could be also formed by the biotin–streptavidin specific binding. Artificial "vesosomes" (encapsulated vesicle aggregates) could be obtained by the preparation in parallel of sized vesicle aggregates displaying biotin and streptavidin with cochleate cylinders. Encapsulation into vesosomes, mediated by the biotin–avidin specific binding, could be obtained upon mixing [44]. Oligonucleotide-directed self-assembly of proteins produced by the combination of the specific binding of complementary DNA hybrid molecules, each conjugated to biotinylated antibody or an enzyme, with the specific biotin–streptavidin binding allowed for the construction of multicomponent arrays on a chip [45].

Thin liquid films on a fluid surface were also employed for the construction of protein arrays [40]. The construction of a tightly chemically bound protein monolayer onto a solid support required detailed systematic study involving careful optimization of reaction conditions and comparison of the efficacy of several alternatives [46].

It should be mentioned here, however, that protein patterning on a solid surface at the nanoscale is still a major challenge: Most available data on protein oriented immobilization and patterning still relate to micro- rather than nanoscale patterning [46,47]. The main reason for this seems to be the common approach to initiate protein patterning and localization by photolithography techniques, e.g., UV irradiation of masked silicon. As in many cases, such techniques provide treated arreas wider than a few tenths of a micron, and the subsequent protein patterns obtained are micron sized [47–49]. The potential use of quatum dots, x-ray lithography, and nanoprinting by STM/AFM is currently considered the most promising next-phase solution for this problem.

VI. PREDESIGNED THREE-DIMENSIONAL SELF-ASSEMBLY OF PROTEINS: "CRYSTAL ENGINEERING"

Development of methodologies for protein self-assembly into a predesigned three-dimensional protein lattice carries potential for the construction of protein scaffolds, offering potential applications such as platforms for the ordered positioning of other proteins via protein fusion or of organic molecules by specific binding.

We recently developed such a methodology [50], employing a binding protein with a known detailed 3D structure as a "building block" and its specific cross-linking by an appropriate "biligand" imposing predetermined relative orientation on the cross-linked protein molecules, leading them into the predesigned lattice configuration. Such imposed orientation is assumed to be affected by the close proximity of the surface of the cross-linked proteins, due to ligand-specific binding and intermolecular interactions.

The feasibility of this approach was recently demonstrated with the nearly tetrahedral lectin concanavalin A as a model "building block." This lectin is a tetramer of a fully characterized 3D structure, presenting four binding sites for α-D-mannopyranoside or α-D-glucopyranoside located in an analogous way to sp^3 carbon atom configuration. Cross-linking of concanavalin A by a dimannoside with an appropriate spacer imposing staggered positioning (Fig. 2a) will lead to the formation of the computer-modeled diamond-like three-dimensional protein lattice described in Figure 2b.

The dimannoside spacer required for this purpose was deduced from calculations of the interaction energies between two concanavalin A–mannose complexes approaching each other on an imaginary line connecting their mass centers, as a function of the distance between the two anomeric oxygens of the complexed mannoses and the dihedral angles. The results of these calculations clearly indicated a minimum of interaction energy for the combination of two-carbon-atom spacer, predicted to effect the desired staggered positioning.

1,2-Ethyl-α-D-mannopyranoside, presenting the predicted dimannoside with a two-carbon-atom spacer, was synthesized, purified, and characterized. Addition of 2:1 molar ratio of this cross-linker to concanavalin A solution effected quantitative and rapid formation of protein crystals. Chemical analysis of the dimannoside content of this precipitate confirmed the anticipated molar ratio of 2:1, supporting the working hypothesis that each molecule of concanavalin A was cross-linked via four biligand molecules to its neighbors, in accord with the envisaged diamond-like model. The crystals thus obtained were stable to environmental changes, including medium substitution by bidistilled water, pH changes, and exposure to high concentrations of a competitive ligand.

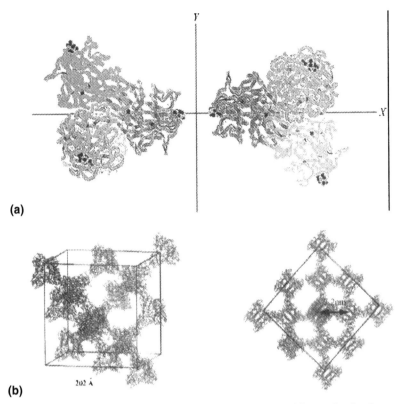

FIG. 2 (a) Model of two concanavalin A molecules approaching each other in a staggered orientation (dihedral angle of 60°; distance shown between anomeric oxygens of complexed sugars in 10 Å). (b) Two views of predicted diamond-like protein lattice formed by cross-linking concanavalin A with dimannoside exhibiting C_2 spacer.

Electron transmission micrographs of negatively stained crystalline precipitates revealed a highly ordered three-dimensional crystalline array with intermolecular distances in good agreement with the predesigned diamond-like model.

X-ray diffraction experiments revealed a pseudo-cubic orthorombic unit cell with cell dimensions similar to the expected cubic F centered arrangement of the predesigned diamond-like crystal.

VII. APPLICATIONS OF NANOSTRUCTURES MADE OF BIOLOGICAL MACROMOLECULES

In some contrast to the progress made in establishing working methodologies for the design and fabrication of nanostructures made of self-assembled biological macromolecules, the realization and feasibility demonstrations of claimed potential applications of these systems are still in their infancy.

A major challenge and important application is "nano-wiring" of electronic circuits mediated by self-assembled DNA or protein structures providing conducting connection between miniaturized electrodes [51,52]. The use of self-assembled DNA for "wiring" two electrodes was recently demonstrated on a several-micron-scale (12–16 μ-wide) electrode gap [53]. The λ-DNA bridge was established by adding suspended λ-DNA onto a chip presenting gold electrodes pretreated with complementary thiol-oligonucleotides capable of hybridizing with the added λ-DNA, thus leading it to place. The DNA bridge thus obtained served as a template for the adsorption of silver ions later converted electrochemically into 12-μ-long and 100-nm-wide silver wire. The conductivity of this wire was, however, relatively poor, raising again the issue of providing good electrical conductivity to biological molecules serving as self-assembled templates.

The "natural" conductivity of DNA and the potential means for its improvement have recently been re-evaluated [54,55]. Use of DNA bridges for nano-wiring appears very attractive in view of the recent great achievements in the design, control, and fabrication of DNA-made nanostructures [5]. It appears, however, that the conversion of DNA wires into effective nano-conductors is still a major problem.

The conversion of protein-made nanostructures, e.g., microtubules, into conducting nano-wires was also recently investigated [11]: "Metallization" of microtubules, by an electrolyte nickel deposition technique, initiated by molecular palladium catalysts, was described.

The development of methodologies for the conversion of DNA or protein template into a conducting nano-wire is in its infancy. Furthermore, there are technical difficulties in measuring and characterizing the currents involved. A promising approach to this problem appears to be the use of a coupled enzymatic system generating measurable currents into the nano-wire connecting it to a biosensor electrode [55].

Recent developments of biosensors as effective analytical tools have generated great interest in the extension of biosensor miniaturization into the nanoscale, involving the use of sensors based on few biocatalyst molecules [52,56–58]. Because most recently developed miniature biosensors are still within a several-microns scale [59], further miniaturization requires the exact location of an enzyme or antibody with accurate oriented immobilization as well as sensitive measurement of signals evolving from the activity of a single or a few molecules. Oriented immobilization has been successfully demonstrated on a micrometer scale (see Section V), but its successful implementation on the level of single molecules is rare [58]. Monitoring of binding reaction of an individual antibody–antigen pair was recently demonstrated [60]. A single-chain antifluorescein antibody fused with a polypeptide terminated by cysteine was prepared, oriented, and immobilized as separated individual molecules on a gold surface. Fluorescein was connected to the tip of AFM through a long linker. Its binding to the gold-anchored individual molecule could hence be monitored by AFM. Analog detection, however, of a single enzyme molecule by visual or amperometric methods is still a major challenge. Recent developments in the immobilization of enzymes on electrodes via mediating conducting polymer layers [61–63] and silicon technology [64,65] may facilitate these efforts.

Several attempts were made to apply nanostructures made of DNA or proteins to the development of alternative computation or computer memory. The concept of *DNA computing* was developed as an alternative computation approach based on information and data stored as sequenced DNA nucleotides and DNA-specific hybridization and elongation as a means to reach the answer or solution to a problem. Available tools of molecular biology were employed to identify and analyze the results [66–68]. This multistage "computation" is based on the assumption that "solutions" can be sought in parallel, thus compensating for the relatively slow processing time.

Oriented bacteriorhodopsin immobilization has been considered as a potential alternative to conventional computer memory [69]. It was suggested that "writing" and "reading" of information on such layers by means of laser beams may be possible. Unlike the previous application, this application has been suggested as a new alternative to conventional memory, to be integrated within existing common computer infrastructures. Other optoelectronic potential applications for bacteriorhodopsin-oriented layers were also suggested [36].

Another alternative prototype of memory array, consisting of data stored as electrostatic charge or molecular dipole in a two-dimensional network of streptavidin cross-linked by biotinylated porphyrin derivative, was also suggested. Information "reading" was expected to be carried out using the electric force mode of the atomic force microscope [70].

The potential application of self-assembled "scaffolds" offering exact positioning of biological macromolecules within micron- to milimeter-size DNA- or protein-made crystals for structural analysis by x-ray crystallography was also suggested [5,50].

The development of "nano-machines" has been the ultimate expectation from nanostructure technology [71–73]. Motion could be affected by temperature change in the environment of synthetic polypeptides [73]. Similarly, raising concentrations of appropriate chemicals isothermally could also result in motion [73]. The relatively low stability of natural proteins was considered a major obstacle to the realization of the self-assembled proteins for the construction of nano-machines [72]. Recently, a DNA-made nano-mechanical prototype device was developed [5,74]. The motion of two DNA "wings" could be effected by the addition of a salt producing a conformational switch of B DNA into Z DNA and vice versa. This effect was made possible following the fabrication of rigid molecular DNA structures providing the "wings" of the device [5]. In view of recent achievements made in the design and demonstration of nano-machines made of small synthetic organic molecules [75], nano-machinery based on biological macromolecules is still in its infancy.

VIII. CONCLUSIONS

The realization of the potential inherent in the self-assembly of biological macromolecules has been promoted mainly at the level of design, fabrication, and control of the self-assembly process. It appears that much can be inspired and learned from natural phenomena within this context. A detailed understanding of the mechanism producing the self-assembly process at the molecular level seems to be essential for the design and fabrication of protein-made nanostructures. Most progress made so far is, however, at an early stage of development or feasibility demonstration. Accumulating experience in protein-made nanostructures may soon pave the way to first-generation applications.

REFERENCES

1. JM Lehn (ed.). Comprehensive Supramolecular Chemistry. Vol 9. Oxford: Pergamon, 1996.
2. D Armspach, PR Ashton, CP Moore, N Spencer, JF Stoddari, TJ Wear, DJ Williams. Angew Chem Int Ed 32:854–858, 1993.
3. CM Niemeyer. Angew Chem Int Ed 36:585–587, 1997.
4. NC Seeman. Ann Rev Biophys Biomed Struct 27:225–248, 1998.
5. NC Seeman. TIBTECH 17:437–443, 1999.
6. KP McGareth, MM Butler. In: KP McGareth, D Kaplan, eds. Protein-Based Materials. Boston: Birkhauser, 1997, pp 251–279.
7. D Pum, UB Sleytr. TIBTECH 17:8–12, 1999.
8. UB Sleytr, M Sara. TIBTECH 15:20–26, 1997.

9. UB Sleytr, H Bayley, M Sara, A Breitweiser, S Kupcu, C Mader, S Weigert, FM Unger, P Messner, B Jahn-Schmid, B Schuster, D Pum, K Douglas, NA Clark, JT Moore, TA Winningham, S Levy, I Frithsen, J Pankovc, P Beale, HP Gillis, DA Choutov, KP Martin. FEMS Microbiol Rev 20:151–175, 1997.

10. Y Engelborghs. Biosens Bioelectron 9:685–689, 1994.

11. R Kirsch, M Mertig, W Pompe, R Wahl, G Sadowski, KJ Bohm, E Unger. Thin Solid Films 305:248–253, 1997.

12. DJ Odde. Biophys J 73:88–96, 1997.

13. DJ Prockop, A Fertala. J Structur Biol 122:111–118, 1998.

14. M Wick. Poultry Sci 78:735–742, 1999.

15. RL Sohn, KL Vikstrom, M Strauss, C Cohen, AG Szent-Gyorgyi, LA Leinward. J Mol Biol 266:317–330, 1997.

16. MJ Kieliszewski, TA Lamport. Plant J 5:157–172, 1994.

17. S Mann. In: JP Sauvage, MW Hosseini, eds. Comprehensive Supramolecular Chemistry. Oxford: Pergamon, 1996, pp 529–564.

18. CS Wright. J Biol Chem 267:14345–14352, 1992.

19. A Dessen, D Gupta, S Sabesan, CF Brewer, JC Sacchettini. Biochemistry 34:4933–4942, 1995.

20. Y Bourne, B Bolgiano, DI Liao, G Strecker, P Cantau, O Herzberg, T Feizi, C Cambillau. Nat Struct Biol 1:863–870, 1994.

21. L Bhattacharyya, J Fant, H Leon, CF Brewer. Biochemistry, 29:7523–7530, 1990.

22. D Gupta, L Bhattacharyya, J Fant, F Macaluso, S Sabesan, CF Brewer. Biochemistry 33:5614–5622, 1994.

23. PW Latham. Nature Biotechnol 17:755–757, 1999.

24. N Voyer. In: FP Schmidtchen, ed. Bioorganic Chemistry—Models and Applications. Berlin: Springer-Verlag, 1997, pp 1–37.

25. WH Landschulz, PF Johnson, SL MacKnight. Science 240:1759–1764, 1988.

26. EK O'shea, R Rutkowski, PS Kim. Cell 68:699–708, 1992.

27. S Zhang, T Holmes, C Lockshin, A Rich. Proc Nat Acad Sci USA 90:3334–3338, 1993.

28. L Brany, W Campbell, K Ohshima, S Fujimoto, M Boros, H Okada. Nature Med 9:894–901, 1995.

29. K Suzuki, H Hiroaki, D Kohda, C Tanaka. Prot Eng 11:1051–1055, 1998.

30. DH Lee, MR Ghadiri. In: JP Sauvage, MW Hosseini, eds. Comrehensive Supramolecular Chemistry. Oxford: Pergamon, 1996, pp 451–481.

31. TM Handel, SA Williams, WF DeGrado. Science 261:879–885, 1993.

32. EJ Leon, N Verma, S Zhang, DA Lauffenburger, RD Kam. J Biomat Sci 9:297–312, 1998.

33. JG Tirrell, DA Tirrell, MJ Fournier, TL Mason, IK McGrath, D Kaplan, eds. Protein-Based Materials. Boston: Birkhauser, 1997, pp 61–99.

34. JM Buriak, MR Ghdiri. Mater Sci Eng C4:207–212, 1997.

35. H Rapaport, HS Kim, K Kjaer, PB Howes, S Cohen, JA Nielsen, MR Ghdiri, L Leiserowitz, M Lahav. J Am Chem Soc 121:1186–1191, 1999.

36. K Koyama, N Yamaguchi, T Miyasaka. Science 265:762–765, 1994.

37. W Frey, WR Schief, DW Pack, CT Chen, A Chilkoti, P Stayton, V Vogel, FH Arnold. Proc Nat Acad Sci USA 93:4937–4941, 1996.

38. MJ Dabrowski, JP Chen, H Shi, WC Chin, WM Atkins. Chem Biol 5:689–697, 1998.

39. P Ringler, W Muller, H Ringsdorf, A Brisson. Chem Eur J 3:620–625, 1997.

40. K Nagayama. Supramolec Sci 3:111–122, 1996.

41. W Muller, H Ringsdorf, E Rump, G Wildburg, X Zhang, L Angermaier, W Knoll, M Liley, J Spinke. Science 262:1706–1708, 1993.

42. M Dontha, WB Nowall, WG Kuhr. Anal Chem 69:2619–2625, 1997.

43. C Bieri, OP Ernst, S Heyese, KP Hofmann, H Vogel. Nature Biotechnol 17:1105–1108.

44. SA Walker, MT Kennedy, JA Zasadzinski. Nature 387:61–64, 1997.

45. CM Niemeyer, T Sano, CL Smith, CR Cantor. Nucl Ac Res 22:5530–5539, 1994.

46. RA Williams, HW Blanch. Biosens Bioelectron 9:159–167, 1994.
47. AS Blawas, WM Reichert. Biomaterials 19:595–609, 1998.
48. M Mrksich, GM Whitesides. TIBTECH 13:228–235, 1995.
49. J Miake, M Hara. Mater Sci Eng C4:213–219, 1997.
50. N Dotan, D Arad, F Frolow, A Freeman. Angew Chem Int Ed 38:2363–2366, 1999.
51. Editorial. Science 286:2442–2444, 1999.
52. W Gopel. Biosens Bioelectron 13:723–728, 1998.
53. E Braun, Y Eichen, U Sivan, G Ben Yosseph. Nature 391:775–778, 1998.
54. Editorial. Chem Eng News 77:43–48, 1999.
55. G Hartwich, DJ Caruana, T deLumley-Woodyear, Y Wu, CN Campbell, A Heller. J Amer Chem Soc 121:10803–10812, 1999.
56. JM Laval, J Chopineau, D Thomas. TIBTECH 13:474–481, 1995.
57. CC Liu, Z Jin. TIBTECH 15:213–216, 1997.
58. L Tiefenauer, C Padeste. Chimia 53:62–66, 1999.
59. DJ Caruana, A Heller. J Am Chem Soc 121:769–774, 1999.
60. R Rose, F Schweisinger, D Anselmetti, M Kubon, R Shafer, A Pluckthum, L Tiefenauer. Proc Natl Acad USA 95:7402–7408, 1998.
61. T deLumley-Woodyear, P Rocca, J Lindsay, Y Dror, A Freeman, A Heller. Anal Chem 67:1332–1338, 1995.
62. T Livache, H Bazin, P Caillat, A Roget. Biosens Bioelectron 13:629–634, 1998.
63. T deLumley-Woodyear, CN Campbell, E Freeman, A Freeman, G Georgiou, A Heller. Anal Chem 71:535–538, 1999.
64. M Adami, M Martini, L Piras. Biosens Bioelectron 10:633–638, 1995.
65. RW Bogue. Biosens Bioelectron 12:27–29, 1997.
66. L Adelman. Science 266:1021–1024, 1994.
67. A Gibbons, M Amos, D Hodgson. Curr Opin Biotechnol 8:103–106, 1997.
68. M Conrad, KP Zauner. Biosystems 45:59–66, 1998.
69. RR Birge. Sci Amer 272(3):66–71, 1995.
70. DM Taylor. Thin Solid Films 331:1–7, 1998.
71. KE Drexler, ed. Nanosystems: Molecular Machinery, Manufacture and Computation. New York: Wiley Interscience, 1992.
72. KE Drexler. Ann Rev Biomed Struct 23:377–405, 1994.
73. DW Urry. Angew Chem Int Ed 32:819 841, 1993.
74. C Mao, W Sun, NC Seeman. Nature 397:144–146, 1999.
75. MD Ward. Chem Ind 2000(1):22–26.

13

Reversed Micelles as Nanometer-Size Solvent Media

VINCENZO TURCO LIVERI   University of Palermo, Palermo, Italy

I. INTRODUCTION

The potential energy of different topological arrangements of an ensemble of molecules is determined by the strength of the intermolecular interactions triggered by the molecular nature, size, shape, and dynamics. In condensed phases, i.e., when this parameter has a key role, the structural and dynamic properties of matter can be considered a consequence of the more or less wide spectrum of the explored topological arrangements. Concerning the nature of the intermolecular interactions and its effects on molecular aggregation, molecules can be fruitfully classified as apolar, polar, or amphiphilic.

Above the intermolecular contact distance, *apolar* molecules create a force field around them that is always attractive, independent of the nature and orientation of the surrounding molecules. This is because the correlation of the electron density fluctuations of neighboring molecules originates, on the average, from favorably oriented instantaneous dipole moments. Thus, an ensemble of apolar molecules is characterized by a wide spectrum of nearly compact molecular arrangements of comparable potential energy that are, quite equally, frequently explored. For this reason apolar molecules display a limited tendency to give, in the liquid state, a long-range ordered molecular arrangement. This also involves the mixture of two apolar components that are similar in size behaving ideally, and each component is statistically distributed in the total volume of the system.

On the other hand, *polar* molecules create a force field around them that is attractive or repulsive, depending on the relative orientation of the neighboring polar molecule. In this case, the spectrum of molecular arrangements actually explored by an ensemble of strongly polar molecules is severely restricted. It follows that these molecules display a more marked tendency to give a dimensionally unlimited ordered molecular arrangement and a limited mutual solubility with apolar solvents.

In the case of *amphiphilic* molecules, characterized by the coexistence of spatially separated apolar (alkyl chains) and polar moieties, both parts cooperate to drive the intermolecular aggregation. This simple but pivotal peculiarity makes amphiphilic molecules soluble in both polar and apolar solvents and able to realize, in suitable conditions, an impressive variety of molecular aggregates characterized by spatially separated apolar and polar domains, local order at short times and fluidity at long times, and differences in size, shape (linear or branched chains, cyclic or globular aggregates, extended fractal-like molecular networks), and lifetime.

Obviously the various transient microstructures realized in a given system and their relative populations are not decided solely by the nature of the amphiphile, but are the result of a delicate equilibrium of many factors, including the nature of the solvent, the presence of additives, composition, and temperature.

II. REVERSED MICELLES

A. Structural and Dynamic Properties of Reversed Micelles

An enormous literature has been produced in recent decades in the field of molecular aggregation of amphiphilic molecules in liquid systems, emphasizing the extremely wide variety of accessible structures and dynamics. Among these molecular aggregates, in this chapter our attention will be restricted to those formed by some amphiphilic molecules (surfactants) in apolar solvents called *reversed micelles* [1].

The structure of these globular aggregates is characterized by a micellar core formed by the hydrophilic heads of the surfactant molecules and a surrounding hydrophobic layer constituted by their opportunely arranged alkyl chains whereas their dynamics are characterized by conformational motions of heads and alkyl chains, frequent exchange of surfactant monomers between bulk solvent and micelle, and structural collapse of the aggregate leading to its dissolution, and vice versa [2–7].

The conformational dynamics of chain segments near the head groups is more restricted than that of those far from the micellar core [8]. Moreover, to avoid the presence of energetically unfavorable void space in the micellar aggregate and as a consequence of the intermolecular interactions, surfactant molecules tend to assume some preferential conformations and a staggered position with respect to the micellar core [9]. A schematic representation of a reversed micelle is shown in Figure 1.

From a thermodynamic point of view, self-aggregation of amphiphilic molecules in apolar solvents is favored by the negative enthalpic term arising from hydrogen bonding and/or dipole–dipole interactions among surfactant head groups and hindered by the negative entropic term arising from the partial loss of molecular translational and rotational degrees of freedom [10].

FIG. 1 Representation of a reversed micelle ("surfactant molecules" are obtained by combining rubber pipette bulbs and magnetic stir bars).

The frequent breaking and reforming of the labile intermolecular interactions stabilizing the reversed micelles maintain in thermodynamic equilibrium a more or less wide spectrum of aggregates differing in size and/or shape whose relative populations are controlled by some internal (nature and shape of the polar group and of the apolar molecular moiety of the amphiphile, nature of the apolar solvent) and external parameters (concentration of the amphiphile, temperature, pressure) [11]. The tendency of the surfactants to form reversed micelles is, obviously, more pronounced in less polar solvents.

Depending on the polydispersity of these aggregates, surfactants can be divided into two main groups. The first, characterized by a low cooperativity of the intermolecular interactions, is well described by a multistep association process, low mean aggregation number, and high polydispersity, whereas the second, characterized by a high cooperativity, is well described by a single monomer/n-mer equilibrium and high aggregation number [11–13].

Highly monodisperse reversed micelles are formed by sodium bis(2-ethylhexyl) sulfosuccinate (AOT) dissolved in hydrocarbons that are in equilibrium with monomers whose concentration (cmc) is 4×10^{-4} M, have a mean aggregation number of about 23, a radius of 15 Å, exchange monomers with the bulk in a time scale of 10^{-6} s, and dissolve completely in a time scale of 10^{-3} s [1,2,4,14]. Other very interesting surfactants able to form reversed micelles in a variety of apolar solvents have been derived from this salt by simple replacing the sodium counterion with many other cations [15,16].

Examples of other frequently used surfactants that able to form reversed micelles without the addition of cosurfactants are didodecyldimethyl ammonium bromide [17], dodecylammonium propionate, benzyldimethylhexadecyl ammonium chloride [18], lecithin [19], tetraethyleneglycol monododecylether ($C_{12}E_4$) [20], decaglycerol dioleate [21], dodecylpyridinium iodide [22], and sodium bis(2-ethylhexyl) phosphate [23].

B. Solubilization in Reversed Micelles

The main peculiarity of solutions of reversed micelles is their ability to solubilize a wide class of ionic, polar, apolar, and amphiphilic substances. This is because in these systems a multiplicity of domains coexist: apolar bulk solvent, the oriented alkyl chains of the surfactant, and the hydrophilic head group region of the reversed micelles. Ionic and polar substances are hosted in the micellar core, apolar substances are solubilized in the bulk apolar solvent, whereas amphiphilic substances are partitioned between the bulk apolar solvent and the domain comprising the alkyl chains and the surfactant polar heads, i.e., the so-called palisade layer [24].

This peculiar kind of solubilization entails that:

Ionic, polar, and amphiphilic solutes display local concentrations very different from the overall.

Amphiphilic molecules solubilized in the palisade layer are forced to assume an oriented arrangement.

Ionic, polar and amphiphilic solubilizates are forced to reside for relatively long times in very small compartments within the micelle (intramicellar confinement, compartmentalization) involving low translational diffusion coefficients and enhancement of correlation times.

Ionic, polar, apolar, and amphiphilic molecules can coexist in the same liquid system, frequently coming in contact as a consequence of the micellar dynamics and of the large interfacial area between different domains (a typical value of the interfacial area is about 100 m^2/cm^3).

The structure and dynamics of the reversed micelle hosting the solubilizate, as well as the physicochemical properties (structure, dynamics, and reactivity) of the solubilizate, are modified.

Moreover, taking into account that reversed micelles coexist with surfactant monomers, in principle, further effects due to the aggregation of polar and amphiphilic solubilizates with surfactant monomers and the shift of the monomer/reversed micelle equilibrium must be also considered [25,26].

Generally, solubilization occurs spontaneously when the pure solubilizate contacts the solution of reversed micelles. Often, vigorous stirring consistently reduces the time necessary to obtain complete solubilization and thermodynamically stable systems.

Investigations of the solubilization capacity of AOT-reversed micelles have shown that, at infinite dilution of the solubilizates, alcohols and diamines are partitioned between the micellar palisade layer and the bulk solvent. The partition constants, obtained by assuming a Nernstian distribution law, decrease with an increase of the alkyl chain length of the solubilizates, i.e., by increasing their lipophilic character, whereas their transfer enthalpy from the bulk solvent to AOT-reversed micelles is negative [25,27–29]. The analysis of these data reveals the occurrence of hydrogen bonding and dipole–dipole interactions between the polar groups of surfactant and solubilizate and a preference for an ordered packing of amphiphilic solubilizates at the micellar palisade layer [30].

In certain cases, solubilization is driven by specific interactions. For example, the formation of a strong bromine–AOT charge-transfer complex has been considered responsible for the solubilization and location of bromine in AOT-reversed micelles [26].

The contemporaneous presence of different solubilizates sometimes involves competition for the micellar binding sites [31]. For instance, from an analysis of the heats of solution of benzene and water in solutions of reversed micelles of tetraethylene glycol dodecyl ether in decane, a competition between water and benzene for the surfactant hydrophilic groups was shown [32].

Incidentally, it must be pointed out that the contemporaneous or sequential solubilization of finite amounts of appropriately chosen substances within the micellar core is an unexplored research field that potentially opens the door to the study of highly complex and intriguing phenomena.

At infinite dilution, 1-pentanol monomers distribute between AOT-reversed micelles and the continuous organic phase, whereas at finite alcohol concentration, given the ability of alcohol to self-assemble in the apolar organic solvent, a coexistence between reversed micelles (solubilizing 1-pentanol) and alcoholic aggregates (incorporating AOT molecules) is realized [25].

Dynamic light-scattering experiments or the analysis of some physicochemical properties have shown that finite amounts of formamide, N-methylformamide, NN-dimethylformamide, ethylene glycol, glycerol, acetonitrile, methanol, and 1,2 propanediol can be entrapped within the micellar core of AOT-reversed micelles [33–36]. The encapsulation of formamide and N-methylformamide nanoclusters in AOT-reversed micelles involves a significant breakage of the H-bond network characterizing their structure in the pure state. Moreover, from solvation dynamics measurements it was deduced that the intramicellar formamide is nearly completely immobilized [34,35].

In spite of the potentialities of reversed micelles entrapping nonaqueous highly polar solvents [34], very few investigations on the solubilization in such systems are reported in the literature. An example is the study of the solubilization of zinc-tetraphenylporphyrin (ZnTPP) in ethylene glycol/AOT/hydrocarbon systems by steady-state and transient

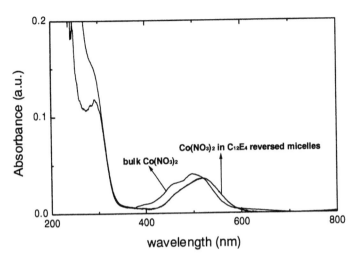

FIG. 2 Comparison between the UV-vis spectra of a $Co(NO_3)_2/C_{12}E_4$/cyclohexane system ($[C_{12}E_4] = 0.094$ mol-kg^{-1}; $[Co(NO_3)_2] = 0.0043$ mol-kg^{-1}) and of a thin film of bulk $Co(NO_3)_2$. The band at about 292 nm is due to the nitrate ion and that at 518 nm to the cobalt ion.

fluorescence techniques, which showed the existence of different spectroscopic ZnTPP species, i.e., species located at different domains of the system [37].

It is also possible to solubilize finite amounts of solid substances within reversed micelles [38–40]. For example, in Figure 2, the UV-vis spectrum of $Co(NO_3)_2$ solubilized in reversed micelles of $C_{12}E_4$ is compared with that of a thin film of bulk $Co(NO_3)_2$. It is interesting to note both similarities and differences between the two spectra. Another example is given by urea, which, as emphasized by the IR spectrum reported in Figure 3, can be

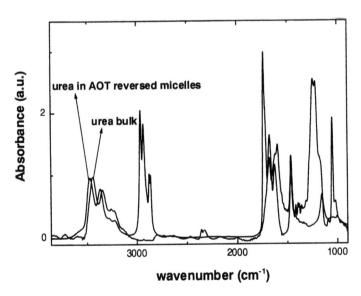

FIG. 3 Comparison between the IR spectra of a urea/AOT/CCl$_4$ system ($[AOT] = 0.1$ mol-kg^{-1}; $[urea]/[AOT] = 0.73$) and of a bulk urea.

solubilized in AOT-reversed micelles dispersed in CCl_4. Both NH stretching band position and shape compared with those of pure urea indicate that it is solubilized in the micellar core, forming strongly perturbed H bonds [41].

Sometimes, solubilization involves marked structural and/or dynamic changes of the reversed micelle solutions. It has been reported that solubilization of stoichiometric amounts of p-cresol, p-ethylphenol, and benzenediols in reversed micelles of AOT in iso-octane leads to dramatic viscosity increases and to gel formation. This has been attributed to the formation of an extended dynamic network constituted of molecules of these substances and AOT linked through hydrogen bonding [42,43]. Given their chemical composition, such highly viscous surfactant-based systems are generally called *organogels*.

Obviously, water, aqueous solutions of salts, and mixtures of highly hydrophilic solvents have also been found to be solubilized in the micellar core [13,44]. The maximum amount of such solubilizates that can be dissolved in reversed micelles varies widely, strongly depending on the nature of the surfactant and the apolar solvent, on the concentrations of surfactant and of additives, and on temperature [24,45–47].

For instance, it has been observed that the addition of additives such as cyclohexane, benzene, and nitrobenzene to water/AOT/isooctane systems considerably increases the maximum amount of solubilized water [48]. The same effect has been observed in the presence of finite amounts of cytochrome c [49].

In some systems containing surfactant mixtures, a synergistic effect on the water solubilization capacity has been observed [50].

In general, solubilization of appropriate substances in a solution of reversed micelles forming a solid, liquid, or gel core within the reversed micelle is the preliminary step to realize systems easily handled and interesting both from the theoretical and the practical points of view.

The effects of the intramicellar confinement of polar and amphiphilic species in nanoscopic domains dispersed in an apolar solvent on their physicochemical properties (electronic structure, density, dielectric constant, phase diagram, reactivity, etc.) have received considerable attention [51,52]. In particular, the properties of water confined in reversed micelles have been widely investigated, since it simulates water hydrating enzymes or encapsulated in biological environments [13,23,53–59].

Studies of reversed micelles dispersed in supercritical fluids have shown their ability to solubilize hydrophilic substances, including biomolecules and dyes, opening the door to many new applications [60,61]. In particular, solutions of reversed micelles in liquid and supercritical carbon dioxide have been suggested as novel media for processes generating a minimum amount of waste and with a low energy requirement [62].

A mimic system of a photosynthetic apparatus was realized by solubilizing C-phycocyanin and zinc phthalocyanine in reversed micelles of tween-80 dispersed in cyclohexane [63].

Dodecylpyridinium iodide–reversed micelles trapping chlorophyll a have been suggested as interesting photochemical model systems [22] and water/AOT/chloroethylene systems as peculiar dry-cleaning solvents [64].

Recently, an electrorheological effect, i.e., an increase in the viscosity and dynamic shear moduli of lecithin/n-decane solutions in the presence of small amounts of polar additives (water or glycerol) when an external electric field is applied to the system, has been observed [65].

Considering that microwaves couple mainly with polar and amphiphilic molecules, it has been suggested that microwave irradiation of these microheterogeneous systems, at appropriately chosen frequencies and within a system-specific time scale, allows a selective heating of only the hydrophilic microdomains. Then intriguing effects on reaction rates and/or mechanisms can be expected, such as selective thermal decomposition of hydrophilic solubilizates, which leaves practically unaltered apolar thermolabile solubilizates [66].

Other applications are based on the use of solutions of reversed micelles as templates. For example, solutions of reversed micelles have been employed as a matrix to control the porosity of cross-linked polymer resins. The pore size of the polymers was controlled by varying the amounts of water in the AOT-reversed micelles [67].

Given their radio-frequency electrical properties and nuclear magnetic resonance chemical shift components, solutions of reversed micelles constituted of water, AOT, and decane have been proposed as suitable systems to test and calibrate the performance of magnetic resonance imagers [68].

III. WATER-CONTAINING REVERSED MICELLES

A. Structural and Dynamic Properties of Water-Containing Reversed Micelles

Solubilization of water or other highly hydrophilic substances within the micellar core involves solvation of the surfactant head group, accompanied by an increase in the effective head group area, a decrease in the staggering of surfactant molecules, conformational changes of the surfactant, micellar swelling, a marked increase in the surfactant aggregation number, changes in the flexibility, penetrability, and packing order of the surfactant layer, change in the intermicellar interactions, and, at constant surfactant concentration, a decrease in the number density of reversed micelles [69]. Moreover, reversed micelles are transformed from short-living aggregates to more stable molecular assemblies, with a greater persistence in size and shape of the entire aggregate [70].

However, large and frequent fluctuations of local properties have to be considered to achieve a realistic view of these systems. The dynamics of water-containing reversed micelles is in fact characterized by a wide variety of processes, such as lateral diffusion of surfactant molecules at the water/surfactant interface, conformational dynamics of its polar and apolar moieties, fast exchange of water molecules between the surface and the center of the hydrophilic core, micellar shape, and charge (in the case of ionic surfactant) fluctuations, exchange of water and surfactant molecules between bulk solvent and micelle, diffusion and rotation of the entire aggregate, intermicellar encounters, breaking/reforming of adhesive bonds between contacting micelles, and intermicellar material exchange [3,5–7,71–78].

In the case of water-containing AOT-reversed micelles, less than 1 in 1000 intermicellar collisions leads to micelle coalescence followed by separation and a material exchange process occurring in the microsecond to millisecond time scale [3,79].

The intermicellar material exchange process can also be assisted by the exocytotic-endocytotic mechanism, i.e., through the gemmation of a minimicelle from a micelle, its diffusion and subsequent coalescence with another micelle [6,69].

The importance of the material exchange process can hardly be overemphasized since it is the mechanism whereby the equilibrium micellar size and polydispersity are reached and maintained, the reversed micelles of ionic surfactants become charged, polar and amphiphilic solubilizates are transported, and hydrophilic reactants can come in contact and react. Besides, it must be pointed out that solutions of water-containing reversed micelles may be considered effectively continuous systems at timescales longer than the characteristic time of the exchange process, whereas only at shorter time scales (i.e., on millisecond–microsecond timescales) may they be considered to be constituted of discrete entities dispersed in the apolar solvent [5,80].

Obviously, factors that determine a decrease in the material exchange rate are expected to extend intermicellar confinement effects to a longer time scale and/or to lead to more marked confinement effects. Enhancement or inhibition of the material exchange rate can be attained through the control of temperature, the nature of surfactant and apolar solvent, composition, and the presence of additives [5,78].

The size and shape of water-containing reversed micelles and their dependence on the water and surfactant concentrations are system specific. The micellar size is regulated mainly by the pronounced tendency of the surfactant to be located opportunely oriented between water and apolar solvent involving a huge value of the water/apolar solvent interface and nanosize reversed micelles. Striking is the peculiar case of didodecyldimethylammonium bromide, which is practically insoluble in either water or alkanes but forms thermodynamically stable systems in a wide composition range in the presence of both [81].

Spherical water-containing reversed micelles, characterized by the smallest micellar surface-to-volume ratio, occur when water/surfactant interactions are less favorable than water/water and/or surfactant/surfactant interactions, while rodlike water-containing reversed micelles, characterized by a greater surface-to-volume ratio, occur when water/surfactant interactions are more favorable than water/water and/or surfactant/surfactant interactions. Since the strength of water/surfactant and surfactant/surfactant interactions is strongly influenced by the geometric properties of the surfactant hydrophilic and hydrophobic moieties, geometric parameters of the surfactant molecules are useful for predicting their aggregational behavior [82,83].

Independent of the nature of the apolar solvent, nearly spherical and monodisperse water-containing reversed micelles are formed by AOT, whose size is quite independent of the surfactant concentration and regulated mainly by the molar ratio R (R = [water]/[surfactant]) [5,84,85].

A simple geometric model, based on the hypothesis that water plus surfactant are subdivided in nanospheres and that their total surface is fixed by the amount of surfactant, can predict the dependence of the micellar radius (r) on R and that of the micellar concentration on R and on the surfactant concentration.

Assuming spherical micelles, the volume V_m and the surface A_m of a micelle are given by

$$V_m = \frac{4\pi r^3}{3} = n_s V_s + n_w V_w \tag{1}$$

$$A_m = 4\pi r^2 = n_s A_s \tag{2}$$

where n_s and n_w are, respectively, the number of moles of surfactant and water in one micelle, V_s and V_w their molar volumes, and A_s the molar interfacial area of the surfactant at the surfactant/apolar solvent interface. Combining Eqs. (1) and (2) and remembering that $R = n_w/n_s$, one obtains

$$r = \frac{3V_s}{A_s} + \frac{3V_w R}{A_s} \tag{3}$$

Experimentally, it has been found that for water-containing AOT-reversed micelles, V_s, V_w, and A_s change with R, becoming nearly constant above $R = 10$ [86,87]. It follows that only above $R = 10$ it can a linear relationship between r and R be expected [88].

The number density (N) of micelles as a function of the volume fraction Φ of the dispersed phase (water plus surfactant) can be calculated by

$$N = \frac{\Phi}{V_m} = \frac{\Phi}{\dfrac{4\pi r^3}{3}} = \frac{\Phi}{36\pi\left(\dfrac{V_s}{A_s} + \dfrac{V_w}{A_s}R\right)^3} \tag{4}$$

Equation 4 shows that, at constant Φ, a change of the external parameter R affects not only the radius but also the concentration of water-containing reversed micelles. It is also of interest that, by increasing R, the fraction of bulklike water molecules located in the core (or the time fraction spent by each water molecule in the core) of spherical reversed micelles increases progressively, whereas the opposite occurs for perturbed water molecules located at the water–surfactant interface, as a consequence of the parallel decrease of the micellar surface-to-volume ratio.

According to current knowledge, the spherical water-containing reversed micelles can be well represented by the onion model (see Fig. 4). In a time scale smaller than that of the molecular exchange between different micellar sites, this model takes into account the presence of two different hydrophilic regions (B and C) and three different solubilization sites (the aqueous core, the hydrated head group region, and the palisade layer), which are more or less greatly affected by a change of R.

The apparent molar volume of interfacial water in AOT-reversed micelles is lower and its refractive index is greater than that of pure water. These findings, together with other experimental evidence, emphasize that these water molecules are destructured, immobilized, and polarized by the ionic head of AOT [2,84,89]. In particular, it has been reported that the

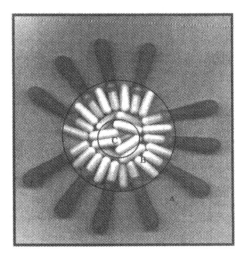

FIG. 4 Onion model of spherical water-containing reversed micelles. Solvent molecules are not represented. A, surfactant alkyl chain domain; B, head group plus hydration water domain; C, "bulk" water domain. (For water-containing AOT-reversed micelles, the approximate thickness of layer A is 1.5 nm, of layer B is 0.4 nm, whereas the radius of C is given by the equation $r_w = 0.17R$ nm.)

characteristic time of orientational relaxation of the water molecules at the micellar interface (2–8 ns) is much slower than that of water molecules in bulk water (0.3 ps) [77].

Sound velocity results showed that water-filled reversed micelles are more floppy than bulk water [70].

The small and positive values of enthalpy of solution of water in AOT-reversed micelles indicate that its energetic state is only slightly changed and that water solubilization (unfavorable from an enthalpic point of view) is driven mainly by a favorable change in entropy (the destructuration of the water at the interface and its dispersion as nanodroplets could be prominent contributions) [87].

The water structure at the water/surfactant interface depends on the nature of the surfactant head group, whereas the hydrophobic interface plays only a secondary role [91–93].

In contrast, thermodynamic as well as spectroscopic properties of "core" water in AOT-reversed micelles are similar to those of pure water. Together with electrostatic considerations, this suggests that the penetration of counterions in the micellar core is negligible and that a relatively small number of water molecules are able to reconstruct the typical extended H-bonded structure of bulk water.

Indeed, the degree of binding of the counterions to the micellar surface, even in the largest aqueous core, is found to be 72% [2,94]. This means that virtually all counterions are confined in a thin shell near the surface (about 4 Å), the concentration of ions in this domain is very high, and a nearly ordered bidimensional spherical lattice of charges is formed at the water/surfactant interface of ionic surfactants.

It has been suggested that this spherical layer of charges effectively screens water from the external electric field, leading to an apparent dielectric constant much lower than that of bulk water [35,91,95,96].

Differential scanning calorimetry measurements have shown a marked cooling/heating cycle hysteresis and that water entrapped in AOT-reversed micelles is only partially freezable. Moreover, the freezable fraction displays strong supercooling behavior as an effect of the very small size of the aqueous micellar core. The nonfreezable water fraction has been recognized as the water located at the water/surfactant interface engaged in solvation of the surfactant head groups [97,98].

Incidentally, it is of interest to note that solutions of water-containing reversed micelles could be employed to study the physicochemical properties of nanosize solid water.

Also, the segmental mobility and preferential conformation of surfactant alkyl chains is perturbed by water addition [92,99].

By ^{1}H and ^{13}C NMR, the occurrence of conformational changes of the AOT molecules with R has been emphasized [13]. An increase of the lateral packing order of the surfactant alkyl chains located in the micellar palisade layer due to water addition has also been pointed out by FT-IR [58,92].

Elongated micelles occur when water/surfactant interactions are more favorable than surfactant/surfactant and water/water interactions involving the unidimensional growth of the micelles with R and the absence of a bulklike water domain within the micellar core (see Figure 5). Lecithin, for example, being able to establish strong hydrogen bonds with water, forms very long rodlike water-containing reversed micelles [58,100–103]. Their length is triggered mainly by the scission energy, i.e., the energy necessary to break a micelle into two parts [104].

The solubilization of water in lecithin-reversed micelles has been found to be an exothermic process. This finding confirms that water interacts with the zwitterionic head group of lecithin, promoting the formation of strong intermolecular H bonds [104].

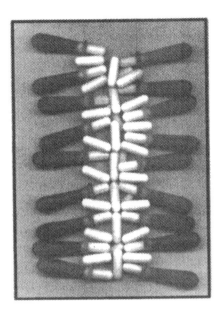

FIG. 5 Representation of rodlike water-containing reversed micelles.

The vibrational dynamics of water solubilized in lecithin-reversed micelles appears to be practically indistinguishable from those in bulk water; i.e., in the micellar core an extensive hydrogen bonded domain is realized, similar, at least from the vibrational point of view, to that occurring in pure water [58]. On the other hand, the reorientational dynamics of the water domain are strongly affected, due to water nanoconfinement and interfacial effects [105,106].

In many cases, under changing experimental conditions, water-containing reversed micelles evolve, exhibiting a wide range of shapes such as disks, rods, lamellas, and reverse-vesicular aggregates [15,107,108]. Nickel and copper bis(2-ethylhexyl) sulfosuccinate and sodium bis(2-ethylhexyl) phosphate, for example, form rod-shaped droplets at low water contents that convert to more spherical aggregates as the water content is increased [23,92,109,110].

Didodecyldimethylammonium bromide forms in the dilute-phase cylindrical reversed micelles in the range $2 < R < 8$, showing an abrupt structural change at $R = 10$ and forming spherical aggregates at $R > 10$ [17].

The cobalt, nickel, and copper bis(2-ethylhexyl) phosphate surfactants dissolved in n-heptane lead to quasi-one-dimensional association microstructures, i.e., rodlike reversed micelles that increase in size via water solubilization [111].

B. Solubilization in Water-Containing Reversed Micelles

Many investigations have been performed on the solubilization of solutes within water-containing reversed micelles to probe micellar structure and dynamics, to study intramicellar and intermicellar confinement effects, to define their partitioning, and to point out mutual modifications due to solute–micelle interactions. The knowledge of the solubilization site of a solute in solution of water-containing reversed micelles and of the interaction forces governing its partitioning between the different microregions is fundamental to rationalize its enhanced solubility and reactivity [112,113].

It is of interest that, as a consequence of the peculiar state of reactants in such systems, reactions rates and equilibrium constants are very often altered by several orders of magnitude as compared with those in homogeneous solution [114,115].

Moreover, solutions of water-containing reversed micelles (also called w/o microemulsions) have been considered useful solvent and reaction media for many technological applications (solvent extractions, lubrication, enhanced solubilization, and stabilization of drugs and cosmetics, photochemical reactions, polymerization processes, nanoparticle synthesis, corrosion inhibition, catalysis, tertiary oil recovery, etc.). Also, some resemblance between these systems and biological environments has been the driving force to employ solutions of reversed micelles to model or to mimic some aspects of biological processes, such as biomineralization and enzyme-mediated synthesis, or to realize pharmaceutical preparations [116,117]. This is because, as in biological systems, it is possible to take advantage of confinement effects to drive and to control processes such as polymerization, precipitation, catalysis, diffusion, and molecular separation.

Since very often the physicochemical properties of solubilizates are modified when they are entrapped in reversed micelles, almost all the experimental techniques can been used, and have been used, to study solubilization [28,31,118–122].

In order to rationalize the experimental results of solubilization investigations at finite solubilizate concentrations, it is fundamental to consider the mean number of solubilizate molecules per micelle, their location within each micelle, and their distribution among the micellar aggregates. Several distribution models of solubilizates in solutions of water-containing reversed micelles have been proposed. In the case of solubilizates at low concentration and that do not significantly influence the micellar structure, a random distribution is generally found. When there is a marked attractive interaction between solubilizate molecules or the changes due to the solubilizate molecule make the micelle more disposed to host other solubilizate molecules, they tend to be located in the same micelle. This effect is dramatically observed during the formation of solid nanoparticles within reversed micelles. On the other hand, the complex interactions responsible for the solubilizate distribution can also lead to a preferential solubilization in empty micelles (repulsive distribution) [44].

As a result of their size and of specific interactions, hydrophilic macromolecules or solid nanoparticles cause strong changes in micellar size and dynamics, and their structural and dynamic properties are strongly affected. In these cases, the distribution among reversed micelles can be only described by ad hoc models [13,123].

Moreover, as a general behavior and indirect effect of solubilizate/micelle interactions, incorporation of additives influences the stability and the phase diagram of the system [48,119,124].

In recent decades, many investigations have been carried out on the solubilization and on the physicochemical characterization of a wide variety of substances confined in water-containing reversed micelles. Even if these studies have not produced a general theory to predict a priori all the effects accompanying the solubilization process, some general aspects nonetheless have been underlined. In the following, the results of some of these investigations, selected to show the extent of some peculiar behaviors, will be reported.

C. Solubilization of Electrolytes

The secondary solubilization of electrolytes, namely, their solubilization into water solubilized in the micelle, has been widely investigated [125–127].

Electrolytes are obviously solubilized only in the aqueous micellar core. Adding electrolytes in water-containing AOT-reversed micelles has an effect that is opposite to that observed for direct micelles, i.e., a decrease in the micellar radius and in the intermicellar attractive interactions is observed. This has been attributed to the stabilization of AOT ions at the water/surfactant interface [128].

Effects on the micellar shape are also induced by electrolyte addition. It has been observed that, in decane, the water-containing AOT-reversed micelles become more spherical upon addition of salt (NaCl, $CaCl_2$) [6].

The addition of salts modifies the composition of the layer of charges at the micellar interface of ionic surfactants, reducing the static dielectric constant of the system [129,130]. Moreover, addition of an electrolyte (NaCl or $CaCl_2$) to water-containing AOT-reversed micelles leads to a marked decrease in the maximal solubility of water, in the viscosity, and in the electrical birefringence relaxation time [131].

Investigations of the solubilization of water and aqueous NaCl solutions in mixed reverse micellar systems formed with AOT and nonionic surfactants in hydrocarbons emphasized the presence of a maximum solubilization capacity of water, occurring at a certain concentration of NaCl, which is significantly influenced by the solvent used [132].

By NMR, it has been observed that the solubilization of various diamagnetic salts in water/AOT/n-octane microemulsions induces a micellar reorganization that is dependent on the electrolyte valence and concentration [133].

Sometimes, the physicochemical properties of ionic species solubilized in the aqueous core of reversed micelles are different from those in bulk water. Changes in the electronic absorption spectra of ionic species (I^-, Co^{2+}, Cu^{2+}) entrapped in AOT-reversed micelles have been observed, attributed to changes in the amount of water available for solvation [2,92,134]. In particular, it has been observed that at low water concentrations cobalt ions are solubilized in the micellar core as a tetrahedral complex, whereas with increasing water concentration there is a gradual conversion to an octahedral complex [135].

D. Solubilization of Small Polar and Amphiphilic Molecules

The different location of polar and amphiphilic molecules within water-containing reversed micelles is depicted in Figure 6. Polar solutes, by increasing the micellar core matter of spherical micelles, induce an increase in the micellar radius, while amphiphilic molecules, being preferentially solubilized in the water/surfactant interface and consequently increasing the interfacial surface, lead to a decrease in the micellar radius [49,136,137]. These effects can easily be embodied in Eqs. (3) and (4), allowing a quantitative evaluation of the mean micellar radius and number density of reversed micelles in the presence of polar and amphiphilic solubilizates. Moreover it must be pointed out that, as a function of the specific distribution law of the solubilizate molecules and on a time scale shorter than that of the material exchange process, the system appears polydisperse and composed of empty and differently occupied reversed micelles [136].

By small-angle neutron scattering experiments on water/AOT/hydrocarbon microemulsions containing various additives, the change of the radius of the micellar core with the addition of small quantities of additives has been investigated. The results are consistent with a model in which amphiphilic molecules such as benzyl alcohol and octanol are preferentially adsorbed into the water/surfactant interfacial region, decreasing the micellar radius, whereas toluene remains predominantly in the bulk hydrocarbon phase. The effect of n-alcohols on the stability of microemulsions has also been reported [119].

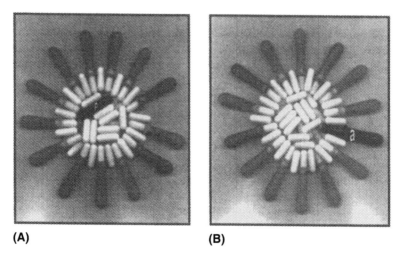

(A)   **(B)**

FIG. 6 Representation of spherical water-containing reversed micelles solubilizing a polar molecule (p) in the micellar core (A) or an amphiphilic molecule (a) in the palisade layer (B).

In addition to the degree of hydrophilicity of the solubilizates, their size and structure, the size of the host microregions, or the occurrence of specific processes must be taken into account in order to rationalize the driving forces of the solubilization process and of the solubilization site within water-containing reversed micelles [25,138,139].

The solubilization of small amphiphilic solutes in water-containing reversed micelles can be consistently treated by a simple Nernstian distribution law, taking into account all the possible solubilization sites [29]. It has been found that the distribution coefficients of alcohols and diamines between aqueous core and bulk apolar solvent decrease by increasing the hydrophobicity of the solubilizate, while the opposite occurs to the distribution coefficients between the palisade layer of AOT-reversed micelles and the bulk apolar solvent. Moreover, at a sufficiently high R value, for the more hydrophilic solubilizates the preferential solubilization site changes from the palisade layer to the micellar core. The enthalpies of transfer from the apolar solvent to the micelles are exothermic (solubilization driven mainly by a favorable enthalpic term), and to their values also contribute changes in the micellar structure and/or preferential location of the solubilizate in the palisade layer [28]. These thermodynamic data also indicate that in the palisade layer the solubilizates are forced to orient so that only one polar group of the diamines interacts with the hydrated surfactant head groups, while this does not occur in the aqueous core, where both polar groups of the diamines can be hydrated, emphasizing the peculiar solvent properties of the water-containing AOT-reversed micelles.

Fluorescence investigations of the partitioning of the aromatic fluorophore Prodan in water/AOT/n-heptane, water/DTAB/n-hexanol/n-heptane, and water/CTAB/n-hexanol/n-heptane microemulsions proved that this molecule, as a consequence of a variety of noncovalent interactions, is distributed in several distinct micellar domains [140].

In the case of Kryptofix 221D, a cryptand able to complex the alkali metal cations [141–143], it has been observed that it is solubilized mainly in the palisade layer of the AOT-reversed micelles. And from an analysis of the enthalpy of transfer of this solubilizate from the organic to the micellar phase it has been established that the driving force of the solubilization is the complexation of the sodium counterion. In addition, the enthalpy values show the peculiar solvation state of sodium counterions and that they are located essentially near the water–AOT interface [138].

A rather unexpected solubilization phenomenon has also been described, i.e., the pressure-induced encapsulation of low-molecular-weight gases in the aqueous micellar core, followed by clathrate hydrate formation [144,145].

Electronic properties of solubilizates are significantly influenced by the micellar microenvironment as a consequence of the abnormal water properties in w/o microemulsions and/or of their specific location, and this effect has been employed to define their solubilization site [46–148]. By UV-vis spectrophotometry, the interaction between melatonin- and water-containing AOT- and lecithin-reversed micelles has been investigated [113]. The experimental results indicate that melatonin strongly prefers to be located at the micellar palisade layer. This, while supporting the hypothesis that melatonin may provide antioxidant protection without the benefit of receptors, suggests that it could easily scavenge aqueous as well as lipophilic radicals.

The fluorescence characteristics of daunomycin, doxorubicin, and other anthracycline drugs solubilized in the water/AOT/n-heptane system were used to monitor their localization [121].

The microenvironment in water-containing AOT-reversed micelles has a marked effect on the spectral properties of fluorescein. The absorption peaks are red-shifted by about 10 nm from the corresponding positions in aqueous solution, the absorption extinction coefficient increases with R, and the fluorescence is more effectively quenched in AOT-reversed micelles than in aqueous solution [149].

By flourescence techniques, it was observed that the fluorescence yield and lifetime of 1,8-anilinonaphthalenesulfonate decrease with an increase in the aqueous core of AOT-reversed micelles, while the position of the emission maximum shifts to longer wavelengths [64]. These changes in the electronic properties were attributed to the peculiar effective polarity and viscosity of the micellar core and to their evolution with R.

It has been reported that while in aqueous solution the lifetime of optically excited nile red is 0.65 ns; inside AOT-reversed micelles it is 3.73 ns, becoming 2.06 ns at $R = 30$ [150].

The fluorescence lifetime of trans-4-[4-(dimethylamino-styryl]-1-methylpyridinium iodide trapped in water-containing AOT-reversed micelles has been found to be markedly influenced by R, implying a significant effect on its excited-state twisting motion [122].

Using the UV-vis absorption spectra of phenol blue and methyl orange solubilized in solutions of AOT-reversed micelles in supercritical ethane, it was ascertained that phenol blue resides in the palisade layer, whereas methyl orange resides in the reversed micelle core [61]. Depending on temperature it has also been observed that methyl orange solubilized in a w/o microemulsion composed of water, potassium oleate, hexanol, and hexadecane is located in the micellar palisade layer as monomer and/or aggregates in the parallel orientation [120].

Also, acridine orange dissolved in an AOT-based w/o microemulsion is located at the micellar palisade layer, forming bidimensional aggregates by increasing the dye concentration [151].

Even molecules highly soluble in water, such as phenols, are preferentially solubilized in the micellar palisade layer. The binding of phenol to AOT-reversed micelles in isooctane shows a decrease of the binding constant followed by a plateau value as the water-to-AOT ratio is increased. This behavior was rationalized in terms of phenol–water competition for interfacial binding sites [152].

The solubilization of amino acids in AOT-reversed micelles has been widely investigated showing the importance of the hydrophobic effect as a driving force in interfacial solubilization [153–157]. Hydrophilic amino acids are solubilized in the aqueous micellar core through electrostatic interactions. The amino acids with strongly hydrophobic groups are incorporated mainly in the interfacial layer. The partition coefficient for tryptophan and micellar shape are affected by the loading ratio of tryptophan to AOT [158].

By IR spectroscopy it was emphasized that the solubilization of amino acids or oligopeptides in water-containing lecithin-reversed micelles involves structural changes in the aqueous micellar core [159].

Some investigations have tested the ability of reversed micelles to act as efficient carriers of molecular species. Solutions of water-containing AOT-reversed micelles have been employed for the selective transport and the efficient separation of the two amino acids tryptophane and p-iodophenylalanine [160].

The transdermal permation of glyceryl trinitrate encapsulated in AOT-reversed micelles was compared with that of an aqueous solution, and an enhancement in permeation was found as well as the absence of skin irritation [161].

E. Solubilization of Macromolecules

The solubilization of enzymes and proteins in water-containing reversed micelles has attracted a great deal of interest for their selective separation, purification, and efficient refolding and for bioreactions involving a wide class of polar, apolar, and amphiphilic reactants and products [13,44,162–164].

Entrapment of enzymes within reversed micelles can be achieved simply by dissolving the biopolymer, pure or solubilized in an appropriate solvent, in a solution of reversed micelles or by extraction from an immiscible liquid phase [13,165,166].

Using a solution of water-containing reversed micelles of di(2-ethylhexyl)phosphorothioic acid in isooctane, hemoglobin was extracted and concentrated. Desolubilization of the protein entrapped in the reversed micelles by weak alkaline solution was realized by adding small amounts of n-octanol [167].

The influence of pH, ionic strength, and protein concentration on the extraction of α-lactalbumin and β-lactoglobulin from an aqueous solution with water/AOT/isooctane microemulsions and their separation has been reported [168].

The entrapment of α-chymotrypsin, lysozyme, and myelin in AOT-reversed micelles is accompanied by an increase in the micellar water content and in the size of the micelle. As a consequence of the redistribution of water among reversed micelles, the micellar solution results in being constituted by large protein-containing micelles and small unfilled ones [169].

The influence of system parameters such as protein charge, size, and concentration, ionic strength, and water content on the sizes of filled and unfilled reversed micelles has been investigated [170].

By adding 1-alkanols to AOT-based w/o microemulsions, some proteins (ribonuclease, lysozyme, alpha-chymotrypsin, pepsin, bovine serum albumin, and catalase) are readily expelled, while the major part of the surfactant remained in solution [171].

Solubilization of biomolecules could induce change in the microemulsion structure. For example, in the presence of the human serum albumin and at low R value, the ternary microemulsion AOT/water/isoctane shows a transition to a bicontinuous microstructure [172].

As a result of the micellar environment, enzymes and proteins acquire novel conformational and/or dynamic properties, which has led to an interesting research perspective from both the biophysical and the biotechnological points of view [173–175]. From the comparison of some properties of catalase and horseradish peroxidase solubilized in water/AOT/n-heptane microemulsions with those in an aqueous solution of AOT it was ascertained that the secondary structure of catalase significantly changes in the presence of an aqueous micellar solution of AOT, whereas in AOT/n-heptane reverse micelles it does not change. On the other hand, AOT has no effect on horseradish peroxidase in aqueous solution, whereas slight changes in the secondary structure of horseradish peroxidase in AOT/n-heptane reverse micelles occur [176].

In the case of myelin proteolipid solubilized in water/tetraethylene glycol monododecyl ether/dodecane microemulsions, its α-helical structure is preserved [176,177].

For many solubilized enzymes the greatest catalytic activity and/or changes in conformation are found at $R < 12$, namely, when the competition for the water in the system between surfactant head groups and biopolymers is strong. This emphasizes the importance of the hydration water surrounding the biopolymer on its reactivity and conformation [13].

It has been reported that enzymes incorporated in the aqueous polar core of the reversed micelles are protected against denaturation and that the distribution of some proteins, such as chymotrypsine, ribonuclease, and cytochrome c, is well described by a Poisson distribution. The protein state and reactivity were found markedly different from those observed in bulk aqueous solution [178,179].

It has been observed that whereas the catalytic activity of malic dehydrogenase in water is not influenced by pressure, in reversed micelles it shows a bell-shaped dependence, suggesting regulation of the enzymatic activity by pressure application, which cannot be realized in aqueous solutions [180].

The activity of α-chymotrypsin was found to be insensitive to the R value, i.e., from the size of the reversed micelles. This was taken as an indication that this enzyme is able to create its own micelles in the hydrocarbon rather than occupy empty ones and that the so-called exclusion effect, i.e., protein larger than the empty micelle cannot be solubilized, is incorrect [181,182].

Even entrapment of entire cells within reversed micelles without loss of their functionality has been achieved. For example, mitochondria and bacteria (*Actinobacter calcoaceticus, Escherichia coli, Corynebacterium equi*) have been successfully solubilized in a microemulsion consisting of isopropyl palmitate, polyoxyethylene sorbitan trioleate [162].

Enhanced hydrogen photoproduction by the bacterium *Rhodopseudomonas sphaeroides* or by the coupled system *Halobacterium halobium* and chloroplasts organelles entrapped inside the aqueous core of reversed micelles with respect to the same cells suspended in normal aqueous medium has been reported [183,184].

The entrapment of nonbiological polymeric materials within water-containing reversed micelles has also been investigated [185]. From the technological point of view, the interest in mixed molecular aggregates composed of polymers and surfactants is due to the peculiar nanostructures that can be realized and consequently to the possibility of obtaining advanced materials for specific applications. Studies on the solubilization of polyacrylic acid and its sodium salt in the rodlike reversed micelles of the cationic surfactant didodecyldimethylammonium bromide indicated the occurrence of conformational change of these polyelectrolytes [186].

In the presence of the polyelectrolyte polyallylamine hydrochloride (PAAN), the formation of a pearl-necklace structure between AOT-reversed micelles and PAAN was characterized by two sizes: the size of each micelle and that of the polymer chain connecting the reversed micelles [187]. The preferential adsorption of AOT-reversed micelles on the polymeric chain of polystyrene in toluene has also been revealed [188].

Solubilization of a graft copolymer comprising a hydrophobic poly(dodecyl-methacrylate) backbone and hydrophilic poly(ethylene glycol) monomethyl ether side chains in water/AOT/cyclohexane w/o microemulsions was rationalized in terms of the backbone dissolved in the continuous apolar phase and the side chains entrapped within the aqueous micellar cores [189].

Effects due to the addition of water-soluble polymers (polyoxyethylene glycol, poly-acrylamide, and polyvinyl alcohol) on water/AOT/decane w/o microemulsions have been reported [190].

In addition to solubilization, entrapment of polymers inside reversed micelles can be achieved by performing in situ suitable polymerization reactions. This methodology has some specific peculiarities, such as easy control of the polymerization degree and synthesis of a distinct variety of polymeric structures. The size and shape of polymers could be modulated by the appropriate selection of the reversed micellar system and of synthesis conditions [31,191]. This kind of control of polymerization could model and/or mimic some aspects of that occurring in biological systems.

By performing in situ the polymerization of acrylamide in water/AOT/toluene microemulsions, clear and stable inverse latexes of water-swollen polyacrylamide particles stabilized by AOT and dispersed in toluene have been found [192–194]. It was shown that the final dispersions consist of two species of particles in equilibrium, surfactant-coated polymer particles (size about 400 Å) with narrow size distribution and small AOT micelles (size about 30 Å).

Nanosize particles of polyacrylic acid were synthesized in w/o microemulsions using azobisisobutyronitrile as lipophilic radical initiator, which were considered suitable for encapsulation of peptides and other hydrophilic drugs [195].

F. Hosting Nanoparticles

The production of systems consisting of reversed micelles entrapping nanoparticles and dispersed in an organic solvent of low polarity is of great interest because of their potential technological and biotechnological applications [196–198]. This is so because in addition to the quantum size effects and the huge surface/volume ratio characterizing the properties of nanoparticles, new distinct features are conferred by their formation and confinement within reversed micelles, adsorption of surfactant molecules at the nanoparticle surface, and dispersion in an apolar medium [199–202]. For example, compared with Mn-doped ZnS materials synthesized through the conventional aqueous reaction, the same particles prepared in w/o microemulsion show a significant enhancement of the photoluminescence [201].

Other advantages of the use of a solution of water-containing reversed micelles as solvent and reaction media are:

Synthesis of nanoparticles can be performed at mild conditions (high temperature or very low pressure are unnecessary).

A wide class of materials (metals, semiconductors, superconductors, biominerals, water-soluble inorganic and organic compounds, etc.) can be produced using these systems [203–206].

The synthesis can easily be modulated to obtain coated nanoparticles, doped nanoparticles, mixed nanoparticles, or onion nanoparticles [207–211].

Since by changing the nature and/or the concentration of the components of the w/o microemulsions it is possible to change size and/or shape (spheres, needles, cubes,
wires, bundles, etc.) of the hydrophilic microregions, in principle, the size, size distribution, and shape of the nanoparticles could easily be modulated [197,202,
212,213].
By simple evaporation of the volatile components of nanoparticle-containing w/o microemulsions, it is possible to obtain very interesting nanoparticle/surfactant composites [214].

Moreover, stable liquid systems made up of nanoparticles coated with a surfactant
monolayer and dispersed in an apolar medium could be employed to catalyze reactions involving both apolar substrates (solubilized in the bulk solvent) and polar and amphiphilic
substrates (preferentially encapsulated within the reversed micelles or located at the surfactant palisade layer) or could be used as antiwear additives for lubricants. For example,
monodisperse nickel boride catalysts were prepared in water/CTAB/hexanol microemulsions and used directly as the catalysts of styrene hydrogenation [215].

Since some structural and dynamic features of w/o microemulsions are similar to
those of cellular membranes, such as dominance of interfacial effects and coexistence of
spatially separated hydrophilic and hydrophobic nanoscopic domains, the formation of
nanoparticles of some inorganic salts in microemulsions could be a very simple and realistic way to model or to mimic some aspects of biomineralization processes [216,217].

In addition, it is of interest to note that investigations of the microscopic processes
leading to nucleation, growth, oriented growth by the surfactant monolayer, and growth inhibition of nanoparticles in reversed micelles and of confinement and adsorption effects on
such phenomena represent an intriguing and quite unexplored research field [218].

Taking into account that the state of nanoparticles is thermodynamically unstable
against an unlimited growth, the physicochemical processes allowing reversed micelles to
lead to stable dispersions and to a size control of nanoparticles are:

The tendency to maintain their spontaneous size and shape
Charging of nanoparticles [219]
Adsorption of surfactant molecules and/or of suitable coating agents at the nanoparticle
surface
Compartimentalization of nanoparticles in spatially distinct domains

Nanoparticles solubilized in w/o microemulsions have been obtained by performing
in situ suitable reactions [196], by dispersion of particles [219,220], or by controlled nano-
precipitation of a solubilizate [221,222].

Generally, nanoparticles entrapped in water-containing reversed micelles are obtained by mixing two w/o microemulsions carrying inside the reversed micelles the appropriate hydrophilic reagents, which, upon coming in contact and reacting, form the precursors of the nanoparticles. The accumulation of these precursors in reversed micelles leads
to the formation of nanoparticles. Then the dispersion of nanoparticles in spatially distinct
domains and/or surfactant adsorption on the nanoparticle surface could prevent unlimited
growth and precipitation [5,217,223].

Some investigations have emphasized the importance of micellar size as a control parameter of nanoparticle size [224]. It has been suggested that other factors also influence
the nanoparticle size, such as the concentration of the reagents, hydration of the surfactant
head group, intermicellar interactions, and the intermicellar exchange rate [198,225–228].

For instance, nanoparticles of silver chloride have been synthesized by mixing two microemulsions, one containing silver ions and the other containing chloride ions. It was shown that the average particle size, the polydispersity and the number of particles formed depend on the intermicellar exchange rate and/or the rigidity of the surfactant shell [228].

It must be pointed out that formation and stabilization of nanoparticles in reversed micelles are the result of a delicate equilibrium among many factors. In addition, lacking a general theory enabling the selection a priori of the optimal conditions for the synthesis of nanoparticles of a given material with the wanted properties, stable nanoparticles containing w/o microemulsions can be achieved only in some system-specific and experimentally selected conditions.

The longtime stability of surfactant-coated Pd nanoparticles in w/o microemulsions has been investigated. It has been proven that under suitable conditions, the use of the functionalized surfactant $Pd(AOT)_2$ allows very stable nanosize Pd particles to be obtained and to finely control their average size [229].

The time evolution of the mean size of CdS and ZnS nanoparticles in water/AOT/n-heptane microemulsions has been investigated by UV-vis spectrophotometry. It was shown that the initial rapid formation of fractal-like nanoparticles is followed by a slow-growing process accompanied by superficial structural changes. The marked protective action of the surfactant monolayer adsorbed on the nanoparticle surface has been also emphasized [230,231].

Another way to obtain, under suitable conditions, stable dispersions of surfactant–stabilized nanoparticles consists in the direct suspension of some materials in w/o microemulsions. The formation of stable dispersions of rutile (size 80–450 nm) and carbon black (200–500 nm) in AOT/p-xylene and of rutile, lead chloride, aluminium, antimony in solutions of calcium soaps in benzene has been reported [219,220].

Nanoparticles of water-soluble compounds can be also obtained by simply solubilizing the solid compound in dry surfactant/apolar solvent solutions. Typical electronic micrographs of nanoparticles of $Co(NO_3)_2$ and urea obtained using this methodology are shown in Figures 7 and 8 [41].

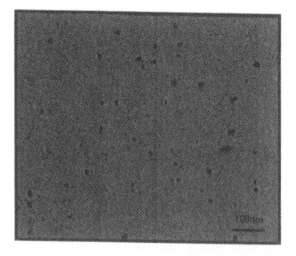

FIG. 7 Typical electronic micrograph of $Co(NO_3)_2$ nanoparticles in deposited film from a $Co(NO_3)_2$/$C_{12}E_4$/cyclohexane system ([$C_{12}E_4$] = 0.094 mol-kg^{-1}; [$Co(NO_3)_2$] = 0.0043 mol-kg^{-1}). Mean size of nanoparticles 19 nm; magnification 300,000.

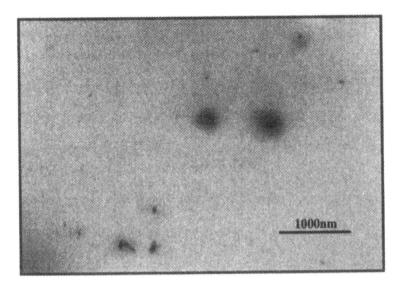

FIG. 8 Typical electronic micrograph of urea nanoparticles in deposited film from a urea/AOT/CCl$_4$ system ([AOT] = 0.1 mol-kg^{-1}; [urea]/[AOT] = 0.73). Mean size of nanoparticles 120 nm; magnification 12,000.

Another method is based on the evaporation of a w/o microemulsion carrying a water–soluble solubilizate inside the micellar core [221,222]. The contemporaneous evaporation of the volatile components (water and organic solvent) leads to an increase in the concentration of micelles and of the solubilizate in the micellar core. Above a threshold value of the solubilizate concentration, it starts to crystallize in confined space. Nanoparticle coalescence could be hindered by surfactant adsorption and nanoparticle dispersion within the surfactant matrix.

G. Hosting Nanogels

The core of reversed micelles can be transformed to a highly viscous domain (nanogel) by entrapping appropriate species, such as viscous solvents and hydrophilic macromolecules, or by performing in situ appropriate polymerization reactions or intramolecular cross-linking of water-soluble polymer chains [232–234].

Solutions of surfactant-stabilized nanogels share both the advantage of gels (drastic reduction of molecular diffusion and of internal dynamics of solubilizates entrapped in the micellar aggregates) and of nonviscous liquids (nanogel-containing reversed micelles diffuse and are dispersed in a macroscopically nonviscous medium). Effects on the lifetime of excited species and on the catalytic activity and stability of immobilized enzymes can be expected.

Finite amounts of glycerol (its viscosity is 945 cP at 25°C) can be dispersed in AOT/heptane or in CTAB/heptane + chloroform systems. The resulting solutions consist of thermodynamically stable, spherical droplets of glycerol stabilized by the surfactant [33,235]. The presence of glycerol within the micellar core results in a reduction of the surfactant mobility [137].

Some potential applications of dispersions of nanodroplets of such highly viscous solvents as novel reaction media for controlled synthesis have been investigated [236].

Insertion of gelatin within water-containing AOT-reversed micelles at $R = 30$ involves an increase in the micellar density, the refractive index, the permittivity at zero frequency, the dielectric relaxation time, and the hypersonic velocity and a decrease in the micellar adiabatic compressibility. All these results point to a picture of rigid, dense, anisotropic AOT-reversed micelles containing water and gelatin, which can surely be viewed as nanogels [233]. The results of scattering experiments on gelatin containing w/o microemulsions near the sol–gel transition indicated the existence of a network of nanogels cross-linked by the polymer [232,237].

By solubilizing very viscous aqueous solutions of polyethylene glycol in AOT/isooctane solutions, it has been observed that the polymer leads to a decrease in the intermicellar interactions and enhances the stability of very large droplets with R values ranging from 55 to 150. The largest reversed micelle may contain up to 200 polymer molecules [238].

Solubilization of vinylpyrrolidone, acrylic acid, and N,N'-methylene-bis-acrylamide in AOT-reversed micelles allowed the synthesis in situ of a cross-linked polymer with narrow size distribution confined in the micellar domain. These particles displayed high entrapment efficiency of small hydrophilic drugs and have been considered interesting drug delivery systems [239].

Nanogels made up of various intramolecularly cross-linked macromolecules have been prepared simply by performing the polymerization of hydrophilic monomers solubilized in the micellar core of reversed micelles, and they represent distinct macromolecular species from those obtained in bulk [191,240].

IV. SOLUTIONS OF WATER-CONTAINING REVERSED MICELLES

A. Intermicellar Interactions in Semidilute Solutions of Water-Containing Reversed Micelles

It has been found that at surfactant concentrations higher than 0.1 M, water-containing reversed micelles of AOT are not randomly dispersed in an isolated state in n-heptane but form clusters through intermicellar flocculation [241,242].

Time-resolved luminescence quenching measurements using the probe Tb(pyridine-2,6-dicarboxylic acid)$_3^{3-}$ and the quencher bromophenol blue show the existence of micellar clusters in AOT-based w/o microemulsions. The fast exchange appearing over several microseconds was attributed to intracluster quenching, whereas the slow exchange on the millisecond time scale was attributed to intercluster exchange [243].

It follows that in spite of the apolar coat surrounding water-containing AOT-reversed micelles and their dispersion in an apolar medium, some microscopic processes are able to establish intermicellar attractive interactions. These intermicellar interactions between AOT-reversed micelles increase with increasing temperature or the chain length of the hydrocarbon solvent molecule, thus leading to the enhancement of the clustering process [244–246], whereas they are reduced in the presence of inorganic salts [131].

It has been proposed that the overlapping of the surfactant hydrocarbon tails is mainly responsible for the micelle–micelle interactions [247]. However, since tail–tail interactions are of the same order of magnitude as tail–apolar solvent interactions, it seems more reasonable to consider the overlapping of the surfactant hydrocarbon tails as an effect rather than the origin of the micelle–micelle interactions.

More recently, in the case of water-containing AOT-reversed micelles, it has been suggested that a pivotal role in the intermicellar interactions is played by the surfactant dissociation leaving hydrated charged heads and counterions at the micellar surface [89,248]. Then the continuous concerted jumping of hydrated AOT^- anions (hopping mechanism) [89,249–251] among neighboring micelles, forming fluctuating oppositely charged micelles, is responsible for the attractive intermicellar interactions leading to the formation of extended clusters of reversed micelles and for the conductometric behavior of water/AOT/hydrocarbon microemulsions [248,243,252].

Moreover, as a consequence of their transient character, a hierarchy of clusters in dynamic equilibrium that may differ in shape and size can be hypothesized [253]. Mass, momentum, and charge transport within a cluster of reversed micelles is expected to be strongly enhanced as compared to that among isolated reversed micelles. It has been shown that the dynamics of a network of interacting reversed micelles is successfully described by a model developed by Cates [35,69,254].

Another mechanism postulated to explain the conductometric behavior of these microemulsions attributes it to the transfer of sodium counterions from a reversed micelle to another through water channels opened by intermicellar coalescence [255–258].

In the case of water-containing lecithin-reversed micelles, an enhancement of hydrogen bonding induced by a temperature decrease or an increase in water and/or micellar concentration has been suggested to account for the huge increase in the intermicellar interactions leading to unidimensional extensive aggregation of lecithin-reversed micelles, i.e., the formation of very long and flexible wormlike aggregates characterized by the continuous breaking and reforming of intermicellar adhesive contacts [69,259]. Then an increase in the temperature determines a destabilization of these aggregates and a decrease in the viscosity of the system [260].

As expected, substitution of water for other polar substances or the presence of additives leads to changes in the intermicellar interactions [261].

By dynamic light scattering it was found that, in surfactant stabilized dispersions of nonaqueous polar solvents (glycerol, ethylene glycol, formamide) in iso-octane, the interactions between reversed micelles are more attractive than the ones observed in w/o microemulsions. Evidence of intermicellar clusters was obtained in all of these systems [262].

Attractive intermicellar interactions become larger by increasing the urea concentration in water/AOT/n-hexane microemulsions at $R = 10$ [263].

The addition of water-soluble polymers, namely, polyoxyethylene glycol, polyacrylamide, and polyvinyl alcohol, to water/AOT/alcohol/decane w/o microemulsions decreases the intermicellar attractive interactions [190].

B. Concentrated Solutions of Water-Containing Reversed Micelles

Interesting phenomena are observed by increasing the concentration of reversed micelles, changing the temperature or pressure, applying high electric fields, or adding suitable solutes. In some conditions, in fact, a dramatic increase in some physicochemical properties has been observed, such as viscosity, conductance, static permittivity, and sound absorption [65,80,173,233,243,249,255,264–269].

For example, the conductivity, on the order of 10^{-18}–10^{-14} S-m^{-1} in pure oil and 10^{-8}–10^{-6} S-m^{-1} in dilute w/o microemulsions of ionic surfactants, in such conditions becomes on the order of 10^{-1} S-m^{-1} [72].

Two system-dependent interpretative pictures have been proposed to rationalize this percolative behavior. One attributes percolation to the formation of a bicontinuous structure [270,271], and the other it to the formation of very large, transient aggregates of reversed micelles [249,263,272]. In both cases, percolation leads to the formation of a network (static or dynamic) extending over all the system and able to enhance mass, momentum, and charge transport through the system. This network could arise from an increase in the intermicellar interactions or for topological reasons. Then all the variations of external parameters, such as temperature and micellar concentration leading to an extensive intermicellar connectivity, are expected to induce percolation [273].

By considering the viscosimetric behavior of water/AOT/n-heptane microemulsions at various R, it was observed that at very low R values or at $R > 10$ these systems behave as suspensions of highly monodisperse hard-sphere particles, whereas at intermediate R values they interact strongly [30,273,274]. This is in agreement with the finding that, in the semidilute region, AOT-reversed micelles form intermicellar aggregates in the range $0 < R < 10$, whereas at $R > 10$ this does not occur [88], and also that the globular structure of water-containing reversed micelles persists even at the higher volume fractions of the dispersed phase and bicontinuous structure never occurs [89,257,275].

Percolated lecithin-based microemulsions (also called *lecithin organogels*) have been widely investigated [19,276]. For these systems, percolation has been attributed to the formation of an infinitely extended dynamic network of very long, entangled, unidimensional aggregates of reversed micelles sustained mainly by topological hindrances [100,104,277,278]. This network is similar to that found in semidilute solutions of living polymers, i.e., linear chain polymers, which can break or recombine along the chain, reversibly.

On the other hand, dodecylmethylbutylammonium bromide– and benzyldymethylheadecylammonium chloride–based w/o microemulsions, which consist of reversed micelles below the percolation threshold, form a bicontinuous structure above the percolation threshold [279].

Effects of additives (electrolytes, surfactants, nonelectrolytes) on the volume fraction and temperature percolation thresholds of a water/AOT/n-heptane system have been investigated [280,281].

It has been reported that the percolation of conductance of water/AOT/n-heptane microemulsions is assisted by sodium cholate and retarded by sodium salicylate [282].

The addition of linear chained alkyl alcohols shifts the percolation of AOT microemulsions to higher temperature, whereas the opposite effect is obtained by adding polyoxyethylene alkyl ethers [261].

It has also been pointed out that percolation is hindered by molecules stiffening the micellar interface, such as cholesterol, whereas it is favored by molecules that make the interface more flexible, such as gramicidin and acrylamide [263].

In sodium bis(2-ethylhexyl) phosphate microemulsions, which are composed of cylindrical micelles in the dilute region, it has been observed that the formation of micellar clusters is characterized by a branched structure as the volume fraction (Φ) of the aggregates increases. At $\Phi > 0.2$, these clusters mutually overlap, forming a network expanded overall [283].

It is worth noting that the picture of clusters of water-containing reversed micelles extending for macroscopic distances makes these systems particularly interesting, since they could allow a spatially ordered distribution of micellar hosts (single molecules, nanoparticles, etc.), or, from the biological point of view, since they can be considered the starting point of a realistic model of some aspects of an assembly of living cells.

The possibility of realizing via percolated w/o microemulsion conductor/insulating composite materials with very large dielectric constant and exotic optical properties has been pointed out [284].

A kinetic study of the basic hydrolysis in a water/AOT/decane system has shown a change in the reactivity of p-nitrophenyl ethyl chloromethyl phosphonate above the percolation threshold. The applicability of the pseudophase model of micellar catalysis, below and above the percolation threshold, was also shown [285].

Under comparable conditions, the reaction rates of the octyl decanoate synthesis by chromobacterium viscosum lipase in AOT-based microemulsions or immobilized in AOT-based organogels were similar [286,287].

Percolated microemulsions composed of biocompatible substances, such as some lecithin-based organogels, have been considered interesting vehicles for the delivery of drugs [288].

It has also been suggested that solubilization of enzymes in organogels allows interesting reaction media to be realized since they facilitate enzyme reuse and easy product separation [281].

V. CONCLUSION

A wide variety of substances can be encapsulated in reversed micelles and water-containing reversed micelles. This allows nanostructures to be obtained that are interesting from the theoretical and technological points of view. The list of the theoretical questions and of the potential and actual applications of solutions of reversed micelles at present is very long, and its length is steeply increasing with time. What appears of utmost importance is that all the investigations performed in this research field could represent the first steps toward the preparation of nanostructures with increasing complexity, both in structure and functionality, trending to that observed in biological systems. In the future a better ability to model or to mimic nature by reversed micelles will give to researchers the possibility of realizing new and exciting living nano-devices.

ACKNOWLEDGMENTS

Financial support from CNR, MURST 60%, and MURST 40% (cofin MURST 97 CFSIB) is gratefully acknowledged.

REFERENCES

1. HF Eicke. In: FL Boschke, ed. Topics in Current Chemistry. Vol 87. New York: Springer-Verlag, 1980, pp 85–145.
2. M Wong, JK Thomas, T Nowak. J Am Chem Soc 99:4730–4736, 1977.
3. PDI Fletcher, BH Robinson. Ber Bunsenges Phys Chem 85:863–867, 1981.
4. M Kotlarchyk, JS Huang, SH Chen. J Phys Chem 89:4382–4386, 1985.
5. PDI Fletcher, AM Howe, BH Robinson. J Chem Soc Faraday Trans I 83:985–1006, 1987.
6. J Lang, A Jada, A Malliaris. J Phys Chem 92:1946–1953, 1988.
7. A D'Aprano, G D'Arrigo, M Goffredi, A Paparelli, V Turco Liveri. J Phys Chem 93: 8367–8370, 1989.
8. M Ueno, H Kishimoto, Y Kyogoku. J Coll Int Sci 63:113–119, 1978.
9. A Maitra. J Phys Chem 88:5122–5125, 1984.
10. FY Lo, BM Escott, EJ Fendler, ET Adams, RD Larsen, PW Smith. J Phys Chem 79: 2609–2621, 1975.

11. N Muller. J Coll Int Sci 63:383–393, 1978.
12. JB Nagy. In: KL Mittal, EJ Fendler, eds. Solution Behavior of Surfactants. Vol 2. New York: Plenum Press, 1982, pp 743–766.
13. PL Luisi, LJ Magid. CRC Critical Rev Biochem 20:4091–474, 1986.
14. F Heatley. J Chem Soc Faraday Trans I 83:517–526, 1987.
15. J Eastoe, TF Towey, BH Robinson, J Williams, RH Heenan. J Phys Chem 97:1459–1463, 1993.
16. J Eastoe, DC Steytler, BH Robinson, RH Heenan, AN North, JC Dore. J Chem Soc Faraday Trans 90:2479–2504, 1994.
17. J Eastoe, RK Heenan. J Chem Soc Faraday Trans I 90:487–492, 1994.
18. R Zana, J Lang, D Canet. J Phys Chem 95:3364–3367, 1991.
19. R Scartazzini, PL Luisi. J Phys Chem 92:829–833, 1988.
20. A Merdas, M Gindre, R Ober, C Nicot, W Urbach, M Waks. J Phys Chem 100:15180–15186, 1996.
21. AB Mandal, B Unni-Nair. J Chem Soc Faraday Trans I 87:133–136, 1991.
22. GR Seely, XC Ma, RA Nieman, D Gust. J Phys Chem 94:1581–1598, 1990.
23. Q Li, S Weng, J Wu, N Zhou. J Phys Chem, 102:3168–3174, 1998.
24. A Kitahara. Adv Coll Int Sci 12:109–140, 1980.
25. A D'Aprano, A Lizzio, V Turco Liveri. J Phys Chem 92:1985–1987, 1988.
26. L Garcia-Rio, JC Mejuto, R Ciri, IB Blagoeva, JR Leis, MF Ruasse. J Phys Chem B 103:4997–5004, 1999.
27. P Menassa, C Sandorfy. Can J Chem 63:3367–3370, 1985.
28. A D'Aprano, ID Donato, F Pinio, V Turco Liveri. J Sol Chem 18:949–955, 1989.
29. G Pitarresi, C Sbriziolo, ML Turco Liveri, V Turco Liveri. J Sol Chem 22:279–287, 1993.
30. V Turco Liveri. In: N Garti, ed. Thermal Behavior of Dispersed Systems. New York: Marcel Dekker, 2001, pp 1–22.
31. NM Correa, EN Durantini, JJ Silber. J Coll Int Sci 208:96–103, 1998.
32. M Nakamura, GL Bertand, SE Friberg. J Colloid Int Sci 91:516–524, 1983.
33. PDI Fletcher, MF Galal, BH Robinson. J Chem Soc Faraday Trans I 80:3307–3314, 1984.
34. RE Riter, JR Kimmel, EP Undikis, NE Levinger. J Phys Chem B 101:8292–8297, 1997.
35. V Arcoleo, F Aliotta, M Goffredi, G La Manna, V Turco Liveri. Mat Sci Eng C5:47–53, 1997.
36. H Shirota, K Horie. J Phys Chem B 103:1437–1443, 1999.
37. P Lopez-Cornejo, SMB Costa. Langmuir 14, 2042–2049, 1998.
38. JP Wilcoxon, RL Williamson, R Baughman. J Chem Phys 98:9933–9950, 1993.
39. JP Wilcoxon, PP Provencio. J Phys Chem B 103:9809–9812, 1999.
40. V Marcianò, A Minore, V Turco Liveri. Coll Polym Sci 278:250–252, 2000.
41. V Turco Liveri. Unpublished data.
42. X Xu, M Ayyagari, M Tata, VT John, GL McPherson. J Phys Chem 97:11350–11353, 1993.
43. M Tata, VT John, YY Waguespack, GL McPherson. J Mol Liq 72:121–135, 1997.
44. PL Luisi, M Giomini, MP Pileni, BH Robinson. Biochim Biophys Acta 947:209–246, 1988.
45. EB Abuin MA Rubio, EA Lissi. J Coll Int Sci 158:129–132, 1993.
46. E Bardez, R Giordano, MP Jannelli, P Migliardo, U Wanderlingh. J Mol Structure 383:183–190, 1996.
47. Y Ikushina, N Saito, M Arai. J Coll Int Sci 186:254–263, 1997.
48. HF Eicke. J Coll Int Sci 68:440–450, 1979.
49. MP Pileni. J Phys Chem 97:6961–6973, 1993.
50. A Bumajdad, J Eastoe, P Griffths, DC Steyler, RK Heenan, JR Lu, P Timmins. Langmuir 15:5271–5278, 1999.
51. EV Votyakov, YK Tovbin, JMD MacElroy, A Roche. Langmuir 15:5713–5721, 1999.
52. S Senapati, A Chandra. J Chem Phys 111:1223–1230, 1999.
53. H MacDonald, B Bedwell, E Gulari. Langmuir 2:704–708, 1986.
54. TK Jain, M Varshney, A Maitra. J Phys Chem 93:7409–7416, 1989.

55. A Maitra, TK Jain, Z Shervani. Colloids Surfaces 47:255–267, 1990.
56. G Giammona, F Goffredi, V Turco Liveri, G Vassallo. J Coll Int Sci 154:411–415, 1992.
57. G Onori, A Santucci. J Phys Chem 97:5430–5434, 1993.
58. G Cavallaro, G La Manna, V Turco Liveri, F Aliotta, ME Fontanella. J Coll Int Sci 176:281–285, 1995.
59. MB Temsamani, M Haeck, I El Hassani, HD Hurwitz. J Phys Chem B 102:3335–3340, 1998.
60. RD Smith, JL Fulton, JP Blitz, JM Tingey. J Phys Chem 94:781–787, 1990.
61. CB Roberts, JB Thomson. J Phys Chem B 102:9074–9080, 1998.
62. ELV Goetheer, MAG Vorstman, JTF Keurentses. Chem Eng Sci 54:1589–1596, 1999.
63. JQ Zhao, XM Ding, LJ Jiang. Sci China B Chem 42:153–158, 1999.
64. M Wong, JK Thomas, M Gratzel. JACS 98:2391–2397, 1976.
65. YA Shchipunov, T Durrschmidt, H Hoffmann. J Coll Int Sci 212:390–401, 1999.
66. P Calandra, E Caponetti, D Chillura, P D'Angelo, A Minore, V Turco Liveri. J Mol Structure 522:165–178, 2000.
67. XX Zhu, K Banana, HY Liu, M Krause, M Yang. Macromolecules 32:277–281, 1999.
68. LJ Schwartz, CL DeCiantis, S Chapman, BK Kelley, JP Hornak. Langmuir 15:5461–5466, 1999.
69. V Turco Liveri. In: JE Desnoyers, F Franks, G. Gritzner, LG Hepler, H Ohtaki, eds. Current Topics in Solution Chemistry. Vol 2. Trivandrum: Research Trends. 1997, pp 143–156.
70. G D'Arrigo, A Paparelli, A D'Aprano, ID Donato, M Goffredi, V Turco Liveri. J Phys Chem 93:8367–8370, 1989.
71. G Carlstrom, B Halle. J Phys Chem 93:3287–3299, 1989.
72. HF Eicke, M Borkovec, B Das-Gupta. J Phys Chem 93:314–317, 1989.
73. DC Hall. J Phys Chem 94:429–430, 1990.
74. AS Bommarius, JF Holzwarth, DIC Wang, TA Hatton. J Phys Chem 94:7232–7239, 1990.
75. J Zhang, FV Bright. J Phys Chem 95:7900–7907, 1991.
76. N Sarkar, K Das, K Bhattacharyya. J Phys Chem 100:10523–10527, 1996.
77. N Sarkar, A Datta, K Das, K Bhattacharyya. J Phys Chem 100:15483–15486, 1996.
78. TK Jain, G Cassin, JP Badiali, MP Pileni. Langmuir 12:2408–2411, 1996.
79. J Clarke, J Nicholson, K Regan. J Chem Soc Faraday Trans I 81:1173–1182, 1985.
80. A D'Aprano, G D'Arrigo, M Goffredi, A Paparelli, V Turco Liveri. J Chem Phys 95:1304–1309, 1991.
81. DF Evans, BW Ninham. J Phys Chem 90:226–234, 1986.
82. JN Israelachvili, DJ Mitchell, BW Ninham. J Chem Soc Faraday Trans II 72:1525–1568, 1976.
83. D Oakenfull. J Chem Soc Faraday Trans I 76:1875–1886, 1980.
84. R Day, BH Robinson, J Clarke, J Doherty. J Chem Soc Faraday Trans I 75:132–139, 1979.
85. JP Blitz, JL Fulton, RD Smith. J Phys Chem 92:2707–2710, 1988.
86. HF Eicke, R Kubik. Faraday Discuss Chem Soc 76:305–315, 1983.
87. A D'Aprano, A Lizzio, V Turco Liveri. J Phys Chem 91:4749–4751, 1987.
88. M Hirai, R Kawai-Hirai, S Yabuki, T Takizawa, T Hirai, K Kobayashi, Y Amemiya, M Oya. J Phys Chem 99:6652–6660, 1995.
89. M Goffredi, V Turco Liveri, G Vassallo. J Solution Chem 22:941–949, 1993.
90. MS Altamirano, CD Borsarelli, JJ Cosa, CM Previtali. J Coll Int Sci 205:390–396, 1998.
91. P Linse. J Chem Phys 90:4992–5004, 1989.
92. V Arcoleo, M Goffredi, V Turco Liveri. J Coll Int Sci 198:216–223, 1998.
93. DE Gragson, GL Richmond. J Phys Chem B 102:3847–3861, 1998.
94. P Karpe, E Ruckenstein. J Coll Int Sci 137:408–424, 1990.
95. A Luzar, D Bratko. J Chem Phys 92:642–648, 1990.
96. M Belletete, M Lachapelle, G Durocher. J Phys Chem 94:5337–5341, 1990.
97. C Boned, J Peyrelasse, M Moha-Ouchane. J Phys Chem 90:634–637, 1986.
98. H Hauser, G Haering, A Pande, PL Luisi. J Phys Chem 93:7869–7887, 1989.

99.   A Yoshino, H Okabayashi, T Yoshida, K Kushida. J Phys Chem 100:9592–9597, 1996.
100. PL Luisi, R Scartazzini, G Hearing, P Schurtenberger. Colloid Polym Sci 268:356–374, 1990.
101. P Schurtenberger, LJ Magid, SM King, P Lindner. J Phys Chem 95:4173–4176, 1991.
102. P Schurtenberger, C Cavaco. Langmuir 10:100–108, 1994.
103. P Schurtenberger, C Cavaco. J Phys Chem 98:5481–5486, 1994.
104. V Arcoleo, M Goffredi, G La Manna, V Turco Liveri, F Aliotta, ME Fontanella. J Thermal Anal 50:823–830, 1997.
105. G La Manna, V Turco Liveri, F Aliotta, ME Fontanella, P Migliardo. Coll Polym Sci 271:1172–1176, 1993.
106. DM Willard, RE Riter, NE Levinger. J Am Chem Soc 120:4151–4160, 1998.
107. JC Ravey, M Buzier, C Picot. J Coll Int Sci 97:9–25, 1984.
108. U Olsson, K Nakamura, H Kuneida, R Strey. Langmuir 12:3045–3054, 1996.
109. F Mantegazza, V De Giorgio, ME Giardini, AL Price, DC Steyler, BH Robinson. Langmuir 14:1–7, 1998.
110. I Lisiecki, P Andre, A Filankembo, C Petit, J Tanori, T Gulik-Krzywicki, BW Ninham, MP Pileni. J Phys Chem B 103:9168–9175, 1999.
111. ZJ Yu; TH Ibrahim; RD Neuman. Solvent Extraction Ion Exchange 16:1437–1463, 1998.
112. FP Cavasino, C Sbriziolo, ML Turco Liveri, V Turco Liveri. J Chem Soc Faraday Trans 90:311–314, 1994.
113. L Cerauolo, M Ferruggia, L Tesoriere, S Segreto, MA Livrea, V Turco Liveri. J Pineal Res 26:108–112, 1999.
114. ML Turco Liveri, V Turco Liveri. J Coll Int Sci 176:101–104, 1995.
115. D Grand. J Phys Chem B 102:4322–4326, 1998.
116. A Derouiche, C Tondre. J Chem Soc Faraday Trans I 85:3301–3308, 1989.
117. PK Das, GV Srilakshimi, A Chandhuri. Langmuir 15:981–987, 1999.
118. E Bardez, E Monnier, B Valeur. J Coll Int Sci 112:200–207, 1986.
119. AM Howe, C Toprakcioglu, JC Dore, BH Robinson. J Chem Soc Faraday Trans I 82:2411–2422, 1986.
120. T Fujieda, K Ohta, N Wakabayashi, S Higuchi. J Coll Int Sci 185:332–334, 1997.
121. KK Karukstis, EHZ Thompson, JA Whiles, RJ Rosenfeld. Biophysical Chem 73:249–263, 1998.
122. J Kim, M Lee. J Phys Chem 103:3378–3382, 1999.
123. PL Luisi. Angew Chem 97:449–460, 1985.
124. M Mangey, AM Bellocq. Langmuir 15:8602–8608, 1999.
125. A Kitahara, K Kon-no. J Phys Chem 70:3394–3398, 1966.
126. M Wentz, WH Smith, AR Martin. J Coll Int Sci 29:36–41, 1969.
127. K Kon-no, A Kitahara. J Coll Int Sci 34:221–227, 1970.
128. B Bedwel, E Gulari. J Coll Int Sci 102:88–100, 1984.
129. HR Rabie, JH Vera. J Phys Chem B 101:10295–10302, 1997.
130. HF Eicke, JCW Shepard. Helv Chim Acta 57:1951–1963, 1974.
131. J Rouviere, JM Couret, A Lindhemeir, B Brun. J Chim Phys 76:297–301, 1979.
132. DJ Liu, JM Ma, HM Cheng, ZG Zhao. Coll Surf A 143:59–68, 1998.
133. SG Frank, YH Shaw, NC Li. J Chem Phys 77:238–241, 1973.
134. J Sunamoto, T Hamada. Bull Chem Soc Jpn 51:3130–3135, 1978.
135. T Handa, M Sakai, N Nakagaki. J Phys Chem 90:3377–3380, 1986.
136. MP Pileni, T Zemb, C Petit. Chem Phys Lett 118:414–420, 1985.
137. PDI Fletcher, BH Robinson, J Tabony. J Chem Soc Faraday Trans I 82:2311–2321, 1986.
138. A D'Aprano, ID Donato, F Pinio, V Turco Liveri. J Sol Chem 19:589–595, 1990.
139. A D'Aprano, ID Donato, V Turco Liveri. J Sol Chem 19:1055–1061, 1990.
140. KK Karustis, AA Frazier, CT Loftus, AS Tware. J Phys Chem B 102:8163–8169, 1998.
141. JM Lehn, JP Sauvage. J Am Chem Soc 75:6700–6707, 1975.
142. BG Cox, N Van Troning, H Schneider. J Am Chem Soc 106:1273–1280, 1984.

143. RM Izatt, JS Bradshaw, K Pawlak, RL Bruening, BJ Tarbet. Chem Rev 92:1261–1354, 1992.
144. HT Nguyen, N Kommareddi, VT John. J Coll Int Sci 155:482–487, 1993.
145. H Nguyen, VT John, WF Reed. J Phys Chem 95:1467–1471, 1991.
146. P Lavallard, M Rosenbauer, T Gacoin. Phys Rev A 54:5450–5453, 1996.
147. S Das, A Datta, K Bhattacharyya. J Phys Chem A 101:3299–3304, 1997.
148. NM Correa, JJ Silber. J Mol Liq 72:163–170, 1997.
149. CY Wang, CY Liu, Y Wang, T Shen. J Coll Int Sci 197:126–132, 1998.
150. A Datta, D Mandal, SK Pal, K Bhattacharyya. J Phys Chem B 101:10221–10225, 1997.
151. O Ortona, V Vitagliano, BH Robinson. J Coll Int Sci 125:271–278, 1988.
152. LJ Magid, K Kon-no, CA Martin. J Phys Chem 85:1434–1439, 1981.
153. EB Leodidis, TA Hatton. J Phys Chem 94:6400–6411, 1990.
154. EB Leodidis, TA Hatton. J Phys Chem 94:6411–6420, 1990.
155. EB Leodidis, AS Bonmarius, TA Hatton. J Phys Chem 95:5943–5956, 1991.
156. EB Leodidis, TA Hatton. J Phys Chem 95:5957–5965, 1991.
157. MM Cardoso, MJ Barradas, KH Kroner, JG Crespo. J Chem Techn Biotechn 74:801–811, 1999.
158. M Addachi, M Harada, A Shioi, Y Sato. J Phys Chem 95:7925–7931, 1991.
159. CA Boicelli, M Giomini, AM Giuliani. Spectrochim Acta 37A:559–560, 1981.
160. M Hebrant, C Tondre. Analytical Sci 14:109–115, 1998.
161. M Varshney, T Khanna, M Changer. Coll Surf B 13:1–11, 1999.
162. ME Leser, G Wei, P Luthi, G Haering, A Hochkoeppler, E Blochliger, PL Luisi. J Chim Physique 84:1113–1118, 1987.
163. E Ruckenstein, P Karpe. J Coll Int Sci 139:408–436, 1990.
164. YX Chen, XZ Zhang, SM Chen, DL You, XX Wu, XC Yang, WZ Guan. Enzyme Microbial Techn 25:310–315, 1999.
165. P Plucinski, W Nitsch. Ber Bunsenges Phys Chem 93:994–997, 1989.
166. T Ono, M Goto. Curr Opinion Coll Int Sci 2:397–401, 1997.
167. L Rong, T Yamane, H Takeuchi. J Chem Eng Jpn 31:434–439, 1998.
168. SR Dungan. J Food Sci 63:601–605, 1998.
169. GG Zampieri, H Jackle, PL Luisi. J Phys Chem 90:1849–1853, 1986.
170. RS Rahaman, TA Hatton. J Phys Chem 95:1799–1811, 1991.
171. DG Hayes, C Marchio. Biotech Bioeng 59:557–566, 1998.
172. M Monduzzi, F Caboi, C Moriconi. Colloid Surfaces A 130:327–338, 1997.
173. G Hearing, PL Luisi. J Phys Chem 90:5892–5895, 1986.
174. GB Strambini, M Gonnelli. J Phys Chem 92:2850–2853, 1988.
175. P Marzola, E Gratton. J Phys Chem 95:9488–9495, 1991.
176. L Gebicka, J Gebicki. Biochem Molecular Biol Int 45:805–811, 1998.
177. A Merdas, M Gindre, JY LeHuerou, C Nicot, R Ober, W Urbach, M Waks. J Phys Chem B 102:528–533, 1998.
178. C Petit, P Brochette, MP Pileni. J Phys Chem, 90:6517–6521, 1986.
179. P Brochette, C Petit, MP Pileni. J Phys Chem 92:3505–3511, 1988.
180. NL Klyachko, PA Levashov, AV Levashov, C Balny. Biochem Biophys Res Commun 254:685–688, 1999.
181. FM Menger, K Yamada. J Am Chem Soc 101:6731–6734, 1979.
182. M Adachi, M Harada. J Phys Chem 97:3631–3640, 1993.
183. A Singh, KP Pandey, RS Dubey. World J Microbiol Biotechnol 15:277–282, 1999.
184. A Singh, KP Pandey, RS Dubey. Int J Hydrogen Energy 24:693–698, 1999.
185. C Gonzales-Blanco, LJ Rodriguez, MM Velasquez. Langmuir 13:1938–1945, 1997.
186. A Shioi, M Harada, M Obika, M Adachi. Langmuir 14:4737–4743, 1998.
187. A Shioi, M Harada, M Obika, M Adachi. Langmuir 14:5790–5794, 1998.
188. S Geiger, M Mandel. J Phys Chem 93:4195–4198, 1989.
189. A Holmenberg, P Hansson, L Piculell, P Linse. J Phys Chem B 103:10807–10815, 1999.

190. MJ Suarez, H Levy, J Lang. J Phys Chem 97:9808–9816, 1993.
191. NB Graham, A Cameron. Pure Appl Chem 70:1271–1275, 1998.
192. YS Leong, F Candau. J Phys Chem 86:2269–2271, 1982.
193. F Candau, YS Leong, G Pouyet, S Candau. J Coll Int Sci 101:167–183, 1984.
194. MT Carver, E Hirsch, JC Wittmann, RM Fitch, F Candau. J Phys Chem 93:4867–4873, 1989.
195. B Kriwet, E Walter, T Kissel. J Controlled Release 56:149–158, 1998.
196. JH Fendler. Chem Rev 87:877–899, 1987.
197. A D'Aprano, F Pinio, V Turco Liveri. J Solution Chem 20:301–306, 1991.
198. MP Pileni. Langmuir 13:3266–3276, 1997.
199. Y Wang, N Herron. J Phys Chem 95:525–532, 1991.
200. MA Lopez-Quintela, J Rivas. J Coll Int Sci 158:446–451, 1993.
201. LM Gan, B Liu, CH Chew, SJ Xu, SJ Chua, GL Loy, GQ Xu. Langmuir 13:6427–6431, 1997.
202. GD Rees, R Evans-Gowing, SJ Hammond, BH Robinson. Langmuir 15:1993–2002, 1999.
203. M Boutonnet, J Kizling, P Stenius, G Maire. Colloids Surfaces 5:209–225, 1982.
204. P Barnickel, A Wokaun, W Sager, HF Eicke. J Coll Int Sci 148:80–90, 1992.
205. E Joselevich, I Willner. J Phys Chem 98:7628–7635, 1994.
206. F Aliotta, V Arcoleo, S Buccoleri, G La Manna, V Turco Liveri. Thermochim Acta 265:15–23, 1995.
207. J Cizeron, MP Pileni. J Phys Chem B 101:8887–8891, 1997.
208. F Parsapour, DF Kelley, RS Williams. J Phys Chem B 102:7971–7977, 1998.
209. CT Seip, CJ O'-Connor. Nanostructured Materials 12:183–186, 1999.
210. CJ O'-Connor, CT Seip, EE Carpenter. Nanostructured Materials 12:65–70, 1999.
211. DW Kim, SG Oh, JD Lee. Langmuir 15:1599–1603, 1999.
212. TS Chneider, M Haase, A Kornowski, M Antonietti. Ber Bunsenges Phys Chem 101:1654–1656, 1997.
213. S Qiu, J Dong, G Chen. J Coll Int Sci 216:230–234, 1999.
214. V Arcoleo, V Turco Liveri. Chem Phys Lett 258:223–227, 1996.
215. H Ma, X Wang, G Li. J Dispersion Sci Tech 19:77–91, 1998.
216. JH Fendler. Membrane Mimetic Chemistry. New York: Wiley, 1982.
217. V Arcoleo, M Goffredi, V Turco Liveri. Thermochim Acta 233:187–197, 1994.
218. R Tang, C Jiang, Z Tai. J Chem Soc Faraday Trans 93:3371–3375, 1997.
219. DNL McGowen, GD Parfitt, E Willis. J Coll Sci 20:650–664, 1965.
220. JL van der Minne, PHJ Hermanie. J Coll Sci 7:600–615, 1952.
221. V Turco Liveri. In: JE Desnoyers, F Franks, G. Gritzner, LG Hepler, H Ohtaki, eds. Current Topics in Colloid and Interface Science. Vol 3. Trivandrum: Research Trends, 1999, pp 65–74.
222. P. Calanolra, A Longo, V Turco Liveri. Coll Polym Sci 279:0000–0000, 2001.
223. V Arcoleo, M Goffredi, V Turco Liveri. J Thermal Anal 51:125–133, 1998.
224. CL Chang, HS Fogler. Langmuir 3:3295–3307, 1997.
225. S Modes, P Lianos. J Phys Chem 93:5854–5859, 1989.
226. JH Clint, IR Collins, JA Williams, BH Robinson, TF Towey, P Cajean, A Khan-Lodhi. Faraday Discuss 95:219–233, 1993.
227. I Lisiencki, MP Pileni. J Phys Chem 99:5077–5082, 1995.
228. RP Bagwe, KC Khilar. Langmuir 13:6432–6438, 1997.
229. V Arcoleo, M Goffredi, A Longo, V Turco Liveri. Mat Sci Eng C 6:7–11, 1998.
230. TF Towey, A Khan-Lodhi, BH Robinson. J Chem Soc Faraday Trans 86:3757–3762, 1990.
231. E Di Dio, M Goffredi, V Turco Liveri. Mat Eng 9:67–82, 1998.
232. C Quellet, HF Eicke, R Gehrke, W Sager. Europhys Lett 9:293–298, 1989.
233. F Aliotta, V Arcoleo, G La Manna, V Turco Liveri. Coll Polym Sci 274:989–994, 1996.
234. P Ulanski, I Janik, JM Rosiak. Radiation Phys Chem 52:289–294, 1998.
235. PDI Fletcher, MF Galal, BH Robinson. J Chem Soc Faraday Trans I 68:2053–2065, 1985.
236. NZ Atay, BH Robinson. Langmuir 15:5056–5064, 1999.
237. C Quellet, HF Eicke, W Sager. J Phys Chem 95:5642–5655, 1991.

238. AM Bellocq. Langmuir 14:3730–3739, 1998.
239. SK Sahoo, TK De, PK Ghosh, A Maitra. J Coll Int Sci 206:361–368, 1998.
240. M Dreja, W Pyckhout-Hintzen, B Tieke. Macromolecules 31:272–280, 1998.
241. M Hasegawa, Y Yamasaka, N Sonta, Y Shindo, T Sugimura, A Kitahara. J Phys Chem 100:15575–15580, 1996.
242. M Hirai, R Kawai-Hirai, M Sanada, H Iwase, S Mitsuya. J Phys Chem B 103:9658–9662, 1999.
243. H Mays. J Phys Chem B 101:10271–10280, 1997.
244. C Robertus, JGH Joosten, YK Levine. J Chem Phys 93:7293–7300, 1990.
245. J Eastoe, WK Young, BH Robinson, DC Steytler. J Chem Soc Faraday Trans I 86:2883–2889, 1990.
246. JM Tingey, JL Fulton, RD Smith. J Phys Chem 94:1997–2004, 1994.
247. B Lemaire, P Bothorel, D Roux. J Phys Chem 87:1023–1028, 1983.
248. V Arcoleo, M Goffredi, V Turco Liveri. J Sol Chem 24:1135–1142, 1995.
249. MW Kim, JS Huang. Phys Rev A 34:719–722, 1986.
250. C Cametti, P Codestefano, P Tartaglia. Ber Bunsenges Phys Chem 94:1499–1503, 1990.
251. C Cametti, P Codestefano, G D'Arrigo, P Tartaglia, J Rouch, SH Chen. Phys Rev A 42:3421–3426, 1990.
252. U Batra, WB Russel, JS Huang. Langmuir 15:3718–3725, 1999.
253. E Dickenson. J Chem Soc Faraday Trans I 90:173–180, 1994.
254. ME Cates. Macromolecules 20:2289–2296, 1987.
255. MA Van Dijk, G Casteleijn, JGH Joosten, Y Levine. J Chem Phys 85:626–631, 1986.
256. A Jada, J Lang, R Zana. J Phys Chem 93:10–12, 1989.
257. A Maitra, C Mathew, M Varshney. J Phys Chem 94:5290–5292, 1990.
258. Y Feldman, N Kozlovich, I Nir, N Garti, V Archipov, Z Idiyatullin, Y Zuev, V Fedotov. J Phys Chem 100:3745–3748, 1996.
259. YA Shchipunov, EV Shumilina. Progr Coll Polym Sci 106:228–231, 1997.
260. F Aliotta, ME Fontanella, G Squadrito, P Migliardo, G La Manna, V Turco Liveri. J Phys Chem 97:6541–6545, 1993.
261. LMM Nazario, TA Hatton, JPSG Grespo. Langmuir 12:6326–6335, 1996.
262. CAT Laia, P Lopez-Cornejo, SMB Costa, J D'Oliveira, JMG Martinho. Langmuir 14:3531–3537, 1998.
263. CL Costa Amaral, R Itri, MJ Politi. Langmuir 12:4638–4643, 1996.
264. RF Berg, MR Moldover, JS Huang. J Chem Phys 87:3687–3691, 1987.
265. D Capitani, AL Segre, G Haering, PL Luisi. J Phys Chem 92:3500–3504, 1988.
266. F Runge, W Rohl, G Ilgenfritz. Ber Bunsenges Phys Chem 95:485–490, 1991.
267. F Runge, L Schlicht, JH Spilgies, G Ilgenfritz. Ber Bunsenges Phys Chem 98:506–508, 1994.
268. E Tekle, ZA Schelly. J Phys Chem 98:7657–7664, 1994.
269. P Alexandridis, JF Holzwarth, TA Hatton. J Phys Chem 99:8222–8232, 1995.
270. PG De Gennes, GJ Taupin. J Phys Chem 86:2294–2304, 1982.
271. M Borkovec, HF Eicke, H Hammerich, B Dasgupta. J Phys Chem 92:206–211, 1988.
272. H Mays, G Ilgenfritz. J Chem Soc Faraday Trans 92:3145–3150, 1996.
273. M Goffredi, V Turco Liveri. Progr Coll Polym Sci 112:109–114, 1999.
274. A D'Aprano, G D'Arrigo, A Paparelli, M Goffredi, V Turco Liveri. J Phys Chem. 97:3614–3618, 1993.
275. SH Chen, JS Wang. Phys Rev Lett 55:1888–1891, 1985.
276. D Capitani, AL Segre, F Dreher, P Walde, PL Luisi. J Phys Chem 100:15211–15217, 1996.
277. P Schurtenberger, R Scartazzini, LJ Magid, ME Leser, PL Luisi. J Phys Chem 94:3695–3701, 1990.
278. D Capitani, E Rossi, AL Segre, M Giustini, PL Luisi. Langmuir 9:685–689, 1993.
279. MS Batista, CD Tran. J Phys Chem B 101:4209–4217, 1997.
280. S Ray, SR Bisal, SP Moulik. J Chem Soc Faraday Trans 89:3277–3282, 1993.

281. SP Moulik, BK Paul. Adv Coll Int Sci 78:99–195, 1998.
282. SP Moulik, GC De, BB Bhowmik, AK Panda. J Phys Chem B 103:7122–7129, 1999.
283. K Kuramada, A Shioi, M Harada. J Phys Chem 98:12382–12389, 1994.
284. DJ Bergman, Y Imry. Phys Rev Lett 39:1222–1225, 1977.
285. LY Zakharova, FG Valeeva, LA Kudryavtseva, NL Zakhartchenko, YF Zuev. Mendeleev Communications 6:224–227, 1998.
286. TRJ Jenta, G Batts, GD Rees, BH Robinson. Biotech Bioeng 54:416–427, 1997.
287. TRJ Jenta, G Batts, GD Rees, BH Robinson. Biotech Bioeng 53:121–131, 1997.
288. MJ Lawrence. Current Opinion Coll Int Sci 1:826–832, 1996.

14

Engineering of Core-Shell Particles and Hollow Capsules

FRANK CARUSO Max-Planck-Institute of Colloids and Interfaces, Potsdam, Germany

I. INTRODUCTION

The construction of nanostructured colloidal materials is an important area of research in modern materials science and of great technological importance. Over the years there has been an immense interest in the design and manufacture of particles with desired surface properties [1–9]. The technology used to achieve the synthesis of particles with well-defined morphologies and compositions is commonly referred to as *particle engineering*. This entails the modification of particle surfaces through the tailoring of their surface properties, which is often accomplished by coating the particles or encapsulating them within a shell of a desired material. Composite particles that contain an inner core covered by a shell (*core-shell particles*) exhibit significantly different properties from those of the core itself, with the surface properties governed by the characteristics of the coating. Surface-engineered colloids are widely exploited in the areas of coatings, electronics, photonics, catalysis, separations, and diagnostics. Besides such technological applications, the creation of core-shell particles with well-defined properties is of interest from a fundamental scientific viewpoint as well. They can be employed to gain valuable information about the properties of concentrated dispersions [10] and as model systems to investigate factors governing colloidal interactions and stabilization [11–13].

An important class of materials that originates from the precursor core-shell particles is hollow capsules. Hollow capsules (or "shells") can be routinely produced upon removal of the core material using chemical and physical methods. Much of the research conducted in the production of uniform-size hollow capsules arises from their scientific and technological interest. Hollow capsules are widely utilized for the encapsulation and controlled release of various substances (e.g., drugs, cosmetics, dyes, and inks), in catalysis and acoustic insulation, in the development of piezoelectric transducers and low-dielectric-constant materials, and for the manufacture of advanced materials [14].

Given the numerous practical applications of both core-shell particles and hollow capsules, and their enormous economical potential, methods to "engineer" these materials with controlled precision have long been sought. A variety of chemical and physicochemical procedures have been employed for their fabrication, many of which have serious disadvantages, thereby limiting the applications and commercialization of the final products [1,15]. The controlled and uniform coating of colloids with organized layers still remains a technical challenge, despite the fact that the advantages of uniformly coated and stable

505

particles have been recognized for decades [16]. Recent methods, however, especially those based on self-assembly and colloidal science, offer new alternatives for the controlled synthesis of novel core-shell particles as well as hollow capsules.

This chapter provides an overview of the various technologies used to fabricate core-shell particles and hollow capsules. Particular emphasis is placed on the utilization of the layer-by-layer (LBL) self-assembly strategy [17], since this is a relatively new method (especially when applied to colloids) and because it offers new opportunities for creating novel core-shell and hollow particles. Although there is also interest in the arrangement of nanostructured colloidal entities into complex functional structures, here only the design and surface modification of colloids, and hollow capsule processing, are discussed and illustrated. In order to exemplify the versatility of the approach, a number of systems, ranging from ordered architectures of polymers, nanoparticles, or proteins on colloids to the encapsulation of (bio)crystals in molecularly engineered polymer multilayer cages, will be presented. The control that can be afforded over the size, shape, composition, and wall thickness of the deposited layers and resulting hollow capsules will also be detailed. The second section is focused on core-shell particles, while the third section is concerned with the design and construction of hollow capsules.

II. CORE-SHELL PARTICLES

A. Engineering of Particle Surfaces

A number of methods have been employed to produce core-shell particles, that is, particles that consist of solid or liquid cores surrounded by shells of organic or inorganic material. These include heteroaggregation (aggregation of oppositely charged particles) [18–20], polymerization processes (e.g., interfacial polymerization in emulsions, photopolymerization of monomers in two-phase aerosol droplets, dispersion/precipitation polymerization) [21–25], and controlled phase separation of polymers within droplets of oil-in-water emulsions [26]. Emulsion polymerization has also been used to coat submicrometer- and micrometer-size organic and inorganic particles [27,28]. Although this method often leads to aggregated particles embedded in a polymer matrix, a recent study has demonstrated the encapsulation of (unaggregated) silver nanoparticles by a uniform and well-defined polymer shell comprising polystyrene and methacrylate via emulsion polymerization [29]. An alternative and promising strategy for the formation of polymer-coated metal (e.g., gold) nanoparticles involves trapping and aligning the nanoparticles in the pores of membranes by vacuum filtration, followed by polymerization of a conducting polymer inside the pores [30–32].

The two main approaches that have been employed to produce shells of various inorganic coatings (silica, yttrium basic carbonate, and zirconium hydrous oxide) on microparticles are those employing direct surface reactions and the controlled precipitation of inorganic molecular precursors from solution [2–9,33–39]. For example, in 1992 Ohmori and Matijevic coated spindle-shaped hematite particles with silica layers by hydrolysis of the alkoxide tetraethylorthosilicate [6]. Submicrometer-size silica spheres have also been coated with titania (submonolayer to 7 nm thick) by hydrolysis of titanium alkoxide precursors [40]. In that work it was found that the ratio of the titanium alkoxide to water and the dilution of the reactant mixture in ethanol control the nature of the coating. More recently, a novel two-step silica-coating process comprising a sol-gel step followed by a dense liquid-coating exposure was used to coat maghemite with silica, affording a magnetic nanocomposite [41]. Dokoutchaev et al. have deposited metal particles of 2- to 4-nm diameter onto polystyrene microspheres by depositing the precursor metal (Pd) oxide or hydroxide onto the spheres, followed by reduction to give fine metal particles on the surface [39]. Even though these inorganic coating methods have been and continue to be widely used, they have a number of disadvantages for the preparation of coated particles: In many cases the formation of uniform and regular shell coatings is not obtained, control of the shell thickness is difficult to achieve, and particle aggregation often occurs.

An alternative approach to the formation of core-shell particles is by using sonochemistry. In sonochemical processes, the chemical effects of ultrasound, which arise from the formation, growth, and implosive collapse of bubbles in liquid (known as *acoustic cavitation*), have been exploited to prepare a variety of metal, oxide, and composite nanoparticles [42–44]. The generation of nanoparticles in the presence of larger colloids has led to the production of core-shell particles [45–47]. Semiconductor nanoparticles (ZnS) on submicrometer-size silica were prepared by the ultrasound irradiation of a slurry of silica, zinc acetate, and thioacetamide in water near room temperature [45]. The ZnS nanoparticles (1–5 nm in diameter) coated the colloidal silica surface as thin layers or nanoclusters, depending on the reactant concentrations. Ultrasound-induced cavitation has also been used to coat nanosize nickel on alumina submicrospheres [46] and cobalt clusters on silica spheres [47], imparting a magnetic function to the particles. Despite the sonochemical deposition technique being an interesting strategy, the type of core-shell particles that can be produced using this method are limited by the type of nanoparticles that can be synthesized sonochemically.

The emergence of self-assembly techniques for film construction has led to a wealth of research on the construction and applications of nanostructured thin-film materials [48]. In contrast, considerably less attention has been paid to the controlled modification of colloidal particle surfaces via classical self-assembly strategies, particularly with respect to the assembly of organized layered materials as a thin shell on the colloids. Recent investigations have exploited the electrostatic self-assembly of preformed organic or inorganic macromolecular species onto colloids in order to produce core-shell particles [39,49–60]. This method, originally introduced for the formation of polymer thin films on planar substrates [17], uses colloidal particles as templates to assemble nanocomposite multilayer shells on their surface. Since the driving force for the multilayer shell formation on the particles is due primarily to the electrostatic attraction between the oppositely charged species that are deposited, the assembly proceeds via the successive (i.e., layer-by-layer, LBL) deposition of a wide range of macromolecules and/or particles of opposite charge. Importantly, an overcompensation of charge occurs with adsorption of each layer, hence facilitating the deposition of subsequent layers.

The technology associated with the application of LBL self-assembly to colloidal templates represents a new and promising approach to engineer the surfaces of particles at the nanometer level. Judging by the impact the LBL method has had on the construction of thin planar films [17], it is anticipated that this technology will rapidly develop into a highly attractive and general approach to construct layered nanocomposite materials on colloids. Although the LBL technology for coating macroscopically flat surfaces is well established, it is still in its infancy when applied to colloidal particles. Therefore, the next section will deal specifically with the recent application of this method to modify particle surfaces and create novel core-shell colloids.

B. Application of Self-Assembly to Colloids

One of the most recent and intriguing developments in the area of colloid science is the surface nanoengineering of colloidal particles in solution. The self-assembly of macromolecular species onto particles via the LBL strategy allows the production of nanostructured colloidal materials with tailored compositions and well-defined morphologies. The concept of depositing particles onto solid substrates in an LBL manner can be traced back to the work of Iler in the mid-1960s [61]. In the early 1990s it was extended by Decher and coworkers to a combination of linear polycations and polyanions [62,63]. The scope of the LBL technique was later expanded to include inorganic nanoparticles, biomolecules, clays, and dyes in polyelectrolyte multilayer assemblies [17,64]. The construction of such multi-layers has, however, almost exclusively been performed on macroscopically flat (two-di-mensional), charged surfaces. It was only recently that the LBL method was applied to colloidal particles, thus permitting the formation of composite core-shell particles [39,49–60,65,66].

In the colloid templating strategy (depicted in Fig. 1), a polymer solution of con-centration sufficient to cause saturation adsorption is added to a colloidal suspen-sion (step 1). A prerequisite is that the added polymer has an opposite charge to that on the colloids, thereby adsorbing through electrostatic interactions and stabilizing the particles via electrostatic as well as steric contributions. At this stage the charge on the surface of the particles is reversed. Subsequent stepwise adsorption of oppositely charged polymer (step 2) or nanoparticles (step 3) (or other oppositely charged macro-molecules [59,60]) results in the deposition of polymer or nanoparticles, respectively. Repeating this sequence allows further deposition of regular layers in a controlled fash-ion. The formation of hollow capsules from core-shell particles will be discussed in the next section.

In addition to our recent work on the coating of colloids [49–53,56–60], several other groups have independently employed the preceding procedure to fabricate nanocomposite particles [39,54,55,65,66]. Keller et al. reported the preparation of alter-nating composite multilayers of exfoliated zirconium phosphate sheets and charged re-dox polymers on (3-aminopropyl)-triethoxysilane-modified silica particles [54], while Chen and Somasunduran deposited nanosize alumina particles in alternation with poly(acrylic acid), which acts as the bridging polymer, on submicrometer-size alumina core particles [55]. In the latter work the evidence for the formation of the composite par-ticles was provided mainly by adsorption studies and zeta-potential measurements, and it was stated that scanning electron microscopy revealed monolayer coverage of the nanoparticles on the core particles. However, in both of these earlier studies, no quanti-tative experimental evidence was provided for stepwise multilayer shell growth of alter-nating inorganic-polymer shells. More recently, Dokoutchaev et al. studied the alternat-ing assembly of metal nanosize particles (Au, Pd, and Pt) and oppositely charged polyelectrolyte onto polystyrene microspheres [39]. Mostly nonuniform nanoparticle coatings were observed, although the nanoparticle loading could be increased by re-peated depositions of nanoparticle and polyelectrolyte in the LBL manner [39]. In sev-eral detailed investigations using the same strategy, Caruso et al. demonstrated that ho-mogeneous and regular multilayer coatings of polymer, nanoparticles, or protein could be fabricated on submicrometer-size particles with nanometer precision [49–53,56–60]. The remainder of this section will deal with the fabrication and characterization of a range of novel core-shell particles using the colloidal templating and self-assembly ap-proach. The first part will focus on the coating of colloids with pure polymer multilayers and the second on nanoparticle/polyelectrolyte coatings. It will be shown that by using this process, important parameters, such as shell thickness, composition, and coating uni-formity, can be readily controlled.

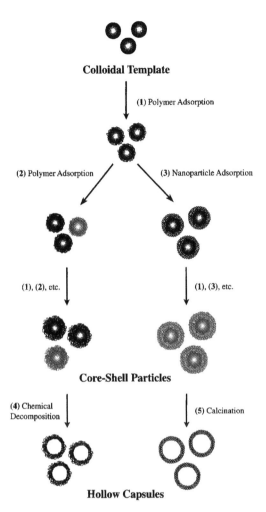

FIG. 1 Schematic illustration of the production of core-shell particles and hollow capsules by self-assembly and colloidal templating. The templates can be of synthetic (polymer latices) or biological origin (proteins, cells) with diameters in the nanometer to micrometer range. The process entails the consecutive deposition of oppositely charged species onto the colloidal particles (steps 1–3), exploiting primarily electrostatic interactions for the multilayer shell buildup. Following the deposition of each layer, unadsorbed polyelectrolyte or nanoparticles are removed by repeated centrifugation or filtration, with intermittent washings. Hollow capsules are obtained by removal of the core from the composite colloids, achieved by either chemical means (step 4) or thermal means (step 5). The method of core removal depends on the final composition of the hollow capsules required. For example, exposure of the coated colloids to acid, dimethylsulfoxide, or highly oxidizing solutions decomposes the core only (step 4), while calcination removes both the colloidal core and bridging polymer (step 5).

1. Polymer Multilayers on Colloids

Electrophoresis measurements provide a qualitative indication of the assembly of polymer multilayers on colloids [49,50]. The ζ-potential as a function of polyelectrolyte layer number for negatively charged polystyrene (PS) particles coated with poly(diallyldimethylammonium chloride) (PDADMAC) and poly(styrenesulfonate) (PSS) are displayed in Figure 2. As expected, the negatively charged (uncoated) PS particles yield a negative ζ-potential, the value being ca. -65 mV. The presence of a single layer of adsorbed PDADMAC on the PS latices causes a reversal in ζ-potential (ca. $+55$ mV). Subsequent deposition of PSS onto the PDADMAC-coated PS particles again changes the ζ-potential to negative values. Further polyelectrolyte depositions cause the ζ-potential to alternate in sign, depending on whether the outermost layer is the polycation (PDADMAC) or the polyanion (PSS). The alternating values qualitatively demonstrate a successful recharging of the particle surface with each polyelectrolyte deposition, suggesting that stepwise polyelectrolyte multilayer growth occurs on the particles. Alternating ζ-potentials were also observed for a range of other polycation/polyanion combinations [49–53].

Single particle light scattering (SPLS) is a sensitive optical technique [67,68] that enables determination of the thickness of adsorbed layers on colloids, as well as the state and degree of aggregation of the coated colloids [49,50]. The SPLS technique involves recording the light scattered from a single particle at a given moment in time. By recording the light scattered from many individual particles, one obtains a histogram of particle number versus scattering intensity (or SPLS intensity distributions). The normalized SPLS intensity distributions for neat PS latices and those coated with one, five, and nine PDADMAC/PSS layers are shown in Figure 3. The deposition of the polyelectrolytes onto the particles is manifested as a shift in the SPLS intensity distribution; there is a systematic shift (in the x-axis direction) with increasing polyelectrolyte layer number, confirming layer

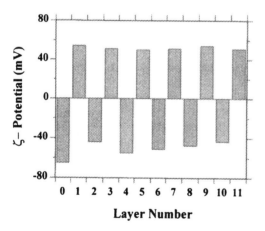

FIG. 2 ζ-potential as a function of layer number for PDADMAC/PSS multilayers on sulfate-stabilized polystyrene (PS) latices. The multilayers were assembled onto the negatively charged PS latices (ζ-potential of ca. -65 mV, layer number = 0) by the consecutive deposition of PDADMAC (odd layers) and PSS (even layers). Positive values are observed for PDADMAC deposition, and negative values for PSS adsorption. The alternating values are characteristic of stepwise growth of multilayer films on colloids.

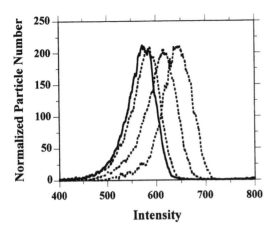

FIG. 3 Normalized SPLS intensity distributions of (from left to right) neat PS latices and PS lat-ices coated with one, five, and nine PDADMAC/PSS multilayers. (From Ref. 50.)

growth, as suggested by the ζ-potential data. The SPLS intensity distributions shown cor-respond to those of single particles. No peaks at higher intensities, which would be charac-teristic of aggregated particles (i.e., doublets, triplets, or higher-order aggregates), were ob-served for the multilayer-coated PS latices. This provides evidence that no significant aggregation of the PS latices occurs as a result of their coating with polymer multilayers. The thickness of the deposited layers can be derived from the SPLS data by using the Rayleigh–Debye–Gans theory and a refractive index for the polymer layers (1.47) [50]. The calculated average layer thickness increases with the number of polyelectrolyte layers deposited. This is depicted in Figure 4 for the PDADMAC/PSS layers on PS particles. The average thickness of each polymer layer is about 1.5 nm (equivalent to approximately 1.0 mg m^{-2}) [50]. Similar values have also been measured for the polyelectrolyte pair poly(al-lylamine hydrochloride) and PSS [50]. These data show that the shell thickness can be con-

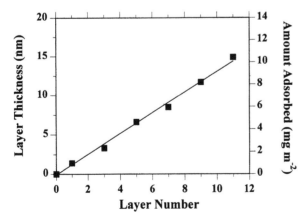

FIG. 4 Thickness of PDADMAC/PSS multilayers assembled on PS latices as a function of layer number. The layers were assembled by the consecutive adsorption of PDADMAC and PSS. The thicknesses were determined from SPLS data.

trolled at the nanometer level (to within 2 nm for the polymer layers). They are also in excellent agreement with those from x-ray reflectivity measurements for similar polymer multilayer films formed on (flat) silicon substrates under the same ionic strength conditions [69]. When the polymer contains strongly absorbing or fluorescing chromophores, evidence for its adsorption onto particles can also be provided by UV-vis or fluorescence measurements, respectively [51–53]. In summary, the foregoing data highlight that regular, stepwise growth of polymer multilayers on colloids occurs using the LBL strategy. Further evidence that multilayers were formed on the colloids is provided in the third section of this chapter, where hollow polymer capsules are derived from the polymer multilayer-coated particles.

The LBL-colloid approach was also extended to encapsulate biologically significant materials in the form of biocolloids [70]. It was demonstrated that the LBL approach could be applied to encapsulate the enzyme catalase via the sequential deposition of alternating polyelectrolytes onto catalase crystal templates. An extremely high enzyme loading in each polymer capsule was obtained, and the activity of the encapsulated enzyme was preserved. Another important and recent investigation concerned the encapsulation of uncharged, water-insoluble low-molecular-weight substances using the same procedure [71]. Crystals of pyrene and fluorescein diacetate (after a precharging step) were coated with polymer multilayers and contained within the polymer cages. These studies further demonstrate that the LBL strategy, when combined with colloidal templating, provides a simple, general, and versatile approach that can potentially be applied to the encapsulation of various crystallized substances for catalysis and drug delivery applications.

2. Nanoparticle Multilayers on Colloids

Nanosize particles (e.g., metals, semiconductors, etc.) are of continuing interest because they possess fascinating catalytic, electronic, and optical properties. Larger particles decorated with smaller nanoparticles on their surface are of interest because of their potential use as heterogeneous catalysts and their relevance in electronic and optical sensor applications as well as surface-enhanced Raman scattering [39,72–75].

Despite several investigations on the coating of larger particles with nanoparticles (see Section II.B), the stepwise and regular formation of uniform nanoparticle layers on colloidal particles was achieved only recently [56,58]. Silica nanoparticles (SiO_2) approximately 30 nm in diameter were alternately assembled with PDADMAC on submicrometer-size PS latices [56,58]. The ζ-potential data revealed an alternating trend (in magnitude) between the SiO_2 (negatively charged) and PDADMAC, again suggesting multilayer buildup. This was confirmed by SPLS, which showed that the thickness of each nanoparticle (SiO_2)/polymer layer is 30–40 nm, corresponding to approximately one monolayer of nanoparticles adsorbed with each deposition step. Importantly, SPLS also revealed that the colloidal core-nanocomposite shell particles prepared in this way exist as single, unaggregated particles in solution.

The morphology of the neat PS latices and the composite particles with PS latices as the core and SiO_2/PDADMAC shell coatings were examined by scanning (SEM) and transmission (TEM) electron microscopy and atomic force microscopy (AFM). Figure 5 shows SEM images of uncoated PS latices (a) and those coated with SiO_2/PDADMAC (b). The uncoated particles are hexagonally packed and exhibit a relatively smooth surface. The coated PS latices are decorated with close-packed SiO_2 nanoparticles on their surface, and an increase in surface roughness is also seen. Further evidence for the stepwise assembly of SiO_2/PDADMAC multilayers on the PS latices was provided by TEM. Figure 6 shows

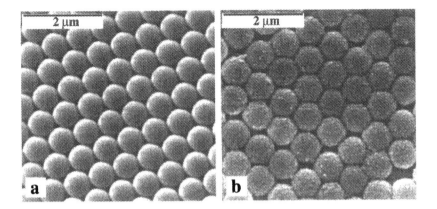

FIG. 5 SEM micrographs of (a) uncoated PS latices and (b) polyelectrolyte-modified PS latices coated with SiO$_2$/PDADMAC. An increase in surface roughness and diameter can be clearly seen for the particles coated with SiO$_2$/PDADMAC (compare b with a). (From Ref. 58.)

TEM micrographs of a neat PS particle (a) and those coated with one to five SiO$_2$/PDAD-MAC multilayers (b–f). The uncoated PS latices are spherical and feature a smooth surface (a). The presence of SiO$_2$/PDADMAC multilayers on the PS latices result in both an increase in surface roughness (due to SiO$_2$) and a systematic increase in the diameter of the PS latices (b–f). Consistent with the SPLS and SEM measurements, the PS particles are homogeneously coated with nanoparticles. The TEM data (Fig. 6) yield an average diame-

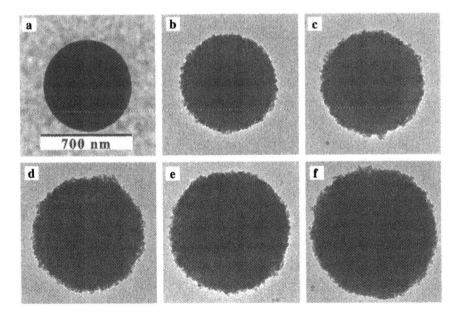

FIG. 6 TEM micrographs of (a) an uncoated PS particle and polyelectrolyte-modified PS latices coated with (b) one, (c) two, (d) three, (e) four, and (f) five SiO$_2$/PDADMAC multilayers. A systematic increase in diameter of the particles with increasing SiO$_2$/PDADMAC layer number is observed, confirming regular, stepwise multilayer growth. The average diameter increase is 65 ± 5 nm. The scale bar corresponds to all six TEM images shown. (From Ref. 58.)

TABLE 1 Thickness of SiO$_2$/PDADMAC Multilayers Assembled onto Negatively Charged PS Particles, as Determined by SPLS, SEM, and TEM

Number of multilayers	Multilayer thickness (nm)[a]		
	SPLS	SEM[b]	TEM[c]
1	24	38	30
2	68	—	63
3	112	122	100
4	155	—	123
5	181	190	155

[a] The error in the values is estimated as 10%.
[b,c] Values are the averages determined from measuring the diameters of coated particles.
Source: Ref. 58.

ter increment of ca. 65 nm, or a layer thickness of approximately 30 nm for each nanoparticle/polymer layer pair. These values are in agreement with those obtained from SPLS and SEM (see Table 1). Furthermore, AFM also revealed a uniform nanoparticle coating and an increase in the diameter of the coated colloids [58].

Micrometer-size composite particles with multilayer arrays of magnetic nanoparticles [57] and luminescent semiconductors [76,77] have also been created by using the LBL method. Functional core-shell particles were prepared by assembling a composite multilayer shell of charged polyelectrolytes and luminescent CdTe(S) nanocrystals (cadmium telluride with a certain content of sulfide) via their consecutive electrostatic adsorption from solution onto latex particles [76,77]. A confocal microscopy image of the polyelectrolyte/CdTe(S) shell assembled on PS spheres is shown in Figure 7. The luminescence from the CdTe(S) nanoparticles produces a ring around the particle core (the rings differ in size due to the focal point of the microscope). This image clearly shows that the CdTe(S) nanoparticles are embedded within the polyelectrolyte multilayer shell.

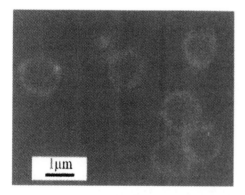

FIG. 7 Confocal laser scanning microscopy image of a four-layer polyelectrolyte/CdTe(S) nanocrystal shell assembled on 1.5-μm-diameter MF particles. The polyelectrolyte film consists of two bilayers of PAH and PSS. (From Ref. 76.)

The preceding examples emphasize the versatility of the LBL strategy used to fabricate composite core-shell particles. The technology has recently been extended to colloidal particles with diameters in the nanometer range [52,53]. Furthermore, biocolloids [70] and microcrystals [71] with diameters in the micrometer range can be successfully templated, thereby achieving their encapsulation. Proteins have also been assembled as multilayer arrays on colloidal particles [59,60]. In summary, the LBL method, when applied with colloids, is a powerful strategy for the engineering of core-shell colloids and is therefore expected to open many new possibilities for the technological application of such novel colloids.

III. HOLLOW CAPSULES

The study and preparation of hollow capsules has attracted considerable attention in recent years. Hollow capsules are of immense interest in a long list of potential applications. These include drug delivery, gene therapy, catalysis, waste removal, acoustic insulation, piezoelectric transducers, and functional materials [14].

A. Hollow Capsule Processing

There are a variety of routes currently utilized to fabricate a wide range of hollow capsules of various compositions. Among the more traditional methods are nozzle reactor processes, emulsion/phase-separation procedures (often combined with sol-gel processing), and sacrificial core techniques [78]. Self-assembly is an elegant and attractive approach for the preparation of hollow capsules. Vesicles [79,80], dendrimers [81,82], and block hollow copolymer spheres [83,84] are all examples of self-assembled hollow containers that are promising for the encapsulation of various materials.

Nozzle and sacrificial core approaches generally yield coarse hollow capsules in the micrometer- to millimeter-size range, while emulsion/sol-gel methods produce hollow spheres of nanometer to micrometer sizes. Hollow polymer, oxide, metal, and glass composite microspheres have been produced via nozzle-reactor methods (spray drying or pyrolysis) [14,85–87]. For example, Bruinsma et al. employed a spray-drying technique to prepare hollow silica spherical particles [85]. Hollow titanium dioxide microspheres with a 50-nm thin shell were prepared by spray-drying a colloidal suspension of exfoliated titanate sheets, followed by heating [87].

The sacrificial core approach entails depositing a coating on the surface of particles by either the controlled surface precipitation of inorganic molecular precursors from solution or by direct surface reactions [2,3,5,6,8,9,33–35,38], followed by removal of the core by thermal or chemical means. Using this approach, micron-size hollow capsules of yttrium compounds [2], silica spheres [38], and monodisperse hollow silica nanoparticles [3,35] have been generated.

Hollow and porous polymer capsules of micrometer size have been fabricated by using emulsion polymerization or through interfacial polymerization strategies [79,83–84, 88–90]. Micron-size, hollow cross-linked polymer capsules were prepared by suspension polymerization of emulsion droplets with polystyrene dissolved in an aqueous solution of poly(vinyl alcohol) [88], while latex capsules with a multihollow structure were processed by seeded emulsion polymerization [89]. Ceramic hollow capsules have also been prepared by emulsion/phase-separation procedures [14,91–96]: For example, hollow silica capsules with diameters of 1–100 micrometers were obtained by interfacial reactions conducted in oil/water emulsions [91].

Self-assembly phenomena can be exploited to create a range of versatile and use-ful hollow capsules. Lipid liposomes and vesicles are a special group of hollow structures that are formed from phospholipids through self-assembly. They consist of closed bilayer aggregate systems, where the bilayer structures separate an aqueous interior from an aqueous exterior, hence allowing water-soluble drugs to be encapsulated within them. Despite the fact that they are widely employed as delivery systems for various com-pounds in the pharmaceutical and cosmetic industries [97], problems exist with their sta-bility and permeability. This has resulted in the use of different methods to effect their stabilization [79,80,98–101]. For example, tough vesicles (termed *polymersomes*) were made from diblock copolymers of polyethyleneoxide-polyethylethylene [80]. These polymersomes were found to be almost an order of magnitude tougher than lipid mem-branes and 10 times less permeable than phospholipid bilayers. Hollow polymer spheres were also obtained by polymerization of hydrophobic monomers in the interior of the surfactant bilayer of vesicles [79]. Copolymers are another example of materials that can self-assemble into hollow structures [80,83–84,102]. The rod-coil copolymer poly(phenylquinoline)-block-polystyrene self-organizes into stable hollow capsules, with diameters of 1.5–10 μm [102]. These hollow capsules have been shown to be suitable for the encapsulation of fullerenes. Wooley et al. have also utilized block copolymers to pro-duce self-assembled (shell) cross-linked polymeric micelles with diameters in the 10- to 100-nm range [83,84]. Nanocapsules have been derived from these polymer micelles (made from poly(isoprene-*b*-acrylic acid)) by chemically removing the polyisoprene core by ozonolysis.

Another method to synthesize hollow nanocapsules involves the use of nanoparticle templates as the core, growing a shell around them, then subsequently removing the core by dissolution [30–32]. Although this approach is reminiscent of the sacrificial core method, the nanoparticles are first trapped and aligned in membrane pores by vacuum fil-tration rather than coated while in aqueous solution. The nanoparticles are employed as templates for polymer nucleation and growth: Polymerization of a conducting polymer around the nanoparticles results in polymer-coated particles and, following dissolution of the core particles, hollow polymer nanocapsules are obtained.

An alternative approach for the production of hollow capsules is that which exploits both self-assembly and colloidal templating (as was introduced in Section II.B; see Fig. 1). The basis of the technology is first to deposit ordered shell architectures on colloidal parti-cles by self-assembly, followed by removal of the decomposable colloidal core, either chemically or thermally [52,70,71,103–106]. The procedure makes use of the semiperme-able nature of the multilayer shell on the colloid: The decomposed core constituents per-meate the shell wall, leaving behind hollow capsules. In this section it will be demonstrated that by using this process, important parameters, such as size, geometry, composition (polymer, inorganic, or composites), wall thickness, uniformity, and the diameter-to-wall-thickness ($d{:}t$) ratio of the hollow capsules formed, can be controlled. The ability to control such parameters is crucial in the application and commercialization of hollow capsules [14], and it overcomes many of the disadvantages associated with current techniques used for their production [14,15]. Further, relatively harsh conditions are employed in a number of the approaches outlined earlier, therefore making them unsuitable for the encapsulation of various sensitive materials. The applicability of the layer-by-layer self-assembly tech-nique to encapsulate biologically significant and sensitive materials will also be verified. It is expected that application of this technology will allow new classes of hollow capsules to be generated, hence extending the potential uses of hollow colloids.

1. Polymer Capsules

The versatility of the approach for creating hollow polymer capsules is demonstrated in Figure 8. This is evidenced by the different size, shell composition, colloidal template, and core-removal procedures employed to produce the polymer capsules. Figure 8a is a TEM image of an air-dried hollow polymer capsule comprising a Fe(II) metallo-supramolecular coordination polyelectrolyte (Fe(II)-MEPE) and PSS, and Figure 8b is a TEM micrograph of a hollow polymer capsule composed of PSS and PAH (also air-dried). The Fe(II)-MEPE/PSS capsule was produced by templating weakly cross-linked, spherical melamine-formaldehyde (MF) particles of 1.7-μm diameter (i.e., coating them with polymer multilayers) and subsequently decomposing the core by exposure of the coated particles to an acidic solution of pH < 1.6 [52]. In contrast, the PSS/PAH capsules were prepared by using biocrystal templates (catalase crystals) of approximately 10-μm size for polymer multilayer deposition, followed by template removal by an oxidizing solution (e.g., deproteinizer) [70]. In both cases, the core was removed by chemical means: For the Fe(II)-MEPE capsules, the acid caused decomposition of the MF particle into its constituent oligomers, and the oligomers were then readily expelled by permeating the polymer multilayer shell [103,104]. The enzyme was decomposed by the deproteinizer treatment (pH ca. 12) and likewise removed by permeating the capsule walls [70]. It should be noted that even the deposition of only three polyelectrolyte layers onto colloids results in the production of polyelectrolyte capsules when the core is removed [103]. This attests to the exceptionally strong electrostatic forces involved in the formation of the polyelectrolyte

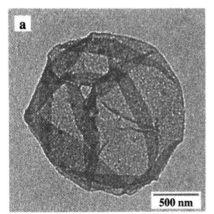
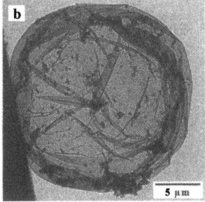

FIG. 8 Microscopy images of air-dried hollow polymer capsules. They were produced by coating polymer- and biocolloids of submicrometer and micrometer size, respectively, with polyelectrolyte multilayers and subsequently removing the templated core. (a) TEM image of a hollow polymer capsule composed of Fe(II)-MEPE and PSS (total of five layers). The multilayers were assembled on MF particles, and the core was decomposed by an acidic solution of pH < 1.6. (b) TEM image of a hollow polymer capsule comprising $(PSS/PAH)_4$ (total of eight polyelectrolyte layers). The capsule was obtained after decomposition of the catalase crystal template (via deproteinizer treatment) on which the polymer layers were deposited. A small amount of undecomposed enzyme can still be seen in the capsule. The dimensions of the hollow polymer capsules are determined by the size of the colloidal template (see text for details). The folds and creases seen in the polymer capsules are a result of the drying process (i.e., evaporation of the aqueous content by air-drying). Some spreading upon drying also occurs.

coating. The polymer multilayers assembled on the enzyme crystals underwent a morphology change from somewhat rectangular to close to spherical upon solubilization of the enzyme crystal [70]. This was attributed to the osmotic pressure built up inside the capsules as a result of solubilization of the enzyme, since it was entrapped inside the capsules.

The latex and biocrystal templates used to produce the capsules in Figure 8 had an inherent surface charge and readily dispersed in water prior to coating. In contrast, the capsule displayed in Figure 9 was obtained from a fluorescein diacetate (FDA) microcrystal template, which is hydrophobic and uncharged. Figure 9 represents a confocal laser scanning microscope (CLSM) image of a hollow polymer capsule in *an aqueous environment* comprising 11 PAH/PSS layers [71]. Here the polymer capsule essentially retains the original shape of the template. The key step in the coating of uncharged crystals is the introduction of a charge to their surface, therefore making them amenable to dispersion in an aqueous solution. This was achieved by exposing the crystals to a solution containing amphiphilic molecules, such as surfactant. The polyelectrolytes were then LBL deposited onto the charged templates, and hollow microcapsules obtained by removal of the FDA template via its solubilization in an ethanol solution. The polymer-multilayer coating of uncharged templates represents a significant step forward in encapsulation technologies, for this process is envisaged to be applicable to a variety of other low-molecular-weight, crystallized (or amorphous) materials, i.e., drugs. In addition, this approach provides a facile approach to the production of hollow polymer capsules, avoiding acid [103] or basic (i.e., deproteinizer) [70] solutions for core removal.

The foregoing examples show that hollow polymer capsules with varying composition and sizes of ca. 2–20 micrometers can be produced, either by templating charged (latex particles and biocrystals) or uncharged (organic microcrystals), and that different core removal procedures can be employed. Nanometer-size polymer capsules have also been produced by employing smaller particle templates [107].

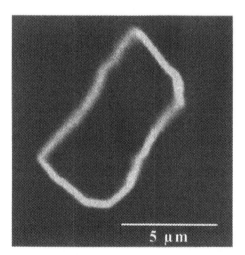

FIG. 9 Confocal laser scanning micrograph of a hollow polymer capsule. The polymer capsule was obtained from polymer multilayer-templated FDA microcrystals after removal of the colloidal core. The FDA microcrystals were coated with SDS and 11 polyelectrolyte layers [(PAH/PSS)$_3$/PAH/ (PSS/PAH-FITC)$_2$]. (PAH-FITC = PAH labeled with fluorescein isothiocyanate.) The microcrystal core was removed by exposure of the coated microcrystals to ethanol, causing solubilization of FDA.

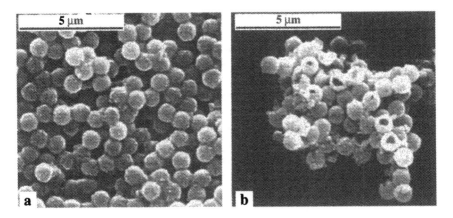

FIG. 10 SEM micrographs of (a) silica nanoparticle/polymer [SiO$_2$/PDADMAC)$_3$]-coated PS latices and (b) hollow silica capsules. The hollow silica capsules were obtained by calcining coated particles as shown in (a). The calcination process removes the PS core and the polymer bridging the silica nanoparticles, while at the same time fusing the silica nanoparticles together. Some of the silica capsules were deliberately broken to demonstrate that they were hollow (b). (From Ref. 106.)

2. Inorganic Capsules

The formation of inorganic capsules via the LBL self-assembly approach requires the uniform deposition of nanoparticles in a close-packed array onto larger-size particles. Subsequent calcination of the nanoparticle-coated microspheres allows the production of hollow inorganic capsules. An example of this is the creation of hollow silica capsules [105,106], which are prepared first by sequentially adsorbing silica nanoparticles and polymer from dilute aqueous solutions onto decomposable micrometer-size polystyrene particles (see Section II.B.2), and second by removing the templated colloidal core and polymer by calcination. SEM images of PS particles coated with three layers of SiO$_2$ and PDADMAC before and after calcination are shown in Figure 10. The calcination process has a dual effect: (1) It removes the organic matter, i.e., colloidal core and bridging polymer (PDADMAC) during heating to 450°C, as confirmed by thermogravimetric analysis, and (2) it causes condensation (i.e., cross-linking) of the silica nanoparticles, hence providing structural integrity for the hollow capsules. Both intact and broken hollow capsules are seen in Figure 10b. The broken spheres were obtained by deliberately crushing them in order to confirm that they were hollow. The diameters of the hollow capsules are about 5–10% smaller than those of the uncalcined nanocomposite particles. Using this approach, complete hollow capsules were produced when the capsule wall consisted of two or more SiO$_2$ nanoparticle layers. The initial spherical shape of the PS latices was retained upon removal of the core.

TEM micrographs of a PS particle coated with two SiO$_2$/PDADMAC layer pairs (a) before calcination and hollow silica capsules obtained after calcination of PS latices coated with (b) one, (c) two or (d) three SiO$_2$/PDADMAC layer pairs are displayed in Figure 11. The PS latices are uniformly coated by the silica nanoparticles (a). Close examination of the TEM images of the hollow silica capsules reveals that individual silica nanoparticles forming the multilayer shell fuse together after calcination. This is due to the high-temperature treatment, which causes condensation of the silica nanoparticles (i.e., fusion). However, cross-linking between the hollow capsules is limited. Individual hollow silica spheres were produced, showing that coalescence of individual SiO$_2$ nanoparticles occurred predominantly within individual capsules rather than between capsules. Figure 11 also reveals the control that is achievable over the capsule wall thickness. The wall thickness and outer sphere diameter increase regularly (by 30 and 60 nm, respectively) with the number of silica nanoparticle layers deposited. The high uniformity of the wall thickness is a further reflection of the regularity of the coating process.

Further evidence that hollow capsules were produced by the LBL approach was provided by ultramicrotoming the samples [106]. A TEM image of an ultrathin (30–50 nm) cross section of the silica capsules is shown in Figure 12. The spherical shape and the regular wall thickness of the hollow capsules can be clearly seen. The average thickness of the silica shell (produced from three SiO_2/PDADMAC layer pairs) is 100 ± 10 nm. This value is in excellent agreement with that expected for three monolayers of SiO_2 nanoparticles. The hollow silica capsules produced are porous: The resin used to set the hollow capsules prior to ultramicrotoming permeated the capsule walls [106]. This is an important aspect with regard to future applications of these hollow capsules.

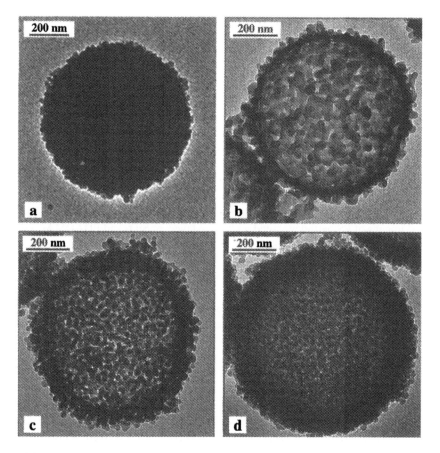

FIG. 11 TEM images of (a) a [(SiO_2/PDADMAC)$_2$]-coated PS particle and hollow silica capsules produced from PS latices coated with (b) one, (c) two, or (d) three SiO_2 layers. The hollow silica capsules maintain the shape of the original PS particle template. Removal of the core by calcination is confirmed by the reduced electron density in the interior of the capsules (compare b–d with a). The images of the hollow silica capsules show the nanoscale control that can be exerted over the wall thickness and their outer diameter. (From Ref. 106.)

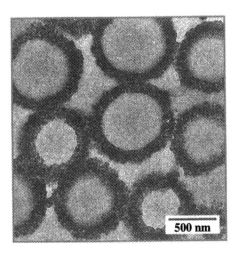

FIG. 12 TEM micrograph of a cross section of hollow silica capsules. The hollow capsules were prepared by calcining PS latices coated with [SiO$_2$/PDADMAC)$_3$]. (From Ref. 110.)

The foregoing results demonstrate that the thickness of the capsule wall can be controlled at the nanometer level by varying the number of deposition cycles, while the shell size and shape are predetermined by the dimensions of the templating colloid employed. This approach has recently been used to produce hollow iron oxide, magnetic, and hetero-composite capsules [108]. The fabrication of these and related capsules is expected to open up new areas of applications, particularly since the technology of self-assembly and colloidal templating allows unprecedented control over the geometry, size, diameter, wall thickness, and composition of the hollow capsules. This provides a means to tailor their properties to meet the criteria of certain applications.

3. Inorganic/Organic Composite Capsules

Hollow inorganic/organic composite spheres can also be produced using the preceding process [105,106]. The key step to their production is the selection of a suitable core-removal procedure, i.e., one that decomposes the templated core but leaves the polymer bridging the nanoparticles in the shell. The most common method of core removal has been the use of solvents: acidic and dimethylsulfoxide decompose melamine-formaldehyde polymer latex core templates [103–105], tetrahydrofuran removes polystyrene cores [105], and highly oxidizing solutions (e.g., deproteinizer) decompose proteinaceous cores [70]. Figure 13 shows a TEM micrograph of a silica nanoparticle/polymer composite capsule produced by the decomposition and removal of an MF latex core from a silica nanoparticle/polymer multilayer shell assembled onto MF particles. The hollow composite capsules assume a flat conformation on the substrate when dried [106], similar to the hollow polymer capsules (see Fig. 8). The capsule is composed of nanoparticles embedded in the polymer matrix. Recently, hollow iron oxide–polymer composite capsules have been produced using this strategy [109], thereby introducing a magnetic function to the capsules. This would be of interest in an application where the capsules need to be displaced under the influence of a magnetic field. Nanoparticle polymer composite hollow structures have also been obtained by coating biocolloids (gluteraldehyde-fixed echinocytes) and removing the core by exposure to deproteinizer [110].

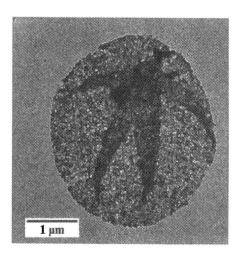

FIG. 13 TEM micrograph of a hollow composite nanoparticle/polymer capsule dried on a carbon grid. The hollow composite capsule was obtained after removal of the MF core from SiO$_2$/PDAD-MAC-coated MF particles by treatment with hydrochloric acid. The "shadowing" seen is a result of collapse and overlapping of the hollow capsule upon drying. (From Ref. 110.)

IV. CONCLUSIONS AND OUTLOOK

The examples presented in this chapter highlight the advantages of combining self-assembly and colloidal templating for the fabrication of a large range of coated colloids and hollow capsules in the submicrometer- to micrometer-size range. Important parameters, such as composition, geometry, diameter, and wall thickness, can be controlled with nanometer-scale precision. This is readily achieved by utilizing colloids of a certain shape and size and by varying the number of coating cycles. Another advantage is the versatility of the approach, as exemplified by the vastly different colloidal core templates (polymer particles, biocolloids, and hydrophobic crystals) that are amenable to the process, as well as the various core-removal processes (chemical or thermal pathways).

The nanoengineered synthetic core-shell particles and hollow capsules described make intriguing new materials available for applications in chemistry, bioscience, and materials science. For example, composite particles can be exploited in the areas of coatings, electronics, photonics, catalysis, separations, and diagnostics, and the hollow capsules can be utilized as drug delivery vehicles, reactor systems, and specific recognition systems (once functionalized). Attaching biocomponents to the surfaces of the composite particles or capsules would yield biofunctionalized colloidal entities for sensor chemistry. For encapsulation and delivery system applications, tailoring the thickness and composition of the walls is an important target in order to achieve selective and switchable release of the encapsulated materials. The technology has already been shown to be suitable for the encapsulation of bioactive [70,110] and low-molecular-weight substances [71]. Chemically reactive sites inside the hollow capsule walls have also been utilized to effect precipitation of various materials [111]. The strategy has also recently been applied to nanoparticle templates [107].

On the whole, the technology utilized to produce the variety of new nanostructured colloidal materials, as outlined in this chapter, is unparalleled in its versatility and simplicity and is therefore foreseen to become widely used in the engineering of colloidal entities for various applications in the physical and life sciences.

ACKNOWLEDGMENTS

The support of colleagues from the Max-Planck-Institute of Colloids and Interfaces, Potsdam, Germany, is gratefully acknowledged.

REFERENCES

1. R Davies, GA Schurr, P Meenan, RD Nelson, HE Bergna, CAS Brevett, RH Goldbaum. Adv Mater 10:1264, 1998.
2. N Kawahashi, E Matijevic. J Colloid Interface Sci 143:103, 1991.
3. M Giersig, T Ung, LM Liz-Marzan, P Mulvaney. Adv Mater 9:570, 1997.
4. H Bamnolker, B Nitzan, S Gura, S Margel. J Mater Sci Lett 16:1412, 1997.
5. D Walsh, S Mann. Nature 377:320, 1995.
6. M Ohmori, E Matijevic. J Colloid Interface Sci 150:594, 1992.
7. MA Correa-Duarte, M Giersig, LM Liz-Marzan. Chem Phys Lett 286:497, 1998.
8. S Margel, E Weisel. J Polym Sci Chem Ed 22:145, 1984.
9. AP Philipse, MPB van Bruggen, C Pathmamanoharan. Langmuir 10:92, 1994.
10. E Bartsch, V Frenz, J Baschnagel, W Schaertl, H Silescu. J Chem Phys 106:3743, 1997.
11. DH Napper. Polymeric Stabilization of Colloidal Dispersions. New York: Academic Press, 1983.
12. DA Antelmi, O Spalla. Langmuir 15:7478, 1999.
13. T Sato, R Ruch. Stabilization of Colloidal Dispersions by Polymer Adsorption (Surfactant Science Series, No. 9). New York: Marcel Dekker, 1980, pp 65–119.
14. DL Wilcox, M Berg, T Bernat, D Kellerman, JK Cochran, eds. Hollow and Solid Spheres and Microspheres: Science and Technology Associated with Their Fabrication and Application. Vol 372. Pitsburgh: Materials Research Society Proceedings, 1995.
15. JK Cochran. Current Opinion Solid State Mater Sci 3:474, 1998.
16. A Rembaum, WJ Dreyer. Science 208:364, 1980.
17. For a review, see: G Decher. Science 277:1232, 1997.
18. S Harley, DW Thompson, B Vincent. Colloids Surf 62:163, 1992.
19. F Dumont, G Ameryckx, A Watillon. Colloids Surf 51:171, 1990.
20. M Okubo, Y He, K Ichikawa. Colloid Polym Sci 269:125, 1991.
21. HK Mahabadi, TH Ng, HS Tan. J Microencapsulation 13:559, 1996.
22. T Dobashi, FJ Yeh, QC Ying, K Ichikawa, B Chu. Langmuir 11:4278, 1995.
23. R Arshady. Polym Eng Sci 30:915, 1990.
24. C Esen, T Kaiser, MA Borchers, G Schweiger. Colloid Polym Sci 275:131, 1997.
25. E Bourgeat-Lami, J Lang. J Colloid Interface Sci 197:293, 1998.
26. A Loxley, B Vincent. J Colloid Interface Sci 208:49, 1998.
27. WD Hergeth, UJ Steinau, HJ Bittrich, K Schmutzler, S Wartewig. Prog Colloid Polym Sci 85:82, 1991.
28. AM van Herk. NATO ASI Ser, Ser E 335:435, 1997.
29. L Quaroni, G Chumanov. J Am Chem Soc 121:10642, 1999.
30. SM Marinakos, LC Brousseau, A Jones, DL Feldheim. Chem Mater 10:1214, 1998.
31. SM Marinakos, DA Shultz, DL Feldheim. Adv Mater 11:34, 1999.
32. SM Marinakos, JP Novak, LC Brousseau, AB House, EM Edeki, JC Feldhaus, DL Feldheim. J Am Chem Soc 121:8518, 1999.
33. A Garg, E Matijevic. J Colloid Interface Sci 126:243, 1998.
34. N Kawahashi, E Matijevic. J Colloid Interface Sci 138:534, 1990.
35. M Giersig, LM Liz-Marzan, T Ung, DS Su, P Mulvaney. Ber Bunsenges Phys Chem 101:1617, 1997.
36. LM Liz-Marzan, M Giersig, P Mulvaney. Langmuir 12:4329, 1996.
37. LM Liz-Marzan, M Giersig, P Mulvaney. J Chem Soc Chem Commun 731, 1996.
38. H Bamnolker, B Nitzan, S Gura, S Margel. J Mater Sci Lett 16:1412, 1997.

39. A Dokoutchaev, JT James, SC Koene, S Pathak, GKS Prakash, ME Thompson. Chem Mater 11:2389, 1999.
40. A Hanprasopwattana, S Srinivasan, AG Sault, AK Datye. Langmuir 12:3173, 1996.
41. Q Liu, Z Xu, JA Finch, R Egerton. Chem Mater 10:3936, 1998.
42. KS Suslick, SB Choe, AA Cichowlas, MW Grinstaff. Nature 353:414, 1991.
43. X Cao, Y Koltypin, G Katabi, I Felner, A Gedanken. J Mater Res 12:405, 1997.
44. N Arul Dhas, A Gedanken. J Phys Chem 101:9495, 1997.
45. N Arul Dhas, A Zaban, A Gedanken. Chem Mater 11:806, 1999.
46. Z Zhong, Y Mastai, Y Koltypin, Y Zhao, A Gedanken. Chem Mater 11:2350, 1999.
47. S Ramesh, Y Cohen, R Prosorov, KVPM Shafi, D Aurbach, A Gedanken. J Phys Chem B 102:10234, 1998.
48. A Ulman. An Introduction to Ultrathin Organic Films: From Langmuir–Blodgett to Self-Assembly. Boston, Academic Press, 1991; JH Fendler. Nanoparticles and Nanostructured Films: Preparation, Characterization and Application. Weinheim, Germany: Wiley VCH, 1998.
49. F Caruso, E Donath, H Möhwald. J Phys Chem B 102:2011, 1998.
50. F Caruso, H Lichtenfeld, E Donath, H Möhwald. Macromolecules 32:2317, 1999.
51. GB Sukhorukov, E Donath, H Lichtenfeld, E Knippel, M Knippel, H Möhwald. Colloids Surf A: Physicochem Eng Aspects 137:253, 1998.
52. F Caruso, C Schüler, DG Kurth. Chem Mater 11:3394, 1999.
53. DG Kurth, F Caruso, C Schüler. Chem Commun 1579, 1999.
54. SW Keller, SA Johnson, ES Brigham, EH Yonemoto, TE Mallouk. J Am Chem Soc 117:12879, 1995.
55. T Chen, P Somasundaran. J Am Ceram Soc 81:140, 1998.
56. F Caruso, H Lichtenfeld, H Möhwald, M Giersig. J Am Chem Soc 120:8523, 1998.
57. F Caruso, AS Susha, M Giersig, H Möhwald. Adv Mater 11:950, 1999.
58. F Caruso, H Möhwald. Langmuir 15:8276, 1999.
59. F Caruso, H Möhwald. J Am Chem Soc 121:6039, 1999.
60. F Caruso, H Fiedler, K Haage. Colloids Surf A: Physicochem Eng Aspects 169:287, 2000.
61. RK Iler. J Colloid Interface Sci 21:569, 1966.
62. G Decher, J-D Hong. Makromol Chem, Macromol Symp 46:321, 1991.
63. G Decher, J-D Hong. Ber Bunsen-Ges Phys Chem 95:1430, 1991.
64. For a review, see: G Decher. In: J-P Sauvage, MW Hosseini, eds. Templating, Self-Assembly and Self-Organization. Vol 9. Oxford: Pergamon Press, 1996, pp 507–528.
65. T Okubo, M Suda. Colloid Polym Sci 277:813, 1999.
66. T Okubo, M Suda. Colloid Polym Sci 278:380, 2000.
67. H Lichtenfeld, L Knapschinsky, H Sonntag, V Shilov. Colloids Surf A: Physicochem Eng Aspects 104:313, 1995.
68. H Lichtenfeld, L Knapschinsky, C Dürr, H Zastrow. Progr Colloid Polym Sci 104:148, 1997.
69. GB Sukhorukov, J Schmitt, G Decher. Ber Bunsenges Phys Chem 100:948, 1996.
70. F Caruso, D Trau, H Möhwald, R Renneberg. Langmuir 16:1485, 2000.
71. F Caruso, W Yang, D Trau, R Renneberg. Langmuir 16:8932, 2000.
72. A Henglein. J Phys Chem 97:5457, 1993.
73. L Armelao, R Bertoncello, MD Dominicus. Adv Mater 9:736, 1997.
74. T Sun, K Seff. Chem Rev 94:857, 1994.
75. M Haruta. Catal Today 36:153, 1997.
76. A Susha, F Caruso, AL Rogach, GB Sukhorukov, A Kornowski, H Möhwald, M Giersig, A Eychmüller, H Weller. Colloids Surfaces A: Physicochem Eng Aspects 163:39, 2000.
77. A Rogach, A Susha, F Caruso, G Sukhorukov, A Kornowski, S Kershaw, H Möhwald, A Eychmüller, H Weller. Adv Mater 12:333, 2000.
78. DL Wilcox, M Berg. In: Hollow and Solid Spheres and Microspheres: Science and Technology Associated with Their Fabrication and Application. Vol 372. Pittsburgh: Materials Research Society Proceedings, 1995, pp 3–13.

79. J Hotz, W Meier. Langmuir 14:1031, 1998.
80. BM Discher, Y-Y Won, DS Ege, JC-M Lee, FS Bates, DE Discher, DA Hammer. Science 284:1143, 1999.
81. M Zhao, L Sun, RM Crooks. J Am Chem Soc 120:4877, 1998.
82. MS Wendland, SC Zimmerman. J Am Chem Soc 121:1389, 1999.
83. KB Thurmond, T Kowalewski, KL Wooley. J Am Chem Soc 119:6656, 1997.
84. KB Thurmond, H Huang, CG Clark Jr, T Kowalewski, KL Wooley. Colloids Surfaces B: Biointerfaces 16:45, 1999.
85. PJ Bruinsma, AY Kim, J Liu, S Baskaran. Chem Mater 9:2507, 1997.
86. Y Lu, H Fan, A Stump, TL Ward, T Rieker, CJ Brinker. Nature 398:223, 1999.
87. M Iida, T Sasaki, M Watanabe. Chem Mater 10:3780, 1998.
88. M Okubo, Y Konishi, H Minami. Colloid Polym Sci 276:638, 1998.
89. XZ Kong, CY Kan, HH Li, DQ Yu, Q Yuan. Polym Adv Technol 8:627, 1997.
90. B Miksa, S Slomkowski. Colloid Polym Sci 273:47, 1995.
91. S Schacht, Q Huo, IG Voigt-Martin, GD Stucky, F Schuth. Science 273:768, 1996.
92. KJ Pekarek, JS Jacob, E Mathiowitz. Nature 367:258, 1994.
93. JG Liu, DL Wilcox. J Mater Res 10:84, 1995.
94. U Kubo, H Tsubakihaara. J Vac Sci Tech A 5:2778, 1987.
95. KH Moh, HG Sowman, TE Wood. US Patent No. 5,077,241, 1991.
96. HG Sowman. US Patent No. 4,349,456, 1982.
97. DD Lasic. Liposomes: From Physics to Applications. Amsterdam: Elsevier, 1993.
98. H Ringsdorf, B Schlarb, J Venzmer. Angew Chem 100:117, 1988.
99. DD Lasic. Angew Chem Int Ed Engl 33:1685, 1994.
100. W Meier, J Hotz, S Günther-Ausborn. Langmuir 12:5028, 1996.
101. D Kippenberger, K Rosenquist, L Odberg, JH Fendler. J Am Chem Soc 105:1129, 1983.
102. SA Jenekhe, XL Chen. Science 279:1903, 1998.
103. E Donath, GB Sukhorukov, F Caruso, SA Davis, H Möhwald. Angew Chem Int Ed 37:2201, 1998.
104. GB Sukhorukov, E Donath, S Davis, H Lichtenfeld, F Caruso, VI Popov, H Möhwald. Polym Adv Technol 9:759, 1998.
105. F Caruso, RA Caruso, H Möhwald. Science 282:1111, 1998.
106. F Caruso, RA Caruso, H Möhwald. Chem Mater 11:3309, 1999.
107. DI Giffins, F Caruso. Adv Mater 12:1947, 2000.
108. F Caruso, M Spasova, A Susha, M Giersig, RA Caruso. Chem Mater 13:109, 2001.
109. A Voigt, H Lichtenfeld, GB Sukhorukov, H Zastrow, E Donath, H Baümler, H Möhwald. Ind Eng Chem Res 38:4037, 1999.
110. F Caruso. Chem Eur J 6:413, 2000.
111. G Sukhorukov, L Dähne, J Hartmann, E Donath, H Möhwald. Adv Mater 12:112, 2000.

15

Electro-Transport in Hydrophilic Nanostructured Materials

BRUCE R. LOCKE   Florida State University, Tallahassee, Florida

I. INTRODUCTION

Electrophoretic transport, or, more generally, electric-field-induced molecular motion, in uncharged hydrophilic materials with nanometer-scale pore structure is important in a number of processes, including biochemical separation and purification methods, electric-field-enhanced drug delivery for biomedical engineering applications, and some biological phenomena. For example, electrophoresis in polymer hydrogels has been extensively used and developed for the separation and purification of biological macromolecules. A wide range of materials, from fibrous membranes to chemically and physically cross-linked gels and polymer solutions, has been utilized for electrophoretic separations. Recent interest has also focused on electrophoresis and chromatography in silicon and other inorganic surfaces (with and without polymeric supports) with etched channels of nanometer-to-micrometer dimension. Many electrophoretic processes have been developed that use these media under various electric field conditions, from constant applied voltages or currents to single or multidimensional pulsed electric fields. In addition, separation methods such as capillary electrophoresis, electrochromatography, and field flow fractionation may utilize hydrophilic polymeric media to increase the resolution of macromolecular separations in situations where the applied electric field is coupled to bulk solution-phase hydrodynamic flow via electro-osmotic flow or pressure-driven flow. Electric fields have also been used to enhance the ultrafiltration of proteins by altering the structure of the protein gel layer formed on the ultrafiltration membrane surface. Other examples where applied electric fields affect macromolecular transport in media consisting of polymeric materials include transport through biological tissue (e.g., biopolymers of proteins or polysaccharides) under the influence of constant or pulsed electric fields. This is of much current interest for applications in the development of transdermal drug delivery processes via iontophoresis and electroporation, for cancer treatment, for wound healing, and for understanding the general effects of electric fields on biological cells, membranes, and tissues.

In the analysis of transport of macromolecules in hydrophilic nanometer-scale media there are four fundamental questions that may be considered. These include:

1. How does the structure of the media, including the size, shape, geometry, and distribution of pores, influence the transport characteristics for a given solute?
2. What is the role of the shape, size, and other molecular and chemical characteristics of the transporting solute on the rate of transport?

527

3. How do the solution conditions including pH, ionic strength, and buffer type, affect the transport, and how can these conditions be controlled and optimized for a particular application?
4. What is the optimal means for applying the electric field to facilitate the motion of the solute?

These four questions are of course interrelated; however, theory and experiments can be, and have been, designed to isolate the effects of the individual factors of media structure, solute properties, electric field conditions, and solvent characteristics. The roles of the media properties and the size and shape of the solute have been considered extensively in the literature on electrophoresis and gel chromatography over the past 20–30 years [75,77,326–329]. The role of the nature of the applied electric field and more extensive consideration of highly elongated or unusually shaped molecular solutes has been considered only within the last 5–10 years with the advent of pulsed-field electrophoresis [263,447], and very recent studies have considered novel methods of controlling the electric field [46]. The effects of solvent conditions have also been considered extensively, especially for nongel applications, in moving-boundary electrophoresis, isotachophoresis, zone electrophoresis, and isoelectric focusing [37,38,76,254,286,287,306,347,352,353]. Therefore, in order to understand the nature of molecular motion under applied electric fields in macromolecular media and to improve the practical applications of this phenomena, a detailed analysis of the preceding four questions is needed.

At the present time, there are several reasons for a detailed re-evaluation and analysis of the literature concerning the study of these questions. These reasons include:

1. The need to develop new materials for electrophoretic analysis and macromolecular separations prompted by the needs of the human genome project and the rapidly advancing fields associated with biotechnology, advances in the development of new analytical instrumentation—especially capillary electrophoresis, and practical limitations of the media currently used for gel electrophoresis [73]
2. Advances in the study of macromolecular structures, e.g., complex and supramolecular fluids, and the synthesis of new materials, e.g., nanostructured media, which may lead to the design of optimal materials for given separations or other applications
3. Advances in the application of pulsed electric fields for the separation and analysis of large macromolecules; use of multidimensional electric fields for enhanced separation of nucleic acids
4. Advances in experimental techniques, including pulsed-field gradient NMR, and theoretical methods, including volume averaging, macrotransport, and variational methods, that may lead to the resolution of a number of the fundamental issues in gel electrophoresis and to improvements in the practical application of electrotransport in polymeric media

The overall objective of this chapter is to review the fundamental issues involved in the transport of macromolecules in hydrophilic media made of synthetic or naturally occurring uncharged polymers with nanometer-scale pore structure when an electric field is applied. The physical and chemical properties and structural features of hydrophilic polymeric materials will be considered first. Although the emphasis will be on classical polymeric gels, discussion of polymeric solutions and nonclassical gels made of, for example, un-cross-linked macromolecular units such as linear polymers and micelles will also be considered in light of recent interest in these materials for a number of applications discussed earlier. Emphasis will be placed on material currently used or under development for gel permeation chromatography and electrophoresis; i.e., primarily polyacrylamide- and agarose-based materials; however, additional discussion of other materials that may be of specific interest for future applications in chromatography and electrophoresis will also be presented. Specific issues related to the development and application of materials with nanometer-scale pore structure will be considered. The discussion will be limited to un-charged media, since a wide range of the materials of interest are not charged or have a neg-ligibly small charge, and the theoretical analysis of transport in charged media is less developed. Furthermore, it will be assumed that the media is not distorted by the applica-tion of the electric field, although future work in this direction is needed.

Diffusive transport in these materials will be considered in detail, since diffusive mo-tion plays a fundamental role in governing molecular motion, even in the presence of an ap-plied electric field, and the methodology used in the analysis of electrophoresis in porous media has much in common with that used for diffusive transport in porous media. A de-tailed review of the literature concerning diffusion in gels and polymeric networks and so-lutions will thus be presented in order to provide the framework for the discussion of elec-trophoresis. The review will emphasize electric-field-induced transport in solutions and gels with *constant* applied fields, including both theoretical methodologies and experi-mental results, and will include comments on where further experimental data or theoreti-cal development is needed. Particular emphasis will be placed on experimental methods us-ing pulsed-field gradient nuclear magnetic resonance (PFGNMR), since this technique may provide the experimental evidence on molecular motion that can be used to test a number of untested or incompletely tested theories. In addition, among the theoretical methods dis-cussed, the emphasis will be on the application of volume-averaging methods.

II. PHYSICAL AND CHEMICAL PROPERTIES OF NANOSTRUCTURED ELECTROPHORESIS MATERIALS

A. Classification of Electrophoresis Media

Since the emphasis of this review is on polymeric hydrogels, it is useful first to consider the defining characteristics of the gel state and to discuss the differences between a classical gel made from chemically or physically cross-linking a polymer into a three-dimensional network and other gels and solutions of un-cross-linked polymers. Flory [119] was the first to develop a rigorous analysis of the gel state by considering the formation of three-di-mensional networks of polymers through condensation chemical reactions whereby "gela-tion occurs when a critical number of intermolecular linkages has been exceeded." He showed that network formation requires trifunctional or higher monomeric units to react, and thus bifunctional monomers cannot form gels since interconnected networks will not form. He considered a gel as an "infinitely large three-dimensional network," or a macro-molecule with an infinite molecular weight that forms once a critical gelation threshold has been exceeded. These gels are essentially viscoelastic solids, and one experimental condi-tion defining the gel state is the state where the viscosity goes to infinity. Figure 1 shows a schematic of the viscosity for a pregel solution formed through chemical polyesterification versus time. The gel point is determined by extrapolation to the time where the viscosity increases without bound. This figure clearly shows that the viscosity increases rapidly near the gel point; therefore, one of the major defining characteristics of a gel is its solidlike me-chanical behavior.

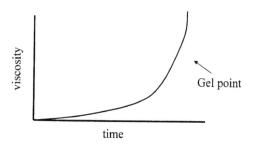

FIG. 1 Viscosity vs. time during gel formation and the gel point. (Based on Ref. 119.)

Flory [120] went on to expand the concept of gels beyond the "chemical gels" just mentioned by classifying gels into the four categories: (1) well-ordered lamellar structures, (2) completely disordered covalent polymeric networks, (3) polymeric networks formed by physical aggregation—mostly disordered but containing regions of local order, and (4) particulate disordered structures. Although the materials included in this classification are very different, Flory examined the unifying concept of "continuity of structure" that links these various materials. He concluded that the "universal characteristic of those systems regarded as gels is that they must possess a continuous structure of some sort, the range of continuity of the structure being of macroscopic dimensions." He further stated, "As a corollary inference, the continuity of structure must possess a degree of permanency—at least for a period of time commensurate with the duration of the experiment." Thus, gels are molecular networks where the individual units are linked primarily by covalent chemical bonds or multiple cooperative hydrogen bonds, although electrostatic and van der Waals interactions may also play a role in stabilizing the network structure. True gels consist of at least two components where the polymer also requires a solvent to mediate the interactions between units of the network.

Flory considered examples of the first category of gels to be soaps, phospholipids, and clays [120]. These materials generally have a more ordered structure than the other three classes of materials discussed by Flory, and indeed are really more akin to liquid crystals [67,82,91]. One may therefore consider broadening this group of ordered materials to include: (1) lyotropic liquid crystal–forming polymers, including surfactants that form micelles, hexagonal cubic phases, bicontinuous cubic phases, and lamellar phases, and (2) ordered arrays of colloidal particles [95]. Table 1 summarizes the variety of media or "gel-like" media that may be considered for electrophoresis studies. All of the materials discussed here are made of macromolecular units or aggregates of macromolecular structures and generally have structural and/or pore features on the nanometer scale. The media are classified into "ordered" and "nonordered" media, reflecting their basic structural features. Some of the materials classified as ordered may have only local order on the micrometer or smaller scale and may be more randomly distributed at the large scale.

Figure 2a shows a schematic phase diagram for lyotropic liquid crystals. This figure shows the formation of micelles, cubic phases, bicontinuous cubic phases, and lamellar phases as the concentration of surfactant increases. Also shown in this figure is a schematic diagram of an ordered bicontinuous cubic phase (Fig. 2b). Another interesting example in

TABLE 1 Classification of Media Types

I. Ordered media
  A. Colloidal particles
    1. Spheres, crystals, "particle gels"
    2. Rigid rods, nematic & smectic phases
    3. Semiflexible chains
  B. Surfactant structures
    1. Micelles, cubic arrays
    2. Hexagonal phases
    3. Bicontinuous cubic phases
    4. Lamellar phases
II. Nonordered media
  A. Type 2: chemical cross-linked gels, polyacrylamide
  B. Type 3: physical cross-linked gels, agarose
  C. Type 4: flocculated or aggregated gels
  D. Microemulsions, ternary microemulsions

this class of ordered materials is a commercially available material made of triblock copolymers of polypropylene oxide and ethylene oxide (Pluronic) that will form a gel-like phase, of face-centered ordered micelles, at high concentration of surfactant [51,175, 211,212,230,231,252,253,304,416,439,443].

Examples of Flory's second category include the polyacrylamide gels that are extensively used in electrophoresis [18] and many other synthetic nonaqueous-based polymers. Examples of Flory's third category include agarose gels, also extensively used for electrophoresis [350], and a number of other important biological polymers, such as protein gels [120]. These gels are of much biological relevance and include the basic structural molecules of tissues and cells, i.e., collagen, hylauronic acid, keratin, and others, in living organisms. Flory's fourth category may include gels made of flocculent precipitates and protein aggregates [120,259]. The last category in Flory's classification shares with types 2 and 3 the characteristic of less order than those in the first group, and many of these materials are polyelectrolytes.

Most of the gels used in chromatography and electrophoresis are of types 2 and 3 in the Flory classification. Current research seeks to explore other types of structures included in the first category as well as hybrid combinations of gels from several of the other categories and linear non-gel-forming polymers. Gels of the fourth category have not been of direct interest in electrophoresis and gel chromatography; however, they are important in the ultrafiltration of proteins [259] and other macromolecules—electric fields have been applied to these protein gel layers to enhance ultrafiltration separations [308]. Other gel-like media of interest include ternary microemulsions that form continuous nonordered biphasic materials [71,177,196–198,248]. Figure 2c shows an example of a nonordered ternary microemulsion. The issue of order vs. nonorder in the biphasic material is a difficult one to resolve, and much extensive structural characterization is required to demonstrate ordering in these materials.

The analysis of gel structure conducted by Flory encompasses primarily category 2 and category 3 gels, which have in common random structures, although some considera-

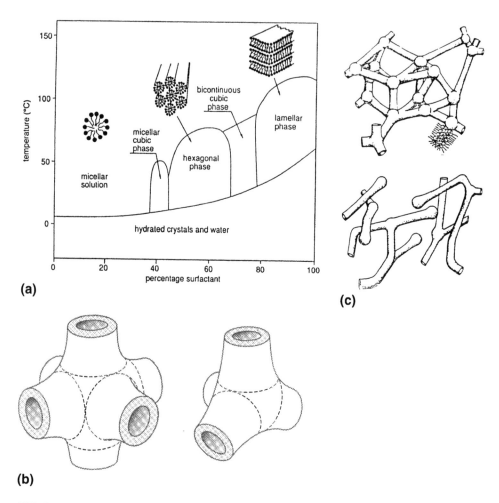

FIG. 2 Example media: (a) Surfactant–water phase diagram. (Reprinted from Ref. 206, Copyright 1991, with permission from Elsevier Science.) (b) Ordered periodic and bicontinuous structures. (Reprinted from Ref. 178 with permission from Academic Press, Ltd.) (c) Nonordered membrane structures from ternary microemulsions. (Reprinted with permission from Ref. 177, Copyright 1989, American Chemical Society.)

tion of type 4 gels is made; type 1 gels are excluded in Flory's analysis due to their regular structures. Reviews of aspects of ordering in supramolecular fluids that would be relevant to the first group of materials are available [67,82,91,178,206].

The physical gels (types 1, 3, and 4 in Flory's classification), i.e., gels not chemically cross-linked, are to a certain degree dynamic structures. They are also referred to as "reversible" gels as opposed to the covalently cross-linked "irreversible" gels. According to Flory, "In a gel structure involving interchain hydrogen bonds, the cross-linkages will be continually forming and dissociating," and "even in the quiescent state, the gel structures will be rearranging on a microscale; networks will be forming and disintegrating in a state of dynamic equilibrium." Flory also considered that physical gels must have some "degree of chain contour cooperativity," and this concept has been considered in detail more recently [333]. This implies that a critical number of overlapping bonds (hydrogen bonds as well as van der Waals interactions may be important) must form between segments of the polymers in order to form a stable structure.

De Gennes [127] further considered the difference between strong gelation and weak gelation. In strong gelation, the cross-links are completely stable and the transition from a solution to a gel is very sharp. In weak gelation, the cross-links are not completely stable, and reversible bond formation can occur. Weak gelation is similar to a glass transition.

De Gennes also emphasized the characteristic of gels whereby they are highly dependent upon the method of preparation. He uses the terminology "preparative ensemble" and "final ensemble" to describe the state of the gel in terms of its preparation method and its final characteristics, respectively. This has been shown to be particularly important for polyacrylamide gels [382,395]. Gelation kinetics also play a major role in determining the final structure of the gel. Other gel properties of importance include mechanical strength, swelling and solvent effects, and phase transitions.

Following the pioneering work of Katzir-Katchalsky [191], who studied phase transitions and mechanochemical coupling in polyelectrolyte gels, Tanaka [381] has extensively studied the phase transitions for gels that give rise to large volume changes in response to changes in solution chemical properties, including solvent composition, pH, and ionic strength, and to physical stimuli, including electric field, light, and temperature. In order for these phase transitions to occur, Tanaka explains, the polymer must interact through repulsive (electrostatic or hydrophobic effects) and attractive interactions. Further aspects of these phase transitions will not be considered in the present review since they are more applicable to charged gels. (For recent advances in polyelectrolyte theory see Ref. 258.)

Hydrogels were originally used for electrophoretic separations because they dampened the thermally induced convection that arises when an electric current is passed through a solution. This convection reduces the resolution of an electrophoretic separation since it causes mixing of the solutes. In addition to its function as a means for dampening hydrodynamic instability, gel structure also plays an important role in improving the separation of a mixture through its sieving properties and, occasionally, adsorption properties. The sieving effect enhances electrophoretic separation through increased retardation in the gel as a function of the molecular weight of the solute. Other work has considered the addition of controlled charged groups into the separation matrix in order to provide preformed pH gradients for isoelectric focusing separations in gels or for deliberately introducing and controlling electro-osmotic flow [70]. Pore gradient electrophoresis [114,331,335] has also been developed whereby the gel is cast in such a manner as to provide a gradient in the mean pore size. Gradients in the buffer composition have also been used to enhance separation in gel electrophorsis [18] and may also be useful in electrochromatography [216]. Proposals have been suggested to capitalize on the electric-field-dependent orientation of certain liquid crystals to create tunable separations media [402].

The desirable properties for media to be used in electrophoresis include: (1) mechanical stability—materials of structural integrity that can easily be handled or that are not disrupted during the course of an experiment; (2) chemical stability—ability to resist chemical degradation by hydrolysis or other means; (3) optical clarity—to allow observation of the transporting species, (4) reproducibility—to allow for clear band formation and little run-to-run variability; (5) ease of preparation; (6) hydrophilic—although some recent work has been performed on electrophoresis in organic solvents; (7) low charge—although some

TABLE 2 Summary of Media Currently Used in Electrophoresis

I. Normal slab electrophoresis
    A. Polyacrylamide, cross-linked
        1. Bis/acrylamide
        2. Variation of cross-linker
        3. Addition of un-cross-linked polymers (polyethylene glycol, glycerol, surfactants, liquid crystals)
        4. Covalent linkage of PA to agarose
    B. Agarose
        1. Normal agarose
        2. Addition of un-cross-linked polymers (polyethylene glycol, glycerol)
II. Nongels
    A. Polymer solutions (used in capillary electrophoresis and slab electrophoresis)
        • linear acrylamide
        • cellulose derivatives—hydroxyethyl, hydroxymethyl cellulose
        • polyethylene oxide
        • dextran
        • liquified agarose
        • polyethylene glycol
    B. Others
        1. Electrophoresis sponges
        2. Porogens
        3. Nonionic micelles (Pluronics)

applications capitalize on controlled charge; and (8) controlled pore size—structures that can be made with a variety of specified or predetermined pore sizes in a wide range of size, from a few nanometers to microns, and with narrow pore size distributions). Table 2 summarizes some of the media currently in use for electrophoresis studies.

B. Example Electrophoresis Media

1. Polyacrylamide Gels

Chemically cross-linked gels can be formed, for example, by condensation polymerization, additive polymerization, or free-radical-induced polymerization [119,394]. A wide range of materials can be made by varying the length of the chains between cross-links. Polyacrylamide (PA), since its introduction in 1959 by Raymond and Weintraub [311] and Ornstein and Davis [278], is perhaps the most extensively used hydrogel in this class. Ornstein [277] provides a historical account of the early work on polyacrylamide gels, and Righetti and Gelfi [317] review more recent work on alternative formulations of polyacrylamide. Extensive reviews of the preparation of polyacrylamide gels are available [18]. The PA gels are synthesized by chemical or photoinduced free-radical polymerization of the acrylamide bifunctional monomer and the tetrafunctional bisacrylamide (Bis) cross-linking agent. Figures 3a and 3b show the chemical composition of polyacrylamide gels and several cross-linking agents used in the preparation of PA gels. It can be noted that the gel consists primarily of carbon, nitrogen, oxygen, and hydrogen. The gels are hydrophilic and have essentially zero charge. The conventional notation for PA gels is to specify the fraction of total monomer (Bis plus acrylamide) as T (defined as mass of Bis plus acrylamide in 100 mL of solution) and the weight fraction of cross-linking agent (Bis only) as C (mass of Bis divided by total mass of Bis plus acrylamide). Specification of T and C along with the solvent and buffer composition and temperature have been considered to completely define a PA gel.

The physical characteristics of PA gels have been studied using a number of techniques that will be discussed in a later section; however, it is useful to note that the pore

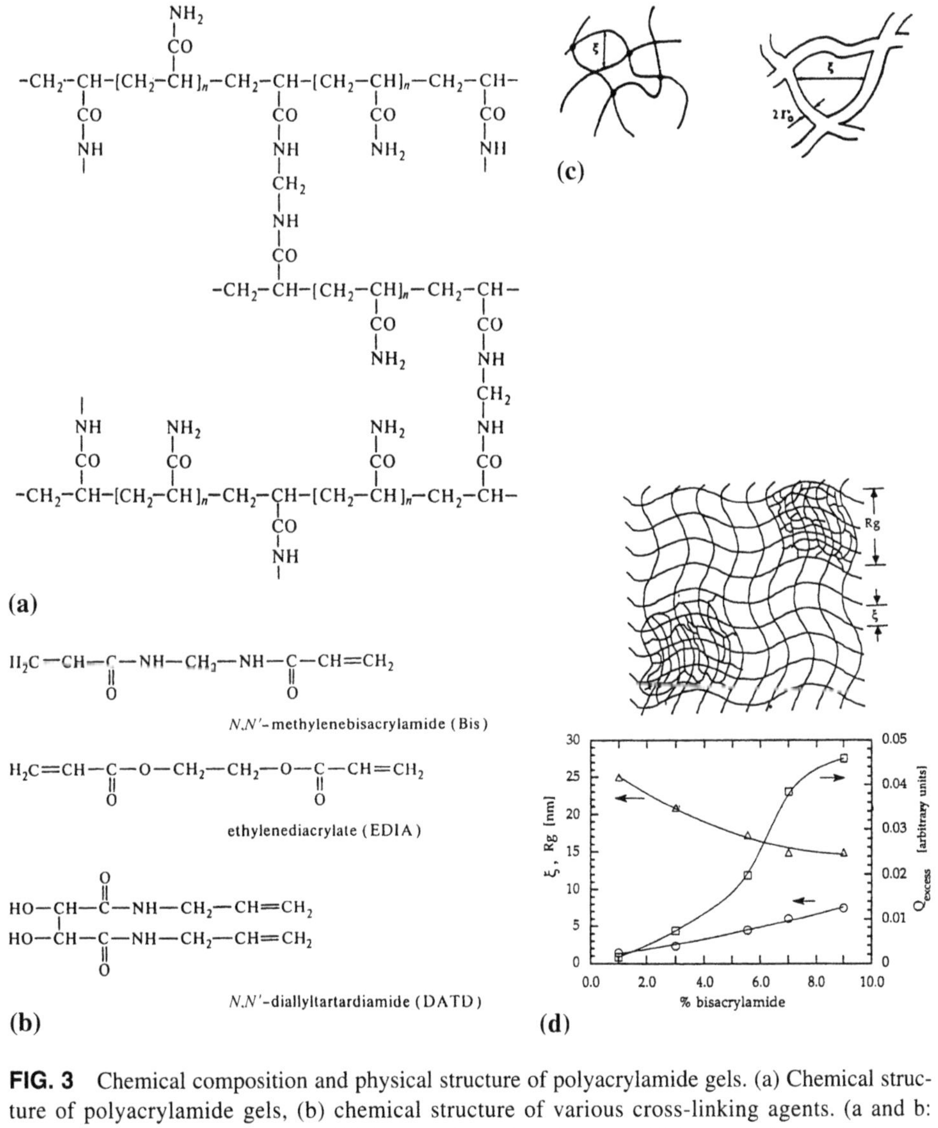

FIG. 3 Chemical composition and physical structure of polyacrylamide gels. (a) Chemical structure of polyacrylamide gels, (b) chemical structure of various cross-linking agents. (a and b: Reprinted by permission of Oxford University Press from Ref. 18, 1986, Oxford University Press.) (c) Model of physical structure as increasing fiber thickness. (Reprinted with permission from Ref. 153, Copyright 1985, American Chemical Society.) (d) Model of physical structure as aggregates of increasing size. (Reprinted by permission of John Wiley & Sons, Inc., from Ref. 80, Copyright 1992, John Wiley & Sons, Inc.)

size of the PA gel generally varies in the range of a few nanometers (at high $\% \ T$) to the tens of nanometers (at low $\% \ T$), and thus they have been employed extensively for the separation of small proteins (<100 kDa) and nucleic acid restriction fragments (<1–2 kpb). The lower limit for the formation of a gel with PA is approximately 2% T, and it has been reported that T4 bacteriophage virus, 75 nm, can be fractionated in this very dilute, almost liquid, gel [350]. Variation of the $\% \ C$ has lead to further insights on the nature of the gel structure. Most of the studies that characterize the gel structure conclude that the gel is heterogeneous on the scale of 50–100 nm [80]; i.e., it contains microdomains that have higher concentrations of Bis. High concentrations of Bis tend to lead to larger overall pore sizes because the Bis, with a solubility limit of around 40%, is much less water soluble than acrylamide, and it therefore appears to "cluster" together in dense aggregates. Figures 3c and 3d also show schematics of two possible structures of polyacrylamide gels. One possibility suggests that the polyacrylamide fibers become thicker as the amount of Bis is increased, and the other possibility suggests that regions of high-concentration Bis form globular domains as the amount of Bis is increased. Further work is needed to distinguish between these alternatives.

Acrylamide gel polymerization, with chemical catalysts rather than by photopolymerization, proceeds rapidly, with approximately 90% conversion in the first hour. The percentage of Bis reacted has recently been found to be higher than the percentage of acrylamide monomer reacted [315] at the gel point. Extensive studies of gel polymerization kinetics (especially photopolymerization with dye-sensitized gels) have been conducted [57,58,125,126,225,315,318] and lead to the concept that at gelation, the gel point, approximately 80% of the double bonds in the Bis have reacted. The remaining reactions occur more slowly over a longer time period but eventually lead to over 99% conversion of the reacting molecules. Chain propagation in free-radical polymerization is generally much more rapid than radical initiation [43], and this has also been used to partially explain some of the heterogeneity observed in polyacrylamide gels.

Conventional PA gels suffer from the disadvantages that they undergo hydrolysis at high pH, are generally not reusable [73], and have fairly low pore-size limits (i.e., they are not useful for very large macromolecules, i.e., DNA size). Righetti and coworkers [317] have found methods to reduce PA hydrolysis through modification of monomeric units and (as further discussed later) to increase the pore size using macromolecular templates.

2. Agarose Gels

Physically cross-linked materials can be made according to De Gennes through (1) formation of helical structures, (2) formation of microcrystals, or (3) formation of nodules with block copolymers. Examples of helical structures include agarose and hyaluronic acid. Figure 4 shows a typical structure of an agarose gel [350]. Agarose is a polysaccharide (shown in Fig. 4) made of six-member rings with various substitutes, such as sulfate, pyruvate, and methoxyl groups, on the basic polysaccharide chain, and it may thus also contain a small residual charge. Agarose will form a double helix; however, this helix is not sufficient to form a gel [350]. Gels are formed as the agarose double helices aggregate to form "superfibers," with radii, observed by small-angle x-ray scattering (SAXS), of 15 nm [96]. The triple helical units, cross-linked by hydrogen bonds and entanglements, are shown in this figure. The pores in agarose are usually very large, on the order of hundreds of nanometers

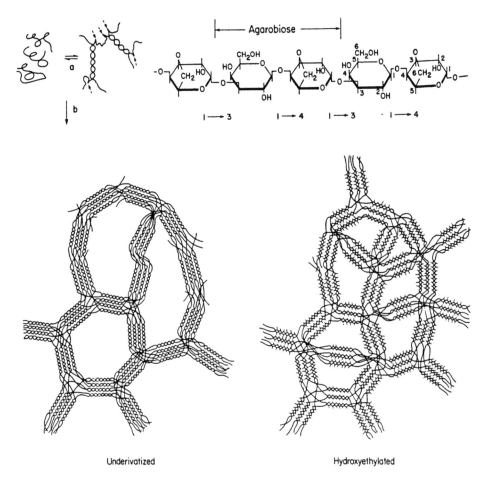

FIG. 4 Chemical composition and physical structure of agarose gels. (Reprinted by permission of Wiley-VCH and P. Serwer from Ref. 350, Copyright 1983, Wiley-VCH.)

[74,226,350], and thus agarose gels have been extensively used for the separation of large proteins, viruses, and, most typically, larger nucleic acids. Agarose gels, due to small residual charges, may lead to small amounts of electro-osmotic flow, nonspecific adsorption, and fiber alignment in electric fields [366], although several commercial suppliers have made advances in reducing the residual charge.

3. Hybrid Gels

Due to the previously mentioned disadvantages and limitations of polyacrylamide and agarose gels, a number of hybrid gel materials have recently been studied [74,317]. Gels that are more resistant to chemical degradation than polyacrylamide and gels that have pore sizes intermediate between polyacrylamide and agarose are of particular interest [74]. Alternative gels can be synthesized using different cross-linking agents with acrylamide and using different monomers. More recent work has considered the addition of other polymers to the acrylamide/bisacrylamide mixture prior to initiation of polymerization [73]. Righetti and coworkers [73] found that the use of a new class of mono- and disubstituted acrylamide monomers lead to hydrogels with enhanced resistance to alkaline hydrolysis. Righetti and coworkers also introduced such polymers as polyethylene glycol [316] and glycerol [70] into a mixture of acrylamide and Bis with the intent to increase the average pore size of an acrylamide gel. Due to the sequestering of water by the PEG, the growing polyacrylamide chains form large "latterly aggregated" "bundles, held together by, preferentially (but perhaps not solely), interchain H bonds." Figure 5 shows a schematic of these laterally aggregated gels where the polymer fibers are assumed to thicken. The macropores created in this gel were found to be of order 500 nm. It was found that these gels had very thick walls between the pore spaces and that during the polymerization, prior to cross-linking, the nascent polymer chains were held together in bundles by hydrogen bonds rather than having a random orientation in the solvent. A similar preparation using agarose gels with the polyethylene glycol or glycerol lead to a bimodal pore distribution, with macropores of 10-μm size and micropores of 100-nm size. Hybrid gels made by covalently linking agarose to poly-

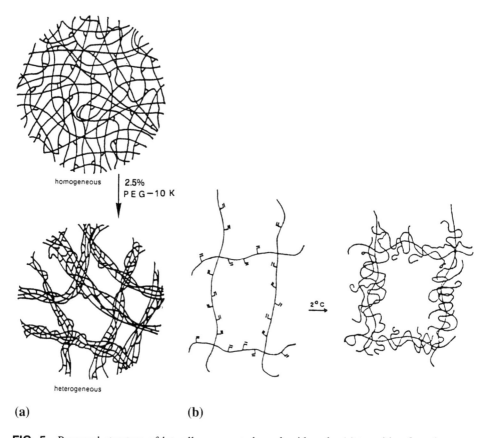

(a)   **(b)**

FIG. 5 Proposed structure of laterally aggregated acrylamide gels: (a) transition from homogeneous to heterogeneous structure by polymerization with large PEG molecules; (b) model of chain bundling at low temperature. (Reprinted by permission of Wiley-VCH and P. Righetti from Ref. 316, Copyright 1992 and 1993, Wiley-VCH.)

acrylamide [72] have also been found to be mechanically stronger and more elastic than conventional polyacrylamide gels and to perform well in DNA electrophoresis. Caglio and Righetti [59] have investigated various means to improve the conversion efficiency of polyacrylamide gels by varying the pH and catalysts.

4. Template Media

One of the earliest reported attempts at creating a porous material using molecular templates was that of Dickey [94], who synthesized silica gel with dye molecules such as methyl orange. After removing the dye from the templated gel it was found that the templated gel had specific affinity up to 20 times higher than gels synthesized without the template. More recently, Mosbach [19,110,201,404,440] and others [299] have developed extensive work on the introduction of templating molecules, termed *molecular imprinting*, into the polymerization process in order to make chromatography material with highly specific affinity for given solutes. Good separation of chiral molecules has been demonstrated using thin films of polymers (made of methacrylic acid and 2-vinyl pyriline as monomers and ethylene dimethacrylate or trimethylol propane trimethacrylate as crosslinker, with toluene as template) for open-tube chromatography and capillary electrochromatography [380].

De Gennes [90] suggested that the incorporation of a hydrophilic polymer into the water layer formed within the lamellar phase of a lipid and water may lead to an interesting and anisotropic gel where "one might cross-link the chains and wash out the lipid with suitable solvents." An anisotropic gel in this context is one where the pores have a highly organized structure with a high degree of orientational order.

The approach of using a template to create defined pores in materials has also been extensively used in the synthesis of inorganic zeolites for catalysis and elsewhere [65,117,118,200,207,249,396,442]. Pores of 5–30 nm were produced in mesoporous silica using triblock copolymers as templates [442], and magnetic field alignment of silicate-surfactant liquid crystals has been shown to lead to anisotropic material. Ordered three-dimensional periodic biphasic media have been synthesized using inorganic and organic anions and cations [174], and silica and titania colloidal particles have also been used to create ordered media [213]. These materials have not been investigated for electrophoresis or for effects on electric-field-induced transport, and any residual charge incorporated in the matrix may have deleterious effects on electrotransport.

The addition of surfactants during gel polymerization has been explored by a number of groups [78,84]. Chu et al. [78] explored complexes of polyelectrolytes and surfactants and found that the ionic surfactants became more concentrated in the charged network and formed micelle-like aggregates within the gel. These surfactant gel complexes, i.e., SDS with gelatin, have been studied by small-angle neutron scattering. These materials, however, have not been investigated for electrophoresis and may not be particularly useful for this application due to the large residual charges from the ionic surfactants.

In order to study the effect of nanometer-scale channels or pores in polyacrylamide gels, Rill et al. [322] have demonstrated that templates such as DNA, xanthan, and spherical micelles of sodium dodecyl sulfate (SDS) and other surfactants can be cast into polyacrylamide gels during polymerization, that these templates can be removed from the resulting gels by soaking or electrophoresis, and that the pores created in these gels upon removal of the template can significantly affect the electrophoretic mobility and gel permeation chromatography characteristics of proteins. The presence of the template was found to have a very small effect on the gelation kinetics, and it was found that over 90% of the template molecules could be removed by soaking in a buffer or by applying a small electric field. The goal of this work was to introduce macromolecular templates, i.e., of the order of nanometers, in order to create pores of defined characteristic shape and size that may alter the selectivity of the templated gel material during electrophoresis. The pores expected in these studies differ substantially from those obtained by Righetti et al. [316]; the use of small macromolecular templates is expected to give pores closer in size to the natural pores of the polyacrylamide matrix rather than the very large pores obtained upon use of such large polymers such as PEG.

Experiments by Rill et al. [322] with templated gels showed that the electrophoretic mobilities of SDS proteins in gels templated with 20% SDS depended strongly on the acrylamide concentration and could not be explained by changes in the overall pore size of the gel. Figure 6 shows plots of the mobility ratios of two types of SDS-templated gels (symbols) and analogous Ferguson plots (plots of logarithm of the mobility versus gel concentration [18,115,154] for normal polyacrylamide gels. If the templated gel behaved as a normal gel with a specific size distribution, the data would fall along a single straight line. It is clear that the data for the templated gels do not fall along a single line corresponding to a normal gel. Indeed the data (for example, the solid squares) indicate that low-molecular-weight proteins were separated as if the gel was made of 18% acrylamide and the high-molecular-weight proteins were separated as if on gels made of 24% acrylamide. The templated gel appears to increase the separation between small and large proteins in relation to normal gels of similar acrylamide concentration.

In a further study, Rill et al. [325] developed a model of gel permeation chromatography that included a bimodal pore structure. The smallest mode in the pore-size distribution represents the basic background polyacrylamide pore structure of about 1-nm mean radius, and the second mode was around 5 nm, i.e., in the range of size of the molecular templates. The introduction of this second pore structure was found to substantially improve the peak resolution for molecules with molecular sizes in the range of the pore size.

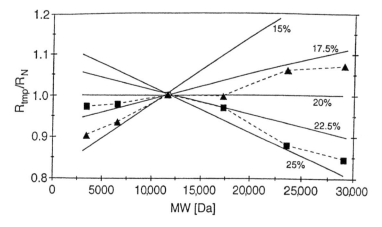

FIG. 6 Comparison of protein electrophoretic mobility ratios as functions of protein molecular weight for SDS-templated gels of various compositions (data points) to Ferguson plots of reference normal gels. (Reprinted by permission of Wiley-VCH from Ref. 322, Copyright 1996, Wiley-VCH.)

Polyacrylamide has also been templated with dodecyltrimethylammonium bromide (DTAB), tetradecyltrimethylammonium bromide (TTAB), hexadecyltrimethylammonium bromide CTAB), polyoxyethylene 4 laauryl ether ($C_{12}EO_4$), polyoxyethylene 10 lauryl ether ($C_{12}EO_{10}$), and Pluronic F127. A wide range of gel permeation experiments have been conducted with these templated gels using a series of proteins and oligonucleotides as test compounds [288]. The results show that PA can be polymerized in the presence of the surfactant templates; the templates can be removed, the resulting gels are mechanically stable in GPC, and pores are formed in the gels with sizes close to those of the surfactant template. The pore sizes in these templated materials were determined using gel permeation chromatography of series of proteins and oligonucleotides.

In a related but different physical/chemical method, porogens, the solvent used during the polymerization process [378], can have a profound effect on the pore structure of a gel. In the case of a nonsolvating solvent (i.e., a solvent in which the resulting polymer is not soluble), as the polymerization proceeds the polymer phase separates from the solvent phase and the monomers also tend to partition into the polymer phase, where they react to form more polymer. This phase separation during the polymerization process leads to local heterogeneity in the final structure, and control of the pore size depends upon a wide range of factors, including solvent type. Svec and Frechet [378] have synthesized poly(glycidyl meth-acrylate-co-ethylene dimethacrylate) monoliths with pore sizes ranging from 100 nm to 3000 nm using dodecanol as the porogenic solvent.

Other microporous materials have been synthesized using the porogen polyethylene glycol in polyethylene oxide-urethane gels [27]. Micropores were formed in the gel, and it was found that the diffusion of larger species, vitamin B12, was enhanced relatively more than that of a smaller species, proxyphylline. This result is in qualitative agreement with that found for electrophoretic transport by Rill et al. [322] discussed earlier, where the mobility of larger species was preferentially enhanced in the templated media.

Electro-optic materials can be made using liquid crystal polymer combinations. In these applications, termed *polymer-stabilized liquid crystals* [83,86], the liquid crystal is not removed after polymerization of the monomer and the resulting polymer network stabilizes the liquid crystal orientation.

5. Gel-Like Media

Triblock copolymers made of chains of poly(ethylene) oxide (PEO) and poly(propylene) oxide (PPO) subunits are currently subjects of intensive interest and research due to their potential as support media for capillary electrophoresis [321,323,324,438,439,441] as well as their many other uses (see, for example, Ref. 442). Chains of $(PEO)_x(PPO)_y(PEO)_x$ are commercially available with a wide range of subunit compositions and go under the trade name of Pluronics. Due to the physical-chemical nature of the oxide subunits, single chains of these molecules will associate to form micelles, with the more hydrophobic PPO units packed within a core surrounded by the more hydrophilic PEO units. Extensive studies have been conducted to determine the phase behavior of a variety of triblock formulations as functions of temperature and polymer concentration in aqueous solutions [211,212,230,416]. Light-scattering methods [51,443], small-angle neutron scattering [253,304], rheological methods [175,304], transmission electron microscopy [253], scanning calorimetry [175], and NMR water self-diffusion [145,231] have been applied to determine the basic structural features of Pluronic micelles, and the general view is that the micelles will pack in a cubic lattice [252,438] to form a gel-like medium under well-defined conditions of concentration and temperature.

F-127, i.e. $(PEO)_{106}(PPO)_{70}(PEO)_{106}$, will form a gel-like phase of associated individual micelles at room temperature and at a concentration of about 20%. Under these conditions the hydrophobic core, consisting of PPO subunits, is approximately 9 nm in diameter and is surrounded by a region of PEO, giving an overall micelle diameter of approximately 13 nm and a center-to-center micelle spacing of approximately 18 nm [323,416]. This medium has been shown to be effective for the separation of a wide range of macromolecules, from double-stranded DNA up to 3000 base pairs by conventional electrophoresis [323] and single-stranded DNA from 4 to 60 nucleotides long and oligonucleotides (poly-thymidines) of 12–24 nucleotides long by capillary electrophoresis [323,441]. These results are very interesting not only for their practical utility, but also for the implications on fundamental aspects of macromolecular transport in complex porous media [144,145].

The structure of these gel-like systems of micelles is very different from that of conventional electrophoresis media made from chemically and physically cross-linked polymers of polyacrylamide and agarose [75]. The absence of chemical or physical cross-links in the Pluronic gel-like phases may allow a larger degree of freedom for macromolecular transport around the obstacles that make up the medium than occurs in conventional electrophoresis media.

6. Nongel Media

Cross-linked gels that contain linear polymers, in addition to the chemically cross-linked matrix, have also been studied. These gels are synthesized by adding the linear polymers to solutions containing the gel-forming monomers. The linear polymer does not react with the gel-forming monomers during polymerization. A wide range of electrophoresis studies with un-cross-linked polymers has also been initiated due to advances in capillary electrophoresis. Water-soluble linear polymers, including linear polyacrylamide [307,317, 418], cellulose derivatives including hydroxyethyl cellulose and hydroxymethyl cellulose, polyethylene oxide [142], dextran, liquefied agarose, and polyethylene glycol, were found to lead to size-dependent separations of proteins [41,42,128] and nucleic acids [194,351]. These linear polymers were found to have sieving properties very similar to cross-linked gels; however, in some cases an anticonvective medium was still required to prevent mixing induced by thermal convection [418]. Butterman et al. [56] explored the use of dynamically cross-linked polymer solutions, which were found to have better behavior than uniform chain polymer solutions.

Harrington et al. [148] have considered the use of "electrophoresis sponges." These materials are porous polymers, not gels, made of polyethylene, polyvinylidene difluoride, and polypropylene, which can have pore sizes in the subnanometer to 100-micron range. In an examination of these materials for protein electrophoresis using isoelectric focusing they found a threefold increase in the speed of electrophoretic transport and improved resolution in the presence of an organic solvent. These materials are percolable, in that they will allow fluid to flow through them under suitable pressure gradients.

There are other, nonhydrogel, new materials for chromatographic and electrophoretic separations [7,8,103,164,199,214,377,407]. For example, Volkmuth and Austin [407] proposed electrophoretic studies in microlithographic arrays of posts and channels etched into silicon wafers. This material may be useful for studying fundamental transport characteristics of macromolecules in defined media, and many recent studies have been conducted to develop chromatography and electrophoresis on silicon wafers with micron-scale channels

[99,100,107,170]. Rossier et al. [332] used UV excimer laser photoablation to cut channels 50 microns deep by 100 microns wide in laminated PET. These channels were filled with PA, and rapid separation of proteins by isoelectric focusing was demonstrated.

A range of flow-through, or perfusion, chromatography media has also been developed in order to overcome some of the mass transport limitations commonly encountered with column packings [7,8,164,199,214,377]. Chromatography particles with macropores of order 600–800 nm and micropores of order 80–150 nm have been synthesized by suspension polymerization of polystyrene divinyl benzene [7,8]. In an extension of this perfusion approach, Hjerten et al. [164] and Svec and Frechet [377,378] have developed chromatography columns with a single continuous porous packing, "plugs," and over 500-micron-diameter spherical particles through polymerization with porogens. These materials may also have use in electrophoresis and electrochromatography separations.

C.   Structural Properties of Electrophoresis Media

Depending upon the type of media considered, with reference to the classification discussed earlier, the structure of a particular media may be one of a wide range of forms. Media made of cross-linked polyacrylamide or of agarose, type 2 and 3 gels, are often considered as beds of interconnected fibers, although, as will be discussed later, this description may be appropriate only for agarose gels and low-cross-link-density polyacrylamide gels. For example, media made of non-cross-linked spherical micelles in ordered cubic packing, i.e., type 1 gels, may be more analogous to beds of spherical packing. Nonordered particle gels made of spherical colloids, type 4 gels, can also form fractal structures or percolation clusters [95].

Analysis of porous media in general has been extensively pursued in the literature. An excellent review by Alder [6] considers convective and diffusive transport in a range of media, with emphasis on fractal media of biological and geological interest, and artificial media. Although Alder's review was not specifically oriented to gels, some common features for understanding the transport in gels and other types of porous media emerge. Alder [6] classified porous media into four categories: (1) spatially periodic media, (2) random media, (3) fractal media, and (4) reconstructed media. Spatially periodic media relate to the concept of a unit cell, whereby the overall structure of the media can be represented by a single unit cell of given geometry. Repetition of these unit cells can be used to model the entire media. Random media cannot be defined as precisely; however, to some degree most random media can be represented as spatially periodic if the unit cell encompasses a large enough representative domain. Fractal media, based on the original definitions of Mandelbrot [232], have the major characteristic of scaling or self-similarity, whereby certain patterns of the structure remain invariant as the size of the region considered varies. Sahimi [341] discusses the relationship between long-range correlations in the physical properties of porous media and their fractal characteristics. Reconstructed media involve the artificial construction of media that more closely approximate real porous media. The media to be considered in this review are primarily in the first three categories.

Other key features in the analysis of pore structure are the length scales associated with the various micro- (nano)-scale obstacles and pores, the possible larger-scale variations in structure, and the averaging domain over which information is needed [6,341,436]. The literature refers to analysis of homogeneous and heterogeneous porous media, where *homogeneous* refers to media with no variation in physical properties (e.g., porosity, diffusion, dispersion, velocity) over the length scale of interest. For example, while typical polyacrylamide gels may be heterogeneous over the nanometer scale, they are usually homogeneous over the micrometer scale.

A number of models have been developed to describe gel structure [326]. Models for gel structure can be classified into: (1) geometrical models of specific uniform pore shapes, (2) statistical models that assume specific size distributions of pores, (3) combinations of the pore-size distributions with unit cells of given geometry, (4) thermodynamic models that describe the swelling behavior of elastic networks, and (5) kinetic models that describe the polymerization kinetics and the subsequent location of intermolecular bonds on some type of lattice. Geometrical models require an accurate description of the geometrical configuration of the media and are mostly appropriate for well-defined rigid media. Size distribution models generally assume some type of pore-size distribution function and have been extensively pursued in the gel chromatography literature. Statistical mechanical models of geometrical structure have also been developed by Giddings et al. [132] and Rikvold and Stell [319,320,365]. Thermodynamic models are based on the underlying phase equilibrium behavior of a multicomponent system. Direct simulations of gel structure in order to describe scattering properties have also been conducted [255]. Examples of these models will be given later in order to illustrate the nature of the structure of gels. The structure of polyacrylamide and agarose gels will be considered in detail, since PA is the most widely used chemically cross-linked gel, agarose is the most widely used physically cross-linked gel, and extensive work has been performed on both polyacrylamide and agarose gels using light scattering, electron microscopy, x-ray diffraction, osmotic swelling, mechanical stress and compression, nuclear magnetic resonance, atomic force microscopy, and gel permeation chromatography to characterize the structure of these gels [80,81,113,153,169, 192,226,229,290,313,314,349].

1. Uniform-Pore Models

Models of regular geometrical pores with rectangular, spherical, cylindrical, and conical shapes have been developed for electrophoresis and gel chromatography media. Figure 7, from Ref. 314, gives samples of these uniform structures. These uniform-pore models have been used more extensively in the analysis of gel filtration chromatography.

Ornstein [276] developed a model for a rigidly organized gel as a cubic lattice, where the lattice elements consist of the polyacrylamide chains and the intersections of the lattice elements represent the cross-links. Figure 7 shows the polymer chains arranged in a cubic lattice as in Ornstein's model and several other uniform pore models for comparison. This model predicted r_p, the pore size, to be proportional to $1/\sqrt{T}$, where T is the concentration of total monomer in the gel, and he found that for a 7.5% T gel the pore size was 5 nm. Although this may be more appropriate for regular media, such as zeolites, this model gives the same functional dependence on T as some other, more complex models.

Squire [364] and Porath [300,301] developed geometrical pore models for gel chromatography media. Squire considered a gel with a set of conical, cylindrical, and rectangular crevices, and found the pore volume, assumed equal to the partition coefficient K_{av}, to vary as

$$K_{av} = \left[1 + a_1 \left(1 - \frac{M^{1/2}}{a_2^{1/3}} \right) \right]^3 \tag{1}$$

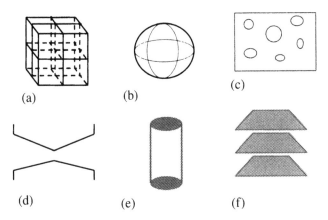

FIG. 7 Uniform-pore models (figure based on that of Ref. 276). (a) Cubic lattice. (From Ref. 276.) (b) Conical lattice. (From Refs. 300 and 301.) (c) Spherical lattice. (From Ref. 297.) (d) Cylindrical lattice. (From Ref. 3.) (e) Circular pores in rectangular sheets. (From Ref. 3.) (f) Rectangular pores. (From Ref. 364.)

where a_1 and a_2 are coefficients to be determined from experimental data and M is the molecular weight of the probe. The coefficient a_2 would correspond to the molecular weight of the smallest molecule that could enter the pore, and the coefficient a_1 reflects the nature of the distribution of the conical, cylindrical, and rectangular pores. K_{av} is defined experimentally for gel filtration chromatography as

$$K_{av} = \frac{V_e - V_0}{V_T - V_0} \tag{2}$$

where V_e is the elution volume, V_0 is the void volume, and V_T is the total volume of the gel chromatography bed. Porath's model for only conical pores leads to a result similar to that of Squire:

$$K_{av} = a_1 \left[1 - a_2 \frac{M^{1/2}}{(a_3 - a_4)^{1/3}} \right]^3 \tag{3}$$

where a_1, a_2, a_3, and a_4 are coefficients to be determined from data.

Models of regular structures, such as zeolites, have been extensively considered in the catalysis literature. Recently, Garces [124] has developed a simple model where the complex pore structure is represented by a single void with a shell formed by n-connected sites forming a net. This model was found to work well for zeolites. Since polymer gels consist of networks of polymers, other approaches, discussed later, have been developed to consider the nature of the structure of the gel.

2. Ogston Theory

Although recent data indicate that the gel structure is more ordered (although not as ordered as the regular geometric models discussed earlier), the vast majority of work pertaining to gel electrophoresis begins with the Ogston model for the distribution of spherical particles in a random suspension of fibers. Ogston [269], using geometrical and statistical considerations, showed that the fraction, f, of the total volume of a suspension of randomly oriented fibers that can accommodate particles of radius r_p, is

$$f = \exp\left[-\left(2\pi v l r_p^2 + \frac{4}{3}\pi v r_p^3 \right) \right] \tag{4}$$

where v is the average number of fiber centers/cm^3 and l is the half-length of a fiber. The first term in the exponential represents particle–fiber tangential contacts and the second term represents particle-to-fiber-end contact. In the original development of the Ogston equation, it was assumed that the fibers were of infinitesimal thickness, that they were rigid nonflexible rods, and that the pore space was randomly distributed. Laurent and Killander [205] have used the Ogston model for analysis of gel filtration. For the case of long fibers, the end-to-end contacts are small compared to tangential contacts, and with nonnegligible, but small, fiber radii, the Ogston equation becomes

$$f = \exp[-(\pi 2 v l (r_p + r_r)^2)] \tag{5}$$

where the fiber radius r_r has been added to the particle radius. Laurent and Killander [205] assumed that the distribution coefficient from gel chromatography, K_{av}, was equal to the fraction, f, obtained from the Ogston model. They verified this approach by correlating K_{av} determined for a variety of proteins in agarose gels with r_p determined from diffusion coefficients via the Stokes equation. The product of parameters (lv) was determined from the best fit of the data to the preceding equations. Early work used gel chromatography of proteins to assess the structure of polyacrylamide and agarose gels; in turn, through calibration of such gels with known proteins, these gels were used to determine molecular properties of unknown proteins. In the early work, the polyacrylamide or agarose continuous slab gels were broken into small pieces and placed in chromatography columns for gel chromatography studies. Calibration required the assumption of particular functional forms for the structural properties of gels, and empirical fitting of the data to the model provided coefficients that were assumed to reflect the structural properties of the gels. Most work has considered the radius of the polymer network fiber to be an adjustable coefficient. By fitting data from chromatography or electrophoresis to Ogston-type models, this coefficient has been obtained and compared to values of fiber thicknesses from other models and experiments. Through extensive studies, a highly reliable database of electrophoresis measurements of proteins, viruses, and subcellular particles in polyarcylamide and agarose gels has been assembled and used to verify the Ogston model [75,391] for gel structure and the Ogston–Morris–Rodbard–Chrambach model for gel electrophoresis (see electrophoresis section, later).

Richards and Temple [313] extended the Ogston theory by considering the properties of an ideal gel. In their approach, an ideal gel is a gel in which all the tetrafunctional cross-linker, e.g., Bis, molecules are connected to four nearest neighbors, and all these cross-linked residues are distributed randomly in space by a Poisson distribution. There are two types of gels in their classification: gels with monomer concentrations below the critical value for an ideal gel—"crumpled gels," and gels with monomer concentrations above the critical value for an ideal gel—"clustered gels." In their scheme, "In a clustered gel there is insufficient acrylamide to connect the Bis residues to their nearest neighbors; if all the Bis residues form knots, the acrylamide chains must either extend beyond their equilibrium mean end-to-end distance or the distribution of Bis molecules must become nonrandom and clustered."

In their analysis Richards and Temple assumed that the end-to-end distance of a growing polymer chain follows a Gaussian distribution, the ratio of the probability of a Bis and an acrylamide residue incorporated in the gel is constant, and the number of chains connected to each knot is four. Thus, they do not take into account the formation of network defects that would arise from topological considerations discussed later. Their work implies that at high Bis content the structure of an acrylamide gel may become more heterogeneous as the Bis "clusters" together; recent evidence from light scattering supports this view [80]. They further stated that "in the case of crumpled gels there is an excess of acrylamide and either the chains must be less extended or they must connect more distant Bis residues." A "crumpled" gel would be expected to "relax" to an ideal gel if it was allowed to swell in a solvent. Through combination of these arguments with the Ogston theory they found

$$f = \exp(-T\,[C(r_p + r_c)^3/6\zeta + k_1(1 - C)(r_p + r_c)^2]) \tag{6}$$

where T and C are the usual monomer and cross-link percentages, ζ is the mean number of Bis residues in a cluster of effective radius r_c, and k_1 is a constant that may depend upon C; however, no explicit relationship for k was provided by Richards and Temple. They found that their model fit gel filtration data and osmotic swelling results for polyacrylamide gels. One of the major results of their model is the clear link between gel structure, including the aggregation of residues, and the composition of the gel, i.e., T and C. In addition, the concept of "clusters" of Bis, which implies a heterogeneous structure for the polyacrylamide gel, has been confirmed more recently by light scattering and a number of other theoretical and experimental approaches. Richards and Temple also noted that at $C = 0$ the fraction f is finite and that this implies that even un-cross-linked PA will have some molecular sieve properties. Recent electrophoresis studies with un-cross-linked polyacrylamide and other polymers confirms the sieving properties of these un-cross-linked solutions.

Figure 8, adapted from Refs. 314, 444, and 445, shows the basic topological structures in cross-linked polymers. A cross-linked gel such as polyacrylamide may have a variety of topological structures, and the nature and distribution of these structures will strongly affect macroscopic properties such as elasticity, swelling behavior, and possibly transport characteristics. Ziabicki [445] classifies the basic topological elements of a network by considering the ways that chains can be connected to cross-links. For a single cross-linker one can consider elements that take up one of the functionalities (e.g., for a tetrafunctional Bis one of the double bond reactions): (1) singlets—chains connected with their two ends to two different cross-links, (2) free-end chains—chains connected with one end only to a given crosslink, and (3) voids—no chains connected to that cross-linker. For elements taking up two of the functionalities of a given cross-linker one can consider: (1) doublets—two chains connecting the same pair of cross-linkers, (2) loops—chains connected with both ends to the same cross-links. Higher-order structures, including triplets, quadruplets, and entanglements, have also been described [444,445]. It is important to note that loops and entanglements will have significant effects on the elastic behavior of the media and no longer lead to the "ideal gels" considered in the framework of Ogston, Richards and Temple, and Flory. Figure 8 also shows Righetti's concept of a variation in Bis content on the structure of polyacrylamide. When no Bis is present, a viscous liquid of linear polyacrylamide is formed, and in gels with high Bis content the cross-links form heterogeneous regions. In the limit of a gel with only cross-linker, a dense network of Bis remains.

Righetti and Caglio [315] have also found that Bis reacts faster than acrylamide and that this leads to the formation of nonhomogeneous regions in the gel. At the critical gel point they found 50% of the acrylamide monomer and 80% of the Bis monomer had reacted. The reaction was found to continue beyond the gel point, with an eventual 99%

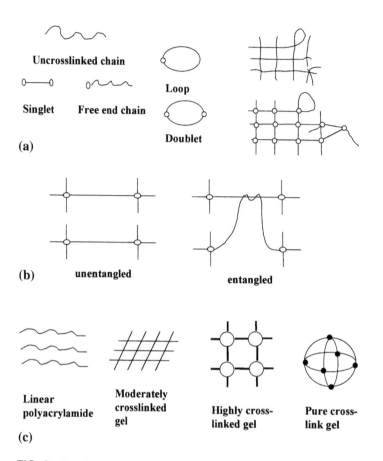

FIG. 8 Topological structures of polymer gels. (a) Structural elements for gels with bifunctional cross-linking. (Redrawn with permission from Ref. 445, Copyright 1978, American Chemical Society.) (b) Entanglement formation. (Redrawn with permission from Ref. 444, Copyright 1979, Elsevier Science.) (c) Proposed gel structure with varying Bis content. (Redrawn by permission of Walter de Gruyter & Co. from Ref. 314.)

conversion of total monomer. According to their work it is the reactions of pendant groups after the gel point that have the strongest effect on the gel structure. They found that a gel polymerized at 50°C was transparent and that a gel polymerized at 2°C was turbid. A gel that was formed at 2°C until the gel point, with its temperature thereafter raised to 50°C for the final reactions of the pendant groups, was found to be clear. They interpreted this result in terms of intrachain hydrogen-bond stabilization of PA chains to form heterogeneous domains at low temperature, while higher temperatures disrupt the hydrogen bonds and lead to the formation of more homogeneous gels. The switch from low temperature to high temperature at the gel point indicates that the cross-linking reactions that stabilize, or "freeze," the heterogeneous structure have not yet occurred at the gel point and that they occur primarily after the gel point.

Some Bis reactions may lead to intramolecular cycles that will not affect the elasticity behavior of the gel, and pendant, unreacted, groups may form where only the cross-linker reacts at one of its double bonds [315]. Tobita and Hamielec [393] found that primary cyclization, i.e., the formation of loop cycles, may consume as much as 80% of pendant double bonds. Primary cyclization does not lead to elastically active gels, in comparison to secondary cyclization, where a cycle is formed between two different chains. It has also been more recently concluded that only a small fraction of the Bis normally produce effective cross-links and that a large amount of Bis leads to intramolecular cyclization of the pendant chains [74].

It is clear that there is a common agreement that nonhomogeneous (tens of nanometers to several hundred nanometers) gels are made during polyacrylamide polymerization and that the degree of this nonhomogeneity varies strongly with the Bis content. However, the finer details of the gel structure are still not clear. In addition to Bis variation, more recent studies to make other modifications to the gel structure also lead to increased nonhomogeneity in the gel and to other structural changes. For gels polymerized in continuous slabs as used in electrophoresis, the pore space has been generally considered to follow Ogston-type behavior; however, gels polymerized by emulsion polymerization or other methods [25,163] to create gel filtration chromatography (i.e., macroscopic particles or beads) media have been assumed to have more uniform pore shapes. In addition, some researchers, as discussed earlier, concluded that the gels used for slab electrophoresis have microdomains of Bis; however, it is not clear whether the geometrical shape of these microdomains is better represented by acrylamide fibers of varying thickness or of spherical clumps linked together by thin strands of PA (see Fig. 3). These finer details of structure, as will be pointed out later, may have a significant affect on the diffusion and electrophoresis properties in the gels.

3. Gel Pore-Size Distributions

Gel filtration chromatography has been extensively used to determine pore-size distributions of polymeric gels (in particle form). These models generally do not consider details of the shape of the pores, but rather they may consider a distribution of effective average pore sizes. Rodbard [326,327] reviews the various models for pore-size distributions. These include the uniform-pore models of Porath, Squire, and Ostrowski discussed earlier, the Gaussian pore distribution and its approximation developed by Ackers and Henn [3,155,156], the log-normal distribution, and the logistic distribution.

Partition coefficients for various gel models are given in Table 3. This table shows the dependence of K_{av} on the assumed structure of the pores. The Ogston model, for example, gives

$$\sqrt{-\ln (K_{av})} = (2lv\pi r_r) + (2lv\pi r_r)\frac{r_p}{r_r} \tag{7}$$

or

$$\sqrt{-\ln (K_{av})} = a_1 + a_2 r_p \tag{8}$$

which, according to Rodbard [326], implies that the log-normal and logistic pore distributions are approximations of Ogston's model. The first derivative of the partition coefficient with respect to the pore size gives the pore-size distribution function.

Polyacrylamide and agarose pore-size distributions have been measured using NMR T_1 relaxation methods by Chul et al. [79] using an extension of the analysis of Brownstein and Tarr [53] (Fig. 9). The distribution of pore sizes was seen to narrow significantly as the amount of acrylamide or agarose increased. This T_1 relaxation methodology requires a priori knowledge of the mean pore size in order to determine the pore distribution. Chul et al. [79] assumed mean pore sizes based upon simple geometrical

TABLE 3 Partition Coefficients

A. Uniform Pores
 1. Porath: $Kd^{1/3}$ vs $M^{1/2}$
 2. Ostrowski: $Kd^{1/3}$ vs $M^{1/3}$
 3. Ornstein
 4. Giddings
B. Gaussian

 1.

$$K_{av} = \frac{1}{\sqrt{2\pi}\sigma} \int_{R_s}^{\infty} \exp\left(-\frac{(x-\mu)^2}{2\sigma^2}\right) dx$$

 2. Ackers' approximation

$$erfc^{-1}(K_{av}) = a + bR_s$$

$$erfc(x) = \frac{1}{\sqrt{2\pi}} \int_{x}^{\infty} \exp(-\mu^2/2)\, d\mu$$

C. Log-normal

$$erfc^{-1}(K_{av}) = a + b \log(M) = a' + b' \log(R_s)$$

D. Logistic

$$\ln\left(\frac{K_{av}}{1-K_{av}}\right) = a'' + b'' \log(M)$$

E. Ogston

$$\sqrt{-\ln(K_{av})} = (2Lv\pi r_r) + (2Lv\pi r_r)r_p/r_r$$

F. Gaussian chain (at theta conditions)

$$K_{av} = \exp\left(-(2L\pi v r_r + 2Lv\pi r_p)^2\right) = \exp\left(-\phi((r_p + r_r)/(r_r))^\varepsilon\right) \qquad \varepsilon = 2$$

$$K_{av} = \exp\left(-\phi((r_p + r_r)/(r_r))^\varepsilon\right) \qquad \varepsilon = 1$$

G. Fractal
H. Random pores

---

considerations, and they concluded that variation of the total acrylamide, keeping Bis constant, had no effect on the overall configuration of the gel structure. One of the major problems with the result reported in this study is that the length scale over which the Brownstein and Tarr [53] NMR method is sensitive (i.e., >20 microns) is much larger than the nanometer-length scales of the pores in the gels reported in this work. It is therefore not clear whether the resulting size distributions from T_1 relaxation are meaningful for PA and agarose gels.

Agarose pore-size distributions, as measured by atomic force microscopy, show a strong dependence on the salt content and indicate that the gels are more homogeneous with lower buffer concentration [226]. Pore size ranged from 200 nm to 800 nm as the agarose concentration varied from 5% to 0.5%. These results are considerably different from those of Chul et al. [79], where the pore size at 5% was closer to 50 nm. Although Chul et al. [79] do not give the buffer conditions of their solutions, for a 1% agarose Maaloum et al. [226]

showed that at 0.001 M ionic strength the pore size was over 300 nm, and it increased as a power law with slope 0.25.

In the case of polyacrylamide gels, Stellwagen [367] found that buffer type (TAE vs TBE) did not affect the apparent pore size (21 nm for 10.5% T/5% C to 200 nm for 4.6% T/2% C), although more extreme variations in salt content and buffer physical properties may very likely strongly affect pore structure in polyacrylamide gels.

Johansson and coworkers [182–184] have analyzed polyacrylamide gel structure via several different approaches. They developed an analytical model of the gel structure using a single cylindrical unit cell coupled with a distribution of unit cells. They considered the distribution of unit cells to be of several types, including: (1) Ogston distribution, (2) Gaussian distribution of chains, and (3) a "fractal" network of pores [182–184]. They [183] used the equilibrium partition coefficient

$$K_{av} = e^{-\phi(\frac{r_p+r_r}{r_r})^\xi} \tag{9}$$

where ϕ is volume fraction of the network, $\xi = 2$ for the Ogston model, $\xi = 1$ for a Gaussian distribution of chains [180], and ξ is given by $3 - d_{eff}$ for the fractal model. Jansons

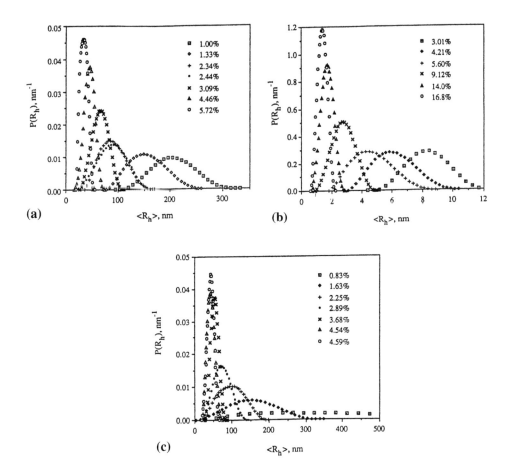

FIG. 9 Polyacrylamide and agarose size distribution. (a) Agar gels, (b) polyacrylamide gels, and (c) agarose gels. (Reprinted with permission from Ref. 79, Copyright 1995, Academic Press.)

and Phillips [180] applied geometric probability theory to determine equilibrium partitioning behavior in polymer networks, and they showed that for a Gaussian distribution of chains $\xi = 1$. For the fractal model d_{eff} was given by

$$d_{\mathrm{eff}} = \frac{d \ln N}{d \ln h_{N=2}} = 2 \; \frac{1 - \dfrac{p}{2r_r\sqrt{6}}\left(1 - e^{-2r_r\frac{\sqrt{6}}{p}}\right)}{1 - e^{-2r_r\frac{\sqrt{6}}{p}}} \tag{10}$$

where p is the persistence length of the chain molecule. This expression was derived by using theoretical results on the mean square end-to-end distance, h^2, for wormlike chains as function of the persistence length, p, the number of units in the chain, N, and the radius of the chain, r_r, given by

$$h^2 = 2\sqrt{6}Nr_rp - 2p^2(1 - \exp(-\sqrt{6}Nr_r/p)) \tag{11}$$

Figure 10 shows a plot of ξ versus the ratio of persistence length to fiber radius. Shown in this figure are points obtained by fitting computer simulations of polymer networks of chains with different flexibility, the solid line represents Eqs. (10) and (11). The limit at low persistence length can be seen to give Ogston behavior at $\xi = 1$, and the limit at very large persistence length gives Gaussian chain behavior $\xi = 2$. To fully test this model, independent measurements of the persistence length of the polymer in the gel would be required.

The effects of various pore-size distributions, including Gaussian, rectangular distributions, and continuous power-law, coupled with an assumption of cylindrical pores and mass transfer resistance on chromatographic behavior, have been developed by Goto and McCoy [139]. This study utilized the method of moments to determine the effects of the various distributions on mean retention and band spreading in size exclusion chromatography.

4. Statistical Mechanical Models

Using the formalism of statistical mechanics, Giddings et al. [135] investigated the effects of molecular shape and pore shape on the equilibrium distribution of solutes in pores. The equilibrium partition coefficient is defined as the ratio of the partition function in the pore

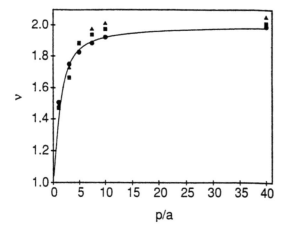

FIG. 10 Scaling parameter for the fractal model. (Reprinted with permission of American Institute of Physics and the authors from Ref. 183, Copyright 1993, American Institute of Physics.)

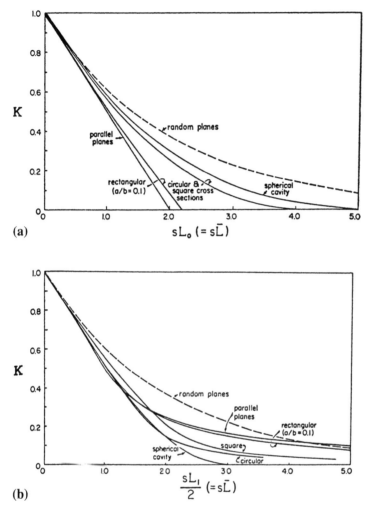

FIG. 11 Pore-size distributions. (a) Spherical molecules in various pore types, (b) thin rods in various pore types. (Reprinted with permission from Ref. 135, Copyright 1968, American Chemical Society.)

to that in the bulk solution. Assuming linear isotherms and rigid molecules and integrating over the space of possible configurations, Giddings et al. [135] found expressions for cases with (1) spherical particles in uniform pore networks of circular pores, infinite rectangular pores, pores made from infinite parallel plates, and pores of elliptical cross-section, (2) rigid molecules with rotational symmetry—i.e., thin rods in the same type of pores as just mentioned, and (3) molecules of any shape in isotropic networks of random planes and random fibers. The last case reproduced the Ogston equation. Figure 11 shows plots of several of these expressions and clearly illustrates a strong relationship between pore shape and partition behavior for various molecular shapes. This is one of the first demonstrations of the importance of the pore and molecule structures on gel chromatography partitioning, and it provides an important motivation for the exploration (experimentally and theoretically) of a much wider variety of structures than those considered before.

Rikvold and Stell [319,320,365] have developed an expression for the partition co-efficient in a random two-phase medium made up of spherical particles. They found the partition coefficient to be essentially an exponential function of the solute radius, which is in qualitative agreement with the Ogston theory.

5. Hydrodynamic Theories

Extensive work has been performed to determine partition and transport behavior in mem-branes and other media using hydrodynamic theories. Deen [89] reviews literature on hindered transport and partitioning in pores filled with liquids. Glandt [136] considered partitioning of solutes in membrane pores of various cross-sectional shapes, including rhomboidal pores and general polygonal shapes. Anderson and coworkers [5,15,16,190] considered a range of different-shaped molecules, including linear, comb-branched, and star-branched polymers as well as short-chain polymers and disklike molecules. Thompson and Glandt [390] have determined the distribution coefficient for flexible linear polymers into a porous solid as functions of the polymer molecular weight and concentration. This work further indicates the importance of the molecule shape and size on the partitioning into well-defined pores.

6. Macroscopic Structures

An additional factor in the comparison of gels made for electrophoresis and gel particles made for gel chromatography is the effect of the macroscopic geometrical structure on the microscopic pore structure. Waldmann-Meyer postulates that the pores in gel particles made for gel filtration chromatography may have a more conical shape due to a higher de-gree of swelling near the surface of the particle than in the center. Waldmann-Meyer, in an extension of Porath's approach, developed expressions for conical pores and applied this approach to controlled pore glass [413] and gels [414]. In the latter case he tested his model with eight proteins in 12 different gels. He concluded that the Ogston model is not appli-cable to agarose particle gels because the agarose radius determined from his data using the Ogston equation did not agree with the radius determined from electron microscopy and x-ray diffraction. This point further underscores the need for experimental–theory compar-ison where all the parameters in the models are obtained from independent experiments. It is also important to consider carefully the differences between gel polymerized in slabs for electrophoresis and gels polymerized in particles for gel chromatography or electrochro-matography.

7. Thermodynamic Models

Elasticity measurements can serve as a measure of the degree of interconnection in gels. Covalently cross-linked networks can be distinguished from physically cross-linked net-works by the use of a technique termed *mechanical spectroscopy* [333]. Compression of gels has also been used to assess the physical structure [28,168,303].

Swelling and solvent effects can be used as an alterative to elasticity measurements to characterize gel networks. Swelling characteristics of uncharged and charged gels have been considered by a number of workers, beginning with Flory and extending to more re-cent work by Prausnitz and coworkers [28,168,303] and others [169]. Within a thermody-namic framework for osmotic swelling, Baker et al. [28] considered the swelling of un-charged and charged polyacrylamide gels. The osmotic swelling pressure at thermodynamic equilibrium is given by

$$\Delta\Pi_{\text{swelling}} \equiv \frac{-\mu_1^{\text{gel}} - \mu_1^{\text{bath}}}{V_1} = \Delta\Pi_{\text{mixing}} + \Delta\Pi_{\text{elastic}} = 0 \qquad (12)$$

where μ_1 is the chemical potential of the solvent, water, and V_1 is the solvent molar volume. The osmotic swelling pressure was divided into contributions due to polymer solvent mixing and elastic effects from network deformation. The Flory–Huggins theory for the mixing effects was used:

$$\Delta\Pi_{\text{mixing}} = -\frac{RT_a}{V_1}[\ln(1 - \phi_2) + \phi_2 + \chi\phi_2^2] \tag{13}$$

where ϕ_2 is the mass swelling ratio, χ is the Flory polymer–solvent interaction energy, R is the universal gas constant, and T_a is the absolute temperature. The contributions due to the elasticity of the network were of two types. An *affine* network is defined as a network where no fluctuations in the cross-links occur. For an idealized affine network, the osmotic pressure is

$$\Delta\Pi_{\text{elastic}}^{\text{affine}} = -2C_cRT_a\left[\left(\frac{\phi_2}{\phi_{2c}}\right)^{1/3} - \frac{1}{2}\left(\frac{\phi_2}{\phi_{2c}}\right)\right] \tag{14}$$

where C_c is the concentration of cross-links in the reference state and ϕ_{2c} is the volume fraction of hydrogel in the reference state at preparation. In the opposite case, a *phantom* network is one where the cross-links are free to move without being affected by neighboring chains. The phantom network model gives for the osmotic pressure

$$\Delta\Pi_{\text{elastic}}^{\text{phan}} = -C_cRT_a\left(\frac{\phi_2}{\phi_{2c}}\right)^{1/3} \tag{15}$$

Baker et al. [28] fit swelling data on polyacrylamide gels to the foregoing model for the phantom elastic model to determine the Flory parameter of 0.48. Figure 12 shows the swelling data from Baker et al. fit to the foregoing model for PA gels in pure water with varying amount of total acrylamide. They concluded that the phantom model gives a better description than the affine model of the gel during swelling as a function of the total acrylamide since it is based on a more realistic description of allowable motion in the network. They also found, however, that for variation of Bis, holding acrylamide constant, a somewhat more complex equation, including both affine and phantom contributions, held [168] for Bis content between about 0.2% and 8.0%. At smaller Bis content the data showed a significant positive deviation from the model fit. Using their concept of an ideal gel and including only the affine elastic contributions and the first term in the mixing contributions to the gel swelling, Richards and Temple [313] found the swelling equilibria for an ideal gel to be represented by

$$Q[2Q^{2/3} - 1] = \bar{v}T/C \tag{16}$$

where Q is the volume expansion factor and $\bar{v}$ is the partial specific volume of the gel matrix. For clustered gels $\bar{v}$ is replaced by $z_{\text{Bis}}\bar{v}$, where z_{Bis} is the number of Bis residues in an aggregate or cluster. Figure 13 shows a plot of swelling data contours as functions of T and C with fixed swelling. The dashed line in this figure shows the model results for an ideal gel obtained from

$$T = 2\frac{M}{N_{\text{av}}}\left(\frac{0.589}{\beta}\right)^3\frac{1}{C}\left(\frac{C}{1 - C}\right)^{3/2} \tag{17}$$

where β is a proportionality coefficient between the root mean end-to-end distance of a growing chain and the number of acrylamide residues in the chain. This expression was ob-

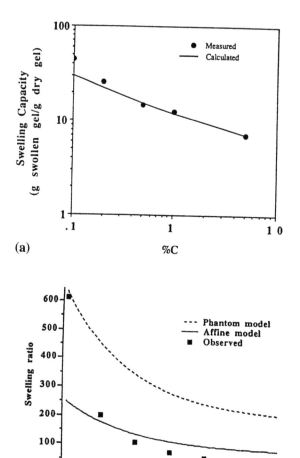

(a)

(b)

FIG. 12 Swelling data. (a) Variation of swelling ratio with % C. (Reprinted with permission from Ref. 168, Copyright 1990, American Chemical Society.) (b) Variation of swelling ratio with % T. (Reprinted with permission from Ref. 28, Copyright 1994, American Chemical Society.)

tained directly from the monomer material balance, the assumed Gaussian distribution of end-to-end distances in the chain, and a Poisson distribution in space of the Bis residues. Comparison of data with the thermodynamic model for gel swelling is shown in Figure 14. Richards and Temple conclude that a detailed quantitative analysis of gel structural data will require more knowledge of the "statistical nature of the clustering of the Bis residues, which in turn depends upon the kinetics of polymerization."

Anseth et al. [20] have reviewed the literature dealing with the mechanical properties of hydrogels and have considered in detail the effects of gel molecular structure, e.g., cross-linking, on bulk mechanical properties using theories of rubber elasticity and viscoelasticity.

8. Kinetic Models

Classical models of gel formation (or sol-gel transitions [4]) by Flory [394], Gordon, Macosko and Miller [228,245], and others (see Ref. 4 for a more complete review) considered

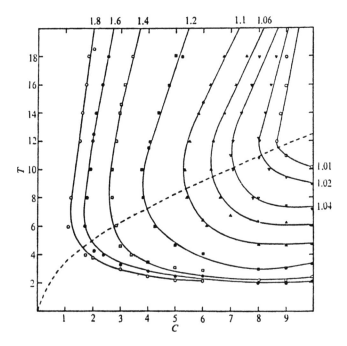

FIG. 13 Gel data contours of constant swelling factor from experimental data. Dashed line indicates the composition of an ideal gel. (Reprinted with permission from Ref. 313, Copyright 1971, Nature.)

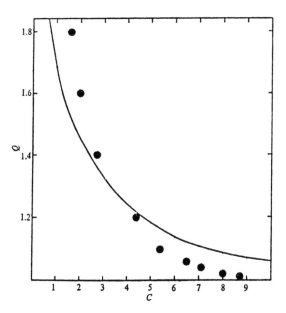

FIG. 14 Swelling data as function of % C. (Reprinted with permission from Ref. 313, Copyright 1971, Nature.)

the formation of treelike branched structures shown in Figure 15. Subsequent analysis, using percolation theory by Hammersley [50,147], De Gennes, and Stauffer allowed consideration of different types of matrix topology. These theories do not strictly consider the kinetics of gelation; rather, they seek to predict critical conditions for gelation, molecular weight distributions, and extent of reaction. More recent work has been performed by Arhabi and Sahimi to consider critical viscoelastic properties of gels using Monte Carlo simulation and scaling theories [21,22,342]. Monte Carlo methods have been very useful for polymer solutions and dilute gels; however, there are no simulations for highly crosslinked gels in which the effects of the gel structure are examined in detail [39].

Kinetic gelation simulations seek to follow the reaction kinetics of monomers and growing chains in space and time using lattice models [43]. In one example, Bowen and Peppas [155] considered homopolymerization of tetrafunctional monomers, decay of initiator molecules, and motion of monomers in the lattice network. Extensive kinetic simulations such as this can provide information on how the structure of the gel and the conversion of monomer change during the course of gelation. Application of this type of model to polyacrylamide gels and comparison to experimental data has not been reported.

Experimental data on polyacrylamide polymerization, in addition to that previously discussed by Righetti and coworkers, has been reported by Nieto et al. [262]. The kinetics of polyacrylamide gelation were studied by Nieto et al. [262] using high-field ^{1}H-NMR spectroscopy. The signal from the —CH_2— bridges in the network was followed with time over the course of 10 hours, and the polymerization was carried out in D_2O to minimize signal from the solvent. Figure 16 shows sample data of acrylamide and Bis conversion as a function of time from their publication. The gel point is clearly seen in this figure, and it is also shown that conversion continues to increase beyond the point of gel formation. Other

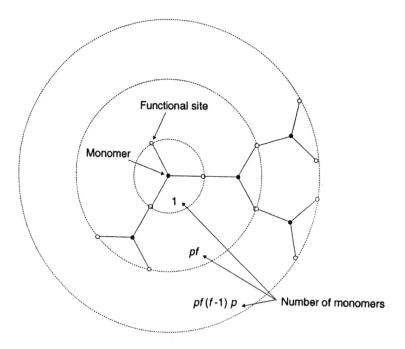

FIG. 15 Gel treelike structure. (Reproduced with permission from Ref. 4, Copyright John Wiley & Sons, Ltd.)

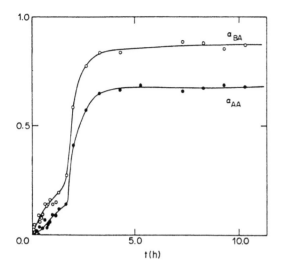

FIG. 16 Gel kinetics from NMR data: conversion of acrylamide and Bis versus time. (Reprinted from Ref. 262, Copyright 1987, with permission from Elsevier Science.)

kinetic studies of gel formation by NMR have been reported by Cohen Addad [81] for different types of polymer gels.

Torkelson and coworkers [274,275] have developed kinetic models to describe the formation of gels in free-radical polymerization. They have incorporated diffusion limitations into the kinetic coefficient for radical termination and have compared their simulations to experimental results on methyl methacrylate polymerization. A basic kinetic model with initiation, propagation, and termination steps, including the diffusion limitations, was found to describe the gelation effect, or time for gel formation, of several samples sets of experimental data.

The effects of diffusion control on cross-link kinetics were investigated by Dusek [102] within the context of polymerization reaction kinetics.

III. GENERAL EQUATIONS FOR TRANSPORT IN ELECTRIC FIELDS

The field of electrohydrodynamics deals with the transport of charged species in fluids under the action of applied electric fields [111,112,203,239]. The general field equations for continuum approaches in single-phase fluids, i.e., gases and liquids, without emphasis on electric fields have been developed and reviewed extensively [40,92,143,360]. Prior to analyzing electric-field-induced transport in gels or heterogeneous materials, the equations required for the analysis of electrohydrodynamic transport in homogeneous multicomponent fluids will be considered.

Under the conditions of a *static* electric field, the Maxwell's field equations reduce to the Poisson equation [111,112,172,261], given by

$$\nabla^2 \Psi^k = -\frac{\sum_{i=1}^{N} c_i^k z_i F}{\varepsilon_d^k} \qquad k = \alpha, \beta, \ldots, \gamma \tag{18}$$

where c_i is the molar concentration of the ith species in the mixture, z_i is the solute charge, F is Faraday's constant, ϵ_d is the dielectric permittivity of the medium [160], N is the total number of charged species in the mixture, Ψ is the electric potential, and the superscript k denotes the phase under consideration. The general field equations for electromagnetic fields (Maxwell's equations) are derived using concepts from irreversible thermodynamics by Kuiken [202], and other derivations are also available [111,112]. The molar species concentration [40] can be determined from the solution of

$$\frac{\partial c_i^k}{\partial t} + \nabla \cdot N_i^k = 0 \qquad k = \alpha, \beta, \ldots, \gamma \tag{19}$$

when no reactions occur. The molar flux N_i^k is usually given by the Nernst–Planck [261] constitutive equation for dilute solutions as

$$N_i^k = c_i^k v^k - D_i^k \nabla c_i^k - z_i u_i' c_i^k \nabla \Psi^k \qquad k = \alpha, \beta, \ldots, \gamma \tag{20}$$

where u_i' is the electrophoretic mobility, D_i is the dilute solution diffusion coefficient, and v is the mass average velocity (assumed equal to the molar average velocity for dilute solutions). Equation (20) is valid only for dilute solutions. In the case of concentrated electrolytes it is necessary to utilize the electrochemical potential in a derivation of a generalized Stefan–Maxwell equation that includes electric interactions [261]. In general, the species continuity equation for all charged species must be considered. However, for a number of specific applications where the solute of interest is in much lower concentration than the current carrying ions and where electroneutrality [261]

$$\sum_{i=1}^{N} c_i^k z_i = 0 \qquad k = \alpha, \beta, \ldots, \gamma \tag{21}$$

can be assumed it is possible to assume a constant electric field and consider only the species continuity equation for the dilute solute of interest [217,218,261]; alternatively, where the species of interest is not in much lower concentration than the current carrying ions, Eq. (21) can be combined with Eqs. (20) and (19) to determine the ion concentrations and the electrostatic potential [261]. Analysis of ion transport in charged nanometer-size pores within biological context, i.e., transport through membrane-spanning proteins, has been conducted by Barcilon, Eisenberg, and Chen [30–32,109]

For a dilute aqueous solution the mass average velocity is determined from the equation of motion for a Newtonian fluid, the Navier–Stokes equation,

$$\rho^k \frac{Dv^k}{Dt} = -\nabla p^k + \rho^k b + \mu^k \nabla^2 v^k + \Sigma \, c_i^k z_i F \nabla \Psi^k \qquad k = \alpha, \beta, \ldots, \gamma \tag{22}$$

where the last term on the right-hand side of the equation represents the effect of the electrical body forces on the fluid motion [239,346]. Whitaker [427,430,435] gives an excellent derivation of the general momentum balance and the specific equations of motion for Newtonian fluids in the absence of electric fields, beginning with the species momentum balance (see also Slattery [360]). Non-Newtonian fluids will also be of interest, especially in the cases of the addition of viscous polymers to the separation mixture. Extensive discussion of the electrical body and surface forces present in the equation of motion are given in the fluid mechanics and colloids literature [160,172,173]. The thermal energy balance, neglecting viscous dissipation, radiative heat transfer, reaction sources, and work terms due to volume expansion, reduces to

$$\rho^k C_p^k \frac{DT^k}{Dt} = k^k \nabla^2 T^k + Q^k \qquad k = \alpha, \beta, \ldots, \gamma \tag{23}$$

where Q^k is the volumetric heat source given by $I_e^2 R_e$. Whitaker [431] again gives an excellent derivation of the thermal energy balance in the absence of electric fields, beginning with the species energy balances. The current I_e [261] is given by the flux of ions by

$$I_e^k = -F^2 \nabla \Psi^k \sum_{i=1}^{N} z_i^2 (u_i')^k c_i^k - F \sum_{i=1}^{N} z_i D_i^k \nabla c_i^k + F v^k \sum_{i=1}^{N} z_i c_i^k \qquad k = \alpha, \beta, \ldots, \gamma \tag{24}$$

The foregoing equations are coupled and are generally nonlinear; no general solution exists. However, these equations serve as a starting point for most of the analysis that is relevant to electrophoretic transport in solutions and gels. Of course, the specific geometry and boundary conditions must be specified in order to solve a given problem. Boundary conditions for the electric field include specification of either (1) constant potential, (2) constant current, or (3) constant power.

It can be noted that other approaches, based on irreversible continuum mechanics, have also been used to study diffusion in polymers [61,224]. This work involves development of the species momentum and continuity equations for the polymer matrix as well as for the solvent and solute of interest. The major difficulty with this approach lies in the determination of the proper constitutive equations for the mixture. Electric-field-induced transport has not been considered within this context.

IV. DIFFUSION

The physical properties of gels, and of porous media in general, are known to significantly influence the diffusion of probe species in gels. The dimensions and concentration of obstacles and the pore size and size distribution are some of the physical factors that affect diffusion in porous media. In addition to these factors, the solute size, shape, and type will influence the diffusion. A number of theories to describe diffusion in porous media have been developed to account for these factors. Models for diffusion in porous media can be classified based upon the method of derivation; for example, they may include: (1) continuum theories, where the solute is much smaller than the obstacles [e.g., volume averaging, exact solutions, variational bounds, perturbation methods, and effective medium approximations], (2) continuum theories that account for hydrodynamic interactions, (3) network theories, (4) free volume theory, (5) molecular dynamics simulations, and (6) statistical theories applicable to a variety of cases where the solute can be of similar magnitude or of very different magnitude to the obstacles—these may include Brownian motion or random walk simulations [171] and Monte Carlo simulations. Alternatively, diffusion theories can be classified based upon the physical factors that they address. These may include the effects of: (1) obstructions and path length and tortuosity, (2) the solute size, (3) hydrodynamic interactions, or (4) variations in solvent properties. The following review is not meant to be an exhaustive review of the literature; the objective is to provide a survey of several models that may be relevant to diffusion in electrophoresis media and to highlight the theories for diffusive transport that have been extended to account for electrophoresis in gels or porous media in general.

A. Comments on Diffusion Coefficients in Single-Phase Media

Before attempting to review the available literature on diffusion in gels, it is important to understand the definitions and nomenclature of diffusion coefficients. The fundamental relationships between transport flux and their driving forces can be obtained using irreversible thermodynamics or molecular-kinetic theory. The fundamentals of irreversible thermodynamics have been well reviewed and discussed in the literature [92,143,202,421]. For the case of diffusive transport the Stephan–Maxwell theory, originally developed for gases and subsequently applied to liquids [202], provides the appropriate framework for obtaining the constitutive flux. Whitaker [434] provides a very clear development of the classical Stephan–Maxwell diffusion theory, without electric or other body forces, by utilizing the species momentum balances.

The relationship between the diffusional flux, i.e., the molar flow rate per unit area, and concentration gradient was first postulated by Fick [116], based upon analogy to heat conduction Fourier [121] and electrical conduction (Ohm), and later extended using a number of different approaches, including irreversible thermodynamics [92] and kinetic theory [162]. Fick's law states that the diffusion flux is proportional to the concentration gradient through

$$\underline{J}_i = -D_{ij}\nabla c_i \tag{25}$$

where $\underline{J}_i$ is the diffusive flux vector of component i and c_j is the molar species concentration per unit volume. Here D_{ij} is the mutual or interdiffusion coefficient and has units of area per unit time.

Fick [116], using his own data and those of Graham [140], analyzed the diffusion of solutes in a two-component mixture (as cited in Cussler [88]). The diffusion coefficients obtained from this type of experiment are called *interdiffusion coefficients*. Interdiffusion coefficients must be distinguished from the *intradiffusion coefficient*, first introduced by Albright and Mills [10]. An intradiffusion coefficient is defined for a multicomponent system in which a fraction of molecules of one component are labeled, or distinguished from other molecules of the same component. These molecules may be labeled or distinguished from others of the same species by isotopic labeling or, for example, nuclear magnetic vector orientation [401]. A self-diffusion coefficient is a special case of intradiffusion for a system containing only one chemically distinguished species. If a trace of labeled diffusant is introduced at some point in space into otherwise homogeneous medium, which may or may not contain the unlabeled species, the resulting diffusion coefficient is termed a *tracer diffusion coefficient*. At low concentrations of labeled species in comparison to the total concentration, the intra- and tracer diffusion coefficients are the same. The intra- and interdiffusion coefficients are equal at infinite dilution of one of the components. For other than limiting concentrations, a number of correlations are available [401]. For the present purpose we will be concerned primarily with the dilute-solution limit of a single solute in a solvent.

B. Comments on Diffusion in Porous Media

A number of different approaches have been taken to describing transport in porous media. The objective here is not to review all approaches, but to present a framework for comparison of various approaches in order to highlight those of particular interest for analysis of diffusion and electrophoresis in gels and other nanoporous materials. General reviews on the fundamental aspects of experiments and theory of diffusion in porous media are given by Karger and Ruthven [189], Alder [6], and Sahimi [341]. The types of models used in the analysis of transport in porous media ranges from models derived using basic steady-state Laplace equations (e.g., thermal and electrical conduction problems) in simple geometries to complex models based upon rigorous derivations in highly complex structures. In order to provide some context to the more advanced models and to illustrate the physical processes without undue mathematical complexity, it will be useful to consider some of the elementary models. The present review shall rely heavily on the rigorous framework developed extensively by Whitaker in a number of publications over the last 30 years [62,337,420,422,423,426–429,433,436].

Before presenting the details of the volume averaging method, it would be useful first to discuss some of the limitations and advantages and disadvantages of the volume averaging method. The volume averaging theory utilizes the basic material, energy, and momentum balances developed for individual species in pure fluids. The effect of obstacles and media structure on transport in these pure fluids is described through boundary conditions. Since most porous media of interest has very complex geometry, it is generally not feasible, nor is it usually desirable, to solve these equations, either in principle or in approximation, throughout the complex structure. It is not desirable since most experiments do not provide that level of detail with regard to concentration, velocity, or temperature distributions in micron or smaller scales. The volume averaging method provides a direct and rigorously defined process to take these point equations and develop averaged expressions that describe more closely the observed behavior on the macroscopic scale. A major advantage of this method is the clear delineation of assumptions made in the formulations and specification of order-of-magnitude estimates necessary to justify various assumptions; many methods for analyzing transport in porous media make hidden and ad hoc assumptions that are not so clearly explained. The method of volume averaging requires the solution of a well-defined closure problem in a representative region that reflects the major features of the media geometry. The closure problem provides the link between the microscopic physical properties and the macroscopic properties.

Generally, the closure problem reflects the idea of a spatially periodic porous media, whereby the entire structure can be described by small portions (averaging volumes) with well-defined geometry. Two limitations of the method are therefore related to how well the overall media can be represented by spatially periodic subunits and the degree of difficulty in solving the closure problem. Not all media can be described as spatially periodic [6,341]. In addition, the solution of the closure problem in a complex domain may not be any easier than solving the original set of partial differential equations for the entire system.

Despite these limitations, the method has been shown to be well suited for a wide class of heat transfer, mass transfer (with and without chemical reactions) of point solutes, and flow problems in a variety of geometries of both heterogeneous and homogeneous porous media. For systems with well-defined geometry and where the closure problem correctly mimics the actual geometry of the system, comparison with experimental data [436] has been quite good. In addition, calculations by volume averaging have been shown to compare vary well with direct simulations using Monte Carlo methods [400]. The method has not been applied to cases of macromolecular transport in porous media, to cases with highly nonlinear transport processes, or to some classes of fractal or other types of porous media with, for example, continuous distributions of porous structures. This is not to say that the method cannot be adapted to those problems; future work should be developed along those lines. However, the general question remains as to whether or not there is an advantage to adapt the method to a given problem or if another approach may prove simpler. The method remains, however, one of the most powerful techniques for deriving, from the basic equations of change for transport in single-phase fluids, transport parameters, e.g., effective diffusion coefficient, thermal conductivity, and flow permeability to reflect macroscopic measurements.

Consider a porous medium as shown in Figure 17. This example medium consists of two phases, α and β, whereby phase β is completely impermeable to the solute of interest. The solute of interest is a point species, with no interactions with the interface bounding the two phases. The species continuity equation for single-phase homogeneous media is given by

$$\frac{\partial c^k}{\partial t} = D^k \nabla^2 c^k \qquad k = \alpha \tag{26}$$

and is valid only in the α phase. Measurement of the flux of solute through or across the composite media, i.e., both α and β phases, will usually provide values for average fluxes, and measurements of the concentration will generally provide the average concentration in the α phase (assuming of course that the other phase is impermeable to the solute of interest). [An illustration of this is the pulsed-field gradient NMR experiment, provided the time for the experiment is long enough so that the distance traveled by the species of interest is much larger than the space between obstacles.] Therefore, Eq. (26), the point species continuity equation, cannot describe properly the overall transport in the porous media.

Previously (e.g., Ref. 344), it has been noted that Eq. (26) will still be valid if the point concentration variable is replaced by the average concentration; however, the diffusion coefficient was found to differ from the molecular diffusion coefficient obtained in the pure fluid. This diffusion coefficient was termed the *effective diffusion coefficient*. The

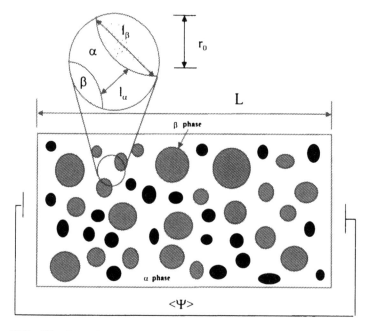

FIG. 17 Generalized model of multiphase media. (Reprinted with permission from Ref. 215, Copyright 1998, American Chemical Society.)

single-phase species continuity equation (26) in this case is replaced by the average equation [423],

$$\frac{\partial \langle c \rangle}{\partial t} = D_{\text{eff}} \nabla^2 \langle c \rangle \tag{27}$$

where $\langle c \rangle$ is the *phase average* concentration in the media, analogous to a superficial velocity and defined rigorously by [422]

$$\langle c \rangle = \frac{1}{V} \int_{V^\alpha} c^\alpha \, dV \tag{28}$$

where V is a suitably defined representative averaging volume [423]. The phase average represents the concentration of the solute of interest over the entire media. It is generally more convenient to work with the intrinsic phase average, defined as

$$\langle c \rangle^\alpha = \frac{1}{V^\alpha} \int_{V^\alpha} c^\alpha \, dV \tag{29}$$

where V^α is a suitably defined representative averaging volume in the α phase. The intrinsic phase average represents the concentration measured inside the α phase or measured by extraction of the fluid from the α phase. This average is analogous to the interstitial velocity from elementary fluid mechanics. For the case of diffusion in a two-phase (isotropic) porous media with one phase impermeable to the solute of interest, the point species continuity equation in the α phase can be averaged to

$$\frac{\partial \langle c \rangle^\alpha}{\partial t} = D_{\text{eff}} \nabla^2 \langle c \rangle^\alpha \tag{30}$$

Note that the intrinsic phase average and the overall phase average are related by the porosity or volume fraction of the phases by [423]

$$\langle c \rangle = \varepsilon \langle c \rangle^\alpha \qquad \varepsilon = \frac{V^\alpha}{V} \tag{31}$$

The average molar flux across some domain in terms of the intrinsic phase average concentration or the phase average concentration is [427]

$$\langle N \rangle = -D_{\text{eff}} \varepsilon \nabla \langle c \rangle^\alpha = -D'_{\text{eff}} \nabla \langle c \rangle^\alpha \tag{32}$$

where we have defined

$$D'_{\text{eff}} = D_{\text{eff}} \varepsilon \tag{33}$$

It is important to point out that care must be taken when comparing effective diffusion coefficients from the literature to distinguish the effects of the porosity as given in Eq. (33). If one measures diffusive transport through use of the point species equations, as, for example, in pulsed-field gradient NMR spectroscopy, the diffusion coefficient determined will be D_{eff}, whereas if one measures average macroscopic flux across some domain using Eq. (32), as in a diffusion cell [26], in terms of the phase average concentration gradient, D'_{eff} will be determined.

The objective of most of the theories of transport in porous media is to derive analytical or numerical functions for the effective diffusion coefficient to use in the preceeding averaged species continuity equations based on the structure of the media and, more recently, the structure of the solute.

C. Obstruction Effects

Many investigators have studied diffusion in systems composed of a stationary porous solid phase and a continuous fluid phase in which the solute diffuses. The effective transport coefficients in porous media have often been estimated using the following expression:

$$\frac{D'_{eff}}{D_0} = \frac{\varepsilon}{\tau} \tag{34}$$

where D_0 is the diffusion coefficient in the continuous phase in the absence of porous solids, ε is the pore volume fraction, and τ is the "tortuosity" factor (here including all factors that affect the diffusion in the porous media—both increased pore path length and constriction effects). This equation was originally postulated as an empirical relationship and much effort was expended in measuring the tortuosity factor for heterogeneous catalysts [344] and other applications.

Simple mathematical expressions for the tortuosity factor have been developed in the literature (see Ref. 344 for a review of some of the early literature related to heterogeneous catalysis). Carman [63] studied the analogous problem of hydrodynamic flow of gases in porous media, and he defined the tortuosity factor as the square of the ratio of an effective path length in the porous media to the shortest distance measured in a given direction ($\tau' = (l_a/l)^2$). This increase in path length due to the "tortuous" nature of the porous media will decrease the effective diffusion coefficient by making a given solute travel a farther distance than it would need to in the absence of the porous solid. In addition to a "pure" "tortuosity" effect, defined here as τ', it has been further noted that the porous media will constrict the motion of a given solute. This constriction factor, θ, has been assumed to be separable from the tortuosity factor through

$$\frac{D'_{eff}}{D_0} = \varepsilon \frac{\theta}{\tau'} \qquad \tau' = \tau\theta \tag{35}$$

In order to illustrate the effects of media structure on diffusive transport, several simple cases will be given here. These cases are also of interest for comparison to the more complex theories developed more recently and will help in illustrating the effects of media on electrophoresis. Consider the media shown in Figure 18, where a two-phase system contains uniform pores imbedded in a matrix of nonporous material. Solution of the one-dimensional point species continuity equation for transport in the pore, i.e., α phase, for the case where the external boundaries are at fixed concentration, c_I and c_{II}, gives an expression for total average flux

$$\langle N \rangle = D_0 \frac{a_2(c_I - c_{II})}{(a_2 + a_1)L} \tag{36}$$

where a_1 and a_2 are the cross-sectional areas of the nonporous material and the open channels, respectively, and L is the length of the channel. It is clear that an effective diffusion coefficient can be defined by

$$D'_{eff} = D_0\varepsilon \qquad \varepsilon = \frac{La_2}{L(a_1 + a_2)} \tag{37}$$

The D'_{eff} across the porous medium for this example is linearly related to the porosity of the path, which is in turn simply the ratio of the open cross-sectional area to the total cross-sectional area. There are no constriction or tortuosity effects in this example; i.e., $\tau = 1$ and $D_{eff} = D_0$.

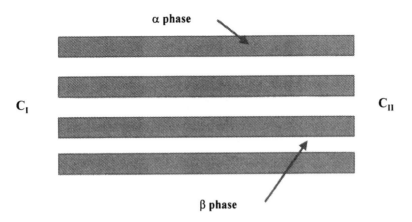

α phase

C_I C_{II}

β phase

FIG. 18 Parallel-pore model of multiphase media.

For the case of a two-phase system with two parallel noninteracting paths that both contribute to the diffusion of the solute and where the diffusion coefficients of the solute of interest are different in the two phases, the solution of the two isolated one-dimensional steady-state diffusion models gives

$$D'_{eff} = D_\alpha \varepsilon + D_\beta (1 - \varepsilon) \qquad D_{eff} = D_\alpha + \frac{1 - \varepsilon}{\varepsilon} D_\beta \qquad \varepsilon = \frac{a_\alpha}{a_\alpha + a_\beta} \tag{38}$$

When a two- or higher-phase system is used with two or more phases permeable to the solute of interest and when interactions between the phases is possible, it would be necessary to apply the principle of local mass equilibrium [427] in order to derive a single effective diffusion coefficient that will be used in a one-equation model for the transport. Extensive justification of the principle of local *thermal* equilibrium has been presented by Whitaker [425,432]. If the transport is in series rather than in parallel, assuming local equilibrium with equilibrium partition coefficients equal to unity, the effective diffusion coefficient is

$$D'_{eff} = \left[\frac{\varepsilon}{D_\alpha} + \frac{1 - \varepsilon}{D_\beta} \right]^{-1} \qquad \varepsilon = \frac{l_1}{L} \tag{39}$$

Combination of these two models for a combined series- and parallel-path system gives

$$D'_{eff} = \left[\frac{\varepsilon_\alpha}{D_\alpha} + \frac{\varepsilon_\beta}{D_\beta} \right]^{-1} + D_\gamma \varepsilon_\gamma \tag{40}$$

where the porosities are defined by

$$\varepsilon_\alpha = \frac{l_1 a_1}{L(a_1 + a_2)} \qquad \varepsilon_\beta = \frac{(L - L_1)a_1}{L(a_1 + a_2)} \qquad \varepsilon_\gamma = \frac{La_2}{L(a_1 + a_2)} \tag{41}$$

Michaels [241], using a model pore shown in Figure 19, considered the constriction effect by solving the steady-state species continuity equations in a model pore consisting of a single constriction. The result for the pore shown in Figure 19 is given by

$$\frac{D'_{eff}}{D_0} = \frac{(1 - A)/\varepsilon}{1 - A\varepsilon + A^2(\varepsilon - 1)} \qquad 1/A < \varepsilon < 1 \tag{42}$$

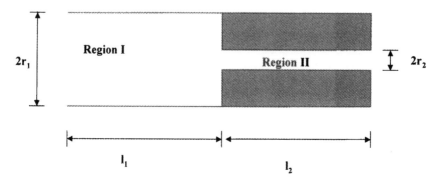

FIG. 19   Model media based on concept proposed by Michaels [241].

or in terms of L and ε by

$$\frac{D'_{\text{eff}}}{D_0} = \frac{(L+1)((L+1)\varepsilon - L)}{\varepsilon(L((L+1)\varepsilon - L) + 1)} \tag{43}$$

where A, L, and ε are defined by

$$A = \frac{a_1}{a_2} \qquad L = \frac{l_1}{l_2} \qquad \varepsilon = \frac{a_1 l_1 + a_2 l_2}{a_1(l_1 + l_2)} = \frac{AL + 1}{A(L+1)} \tag{44}$$

This example will be of particular interest in our consideration of electrophoresis, and it is also of interest from the point of view of introducing anisotropy into the media structure. Figure 20 shows plots of the effective diffusion coefficient versus porosity for various

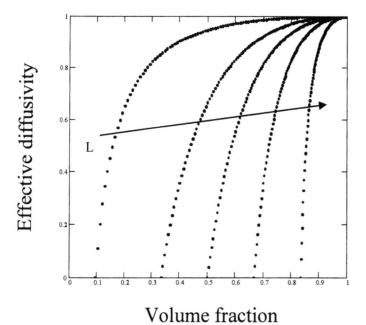

Volume fraction

FIG. 20   Effective diffusion coefficients using Michaels' model [241], Eq. (43), versus porosity for various ratios of pore lengths.

values of the length scale ratio L. It is clear that certain values of porosity are inadmissible and that the range of validity of porosity for this model depends upon the value of L. The lower limit for the value of porosity for a given value of L is given by $L/(L + 1)$. This reflects the fact that for certain length scales the geometry becomes closed, i.e., percolation occurs. Percolation concepts have been extensively considered in the literature on transport in porous media [6,341]. Various pores with different geometries, including sinusoidally varying pores, have also been considered in the literature [292].

Wakao and Smith [411] examined diffusion of gas molecules in porous media composed of micropores (free space in the solid particles) and macropores (space between the solid particles). The diffusing molecule was considered to be moving through three possible paths: (1) through the macropores, (2) through the micropores, and (3) through micropores and macropores in series. In order to solve for the diffusion coefficients, they used a general equation for the diffusion flux of gas A in a circular capillary at constant pressure in a binary system of gases A and B and applied it to all three diffusion mechanisms. The resulting diffusion coefficient for the aggregation of fluxes corresponding to all three diffusion mechanisms was

$$D'_{eff} = \varepsilon^2 D_a + (1 - \varepsilon)^2 D_i + 2\varepsilon(1 - \varepsilon) \frac{2}{\dfrac{1}{D_a} + \dfrac{1}{D_i}} \tag{45}$$

where the first two terms are the contributions from parallel diffusion in macropores and micropores, respectively, and the last term is the contribution from macro-/micropores in series. It was assumed in this derivation that the pores are randomly distributed. The diffusion coefficient in the micropores is given by

$$D_i = \frac{\varepsilon_i^2/(1 - \varepsilon_i)^2}{(1 - \alpha y_A)/D_{AB} + 1/D_{ki}} \tag{46}$$

where y_A is the mole fraction of species A, α is a coefficient given by 1 plus the ratio of fluxes of species A and B, and the diffusion coefficient in the macropores is given by

$$D_a = \frac{1}{(1 - \alpha y_A)/D_{AB} + \dfrac{1}{D_{ka}}} \tag{47}$$

The contributions of the individual fluxes were based upon the volume fractions of the respective areas available for diffusion, as would be expected from the simple models given earlier. For low-density pellets only the macropore contribution is significant, and if the macropores are large enough, or the pressure high enough, the Knudsen part of the diffusion in the macropores can be neglected. Under these conditions their expression simplifies to

$$\frac{D'_{eff}}{D_0} = \varepsilon^2 \tag{48}$$

It is clear that by making more and more complex structures and including nonunity partition coefficients the effective diffusion coefficient can be derived using the simple straightforward approach outlined earlier. This type of approach has been considered in a number of models of interest in biomedical engineering and chemical reaction engineering. For example, Michaels et al. [242] have developed a more extensive model of drug permeation across skin using this methodology. However, it is important to recognize that there are a number of shortcomings with this type of approach. First, it is not strictly valid to ignore the connections between the constriction effects and the tortuous-path effects. Second, simple one-dimensional models will fail to properly account for effects due to two- and three-dimensional transport. Third, it may be difficult or impossible to extend these models to more complex media.

One approach to extend such theories to more complex media is network theory. This approach utilizes solutions for transport in single pores, usually in one dimension, and couples these solutions through a network of nodes to mimic the general structure of the porous media [341]. The complete set of equations for all pores and nodes is then solved to determine overall transport behavior. Such models are computationally intense and are somewhat heuristic in nature.

One must be very careful in reviewing the older, and some more recent, literature in consideration of the tortuosity and constriction factors; some work has attempted to separate these two factors; however, more modern developments show that they cannot be strictly decoupled. This aspect will be particularly important when reviewing the barrier and tortuous-path theories of electrophoresis, as discussed later.

In order to overcome these problems, one approach, originally developed by Whitaker [420], Slattery [359], and Anderson and Jackson [17], involves the method of volume averaging. Using volume averaging theory, Whitaker and coworkers [193,264,268,337,436] found the effective diffusion *tensor* for a two-phase system to be given by

$$\frac{\underline{D}_{\text{eff}}}{D} = \underline{I} + \frac{1}{V^\alpha} \int_{A_{\alpha\beta}} \frac{1}{2}(\underline{n}_{\alpha\beta} \, \underline{f} + \underline{f}\underline{n}_{\alpha\beta}) \, dA \tag{49}$$

and the $\underline{f}$ vector field is determined by solving the closure problem in a unit cell. This expression is valid for anisotropic as well as isotropic media. Ochoa [193,268], Nozad [264], Ryan et al. [337], and Saez et al. [340] have developed extensive analysis and computations of the effective diffusion coefficient in model porous media using the volume averaging method. Whitaker [427] and Kim et al. [193] compared the results from volume averaging for an isotropic media (Figure 21a) with those of Maxwell, Wakao and Smith, and Weissberg. They found the Wakao and Smith result to underpredict the diffusion coefficient at low porosities and the Weissberg results to slightly overpredict the volume averaging results. As expected, the Maxwell model also gives an upper limit to the range of diffusion coefficients. The results of Kim et al. [193] for a unit cell analogous to the constriction model developed by Michaels, discussed earlier, are particularly interesting. The volume averaging method gives both diagonal components of the diffusion tensor, and, as seen in Figure 21b, a percolation limit is observed in a fashion analogous to that seen in the simple one-dimensional model. Figure 21b shows volume averaging results for various length scales of the unit cell. This plot can be compared to Figure 20, developed using the analysis of Michaels. The more rigorous analysis using volume averaging methods of course give the two-dimensional results (and three-dimensional results can also be obtained with this method); i.e., both D_x and D_y and that of the one-dimensional model of Michaels can only give a one-dimensional result. It is interesting to note that the simple one-dimensional analysis gives percolation limits in an analogous fashion to that given by the multidimensional model, however, significant quantitative differences exist. Saez et al. [340] and Quintard [305] performed more extensive calculations for three-dimensional media and found some degree of improvement in the comparison with the experimental results for anisotropic media. Trinh et al. [398,400] showed good agreement between Monte Carlo

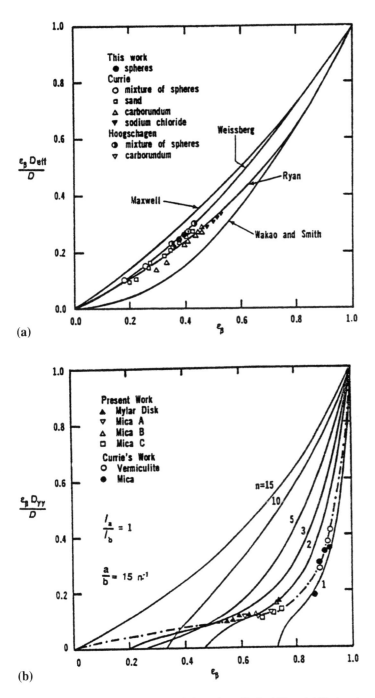

FIG. 21 Effective diffusion coefficients from Refs. 337 and 193 showing comparison of volume average results (Ryan) with models of Maxwell, Weisberg, Wakao, and Smith for isotropic systems (a), and volume averaging calculations (solid lines) and comparison with data for anisotropic systems (b). (Reproduced with kind permission of Kluwer Academic Publishers from Ref. 193, Fig. 3 and 12, Copyright Kluwer Academic Publishers.)

simulations and calculations using volume averaging for these two-dimensional isotropic and anisotropic media. Figure 22 shows the excellent agreement of Monte Carlo simulations with volume averaging results for isotropic systems.

For the case of a three-phase problem, where the solute is accessible to the α, β, and γ phases, Whitaker [427] finds the overall average phase concentration for the case of local mass equilibrium given by

$$(\varepsilon_\alpha + \varepsilon_\beta \varepsilon_\gamma) \frac{\partial \langle c \rangle}{\partial t} = \nabla \cdot (\varepsilon_\alpha + \varepsilon_\beta \varepsilon_\gamma) \underline{\underline{D}}_{\text{eff}} \cdot \nabla \langle c \rangle \tag{50}$$

with the effective diffusion tensor given by

$$\underline{\underline{D}}_{\text{eff}} = \frac{\varepsilon_\alpha D_\alpha + \varepsilon_\beta \varepsilon_\gamma D_\gamma}{(\varepsilon_\alpha + \varepsilon_\beta \varepsilon_\gamma)} \underline{I} + \left(\frac{D_\alpha - \varepsilon_\beta D_\gamma}{(V \varepsilon_\alpha \varepsilon_\beta \varepsilon_\gamma)} \right) \int_{A_{\beta\gamma}} \underline{n}_{\alpha\beta} \underline{f} \, dA \tag{51}$$

and the overall intrinsic phase average concentration was defined by

$$\langle c \rangle = \frac{\varepsilon_\alpha \langle c_\alpha \rangle^\alpha + \varepsilon_\beta \varepsilon_\gamma \langle c_\gamma \rangle^\gamma}{} \tag{52}$$

In addition to the volume averaging methods and the simple heuristic arguments given earlier, a number of other techniques, including variational theory [149], macrotransport [48], effective medium [341], and others, have been applied to determining transport coefficients in porous media. Many theories originally developed for estimating thermal conductivity [34], electrical conductivity [23], or magnetic permeability [149] can be applied to diffusion, since the mathematical treatment of the problem is essentially common to all, at least for the dilute-solution limit of mass transfer. Some of these theories, along with equations originally developed for diffusion processes, are reported later. The derived expressions are reported here in terms of diffusion coefficients and contain the volume fraction of one of the two phases as the structural parameter.

Neal and Nader [260] considered diffusion in homogeneous isotropic medium composed of randomly placed impermeable spherical particles. They solved steady-state diffusion problems in a unit cell consisting of a spherical particle placed in a concentric shell and the exterior of the unit cell modeled as a homogeneous media characterized by one parameter, the porosity. By equating the fluxes in the unit cell and at the exterior and applying the definition of porosity, they obtained

$$\frac{D'_{\text{eff}}}{D_0} = \frac{2\varepsilon}{3 - \varepsilon} \tag{53}$$

This equation is identical to the Maxwell [236,237] solution originally derived for electrical conductivity in a dilute suspension of spheres. Hashin and Shtrikman [149] using variational theory showed that Maxwell's equation is in fact an upper bound for the relative diffusion coefficients in isotropic medium for any concentration of suspended spheres and even for cases where the solid portions of the medium are not spheres. However, they also noted that a reduced upper bound may be obtained if one includes additional statistical descriptions of the medium other than the void fraction. Weissberg [419] demonstrated that this was indeed true when additional geometrical parameters are included in the calculations. Batchelor and O'Brien [34] further extended the Maxwell approach.

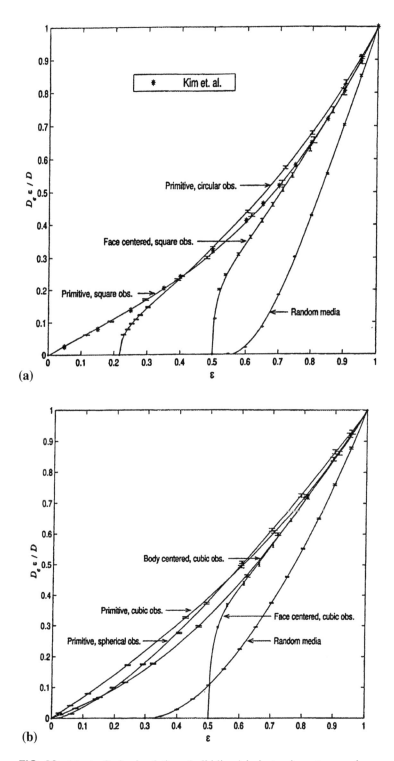

FIG. 22 Monte Carlo simulations (solid lines) in isotropic systems and comparison to results of volume averaging theory (horizontal bars). (Reproduced with kind permission of Kluwer Academic Publishers from Ref. 398, Fig. 6 and 8, Copyright Kluwer Academic Publishers.)

Lord Rayleigh [310] modeled transport in a homogeneous suspension of spheres placed in a square lattice. In terms of diffusion coefficients, his solution was

$$\frac{D'_{\text{eff}}}{D_0} = \frac{2\varepsilon - 0.39381(1-\varepsilon)^{10/3}}{3 - \varepsilon - 0.3938(1-\varepsilon)^{10/3}} \tag{54}$$

A number of researchers, such as Perrins et al. [291], have followed the method first described by Lord Rayleigh to include transport in other geometries.

Jeffrey [181] estimated the effective conductivity in a dilute suspension of spherical particles, and in terms of diffusion coefficient his solution was

$$\frac{D'_{\text{eff}}}{D_0} = 1 - 1.5(1-\varepsilon) - 0.88(1-\varepsilon)^2 \qquad \varepsilon \to 1 \tag{55}$$

Meredith and Tobias [240] found

$$\frac{D'_{\text{eff}}}{D_0} = \frac{2\varepsilon + 1.227(1-\varepsilon)^{7/3} - 1.600(1-\varepsilon)^{10/3}}{3 - \varepsilon + 1.227(1-\varepsilon)^{7/3} - 1.600(1-\varepsilon)^{10/3}} \tag{56}$$

Archie [23] examined electrical resistivity of various sand formations having pore spaces filled with saline solutions of different salt concentrations. Based upon his own experimental results, he obtained a simple relationship for the conductivity of beds of sand (assuming the sand itself is nonconductive) containing saline solution in terms of the porosity. In terms of diffusion coefficients his expression is

$$\frac{D'_{\text{eff}}}{D_0} = \varepsilon^m \tag{57}$$

where m is an empirically determined coefficient that varies with the degree of consolidation of sand. Using concepts from fractal media, Alder [6] was able to derive Archie's law. Bruggeman [54] in 1935 extended work by Maxwell [236,237] and Lorentz [221,222] to conductivities in dispersions composed of concentrated random-size spherical particles. When the dispersed phase is nonconducting, the coefficient m in Archie's equation is equal to 3/2. Meredith and Tobias [240] noted that neither Maxwell's nor Bruggemann's equations yield satisfactory results for their experimental data on conductivities in emulsions at high particle concentrations. Accounting for the interactions of the fields around the particles of the dispersed phase, they proposed a simple approximation valid for elliptical or spherical particles. If the dispersed phase is nonconducting, their equation in terms of the diffusion coefficients is

$$\frac{D'_{\text{eff}}}{D_0} = \frac{1+\varepsilon}{2 + (W-1)(1-\varepsilon)} \frac{2\varepsilon}{2\varepsilon + W(1-\varepsilon)} \tag{58}$$

where W is a function of particle geometry; for spheres, W is 3/2. This equation proved to be much more successful in predicting conductivities in water–propylene carbonate emulsions at high particle concentrations as reported by Meredith and Tobias [240].

Prager [302] examined diffusion in concentrated suspensions using the variational approach. (A discussion of the basic principles in variational theory is given in Ref. 6.) Prager's result is applicable to a very general class of isotropic porous media. Prager's solution for a limiting case of a dilute suspension of particles was

$$\frac{D'_{\text{eff}}}{D_0} = \frac{\varepsilon}{2}(\varepsilon + 1) \qquad \varepsilon \to 1 \tag{59}$$

Akanni et al. [9] employed Monte Carlo simulations for diffusion in porous solids consisting of overlapping randomly placed spheres in a near-cubical volume and compared their results to all of the preceding correlations. They concluded that the correlations proposed by Maxwell [236,237], Bruggeman [54], Prager [302], and Weissberg [183,376] agree well with their results. These models are appropriate for weakly consolidated porous media. And although Akanni et al. [9] intended their calculations for intermediate degree of consolidation, their assumptions involving the structure of the porous media are closer to weakly rather than strongly consolidated media, thus explaining the particular match between the results. Weissberg reformulated the calculations by Prager so as to apply the variational theory specifically to a bed of spherical particles. The calculations were simplified by considering an idealized bed in which centers are randomly situated without restricting the spheres to nonoverlapping locations. Other investigators [187,188] have further presented improved derivations based upon the variational theory approach first formulated by Hashin and Shtrikman [149].

Perrins et al. [291] model for a square lattice of solid cylindrical fibers was

$$\frac{D'_{eff}}{D_0} = \frac{1 - 2\phi}{1 + \phi - \dfrac{0.305827\phi^4}{1 - 1.402958\phi^8} - 0.013362\phi^8} \tag{60}$$

where ϕ is the volume fraction of fibers $(1 - \varepsilon)$. The results obtained from this equation were found to be in good agreement with the experimental data obtained from measuring the conductivities in unit cells constructed according to the specified geometry. It is also interesting to note that the numerical results of this equation match very well those of volume averaging. Since most of the analytical solutions are applicable to particle media exhibiting insignificant overlap, the need to account for overlapping was stressed by Milton et al. [246], who extended the work of Perrins et al. [291] to cylinders placed in square arrays. Tomadakis and Sotirchos [397] used random walk simulation to derive transport coefficients in media composed of random arrays of freely overlapping cylinders of various orientation distributions. They found their results to be in good agreement with the analytical solution that Perrins et al. [291] derived for circular cylinders placed in square and hexagonal arrays.

Other expressions, such as

$$\tau = 1 - \frac{1}{2}\ln \varepsilon \tag{61}$$

$$\tau = \frac{1}{\varepsilon^2} \tag{62}$$

have been reported by Weissberg [343], who applied the variational approach of Prager [302] to a bed of randomly overlapping spheres, and Dullien [55], respectively.

Figure 23 shows some of the one-parameter theories just discussed as functions of the porosity. It must be noted that all of the theories have been derived using different assumptions, so care must be used in applying them to experimental results. The model of Maxwell appears to give an upper limit, as was predicted from variational theory. Many of the models shown in this figure are very close, and it may be difficult to measure diffusion coefficients accurately enough to distinguish between some of these theoretical results. The Mackie–Mears model, discussed next for the case of a solute with magnitude of order of the obstacle size, predicts much lower effective diffusion coefficients than all of the other models in this figure.

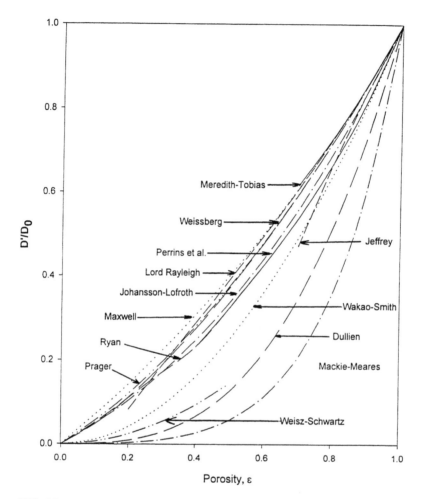

FIG. 23 Comparison of various one-parameter diffusion models. (Reproduced with permission from Ref. 448, "Analysis of Diffusion and Structure in Polyacrylamide Gels by Nuclear Magnetic Resonance," M.S. Thesis, Florida State University, Copyright 1997, Brigita Penke.)

For the case where the solute of interest is of a similar size to the obstructions, a stochastic approach was employed by Mackie and Mears [227], who considered the diffusion of electrolytes in a cation-exchange resin membrane. They modeled the resin phase as a collection of impenetrable objects organized in cubic lattice with one axis parallel to the direction of diffusion. The resin is an obstacle to the diffusing molecule, thus increasing the path length over which the molecule must travel. The lattice spacing is considered comparable to the average size of the diffusion jump. Based upon the probabilistic arguments, they derived an expression for tortuosity:

$$\tau = \frac{1 + \phi}{1 - \phi} \tag{63}$$

They also derived an expression relating mobility to tortuosity, so

$$\frac{D_{\text{eff}}}{D_0} = \frac{1}{\tau^2} \tag{64}$$

By combining these two expressions, one can relate the diffusion coefficients to the polymer volume fraction:

$$\frac{D_{eff}}{D_0} = \left(\frac{1 - \phi}{1 + \phi}\right)^2 \tag{65}$$

The Mackie–Mears expression has been extensively used in the analysis of diffusion in polymers where it is assumed that the obstacles, i.e., the polymer fibers, are of the same order of magnitude as the radius of the solute.

D. Hydration Effects

Wang [415] considered diffusion of water molecules in elliptical-shaped protein medium. The much larger protein molecules obstruct the path of the water molecule, thus increasing its effective diffusional path. Wang derived effective diffusion coefficients by solving the steady-state diffusion problem and obtaining the diffusional flux. He also considered hydration effects caused when water molecules attached to the proteins do not contribute to the diffusional flux. Accounting for this effect is more difficult, because of the time factor introduced by the rate of exchange of labeled water molecules between the bound and free water. In order to incorporate this effect into the water diffusion equation, Wang suggested solving the transient diffusion problem. Furthermore, he concluded, based upon his own experimental evidence, that the rate of exchange of labeled molecules between the bound and free water states can be considered as instantaneously fast. The steady-state solution of Wang's equation was

$$\frac{D'_{eff}}{D_0} = 1 - \alpha\phi \tag{66}$$

where α is the mean value of the structural parameters of the ellipsoid. Muhr et al. [256] noted that the equation of Wang [415] can be derived as a special case of Fricke's treatment, published 30 years earlier. Fricke [123] investigated the conductivity of a suspension of homogeneous spheroids and derived an expression relating the conductivity of the suspension, the suspending medium, and the suspended spheroids to the geometry and concentration of the spheroids. The result for the case when the conductivity of the spheroids is zero can be related to the diffusion in a media consisting of suspended stationary spheroids:

$$\frac{D'_{eff}}{D_0} = \frac{1 - \phi}{1 + \phi/x} \tag{67}$$

where ϕ is the volume fraction of the spheroids, $x = 2$ for spheres, x approaches 0 in the limit for oblate spheroids, and x approaches 3/2 in the limit for prolate spheroids. When $x = 2$, this equation becomes equal to Maxwell's equation, as x goes to infinity, this approaches Wang's equation. Fricke compared his analytical solution to the experimental results for the conductivity of the blood of dog and found an excellent agreement for concentration from 10 to 90%.

Derbyshire and Duff [93] used NMR self-diffusion measurements of water in agarose gels to determine the amount of hydrated water. Using Wang's theory they found hydration numbers of about 0.35 g of water per gram of agarose. Penke et al. [290] utilized the volume averaging method, including the effects of water binding, to analyze NMR experimental data on water self-diffusion and longitudinal relaxation in polyacrylamide gels. Increases in the concentration of cross-linker did not affect the diffusion coefficient of water in the gels; however, it had a substantial effect on the relaxation properties. This effect could be interpreted by the fact that increases in the more hydrophobic Bis cross-linker causes less water to interact with the gel. The effective diffusion tensor in this analysis was found to be given by

$$
\underline{D}_{\text{eff}} = \frac{D_\alpha}{\varepsilon_\alpha + K^{\text{eq}}\varepsilon_\beta}\left[(\varepsilon_\alpha + \kappa\varepsilon_\beta)\underline{I} + \frac{1}{V_T}\int_{A_{\beta\gamma}} \right.
$$

$$
\left. \times (1/2)(\underline{n}_{\alpha\beta}\underline{g} + \underline{gn}_{\alpha\beta})\,dA + \frac{\kappa}{V_T}\int_{A_{\beta\gamma}} (1/2)(\underline{n}_{\alpha\beta}\underline{h} + \underline{hn}_{\alpha\beta})\,dA \right] \tag{68}
$$

where K^{eq} represents the effects of equilibrium partitioning between phases. The advantage of the approach reported by Penke et al. [290] is that both diffusion and T_1 relaxation could be accounted for within one self-consistent theoretical framework of the volume averaging approach.

E.   Effect of Solute Size

For any diffusing molecule larger than a point, the volume fraction inaccessible to the solute is larger than the volume fraction of the polymer, since the diffusing molecules occupy a certain volume in the continuous phase. As mentioned in a previous section, Ogston [269] derived an expression for the space available to the diffusing molecule in a network consisting of randomly oriented, overlapping straight fibers using purely geometrical and statistical arguments. Using the Ogston expression developed for partitioning and probabilistic arguments on diffusive transport in porous media, Ogston et al. [270] derived an equation for diffusion and sedimentation of large particles in polymer solutions (or gels) with a stochastic method. The result was

$$
\frac{D_{\text{eff}}}{D_0} = \exp(-\phi^{0.5}(r_c + r_p)/r_p) \tag{69}
$$

This equation was verified experimentally by comparing the radii of the polymer chains obtained from migration data with those predicted by the equation. The results were found to be consistent with the values deduced from the equilibrium measurements of proteins between free solutions and cross-linked gels for hyaluronic acid [205], dextran [271], and polyacrylamide [251]. In addition, the underlying idea of spaces in a uniform random suspension of fibers has been supported by statistical mechanical calculations and Monte Carlo simulations of Limbach et al. [209] and Giddings et al. [135]. It can be noted that Eq. (69) does not predict that $D_{\text{eff}}/D_0 = f$, where f is given by the volume fraction of gel available for the given solute [see, for example, Eqs. (4) and (5)].

Many investigators have used the Ogston model and its fundamental idea as a basis for their models. Most recently, Johansson and Elvingson [182] obtained the probability distribution $g(r)$ for spaces in a random suspension of fibers; i.e., the probability that a randomly chosen point in a network of fibers is found at a radial distance r to the fiber of closest approach. For a cylindrical cell (CC) model, which consists of an infinite cylindrical cell, containing solvent and polymer, with the polymer represented as a rod centered in the cell, they obtained $g(r)$ for one cylindrical cell as

$$
g^{CC}(r) = \frac{2(r + r_p)}{R_c^2}\,S(r - (R_c - r_p)) \tag{70}
$$

where r_p is the polymer radius and R_c is the cell radius, which is a function of the local polymer concentration. $S(r - (R_c - r_p))$ is a step function, which is equal to 1 if $r \leq (R_c - r_p)$ and 0 if $r > (R_c - r_p)$. The total probability distribution is obtained by integrating over all cells with different radii:

$$g(r) = \int_{r_p}^{\infty} f(R_c)g^{CC}(r)\, dR_c \tag{71}$$

where $f(R_c)$ is the weighting function for each cell. If $g(r)$ is known by measurement or by assuming a pore distribution, $f(R_c)$ can be calculated by deconvolution. Once the frequency function is known, the effective diffusion coefficients can be calculated from the local diffusion coefficients given by the Fick's first law in the cylindrical cell models. An analytical solution for cylindrical cell model assuming Ogston's expression for the probability distribution of straight polymer chains was obtained as

$$\frac{D_{eff}}{D_0} = e^{-\alpha} + \alpha^2 e^{\alpha}E_1(2\alpha) \tag{72}$$

and α and E_1 are given by the following expressions:

$$\alpha = \varphi(r_s + r_p)^2/r_p^2$$

$$E_1 = \int_x^{\infty} e^{-u}/u\, du$$

where r_s is the radius of the solute. A simplified expression for their theory was later found by fitting the data from Monte Carlo simulations to an exponential curve:

$$\frac{D_{eff}}{D_0} = \exp\left(-0.84\alpha^{1.09}\right) \tag{73}$$

Subsequent work by Johansson and Lofroth [183] compared this result with those obtained from Brownian dynamics simulation of hard-sphere diffusion in polymer networks of wormlike chains. They concluded that their theory gave excellent agreement for small particles. For larger particles, the theory predicted a faster diffusion than was observed. They have also compared the diffusion coefficients from Eq. (73) to the experimental values [182] for diffusion of poly(ethylene glycol) in k-carrageenan gels and solutions. It was found that their theory can successfully predict the diffusion of solutes in both flexible and stiff polymer systems. Equation (73) is an example of the so-called stretched exponential function discussed further later.

Diffusion of flexible macromolecules in solutions and gel media has also been studied extensively [35,97]. The Zimm model for diffusion of flexible chains in polymer melts predicts that the diffusion coefficient of a flexible polymer in solution depends on polymer length to the $-1/2$ power, $D \sim N^{-1/2}$. This theoretical result has also been confirmed by experimental data [97,122]. The reptation theory for diffusion of flexible polymers in highly restricted environments predicts a dependence $D \sim N^{-2}$ [97,122,127]. Results of various MC simulations and semianalytical theories for diffusion of flexible polymers in random porous media, which have been summarized [35], indicate that the diffusion coefficient in random three-dimensional media follows the Rouse behavior ($D \sim N^{-1}$ dependence) at short times, and approaches the reptation limit ($D \sim N^{-2}$ dependence) for long times. By contrast, the diffusion coefficient follows the reptation limit for a highly ordered media made from infinitely long rectangular rods connected at right angles in three-dimensional space (like a 3D grid).

F. Hydrodynamic Interactions

The hydrodynamic drag experienced by the diffusing molecule is caused by interactions with the surrounding fluid and the surfaces of the gel fibers. This effect is expected to be significant for large and medium-size molecules. Einstein [108] used arguments from the random Brownian motion of particles to find that the diffusion coefficient for a single molecule in a fluid is proportional to the temperature and inversely proportional to the frictional coefficient by

$$D = \frac{kT_a}{f} \tag{74}$$

where k is Boltzmann's constant, T_a is the temperature, and f is the friction coefficient. (See also Refs. 66, 189, 238, and 401 for derivations and further discussion of Brownian motion.) Strictly speaking, this equation is applicable only in the limit of infinite dilution, since no account has been taken of any correlation of the motion of one particle with another. For a sphere moving in a continuum or a large spherical particle in a solvent of low relative molecular mass [363], the frictional coefficient based upon Stokes' [375] solution of the equation of motion for a Newtoniam fluid is given by

$$f = 6\pi\mu r_s \tag{75}$$

where μ is the viscosity coefficient and r_s is the radius of the sphere. Combining these equations, one arrives at the well-known Stokes–Einstein equation:

$$D = \frac{kT_a}{6\pi\mu r_s} \tag{76}$$

It was suggested by Sutherland [363] that for a sphere diffusing through a medium consisting of molecules of comparable size to the diffusant, the friction coefficient is equal to

$$f = 4\pi\mu r_s \tag{77}$$

The Stokes–Einstein equation can be successfully used to explain diffusion under the following conditions [401], where (a) the diffusing molecule is large with respect to the molecules defining the medium, (b) the medium has a very low viscosity, and (c) no solute–solvent interactions occur.

Thus, the Stokes–Einstein equation is expected to be valid for colloidal particles and suspensions of large spherical particles. Experimental evidence supports these assumptions [101], and this equation has occasionally been used for much smaller species.

The Stokes–Einstein equation predicts that $D\mu/T_a$ is independent of the solvent; however, for real solutions, it has long been known that the product of limiting interdiffusion coefficient D_{12} for solutes and the solvent viscosity decreases with increasing solute molar volume [401]. Based upon a large number of experimental results, Wilke and Chang [437] proposed a semiempirical equation,

$$D = 7.4 \times 10^{-8} \frac{\beta M T_a}{\mu V^{0.6}} \tag{78}$$

where M is the molecular weight of the solvent (g/mol) and V is the molar volume (mL/mol) of the diffusant at the normal boiling point. The association factor β is equal to 2.6 if the solvent is water and unity for nonassociated solvents.

Cukier [87], using an effective medium-type approach, analyzed the diffusion of Brownian spheres in two semidilute polymer solutions: The first was composed of long rodlike polymers, and the second was a random coil solution. He obtained an expression for the normalized diffusion coefficient as an exponential function of the screening constant and solute radius,

$$\frac{D_{\text{eff}}}{D_0} = \exp\left(-kr_s\right) \tag{79}$$

where k is the screening constant of the semidilute polymer solution; $k = (fn/\mu_0)^{0.5}$, f is the friction coefficient, and n is the number density of the polymer rods or monomers. The hydrodynamic interactions were examined using the Navier–Stokes equation in the presence of polymer networks. Since the radius of the rod or monomer appears in the friction parameter, Cukier's solution can be written as

$$\frac{D_{\text{eff}}}{D_0} = \exp\left(-Kr_s c^{1/2}\right) \tag{80}$$

where K is a constant and c is the concentration of the polymer rods in g/mL. Although Cukier based his derivation on hydrodynamic interactions, the expression for rodlike media has the same form as the Ogston et al. [270] model. Johansson et al. [184] compared their experimental results for diffusion in rodlike and coil solutions with the Cukier model that predicts that diffusion is faster in a solution of rods than in a solution of random coils. They found faster diffusion in random coil solutions, in contradiction to Cukier's results.

Phillies [296] observed that several authors [87,270] using different underlying arguments have obtained stretched exponential form for diffusion in presence of polymer matrix:

$$\frac{D_{\text{eff}}}{D_0} = \exp\left(-\alpha\phi^\xi\right) \tag{81}$$

where α and ξ are constants. Based upon the hydrodynamic interactions between the diffusing molecule and the polymer and on scaling concepts of polymer chemistry, de Gennes [127] suggests $\xi = 1$ for $M_p < 50,000$, $\xi \propto M_p^{-0.25}$ for $50,000 < M_p < 500,000$, and $\xi = 0.5$ for $M_p > 500,000$, and $\alpha \propto (M_a M_p)^{0.5}$. Phillies [293] compared the stretched exponential function to a wide range of literature data on diffusion of particles and molecules in polymer gels and solutions. He found that the data generally fit the model with ξ in the range of 0.54–1.86 and that this coefficient varied with molecular weight to the -0.25 power. He further developed methods to derive the form of Eq. (81) and to account for both the concentration of the polymer and the size of the diffusing molecule [210,294,295].

It is interesting to note that Johansson and Lofroth [183] found ξ to equal 1.09 for the diffusion analysis, but the partition coefficient followed a streched exponential with varying ξ from 1 to 2, indicating that the ratio of diffusion coefficient is not equal to the accessible volume fraction for large molecules, as is assumed in the ORMC model for gel electrophoresis.

Amsden [14] used the Ogston model coupled with probability argument to find

$$\frac{D_{\text{eff}}}{D_0} = \exp\left(-\pi\left[\frac{r_p + r_r}{\alpha\phi^{-.75}C_\infty^{-.25}(1 - 2\chi)^{-.75} + 2r_r}\right]^2\right) \tag{82}$$

where χ is the Flory–Huggins interaction parameter and C_∞ is the characteristic ratio of the polymer. This equation is restricted to good solvents (i.e., water for hydrogels), larger distances between cross-links than the polymer persistence length, and fully swollen gels. He showed that this model worked well for a range of solutes from 1.9 to 53.5 Å in diameter in polyethylene oxide, polyvinyl alcohol, and dextran gels.

Altenberger and Tirrell [11] utilized the Langevin equation for particle motion coupled with hydrodynamics described by the Navier–Stokes equation to determine particle diffusion coefficients in porous media given by

$$\frac{D_{eff}}{D_0} = 1 - A\phi^{1/2} - B\phi \tag{83}$$

where A and B are function of the solute and obstacle sizes. This equation is valid for small volume fractions, ϕ, of the porous media. This group extended their results for diffusion and developed results for suspension viscosity using mean-field theory [12,13].

For heterogeneous media composed of solvent and fibers, it was proposed to treat the fiber array as an effective medium, where the hydrodynamic drag is characterized by only one parameter, i.e., Darcy's permeability. This hydrodynamic parameter can be experimentally determined or estimated based upon the structural details of the network [297]. Using Brinkman's equation [49] to compute the drag on a sphere, and combining it with Einstein's equation relating the diffusion and friction coefficients, the following expression was obtained:

$$\frac{D_{eff}}{D_0} = \frac{1}{1 + \dfrac{r_s}{\sqrt{k}} + \dfrac{1}{3}\left(\dfrac{r_s}{\sqrt{k}}\right)^2} \tag{84}$$

Brinkman's equation is a semiempirical modification of the Navier–Stokes equation whereby an additional term to account for the forces exerted by the porous media is added to the forces acting on the fluid. (Note that Whitaker showed that this correction could be derived through the application of volume averaging methods to the equation of motion in porous media [424,436]). Phillips et al. [298] compared this result with more rigorous calculations using a generalized Taylor dispersion model. They concluded that the agreement between the results is best when the fiber to particle radii are of the same dimensions. Poor agreement was observed for other solute–fiber radii ratios and less uniform fiber arrangements.

On the basis of hydrodynamic arguments first suggested by Brady, Johnson et al. [186] proposed that the hydrodynamic and steric effects that influence the diffusivity of a macromolecule in a fibrous medium can be separated into two multiplicative factors. The overall functional dependence is of the form

$$\frac{D_{eff}}{D_0} = F(r_s/\sqrt{k})S(f) \tag{85}$$

where $F(r_s/\sqrt{k})$ is the contribution from the hydrodynamic drag, k is Darcy's permeability, and $S(f)$ is the steric effect. The hydrodynamic effect is approximated here using Brinkman's result, although other theories could be used as well. The other factor is a steric tortuosity effect, or obstruction effect, discussed previously. The steric factor S can be obtained using the models of Perrins et al. [291], Johansson and Lofroth [183,184], volume averaging [436], or other methods mentioned earlier. Johnson et al. [186] noted that use of Perrins model in Eq. (85) agrees with the Phillips et al. [298] rigorous calculations.

It is also useful to note that other approaches to describe diffusion in solvent–polymer systems have been developed using free-volume theory [408–410].

G. Experimental Methods and Comparison to Theories

Several studies have been performed to measure diffusion of a range of molecules in poly-acrylamide [131,289,290,394], agarose [47,138,186], and other gels [29,130,145,176]. Pavesi and Rigamonti [289] used pulsed-field gradient NMR to measure water self-diffusion in a number of polyacrylamide gels with different concentrations of acrylamide and Bis. They did not compare their data to any theoretical models; however, they did observe a log-linear decrease in water self-diffusion as the amount of cross-linker increased and a decrease in diffusion coefficient with the diffusion time of the NMR experiment. Figure 24 shows data on water self-diffusion in polyacylamide gels and solutions of un-cross-linked monomers obtained by Penke et al. [290]. This figure clearly shows that the water self-diffusion in the gels is slower than the solution of un-cross-linked monomers and that the diffusion coefficients decrease with increasing total monomer. Not shown in this figure is the result that variation of Bis content has no effect on the water self-diffusion in the cross-linked gels and the fact that no time dependence on diffusion coefficients was observed. The theoretical models shown for reference in this figure indicate that the Maxwell relationship is an upper limit, and the Mackie–Meares relationship is a lower limit. Both the

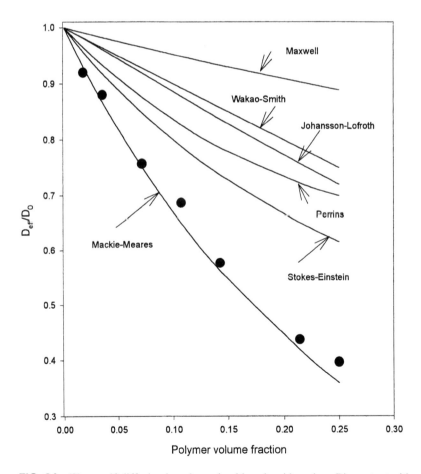

FIG. 24 Water self-diffusion in polyacrylamide gels with various Bis content with comparison to various models from the literature. (Reprinted from Ref. 290, Copyright 1998, Academic Press.)

Perrin model and the Johansson and Elvingston model fall above the experimental data. Also shown in this figure is the prediction from the Stokes–Einstein–Smoluchowski expression, whereby the Stokes–Einstein expression is modified with the inclusion of the Einstein–Smoluchowski expression for the effect of solute on viscosity. Penke et al. [290] found that the Mackie–Meares equation fit the water diffusion data; however, upon consideration of water interactions with the polymer gel, through measurements of longitudinal relaxation, adsorption interactions incorporated within the volume averaging theory also well described the experimental results. The volume averaging theory had the advantage that it could describe the effect of Bis on the relaxation within the same framework as the description of the diffusion coefficient.

Tokita et al. [394] measured diffusion of different molecules with molecular weights varying from 18 to 342 in polyacrylamide gels with constant percentage cross-linker and varying total acrylamide concentration. They found the data to be in good agreement with the stretched exponential of the form

$$\frac{D_{\text{eff}}}{D_0} \propto \exp(-M^{1/3}\phi^{3/4}) \tag{86}$$

Johnson et al. [186] measured diffusion of fluorescein-labeled macromolecules in agarose gels. Their data agreed well with Eq. (85), which combined the hydrodynamic effects with the steric hindrance factors. Gibbs and Johnson [131] measured diffusion of proteins and smaller molecules in polyacrylamide gels using pulsed-field gradient NMR methods and found their data to fit the stretched exponential form

$$\frac{D_{\text{eff}}}{D_0} = \exp\left(-3.2r_p^{0.53}T\right) \tag{87}$$

where T is the concentration of acrylamide. Brown and Stilbs [52] measured diffusion coefficients of ethylene glycol and crown ether in cellulose. They found a first-order decay in diffusion coefficient as the cellulose content increased in the range of 20–55% (w/w). Boyer and Hsu [47] measured protein diffusion (molecular weights 16 kDa to 225 kDa) in Sepharose gel particles using a moment analysis of chromatographic pulses. They found that their results follow generally the stretched exponential of Cukier. The stretched exponential equation therefore represents diffusion behavior for a wide range of different-size molecules in polymer gels with different total monomer concentration.

Balcom et al. [29] have developed a method to use one-dimensional nuclear magnetic resonance imaging to measure mutual diffusion coefficients of paramagnetic species in aqueous gels. This method is different from the pulsed-field gradient method employed by others [130,131,289,394], in that the spatial variation in concentration is determined by measurement of a spatial variation in T_1 relaxation time within the gel. This method may be useful to study spatial distributions of diffusing molecules in the gels; however, the method is limited by MRI resolution of at best 20–50 micron pixels. Therefore, for systems with heterogeneity below this scale, average diffusion coefficients will be obtained.

Another important factor in diffusion measurements that is often encountered in NMR experiments is the effect of time on diffusion coefficients. For example, Kinsey et al. [195] found water diffusion coefficients in muscles to be time dependent. The effects of diffusion time can be described by transient closure problems within the framework of the volume averaging method [195,285]. Other methods also account for time effects [204,247,341].

V. ELECTROPHORESIS

The observation that a charged particle or macromolecule will move in an electric field, a phenomenon called *electrophoresis*, has been well known since the experiments of Reuss in 1809 [312]. Since this beginning, a plethora of experimental and theoretical approaches have been taken to utilize and understand electrophoresis. For example, the molecular size, structure, and charges of proteins have been studied using electrophoresis, and electrophoresis has been used to analyze, isolate, purify, and separate proteins from mixtures [18]. Abramson [1] reviewed the early literature on protein electrophoresis, with emphasis on the early experimental work on electrophoresis in free solution, i.e., without any solid matrices. Andrews [18] reviewed the current literature on electrophoresis, with emphasis on electrophoresis in polyacrylamide and agarose gels. More recent work on the electrophoresis of elongated macromolecules, primarily nucleic acids, is reviewed by Zimm and Levine [447]. This section will consist of a review of the general approaches taken to describe electrophoretic motion in free solution and in gels or other structured media. The emphasis will be on the transport of biological (primarily proteins) macromolecules; however, general principles will also be discussed in order to elucidate the relationships among gel structure, macromolecular properties, and electric field conditions that govern the molecular motion in the gel environment under the influence of an applied electric field.

A. Electrophoresis in Solution

Much of the early work focused on the experimental and theoretical analysis of colloidal particle electrophoresis in free, unbounded solutions. For general reviews in this area see Refs. 141,172,220, and 281–284. Table 4 summarizes various expressions developed for the electrophoretic mobility of a spherical colloidal particle with a uniform distribution of charge spread over the surface of the colloid. *Electrophoretic mobility* is defined as the velocity of the particle due to an applied electric field divided by the electric field strength. For a single isolated charged particle, the electrophoretic mobility is determined by the resultant of the Stoke's drag force and the coloumbic force of the electric field. The mobility is directly proportional to the particle's intrinsic charge and inversely proportional to its radius and the viscosity of the solution. Through combination of the Stoke's–Einstein relationship and the foregoing description of the electrophoretic mobility, the Nernst–Einstein equation [261] gives the relationship between electrophoretic mobility and diffusion coefficient for dilute solution and small molecules:

$$D = RT_a u \tag{88}$$

where R is the universal gas constant, T_a is the absolute temperature, and u is the electrophoretic mobility. This relationship is very important for understanding some of the early arguments for electrophoretic mobilities in gels, as discussed in a later section.

When a charged particle is placed in aqueous media, however, the mobility may no longer be proportional to the intrinsic particle charge, since free counterions in solution will associate and move with the particle and thereby alter the net force exerted on the particle by the electric and fluid flow fields. The region of free or mobile counterions surrounding the particle has been termed the *electrical double layer* or *ionic atmosphere*.

In order to describe the effects of the double layer on the particle motion, the Poisson equation is used. The Poisson equation relates the electrostatic potential field to the charge density in the double layer, and this gives rise to the concepts of zeta-potential and surface of shear. Using extensions of the double-layer theory, Debye and Huckel, Smoluchowski,

TABLE 4   Theoretical Equations for the Transport of a Single
Spherical Particle in a Constant Electric Field

Stokes Law: no effect of ionic atmosphere

$$u = \frac{ze}{6\pi\eta a} = \frac{ze}{kT_a/D}$$

Helmholtz-Smoluchowski: $\kappa a > 100$

$$u = \frac{\varepsilon\xi}{\eta} \qquad q = 4\pi\varepsilon\xi a(1 + \kappa a)$$

Huckel: $\kappa a < 0.1$

$$u = \frac{2\varepsilon\xi}{3\eta} \qquad q = 4\pi\xi a\varepsilon(1 + \kappa a)$$

Henry: entire range of κa, $\Psi e/kT_a < 1$, $\Psi < 25$ mV

$$u = \frac{2\varepsilon\xi}{3\eta} f_1(\kappa a) \qquad q = 4\,\pi\varepsilon\xi a(1 + \kappa a)$$

$$f_1(\kappa a) = 1 + \frac{(\kappa a)^2}{16} - \frac{5(\kappa a)^3}{48} - \frac{(\kappa a)^4}{96} + \frac{(\kappa a)^5}{96}$$

$$- \left[\frac{(\kappa a)^4}{8} - \frac{(\kappa a)^6}{96}\right] \exp(\kappa a) \int_{\infty}^{\kappa a} \frac{\exp(-t)}{t}\, dt$$

Michov's simplification of Henry's function

$$f_1(\kappa a) = \frac{1 + \kappa a + (\kappa a)^2}{1 + \kappa a}$$

Overbeek and Booth: includes relaxation effect

$$u = \frac{2\varepsilon\xi}{3\eta}[f_1(\kappa a) + cub2\xi + c_3\xi^2 + c_4\xi^3 + \cdots] \qquad q = 4\pi\varepsilon\xi a(1 + \kappa a)$$

Gorin: includes effect of counterion size

$$u = \frac{2\varepsilon\xi}{3\eta} f_1(\kappa a) \qquad q = \frac{4\pi\varepsilon\xi a(1 + \kappa a + \kappa a_{ion})}{1 + \kappa a_{ion}}$$

and Henry have found the mobility to be inversely proportional to the viscosity and directly proportional to the potential at the surface of shear; i.e., the zeta-potential [157,284].

The Huckel equation describes the case where the double layers are large in relation to the particle size; in this case the surface of shear lies inside the double layer. The Helmholtz–Smoluchowski equation describes the case of small double layers in relationship to particle size; in this case the surface of shear is approximately at the particle surface. In both cases the net particle charge is related to the zeta-potential by integrating the Poisson equation from the particle surface to the surface of shear; the net charge therefore accounts for all the ions in the double layer and not just the intrinsic charge of the c olloid surface. The Huckel and Helmholtz–Smoluchowski equations in Table 4 represent limiting cases of the more general Henry model for very small particles and very large particles, respectively.

Henry [157] solved the steady-flow continuity and Navier–Stokes equations in spherical geometry, neglecting inertial terms but including pressure and electrical force terms, coupled with Poisson's equation. The electrical force term in Henry's analysis consisted of the sum of the externally applied electric field and the field due to the double layers. His major assumptions are low surface potential (i.e., potentials less than approximately 25 mV) and undistorted double layers. The additional parameter κa appearing in the Henry model consists of a, the particle geometric radius, and κ, the inverse of the radius of the surface of shear. It should be noted that κ is a strong function of ionic strength. At high ionic strength the double layer around the colloid is small (Helmholtz-Smoluchowski limit) and hence κ is large, and at low ionic strength the double layer is large (Huckel limit) and hence κ is small. Henry derived a correction term that gives the proper limits of the Huckel and Helmholtz–Smoluchowski theories. This term varies from 1.0 to 1.5, and Michov [243,244] gives a simplified equation for the complex term derived by Henry, as shown in Table 4.

Overbeek and Booth [284] have extended the Henry model to include the effects of double-layer distortion by the relaxation effect. Since the double-layer charge is opposite to the particle charge, the fluid in the layer tends to move in the direction opposite to the particle. This distorts the symmetry of the flow and concentration profiles around the particle. Diffusion and electrical conductance tend to restore this symmetry; however, it takes time for this to occur. This is known as the *relaxation effect*. The relaxation effect is not significant for zeta-potentials of less than 25 mV; i.e., the Overbeek and Booth equations reduce to the Henry equation for zeta-potentials less than 25 mV [284]. For an electrophoretic mobility of approximately 10×10^{-4} cm^2/V-sec, the corresponding zeta potential is 20 mV at 25°C. Mobilities of up to 20×10^{-4} cm^2/V-s, i.e., zeta-potentials of 40 mV, are not uncommon for proteins at temperatures of 20–30°C, and thus relaxation may be important for some proteins.

Gorin has extended this analysis to include: (1) the effects of the finite size of the counterions in the double layer of spherical particles [137], and (2) the effects of geometry, i.e. for cylindrical particles [2]. The former is known as the Debye–Huckel–Henry–Gorin (DHHG) model. Stigter and coworkers [348,369–374] considered the electrophoretic mobility of polyelectrolytes with applications to the determination of the mobility of nucleic acids.

Many other workers have expended considerable effort in extending the theoretical treatment for single-particle motion. For example, Weirsema et al. [417] extended the Debye–Huckel analysis for spherical colloids to potentials of 125 mV by solving the Poisson–Boltzmann equation numerically. O'Brien and White [265] developed a more general numerical scheme that would work with the highest experimentally observable potentials of 250 mV. They found that the mobility reaches a maximum with increasing potential. Ohshima et al. [272] developed a semiempirical formula for similar cases valid for κa greater than 10. Hermans [158] derived an expression for the mobility of a porous sphere. Ohshima and Kondo [273] found an expression for the case where the colloid has fixed uniform charge distribution within a surface layer and where the ions from solution could penetrate this layer. Ivory [179] developed a model for the transient response of a dielectric sphere with a thin double layer to step changes in the electric field. Volkel and Noolandi [405,406] and Muthukumar [257] have developed expressions for mobilities of flexible and stiff polyelectrolytes. Thus, much theory has been developed, but correspondingly little experimental work has been done to test the various theories. This is particularly true for testing the theories with data on proteins.

An important reason for this lack of experimental work is that the zeta-potential cannot be easily determined independent of the electrophoretic mobility [284]; however, in the case of proteins (as well as some other charged colloids), the intrinsic charge obtained by titration is a parameter that can be measured independent of the electrophoretic mobility. The charge obtained from electrophoretic measurements (i.e., the net charge) via the preceding theories is generally not the same as the charge obtained from titration (i.e., the intrinsic charge). This has commonly been attributed to the binding of counterions to the protein's surface; titration measures only proton, or hydronium ion, equilibria that arise from the charged amino acid groups on the protein surface. Metals, other cations, and many anions are known to bind to a wide range of proteins, and they will affect the net charge and thus the electrophoretic motion of the protein.

For example, Barlow and Margoliash [33] showed that phosphate, chloride, iodide, and sulfate, in decreasing order of effect, reduced the electrophoretic mobility of human cytochrome c at pH 6.0 by up to a factor of 2. The cations lithium, sodium, potassium, and calcium had no effect. It is possible to account for the binding equilibria of these counterions so that the titration and electrophoresis results can be compared; however, in many of the early electrophoresis experiments these data were not available and relevant conditions were not recorded or controlled. For general discussions on the extensive field of ligand binding to proteins, see Cantor and Schimmel [60] and van Holde [403].

The charge of a number of proteins has been measured by titration. The early experimental work focused on the determination of charge as a function of pH; later work focused on comparing the experimental and theoretical results; the latter obtained from the extensions of the Tanford–Kirkwood models on the electrostatic behavior of proteins. Edsall and Wyman [104] discuss the early work on the electrostatics of polar molecules and ions in solution, considering fundamental coulombic interactions and accounting for the dielectric properties of the media. Tanford [383,384], and Tanford and Kirkwood [387] describe the development of the Tanford–Kirkwood theories of protein electrostatics. For more recent work on protein electrostatics see Lenhoff and coworkers [64,146,334].

Example studies of protein charge include: ribonuclease [385,386], horse hemoglobin [279,280], lysozyme [388], a range of marine mammals and horse myoglobin [234,354,355], bovine serum albumin [389], and human hemoglobin [234,235]. However, only in a few cases have the titration data been directly compared to the electrophoretic measurements in order to directly test the theories of electrophoresis. Ovalbumin and lysozyme [384] are two cases where this has been done. Qualitative, but not quantitative, agreement between titration and electrophoresis was observed for these two proteins. According to Tanford, the data on the binding of counterions were sparse; however, even when the data were available, he concluded that the buffer ion binding was not the only reason for the lack of quantitative agreement. Figure 25 shows an example comparison of mobility and titration data for a range of pH values [98].

Waldmann-Meyer [412] has proposed a useful method for evaluating the net charge on a protein that eliminates the need for titration data. By determining the change in electrophoretic mobility for a protein with charged counterions, he was able to calculate the net protein charge and ligand binding constant. For the case of human serum albumin with cadmium ion binding equilibria at pH 5.95, he found a difference of less than 2% between his measured ratio of charge to mobility and the ratio calculated using the DHHG model. Although Waldmann-Meyer suggested that this demonstrates the validity of the DHHG model for proteins, this approach has yet to be applied to a wide range of proteins or even to a single protein over a wide range of pH values.

Douglas et al. [98] have measured protein (serum albumin, ovalbumin, and hemoglobin) mobilities over a range of pH values using a free-flow electrophoresis apparatus and a particle electrophoresis apparatus. They found good agreement between the two measurements; however, they also found some differences between their measurements and those reported in the older literature. They attributed the differences to the use of moving-boundary electrophoresis methods in the early experimental work and to differences in

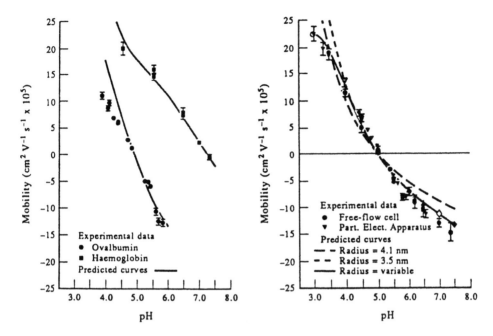

FIG. 25 Titration theory vs mobility charge. (Reprinted from Ref. 98, Copyright 1995, with permission from Elsevier Science.)

buffer and salts. Using titration data from the older literature to determine protein charge and a correction factor for the binding of chloride ions, they calculated the protein mobility using Henry's equation for electrophoretic mobility. They found the mobility to agree with their experimental data over a range of pH values for all three proteins. Figure 25 shows mobility versus pH measurements and comparison with theory for two proteins from the work of Douglas et al. [98].

Chae and Lenhoff [64] have developed a method to determine the free solution mobility of proteins taking into account the details of the protein shape and charge distribution. Using boundary integral formulation, including the velocity distribution, the equilibrium electrostatic potential around the molecule, and the potential distribution due to the applied electric field, they found good agreement between the theoretical predictions and the data for lysozyme and ribonuclease A. Extension of this approach to a wide range of pH values and protein types for comparison of mobility-charge data would be very useful and interesting.

For further discussion of experimental methods for determination of electrophoretic titration curves of proteins, see the recent study by Gianazza et al. [129]. For discussion of the free solution mobility of DNA see Stellwagen et al. [368].

B. Electrophoresis of Noninteracting Solutes in Gels

1. Ogston–Morris–Rodbard–Chrambach Theory

Morris [250] extended the Ogston model to gel electrophoresis. He found empirically that the product of parameters (lv) for gel filtration was proportional to the total monomer concentration, T, and therefore

$$\ln (K_{av}) = -k_o T \tag{89}$$

where k_o can be determined from the Ogston equation; however, it is usually used as an empirical constant.

The electrophoretic mobilities in gels are generally related to the mobilities in free solution by the empirical logarithmic Ferguson equation [18,115,154]

$$\ln \frac{u}{u_0} = -K_R T \tag{90}$$

where u is the mobility in the gel, u_0 is the mobility in free solution, i.e., at zero concentration of the gel, and K_R is the retardation constant. The retardation constant has been found to depend upon pH, ionic strength, buffer type, degree of cross-linking, and polymerization [18]. It was originally found empirically that the retardation constant was linearly related to the molecular weight; however, other work subsequently indicated that the square root of the retardation constant is proportional to the cubic root of the molecular weight and is also proportional to the Stokes radius of the protein [329]. This has been widely used to determine the molecular weights or Stoke's radii for unknown proteins from a set of known standard proteins.

Because of the similarity in form between the Eqs. (89) and (90), and since diffusional transport occurs in both electrophoresis and gel filtration, Morris [250] reasoned that

$$K_{av} = \frac{u}{u_0} = \frac{D}{D_0} \tag{91}$$

The mobility ratio equal to the diffusion ratio in this equation would naturally follow from application of the Nernst–Einstein equation, Eq. (88), to transport gels. Since the Nernst–Einstein equation is valid for low-concentration solutes in unbounded solution, one would expect that this equation may hold for dilute gels; however, it is necessary to establish the validity of this equation using a more fundamental approach [215,219]. (See a later discussion.) Morris used a linear expression to fit the experimental data for mobility [251]

$$\frac{u}{u_0} = aK_{av} + m \tag{92}$$

where a and m are adjustable coefficients. Morris and Morris [251] studied the gel filtration and gel electrophoresis of eight proteins in polyacrylamide gels at constant temperature, pH, and ionic strength. They found a to vary from 0.76 to 0.90 and m to be very small. The constant a is a function of the amount of cross-linker in the gel; for cross-linking above 5%, a was close to 0.9. The intercept m was found to be related to the water content of the gel and thus also related to the degree of cross-linking. They did not investigate the effect of electric field on Eq. (92).

Rodbard and Chrambach [328] suggested using the equation

$$\frac{u}{u_0} = K_{av}^a \tag{93}$$

where a is an empirically determined parameter. They found the constant a in Eq. (93) to be 0.7 for a 5% cross-linked polyacrylamide gel [77,329]. According to Rodbard and

Chrambach this indicates that the effective protein size for gel filtration is larger than the effective size for gel electrophoresis. They concluded that this could not be accounted for by gel swelling, pH, or ionic strength effects. Biefer and Mason [36] found the constant a in Eq. (93) to be 0.93. They measured the conductance of cellulose acetate filter pads with porosities from 0.5 to 0.9 in solutions of 10^{-2} M KCl.

Rodbard and Chrambach [77,329] developed a computer program that allows the determination of molecular parameters, i.e., free mobility, molecular radii, molecular weight, and charge or valence, from measured electrophoretic mobilities in gels with different monomer concentrations. For a set of mobility versus gel concentration data they used the Ferguson [18,115,154] equation to obtain the retardation constant from the negative slope and the free mobility from the extrapolated intercept. From the retardation constant they determined the molecular radius using

$$(K_R)^{1/2} = \alpha(r_p + r_r) \tag{94}$$

where α has been determined by fitting a set of known standard proteins. Equation (94) follows directly from the Ferguson equation (90), the definition of K_{av}, and Eq. (5). The molecular weight was determined by

$$M_{wt} = 4/3 r_p^3 \pi N_{av}/v_p \tag{95}$$

where N_{av} is Avogadro's number and v_p is an assumed molecular volume. Rodbard [326,327] and Chrambach [75] give comprehensive reviews of these and other methods for protein molecular weight determination. In addition to molecular weight, the other major property of proteins is charge. Extensive studies and applications of this approach have been reported by Chrambach [75] and Tietz [391].

To determine the charge, the Gorin version of the modified Henry equation was used. Rodbard and Chrambach [330] suggested that due to the assumptions and limitations in the calculations, the determined charge could be in error by as much as a factor of 2, and since the Gorin model had not been experimentally justified, they remained skeptical about charge predictions. Their original article stated that the charges determined for bromophenol blue and methyl green were in good agreement with their chemical structures; however, data for proteins were not cited. Rodbard and Chrambach [330] also noted that extreme caution should be used in comparing the free mobilities from their program, which relies on data from moving-boundary electrophoresis in gels, to data from free-solution measurements.

Mobility data versus monomer concentration have been found to deviate from the form implied by the Ferguson equation [56]. The mobility of proteins and other spherical subcellular particles in agarose gels were found to deviate from the Ferguson equation at low values of monomer concentration. These deviations were attributed to changes in gel fiber properties with gel concentration [56,392]. Additional studies have also concluded that the fiber properties of polyacrylamide also change with both monomer concentration and degree of cross-linking of the gel. This implies that the determination of free electrophoretic mobilities by extrapolating to zero gel concentration can lead to serious errors.

The OMRC does not account for the orientation or structure of the probe species, nor does it account for the interconnectedness (i.e., the possibility of percolation in the gel whereby certain regions of the gel may be inaccessible to a particular species) of the matrix [361]. The OMRC also assumes a uniform electric field and does not consider the effects of the gel on the electric field.

2. Tortuous-Path and Barrier Theories

The tortuous-path and barrier theories consider the effects of the media on the electrophoretic mobility in a way similar to the effect of media on diffusion coefficients discussed in a previous section of this chapter. The tortuous-path theory seeks to determine the effect of increased path length on electrophoretic mobility. The barrier theory considers the effects of the barrier or media conductivity on the electrophoretic mobility.

Giddings and coworkers [44,45,134] expanded the barrier and tortuous-path theories for electrophoresis in porous membranes originally developed by Tsieluis and Synge [379]. The barrier theory applies to electrophoretic transport in a medium where all the obstructions are conductors; i.e., to the case where the barriers are completely permeable to the background electrolyte and completely impermeable to the migrant [44]. In the barrier theory the electric field was found to have an effect on the mobility ratio. This was the first theoretical case in the literature where the electric field was observed to affect the mobility ratio. This effect could be relatively large for the extreme case assumed in barrier theory; however, in the more realistic intermediate case between the barrier theory and the tortuous-path theory, this effect is much smaller [134]. The tortuous-path theory applied to transport in a medium where all the obstructions are complete insulators [45]. Boyack and Giddings [44] applied this approach to the data of Biefer and Mason [36], with good qualitative agreement. Giddings and Boyack [134] also extended these concepts to the cases intermediate to the two extremes.

Boyack and Giddings [44,45,133,134] considered several cases of electrophoresis in porous media that are of interest in the present context. In the case of a single binary equilibrium between the two species

$$A_1 \rightleftharpoons A_2 \qquad K = \frac{k_{12}}{k_{21}} \tag{96}$$

where k_1 and k_2 are the kinetic constants for the reversible reaction and A_1 and A_2 have different electrophoretic mobilities given by $m_i = z_i D_i / RT_a$. An effective diffusion coefficient for the mixture was found by simple arguments to be

$$\frac{D'_{\text{eff}}}{D_0} = \left[\frac{E}{kT_a}\right]^2 \frac{(z_1 D_1 - z_2 D_2)^2}{k_{12} K (1 + 1/K)^3 D_0} \tag{97}$$

with $D_0 = x_1 D_1 + x_2 D_2$. An order of magnitude analysis leads to

$$\frac{D'_{\text{eff}}}{D_0} = [200 E^2 / k_{12}] D_0 \tag{98}$$

For cases of univalent ions, the magnitude of this quantity was found to be small; however, for large macromolecules with high surface charge, there could be significant electric-field-induced dispersion.

In a more extensive development of the tortuous-path and barrier theories, Boyack and Giddings [45] considered the transport of solute in a simple geometrical system similar to that used in the diffusion analysis of Michaels [241] but with added tortuosity effects. The effective mobility in this system was found to be

$$\frac{u'_{\text{eff}}}{u_0} = \frac{u_{\text{eff}} \varepsilon}{u_0} = \frac{A(L + 1)}{(A + L)(AL + 1)} \left(\frac{L}{L_e}\right)^2 = \frac{\theta}{\tau'} \qquad L = l_1/l_2, A = a_1/a_2 \tag{99}$$

Boyack and Giddings [45] considered the tortuosity effects to be separable from the constriction effects. In their derivation they assumed that the electric field in the constricted channel decreased proportionally to the decrease in cross-sectional area and changes in path length. The field was assumed not to penetrate the barrier. The effect of constriction can be written in terms of porosity and L as

$$\theta = \frac{1}{\varepsilon} \frac{\varepsilon(L + 1) - L}{L[(L + 1)\varepsilon - L] + 1} \tag{100}$$

Figure 26 shows plots of this constriction factor verses porosity for various values of L. As observed in Figure 21 for the Michaels' model of diffusion in such a cell, the percolation limits are seen where the constriction factor goes to zero at $\varepsilon = L/(1 + L)$.

Trinh et al. [399] derived a number of similar expressions for mobility and diffusion coefficients in a similar unit cell. The cases considered by Trinh et al. were: (1) electrophoretic transport with the same uniform electric field in the large pore and in the constriction, (2) hindered electrophoretic transport in the pore with uniform electric fields, (3) hydrodynamic flow in the pore, where the velocity in the second pore was related to the velocity in the first pore by the overall mass continuity equation, and (4) hindered hydrodynamic flow. All of these four cases were investigated with two different boundary condi-

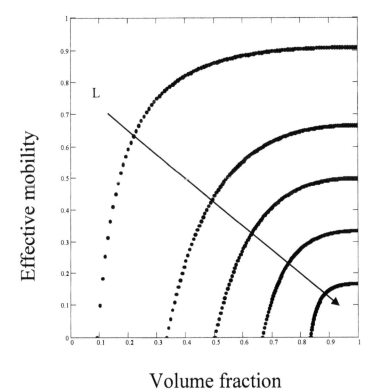

Volume fraction

FIG. 26 Effective mobility versus porosity for various length ratios using Eq. (100). (Based on Ref. 45.)

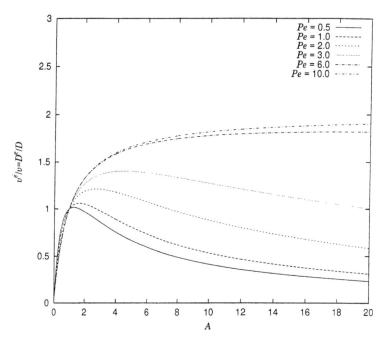

FIG. 27 Effective transport coefficient versus porosity from the model of Trinh et al. (Reproduced with permission from Ref. 399.)

tions at the external boundaries of the model system. For the case of electrophoretic transport they found

$$\frac{u_{\text{eff}}}{u} = \frac{D_{\text{eff}}}{D} = \frac{A(L+1)}{AL+1} \frac{1 - \exp\left(-Pe(1+1/L)\right)}{1 - \exp\left(-Pe\right) + A\exp\left(-Pe\right) * \left[1 - \exp(-Pe/L)\right]} \quad (101)$$

where $Pe = ul_1/D$ is a measure of the strength of the applied electric field. Figure 27 shows plots of the mobility ratio as a function of porosity for given L and Pe. These curves differ from those of Giddings and Boyack and those of Michaels in that a maximum is observed in the effective transport coefficient at a given porosity, and this maximum becomes more pronounced as the parameter L increases, and there is an obvious dependence upon the electric field through the Pe.

Combining hindered diffusion theory with the diffusion/convection problem in the model pore, Trinh et al. [399] showed how the effective transport coefficients depend upon the ratio of the solute to pore size. Figure 28 shows that as the ratio of solute to pore size approaches unity, the effective mobility function becomes very steep, thus indicating that the resolution in the separation will be enhanced for molecules with size close to the size of the pore. Similar results were found for the effective dispersion, and the implications for the separation of various sizes of molecules were discussed by Trinh et al. [399].

Giddings and Boyack further extended their analysis to account for different geometries [45] and matrices that were able to conduct the current and partially affect the solute transport [134]. They assumed that the overall tortuosity could be decoupled into the purely constriction effects and the purely tortuosity effects through

$$\frac{u'_{\text{eff}}}{u_0} = \frac{\theta}{\tau'} \tag{102}$$

For a system on nonconducting spherical particles of radius r, Boyack and Giddings found that the constriction and tortuosity factors are given by

$$\theta = (\varepsilon)^{-1}\left[1 - \delta^{1/3} + \frac{4\pi}{\sqrt{4/\pi - \delta^{2/3}}} \tan^{-1} \frac{\delta^{1/3}}{\sqrt{4/\pi - \delta^{2/3}}}\right]^{-1} \qquad \delta = 6(1 - \varepsilon)/\pi \tag{103}$$

and

$$\tau' = [1 + 0.178(1 - \varepsilon)]^2 \tag{104}$$

Similar expressions for cubic obstacles and cylindrical fibers were also determined, and Figure 29 shows the results for beds of fibers [45,134].

3. Volume Averaging

The volume averaging approach discussed in the section on diffusive transport can also be extended to account for electrophoresis [215] and hydrodynamic flow [215,436]. Locke [215] considered the application of volume averaging to the determination of the effective

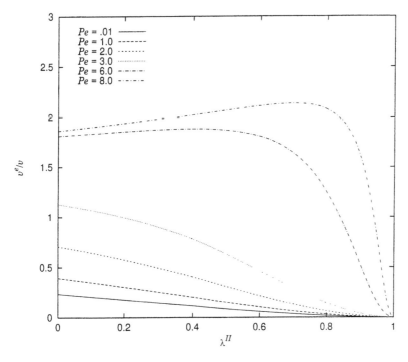

FIG. 28 Effective electrophoretic mobility in the case of hindered diffusion. (Reproduced with permission from Ref. 399.)

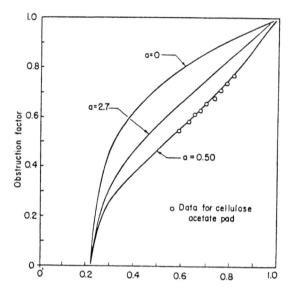

FIG. 29 Obstruction factor from barrier theory of Boyack and Giddings for fibers. (Reprinted with permission Ref. 45, Copyright 1963, Academic Press.)

mobility and dispersion coefficients in uncharged porous media upon the application of an electric field. The following discussion outlines the basic results from that study. Consider a two-phase medium consisting of an α phase and a β phase as shown in Figure 30. The point species molar continuity equation, including the Nernst–Planck expression for the molar flux for the α phase, is given by

$$\frac{\partial c_{i\alpha}}{\partial t} + \nabla \cdot c_{i\alpha}\underline{v}_\alpha = \nabla \cdot (D_{i\alpha}\nabla c_{i\alpha} + u_{i\alpha}c_{i\alpha}\nabla\Psi_\alpha) \tag{105}$$

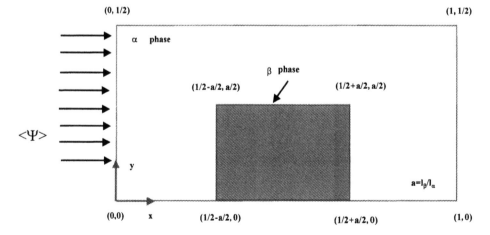

FIG. 30 Unit cell for transport in anisotropic media. (Reprinted with permission from Ref. 215, Copyright 1998, American Chemical Society.)

Examples of the electrophoretic mobility, u_i, as functions of the molecular properties, e.g., solute size, charge, and shape, and solution conditions, were discussed in a previous section. For the two-phase system considered in Figure 30, the flux and equilibrium boundary conditions at the interface between the α and β phases are given by

$$c_{i\alpha} = c_{i\beta} \qquad \underline{n}_{\alpha\beta} \cdot \underline{N}_{i\alpha} = \underline{n}_{\alpha\beta} \cdot \underline{N}_{i\beta} \qquad \text{on } A_{\alpha\beta} \tag{106}$$

where we have assumed for simplicity a unit partition coefficient and have neglected mass transfer resistance across the interface between the two phases [435]. In order to solve for the concentration field, the velocity and the electrostatic potential must be known. In general the velocity, the electrostatic potential, and the species concentration are coupled through the equation of motion, Eq. (5), with body forces due to the applied electric field, and the Poisson equation, for the electrostatic potential. In order to illustrate the methodology and to apply the volume averaging method to cases where the medium is uncharged, the solute of interest is considered to be in the dilute-solution limit, the current-carrying ions are assumed to be uniformly distributed [217], and the electric field has negligible effects on the hydrodynamics. A single solute is thus considered, and therefore all species indices will be dropped in further equations. These assumptions will allow for a sequential coupling of the equation of motion for the velocity field and the Poisson equation, reduced to the Laplace form, for the electrostatic potential to the species continuity equation in a fashion similar to that used by Sauer et al. [345] for modeling electrokinetic transport in a single-phase system, i.e., a capillary tube, with parallel and orthogonal applied fields by area averaging in the tube.

(a) Electric Field. The electrostatic potential in the system, assuming electroneutrality, neglecting surface charges on the medium at the α–β interphase and any contributions of the solute of interest to the electric field, can be determined following the methodology developed for the analogous heat conduction problem by Nozad et al. [264]. The point values of the electrostatic potentials in the two phases are given by

$$\nabla^2 \Psi_\beta = 0 \qquad \nabla^2 \Psi_\alpha = 0 \tag{107}$$

with boundary conditions

$$\Psi_\beta = \Psi_\alpha, \qquad k_\beta \underline{n}_{\alpha\beta} \cdot \nabla \Psi_\beta = k_\alpha \underline{n}_{\alpha\beta} \cdot \nabla \Psi_\alpha \qquad \text{on } A_{\beta\alpha} \tag{108}$$

$$\Psi_\beta = W(t) \quad \text{on } A_{\beta e} \qquad \Psi_\alpha = Y(t) \quad \text{on } A_{\alpha e} \tag{109}$$

where k_α and k_β are the electrical conductivities of the two phases, $A_{\alpha e}$ and $A_{\beta e}$ represent the areas of exits and entrances to the two phases, $A_{\beta\alpha}$ represents the bounding surface between the α and β phases, and $W(t)$ and $Y(t)$ are prescribed functions at the entrances and exits. As noted by Nozad et al. [264], use of the boundary conditions [Eq. (109)] implies that the radius of the averaging volume, r_0, is bounded by the length scale of the macroscopic domain, L, and the length scale of each phase, l_α or l_β, by $l_\alpha, l_\beta \ll r_0 \ll L$. A one-equation model for the overall average potential was shown [264] to be given by

$$\underline{\underline{K}}_{\text{eff}} : \nabla \nabla \langle \Psi \rangle = 0 \qquad \langle \Psi \rangle = \varepsilon_\beta \langle \Psi_\beta \rangle^\beta + \varepsilon_\alpha \langle \Psi_\alpha \rangle^\alpha \tag{110}$$

The effective conductivity tensor can be determined from

$$\frac{\underline{\underline{K}}_{\text{eff}}}{k_\beta} = (\varepsilon_\beta + \varepsilon_\alpha \kappa) \underline{\underline{I}} + \frac{1 - \kappa}{V} \int_{A_{\beta\alpha}} \frac{1}{2} (\underline{n}_{\beta\alpha} \underline{f} + \underline{f} \underline{n}_{\beta\alpha}) \, dA \tag{111}$$

where κ is the ratio of electrical conductivities in the two phases. The functions $\underline{f}$ and $\underline{g}$ are vectors determined from the solutions in a suitably defined unit cell representation of the porous medium of the closure problem given by

$$\nabla^2 \underline{f} = 0 \in V_\beta, \qquad \nabla^2 \underline{g} = 0 \in V_\alpha \tag{112}$$

with boundary conditions given by

$$\underline{f} = \underline{g} n_{\beta\alpha} \cdot \nabla \underline{f} = \kappa \underline{n}_{\beta\alpha} \cdot \nabla \underline{g} + (\kappa - 1)\underline{n}_{\beta\alpha} \qquad \text{on } A_{\beta\alpha} \tag{113}$$

$$\underline{f}(\underline{r} + \underline{l}_i) = \underline{f}(r) \qquad i = 1, 2, 3, \ldots$$

$$\underline{g}(\underline{r} + \underline{l}_i) = \underline{g}(\underline{r}) \qquad i = 1, 2, 3, \ldots \tag{114}$$

and for spatially periodic media

Analytical solutions for the closure problem in particular unit cells made of two concentric circles have been developed by Chang [68,69] and extended by Hadden et al. [145]. In order to use the solution of the potential equation in the determination of the effective transport parameters for the species continuity equation, the deviations of the potential in the unit cell, defined by

$$\Psi_\alpha = \tilde{\Psi}_\alpha + \langle \Psi_\alpha \rangle \qquad \Psi_\beta = \tilde{\Psi}_\beta + \langle \Psi_\beta \rangle \tag{115}$$

and the average potential over the macroscopic region, determined through solution of the preceding equations, are needed. The deviations can be determined from the $\underline{f}$ and $\underline{g}$ fields and the average potential by

$$\tilde{\Psi}_\beta = \underline{f} \cdot \nabla \langle \Psi \rangle \qquad \tilde{\Psi}_\alpha = \underline{g} \cdot \nabla \langle \Psi \rangle \tag{116}$$

Note that for a macroscopically isotropic medium, the tensor given by Eq. (111) has equal elements along the diagonal, and therefore Eq. (110) is equivalent to

$$\nabla^2 \langle \Psi \rangle = 0 \tag{117}$$

Extensive studies of anisotropic diffusion have also been reported [193,266,267].

Determination of the effective transport coefficients, i.e., dispersion coefficient and electrophoretic mobility, as functions of the geometry of the unit cell requires an analogous averaging of the species continuity equation. Locke [215] showed that for this case the closure problem is given by the following *local* problems:

$$D_\alpha \nabla^2 \underline{G}_1 + u_\alpha(\underline{I} + \nabla \underline{g}) \cdot \nabla \langle \Psi \rangle \cdot \nabla \underline{G}_1 = 0$$

$$D_\alpha \nabla^2 \underline{G}_2 + u_\alpha(\underline{I} + \nabla \underline{g}) \cdot \nabla \langle \Psi \rangle \cdot \nabla \underline{G}_2 = 0 \tag{118}$$

Analogous equations can be derived for the $\underline{F}_1$ and F_2 fields. The boundary conditions are

$$\underline{F}_1 = \underline{G}_1 \qquad F_2 = G_2 \qquad \text{on } A_{\alpha\beta} \tag{119}$$

and

$$\underline{n}_{\beta\alpha} \cdot (D_\beta \nabla \underline{F}_1 + u_\beta(\underline{I} + \nabla \underline{f}) \cdot \nabla \langle \Psi \rangle \underline{F}_1)$$

$$= n_{\beta\alpha} \cdot (D_\alpha \nabla \underline{G}_1 + u_\alpha(\underline{I} + \nabla \underline{g}) \cdot \nabla \langle \Psi \rangle \underline{G}_1) + (D_\alpha - D_\beta)\underline{n}_{\beta\alpha} \tag{120}$$

$$\underline{n}_{\beta\alpha} \cdot [D_\beta \nabla F_2 - \underline{u}_\alpha F_2(\underline{I} + \nabla \underline{g}) \cdot \nabla \langle \Psi \rangle]$$

$$= \underline{n}_{\beta\alpha} \cdot [D_\alpha \nabla G_2] + \underline{n}_{\beta\alpha} - [u_\alpha(\underline{I} + \nabla \underline{g}) - u_\beta(\underline{I} + \nabla \underline{f})] \cdot \nabla \langle \Psi \rangle \tag{121}$$

It is important to note that the closure problem for the species continuity equation requires solutions for the deviations of the potential, i.e., the $\underline{f}$ and $\underline{g}$ fields, *and* knowledge of the average potential $\langle \Psi \rangle$. This result is very similar to that found by the area averaging method in Sauer et al. [345]. Utilizing the closure expressions the average species continuity equation becomes

$$\frac{\partial \langle c \rangle}{\partial t} = \underline{\underline{D}}_{\text{eff}} : \nabla \nabla \langle c \rangle + \underline{\underline{U}}_{\text{eff}} : \nabla \langle c \rangle \nabla \langle \Psi \rangle \tag{122}$$

where the effective transport coefficients are given by the tensors

$$\underline{\underline{U}}_{\text{eff}} = u_\alpha (\underline{\underline{I}} \varepsilon_\alpha + \langle G_2 \nabla \underline{g} \rangle) + u_\beta (\underline{\underline{I}} \varepsilon_\beta + \langle F_2 \nabla \underline{f} \rangle) \tag{123}$$

and

$$\underline{\underline{D}}_{\text{eff}} = (\varepsilon_\alpha D_\alpha + \varepsilon_\beta D_\beta) \underline{\underline{I}} + \frac{D_\alpha - D_\beta}{}\int_{A_{\alpha\beta}} [1/2 \underline{n}_{\alpha\beta} \underline{G}_1$$
$$+ 1/2 \underline{G}_1 \underline{n}_{\alpha\beta}] \, dA + \nabla \langle \Psi \rangle \cdot (\langle \nabla \underline{g} \cdot \underline{G}_1 \rangle u_\alpha + \langle \nabla \underline{f} \cdot F_1 \rangle u_\beta) \underline{\underline{I}} \tag{124}$$

It can be noted that in general this result predicts that the ratio of the dispersion coefficient to the free-solution diffusion coefficient is different from the ratio of the effective mobility to the free-solution mobility. In the case of gel electrophoresis, where it is expected that the β phase is impermeable (i.e., the gel fibers), the medium is isotropic, and the α phase is the space between fibers, the transport coefficients reduce to

$$\underline{\underline{u}}_{\text{eff}} = u_\alpha (\varepsilon_\alpha \underline{\underline{I}} + \langle G_2 \nabla \underline{g} \rangle)$$

$$\underline{\underline{D}}_{\text{eff}} = \underline{\underline{I}} \varepsilon_\alpha D_\alpha + \frac{D_\alpha}{V} \int_{A_{\alpha\beta}} \underline{n}_{\alpha\beta} \underline{G}_1 dA + \nabla \langle \Psi \rangle \cdot \langle \nabla \underline{g} \cdot \underline{G}_1 \rangle u_\alpha \underline{\underline{I}} \tag{125}$$

The standard Rodbard–Ogston–Morris–Killander [326,327] model of electrophoresis which assumes that $u_{\text{eff}}/u_\alpha = D_{\text{eff}}/D_\alpha$ is obtained only for special circumstances. See also Locke and Trinh [219] for further discussion of this relationship. With low electric fields the effective mobility equals the volume fraction. However, the dispersion coefficient reduces to the effective diffusion coefficient, as determined by Ryan et al. [337], which reduces to the volume fraction at low gel concentration but is not, in general, equal to the porosity for high gel concentrations. If no electrophoresis occurs, i.e., u_α and u_β equal zero, the results reduce to the analysis of Nozad [264]. If the electrophoretic mobility is assumed to be much larger than the diffusion coefficients, the results reduce to that given by Locke and Carbonell [218].

Figure 31 illustrates the effect of the Pe ($= (\langle \Psi \rangle /L) u_\alpha 1_\alpha /D_\alpha$) in the range of 1 to 100 on the *y*-component of the effective dispersion coefficient tensor for porosities ranging from 0 to 1 in a porous media consisting of nonconducting obstacles. It is clear that the effective dispersion coefficients increase with increasing electric field; however, the magnitude of the increase is not large. The largest effect appears in the intermediate porosities between 0.5 and 0.8, and at porosities of 0 and 1 the $Pe = 0$ limits are recovered. At $Pe = 0$, the dispersion coefficients reduce to the effective diffusion coefficients for the no-flow case reported by Ryan et al. [337]. The mobility was found to be independent of the Pe and dependent upon only the porosity of the medium. Since, in this case the medium perturbs the lines of constant electrostatic potential in a manner analogous to potential flow, the mobility is reduced only by the reduction in area, or volume, available for transport. Of course, with different type of media and electric fields, especially media that allow the electric field

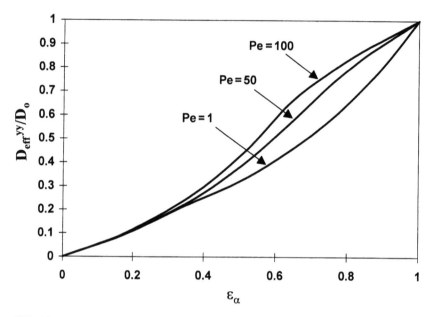

FIG. 31 Effective transport coefficients for unit cell given in Figure 29. (Reprinted with permission from Ref. 215, Copyright 1998, American Chemical Society.)

to propagate through the obstacles, the effective mobility will be a function of the electric field.

4. Hydrodynamic Interactions

Lumpkin [223] extended Hermann's analysis for electrophoresis of porous spheres in free solution to the case of electrophoresis of porous spheres in gel-like media. He applied the Ogston model for the gel structure together with Hermann's hydrodynamic model for the sphere's mobility in order to determine the Ferguson equation as a special case of his more complicated model. He compared his result to only one experimental value and found the predicted value to be below the experimental. He attributed this to an incomplete description of the gel structure. The logarithm of the mobility was found to vary linearly with the concentration of obstacles in the limit of large obstacles; however, at low obstacle concentration, the relationship was convex, with the magnitude of curvature dependent on the Debye–Huckel screening parameter. Figure 32 shows plots of the mobility versus a parameter, σ, which reflects the density of obstacles in the matrix.

5. Macrotransport Theory

Edwards [105] has extended the macrotransport method, originally developed by Brenner [48] and based upon a generalization of Taylor–Aris dispersion theory, to the analysis of electrokinetic transport in spatially periodic porous media. Edwards and Langer [106] applied this methodology to transdermal drug delivery by iontophoresis and electroporation.

6. Monte Carlo Simulations

Monte Carlo simulations of transport by electrophoresis have also been performed. Recent simulations by Slater and Guo [356] have tested the fundamental assumption used in electrophoresis given by Eq. (91). Using Monte Carlo simulations in a two-dimensional periodic gel and a random gel, they concluded that the Ferguson plot is intrinsically non-linear and that the experimental observations of the curvature of this plot are related to the intensity of the electric field and the randomness of the gel fibers. Their simulation shows that the semilog plot of the relative electrophoretic mobility is concave at high gel concentration for the random gel and that this plot is convex for the periodic gel. In both cases the Ferguson equation is valid only in the asymptotic limit at very low gel concentration. This appears to be the reverse of the experimental observations seen by Butterman et al. [56], where the convex deviations from the Ferguson equation were seen at the low arcylamide concentrations.

The electrophoretic mobilities of flexible macromolecules (e.g., DNA, oligonucleotides, and other polymers) in gel media have also been extensively studied by a number of methods, including Monte Carlo simulations [159,165,208,357,358,361,362,447]. In general, the mobility is expected to vary with the length of the polymer to the -1 power ($\mu \sim N^{-1}$); however, there are complicating effects of the applied electric field as well as the

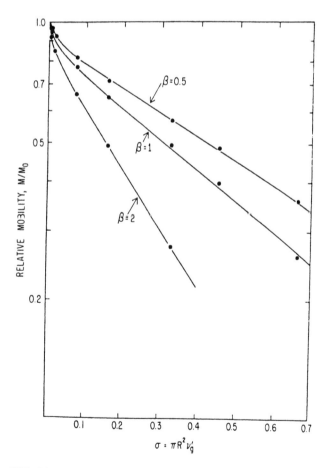

FIG. 32 Effective mobility. (Reprinted with permission of the American Institute of Physics and O. Lumpkin from Ref. 223, Copyright 1984, American Institute of Physics.)

size of the molecules relative to the pores. Biased reptation theory leads to size-independent mobility for very large fields in "tight" gels. Monte Carlo simulations reported for electrophoresis in gels have assumed uniform electric fields unaffected by the presence of the media to any significant degree. Baumgartner and Muthakumar [35] utilized a Debye–Huckel potential on each bead of a chain in combination with excluded volume interactions and the kink-jump technique in Monte Carlo simulations of flexible macromolecules in random media. Effects of media order and nonuniformities in the electric field have not been considered for the transport of these flexible macromolecules.

C. Electrochromatography

Electrochromatography utilizes an electric field applied cocurrently, i.e., down the axial dimension, with the convective flow in a column that is packed with chromatographic media [218,336]. The polarity of the electric field is usually oriented so that the solute(s) of interest are retarded relative to the motion of the hydrodynamic flow. This process is of interest in the current review because the separation process is strongly governed by the type of packing media used. Studies with agarose, polyacrylamide, and other gel particles have been reported and applications to the separation of proteins and small molecular species have been considered. The fundamentals of the transport in such a system are similar to those discussed in a previous section of this review, and further analysis can be carried out using the volume averaging methods introduced earlier. The convective flow can be accounted for using the solution obtained by Whitaker [424] for the average velocity and the deviation of the average velocity in the averaging and closure problems of the species continuity equation.

For the case of a flowing fluid in the α phase, Locke [215] showed that the effective transport equation is given by

$$\frac{\partial \langle c \rangle}{\partial t} + \underline{V}_{\text{eff}} : \nabla \langle c \rangle \langle \underline{v}_\beta \rangle^\beta = \underline{D}_{\text{eff}} : \nabla \nabla \langle c \rangle + \underline{U}_{\text{eff}} : \nabla \langle c \rangle \nabla \langle \Psi \rangle \tag{126}$$

where the effective transport coefficients are given by

$$\underline{D}_{\text{eff}} = (\varepsilon_\alpha D_\alpha + \varepsilon_\beta D_\beta)\underline{I} + \frac{D_\alpha - D_\beta}{V} \int_{A_{\alpha\beta}} [1/2\underline{n}_{\alpha\beta}\underline{G}_1 + 1/2\underline{G}_1\underline{n}_{\alpha\beta}]$$

$$+ \nabla \langle \Psi \rangle \cdot (\langle \nabla \underline{g} \cdot \underline{G}_1 \rangle u_\alpha + \langle \nabla \underline{f} \cdot \underline{F}_1 \rangle u_\beta)\underline{I} \tag{127}$$

$$- \langle G_3 \underline{B} \rangle (\langle v_\alpha \rangle^\alpha)^2 - \langle \underline{B} \cdot \underline{G}_1 \rangle \cdot \langle \underline{v}_\alpha \rangle^\alpha \underline{I}$$

where the last two terms include convective dispersion (Taylor–Aris dispersion [24]) and the effect of partitioning into the α phase, respectively, and

$$\underline{V}_{\text{eff}} = (\underline{I} + \langle \underline{B} F_2 \rangle) \tag{128}$$

and the closure problem for G_3 is given by

$$D_\alpha \nabla^2 G_3 + u_\alpha \nabla G_3 \cdot (\underline{I} + \nabla \underline{g}) \cdot \nabla \langle \Psi \rangle = (1 + \underline{B} : \underline{I}) \tag{129}$$

with boundary conditions

$$\underline{n}_{\beta\alpha} \cdot [D_\alpha \nabla G_3 + u_\alpha G_3(\underline{I} + \nabla \underline{g}) \cdot \nabla \langle \Psi \rangle] = 0 \in A_{\alpha\beta} \tag{130}$$

Equation (130) for the G_3 field can be considered a three-dimensional generalization of the results of Sauer et al. [345] for flow in a tube. For zero electric field this term will recover the Taylor–Aris dispersion result.

The form of the effective mobility tensor remains unchanged as in Eq. (125), which implies that the fluid flow does not affect the mobility terms. This is reasonable for an uncharged medium, where there is no interaction between the electric field and the convective flow field. However, the hydrodynamic term, Eq. (128), is affected by the electric field, since electroconvective flux at the boundary between the two phases causes solute to transport from one phase to the other, which can change the mean effective velocity through the system. One can also note that even if no electric field is applied, the mean velocity is affected by the diffusive transport into the stationary phase. Paine et al. [285] developed expressions to show that reversible adsorption and heterogeneous reaction affected the effective dispersion terms for flow in a capillary tube; the present problem shows how partitioning, driven both by electrophoresis and diffusion, into the second phase will affect the overall dispersion and mean velocity terms.

D. Electrophoretic NMR Spectroscopy

The study of electrophoresis by NMR spectroscopy is a relatively new technique that allows for direct in situ determination of electrophoretic transport parameters. This method can be applied to multicomponent systems [166] consisting of a number of molecular and macromolecular species in a wide range of media. In addition, the study of electro-osmosis [233] and other electric-field-induced molecular motion may be performed using NMR instrumentation. Holz [166] reviews some of the early experiments on electrophoretic NMR and discusses the difficulties arising from interactions of the electrophoretic field with the NMR magnetic field as well as from electric heating and bubble formation at the electrode. A number of different types of NMR probes have been investigated in order to overcome these problems, and recent success by several research groups [85,131,150–152,161,166,167,185,233,338,339] has shown that PFGNMR spectroscopy can indeed be used with high accuracy to determine the electrophoretic mobilities of a wide range of species. Nuclei that may be observed using electrophoretic NMR include ^1H, ^7Li, ^{13}C, ^{19}F, ^{23}Na, ^{29}Si, ^{31}P, ^{87}Rb, and ^{205}Tl [166]. Most work to date has focused on protons and fluorine-labeled species.

Johnson's research group [131,152,161,185,338] has developed high-resolution electrophoretic NMR to determine the mobilities of ions in mixtures and to evaluate the effects of electro-osmosis. In addition, extension to two-dimensional electrophoretic NMR allows for the determination of information on the distribution of diffusion and electrophoresis transport parameters. Gibbs and Johnson [131] have measured electrophoretic mobilities (and diffusion coefficients) for small molecules with different molecular weights and diameters ranging from 0.1 to 5.0 nm in polyacrylamide gels containing different amount of total monomer with fixed percentage of Bis. They found the mobilities (and diffusion coefficients) of these species to follow the stretched exponential function given by Eq. (87). In addition, they also found that the ratio of electrophoretic mobility in the gel to that in free solution equaled the ratio of diffusion coefficient in the gel to that in free solution for a range of polyacrylamide concentrations with fixed Bis content for several molecular species. This is the first rigorous experimental test of Eq. (91) (at least the mobility and diffusion part), and it appears to justify this equation. Further work using pulsed-field NMR to measure both diffusion coefficients and electrophoretic mobilities to test these predictions is needed. Radko and

Chrambach [307] also used the stretched exponential to describe electrophoresis of spherical latex particles in polymer solutions.

Pulsed-field gradient nuclear magnetic resonance experiments [145] have shown that the diffusion coefficients of oligonucleotides (single thymidines [T] to T-30) in Pluronic F127 gels and in free solution decrease with molecular size raised to the -0.5 power, following Zimm [97,446], behavior that is expected for free-solution conditions. This result is significant because it demonstrates that these small to medium-size molecules do not diffuse by reptation in the Pluronic gels. The ratios of the diffusion coefficients in the gels to those in free solution were independent of molecular size and were equal to 0.5. This factor could be accounted for using geometrical arguments from volume averaging theory coupled to experimental results on water self-diffusion in the same system. Volume averaging methods were used to determine effective diffusion coefficients of small molecules in systems of spherical geometry consisting of hydrophobic impermeable cores surrounded by hydrophilic permeable outer regions of the sphere, which, in turn, were surrounded by free bulk water. This theory could describe well the effects of Pluronic concentration on water self-diffusion coefficients. An additional very important and significant result of this work was the fact that the diffusion coefficient ratio, in the molecular size range studied (i.e., 300–9000 daltons), was not equal to the electrophoretic mobility ratio, obtained by capillary electrophoresis [309]. The electrophoretic mobility ratio was, in contrast, a function of the molecular size of the oligonucleotides, and this functional dependence could be described by the semiempirical "stretched exponential" equation [131,145]. This molecular weight dependence of electrophoretic mobility and diffusion coefficient is significantly different from that seen in polyacrylamide gels [131], where the dependence of *both* diffusion and electrophoresis was found to fit the stretched exponential, albeit for different molecular species of smaller molecular weight (19–390) and for one protein (bovine serum albumin).

VI. CONCLUSIONS

Electric-field-driven transport in media made of hydrophilic polymers with nanometer-size pores is of much current interest for applications in separation processes. Recent advances in the synthesis of novel media, in experimental methods to study electrophoresis, and in theoretical methodology to study electrophoretic transport lead to the possibility for improvement of our understanding of the fundamentals of macromolecular transport in gels and gel-like media and to the development of new materials and applications for electric-field-driven macromolecular transport. Specific conclusions concerning electrodiffusive transport in polymer hydrogels include the following.

1.   A wide range of hydrogels with various types of pore structure and size (down to the nanometer scale) have been developed. Many of these materials are based upon polyacrylamide and agarose, and these are the primary materials that have been used in most applications to electrophoresis and, to a lesser degree, chromatography. Molecular templating during hydrogel synthesis can lead to materials with nanometer-scale pores that have been demonstrated to be effective at improving electrophoretic and chromatographic separations.

2.   Experimental data on diffusion of a wide range of molecular size species in hydrogels generally fit stretched exponential functions; however, rigorous justification of these models using nonfitted or independently determined structural parameters of the gels has not been performed. The stretched exponential remains a useful tool for fitting experimental data. Volume averaging methods have been used successfully to analyze the diffu-

sion of small to medium-size molecules in several types of gels, including polyacrylamide and Pluronics.

3.   Much of the experimental data on electrophoresis in gels generally appear to follow the semilogarithmic behavior of the Ferguson equation; however, there are important exceptions for the cases of larger macromolecules and at the limits of more concentrated gels. The method of volume averaging has been used to define the conditions under which the ratio of the diffusion coefficient in the gel to that in solution equals the electrophoretic mobility ratio and the volume fraction of voids.

NOTATION

a_1, a_2, a_3, a_4 = model coefficients, also used as cross-sectional areas [Eq. (36)]

A = area ratio, Eq. (44)

b = external nonelectrical body force in equation of motion, Eq. (22)

C = mass of bis/total monomer, Eq. (6)

C_c = concentration of cross-links in reference state

C_p^k = heat capacity in k phase

c = species concentration

c_I, c_{II} = boundary condition concentrations in Eq. (36)

$\langle c \rangle$ = average species concentration, Eq. (28)

d_{eff} = fractal parameter, Eq. (10)

D = diffusion coefficient

D_0 = diffusion coefficient in solution

D_{eff} = effective diffusion coefficient

$D'_{\text{eff}} = D_{\text{eff}}\,\varepsilon$, Eq. (33)

f = fraction of space available in Ogston model Eq. (4), friction coefficient in Eqs. (74) and (75), closure field in Eq. (49)

F = Faraday's constant

g, h = closure functions in volume averaging method, Eq. (68)

h^2 = mean square end-to-end distance, Eq. (11)

I = unit tensor

I_e^k = current, Eq. (24)

J_i = molar diffusive flux of species i, Eq. (25)

K_{av} = gel chromatography partition coefficient, Eq. (2)

K^{eq} = equilibrium distribution coefficient, Eq. (68)

K_R = retardation coefficient in Ferguson equation

k_0, k_1 = model coefficients

k = Boltzmann constant

k^k = thermal conductivity in k phase

l = 1/2 of fiber length in Ogston model

L = ratio of pore lengths, Eq. (44), or channel length, Eq. (36)

M = molecular weights

$n_{\alpha\beta}$ = unit normal vector, Eq. (49)

N = number of units in chain Eq. (11), number of charged species, Eq. (18)

N_{av} = Avogadro's number

N_i = molar flux of species i, Eqs. (19) and (20)

p = persistence length

p^k = pressure in k phase

Pe = Pectlet number, Eq. (101)

Q = swelling ratio, Eq. (16)

Q^k = heat source in k phase

r_c = effective radius of Bis clusters

r_p = solute particle radius

r_r = radius of polymer fiber

R = gas constant

R_e = electrical resistance

R_c = unit cell radius

t = time

T = total acrylamide content, g/100 mL, Eq. (6)

T_a = absolute temperature

u = electrophoretic mobility

u_{eff} = effective electrophoretic mobility

v = mass average velocity, Eq. (20)

V = volume

V_e = elution volume

V_0 = interstitial volume

V_T = total bed volume

x = model parameter, Eq. (67)

z = charge on solute

z_{BIS} = number of Bis residues in a cluster, following Eq. (16)

Greek

α = model coefficient, Eq. (66)

β = constant, Eq. (17)

v = density of fibers per unit volume

$\bar{v}$ = partial specific volume, Eq. (16)

μ_i = chemical potential of species i, Eq. (12)

μ = viscosity

θ = constriction factor, Eq. (100)

ε = porosity, Eq. (31)

ε_d = dielectric permittivity of the medium, Eq. (18)

ζ = mean number of Bis residues in a cluster, Eq. (6)

κ = Debye screening length

ξ = model parameter

Π = osmotic pressure

ρ = density

τ = tortuosity factor

Φ = swelling ratios

ϕ = network void fraction Eq. (9), volume faction of media ($= 1 - \varepsilon$), Eq. (60)

ϕ_2 = swelling ratio, Eq. (13)

χ = Flory model parameter, Eq. (13)

Ψ = electrostatic potential

ACKNOWLEDGMENTS

Support from the National Science Foundation (BCS-9311901 and BES 9521381) and the National Aeronautics and Space Administration (NAG8-1163) is gratefully acknowledged for work on gel electrophoresis in templated media and NMR analysis of diffusion and electrophoresis, respectively. Support from the Whitaker Foundation for NMR studies of transdermal drug delivery by iontophoresis and electroporation is also gratefully acknowledged. Collaboration with Dr. Randolph Rill and Dr. David Van Winkle as well as their students Dr. Yingjie Liu, Dr. Brian Patterson, and Brian Ramey to study electrophoresis in templated media and Pluronics liquid crystals has also been very successful. Dr. Timothy Moerland and Dr. Stephen Gibbs have been very valuable collaborators in the areas of diffusion and electrophoresis studies in gels and biological media using pulsed-field gradient NMR methods. In addition, I would like to acknowledge Dr. Rubin Carbonell for introducing me to volume averaging methods and guidance during graduate studies and Dr. Stephen Whitaker for discussions on volume averaging methods. My graduate students Ms. Brigita Penke, Mr. Darren Hadden, Mr. John Caban, Mr. Mike Acton, and Dr. Craig Galban and postdoctoral associates Dr. Stephen Kinsey and Dr. Sinh Trinh have contributed invaluably to this work.

REFERENCES

1. Abramson, HA, Electrophoresis of Proteins; Hafner: New York, 1964.
2. Abramson, HA; Moyer; Gorin, MA, Electrophoresis of Proteins, Reinhold: New York, 1942.
3. Ackers, GK, A New Calibration Procedure for Gel Filtration Columns, Journal of Biological Chemistry 242, 3237, 1967.
4. Adam, M; Lairez, D, Sol-gel transition. In: Physical Properties of Polymeric Gels; Cohen Addad, JP, ed.; Wiley: Chichester, UK, 1996; 88.
5. Adamski, RP; Anderson, JL, Configurational Effects on Polystyrene Rejection from Microporosou Membranes, Journal of Polymer Science: Part B: Polymer Physics 25, 765, 1987.
6. Adler, PM, Porous Media, Geometry and Transports; Butterworth-Heinemann: Boston, 1992.
7. Afeyan, NB; Fulton, SP; Regnier, FE, Perfusion Chromatography Packing Materials for Proteins and Peptides, Journal of Chromatography 544, 267, 1991.
8. Afeyan, NB; Gordon, NF; Mazsaroff, I; Varady, L; Fulton, SP; Yang, YB; Regnier, FE, Flow-through Particles for the High-performance Liquid Chromatographic Separation of Biomolecules: Perfusion Chromatography, Journal of Chromatography 519, 1, 1990.
9. Akanni, KA; Evans, JW; Abramson, IS, Effective Transport Coefficients in Heterogeneous Media, Chemical Engineering Science 42, 1945, 1987.
10. Albright, JG; Mills, R, Journal of Physical Chemistry 69, 3120, 1965.
11. Altenberger, AR; Tirrell, M, On the Theory of Self-Diffusion in a Polymer Gel, Journal of Chemical Physics 80, 2208, 1984.
12. Altenberger, AR; Tirrell, M; Dahler, JS, Hydrodynamic Screening and Particle Dynamics in Porous Media, Semidilute Polymer Solutions and Polymer Gels, Journal of Chemical Physics 84, 5122, 1986.
13. Altenberger, FR; Dahler, JS; Tirrell, MV, A Mean-Field Theory of Suspension Viscosity, Macromolecules 18, 2752, 1985.
14. Amsden, B, An Obstruction-Scaling Model for Diffusion in Homogeneous Hydrogels, Macromolecules 32, 874, 1999.
15. Anderson, JL; Kathawalla, IA; Lindsey, JS, Configurational Effects on Hindered Diffusion in Micropores, AIChE Symposium Series 84, 35, 1988.
16. Anderson, JL; Quinn, JA, Restricted Transport in Small Pores, a Model for Steric Exclusion and Hindered Particle Motion, Biophysical Journal 14, 130, 1974.
17. Anderson, TB; Jackson, R, A Fluid Mechanical Description of Fluidized Beds, Industrial and Engineering Chemistry Fundamentals 6, 527, 1967.
18. Andrews, AT, Eletrophoresis Theory, Techniques, and Biochemical and Clinical Applications, 2nd ed.; Clarendon Press: Oxford, 1986.

19. Ansell, RJ; Mosbach, K, Magnetic Molecularly Imprinted Polymer Beads for Drug Radioligand Binding Assay, Analyst 123, 1611, 1998.
20. Anseth, KS; Bowman, CN; Brannon-Peppas, L, Mechanical Properties of Hydrogels and Their Experimental Determination, Biomaterials 17, 1647, 1996.
21. Arbabi, S; Sahimi, M, Elastic Properties of Three-Dimensional Percolation Networks with Stretching and Bond-Bending Forces, Physical Review B 38, 7173, 1988.
22. Arbabi, S; Sahimi, M, Critical Properties of Viscoelasticity of Gels and Elastic Percolation Networks, Physical Review Letters 65, 725, 1990.
23. Archie, GE, The Electrical Resistivity Log as an Aid in Determining Some Reservoir Characteristics, Transactions of the American Institute of Mining Engineers 146, 54, 1942.
24. Aris, R, On the Dispersion of a Solute in a Fluid Flowing Through a Tube, Proceedings of the Royal Society of London Series A 235, 67, 1956.
25. Arshady, R, Beaded Polymer Supports and Gels I. Manufacturing Techniques, Journal of Chromatography 586, 181, 1991.
26. Axelsson, A; Westrin, B; Loyd, D, Application of the diffusion cell for the measurement of diffusion in gels, Chemical Engineering Science 46, 913, 1991.
27. Badiger, MV; McNeill, ME; Graham, NB, Porogens in the Preparation of Microporous Hydrogels based on Poly(ethylene oxides), Biomaterials 14, 1059, 1993.
28. Baker, JP; Hong, LH; Blanch, HW; Prausnitz, JM, Effect of Initial Total Monomer Concentration on the Swelling Behavior of Cationic Acrylamide-Based Hydrogels, Macromolecules 27, 1446, 1994.
29. Balcom, B; Fischer, A; Carpenter, T; Hall, L, Diffusion in Aqueous Gels. Mutual Diffusion Coefficients Measured by One-Dimensional Nuclear Magnetic Resonance Imaging, Journal of the American Chemical Society 115, 3300, 1993.
30. Barcilon, V, Ion Flow Through Narrow Membrane Channels: Part I, SIAM Journal of Applied Mathematics 52, 1391, 1992.
31. Barcilon, V, Singular Perturbation Analysis of the Fokker–Planck Equation: Kramer's Underdamped Problem, SIAM Journal of Applied Mathematics 56, 446, 1996.
32. Barcilon, V; Chen, DP; Eisenberg, RS, Ion Flow Through Narrow Membrane Channels: Part II, SIAM Journal of Applied Mathematics 52, 1405, 1992.
33. Barlow, GH; Margoliash, E, Electrophoretic Behavior of Mammalian-Type Cytochromes c, The Journal of Biological Chemistry 241, 1473, 1966.
34. Batchelor, GK; O'Brien, RW, Thermal or Electrical Conduction Through a Granular Material, Proceedings of the Royal Society of London Series A 355, 313, 1977.
35. Baumgartner, A; Muthukumar, M, Polymers in Disordered Media, Advances in Chemical Physics 94, 625, 1996.
36. Biefer, GJ; Mason, SG, Electrokinetic Streaming, Viscous Flow and Electrical Conduction in Inter-Fiber Networks, The Pore Orientation Factor, Transactions of the Faraday Society 55, 1239, 1959.
37. Bier, M; Mosher, RA; Palusinski, OA, Computer Simulation and Experimental Validation of Isoelectric Focusing in Ampholine-Free Systems, Journal of Chromatography 211, 313, 1981.
38. Bier, M; Palusinski, OA; Mosher, RA; Saville, DA, Electrophoresis: Mathematical Modeling and Computer Simulation, Science 219, 1281, 1983.
39. Binder, K, Monte Carlo and Molecular Dynamics Simulations in Polymer Science; Oxford University Press: New York, 1995.
40. Bird, RB; Stewart, WE; Lightfoot, EN, Transport Phenomena; Wiley: New York, 1960.
41. Bode, HJ, SDS-Polyethyleneglycol Electrophoresis: A Possible Alternative to SDS-Polyacrylamide Gel Electrophoresis, FEBS Letters 65, 56, 1976.
42. Bode, H-J, The Use of Liquid Polyacrylamide in Electrophoresis: III. Properties of Liquid Polyacrylamide in the Presence of Cellulose Acetate, Analytical Biochemistry 92, 99, 1979.
43. Bowman, CN; Peppas, NA, A Kinetic Gelation Method for the Simulation of Free-Radical Polymerizations, Chemical Engineering Science 47, 1411, 1992.
44. Boyack, JR; Giddings, JC, Zone and Boundary Diffusion in Electrophoresis, The Journal of Biological Chemistry 235, 1970, 1960.

45. Boyack, JR; Giddings, JC, Theory of Electrophoretic Mobility in Stabilized Media, Archives of Biochemistry and Biophysics 100, 16, 1963.
46. Boyd, BM; Prausnitz, JM; Blanch, HW, High-Frequency Alternating-Cross-Field Gel Electrophoresis with Neutral or Slightly Charged Interpenetrating Networks to Improve DNA Separation, Electrophoresis 19, 3137, 1998.
47. Boyer, PM; Hsu, JT, Experimental Studies of Restricted Protein Diffusion in an Agarose Matrix, AIChE Journal 38, 259, 1992.
48. Brenner, H; Edwards, DA, Macrotransport Processes; Butterworth-Heinemann: Boston, 1993.
49. Brinkman, HC, Applied Science Research A 1, 27, 1947.
50. Broadbent, SR; Hammersley, JM, Percolation Processes I. Crystals and Mazes, Proceedings of the Cambridge Philosophical Society 53, 629, 1957.
51. Brown, W; Schillen, K; Hvidt, S, Triblock Copolymers in Aqueous Solution Studied by Static and Dynamic Light Scattering and Oscillatory Shear Measurements. Influence of Relative Block Sizes, Journal of Physical Chemistry 96, 038, 1992.
52. Brown, W; Stilbs, P; Lindstrom, T, Self-Diffusion of Small Molecules in Cellulose Gels using FT-Pulsed Field Gradient NMR, Journal of Applied Polymer Science 29, 823, 1984.
53. Brownstein, KR; Tarr, CE, Importance of Classical Diffusion in NMR Studies of Water in Biological Cells, Physical Review A 19, 2446, 1979.
54. Bruggeman, VDAG, Berechnung verschiedener physikalisher Konstanten von heterogenen Substanze 1. Dielektrizitatskonstanten und Leitfahigkeiten der Mischkorper aus isotropen Substanzen, Annalen der Physik 5, 636, 1935.
55. Bullien, FAL, Porous Media; Academic Press: New York, 1979.
56. Butterman, M; Tietz, D; Orban, L; Chrambach, A, Ferguson Plots Based on Absolute Mobilities in Polyarcylamide Gel Electrophoresis: Dependence of Linearity of Polymerization Conditions and Application on the Determination of Free Mobility, Electrophoresis 9, 293, 1988.
57. Caglio, S; Chiari, M; Righetti, PG, Gel Polymerization in Detergents: Conversion Efficiency of Methylene Blue vs. Persulfate Catalysis, as Investigated by Capillary Zone Electrophoresis, Electrophoresis 15, 209, 1994.
58. Caglio, S; Righetti, PG, On the Efficiency of Methylene Blue versus Persulfate Catalysis of Polyacrylamide Gels, as Investigated by Capillary Zone Electrophoresis, Electrophoresis 14, 997, 1993.
59. Caglio, S; Righetti, PG, On the pH Dependence of Polymerization Efficiency, as Investigated by Capillary Zone Electrophoresis, Electrophoresis 14, 554, 1993.
60. Cantor, CR; Schimmel, PR, Biophysical Chemistry Part II: Techniques for the Study of Biological Structure and Function; WH Freeman: San Francisco, 1980.
61. Carbonell, RG; Sarti, GC, Coupled Deformation and Mass-Transport Processes in Solid Polymers, Industrial and Engineering Chemistry Research 29, 1194, 1990.
62. Carbonell, RG; Whitaker, S, Heat and Mass Transfer in Porous Media. In: Fundamentals of Transport Phenomena in Porous Media, Nato ASI Series, Series E: Applied Sciences—No. 82 ed.; Bear, J; Corapcioplu, MY, eds.; Marinus Nijhoff: Dordrecht, The Netherlands, 1984; 121.
63. Carman, PC, Flow of Gases Through Porous Media; Academic Press: New York, 1956.
64. Chae, KS; Lenhoff, AM, Computation of the Electrophoretic Mobility of Proteins, Biophysical Journal 68, 1120, 1995.
65. Chan, VZH; Hoffman, J; Lee, VY; Iatrou, H; Avgeropoulos, A; Hadjichristidis, N; Miller, RD; Thomas, EL, Ordered Bicontinuous Nanoporous and Nanorelief Ceramic Films from Self-Assembling Polymer Precursors, Science 286, 1716, 1999.
66. Chandrasekhar, S, Stochastic Problems in Physics and Astronomy, Reviews in Modern Physics 15, 1, 1943.
67. Chandrasekhar, S, Liquid Crystals, 2nd ed.; Cambridge University Press: Cambridge, 1992.
68. Chang, H-C, Multi-Scale Analysis of Effective Transport in Periodic Heterogeneous Media, Chemical Engineering Communications 15, 83, 1982.
69. Chang, H-C, Effective Diffusion and Conduction in Two-Phase Media: A Unified Approach, AIChE Journal 29, 846, 1983.

70. Charlionet, R; Levasseur, L; Malandain, JJ, Eliciting Macroporosity in Polyacrylamide and Agarose Gels with Polyethylene Glycol, Electrophoresis 17, 58, 1996.

71. Chen, SJ; Evans, DF; Ninham, BW, Properties and Structure of Three-Component Ionic Microemulsions, Journal of Physical Chemistry 88, 1631, 1984.

72. Chiari, M; D'Alesio, L; Consonni, R; Righetti, PG, New Types of Large-Pore Polyacrylamide-Agarose Mixed-Bed Matrices for DNA Electrophoresis: Pore Size Estimation from Ferguson Plots of DNA Fragments, Electrophoresis 16, 1337, 1995.

73. Chiari, M; Micheletti, C; Nesi, M; Fazio, M; Righetti, PG, Towards New Formulations for Polyacrylamide Matrices: N-Acryloylaminoethoxyethanol, a Novel Monomer Combining High Hydrophilicity with Extreme Hydrolytic Stability, Electrophoresis 15, 177, 1994.

74. Chiari, M; Righetti, PG, New Types of Separation Matrices for Electrophoresis, Electrophoresis 16, 1815, 1995.

75. Chrambach, A, The Practice of Quantitative Gel Electrophoresis; VCH: Weinheim, Germany, 1985.

76. Chrambach, A, Unified View of Moving Boundary Electrophoresis: Practical Implications, Journal of Chromatography 320, 1, 1985.

77. Chrambach, A; Rodbard, D, Polyacrylamide Gel Electrophoresis, Science 172, 440, 1971.

78. Chu, B; Yeh, F; Sokolov, EL; Starodoubtsev, SG; Khokhlov, AR, Interaction of Slightly Cross-linked Gels of Poly(diallyldimethylammonium chloride) with Surfactants, Macromolecules 28, 8447, 1995.

79. Chui, M; Phillips, R; McCarthy, M, Measurement of the Porous Microstructure of Hydrogels by Nuclear Magnetic Resonance, Journal of Colloid and Interface Science 174, 336, 1995.

80. Cohen, Y; Ramon, O; Kopelman, IJ; Mizrahi, S, Characterization of Inhomogeneous Polyacrylamide Hydrogels, Journal of Polymer Science: Part B: Polymer Physics 30, 1055, 1992.

81. Cohen Addad, JP, NMR and Statistical Structures of Gels. In: The Physical Properties of Polymeric Gels; Cohen Addad, JP, ed.; Wiley: Chichester, UK, 1996; 39.

82. Collings, PJ, Liquid Crystals, Nature's Delicate Phase of Matter; Princeton University Press: Princeton, NJ, 1990.

83. Corvazier, L; Zhao, Y, Induction of Liquid Crystal Orientation through Azobenzene-Containing Polymer Networks, Macromolecules 32, 3195, 1999.

84. Cosgrove, T; White, SJ; Zarbakhsh, A, Small-Angle Scattering Studies of Sodium Dodecyl Sulfate Interactions with Gelatin. 1, Langmuir 11, 744, 1995.

85. Coveney, FM; Strange, JH; Smith, AL; Smith, EG, NMR Studies of Electrophoretic Mobility in Surfactant Systems, Colloids and Surfaces 36, 193, 1989.

86. Crawford, GP; Zumer, S, eds. Liquid Crystals in Complex Geometries, Formed by Polymer and Porous Networks; Taylor and Francis: London, 1996.

87. Cukier, RI, Diffusion of Brownian Spheres in Semidilute Polymer Solutions, Macromolecules 17, 252, 1984.

88. Cussler, EL, Diffusion, Mass Transfer in Fluid Systems; Cambridge University Press: Cambridge, UK, 1984.

89. Deen, WM, Hindered Transport of Large Molecules in Liquid-Filled Pores, AIChE Journal 33, 1409, 1987.

90. De Gennes, PG, Possibilities Offertes Par La Reticulation De Polymeres en Presence D'un Cristal Liquide, Physics Letters 28A, 725, 1969.

91. De Gennes, PG; Prost, J, The Physics of Liquid Crystals, 2nd ed.; Oxford Science Publications: Oxford, 1993.

92. de Groot, SR; Mazur, P, Non-Equilibrium Thermodynamics; Dover: New York, 1984.

93. Derbyshire, W; Duff, ID, NMR of Agarose Gels, Faraday Discussions of the Chemical Society 57, 243, 1974.

94. Dickey, FH, The Preparation of Specific Adsorbents, Proceedings of the National Academy of Science, USA 35, 227, 1949.

95. Dickinson, E, Particle Gels, Chemtech November, 665, 1991.

96. Djabourov, M; Clark, AH; Rowlands, DW; Ross-Murphy, SB, Small-Angle X-Ray Scattering Characterization of Agarose Sols and Gels, Macromolecules 22, 180, 1989.

97. Doi, M; Edwards, SF, The Theory of Polymer Dynamics; Oxford University Press: Oxford, 1986.

98. Douglas, NG; Humffray, AA; Pratt, HRC; Stevens, GW, Electrophoretic Mobilities of Proteins and Protein Mixtures, Chemical Engineering Science 50, 743, 1995.

99. Duke, TAJ; Austin, RH, Microfabricated Sieve for the Continuous Sorting of Macromolecules, Physical Review Letters 80, 1552, 1998.

100. Duke, TAJ; Austin, RH; Cox, EC; Chan, SS, Pulsed-field Electrophoresis in Microlighographic Arrays, Electrophoresis 17, 1075, 1996.

101. Dunning, JW; Angus, JC, Particle-Size Measurement by Doppler-Shifted Laser Light, a Test of the Stokes–Einstein Relation, Journal of Applied Physics 39, 2479, 1968.

102. Dusek, K, Diffusion Control in the Kinetics of Cross-Linking, Polymer Gels and Networks 4, 383, 1996.

103. Edgington, SM, Biotech's New Nanotools, Developing Nanoscale Approaches to Biological Problems Is Now Within the Reach of Every Lab, Bio/Technology 12, 468, 1994.

104. Edsall, JT; Wyman, J, Biophysical Chemistry; Academic Press: New York, 1958; Vol. 1.

105. Edwards, DA, Charge Transport Through a Spatially Periodic Porous Medium: Electrokinetic and Convective Dispersion Phenomena, Philosophical Transactions of the Royal Society of London A 353, 205, 1995.

106. Edwards, DA; Langer, R, A Linear Theory of Transdermal Transport Phenomena, Journal of Pharmaceutical Sciences 83, 1315, 1994.

107. Effenhauser, CS; Bruin, GJM; Paulus, A, Integrated Chip-Based Capillary Electrophoresis, Electrophoresis 18, 2203, 1997.

108. Einstein, A, Investigations on the Theory of the Brownian Movement; Dover: New York, 1956.

109. Eisenberg, RS, Computing the Field in Proteins and Channels, Journal of Membrane Biology 150, 1, 1996.

110. Ekberg, B; Mosbach, K, Molecular Imprinting: A Technique for Producing Specific Separation Materials, Trends in Biotechnology 7, 92, 1989.

111. Eringen, AC; Maugin, GA, Electrodynamics of Continua II, Fluids and Complex Media; Springer-Verlag: New York, 1990; Vol. 2.

112. Eringen, AC; Maugin, GA, Electrodynamics of Continua I, Foundations and Solid Media; Springer-Verlag: New York, 1990; Vol. 1.

113. Fang, L; Brown, W, Decay Time Distributions from Dynamic Light Scattering for Aqueous Poly(vinyl alcohol) Gels and Semidilute Solutions, Macromolecules 23, 3284, 1990.

114. Fawcett, JS; Sullivan, SV; Chrambach, A, Toward a Steady-State Pore Limit Electrophoresis Dimension for Native Proteins in Two-Dimensional Polyacrylamide Gel Electrophoresis, Electrophoresis 10, 182, 1989.

115. Ferguson, KA, Starch-Gel Electrophoresis—Application to the Classification of Pituitary Proteins and Polypeptides, Metabolism 13, 985, 1964.

116. Fick, AE, Philosophical Magazine 10, 30, 1855.

117. Firouzi, A; Atef, F; Oertli, AG; Stucky, GD; Chmelka, BF, Alkaline Lyotropic Silicate-Surfactant Liquid Crystals, Journal of the American Chemical Society 119, 3596, 1997.

118. Firouzi, A; Schaefer, DH; Tolbert, SH; Stucky, GD; Chmelka, BF, Magnetic-Field-Induced Orientational Ordering of Alkane Lyotropic Silicate-Surfactant Liquid Crystals, Journal of the American Chemical Society 119, 9466, 1997.

119. Flory, P Principles of Polymer Chemistry; Cornell University Press: Ithaca, 1953.

120. Flory, PJ, Introductory Lecture, Faraday Discussions of the Chemical Society 57, 7, 1974.

121. Fourier, JBJ, Theorie analytique de la chaleur; The University Press, Cambridge, England, 1878.

122. Freed, KF; Muthukumar, M, On the Stokes Problem for a Suspension of Spheres at Finite Concentrations, Journal of Chemical Physics 68, 2088, 1978.

123. Fricke, H, A Mathematical Treatment of the Electric Conductivity and Capacity of Disperse Systems, Physical Review 24, 575, 1924.

124. Garces, JM, On Void Fraction and the Nature of Porous Solids, Advanced Materials 8, 434, 1996.

125. Gelfi, C; Righetti, PG, Polymerization Kinetics of Polyacrylamide Gels I. Effect of Different Cross-Linkers, Electrophoresis 2, 213, 1981.
126. Gelfi, C; Righetti, PG, Polymerization Kinetics of Polyacrylamide Gels II. Effect of Temperature, Electrophoresis 2, 220, 1981.
127. De Gennes, P-G Scaling Concepts in Polymer Physics; Cornell University Press: Ithaca, NY, 1979.
128. Gersten, DM; Kimball, H; Bijwaard, KE, Gel Electrophoresis in the Presence of Soluble, Aqueous Polymers: Horizontal Sodium dodecyl Sulfate-Polyacrylamdie Gels, Analytical Biochemistry 197, 59, 1991.
129. Gianazza, E; Miller, I; Eberini, I; Castiglioni, S, Low-Tech Electrophoresis, Small but Beautiful, and Effective: Electrophoretic Titration Curves of Proteins, Electrophoresis 20, 1325, 1999.
130. Gibbs, SJ; Chu, AS; Lightfoot, EN; Root, TW, Ovalbumin Diffusion at Low Ionic Strength, Journal of Physical Chemistry 95, 467, 1991.
131. Gibbs, SJ; Johnson, CS, Pulsed Field Gradient NMR Study of Probe Motion in Polyacrylamide Gels, Macromolecules 24, 6110, 1991.
132. Giddings, JC; Bowman, LM; Myers, MN, Isolation of Peak Broadening Factors in Exclusion Chromatography, Macromolecules 10, 442, 1977.
133. Giddings, JC; Boyack, JR, Effect of Stabilizing Media on Zone Mobility and Spreading in Electrophoresis, Journal of Theoretical Biology 2, 1, 1962.
134. Giddings, JC; Boyack, JR, Mechanism of Electrophoretic Migration in Paper, Analytical Chemistry 36, 1229, 1964.
135. Giddings, JC; Kucera, E; Russell, CP; Myers, MN, Statistical Theory for the Equilibrium Distribution of Rigid Molecules in Inert Porous Networks. Exclusion Chromatography, Journal of Physical Chemistry 72, 4397, 1968.
136. Glandt, ED, Noncircular Pores in Model Membranes: A Calculation of the Effect of Pore Geometry on the Partition of a Solute, Journal of Membrane Science 8, 331, 1981.
137. Gorin, MA, An Equilibrium Theory of Ionic Conductance, Journal of Chemical Physics 7, 405, 1939.
138. Gosnell, DL; Zimm, BH, Measurement of Diffusion Coefficients of DNA in Agarose Gels, Macromolecules 26, 1304, 1993.
139. Goto, M; McCoy, BJ, Inverse Size-Exclusion Chromatography for Distributed Pore and Solute Sizes, Chemical Engineering Science 55, 723, 2000.
140. Grahman, T, Philosophical Transactions of the Royal Society of London A 140, 1, 1850.
141. Gray, WG, A Derivation of the Equations for Multiphase Transport, Chemical Engineering Science 30, 229, 1975.
142. Guttman, A, On the Separation Mechanism of Capillary Sodium Dodecyl Sulfate-Gel Electrophoresis of Proteins, Electrophoresis 16, 611, 1995.
143. Haase, R. Thermodynamics of Irreversible Processes; Dover: New York, 1969.
144. Hadden, DA, Master of Science Thesis, Florida State University, Tallahassee, FL, 1999.
145. Hadden, D; Rill, RL; McFadden, L; Locke, BR, Oligonucleotide and Water Self-Diffusion in Pluronic Triblock Copolymer Gels and Solutions by Pulsed Field Gradient Nuclear Magnetic Resonance, Macromolecules 33, 4235, 2000.
146. Haggerty, L; Lenhoff, AM, Relation of Protein Electrostatics Computations to Ion-Exchange and Electrophoretic Behavior, Journal of Physical Chemistry 95, 1472, 1991.
147. Hammersley, JM, Percolation Processes. II. The Connective Constant, Proceedings of the Cambridge Philosophical Society 53, 642, 1957.
148. Harrington, MG; Lee, KH; Bailey, JE; Hood, LE, Sponge-Like Electrophoresis Media: Mechanically Strong Materials Compatible with Organic Solvents, Polymer Solutions and Two-Dimensional Electrophoresis, Electrophoresis 15, 187, 1994.
149. Hashin, Z; Shtrikman, S, A Variational Approach to the Theory of Effective Magnetic Permeability of Multiphase Materials, Journal of Applied Physics 33, 3125, 1962.

150. He, Q; Hinton, DP; Johnson, CS, Measurement of Mobility Distributions for Vesicles by Electrophoretic NMR, Journal of Magnetic Resonance 91, 654, 1991.

151. He, Q; Johnson, CS, Stimulated Echo Electrophoretic NMR, Journal of Magnetic Resonance 85, 181, 1989.

152. He, Q; Johnson, CS, Two-Dimensional Electrophoretic NMR for the Measurement of Mobilities and Diffusion in Mixtures, Journal of Magnetic Resonance 81, 435, 1989.

153. Hecht, A-M; Duplessix, R; Geissler, E, Structural Inhomogeneities in the Range 2.5–2000 Angstroms in Polyacrylamide Gels, Macromolecules 18, 2167, 1985.

154. Hedrick, JL; Smith, AJ, Size and Charge Isomer Separation and Estimation of Molecular Weights of Proteins by Disc Gel Electrophoresis, Archives of Biochemistry and Biophysics 126, 155, 1968.

155. Henn, SW; Ackers, GK, Molecular Sieve Studies of Interacting Protein Systems, The Journal of Biological Chemistry 244, 465, 1969.

156. Henn, SW; Ackers, GK, Molecular Sieve Studies of Interacting Protein Systems. V. Association of Subunits of D-Amino Acid Oxidase Apoenzyme, Biochemistry 8, 3829, 1969.

157. Henry, DC, The Cataphoresis of Suspended Particles. Part I. The Equation of Cataphoresis, Proceedings of the Royal Society of London Series A 133, 106, 1931.

158. Hermans, JJ, Sedimentation and Electrophoresis of Porous Spheres, Journal of Polymer Science 18, 527, 1955.

159. Hervet, H; Bean, CP, Electrophoretic Mobility of Lambda Phage HIND III and HAE III DNA Fragments in Agarose Gels: A Detailed Study, Biopolymers 26, 727, 1987.

160. Hiemenz, PC, Principles of Colloid and Surface Chemistry, 2nd ed.; Marcel Dekker: New York, 1986.

161. Hinton, DP; Johnson, CS, Diffusion Coefficients, Electrophoretic Mobilities, and Morphologies of Charged Phospholipid Vesicles by Pulsed Field Gradient NMR and Electron Microscopy, Journal of Colloid and Interface Science 173, 364, 1995.

162. Hirschfelder, JO; Curtiss, CF; Bird, RB, Molecular Theory of Gases and Liquids; Wiley: New York, 1954.

163. Hjerten, S, The Preparation of Agarose Spheres for Chromatography of Molecules and Particles, Biochimica et Biophysica Acta 79, 393, 1964.

164. Hjerten, S; Liao, JL; Zhang, R, High-Performance Liquid Chromatography on Continuous Polymer Beds, Journal of Chromatography 473, 273, 1989.

165. Hoagland, DA; Smisek, DL; Chen, DY, Gel and Free Solution Electrophoresis of Variably Charged Polymers, Electrophoresis 17, 1151, 1996.

166. Holz, M, Electrophoretic NMR, Chemical Society Reviews, 165, 1994.

167. Holz, M; Lucas, O; Muller, C, NMR in the Presence of an Electric Current, Simultaneous Measurements of Ionic Mobilities, Transference Numbers, and Self-Diffusion Coefficients Using an NMR Pulsed-Gradient Experiment, Journal of Magnetic Resonance 58, 294, 1984.

168. Hooper, HH; Baker, JP; Blanch, HW; Prausnitz, JM, Swelling Equilibria for Positively Ionized Polyacrylamide Hydrogels, Macromolecules 23, 1096, 1990.

169. Horkay, F; Hecht, AM; Zrinyi, M; Geissler, E, Effect of Cross-Links on the Structure of Polymer Gels, Polymer Gels and Networks 4, 451, 1996.

170. Hoyt, JJ; Wolfer, WG, Boundary Element Modeling of Electrokinetically Driven Fluid Flow in Two-Dimensional Microchannels, Electrophoresis 19, 2432, 1998.

171. Hughes, BD, Random Walks and Random Environments Volume I: Random Walks; Clarendon Press: Oxford, 1995.

172. Hunter, RJ, Zeta-Potential in Colloid Science Principles and Applications; Academic Press: London, 1981.

173. Hunter, RJ, Foundations of Colloid Science; Clarendon Press: Oxford, 1986; Vol. I.

174. Huo, Q; Margolese, DI; Ciesla, U; Demuth, DG; Feng, P; Gier, TE; Sieger, P; Firouzi, A; Chmelka, BF; Schuth, F; Stucky, GD, Organization of Organic Molecules with Inorganic Molecular Species into Nanocomposite Biphase Arrays, Chemistry of Materials 6, 1176, 1994.

175. Hvidt, S; Jorgensen, EB; Brown, W; Schillen, K, Micellization and Gelation of Aqueous Solutions of a Triblock Copolymer Studied by Rheological Techniques and Scanning Calorimetry, Journal of Physical Chemistry 98, 12320, 1994.
176. Hyde, PD; Ediger, MD, NMR Imaging of Diffusion of Small Organic Molecules in Silk Fibroin Gel, Macromolecules 24, 620, 1991.
177. Hyde, ST; Ninham, BW; Zemb, T, Phase Boundaries for Ternary Microemulsions. Predictions of a Geometric Model, Journal of Physical Chemistry 93, 1464, 1989.
178. Israelachvili, J, Intermolecular and Surface Forces, 2nd ed.; Academic Press: London, 1992.
179. Ivory, CF, Transient Electrophoresis of a Dielectric Sphere, Journal of Colloid and Interface Science 100, 239, 1984.
180. Jansons, KM; Phillips, CG, On the Application of Geometric Probability Theory to Polymer Networks and Suspensions, I, Journal of Colloid and Interface Science 137, 75, 1990.
181. Jeffrey, DJ, Conduction through a Random Suspension of Spheres, Proceedings of the Royal Society of London Series A 335, 355, 1973.
182. Johansson, L; Elvingson, C; Lofroth, JE, Diffusion and Interaction in Gels and solutions. 3. Theoretical Results on the Obstruction Effect, Macromolecules 24, 6024, 1991.
183. Johansson, L; Lofroth, J-E, Diffusion and Interaction in Gels and Solutions. 4 Hard Sphere Brownian Dynamics Simulations, Journal of Chemical Physics 98, 7471, 1993.
184. Johansson, L; Skantze, U; Lofroth, J-E, Diffusion and Interaction in Gels and Solutions. 2. Experimental Results on the Obstruction Effect, Macromolecules 24, 6019, 1991.
185. Johnson, CS; He, Q, Electrophoretic Nuclear Magnetic Resonance, Advances in Magnetic Resonance 13, 131, 1989.
186. Johnson, EM; Berk, DA; Jain, RK; Deen, WM, Hindered Diffusion in Agarose Gels: Test of Effective Medium Model, Biophysical Journal 70, 1017, 1996.
187. Joslin, CG; Stell, G, Bounds on the Properties of Fiber-Reinforced Composites, Journal of Applied Physics 60, 1607, 1986.
188. Joslin, CG; Stell, G, Effective Properties of Fiber-Reinforced Composites: Effects of Polydispersity in Fiber Diameter, Journal of Applied Physics 60, 1611, 1986.
189. Karger, J; Ruthven, DM, Diffusion in Zeolites and Other Microporous Solids; Wiley: New York, 1992.
190. Kathawalla, IA; Anderson, JL; Lindsey, JS, Hindered Diffusion of Porphyrins and Short-Chain Polystyrene in Small Pores, Macromolecules 22, 1215, 1989.
191. Katzir-Katchalsky, A, Biophysics and Other Topics, Selected Papers by Aharon Katzir-Katchalsky; Academic Press: New York, 1976.
192. Key, PY; Sellen DB, A Laser Light-Scattering Study of the Structure of Agarose Gels, Journal of Polymer Science 20, 659, 1982.
193. Kim, JH; Ochoa, JA; Whitaker, S, Diffusion in Anisotropic Porous Media, Transport in Porous Media 2, 327, 1987.
194. Kim, Y; Morris, MD, Pulsed Field Capillary Electrophoresis of Multikilobase Length Nucleic Acids in Dilute Methyl Cellulose Solutions, Analytical Chemistry 66, 3081, 1994.
195. Kinsey, ST; Locke, BR; Penke, B; Moerland, TS, Diffusional Anisotropy Is Induced by Subcellular Barriers in Skeletal Muscle, NMR in Biomedicine 12, 1, 1999.
196. Knackstedt, MA; Ninham, BW, Model Disordered Media Provided by Ternary Microemulsions, Physical Review E 50, 2839, 1994.
197. Knackstedt, MA; Ninham, BW, Ternary Microemulsions as Model Disordered Media, AIChE Journal 41, 1295, 1995.
198. Knackstedt, MA; Zhang, X, Direct Evaluation of Length Scales and Structural Parameters Associated with Flow in Porous Media, Physical Review E 50, 2134, 1994.
199. Kokufuta, E; Jinbo, E, A Hydrogel Capable of Facilitating Polymer Diffusion through the Gel Porosity and Its Application in Enzyme Immobilization, Macromolecules 25, 3549, 1992.
200. Kresge, CT; Leonowicz, ME; Roth, WJ; Vartuli, JC; Beck, JS, Ordered Mesoporous Molecular Sieves Synthesized by a Liquid-Crystal Template Mechanism, Nature 359, 710, 1992.

201. Kriz, D; Ramstrom, O; Mosbach, K, Molecular Imprinting, New Possibilities for Sensor Technology, Analytical Chemistry June 1, 345A, 1997.
202. Kuiken, GDC, Thermodynamics of Irreversible Processes; Wiley: Chichester, UK, 1994.
203. Landau, LD; Lifshitz, EM; Pitaevskii, LP, Electrodynamics of Continuous Media, 2nd ed.; Pergamon Press: Oxford, 1984.
204. Latour, LL; Kleinberg, RL; Mitra, PP; Sotak, CH, Pore-Size Distributions and Tortuosity in Heterogeneous Porous Media, Journal of Magnetic Resonance Series A 112, 83, 1995.
205. Laurent, TC; Killander, J, A Theory of Gel Filtration and Its Experimental Verification, Journal of Chromatography 14, 317, 1964.
206. Lekkerkerker, HNW, Ordering in Supramolecular Fluids, Physica A 176, 1, 1991.
207. Leonowicz, M; Lawton, JA; Lawton, SL; Rubin, MK, MCM-22: A Molecular Sieve with Two Independent Multidimensional Channel Systems, Science 264, 1910, 1994.
208. Lim, WA; Slater, GW; Noolandi, J, A Model of the DNA Transient Orientation Overshoot During Gel Electrophoresis, Journal of Chemical Physics 92, 709, 1990.
209. Limbach, KW; Nitsche, JM; Wei, J, Partitioning of Nonspherical Molecules Between Bulk Solution and Porous Solids, AIChE Journal 35, 42, 1989.
210. Lin, TH; Phillies, GDJ, Probe Diffusion in Polyacrylic Acid: Water—Effect of Polymer Molecular Weight, Journal of Colloid and Interface Science 100, 82, 1984.
211. Linse, P, Phase Behavior of Poly(ethylene oxide)-Poly(propylene oxide) Block Copolymers in Aqueous Solutions, Journal of Physical Chemistry 97, 13896, 1993.
212. Linse, P; Malmsten, M, Temperature-Dependent Micellization in Aqueous Block Copolymer Solutions, Macromolecules 25, 5434, 1992.
213. Liu, J; Kim, AY; Virden, JW; Bunker, BC, Effect of Colloidal Particles on the Formation of Ordered Mesoporous Materials, Langmuir 11, 682, 1995.
214. Lloyd, L; Warner, FP, Preparative High-Performance Liquid Chromatography on a Unique High-Speed Macroporous Resin, Journal of Chromatography 512, 365, 1990.
215. Locke, BR, Electrophoretic Transport in Porous Media: A Volume Averaging Approach, Industrial and Engineering Chemistry Research 37, 615, 1998.
216. Locke, BR; Arce, P, Modeling Electrophoretic Transport of Polyelectrolytes in Beds of Nonporous Spheres, Separation Technology 3, 111, 1993.
217. Locke, BR; Arce, P; Park, Y, Applications of Self-Adjoint Operators to Electrophoretic Transport, Enzyme Reactions, and Microwave Heating Problems in Composite Media—II. Electrophoretic Transport in Layered Membranes, Chemical Engineering Science 48, 4007, 1993.
218. Locke, BR; Carbonell, RG, A Theoretical and Experimental Study of Counteracting Chromatographic Electrophoresis, Separation and Purification Methods 18, 1, 1989.
219. Locke, BR; Trinh, SH, When Can the Ogston–Morris–Rodbard–Chrambach Model be Applied to Gel Electrophoresis, Electrophoresis 20, 00, 1999.
220. Loeb, AL; Overbeek, JTG; Wiersema, PH, The Electrical Double Layer Around a Spherical Colloid Particle, Computation of the Potential, Charge Density, and Free Energy of the Electrical Double Layer Around a sperical Colloid Particle; M.I.T. Press: Cambridge, MA, 1961.
221. Lorentz, HA, Wied, Ann. 11, 70, 1880.
222. Lorentz, L, Wied. Ann. 9, 641, 1880.
223. Lumpkin, O, Electrophoretic Mobility of a Porous Sphere Through Gel-Like Obstacles: Hydrodnamic Interactions, Journal of Chemical Physics 81, 5201, 1984.
224. Lustig, SR; Caruthers, JM; Peppas, NA, Continuum Thermodynamics and Transport Theory for Polymer-Fluid Mixtures, Chemical Engineering Science 12, 3037, 1992.
225. Lyubimova, T; Righetti, PG, On the Kinetics of Photopolymerization: A Theoretical Study, Electrophoresis 14, 191, 1993.
226. Maaloum, M; Pernodet, N; Tinland, B, Agarose Gel Structure Using Atomic Force Microscopy: Gel Concentration and Ionic Strength Effects, Electrophoresis 19, 1606, 1998.
227. Mackie, JS; Meares, P, The Diffusion of Electrolytes in a Cation-Exchange Resin Membrane I. Theortical, Proceedings of the Royal Society of London Series A 232, 498, 1955.

228. Macosko, CW; Miller, DR, A New Derivation of Average Molecular Weights of Nonlinear Polymers, Macromolecules 9, 199, 1976.

229. Mallam, S; Horkay, F; Hecht, A-M; Geissler, E, Scattering and Swelling Properties of Inhomogeneous Polyacrylamide Gels, Macromolecules 22, 3356, 1989.

230. Malmsten, M; Lindman, B, Self-Assembly in Aqueous Block Copolymer Solutions, Macromolecules 25, 5440, 1992.

231. Malmsten, M; Lindman, B, Water Self-Diffusion in Aqueous Block Copolymer Solutions, Macromolecules 25, 5446, 1992.

232. Mandelbrot, BB, The Fractal Geometry of Nature; WH Freeman: San Francisco, 1983.

233. Manz, B; Stilbs, P; Jonsson, B; Soderman, O; Callaghan, PT, NMR Imaging of the Time Evolution of Electroosmotic Flow in a Capillary, Journal of Physical Chemistry 99, 11297, 1995.

234. Matthew, JB; Hanania, GIH; Gurd, FRN, Electrostatic Effects in Hemoglobin: Bohr Effect and Ionic Strength Dependence of Individual Groups, Biochemistry 18, 1928, 1979.

235. Matthew, JB; Hanania, GIH; Gurd, FRN, Electrostatic Effects in Hemoglobin: Hydrogen Ion Equilibria in Human Deoxy- and Oxyhemoglobin A, Biochemistry 18, 1919, 1979.

236. Maxwell, JC, A Treatise on Electricity and Magnetism, 3rd ed.; Dover: New York, 1981; Vol. 1.

237. Maxwell, JC, A Treatise on Electricity and Magnetism, 3rd ed.; Dover: New York, 1981; Vol. 2.

238. McQuarrie, DA, Statistical Mechanics; Harper Collins: New York, 1976.

239. Melcher, JR; Taylor, GI, Electrohydrodynamics: A Review of the Role of Interfacial Shear Stresses, Annual Review of Fluid Mechanics 1, 111, 1969.

240. Meredith, RE; Tobias, CW, Conductivities in Emulsions, Journal of the Electrochemical Society 108, 286, 1961.

241. Michaels, AS, Diffusion in a Pore of Irregular Cross Section—a Simplified Treatment, AIChE Journal 5, 270, 1958.

242. Michaels, AS; Chandrasekaran, SK; Shaw, JE, Drug Permeation Through Human Skin: Theory and in vitro Experimental Measurement, AIChE Journal 21, 985, 1975.

243. Michov, BM, A Quantitative Theory of the Stern Electric Double Layer, Electrophoresis 9, 201, 1988.

244. Michov, BM, Radically Simplifying the Henry Function, Electrophoresis 9, 199, 1988.

245. Miller, DR; Macosko, CW, A New Derivation of Post Gel Properties of Network Polymers, Macromolecules 9, 206, 1976.

246. Milton, GW; McPhedran, RC; McKenzie, DR, Transport Properties of Arrays of Intersecting Cylinders, Applied Physics 25, 23, 1981.

247. Mitra, PP; Sen, PN; Schwartz, LM; Le Doussal, P, Diffusion Propagator as a Probe of the Structure of Porous Media, Physical Review Letters 68, 3555, 1992.

248. Monduzzi, M; Knacksted, MA; Ninham, BW, Microstructure of Perfluoropolyether Water/Oil Microemulsions, Journal of Physical Chemistry 99, 17772, 1995.

249. Monnier, A; Schuth, F; Huo, Q; Kumar, D; Margolese, D; Maxwell, RS; Stucky, GD; Krishnamurty, M; Petroff, P; Firouzi, A; Janicke, M; Chmelka, BF, Cooperative Formation of Inorganic–Organic Interfaces in the Synthesis of Silicate Mesostructures, Science 261, 1299, 1993.

250. Morris, CJOR, Gel Filtration and Gel Electrophoresis, Protides of the Biological Fluids 14, 543, 1966.

251. Morris, CJOR; Morris, P, Molecular-Sieve Chromatography and Electrophoresis in Polyacrylamide Gels, Biochemical Journal 124, 517, 1971.

252. Mortensen, K; Brown, W; Norden, B, Inverse Melting Transition and Evidence of Three-Dimensional Cubatic Structure in a Block-Copolymer Micellar System, Physical Review Letters 68, 2340, 1992.

253. Mortensen, K; Talmon, Y, Cryo-TEM and Sans Microstructural Study of Pluronic Polymer Solutions, Macromolecules 28, 8829, 1995.

254. Mosher, RA; Saville, DA; Thorman, W The Dynamics of Electrophoresis; VCH: Weinheim, 1992.

255. Moussaid, A; Pusey, PN; Slot, JJM; Joosten, JGH, Simulation of Scattering Properties of Gels, Macromolecules 32, 3774, 1999.
256. Muhr, AH; Blanshard, JMV, Diffusion in Gels, Polymer 23, 1012, 1982.
257. Muthukumar, M, Theory of Electrophoretic Mobility of Polyelectrolyte Chains, Macromolecular Theory and Simulations 3, 61, 1994.
258. Muthukumar, M, Dynamics of Polyelectrolyte Solutions, Journal of Chemical Physics 107, 2619, 1997.
259. Nakao, S-I; Nomura, T; Kumura, S, Characteristics of Macromolecular Gel Layer Formed on Ultrafiltration Tubular Membrane, AIChE Journal 25, 615, 1979.
260. Neale, GH; Nader, WK, Prediction of Transport Processes Within Porous Media: Diffusive Flow Processes Within a Homogeneous Swarm of Spherical Particles, AIChE Journal 19, 112, 1973.
261. Newman, J, Electrochemical Systems; Prentice Hall: Englewood Cliffs, NJ, 1973.
262. Nieto, JL; Baselga, J; Hernandez-Fuentes, I; Llorente, MA; Pierola, IF, Polyacrylamide Networks. Kinetic and Structural Studies by High Field H1-NMR with Polymerization in Situ, European Journal of Polymers 23, 551, 1987.
263. Norden, B; Elvingson, C; Jonsson, M; Akerman, B, Microscopic Behavior of DNA Duing Electrophoresis: Electrophoretic Orientation, Quarterly Reviews of Biophysics 24, 103, 1991.
264. Nozad, I; Carbonell, RG; Whitaker, S, Heat Conduction in Multiphase Systems—I Theory and Experiment for Two-Phase Systems, Chemical Engineering Science 40, 843, 1985.
265. O'Brien, RW; White, LR, Electrophoretic Mobility of a Spherical Colloidal Particle, Journal of the Chemical Society, Faraday Transactions 74, 1607, 1978.
266. Ochoa, JA Ph.D., Dissertation, University of California, Davis, 1988.
267. Ochoa, JA; Stroeve, P; Whitaker, S, Diffusion and Reaction in Cellular Media, Chemical Engineering Science 41, 2999, 1986.
268. Ochoa-Tapia, JA; Stroeve, P; Whitaker, S, Diffusive Transport in Two-Phase Media: Spatially Periodic Modles and Maxwell's Theory for Isotropic and Anisotropic Systems, Chemical Engineering Science 49, 709, 1994.
269. Ogston, AG, The Spaces in a Uniform Random Suspension of Fibers, Transactions of the Faraday Society 54, 1754, 1958.
270. Ogston, AG; Preston, BN; Wells, JD, On The Transport of Compact Particles Through Solutions of Chain-Polymers, Proceedings of the Royal Society of London Series A 333, 297, 1973.
271. Ogston, AG; Silpananta, P, The Thermodynamics of Interaction between Sephadex and Penetrating Solutes, Biochemical Journal 116, 171, 1970.
272. Ohshima, H; Healy, T; White, LR, Approximate Analytic Expressions for the Electrophoretic Mobility of Spherical Colloidal Particles and the Conductivity of their Dilute Suspensions, Journal of the Chemical Society, Faraday Transactions 79, 1613, 1983.
273. Ohshima, H; Kondo, T, Electrophoretic Mobility and Donnan Potential of a Large Colloidal Particle with a Surface Charge Layer, Journal of Colloid and Interface Science 116, 305, 1987.
274. O'Neil, GA; Torkelson, JM, Modeling Insight into the Diffusion-Limited Cause of the Gel Effect in Free Radical Polymerization, Macromolecules 32, 411, 1999.
275. O'Neil, GA; Wisnudel, MB; Torkelson, JM, Gel Effect in Free Radical Polymerization: Model Discrimination of Its Cause, AIChE Journal 44, 1226, 1998.
276. Ornstein, L, Disc Electrophoresis—I Background and Theory, Annals of the New York Academy of Sciences 121, 321, 1969.
277. Ornstein, L, Tenuous but Contingent Connections, Electrophoresis 8, 3, 1987.
278. Ornstein, L; Davis, BJ, Disc Electrophoresis; Distillation Products Div., Eastman Kodak Co.: Rochester, 1962.
279. Orttung, WH, Interpretation of the Titration Curve of Oxyhemoglobin. Detailed Consideration of Coulomb Interactions at Low Ionic Strength, Journal of the American Chemical Society 91, 162, 1969.
280. Orttung, WH, Proton Binding and Dipole Moment of Hemoglobin. Refined Calculations, Biochemistry 9, 2394, 1970.

281. Overbeek, JTG, Quantitative Interpretation of the Electrophoretic Velocity of Colloids. In: Advances in Colloid Science; Mark, H; Verwey, EJW, eds.; Interscience: New York, 1950; Vol. 3, p 97.

282. Overbeek, JTG; Bijsterbosch, BH, The Electrical Double Layer and the Theory of Electrophoresis. In: Electrokinetic Separation Methods; Righetti, PG; van Oss, CJ; Vanderhoff, JW, eds.; Elsevier/North-Holland Biomedical Press:, 1979; 1.

283. Overbeek, JTG; Lijklema, J, Electric Potentials in Colloidal Systems. In: Electrophoresis, Theory, Method and Applications; Bier, M, ed.; Academic Press: New York, 1959; 1.

284. Overbeek, JTG; Wiersema, PH, The Interpretation of Electrophoretic Mobilities. In: Electrophoresis, Theory, Methods and Applications; Bier, M, ed.; Academic Press: New York, 1967; 1.

285. Paine, MA; Carbonell, RG; Whitaker, S, Dispersion in Pulsed Systems—I. Heterogeneous Reaction and Reversible Adsorption in Capillary Tubes, Chemical Engineering Science 38, 1781, 1983.

286. Palusinski, OA; Allgyer, TT; Mosher, RA; Bier, M; Saville, DA, Mathematical Modeling and Computer Simulation of Isoelectric Focusing with Electrochemically Defined Ampholytes, Biophysical Chemistry 13, 193, 1981.

287. Palusinski, OA; Bier, M; Saville, DA, Mathematical Model for Transient Isoelectric Focusing of Simple Ampholytes, Biophysical Chemistry 14, 389, 1981.

288. Patternson, B Ph.D., Dissertation, Florida State University, Tallahassee, FL, 2000.

289. Pavesi, L; Rigamonti, A, Diffusion Constants in Polyacrylamide Gels, Physical Review E 51, 3318, 1995.

290. Penke, B; Kinsey, S; Gibbs, SJ; Moerland, TS; Locke, BR, Proton Diffusion and T1 Relaxation in Polyacrylamide Gels: A Unified Approach Using Volume Averaging, Journal of Magnetic Resonance 132, 240, 1998.

291. Perrins, WT; McKenzie, DR; McPhedran, RC, Transport Properties of Regular Arrays of Cylinders, Proceedings of the Royal Society of London Series A 369, 207, 1979.

292. Petersen, EE, Diffusion in a Pore of Varying Cross Section, AIChE Journal 4, 343, 1958.

293. Phillies, GDJ, Universal Scaling Equation for Self-Diffusion by Macromolecules in Solution, Macromolecules 19, 2367, 1986.

294. Phillies, GDJ, Dynamics of Polymers in Concentrated Solutions: The Universal Scaling Equation Derived, Macromolecules 20, 558, 1987.

295. Phillies, GDJ, Quantitative Prediction of alpha in the Scaling Law for Self-Diffusion, Macromolecules 21, 3101, 1988.

296. Phillies, GDJ, The Hydrodynamic Scaling Model for Polymer Self-Diffusion, Journal of Physical Chemistry 93, 5029, 1989.

297. Phillips, RJ; Deen, WM; Brady, JF, Hindered Transport of Spherical Macromolecules in Fibrous Membranes and Gels, AIChE Journal 35, 1761, 1989.

298. Phillips, RJ; Deen, WM; Brady, JF, Hindered Transport in Fibrous Membranes and Gels: Effect of Solute Size and Fiber Configuration, Journal of Colloid and Interface Science 139, 363, 1990.

299. Piletsky, SA; Andersson, HS; Nicholls, IA, Combined Hydrophobic and Electrostatic Interaction-Based Recognition in Molecularly Imprinted Polymers, Macromolecules 32, 633, 1999.

300. Porath, J, Some Recently Developed Fractionation Procedures and Their Application to Peptide and Protein Hormones, Pure and Applied Chemistry 6, 233, 1963.

301. Porath, J, Some Recent Developments in Preparative Electrophoresis and Gel Filtration, Metabolism 13, 1004, 1964.

302. Prager, S, Diffusion and Viscous Flow in Concentrated Suspensions, Physica 29, 129, 1963.

303. Prange, MM; Hooper, HH; Prausnitz, JM, Thermodynamics of Aqueous Systems Containing Hydrophilic Polymers or Gels, AIChE Journal 35, 803, 1989.

304. Prud'homme, RK; Wu, G; Schneider, DK, Structure and Rheology Studies of Poly(oxyethylene-oxypropylene-oxyethylene) Aqueous Solution, Langmuir 12, 4651, 1996.

305. Quintard, M, Diffusion in Isotropic and Anisotropic Porous Systems: Three-Dimensional Calculations, Transport in Porous Media 11, 187, 1993.

306. Radi, P; Schumacher, E, Numerical Simulation of Electrophoresis: The Complete Solution for Three Isotachophoretic Systems, Electrophoresis 6, 195, 1985.

307. Radko, SP; Chrambach, A, Electrophoretic Migration of Submicron Polystyrene Latex Spheres in Solutions of Linear Polyacrylamide, Macromolecules 32, 2617, 1999.

308. Radovich, JM; Behnam, B; Mullon, C, Steady-State Modeling of Electro-Ultrafiltration at Constant Concentration, Separation Science and Technology 20, 315, 1985.

309. Ramey, BA, Master of Science Thesis, Florida State University, Tallahassee, FL, 1998.

310. Rayleigh, L, On the Influence of Obstacles arranged in Rectangular Order upon the Properties of a Medium, Philosophical Magazine 34, 481, 1892.

311. Raymond, S; Weintraub, LS, Acrylamide Gel as a Supporting Medium for Zone Electrophoresis, Science 130, 7111, 1959.

312. Reiner, M, History of Electrophoresis. In: CRC Handbook of Electrophoresis; Lewis, LA; Opplt, JJ, eds.; CRC Press: Boca Raton, FL, 1980; 3.

313. Richards, EG; Temple, CJ, Some Properties of Polyacrylamide Gels, Nature 230, 92, 1971.

314. Righetti, PG, On the Pore Size and Shape of Hydrophilic Gels for Electrophoretic Analysis In Electrophoresis '81; Walter de Gruyter: Berlin, 1981; 3.

315. Righetti, PG; Caglio, S, On the Kinetics of Monomer Incorporation into Polyacrylamide Gels, as Investigated by Capillary Zone Electrophoresis, Electrophoresis 14, 573, 1993.

316. Righetti, PG; Caglio, S; Saracchi, M; Quaroni, S, "Laterally Aggregated" Polyacrylamide Gels for Electrophoresis, Electrophoresis 13, 587, 1992.

317. Righetti, PG; Gelfi, C, Electrophoresis Gel Media: The State of the Art, Journal of Chromatography B 699, 63, 1996.

318. Righetti, PG; Gelfi, C; Bosisio, AB, Polymerization Kinetics of Polyacrylamide Gels III. Effect of Catalysts, Electrophoresis 2, 291, 1981.

319. Rikvold, PA; Stell, G, D-Dimensional Interpenetrable-Sphere Models of Random Two-Phase Media: Microstructure and an Application to Chromatography, Journal of Colloid and Interface Science 108, 158, 1985.

320. Rikvold, PA; Stell, G, Porosity and Specific Surface for Interpenetrable-Sphere Models of Two-Phase Random Media, Journal of Chemical Physics 82, 1014, 1985.

321. Rill, RL; Liu, Y; Van Winkle, DH; Locke, BR, Pluronic Copolymer Liquid Crystals: Unique, Replaceable Media for Capillary Gel Electrophoresis, Journal of Chromatography A817, 287, 1998.

322. Rill, RL; Locke, BR; Liu, Y; Dharia, J; Van Winkle, D, Protein Electrophoresis in Polyacrylamide Gels with Templated Pores, Electrophoresis 17, 1304, 1996.

323. Rill, RL; Locke, BR; Liu, Y; Van Winkle, DH, Electrophoresis in Lyotropic Polymer Liquid Crystals, Proceedings of the National Academy of Science, USA 95, 1534, 1998.

324. Rill, RL; Ramey, BA; Van Winkle, DH; Locke, BR, Capillary Gel Electrophoresis of Nucleic Acids in Pluronic F127 Copolymer Liquid Crystals, Chromatographia Supplement I, Vol 49, S65, 1999.

325. Rill, RL; Van Winkle, DH; Locke, BR, Templated Pores in Hydrogels for Improved Size Selectivity in Gel Permeation Chromatography, Analytical Chemistry 70, 2433, 1998.

326. Rodbard, D, Estimation of Molecular Weight by Gel Filtration and Gel Electrophoresis I. Mathematical Principles. In Methods of Protein Separation; Catsimpoolas, N. ed.; Plenum Press: New York, 1976; Vol. 2, p 145.

327. Rodbard, D, Estimation of Molecular Weight by Gel Filtration and Gel Electrophoresis II. Statistical and Computational Considerations. In Methods of Protein Separation; Catsimpoolas, ed.; Plenum Press: New York, 1976; Vol. 2, p 181.

328. Rodbard, D; Chrambach, A, Unified Theory of Gel Electrophoresis and Gel Filtration, Proceedings of the National Academy of Science, USA 65, 970, 1970.

329. Rodbard, D; Chrambach, A, Estimation of Molecular Radius, Free Mobility, and Valence Using Polyacrylamide Gel Electrophoresis, Analytical Biochemistry 40, 95, 1971.

330. Rodbard, D; Chrambach, A, In: Electrophoresis and Isoelectric Focusing in Polyacrylamide Gels; Allen, RC; Mauer, HR, eds.; Walter E Gruyter: New York, 1974; 28.

331. Rodbard, D; Kapadia, G; Chrambach, A, Pore Gradient Electrophoresis, Analytical Biochemistry 40, 135, 1971.

332. Rossier, JS; Schwarz, A; Reymond, F; Ferrigno, R; Bianchi, F; Girault, HH, Microchannel Networks for Electrophoretic Separations, Electrophoresis 20, 727, 1999.

333. Ross-Murphy, SB, Physical Gelation of Synthetic and Biological Macromolecules. In: Polymer Gels, Fundamentals and Biomedical Applications; DeRossi, D; Kajiwara, K; Osada, Y; Yamauchi, A, eds.; Plenum Press: New York, 1991; 21.

334. Roth, CM; Lenhoff, AM, Electrostatic and van der Waals Contributions to Protein Adsorption: Comparison of Theory and Experiment, Langmuir 11, 3500, 1995.

335. Rothe, GM, Determination of Molecular Mass, Stoke' radius, Frictional Coefficient and Isomer-Type of Non-denatured Proteins by Time-Dependent Pore Gradient Gel Electrophoresis, Electrophoresis 9, 307, 1988.

336. Rudge, SR; Ladisch, MR, Electrochromatography, Biotechnology Progress 4, 123, 1988.

337. Ryan, DJ; Carbonell, RG; Whitaker, S, A Theory of Diffusion and Reaction in Porous Media, AIChE Symposium Series 71, 46, 1981.

338. Saarinen, TR; Johnson, CS, High-Resolution Electrophoretic NMR, Journal of the American Chemical Society 110, 3332, 1988.

339. Saarinen, TR; Woodward, WS, Computer-Controlled Pulsed Magnetic Field Gradient NMR System for Electrophoretic Mobility Measurements, Reviews of Scientific Instruments 59, 761, 1988.

340. Saez, AE; Perfetti, JC; Rusinek, I, Prediction of Effective Diffusivities in Porous Media Using Spatially Periodic Models, Transport in Porous Media 6, 143, 1991.

341. Sahimi, M, Flow and Transport in Porous Media and Fractured Rock; VCH: Weinheim, Germany, 1995.

342. Sahimi, M; Arbabi, S, Force Distribution, Multiscaling, and Fluctuations in Disordered Elastic Media, Physical Review B 40, 4975, 1989.

343. Sahimi, M; Jue, VL, Diffusion of Large Molecules in Porous Media, Physical Review Letters 62, 629, 1989.

344. Satterfield, CN, Mass Transfer in Heterogeneous Catalysis; Robert E. Krieger: Melbourne, FL, 1970.

345. Sauer, SG; Locke, BR; Arce, P, Effects of Axial and Orthogonal Applied Electric Fields on Solute Transport in Poiseuille Flows. An Area Averaging Approach, Industrial and Engineering Chemistry Research 34, 886, 1995.

346. Saville, DA, The Fluid Mechanics of Continuous Flow Electrophoresis in Perspective, Physicochemical Hydrodynamics 1, 297, 1980.

347. Saville, D; Palusinski, OA, Theory of Electrophoretic Separations Part I: Formulation of a Mathematical Model, AIChE Journal 32, 207, 1986.

348. Schellman, JA; Stigter, D, Electrical Double Layer, Zeta Potential, and Electrophoretic Charge of Double-Stranded DNA, Biopolymers 16, 1415, 1977.

349. Sellen, DB, Laser Light Scattering Study of Polyacrylamide Gels, Journal of Polymer Science: Part B: Polymer Physics 25, 699, 1987.

350. Serwer, P, Agarose Gels: Properties and Use for Electrophoresis, Electrophoresis 4, 375, 1983.

351. Shi, X; Hammond, RW; Morris, MD, DNA Conformational Dynamics in Polymer Solutions Above and Below the Entanglement Limit, Analytical Chemistry 67, 1132, 1995.

352. Shimao, K, Mathematical Simulation of Isotachophoresis Boundary Between Protein and Weak Acid, Electrophoresis 7, 297, 1986.

353. Shimao, K, Mathematical Simulation of Steady State Isoelectric Focusing of Proteins using Carrier Ampholytes, Electrophoresis 8, 14, 1987.

354. Shire, SJ; Hanania, GIH; Gurd, FRN, Electrostatic Effects in Myoglobin. Hydrogen Ion Equilibria in Sperm Whale Ferrimyoglobin, Biochemistry 13, 2967, 1974.

355. Shire, SJ; Hanania, GIH; Gurd, FRN, Electrostatic Effects in Myoglobin. Application of the Modified Tanford–Kirkwood Theory to Myoglobins from Horse, California Grey Whale, Harbor Seal, and California Sea Lion, Biochemistry 14, 1352, 1975.

356. Slater, GW; Guo, HL, Ogston Gel Electrophoretic Sieving: How Is the Fractional Volume Available to a Particle Related to Its Mobility and Diffusion Coefficient(s)?, Electrophoresis 16, 11, 1995.

357. Slater, GW; Rousseau, J; Noolandi, J; Turmel, C; Lalande, M, Quantitative Analysis of the Three Regimes of DNA Electrophoresis in Agarose Gels, Biopolymers 27, 509, 1988.

358. Slater, GW; Turmel, C; Lalande, M; Noolandi, J, DNA Gel Electrophoresis: Effect of Field Intensity and Agarose Concentration on Band Inversion, Biopolymers 28, 1793, 1989.

359. Slattery, J, Flow of Viscoelastic Fluids Through Porous Media, AIChE Journal 13, 1066, 1967.

360. Slattery, JC, Momentum, Energy, and Mass Transfer in Continua; Robert E. Krieger: Melbourne, FL, 1981.

361. Smisek, DL, Capillary Electrophoresis with Polymeric Separation Media: Considerations for Theory, Electrophoresis 16, 2094, 1995.

362. Smisek, DL; Hoagland, DA, Electrophoresis of Flexible Macromolecules: Evidence for a New Mode of Transport in Gels, Science 248, 1221, 1990.

363. Southerland, W, A Dynamical Theory of Diffusion for Non-Electrolytes and the Molecular Mass of Albumin, Philosophical Magazine 9, 781, 1905.

364. Squire, PG, A Relationship Between the Molecular Weights of Macromolecules and their Elution Volumes Based on a Model for Sephadex Gel Filtration, Archives of Biochemistry and Biophysics 107, 471, 1964.

365. Stell, G; Rikvold, PA, Polydispersity in Fluids, Dispersion, and Composites; Some Theoretical Results, Chemical Engineering Communications 51, 233, 1987.

366. Stellwagen, NC, Effect of Pulsed and Reversing Electric Fields on the Orientation of Linear and Supercoiled DNA Molecules in Agarose Gels, Biochemistry 27, 6417, 1988.

367. Stellwagen, NC, Apparent Pore Size of Polyacrylamide Gels: Comparison of Gels Cast and Run in Tris-acetate-EDTA and Tris-borate-EDTA Buffers, Electrophoresis 19, 1542, 1998.

368. Stellwagen, NC; Gelfi, C; Righetti, PG, The Free Solution Mobility of DNA, Biopolymers 42, 687, 1997.

369. Stigter, D, The Charged Colloidal Cylinder with a Gouy Double Layer, Journal of Colloid and Interface Science 53, 296, 1975.

370. Stigter, D, A Comparison of Manning's Polyelectrolyte Theory with the Cylindrical Gouy Model, Journal of Physical Chemistry 82, 1603, 1978

371. Stigter, D, Electrophoresis of Highly Charged Colloidal Cylinders in Univalent Salt Solutions. 2. Random Orientation in External Field and Application to Polyelectrolytes, Journal of Physical Chemistry 82, 1424, 1978.

372. Stigter, D, Electrophoresis of Highly Charged Colloidal Cylinders in Univalent Salt Solutions. 1. Mobility in Transverse Field, Journal of Physical Chemistry 82, 1417, 1978.

373. Stigter, D, Kinetic Charge of Colloidal Electrolytes from Conductance and Electrophoresis. Detergent Micelles, Poly(methacrylates), and DNA in Univalent Salt Solutions, Journal of Physical Chemistry 83, 1670, 1979.

374. Stigter, D, Ionic Charge Effects on the Sedimentation Rate and Intrinsic Viscosity of Polyelectrolytes. T7 DNA, Macromolecules 18, 1619, 1985.

375. Stokes, GG, Mathematical and Physical Papers; Cambridge University Press: Cambridge, 1922; Vol. III.

376. Suzuki, Y; Nishio, I, Quasielastic-Light-Scattering Study of the Movement of Particles in Gels: Topological Structure of Pores in Gels, Physical Review B 45, 4614, 1992.

377. Svec, F; Frechet, JMJ, Continuous Rods of Macroporous Polymer as High-Performance Liquid Chromatography Separation Media, Analytical Chemistry 64, 820, 1992.

378. Svec, F; Frechet, JMJ, Molded Rigid Monolithic Porous Polymers: An Inexpensive, Efficient, and Versatile Alternative to Beads for the Design of Materials for Numerous Applications, Industrial and Engineering Chemistry Research 38, 34, 1999.

379. Synge, RLM, Experiments on Electrical Migration of Peptides and Proteins Inside Porous Membranes: Influences of Adsorption, Diffusion, and Pore Dimensions, Biochemical Journal 65, 266, 1957.

380. Tan, ZJ; Remcho, VT, Molecular Imprint Polymers as Highly Selective Stationary Phases for Open Tubular Liquid Chromatography and Capillary Electrophoresis, Electrophoresis 19, 2055, 1998.

381. Tanaka, T, Phase Transitions of Gels. In: Polyelectrolyte Gels, Properties, Preparation, and Applications, ACS Symposium Series; Harland, RS; Prud'homme, RK, eds.; American Chemical Society: Washington, DC, 1992; Vol. 480, p 1.

382. Tanaka, T; Nishio, I; Sun, S-T; Uneo-Nishio, S, Collapse of Gels in an Electric Field, Science 218, 467, 1982.

383. Tanford, C, Theory of Protein Titration Curves. II. Calculations for Simple Models at Low Ionic Strength, Journal of the American Chemical Society 79, 5340, 1957.

384. Tanford, C, Physical Chemistry of Macromolecules; Wiley: New York, 1961.

385. Tanford, C; Haueinstein, JD, Phenolic Hydroxyl Ionization in Proteins. II. Ribonuclease, Journal of the American Chemical Society 78, 5287, 1956.

386. Tanford, C; Haueinstein, JD; Rands, DG, Hydrogen Ion Equilibria of Ribonuclease, Journal of the American Chemical Society 77, 6409, 1955.

387. Tanford, C; Kirkwood, JG, Theory of Protein Titration Curves. I. General Equations for Impenetrable Spheres, Journal of the American Chemical Society 79, 5333, 1957.

388. Tanford, C; Roxy, Interpretation of Protein Titration Curves Application to Lysozyme, Biochemistry 11, 2192, 1972.

389. Tanford, C; Swanson, SA; Shore, WS, Hydrogen Ion Equilibria of Bovine Serum Albumin, Journal of the American Chemical Society 77, 6414, 1955.

390. Thompson, AP; Glandt, ED, Polymers in Random Porous Materials: Structure, Thermodynamics, and Concentration Effects, Macromolecules 29, 4313, 1996.

391. Tietz, D, Evaluation of Mobility Data Obtained from Gel Electrophoresis: Strategies in the Computation of Particle and Gel Properties on the Basis of the Extended Ogston Model, Advances in Electrophoresis 2, 109, 1988.

392. Tietz, D; Gottlieb, MH; Fawcett, JS; Chrambach, A, Electrophoresis on Uncrosslinked Polyacrylamide: Molecular Sieving and Its Potential Applications, Electrophoresis 7, 217, 1986.

393. Tobita, H; Hamielec, AE, Crosslinking Kinetics in Polyacrylamide Networks, Polymer 31, 1546, 1990.

394. Tokita, M; Miyoshi, T; Takegoshi, K; Hikichi, K, Probe Diffusion in Gels, Physical Review E 53, 1823, 1996.

395. Tokita, M; Tanaka, T, Reversible Decrease of Gel-Solvent Friction, Science 253, 1121, 1991.

396. Tolbert, SH; Firouzi, A; Stucky, GD; Chmelka, BF, Magnetic Field Alignment of Ordered Silicate-Surfactant Composites and Mesoporous Silica, Science 278, 264, 1997.

397. Tomadakis, MM; Sotirchos, SV, Transport Properties of Random Arrays of Freely Overlapping Cylinders with Various Orientation Distributions, Journal of Chemical Physics 98, 616, 1993.

398. Trinh, SH; Arce, P; Locke, BR, Effective Diffusivities of Point-Like Molecules in Isotropic Porous Media by Monte Carlo Simulation, Transport in Porous Media 38, 241, 2000.

399. Trinh, S; Locke, BR; Arce, P, Diffusive-Convective and Diffusive-Electroconvective Transport in Non-Uniform Channels with Application to Macromolecular Separations, Separation and Purification Technology 15, 255, 1999.

400. Trinh, S; Locke, BR; Arce, P, Effective Diffusivity Tensors of Point-Like Molecules in Anisotropic Media by Monte Carlo Simulation, Transport in Porous Media in review, 2001.

401. Tyrrell, HJV; Harris, KR, Diffusion in Liquids; Butterworths: London, 1984.

402. Vaidya, D; Diamond, SL; Nitsche, JM; Kofke, DA, Potential for Use of Liquid Crystals as Dynamically Tunable Electrophoretic Media, AIChE Journal 43, 1366, 1997.

403. van Holde, KE, Physical Biochemistry, 2nd ed.; Prentice-Hall, Englewood Cliffs, NJ, 1985.

404. Vlatakis, G; Andersson, LI; Muller, R; Mosbach, K, Drug Assay using Antibody Mimics Made by Molecular Imprinting, Nature 361, 645, 1993.
405. Volkel, AR; Noolandi, J, Mobilities of Labeled and Unlabeled Single-Stranded DNA in Free Solution Electrophoresis, Macromolecules 28, 8182, 1995.
406. Volkel, AR; Noolandi, J, On the Mobility of Stiff Polyelectrolytes, Journal of Chemical Physics 102, 5506, 1995.
407. Volkmuth, WD; Austin, RH, DNA Electrophoresis in Microlithographic Arrays, Nature 358, 600, 1992.
408. Vrentas, JS; Duda, JL, Diffusion in Polymer-Solvent Systems. I. Reexamination of the Free-Volume Theory, Journal of Polymer Science: Polymer Physics Edition 15, 403, 1977.
409. Vrentas, JS; Duda, JL, Diffusion in Polymer-Solvent Systems. II. A Predictive Theory for the Dependence of Diffusion Coefficients on Temperature, Concentration, and Molecular Weight, Journal of Polymer Science: Polymer Physics Edition 15, 417, 1977.
410. Vrentas, JS; Duda, JL, Diffusion in Polymer-Solvent Systems. III. Construction of Deborah Number Diagrams, Journal of Polymer Science: Polymer Physics Edition 15, 441, 1977.
411. Wakao, N; Smith, JM, Diffusion in Catalyst Pellets, Chemical Engineering Science 17, 825, 1962.
412. Waldmann-Meyer, HK, Protein Ion Equilibria, Total Evaluation of Binding Parameters and Net Charge from the Electrophoretic Mobility as a Function of Ligand Concentration. In: Recent Developments in Chromatography and Electrophoresis; Frigerio, A; McCamish, M, eds.; Elsevier Scientific: Amsterdam, 1980; Vol. 10, p 125.
413. Waldmann-Meyer, H, Structure Parameters of Molecules and Media Evaluated by Chromatographic Partition I. Controlled-Pore Glasses, Journal of Chromatography 350, 1, 1985.
414. Waldmann-Meyer, H, Structure Parameters of Molecules and Media Evaluated by Chromatographic Partition II. Geometrical Exclusion in Gels, Journal of Chromatography 410, 233, 1987.
415. Wang, JH, Theory of the Self-Diffusion of Water in Protein Solutions. A New Method for Studying the Hydration and Shape of Protein Molecules, Journal of the American Chemical Society 76, 4755, 1954.
416. Wanka, G; Hoffman, H; Ulbricht, W, Phase Diagrams and Aggregation Behavior of Poly (oxyethylene)-Poly(oxypropylene)-Poly(exyethylene) Triblock copolymers in Aqueous Solutions, Macromolecules 27, 4145, 1994.
417. Weirsema, PH; Loeb, AL; Overbeek, JTG, Calculation of the Electrophoretic Mobility of a Spherical Colloid Pariole, Journal of Colloid and Interface Science 22, 78, 1966.
418. Weiss, GH; Garner, M; Yarmola, E; Bocek, P; Chrambach, A, A Comparison of Resolution of DNA Fragments Between Agarose Gel and Capillary Zone Electrophoresis in Agarose Solutions, Electrophoresis 16, 1345, 1995.
419. Weissberg, HL, Effective Diffusion Coefficient in Porous Media, Journal of Applied Physics 34, 2636, 1963.
420. Whitaker, S, Diffusion and Dispersion in Porous Media, AIChE Journal 13, 420, 1967.
421. Whitaker, S, Introduction to Fluid Mechanics; Kreiger: Malabar, FL 1968.
422. Whitaker, S, The Transport Equations for Multi-Phase Systems, Chemical Engineering Science 28, 139, 1973.
423. Whitaker, S, A Simple Geometrical Derivation of the Spatial Averaging Theorem, Chemical Engineering Education Winter, 18, 1985.
424. Whitaker, S, Flow in Porous Media I: A Theoretical Derivation of Darcy's Law, Transport in Porous Media 1, 3, 1986.
425. Whitaker, S, Local Thermal Equilibrium: An Application to Packed Bed Catalytic Reactor Design, Chemical Engineering Science 41, 2029, 1986.
426. Whitaker, S, Transient Diffusion, Adsorption and Reaction in Porous Catalysts: The Reaction Controlled, Quasi-Steady Catalytic Surface, Chemical Engineering Science 41, 3015, 1986.

427. Whitaker, S, Transport Processes with Heterogeneous Reaction. In: Concepts and Design of Chemical Reactors; Whitaker, S; Cassano, AE, eds.; Gordon and Breach: Newark, NJ 1986; 1.
428. Whitaker, S, Mass Transport and Reaction in Catalyst Pellets, Transport in Porous Media 2, 269, 1987.
429. Whitaker, S, Diffusion in Packed Beds of Porous Particles, AIChE Journal 34, 679, 1988.
430. Whitaker, S, The Development of Fluid Mechanics in Chemical Engineering. In: One Hundred Years of Cheical Engineering; Peppas, NA, ed.; Kluwer Academic: Dordrecht The Netherlands, 1989; 47.
431. Whitaker, S, Heat Transfer in Catalytic Packed Bed Reactors. In: Handbook of Heat and Mass Transfer; Cheremisinoff, NP, ed.; Gulf: Matawan, NJ, 1989; Vol. 3, p 361.
432. Whitaker, S, Improved Constraints for the Principle of Local Thermal Equilibrium, Industrial and Engineering Chemistry Research 29, 983, 1991.
433. Whitaker, S, The Method of Volume Averaging: An Application to Diffusion and Reaction in Porous Catalysts. In Proceedings of the National Science Council, Part A: Physical Science and Engineering; National Science Council: Taipei, Taiwan, Republic of China, 1991; Vol. 15, p 465.
434. Whitaker, S, Role of the Species Momentum Equation in the Analysis of the Stefan Diffusion Tube, Industrial and Engineering Chemistry Research 29, 978, 1991.
435. Whitaker, S, The Species Mass Jump Condition at a Singular Surface, Chemical Engineering Science 47, 1677, 1992.
436. Whitaker, S, The Method of Volume Averaging; Kluwer Academic: Dordrecht, The Netherlands, 1998.
437. Wilke, CR; Chang, P, Correlation of Diffusion Coefficients in Dilute Solutions, AIChE Journal 1, 264, 1955.
438. Wu, C; Liu, T; Chu, B; Schneider, D; Graziano, V, Characterization of the PEO-PPO-PEO Triblock Copolymer and Its Application as a Separation Medium in Capillary Electrophoresis, Macromolecules 30, 4574, 1997.
439. Wu, G; Chu, B; Schneider, DK, SANS Study of the Micellar Structure of PEO/PPO/PEO Aqueous Solutions, Journal of Physical Chemistry 99, 5094, 1995.
440. Ye, L; Ramstrom, O; Mansson, MO; Mosbach, K, A New Application of Molecularly Imprinted Materials, Journal of Molecular Recognition 11, 75, 1998.
441. Yiu, Y; Locke, BR; Van Winkle, DH; Rill, RL, Optimizing Capillary Gel Electrophoretic Separations of Oligonucleotides in Liquid Crystalline Pluronic F127, Journal of Chromatography A817, 367, 1998.
442. Zhao, D; Feng, J; Hou, Q; Melosh, N; Gredirckson, GH; Chmelka, BF; Stucky, GD, Triblock Copolymer Syntheses of Mesoporous Silica with Periodic 50- to 300-Angstrom Pores, Science 279, 548, 1998.
443. Zhou, Z; Chu, B, Light-Scattering Study on the Association Behavior of Triblock Polymers of Ethylene Oxide and Propylene Oxide in Aqueous Solution, Journal of Colloid and Interface Science 126, 171, 1988.
444. Ziabicki, A, Topological Structure and Macroscopic Behavior of Permanently Crosslinked Polymer Systems, Polymer 20, 1373, 1979.
445. Ziabicki, A; Walasek, J, Topological Structure and Physical Properties of Permanently Cross-Linked Systems. 1. s-Functional, Homogenerous, Gaussian Systems, Macromolecules 11, 471, 1978.
446. Zimm, BH, Dynamics of Polymer Molecules in Dilute Solution: Viscoelasticity, Flow Birefringence and Dielectric Loss, Journal of Chemical Physics 24, 269, 1956.
447. Zimm, BH; Levene, SD, Problems and Prospects in the Theory of Gel Electrophoresis of DNA, Quarterly Reviews of Biophysics 25, 171, 1992.
448. Penke, B, Analysis of Diffusion and Structure in Polyacryalamide Gels by Nuclear Magnetic Resonance, M.S. Thesis, Florida State University, 1997.

16
Electrolytes in Nanostructures

KWONG-YU CHAN   The University of Hong Kong, Hong Kong SAR, China

I. INTRODUCTION

The solid–electrolyte interface takes a variety of geometrical expressions in the nanometer scale. Figure 1 illustrates some examples of these "solid"–electrolyte interfaces. The classical picture of Figure 1(a) is a charged planar rigid surface meeting a semi-infinite region of electrolyte solution. Induced by fixed charges of the solid surface, a differential distribution of ions forms in the solution layer adjacent to the interface. This electrochemical double layer has been a focus of investigations in electrochemistry and related fields such as colloids, materials science, and biology. Classical theories of Gouy–Chapman [1,2] and Stern [3] for the electrochemical double layer have formed the basis for intelligible interpretation of experiments and for useful applications in electrochemical and materials technologies. An electrolyte confined between two surfaces separated by nanometer dimensions behaves differently compared to the case of an open semi-infinite boundary. This confined situation can be found in colloids, as shown in Figure 1(b), or in a porous solid or membranes, as shown in Figure 1(c). The confining boundary or internal pores can be with or without fixed wall charges. Complexity increases with the dynamics of the fixed charges or solid structures in the molecular scale. In Figure 1(d), the dynamic surface structure of a polyelectrolyte (or macroion) has an interdependent relationship with other electrolyte and polyelectrolyte molecules in the solution. An important area is the study of protein folding, where, at the moment, consideration of interaction with explicit electrolyte molecules is rare. One class of these polyelectrolytes is ionic surfactants. Different levels of ordering in ionic or nonionic amphiphilic molecules can lead to a monolayer structure or a microemulsion, as shown in Figure 1(e), and a bilayer membrane with pores, as shown in Figure 1(f). Modern technology has exploited the nanoscopic domains, and important applications can be found in these nano-electrolyte interfaces in Figure 1. One example is the application of scanning probe microscopy in a solution environment. The scanning tip interacts closely with the solid surface and the electrolyte. The scanning tip can be that of an atomic force microscope (AFM), a scanning tunneling microscope (STM), or a near-field scanning optical microscope (NSOM).

The role of electrolyte is critical in these nanoscopic interfaces, but is difficult to predict and quantify. For sufficiently large rigid interfacial structures, one can apply the model of electrolyte interaction with a single charged surface in Figure 1(a). The double-layer theories or the recent integral-equation theories have been applied. Reviews of this subject are available in the literature [4,5]. For electrolytes in a nanostructure, the double layers from two surfaces overlap and behave differently from the case of a single surface. Ad-

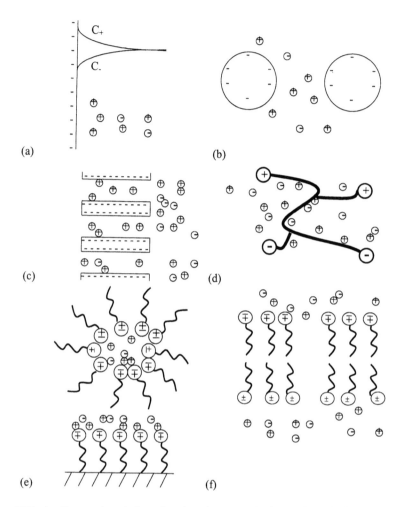

FIG. 1 Geometries of electrolyte interfaces. (a) A planar electrode immersed in a solution with ions, and with the ion distribution in the double layer. (b) Particles with permanent charges or adsorbed surface charges. (c) A porous electrode or membrane with internal structures. (d) A polyelectrolyte with flexible and dynamic structure in solution. (e) Organized amphophilic molecules, e.g., Langmuir–Blodgett film and microemulsion. (f) Organized polyelectrolytes with internal structures, e.g., membranes and vesicles.

vances have been made in experimental studies of nanoscopic phenomena, but studies of electrolytes confined in nanostructures are not common, due partly to the difficult characterization of the charged nanosurface and partly to the dynamics in a solution environment. Molecular modeling provides a useful tool for understanding electrolytes in nanostructures and near nano-surfaces. This chapter attempts to review some of the latest developments in the molecular modeling of electrolytes confined in nanostructures. We will focus our attention on the electrolyte solution and mostly ignore the molecular details and dynamics of the solid structure. Some pertinent questions of interest are: What are the equilibrium concentrations of ions in nanopores? What is the selectivity? What is the force between charged surfaces immersed in an electrolyte solution? How does nano-confinement affect ion diffusion and migration? Because of limitations in analytical theories, mainly computer simulations results will be presented. Section II introduces some models of electrolyte–solid interface studies. Equilibrium properties of adsorption, neutrality, solvent effect, and selectivity will be discussed in Section III. Electrolyte-mediated forces between charged surfaces will be discussed in Section IV, and transport properties will be discussed in Section V. Relevant experimental studies will be discussed, especially those made by recent advances in the nanometer scale.

II. MODELS

A typical biological example of electrolytes confined in a nanostructure is the gramacidin channel in a molecular electrolyte environment. A complete atomistic description of the electrolyte–nano-surface interactions is desirable, for it can account for all the degrees of freedom at the atomic level. Analytical solutions and theories of such a model are difficult, if not impossible, and have not been developed. Computer simulation techniques of molecular dynamics and Monte Carlo simulations have been applied to various extents to the gramacidin channel [6–14]. The computational demand for a complete atomistic model of channel and electrolyte, however, exceeds the present power of computers and restricts the time scale, length scale, and concentration range of electrolytes to be modeled. Only short-time dynamics and short-range structural information can be probed. Capturing the linkage of electrolytes from the nanostructure to the bulk state is difficult. For long-time dynamics, equilibrium concentrations, and profiles of electrolytes from within the electrolyte to the bulk solution, simplifications and approximations have to be made in the model. Table 1 lists some common approaches to describe the surface, the solvent, and the ions.

A. Electrolytes

1. Solvent Primitive Model

The simplest way to treat the solvent molecules of an electrolyte explicitly is to represent them as hard spheres, whereas the electrostatic contribution of the solvent is expressed implicitly by a uniform dielectric medium in which charged hard-sphere ions interact. A schematic representation is shown in Figure 2(a) for the case of an idealized situation in which the cations, anions, and solvent have the same diameters. This is the solvent primitive model (SPM), first named by Davis and coworkers [15,16] but appearing earlier in other studies [17]. As shown in Figure 2(b), the interaction potential of a pair of particles (ions or solvent molecule), i and j, in the SPM are:

TABLE 1   Models of Electrolytes in a Nanostructure

Surface	Ion	Solvent
Hard wall	Point ions (Debye–Huckel)	Constant dielectric background
Soft and smooth wall	Charged hard spheres	Low dielectric layer
Discrete sites with axial and/or radial variations	Charged soft spheres (LJ)	Neutral hard spheres and constant dielectric background
Atom dynamics	Specific adsorption	Dipolar hard sphere
Group contribution and rigid bonds/angels		SPC, ST2, TIPS
		Polarizable; H Bonds

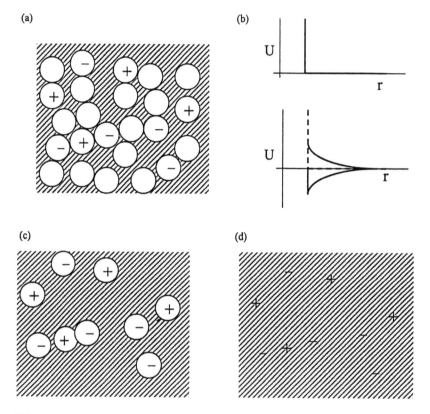

FIG. 2 (a) Solvent primitive model, with charged hard spheres representing the ions, neutral hard spheres as solvent, and a dielectric background. (b) The pair potentials in the SPM model. (c) The primitive model with no explicit presence of solvent molecules. (d) The point-ion model.

$$u_{ij}(r_{ij}) = \left[\begin{cases} \infty & \text{all pairs} & r_{ij} \le d_{ij} \\ \dfrac{z_i z_j e^2}{4\pi\varepsilon_0\varepsilon r_{ij}} & \text{ion–ion} \\ 0 & \text{solvent–solvent, solvent–ion} \end{cases} \right\} \quad r_{ij} > d_{ij} \qquad (1)$$

where r_{ij} is the distance between two particles i and j, $d_{ij} = (d_{ii} + d_{jj})/2$, and d_{ii} is the diameter of the i species and i can be $+$, $-$, or s, representing either the cation, the anion, or the solvent, respectively. Here, e is the electronic charge of 1.6×10^{-19} C, z_i is the charge valency of the ion, ϵ_0 is the permitivity in vacuum, and is ϵ the dielectric constant. The molecular packing will depend on the number density of the solvent ρ_s and the number density of the ions ρ_+ and ρ_-. The concentration of a symmetric electrolyte in moles/liter will be $\rho_+/(1000N_{Av})$, where N_{Av} is Avogadro's number and ρ_+ is in molecules/m³. For an asymmetric electrolyte, the smaller of ρ_+ and ρ_- will be used to calculate the molar concentration. The more popular restricted primitive model (RPM)[18–30] and point ions (Debye–Hückel) model [32–34] can be treated as simplified and degenerate cases of the SPM. More realistic models of solvent with a hard-sphere core, such as the dipolar hard sphere (DHS) model [35–39] and simple point charge (SPC) model [43,44], can be treated as an extension of the SPM by adding extra solvent–solvent and solvent–ion interactions.

2. Primitive Model

The popular and well-studied primitive model is a degenerate case of the SPM with $\rho_s = 0$, shown schematically in Figure (c). The restricted primitive model (RPM) refers to the case when the ions are of equal diameter. This model can realistically represent the packing of a molten salt in which no solvent is present. For an aqueous electrolyte, the primitive model does not treat the solvent molecules explicitly and the number density of the electrolyte is unrealistically low. For modeling nano-surface interactions, short-range interactions are important and the primitive model is expected not to give adequate account of confinement effects. For its simplicity, however, many theories [18–22] and simulation studies [23–25] have been made based on the primitive model for the bulk electrolyte. Applications to electrolyte interfaces have also been widely reported [26–30].

3. Point-Ions Model

Further simplification of the SPM and RPM is to assume the ions are point charges with no hard-core correlations, i.e., $d_{ii} = 0$. This is called the Debye–Hückel (DH) level of treatment, and an early Nobel prize was awarded to the theory of electrolytes in the infinite-dilution limit [31]. This model can capture the long-range electrostatic interactions and is expected to be valid only for dilute solutions. An analytical solution is available by solving the Poisson–Boltzmann (PB) equation for the distribution of ions (charges). The PB equation is

$$\nabla^2 \phi = -\frac{1}{\varepsilon \varepsilon_0} \Sigma \, z_i e \rho_i \, exp\left(-\frac{z_i e \phi}{kT}\right) \tag{2}$$

where k is the Boltzmann factor, T is the absolute temperature, ϕ is the electric potential, and other symbols are the same as in Eq. (1). At low surface charges and dilution concentration, the DH model has been applied successfully to a single isolated double layer to yield the Gouy–Chapman theory [1,2]. The application of the DH model to other interfaces [32–34] often requires numerical solution of the PB equation.

4. Soft-Core Models

Instead of the hard-sphere model, the Lennard–Jones (LJ) interaction pair potential can be used to describe soft-core repulsion and dispersion forces. The LJ interaction potential is

$$u_{LJ}(r_{ij}) = 4\varepsilon_{LJ}\left[\left(\frac{d_{ij}}{r_{ij}}\right)^{12} - \left(\frac{d_{ij}}{r_{ij}}\right)^{6}\right] \tag{3}$$

where ε_{LJ} is the well depth of the attractive dispersion interaction potential, located at $r_{ij} = 2^{1/6} d_{ij}$. The soft core is therefore the cut-and-shift version of the LJ core potential. If only the soft-core part is modeled, then a cut-and-shift version of the LJ potential can be used to replace the hard core in the primitive model and solvent primitive model, as described by

$$u(r_{ij}) = \begin{cases} u_{LJ}(r_{ij}) + \varepsilon + \dfrac{q_i q_j}{4\pi\varepsilon\varepsilon_0 r_{ij}} & r_{ij} < 2^{1/6} d_{ij} \\[2ex] \dfrac{q_i q_j}{4\pi\varepsilon\varepsilon_0 r_{ij}} & r_{ij} > 2^{1/6} d_i \end{cases} \tag{4}$$

The soft-core model may be more convenient in molecular dynamics simulation, since a continuously differentiable potential is available to calculate the force. In the case of a hard-core potential, collision times of all atom pairs have to be monitored and used to control the time step.

5. Water Solvent Model

More realistic treatment of the electrostatic interactions of the solvent can be made. The dipolar hard-sphere model is a simple representation of the polar nature of the solvent and has been adopted in studies of bulk electrolyte and electrolyte interfaces [35–39]. Recently, it was found that this model gives rise to phase behavior that does not exist in experiments [40,41] and that the Stockmeyer potential [41,42] with soft cores should be better to avoid artifacts. Representation of higher-order multipoles are given in several popular models of water, namely, the simple point charge (SPC) model [43] and its extension (SPC/E) [44], the transferable interaction potential (TIPS)[45], and other central force models [46–48]. Models have also been proposed to treat the polarizability of water [49].

B. Nano-Surface

1. Hard Wall

The simplest description of a charged nano-surface is a hard impenetrable wall with electric charges uniformly distributed and localized at the surface, as shown schematically in Figure 1(a). For a smooth planar hard wall, the ion–wall interaction potential is described mathematically as

$$u_{\text{wall}}(x) = \begin{cases} \infty & x < d_{ij}/2 \\ z_i e v_{qw}(x) & x \geq d_{ii}/2 \end{cases} \tag{5}$$

where

$$v_{qw}(x) = -\frac{\sigma_w}{4\pi\varepsilon\varepsilon_0} x$$

is the electrostatic interaction of the charged wall with a charge away from the surface and σ_w is the surface charge density. The model of point ions interacting with a charged hard wall is the classical Gouy–Chapman (GC) theory [1,2]. The modified Gouy-Chapman (MGC) model can treat the ion–wall interaction with a hard core, but still ignores the ion–ion core interaction. The MGC theory can adequately describe the ion distribution near the charged surface when the Debye length is large. For electrolytes confined in small length scales, the double layers of different surfaces will interact, and at the same time, the ion–ion correlation becomes important. Also, numerical solution of the point-ion model is needed for the ion-pore model with a finite boundary [34,50]. Using integral equation theories and grand canonical Monte Carlo (GCMC) simulations, Vlachy, Haymet, and coworkers have shown that the point-ion approach is still good for certain cases of 1:1 electrolyte but fails for 2:2 electrolytes [51,52]. They have quantified salt exclusion behavior in charged micropores [53].

2. Soft Wall

To represent the elasticity and dispersion forces of the surface, an approach similar to that of Eqs. (3) and (4) can be taken. The wall molecules can be assumed to be smeared out. And after performing the necessary integration over the surface and over layers of molecules within the surface, a 10–4 or 9–3 version of the potential can be obtained [54,55]. Discrete representation of a hexagonal lattice of wall molecules is also possible by the Steele potential [56]. The potential is essentially one dimensional, depending on the distance from the wall, but with periodic variations according to lateral displacement from the lattice molecules. Such a representation, however, has not been developed in the cylindrical pore wall geometry except for a smeared-out version of the lattice [57]. The alternate approach, as adopted by Gubbins and coworkers [58], is to predetermine the potential distribution by calculating the values in a fine grid.

3. Electrostatics of the Nano-Surface

The technical difficulty in calculating long-range electrostatic interactions is the major stumbling block in simulation of the electrolyte interface. For a bulk electrolyte, the three-dimensional symmetry allows special techniques of Ewald summation [59] and reaction field method [60] to be deployed. The loss of three-dimensional symmetry in an inhomogeneous interface requires modifications of these techniques [61–63] or use of other techniques. In a planar geometry, Torrie and Valleau [27] have used uniform charged sheets to represent the periodic charge images in the lateral dimensions and a closed-form expression was obtained. Boda et al. [39] have extended the method and used one charged sheet per real charge in the simulation cell in order to achieve a better Markov chain in Monte Carlo procedures. Lee and Chan [64] have applied this method to electrolytes confined in a slit pore. A schematic diagram is shown in Figure 3(a) for two infinitely charged planes and the periodic boundary conditions. This method has not been applied to molecular dynamics simulation. For a cylindrical geometry, i.e., a nanopore, a closed-form expression of the integration over charged cylindrical sheets cannot be obtained. Charged lines may be

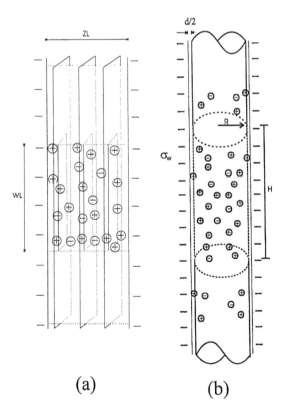

(a)   (b)

FIG. 3 Setup of simulation cell of confined electrolyte with periodic boundary conditions. (a) Electrolyte bound by two infinitely long charged plates, representing a slit pore. (b) Electrolyte in a cylindrical nanopore.

used instead to represent the images. In the simulations of Lo and Chan [65] and Lee et al. [66], a simple cutoff was used. Their justification was that with a sufficiently long simulation cell, the truncated tail was negligible. The electrostatic interaction of a given ion to the cylindrical charged pore wall is given as

$$v_{qw} = \frac{\sigma_w R}{\varepsilon_0 \varepsilon_r} \left[ln \frac{H}{2R} + \ln \left(1 + \sqrt{1 + \left(\frac{2R}{H} \right)^2} \right) \right] \tag{6}$$

where R is the radius of the pore and H is the length of the simulation cell, which must be more than $10R$ for the cutoff to be reasonable. A schematic diagram of the simulation setup is shown in Figure 3(b).

III. ELECTROLYTE ADSORPTION

A question of practical interest is the amount of electrolyte adsorbed into nanostructures and how this depends on various surface and solution parameters. The equilibrium concentration of ions inside porous structures will affect the applications, such as ion exchange resins and membranes, containment of nuclear wastes [67], and battery materials [68]. Experimental studies of electrosorption studies on a single planar electrode were reported [69]. Studies on porous structures are difficult, since most structures are ill defined with a wide distribution of pore sizes and surface charges. Only rough estimates of the average number of fixed charges and pore sizes were reported [70–73]. Molecular simulations of nonelectrolyte adsorption into nanopores were widely reported [58]. The confinement effect can lead to abnormalities of lowered critical points and compressed two-phase envelope [74].

The theoretical and simulation studies of electrolyte adsorption in nanopores were reported using the restricted primitive model [51–53,64–66,75–77]. The main questions are how the equilibrium concentration of counterion and coion can be affected by the confinement and the electrostatics of the charged surfaces. Based on the RPM model, Haymet et al. reported that for 1:1 electrolyte, there is good agreement between integral equation theory, PB theory, and Monte Carlo results but that for 2:2 electrolyte the theories break down [51]. The salt exclusion effect based on coion concentration were reported for a number of pore geometries and charges [52,53]. In experimental studies of the ion exchange capacity, it was conveniently assumed that (1) all coions will be excluded and (2) the number of counterions will be equal to the number of fixed charges. The activity of ions in solution, however, will mean that some coions can enter, and both assumptions will not be valid. Some experimental studies of ion exchange membranes have confirmed this [78,79]. On the other hand, for larger porous structures, it may be assumed that no coions are excluded, and its concentration is the same as that in the bulk, whereas the concentration of the counterion will be in excess to balance the wall charges. In the theoretical approaches, it is usually assumed that the excess charges in the pore solution (i.e., the number of counterions minus the number of coions) should balance those of the fixed charges. This is the electroneutrality assumption. The concentration of coion, however, will be determined by its activity inside the pore and the confinement effect can lead to its exclusion.

A. Charge-Induced Concentration Profiles

In the theoretical approaches of Poisson–Boltzmann, modified Gouy-Chapman (MGC), and integral equation theories such as HNC/MSA, concentration or density profiles of counterions and coions are calculated with consideration of the ion–wall and ion–ion interactions. The amount of ions or charges adsorbed in an idealized single pore can be obtained by integrating the local density over the cross section of the pore. A similar approach is taken in molecular simulation in the canonical ensemble (constraints of constant temperature and of volume and number of particles), but the exact number of counterions and coions can also be counted throughout the simulation. In the grand canonical ensemble Monte Carlo (GCMC) simulation, however, a more direct link between the pore fluid and an equilibrium external fluid can be made by specifying the chemical potential and allowing the number of ions to fluctuate until equilibrium is reached. The chemical potential can be expressed by Widom's equation [80,81],

$$\beta\mu_i(r) = \ln \rho_i(r_i) + 3 \ln \Lambda_i - \ln[\langle exp(-\beta\{u(r_{ij}) + u_{wall}(r_i)\})\rangle] - z_i e\psi \tag{7}$$

where $\rho_i(r_i)$ is the local number density of ion i, Λ_i is the thermal de Broglie wavelength, $\langle\ \rangle$ is the ensemble average, $\beta = 1/kT$ and ψ is the external potential. In Eq. (7), the first two terms account for the ideal activity of the ideal gas, the second term accounts for interaction with the wall and other particles, and the last term accounts for interaction with an external field.

Figures 4 and 5 show the concentration profiles determined by theory and simulation for a 1:1 electrolyte inside an uncharged and a charged cylindrical pore, respectively, with radius five times the diameter of the ion. Results for the RPM and SPM models are presented for the simulation. In an uncharged pore, the profiles of cations and anions in a symmetric electrolyte are the same. The concentration is higher at contact with the wall, and with a solvent, the packing effect is stronger and the profiles more structured, as shown in the curves of the SPM model. The normalized profiles are the same for the neutral hard-sphere solvent and the charged ions. These results are identical to hard spheres inside a pore, for which the integral equation theory predictions, such as HNC, agree with the simulation results. For the pore with a surface charge of -0.05 C/m^2, the induced profiles of counterion and anion are markedly different, especially near contact with the wall. For a 1:1 electrolyte, the Poisson–Boltzmann prediction is roughly correct for the

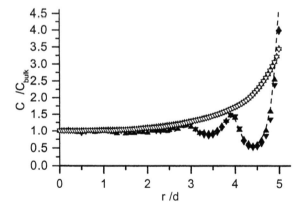

FIG. 4   Normalized concentration distribution of a 0.1 molar 1:1 electrolyte in an uncharged cylindrical pore of radius five times the diameter of the ions. The dashed line, solid up-triangles, and solid down-triangles are the neutral solvent particles, cations, and anions, respectively, in an SPM model with 0.3 solvent packing fraction. The open symbols are for the cations and anions in the RPM model.

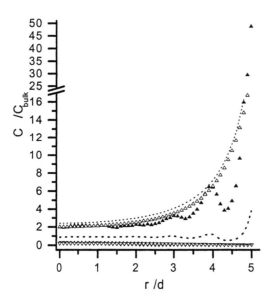

FIG. 5 Normalized concentration distribution in the pore of Figure 4 but charged with -0.05 C/m^2. The symbols are the same as in Figure 4, with the cations being the counterions. The anions (coions) of the RPM and SPM model are not distinguishable on the present scale. The dotted line is the prediction of the modified Gouy–Chapman theory and approximates the simulation results of the RPM.

RPM model, but theories for the SPM model electrolyte inside a nanopore have not been reported. It is noticed that everywhere in the pore, the concentration of counterion is higher than the bulk concentration, also predicted by the PB solution. However, neutrality is assumed in the PB solution but is violated in the single-ion GCMC simulation, since the simulation result of the counterion in the RPM model is everywhere below the PB result. There is exclusion of coion, for its concentration is below the bulk value throughout the pore. Only the solvent profile in the SPM model has the bulk value in the center of the pore.

B. Reduced Ion Exchange Capacity/Nonneutrality

A key question in equilibrium adsorption of ions is whether fixed charges in the pore are balanced by an equivalent amount of excess counterions. In theories and simulations, it was convenient to assume the electroneutrality condition in a nanopore. In the earlier simulations in the GCMC ensemble [51–53], ions were inserted and deleted one pair at a time and electroneutrality was preserved. In their molecular dynamics (MD) simulations, Lo et al. [50] notice that the chemical potentials of counterions and coions in a symmetric RPM electrolyte are not equal unless electroneutrality is violated inside the pore. Grand canonical Monte Carlo (GCMC) simulation works were reported [64–66,75] in which single ionic activities are specified and the number of individual ions can be varied without the necessity of an overall neutral pore. This nonneutrality effect has been reported in spherical pore [75], cylindrical pore [65,66], and slit pore [64]. Lozada–Cassou and coworkers have also shown the excess charges in a slit pore using integral equation theories [82–85]. The dependence of excess charges on pore size obtained from the GCMC is shown in Figure 6. The excess surface charge per unit pore area is defined as

$$\sigma_{ex} = \frac{(N_+ - N_-)}{A} + \sigma_w \qquad (8)$$

where N_+ is the number of cations in the pore, N_- is the number of anions, A is the area of the pore wall, and σ_w is the surface charge density of the pore wall. Without loss of generality, it can be assumed that the wall is negatively charged and cations will be the counterions. From Figure 6 it can be seen that in the limit of large pores, σ_{ex} is basically zero and electroneutrality is obeyed. In the limit of very small pores, no ions can enter and $\sigma_{ex} = \sigma_w$; i.e., the excess unbalanced charges will be the same as the fixed-wall charges. In between the two limits, there will be a gradual variation of counterion exclusion. The exclusion of counterion expressed in terms of excess charges is successively one order of magnitude higher in going from a slit pore to a cylindrical pore and then to a spherical pore. This is due to the successive reduction in the degrees of freedom. The results in Figure 6 are from the primitive model. Anticipating a more severe confinement effect in the more densely packed SPM model, it was surprising, however, to find that the exclusion of counterions is nearly the same as compared to the RPM results in Figure 7. It could be interpreted that the electrostatics effects are similar in the two models, and the exclusion effect depends on the difference of chemical potential between the bulk and the pore fluid. Though the packing is dense in the SPM pore fluid, it is also higher in the bulk, and therefore no extra exclusion is experienced in the SPM model, compared to the RPM model electrolyte.

C. Salt Exclusion

Since there is normally adsorption of counterion, the exclusion of electrolytes has been conventionally defined based on the exclusion of the coion. The exclusion coefficient is defined as

$$\Gamma = \frac{[C_{bulk} - \langle C_{coion} \rangle]}{C_{bulk}} \qquad (9)$$

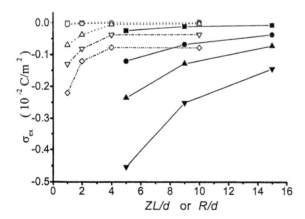

FIG. 6 The unbalanced surface charge in a nanopore with different size and charges and in equilibrium with a 0.1 molar 1:1 RPM model electrolyte. The solid squares, circles, up-triangles, and down-triangles represent original surface charges of −0.001, −0.005, −0.01, and −0.2 C/m², respectively, in a cylindrical pore. The open squares, circles, up-triangles, down-triangles, and diamonds represent original surface charges of −0.05, −0.1, −0.2, −0.25, and −0.3 C/m², respectively, in a slit pore.

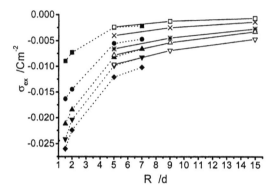

FIG. 7 Unbalanced surface charges in a cylindrical pore in equilibrium with a 0.1 molar 1:1 electrolyte. The solid squares, circles, up-triangles, down-triangles, and diamonds represent original surface charges of -0.01, -0.03, -0.05, -0.05, and -0.1 C/m^2, respectively, from the GCMC results of an SPM model with an 0.3 solvent packing fraction. The open squares, crosses, stars, open up-triangles, and open down-triangles represent original surface charges of -0.01, -0.02, -0.04, -0.05, and -0.07 C/m^2, respectively, from the results of an RPM model electrolyte.

where $\langle C_{coion} \rangle$ is the statistical average of the coion concentration and C_{bulk} is the concentration of the coion in the bulk solution. Figure 8 shows the exclusion coefficient for a 1:1 electrolyte inside a cylindrical pore in the RPM and SPM models, obtained by GCMC simulation. The RPM shows a higher exclusion effect for the coion, whereas there is little difference in the amount of unbalanced charges between the SPM and RPM models. It means that exclusion of the counterion is also higher in the RPM model. While the packing is less dense in the RPM model, this result could be due to osmosis effect of the solvent. The presence of solvent molecules may enhance the adsorption of the ions into the nanopore. Comparing results for different pore sizes, it is also seen that the exclusion decreases rapidly

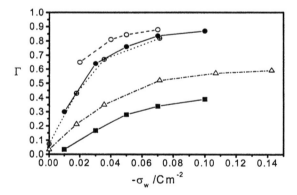

FIG. 8 Salt exclusion as a function of surface charge in a cylindrical pore in equilibrium with a 0.1 molar electrolyte. The open circles are GCMC results for 1:1 RPM electrolyte in a pore of $R = 5d$. The circles with a centered cross are results for a 2:1 electrolyte in a pore of $R = 5d$. The up-triangles are results for a 2:1 electrolyte in a pore of $R = 10d$. The solid circles are results for a 1:1 SPM model with 0.3 solvent packing fraction in a pore of $R = 5d$. The solid squares are the same results for a pore of $R = 7d$.

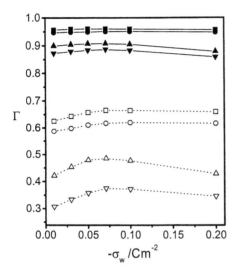

FIG. 9 Salt exclusion as a function of surface charge in a cylindrical pore in equilibrium with an SPM electrolyte of 0.1 solvent packing fraction. The open squares and circles are results of a 0.4 molar 1:2 electrolyte in a pore of $R = 5.0d$ and $R = 7.0d$, respectively. The open up-triangles and down-triangles are results of a 0.4 molar 2:2 electrolyte in a pore of $R = 5.0d$ and $R = 7.0d$, respectively. The corresponding solid symbols are results under the same conditions, but with a 0.1 molar electrolyte.

with pore size. The exclusion shows a monotone increase with surface charge, but for a 2:2 electrolyte it was found that maximum exclusion occurs at a particular surface charge, as shown in Figure 9. Again, the decrease of coion exclusion at a higher surface charge may be due to electrosmosis since the drop in concentration is enhanced for a divalent coion.

D. Donnan Potential and External Field

With a membrane that selectively excludes one type of ions but not the counterions, an electric potential is built up between the pore fluid and the bulk fluid and defined as the Donnan potential. This is potential counterbalance to the nonneutrality created by the exclusion of one type of charges. A similar situation is present with the nonneutrality in a charged nanopore, even for a symmetric electrolyte where the counterion and coion have the same size. Lo [86] has explored the effect of an external potential on the electrolyte equilibrium in a charged nanopore. Figure 10 shows the change of excess charges as a function of external potential and the surface charges. Electroneutrality is retained with a specific external potential equivalent to the Donnan potential [87]. Figure 11 shows the potential variation from a confined fluid to the bulk solution. With the Donnan potential, the transition of electric potential from the internal to the external fluid is smooth. The Donnan potential varies linearly with low-surface-charge density, as shown in Figure 12.

IV. FORCES BETWEEN CHARGED SURFACES

The force between charged objects and surfaces in an electrolyte solution is mediated by the ions and the solvent. This is referred to as solvation forces or electrolyte-mediated

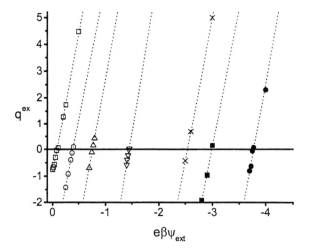

FIG. 10 Normalized unbalanced surface charge in a cylindrical pore with $R = 5d$ in the presence of an external potential Ψ. The results, from left to right, are for original surface charge densities of -0.001, -0.005, -0.01, -0.02, -0.04, -0.05, -0.07123 C/m^2, respectively. The x-intercepts are values of the corresponding equilibrium Donnan potentials.

forces between charged surfaces. The charged surfaces can be colloidal particles, nanoparticles, biological cells, or the tip of an atomic microscope (AFM) operating in solution. Experimental studies of electrolyte-mediated forces were made to understand the stability of suspended colloids, monodispersity of composite nanomaterial, attachment of biological cells, and resolution of the AFM image.

A. Theories and Experiments

The DLVO theory [88,89], a landmark in the study of colloids, interprets stability as dependent on the competition between the long-range repulsion forces of similarly charged

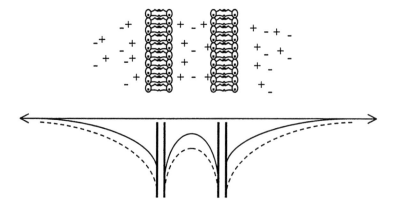

FIG. 11 Schematic illustration of the electric potential profiles inside and outside a nanopore with lipid bilayer membranes separating the internal and external electrolyte solutions. The dotted line is a junction potential representation where the internal potential is shifted.

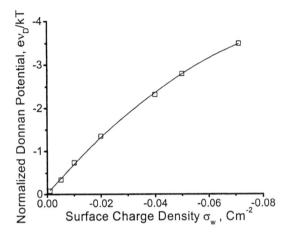

FIG. 12   Donnan potential as a function of surface charge density obtained from Figure 10.

objects and the short-range dispersion attractive forces. The electrolyte simply plays the role of a dielectric continuum to mediate electrostatic forces. Using a surface force apparatus, Israelachvili [90] measured the force between mica surfaces with a liquid between the surfaces. He found oscillating forces with periods equal to the diameter of the solvent. By HNC/MSA theory, Henderson and Lozada-Cassou [91] were able to show similar oscillating forces, logically interpreted as due to successive displacement of layers of solvent molecules. Similar force measurements were extended to charged objects in electrolyte using an AFM [92–95]. Different reports of the existence of oscillating forces and the extent of long-range repulsion were given. Rotsch and Radmacher [96] reported reduction of repulsion forces in changing the electrolyte from univalent to divalent electrolyte. Muller et al. showed that with careful adjustment of electrolyte concentrations, the electrolyte force can be eliminated to give a better AFM resolution [97].

B.   Monte Carlo Simulations

Monte Carlo simulations to calculate the electrolyte force between colloids have been reported [98–100]. Valleau and coworkers[98] revealed the components of forces between two charged surfaces in the RPM model. As shown in Figure 13, the overall pressure (force) has a mild oscillatory behavior with an attractive region between 1.3–1.8 times the diameter of the ions. The electrostatic part is always attractive, indicating favorable electrostatic mediation of the ions. Excess counterions between the surfaces screen the repulsion between them and contribute to an overall minimum-energy configuration, analogous to ionic bonding in a salt crystal. The collisional contribution represents hard-sphere interaction among the ions and is always positive (repulsive). The kinetic part is collision of particles with the wall and is equivalent to the ideal gas pressure. The results of Valleau et al. [98] are obtained in the canonical ensemble with assumption of electroneutrality. The GCMC work of Lee and Chan [101] show that with an open system and single ion equilibrium, the total force for RPM electrolyte is similar to the results in the canonical ensemble, as shown in Figure 14. An attractive region is found only for 2:1 and 2:2 electrolyte, i.e., divalent counterions. For 1:1 RPM electrolyte, the total force is always positive. The results for the SPM model electrolyte, however, show no attractive

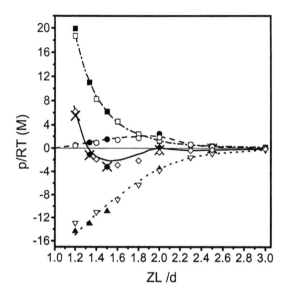

FIG. 13 Contributions to the pressure between two -0.244 C/m^2 charged planar surfaces separated by a 0.1 molar 2:2 RPM electrolyte. The open squares, circles, down-triangles, and diamonds are the kinetic, collision, electrostatic, and total pressures, respectively, from results of Valleau et al. [98]. The corresponding solid symbols are unpublished results of Lee and Chan. The lines are calculations by the hypernatted-chain (HNC) equation.

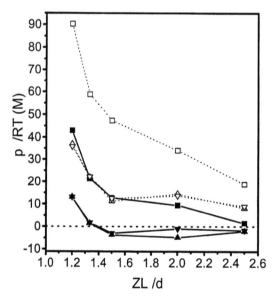

FIG. 14 Comparison of total pressure between two charged surfaces in 1:1, 2:1, and 2:2 electrolytes. The open squares, up-triangles, and down-triangles are results of 1:1, 2:1, and 2:2, respectively, for the SPM model of 0.3 packing fraction. The corresponding solid symbols are for the RPM model.

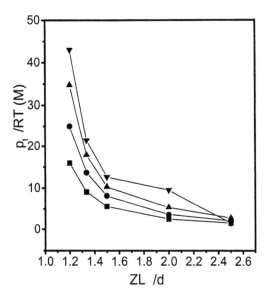

FIG. 15 Total pressure between two charged surfaces in a 0.1 molar 1:1 RPM electrolyte. The down-triangles, up-triangles, circles, and squares are results for surface charges of -0.244, -0.18, -0.12, and -0.07 C/m^2, respectively.

region, even for divalent counterions. The collision contribution from the solvent appeared to play a major role here. The damping of repulsion forces in going from monovalent to divalent ions is in agreement with the experimental observation [96]. One explanation can be offered in comparing the components of pressure. Since fewer ions are needed for charge balance in the divalent electrolytes, the repulsive-collision term is smaller. The effect of surface charge on total force is shown in Figure 15 for the 1:1 RPM model. A large repulsion is shown at higher surface charge, due to an increase in collisions from the increased number of counterions.

Theories and simulation of the operation of AFM in liquid have been attempted [102–104]. In principle, molecular dynamics or NEMD may be a suitable method to mimic the operation of a scanning tip. The time scale, however, precludes simulating a long-enough scan to see a complete atom. Most studies, therefore, were made with equilibrium conditions and a fixed position of the AFM tip. Explicit consideration of electrolytes and electrostatic effects has not been modeled.

V. ION TRANSPORT IN NANOPORES

A. Continuum Theory of Ion Transport

A combination of continuum transport theory and the Poisson distribution of solution charges has been popular in interpreting transport of ions or conductivity of electrolytes. Assuming zero gradient in pressure and concentration of other species, the flux of an ion depends on the concentration gradient, the electrical potential gradient, and a convection term, expressed as

$$J_i = -D_i \frac{dC}{dx} - \frac{z_i e D_i C_i}{RT} \frac{d\phi}{dx} + C_i v_x \qquad (10)$$

where v_x is the fluid velocity in the x-direction. This is the Nernst–Planck (NP) theory. Where there is pressure gradient and gradients in concentration of other species, osmotic flow or electro-osmotic flow exists, and it will be proper to replace the concentration gradient by chemical potential gradient. Equation (10) is a one-dimensional equation, and extension to three-dimensional expression can be made. This is a phenomenological equation, without accounting for interactions between ions and solvents, ions and a confining wall, or ions with other ions. Equation (10) is for a bulk solution without specific boundary conditions. Dresner [105] and Osterle and coworkers [106] have extended the model for ion transport in a charged capillary. A recent review of the theoretical applications to microporous membranes is given by Yaroshchuk [107]. The Gouy–Chapman double-layer theory is used to describe the uneven charge distribution of ions inside the pore. The Navier–Stokes equation, modified with consideration of electric potential gradient, is used to consider the viscous-flow effects. Numerical solution is necessary with the boundary conditions of a long capillary. While the conductivity decreases linearly with the size of the capillary, it reached a constant value when double-layer effects dominate. Westermann-Clark and Anderson [108] have experimentally confirmed the limiting behavior using track-etched mica membrane. The Dresner–Osterle model is two-dimensional and assumes uniform distribution of charges on the pore wall. To model transport of ions through a biological ion channel, Eisenberg and coworkers [109,110] proposed to incorporate an uneven distribution of fixed charges inside the protein channel and solved the one-dimensional Poisson equation within the channel. A numerical iteration based on the Gummel algorithm [111] is used for this Poisson–Nernst–Planck (PNP) approach. It is clear that for nanometer-scale pores, molecular interactions dominate the transport of ions. Analytical theories, however, have not been developed to account for the molecular details and incorporate the parameters of ion size and interaction potential. In the spirit of the MSA, Nonner et al. [112] have proposed modifications to the PNP model by adding an adjustable parameter to account for a specific ion's activity.

B. Experimental Work

Experimental determination of ionic transport through nanoscale pores has met with difficulties of ill-characterized pores and the ability to measure potential and small current localized to the nanometer scale. Various experimental attempts have been made in conductivity measurements of membranes in diverse fields, reflecting the corresponding applications and interests. Alternating current impedance measurements have been made for perfluorosulfonic acid membranes [113–115]. Conductivity was found to increase with increasing water content in the pore. There is a distribution of pore sizes, and the pore size cannot be characterized precisely and changes with the amount of water adsorbed. Gierke and Hsu [70,71] estimated the diameters of pores to be 0.7–1.6 nm, while from x-ray diffraction analyses, a mean diameter of 6 nm was estimated [72,73]. It was not clean whether the increase in conductivity is due to a hydration effect, dilution effect, electosmotic flow, or reduced confinement due to swelling of the pores. In addition, the conductivity varies with the size and charge of the cation, similar to the conductivity in the bulk solution. Using track-etched mica membranes, Westermann-Clark and Anderson [108] measured conductivity through the membrane with various electrolytes, pore sizes, and surface charge densities. The pores are stable and well defined, with a narrow distribution of sizes, but the surface charge density can be determined only indirectly by streaming potential measurement and is assumed to be uniformly distributed over the pore wall. Recent advances in scanning probe microscopy have made nano-manipulation and nano-characterization possible. Hansma et al. [116] have developed a scanning ion-conductance microscope (SICM) to image a surface by monitoring the local conductance. Using a bent micropipette, the topography can be obtained by AFM in contact mode and tapping mode while the conductance can be obtained through a electrochemical circuit through the electrolyte inside the pipette [117]. A single pore of 200 nm can be imaged, and pore current 0–150 pA was measured. By a similar principle, the scanning electrochemical microscope (SECM) has been applied to image porous membrane [118]. A redox couple is used and the faradiac current was measured. The diameter of the platinum electrode tip in the SECM is 1–10 μm, compared to 50 nm for the micropipette in SICM. The reported resolution of SECM is therefore not as good at the moment. While both of these techniques have been applied to image individual pores in membranes under solution, no studies of correlation of conductance with pore sizes, electrolyte concentration, and surface charges have been reported.

In a different context, a micropipette has been applied to monitor the current through a single-ion channel in a biological membrane. The patch-clamp technique invented by Sackmann and Neher [119] led to their Nobel Prize in medicine. The variations in channel current with voltage, concentration, type of ions, and type of channels have been explored. While the functions of specific channels, in particular their ionic selectivity, have been well known, only a handful of channels have the internal geometry and charge distribution determined. The development of a theory to interpret the mass of channel data and to predict channel action is still lacking.

C. Molecular Dynamics Simulation

The Nernst–Planck equation is based on a continuum model and does not capture molecular details of ion–ion, ion–solvent, and ion–wall interactions. Modeling at the molecular level is needed, but molecular theories that yield analytical or simple numerical solutions are still difficult, given the many degrees of freedom. Computer simulation is necessary to compute the many-body interactions and yield results that can be linked to measurable experimental quantities. Monte Carlo simulations and energetic calculations have yielded understanding of binding energies, activation energies, and favorable structures and configurations of electrolytes in ion channels [8–10]. Molecular dynamics simulation, however, is preferred for the direct computation of time-dependent phenomena and for yielding the usual results, such as transport properties. Newton's equations of motion for all particles are solved at each femtosecond time step. The diffusion coefficient can be related to the mean square displacement, or velocity auto-correlation function. In nanopores, mobility is nonisotropic. Mean square displacement in the axial and radial directions have to be separately accumulated in the simulations. Under equilibrium conditions, the corresponding self-diffusion coefficients can be related to the long-time limit of the time derivative of the mean square displacement in each direction,

$$D = \lim_{t \to \infty} \frac{\langle |z_i(t) - z_i(0)|^2 \rangle}{2t} \tag{11}$$

according to the Einstein relation. Under an external field, in a mixture, nonequilibrium effects and osmotic flow or the darken coefficient have to be considered.

1. Biological Ion channel

There are few simulations of electrolyte transport in nanopores with a full atomic model of electrolyte and pore wall. As discussed in Section II, simplifications and assumptions are made to make simulation feasible. In the biophysics area, molecular dynamics of various ion channels have been reported [120–125], the Gramicidin channel is a common structure to study, since its structure is simple and better known. The early simulations focused on the dynamics of the channel in vacuum [120–123]. Recently, simulations of water and different ions have been reported. The time scale of permeation of one ion through an ion channel is sufficiently long (~1 μs) and is inaccessible by full atomic simulations with today's computers. One strategy is to investigate short-time (<1 ns) phenomena and to apply the resulting diffusion coefficient to cruder models or algorithms. The other limitations are that simulations reported for ion channels so far have assumed infinite dilution (i.e., only one ion), the absence of external field, and external chemical or electrochemical potential. Nevertheless, many interesting and qualitative results have been provided. The local mobility of an ion and its correlation to channel structure have been reported [124,125]. It was also known that proton permeability, although lower than that of bulk water, is many times higher than for other cations [13,14]. Proton permeates via a hopping mechanism through a water wire or hydronium wire [11], and the main bottleneck is due to reorientation of the water molecules that hydrate the proton inside the channel [12].

2. Equilibrium Simulations in a Uniform Nanopore

Many molecular dynamics simulations have focused on the electrolyte solution factors and ignored the atomic features of the pore wall. The assumptions of these simulations may match more closely the experiments of inorganic channels, such as track-etched nu-

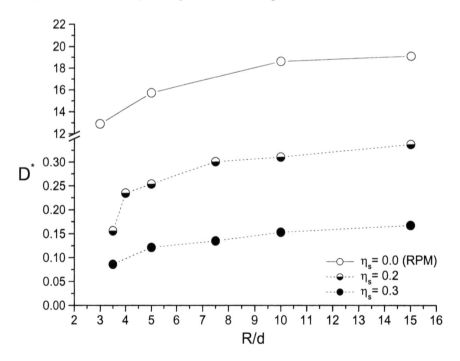

FIG. 16 Reduced self-diffusion coefficients of SPM model ions in pores of different sizes. The zero solvent packing represents the RPM model.

cleopores [108]. The diffusion coefficients of 0.1 molar 1:1 electrolyte in an uncharged, infinitely long, and smooth nanopore have recently been studied for the RPM [126] and SPM models [127] by equilibrium molecular dynamics (EMD) simulations. As shown in Figure 16, it is demonstrated that confinement reduces the diffusion and mobility of the ions in the RPM as well as the SPM model, although the diffusion coefficient of the RPM model is three orders of magnitude higher. The mobility of the SPM solvent is similarly reduced by confinement effects. In the SPM model, the mobility of the ion is lower than that of the solvent, due to a solvation cage effect. The mobility of different monovalent ions inside uniform channels of 1.5–5.5 Å has been investigated by Lynden-Bell and Rasaiah [128]. Using an SPC/E model for water and Lennard-Jones ions, the diffusion coefficients from mean square displacement is about 50% to 10% of the bulk values. The relative mobilities of different ions are similar to those in the bulk. The effect of pore size on mobility is investigated for the sodium ion and correlates with the solvation effects.

The previously mentioned MD simulations are for uncharged pore. For a charged pore, the effect of the wall charges on the axial and radial field is zero, due to symmetry. The potential in the pore, however, will be different and therefore will affect the chemical potential and the equilibrium with external bulk electrolyte. Lo et al. [76,86] have reported some simulation results for a 0.1 molar, 1:1 restricted primitive model in charged nanopores. While the surface wall charges do not have any contribution to the axial and radial fields, the balance of the number of counterions versus coions is affected. An exact balance of surface charges by excess counterions can be assumed, but it was found that the resulting chemical potentials of cation and anion are incorrect, unless an external Donnan potential is assumed to exist. Since the Donnan potential is uniform and does not give any field, there is no effect on the motion of the ions. Rather than the MC method discussed earlier in Section III.D, the value of the Donnan potential can be determined in the MD simulation simply by the arithmetic mean of the individual chemical potentials determined by Widom's method. Therefore,

$$zev_D = \frac{1}{2} \left[\mu_-(r) - \mu_+(r) \right] \tag{12}$$

The axial diffusion coefficients computed for the 1:1 RPM electroyte are of the order of 10^{-6} m^{2}s^{-1} in an uncharged pore and are similar to the values in the bulk. The diffusion coefficient of the counterion decreases with the surface charge density of a charged pore, whereas that of the coion remains unchanged, as shown in Figure 17. This difference is due to the densities of the ions and their distribution. At higher surface charge, there is a higher density of counterions, which are distributed mainly near the pore wall, hence, mobility is reduced. The reduction of mobility here is due mainly to double-layer effects rather than confinement of a small pore. Similar EMD simulation studies have not been reported for the SPM or other molecular solvent models.

3. Nonequilibrium Molecular Dynamics Simulations

The EMD studies are performed without any external electric field. The applicability of the EMD results to useful situations is based on the validity of the Nernst–Planck equation, Eq. (10). From Eq. (10), the current can be computed from the diffusion coefficient obtained from EMD simulations. It is well known that Eq. (10) is valid only for a dilute concentration of ions, in the absence of significant ion–ion interactions, and a macroscopic theory can apply. Intuitively, the Nernst–Planck theory can be expected to fail when there is a significant confinement effect or ion–wall interaction and at high electric

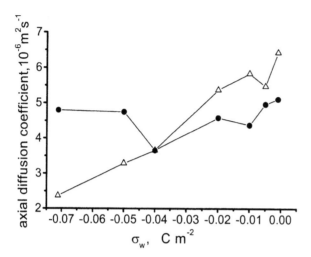

FIG. 17 Diffusion coefficients of the counterions and coions of a 1:1 RPM model electrolyte in a cylindrical nanopore of $R = 10d$. The circles and triangles represent the results of coions and counterions, respectively.

or concentration field. It will, therefore, be necessary to calculate the current directly by performing nonequilibrium molecular dynamics (NEMD) simulations in the presence of an external electric field. The validity of the Nernst–Planck theory can then be assessed by comparing the NEMD current and the current obtained via Eq. (10) using EMD diffusion coefficients.

The technique of NEMD is an active area of research [129], and application to bulk RPM electrolytes has been reported [130,131]. Normally, a concentration gradient or a chemical potential gradient is imposed to induce the flux of a molecular species. For uncharged species, the middle term of Eq. (10) will be absent and the current flux will then be a flux of the uncharged species i. To maintain the two different concentrations at the source and the drain, a dual-control-volume grand canonical molecular dynamics (DCVGCMD) method has been introduced [132]. This method required separate grand canonical simulations at the two different, opposite-end reservoirs. For NEMD simulation of electrolytes, the DCVGCMD method can be avoided, since the ions can be moved with zero concentration gradient, provided an electric field is imposed. The constant-concentration profile, a finite electric field, and the recycling of ions can all be achieved in an NEMD simulation of a cylindrical geometry, as shown in Figure 18. The results of such NEMD simulations are reported [126] for a 0.1 molar 1:1 RPM electrolyte for different radii of uncharged cylindrical pores. The conductivity can be computed by making NEMD simulations at several electric fields and extrapolating to zero fields, assuming the validity of NEMD in the linear response region. The ohmic heat generated has to be properly treated by a Gaussian thermostat. The zero-field conductivity σ_E is given by

$$\sigma = \lim_{E_z \to 0} \frac{J_z}{E_z} \tag{13}$$

where J_z is the axial current density at an electric field strength of E_z.

Surprisingly, an enhanced conductivity is shown for a certain range of confinement, as seen in Figure 19. This can be explained by the decrease in coulombic attractive interaction among the ions in a one-dimensional configuration. But when the pore diameter is

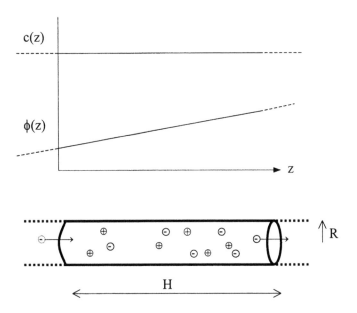

FIG. 18 Setup of the cylindrical simulation cell with periodic boundary condition in the axial (z) direction. In the NEMD simulation, a constant gradient in the potential $\phi(z)$ is applied, and concentration $c(z)$ is maintained constant by recycling ions.

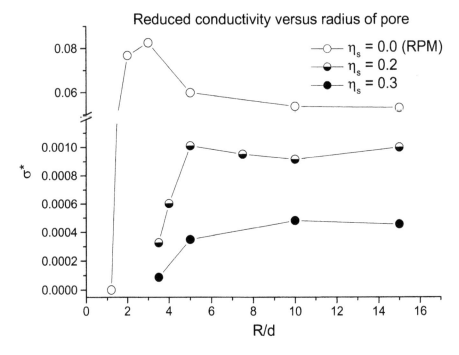

FIG. 19 Extrapolated zero-field conductivity versus pore radius for 0.1 M 1:1 RPM and SPM electrolytes at two solvent packing fractions, 0.1 and 0.2. The conductivity in a pore $R = 1.2d$ is essentially zero on the scale.

about two times the diameter of the ion, significant collisions occur between oppositely traveling anions and cations, and conductivity decreases rapidly. This maximum conductivity in a narrow nanopore is found only for the RPM model, where a solvent viscous effect is absent and changes in coulombic interactions are important. In similar NEMD simulations for the SPM model [127], the conductivity monotonically decreases with pore size. A similar NEMD study on finite-length cylindrical pore with SPC/E solvent electrolytes is also reported [133].

One attraction of MD simulation is the possibility of computer animation. The mobility of ions inside a charged cylindrical pore can be visualized. Some movie clips of EMD and NEMD are downloadable at http://chem.hku.hk/~kyc/movies/*.mpg. Some features that escape statistical averages can be learned in watching the animation. While the coions are present mainly in the center of the pore, occasional collisions with the wall do occur, as observed in the movie. The time scale of a coion staying near the wall is of the order of 1 ps, compared to 10 ps for the counterion. While the averaged equilibrium distributions indicate an infinitesimal concentration of coion at the wall, reaction of coion with the wall can occur within a time scale of 1 ps. From the video, it can also be observed that the radial mobility of the counterion is more significant compared to the coion's and compared to the axial mobility. It is consistent with the statistical results.

VI. SUMMARY AND OUTLOOK

Accurate prediction of the thermodynamics, transport, and structural properties of bulk electrolytes is still technically difficult, even after decades of theoretical and simulation studies at the molecular level. The physical concepts of how molecular properties affect bulk electrolyte behavior, however, are fairly clear and well established. Electrolytes in nanostructures, however, are less understood. The classical theories of Gouy–Chapman and DLVO only help to understand the limiting cases of smooth, rigid, weakly charged, single isolated surfaces and in dilution electrolyte solution. The confinement effects in nanostructures, ion–surface interactions at the nanometer scale, the overlapping of double layers, the disruption of solvation structures, and the indirect effects on charge balance and osmosis effects all give rise to rich possibilities of phenomena not found in the bulk solution. The questions of electroadsorption, electroneutrality, surface forces, and diffusion have been discussed here. Other interesting subjects, e.g., phase transition, dynamic surface, critical phenomena, migration, electrokinetic phenomena in nanostructures, have been little explored. Studies of confined electrolytes span a wide array of subject areas, from the old discipline of colloidal science to modern electrophysiology. Technology development in many areas provides both the ends and the means of these investigations. Exciting progress is possible with advances in experimental techniques and computation techniques. Atomic structures been determined for an increasing number of ion channels in biological membranes. Rapid development has been made in the synthesis of molecular and nanoscopic structures with well-defined geometries. New optical, spectroscopic, and microscopy techniques are increasing their resolutions to nanometer scale and the single molecule level. On the other hand, computational power and techniques can provide new understanding via theoretical approaches. On one hand, the loss of three-dimensional symmetry precludes the use of standard treatments of long-range electrostatic interaction. On the other hand, electrostatic interaction is only long range in one or two dimensions in a confined geometry. A better and more efficient model for water is vital to progress further in computational approaches. For the study of dynamic phenomena, the latest techniques in nonequilibrium molecular dynamics will be needed. Before comprehensive progress on different fronts, judicious assumptions and simplifications will be deployed for progress in modeling and simulation studies.

ACKNOWLEDGMENTS

Financial support from the CRCG and the Outstanding Researcher Award of The University of Hong Kong and the Research Grants Council of Hong Kong are acknowledged.

REFERENCES

1. G Gouy, J. Phys. 9 (1910) 457.
2. DL Chapman, Phil. Mag. 25 (1913) 475.
3. O Stern, Z, Electrochem. 30 (1924) 508.
4. SL Carnie and GM Torrie, Adv. Chem. Phys. 56 (1984) 141.
5. W Schmickler and D Henderson, Progress Surf. Sci. 22(4) (1986) 323–420.
6. DHJ Mackay, PH Berens, and KR Wilson, Biophys. J. 46 (1984) 229–248.
7. PC Jordan, J. Phys. Chem. 91 (1987) 6582–6591.
8. J Xing and HL Scott, Biochem. Biophys. Res. Comm. 165 (1989) 1–6.
9. V Dorman, MB Partenskii, and PC Jordan, Biophys. J. 70 (1996) 121–134.
10. B Roux, Biophys. J. 71 (1996) 3177–3185.
11. DE Sagnella and GA Voth, Biophys. J. 70 (1996) 2043–2051.
12. R Pomes and B Roux, Biophys. J. 71 (1996) 19–39.
13. M Akeson and DW Deamer, Biophys. J. 60 (1991) 101–109.
14. DW Deamer, Biophys. J. 71 (1996) 5.
15. Z Tang, LE Scriven, and HT Davis, J. Chem. Phys. 97(1) (1992) 494; 100(6) (1994) 4527.
16. L Zhang, HT Davis, and HS White, J. Chem. Phys. 98(7) (1993) 5793.
17. MJ Grimson and G Rickayzen, Chem. Phys. Lett. 86 (1982) 71.
18. Wertheim, Phys. Rev. Lett. 8 (1963) 321.
19. E Thiele, J. Chem. Phys. 38 (1963) 1959.
20. JL Lebowitz, Phys. Rev. 133 (1964) 895A.
21. E Waisman and JL Lebowitz, J. Chem. Phys. 56(6) (1972) 3086.
22. KY Chan, J. Phys. Chem. 94 (1990) 8472–8477.
23. DN Card, JP Valleau, J. Chem. Phys. 12 (1970) 6232; JP Valleau, DN Card, J. Chem. Phys. 57 (1972) 5457.
24. WJ Van Megen and IK Snook, J. Chem. Phys. 73 (1980) 4656; 75 (1981) 4104.
25. JP Valleau, LK Cohen, and DN Card, J. Chem. Phys. 72 (1980) 5942.
26. L Blum, J. Phys. Chem. 81 (1977) 136.
27. GM Torrie and JP Valleau, J. Chem. Phys. 73(11) (1980) 5807–5816.
28. L Zhang, HS White, and HT Davis, Molecular Simulation 9 (1992) 247.
29. D Boda, D Henderson, K-Y Chan, DT Wasan, Chem. Phys. Letts. 308 (1999) 473–478.
30. D Boda, D, Henderson, K-Y Chan, J. Chem. Phys. 110 (1999) 5346–5350.
31. P Debye and E Huckel, Z, Physik 24 (1923) 133–305.
32. S Levine, JR Mariott, G Neale, and N Epstein, J. Colloid Interface Sci. 52 (1975) 136.
33. V Vlachy and DA McQuarrie, J. Phys. Chem. 90 (1986) 3248–3250.
34. WY Lo and KY Chan, J. Chem. Phys. 101 (1994) 1431.
35. G Stell and JL Lebowitz, J. Chem. Phys. 48 (1968) 3706.
36. K-Y Chan, KE Gubbins, D Henderson, and L Blum, Mol. Phys. (1989) 66 299.
37. K-Y Chan, J. Phys. Chem. 95 (1991) 7465–7471.
38. L Blum and D Henderson, J. Chem. Phys. 74 (1981) 1902.
39. D Boda, K-Y Chan, and DJ Henderson, J. Chem. Phys. 109 (1998) 7362–7371.
40. JJ Weis and D Levesque, Phys. Rev. Lett. 71 (1993) 2729.
41. I Szalai, D Henderson, and K-Y Chan, J. Chem. Phys. 111 (1999) 337–344.

42. ME van Leeuvan, B Smit, Phys. Rev. Lett. 71 (1993) 3991.
43. HJC Berendsen, JPM Postma, WF van Gunsteren, and J Hermans, in Intermolecular Forces (Eds. B Pullman and D Reidel). Dordrecht, The Netherlands, (1981), pp 331–342.
44. HJC Berendsen, JR Grigera, and TP Straatsma, J. Phys. Chem. 91 (1987) 6269.
45. WL Jorgensen, J. Am. Chem. Soc. 103 (1981) 335.
46. A Rahman and FH Stillinger, J. Chem. Phys. 55 (1971) 3336.
47. FH Stillinger and A Rahman, J. Chem. Phys. 60 (1974) 1545.
48. K Watanabe and M Klein, Chem. Phys. 131 (1989) 157–167.
49. J Caldwell, LX Dang, and PA Kollman, J. Am. Chem. Soc. 112 (1990) 9144.
50. WY Lo, KY Chan, and KL Mok, J. Phys. Condens. Matter 6 (1994) A145.
51. V Vlachy and ADJ Haymet, J. Am. Chem. Soc. 111 (1989) 477.
52. V Vlachy and ADJ Haymet, J. Electroanal. Chem. 283 (1990) 77.
53. B Jamnik and V Vlachy, J. Am. Chem. Soc. 115 (1993) 660.
54. AD Crowell, J. Chem. Phys. 29 (1958) 446.
55. JE Lane and TH Spurling, Aust. J. Chem. 33 (1980) 231.
56. WA Steele, Surf. Sci. 36 (1973) 317.
57. GJ Tjatjopoulos, DL Feke, and JA Mann, Jr., J. Phys. Chem. 92 (1988) 4006.
58. BK Peterson, KE Gubbins, and F van Swol, J. Chem. Phys. 93 (1990) 679.
59. P Ewald, Ann. Phys. 64 (1921) 253.
60. JA Barker and RO Watts, Mol. Phys. 26 (1973) 789.
61. GT Gao, XC Zeng, and W Wang, J. Chem. Phys. 106 (1997) 3311–3317.
62. X Din and EE Michaelides, J. Phys. Chem. 101 (1997) 4323–4331.
63. J Hautman and ML Klein, Mol. Phys. 76 (1992) 379–395.
64. M Lee and KY Chan, Chem. Phys. Lett. 275 (1997) 56.
65. WY Lo and KY Chan, Molec. Phys. 86(4) (1995) 745.
66. M Lee, KY Chan, D Nicholson, and S Zara, Chem. Phys. Lett. 307 (1999) 89.
67. MD Neville, CP Jones, and AD Turner, Progress Nuclear Energy, 32 (1998) 397–401.
68. BE Conway, V Birss, and J Wojtowicz, J. Power Sources 66 (1997) 1–14.
69. NM Markovic, CA Lucas, HA Gasteiger, and PN Ross, Surf. Sci. 365 (1996) 229–240.
70. TD Gierke and WY Hsu, in Perfluorinated Ionomer Membranes, eds. A Eisenberg and HL Yeager, ACS Symposium Series 180 (1982) Ch. 13.
71. WY Hsu and TD Gierke, J. Membrane Sci. 13 (1983) 307.
72. MV Verbrugge and RF Hill, J. Electrochem. Soc. 137 (1990) 886, 893.
73. EH Cwirko and RB Carbonell, J. Membrane Sci. 67 (1992) 211, 227.
74. GS Heffelfinger, F van Swol, and KE Gubbins, Mol. Phys. 61 (1987) 1381.
75. S Rivera and TS Sorenson, Molecular Simulation 13 (1994) 115; TS Sorensen and P Sloth, J. Chem. Soc. Faraday Trans. 88(4) (1992) 571.
76. WY Lo, KY Chan, M Lee, and KL Mok, J. Electroan. Chem. 450 (1998) 265.
77. L Yeomans, SE Feller, E Sanchez, and M Lozada-Cassou, J. Chem. Phys. 98 (1993) 1436–1450.
78. K Kimoto, J. Electrochem. Soc. 130 (1983) 334.
79. A Herra and HL Yeager, J. Electrochem. Soc. 134 (1987) 2446.
80. B Widom, J. Chem. Phys. 39 (1963) 2808.
81. B Widom, J. Stat. Phys. 19 (1978) 563.
82. M Lozada-Cassou, W Olivares, and B Sulbaran, Phys. Rev. E, 53(1) (1996) 522.
83. M Lozada-Cassou, and J Yu, Phys. Rev. Lett. 77 (1996) 4019–4022.
84. M Lozada-Cassou, and J Yu, Phys. Rev. E. 56 (1997) 2958–2965.
85. M Lozada-Cassou, W Olivares, B Sulbaran, and Y Jiang, Physica A 231 (1996) 197–206.
86. WY Lo, Ph.D., Thesis, University of Hong Kong (1995).
87. FG Donnan, Z, Elektrochem. 17 (1911) 572.
88. BV Derjaguin and L Landau, Acta Phys. Chem. USSR XIV (1941) 633.
89. EJV Verwey and J Th G Overbeek, Theory of the Stability of Lyophobic Colloids, Elsevier, Amsterdam, 1948.
90. JN Israelachvili, Intermolecular and Surface Forces, Academic Press, London (1992).

91. D Henderson and M Lozada-Cassou, J. Colloid Interface Sci. 114 (1986) 180.
92. R Kjellnader, S Marcelja, RM Pashley, and JP Quirk, J. Phys. Chem. 92 (1988) 6489; J. Chem. Phys. 92 (1990) 4399.
93. R Kjellander and S Marcelja, Chem. Phys. Lett. 112 (1984) 49; 114 (1985) 124(E); 127 (1986) 402; 142 (1987) 485; J. Phys. Chem. 1230 90 (1985); J. Chem. Phys. 82 (1985) 2122; 88 (1988) 7138.
94. R Kjellander, J. Phys. Chem. 88 (1988) 7129.
95. G Toikka and RA Hayes, J. Colloid Interface Sci. 191 (1997) 102–109.
96. C Rotsch and M Radmacher, Langmuir 13 (1997) 2825–2832.
97. DJ Muller, D Fotiadis, S Scheuring, SA Muller, and A Engel, Biophys. J. 76 (1999) 1101–1111.
98. JP Valleau, R Ivkov, and GM Torrie, J. Chem. Phys. 95 (1991) 520.
99. L Gulbrand, B Jonsson, H Wennerstrom, and P Linse, J. Chem. Phys. 80 (1984) 2221.
100. B Svensson and B Jonsson, Chem. Phys. Lett. 108 (1984) 580.
101. M Lee and K-Y Chan, unpublished results.
102. K Koga and XC Zeng, Phys. Rev. Letts, 79 (1997) 853.
103. LD Gelb and RM Lynden-Bell, Chem. Phys. Lett. 211 (1993) 328.
104. M Callaway, DJ Tildesley, and N Quirke, Langmuir 10 (1994) 3350.
105. L Dresner, J. Phys. Chem. 67 (1963) 1635.
106. RJ Cross and JF Osterle, J. Chem. Phys. 49 (1968) 228; FA Morrison Jr., JF Osterle. J. Chem. Phys. 43 (1965) 2111.
107. AE Yaroshchuk, Ad. Colloid Interface. Sci. 60 (1995) 1–93.
108. GB Westermann-Clark and JL Anderson, J. Electrochem. Soc. 130 (1983) 839.
109. RS Eisenberg, J. Membr. Biol. 150 (1996) 1.
110. RS Eisenberg, Accounts Chem. Res. 31 (1998) 117.
111. HK Gummel, IEEE Trans. on Electron Devices (1964) 455.
112. W Nonner, L Catacuzzeno, and B Eisenberg, Biophys. J. 79 (2000) 1976.
113. G Pucelly, A Oikinomou, C Gavach, and HD Hurwitz, J. Electroanal. Chem., 287 (1990) 43.
114. C Gavach, G Pamboutzoglou, M Nedyalkov, and G Poucelly, J. Membrane Sci. 45 (1989) 37.
115. A Steck and HL Yeager, J. Electrochem. Soc. 130 (1983) 1297.
116. PK Hansma, B Drake, O Marti, SAC Gould, and CB Prater, Science 243 (1989) 641.
117. R Proksch, R Lal, PK Hansma, D Morse, and G Stucky, Biophys. J. 71 (1996) 2155.
118. A Bath, RD Lee, HS White, and ER Scott, Anal. Chem. 65 (1993) 1537; 70 (1998) 1047.
119. B Sackmann and E Neher, Single Channel Recording, Plenum, New York (1995).
120. S-W Chiu, JA Novotny, and E Jakobsson, Biophys. J. 64 (1993) 98.
121. S-W Chiu and E Jakobsson, Biophys. J. 55 (1989) 147.
122. E Jakobsson and S-W Chiu, Biophys. J. 52 (1987) 33.
123. B Roux and M Karplus, JACS, 115 (1993); J. Phys. Chem. 95 (1991) 4856; Biophys. J. 59 (1991) 961.
124. GR Smith and MSP Sansom, Biophys. Chem. 79 (1999) 129.
125. GR Smith and MSP Sansom, Biophys. J. 73 (1997) 1364.
126. YW Tang, I Szalai, and K-Y Chan, Molec. Phys. 99 (2001) 309.
127. YW Tang, I Szalai, and K-Y Chan, J. Phys. Chem. (2001) under revision.
128. RM Lynden-Bell and JC Rasaiah, J. Chem. Phys. 105(20) (1996) 9266.
129. DJ Evans and GP Morriss, Statistical Mechanics of Non-equilibrium Liquids, Academic Press, London, 1990.
130. IM Svishchev and PG Kusalik, Physica A 192 (1993) 628.
131. IM Svishchev and PG Kusalik, Phys. Chem. Liquids 26 (1994) 237.
132. GS Heffelfinger and F van Swol, J. Chem. Phys. 100 (1994) 7548.
133. PS Crozier, RL Rowley, NB Holladay, D Henderson, and DD Busath, Phys. Rev. Letts. 86 (2001) 2467.

17

Polymer–Clay Nanocomposites: Synthesis and Properties

SYED QUTUBUDDIN and XIAOAN FU

Case Western Reserve University, Cleveland, Ohio

I. INTRODUCTION

Conventional polymer composites are widely used in diverse applications, such as construction, transportation, electronics, and consumer products. Composites offer improved properties, including higher strength and stiffness, compared to pristine polymers. The properties of polymer composites are greatly affected by the dimension and microstructure of the dispersed phase. Nanocomposites are a new class of composites that have a dispersed phase with at least one ultrafine dimension, typically a few nanometers [1–3]. Nanocomposites possess special properties not shared by conventional composites, due primarily to large interfacial area per unit volume or weight of the dispersed phase (e.g., 750 m^2/g). Clay layers dispersed at the nanoscale in a polymer matrix act as a reinforcing phase to form polymer–clay nanocomposites, an important class of organic–inorganic nanocomposites. These nanocomposites are also referred to as polymer–silicate nanocomposites and organic–inorganic hybrids. Polymer–clay nanocomposites can drastically improve mechanical reinforcement and high-temperature durability, provide enhanced barrier properties, and reduce flammability [4–6]. Clays that have a high aspect ratio of silicate nanolayers are desirable for polymer reinforcement.

Colloid and surface chemistry play important roles in the synthesis of polymer–clay nanocomposites. Dispersion of clay layers in polymers is hindered by the inherent tendency to form face-to-face stacks in agglomerated tactoids due to high interlayer cohesive energy. Nanoscale dispersion of the clay tactoids into individual nanolayers is known as exfoliation or delamination. Exfoliation is further prevented by the incompatibility between hydrophilic clay and hydrophobic polymers. Treatment or functionalization of clay by adsorption of organic molecules weakens the interlayer cohesive energy. Intercalation, i.e., penetration of organic molecules into the clay interlayers, increases the compatibility between clay and polymer matrix. Due to the negative charge on the clay surface, cationic surfactants and polymers are commonly used for intercalation. The ion exchange of inorganic cations in clay galleries by organic cations renders the clay organophilic. Such organoclays have found large-scale applications for decades in cosmetics, drilling mud, paints, coatings, inks, and wastewater treatment [7]. There is a growing interest in the surface chemistry of clays in pursuit of nanocomposite synthesis using specific monomers, prepolymers, and polymer melts. This chapter provides a review of recent developments in the synthesis and properties of modified clay and polymer–clay nanocomposites.

653

II. CLAY STRUCTURE AND DISPERSION IN POLYMER

Clay consists of small crystalline particles made up of aluminosilicates of various compositions, with possible iron and magnesium substitutions by alkalis and alkaline earth elements [8–12]. The basic silicon-oxygen unit is a tetrahedron, with four oxygen atoms surrounding the central silicon. The tetrahedra are linked to form hexagonal rings. This pattern repeats in two dimensions to form a sheet. Aluminum, in combination with oxygen, forms an octahedron, with the aluminum at the center, and the octahedra link to form a more closely packed two-dimensional sheet. There are two basic types of clay structures (1:1 and 2:1). Kaolinite is 1:1 type of nonswelling dioctahedral clay. The kaolinite crystal is a sheet of alumina octahedra sitting on top of a sheet of silica tetrahedra. The apical oxygen atoms from the silica are shared with the aluminum atoms of the upper layer. The other basic type of clay is of the 2:1 type (i.e., two sheets of silica to one of alumina or two sheets of silica to one of magnesium oxide). The two parent materials are pyrophyllite and talc, with alumina and magnesia, respectively, in the central layer.

Clays used in preparing polymer–clay nanocomposites belong to the 2:1 layered structure type. A member of the 2:1 family, montmorillonite is one of the most interesting and widely investigated clays for polymer nanocomposites. The structure of montmorillonite consists of layers made up of one octahedral alumina sheet sandwiched between two tetrahedral silica sheets, as shown in Figure 1 [8]. Stacking of the silicate layers leads to a regular van der Waals gap between the layers. Approximately one in six of the aluminum ions in the octahedral layers of montmorillonite is isomorphously substituted by magne-

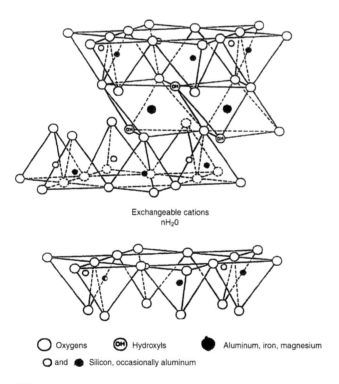

Exchangeable cations
nH_2O

○ Oxygens   (OH) Hydroxyls   ● Aluminum, iron, magnesium

○ and ● Silicon, occasionally aluminum

FIG. 1 Idealized structure of a montmorillonite layer showing two tetrahedral-site sheets fused to an octahedral-site sheet (2:1 type). (From Ref. 8.)

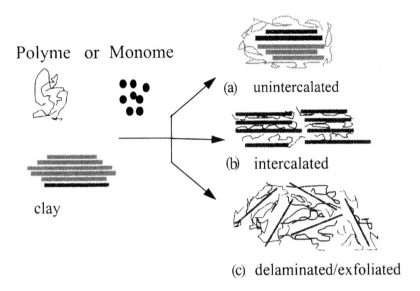

Polyme or Monome

(a) unintercalated

(b) intercalated

clay

(c) delaminated/exfoliated

FIG. 2 Schematic illustration of three types of polymer–clay composites.

sium or other divalent ions. The isomorphic substitution renders negative charges that are counterbalanced by cations residing in the interlayer. Pristine clay usually contains hydrated inorganic cations such as Na^+, K^+, and Ca^{2+}. When the inorganic cations are exchanged by organic cations, such as from surfactants and polyelectrolytes, the clay surface changes from hydrophilic to hydrophobic or organophilic [13,14]. The organic cations lower the surface energy and decrease the cohesive energy by expanding the interlayer distance, thus facilitating the wetting and intercalation of monomer or polymer. In addition, the organic cations may contain various functional groups that react with monomer or polymer resin to improve interfacial adhesion between clay nanolayers and polymer matrix.

Complete dispersion or exfoliation of clay tactoids in a monomer or polymer matrix may involve three steps similar to the dispersion of powders in liquids, as identified by Parfitt [15]. The first step is wetting the surface of clay tactoids by monomer or polymer molecules. The second step is intercalation or infiltration of the monomer or polymer into the clay galleries, and the third step is exfoliation of clay layers. The first and second steps are determined by thermodynamics, while the third step is controlled by mechanical and reaction driving forces. The dispersion of clay tactoids in a polymer matrix can result in the formation of three types of composites, as shown in Figure 2. The first type is a conventional composite that contains clay tactoids with the nanolayers aggregated in un-intercalated face-to-face form. In this case, the clay tactoids are dispersed simply as a segregated phase, resulting in poor mechanical properties of the composite. The second type is intercalated polymer–clay nanocomposite, which is formed by the infiltration of one or more molecular layers of polymer into the clay host galleries. The last type is exfoliated polymer–clay nanocomposites, characterized by a low clay content, a monolithic structure, and a separation between clay layers that depends on the polymer content of the composite. Exfoliation is particularly desirable for improving specific properties that are affected by the degree of dispersion and resulting interfacial area between polymer and clay nanolayers.

Homogeneous dispersion of clay nanolayers in a polymer matrix provides maximum reinforcement via distribution of stress and deflection of cracks resulting from an applied load. Interactions between exfoliated nanolayers with large interfacial area and surrounding polymer matrix lead to higher tensile strength, modulus, and thermal stability [4–6]. Conventional polymer–filler composites containing micron-size aggregated tactoids also improve stiffness, but at the expense of strength, elongation, and toughness. However, exfoliated clay nanocomposites of Nylon-6 and epoxy have shown improvements in all aspects of thermomechanical behavior. Exfoliation of silicate nanolayers with high aspect ratio also provides other performance enhancements that are not achievable with conventional particulate composites. The impermeable clay nanolayers provide a tortuous pathway for a permeant to diffuse through the nanocomposite. The hindered diffusion in nanocomposites leads to enhanced barrier property, reduced swelling by solvent, and improvements in chemical stability and flame retardance.

III. CATION EXCHANGE OF CLAY WITH SURFACTANTS

Industrial applications of organoclays [7,16] have stimulated scientific efforts to understand the mechanism of surfactant ion exchange and adsorption. The adsorption of cationic surfactants onto a homoionic montmorillonite dispersed in water was found to be independent of the size of hydrophilic head group of cationic surfactants at its natural pH [17]. The amount of cationic surfactant adsorbed as a monolayer is almost the same as the cationic exchange capacity, CEC. The critical coagulation concentrations are also close to CEC. The completeness of the exchange of inorganic cations by cationic surfactants and the chemical adsorption stability of surfactant–clay complexes greatly affect the application of organoclays [18,19]. The structure of the adsorption layer of cationic surfactants in the galleries of swelling clays depends strongly on the initial degree of clay dispersion. Initial conditions that correspond to a homogeneous dispersion of swelling clay (e.g., Na-saturated clay and low ionic strength) result in random organic/inorganic cation distribution in the interlayers. Also, the adsorbed surfactant layer has a loose structure at low organic cation concentration.

Adsorption isotherms and precise calorimetric experiments were used to identify different types of adsorption and quantify the interactions between cationic surfactants and clay [20,21]. The type of clay and the alkyl chain length of surfactant significantly affect the amount and the enthalpy of adsorption, which is exothermic. Several models have been proposed to account for surfactant adsorption on solid surfaces. However, these models rely on assumptions about the structure of the adsorbed surfactant layer [22–24]. Bohmer and Koopal [25,26] investigated surfactant adsorption on nonswelling clay surfaces using a self-consistent lattice model. The model predicts the structure of the adsorbed layers and a gradual increase in surfactant adsorption with surfactant concentration. Most experimental data supporting the foregoing models were obtained with nonswelling solids. Swelling layered clays certainly exhibit some differences in adsorption behavior, because the structure of the adsorbed surfactant is quite different from that on nonswelling clays. The intercalation of surfactant in swelling clays is discussed in the next section.

IV. INTERCALATION OF CATIONIC SURFACTANTS IN CLAY GALLERIES

The exchange of inorganic cations by organic surfactant ions in the clay galleries not only makes the organoclay surface compatible with monomer or polymer matrix, but also de-

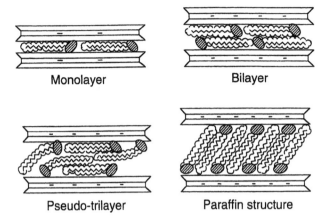

Monolayer   Bilayer

Pseudo-trilayer   Paraffin structure

FIG. 3 Orientations of alkylammonium ions in the galleries of clay layers with different layer charge densities. (From Ref. 28.)

creases the interlayer cohesive energy by expanding the d-spacing. The orientation of the surfactant in the galleries depends on its chemical structure and the charge density of the clay. Increasing the surfactant chain length or the charge density of the clay leads to larger d-spacing and interlayer volume. The adsorbed organic cations in swelling clays such as montmorillonites and vermiculites may adopt (after drying) several configurations in the interlayers. Fourier transform infrared spectroscopy (FTIR), x-ray diffraction (XRD), and differential scanning calorimetry (DSC) are some techniques used to probe the interlayer structure and packing of intercalated cationic surfactants [27–29]. Some possible configurations, such as flat monolayer, bilayer, pseudo-trilayer, and inclined paraffin structure, are shown in Figure 3 [27]. Thus, the molecular environment varies from solidlike to liquidlike. The surfactant chains adopt a more disordered, liquidlike structure with decreasing packing density or chain length and with higher temperature. When the surface area per molecule is within an intermediate range, the chains are not completely random, but retain some orientation, as in a liquid crystalline state.

Recent modeling has provided further insights into the packing characteristics of the alkyl chains in intercalated clay. Hackett et al. [30] used molecular dynamics (MD) simulations to investigate properties such as density profiles, pressure, chain configurations, and trans–gauche conformer ratios. The internal gallery pressure determines the d-spacing of an organoclay, as shown in Figure 4 for three different clays with varying surfactant length. A random liquidlike arrangement of chains was preferred for mono-, bis, and psuedo-trilayers with d-spacings of 1.32, 1.8, and 2.27 nm, respectively. The MD simulations agreed well with experimental XRD and FTIR data for intercalated surfactants with chain length less than 15.

V. SYNTHESIS AND PROPERTIES OF POLYMER–CLAY NANOCOMPOSITES

There are three general approaches to the synthesis of polymer–clay nanocomposites. In the first approach, a monomer or precursor is mixed with organophilic clay and followed by polymerization. This in situ polymerization technique was first developed by the

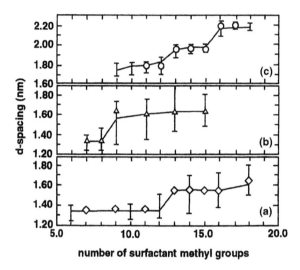

FIG. 4 Experimental (vertical bars) and simulated (symbols) values of the d-spacings for alky-lammonium-exchanged clay at three different cation exchange capacities (CECs): (a) SWy2 mont-morillonite, CEC = 0.8 meq/g; (b) AMS montmorillonite (Nanocor), CEC = 1.0 meq/g; (c) fluoro-hectorite (Dow-Corning), CEC = 1.5 meq/g. (From Ref. 30.)

Toyota group to make Nylon-6 nanocomposites from caprolactam monomer [31–33]. It has been applied to several other systems, including epoxies [34–43] and styrene [44,45]. The monomer intercalates into the galleries and swells the organoclay. For clays modified with a long-chain surfactant, the galleries swollen by the monomer or precursor show a d-spacing indicative of a paraffin monolayer arrangement, as illustrated in Figure 5. Upon polymerization, the clay nanolayers are forced apart and no longer interact through the surfactant chains. Thus, highly exfoliated nanocomposites are formed. In terms of both experimental results and thermodynamic considerations, this method is most promising for the synthesis of highly exfoliated nanocomposites. The synthesis and characteristics of nanocomposites prepared via in situ polymerization of various monomers and precursors are discussed in Sections V.A and V.B respectively.

The second method of nanocomposite synthesis involves dissolving a polymer in a solvent, mixing with organophilic clay, and then removing the solvent [14,46–48]. Some

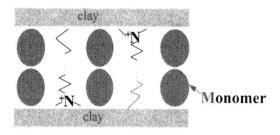

FIG. 5 Model of alkylammonium-exchanged clay swollen by monomer or polymer precursors such as styrene, ε-caprolactam, and epoxide.

examples of this approach are provided in Section V.C. The third approach is melt intercalation and involves heating a mixture of polymer and organophilic clay above the glass transition or melting temperature [5]. Nylon-6–clay [49] and polypropylene–clay [50] nanocomposites were also prepared via this approach. The properties of nanocomposites depend on the compatibility or interaction between the polymer and organophilic clay. A major difficulty in nanocomposite synthesis is that nonpolar polymers such as polypropylene do not easily intercalate into clay galleries. The synthesis and properties of various polymer–clay nanocomposites via melt intercalation are briefly described in Section V.D.

A. Polymer–Clay Nanocomposites Synthesized from Monomers

1. Nylon-6–Clay Nanocomposites

Polymer–clay nanocomposites from monomers were first synthesized by the Toyota group [31–33]. Inorganic cations were replaced by alkylammonium acids to make the clay surface compatible with the monomer, ε-caprolactam. The monomer was polymerized in the interlayer galleries of modified montmorillonite (MMT) to form Nylon-6–clay hybrids or nanocomposites. The protonated alkylammonium acidic cations catalyze the intragallery polymerization of caprolactam, thereby providing a driving force for nanolayer exfoliation in the resulting composite. Individual silicate layers (less than 1-nm thickness) of montmorillonite were completely exfoliated and homogeneously dispersed in Nylon-6 matrix, as revealed by XRD and transmission electron microscopy (TEM). There were significant improvements in the properties of Nylon-6–clay nanocomposite containing 4.2 wt.% clay compared with pure Nylon-6, as shown in Table 1 [33]. The strength increased more than 50%, the modulus doubled, and the heat distortion temperature increased by 80°C compared to pristine Nylon-6. Exfoliated Nylon-6 nanocomposites also demonstrated significant improvements in dimensional stability, barrier properties, and flame retardant properties [51,52]. The enhanced ablative performance of Nylon-6–clay nanocomposites was also studied [53]. α and γ crystal forms and amorphous region of Nylon-6–clay nanocomposites were observed by solid-state nuclear magnetic resonance [54]. The mechanisms behind drastic improvements in the performance of nanocomposites such as Nylon-6 are still unclear. More significantly, Nylon-6 nanocomposites are being used in under-the-hood applications in the automobile industry [55]. Nylon-6–clay nanocomposites were also prepared by melt intercalation [49], as discussed later.

TABLE 1 Mechanical and Thermal Properties of Nylon-6 and Nylon-6–Clay Nanocomposites

Property	Nanocomposite	Nylon-6
Tensile modulus (GPa)	2.1	1.1
Tensile strength (MPa)	107	69
Heat distortion temperature (°C)	145	65
Impact strength (kJ/m^2)	2.8	2.3
Water adsorption (%)	0.51	0.87
Coefficient of thermal expansion (x,y)	6.3×10^{-5}	13×10^{-5}

Source: Ref. 33.

2. Poly(ε-caprolactone)–Clay Nanocomposites

Poly(ε-caprolactone)–clay nanocomposites were prepared via in situ polymerization of ε-caprolactone in the presence of treated clay containing chromium [56] and protonated 12-aminododecanoic acid [57] ions. Poor intercalation was observed with the inorganic cation. However, the surfactant cation allowed swelling of the galleries at 170°C and also catalyzed the polymerization in the layers. The barrier properties of poly(ε-caprolactone)–clay film prepared by casting were studied by monitoring the adsorption of water. The permeability decreased linearly with clay loading, as shown in Figure 6 [57]. The relative permeability of the nanocomposite film was reduced to 0.2 with less than 5% clay by volume.

3. Polystyrene–Clay Nanocomposites

Several attempts to prepare polystyrene–clay nanocomposites are reported in the literature. The primary technique used is impregnating clay in styrene monomer, followed by polymerization. The hydrophilic nature of untreated clay impedes its homogeneous dispersion in styrene. Either no change or a slight expansion of the d-spacing of clay galleries was reported in early studies by Friedlander and Grink [58] and Blumstein [59]. Recently, Cu^{++}-exchanged hectorite was used to catalyze the oxidation of styrene in clay [60]. However, the technique did not work, probably due to the lack of styrene intercalation into clay. The intercalation of polystyrene (PS) was achieved in stearyltrimethyl ammonium cation–exchanged MMT by Kato et al. [61].

Akelah and Moet [62] followed a modified approach to prepare PS–clay intercalated nanocomposites using a solvent to facilitate intercalation. MMT was ion exchanged using a polymerizable surfactant, vinylbenzyltrimethylammonium chloride. Acetonitrile was found to be the most effective solvent, producing a d-spacing of 2.45 nm versus 2.22 and

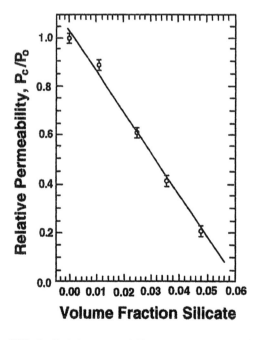

FIG. 6   Relative permeability (P_c/P_o) vs volume fraction of clay for poly(ε-caprolactone)–montmorillonite nanocomposites. (From Ref. 57.)

FIG. 7   TEM of polystyrene–montmorillonite nanocomposite. (From Ref. 45.)

18.1 nm for acetonitrile–THF and acetonitrile–toluene mixtures, respectively. Doh and Cho [63] prepared PS-MMT intercalated nanocomposites by directly mixing styrene with organoclay followed by in situ polymerization. The resulting nanocomposites exhibited higher thermal stability compared with virgin PS or PS/pristine-MMT microcomposite. The PS nanocomposite obtained with organoclay containing benzyl-units exhibited the best thermal stability. Weimer et al. [64] prepared PS–clay nanocomposites by anchoring a living free-radical polymerization initiator inside the clay galleries, followed by intercalation and polymerization of styrene. Noh and Lee [65] used an emulsion polymerization technique to prepare PS–clay nanocomposites with enhanced thermal properties.

Fu and Qutubuddin [44,45] synthesized highly exfoliated PS–clay nanocomposites via in situ polymerization of styrene and reactive organoclay. A representative TEM micrograph is shown in Figure 7 [45]. The organoclay was prepared by cationic exchange with vinylbenzyldimethyldodecylammonium chloride (VDAC). VDAC-functionalized clay has a d-spacing of 1.92 nm, compared to 0.99 nm for pristine MMT. VDAC-MMT swells in styrene more than other functionalized clays [44]. Thus, gelation was observed when only 3 wt% of organoclay was dispersed in styrene. The storage modulus of the nanocomposite was higher, depending on the loading of clay, as illustrated in Figure 8 [44]. Exfoliated PS nanocomposites also exhibit a higher thermal degradation temperature than pristine PS. VDAC can copolymerize with styrene; thus the covalent bonding between PS and clay improves the interfacial strength.

B.   Polymer–Clay Nanocomposites Synthesized from Precursors

1.   Epoxy–Clay Nanocomposites

Epoxy–clay nanocomposites from epoxide precursors have been investigated by research groups at Michigan State University [34–40], Cornell University [41], and Case Western Reserve University [42,43]. In general, the synthesis is similar to that of Nylon-6 and PS

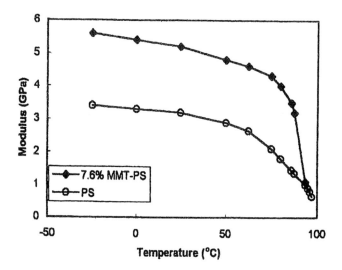

FIG. 8 DMA scans of polystyrene and 7.6 wt% VDAC-MMT polystyrene nanocomposites. (From Ref. 45.)

described earlier, but with the additional need for a curing agent. First, the clay is rendered hydrophobic by cationic exchange with appropriate surfactant molecules. Next, the organo-clay is dispersed in a mixture of epoxy resin and curing agent. Finally, the temperature is increased to cure the resin. Acidic onium ions catalyze intragallery polymerization at a rate that is comparable to extragallery polymerization. The relative rates of reagent intercala-tion, chain formation, and network cross-linking have an important effect on the initial gelation and final curing of epoxy–clay exfoliated nanocomposite [39]. Aliphatic amine, aromatic amine, anhydride, and catalytic curing agents have been investigated to form an epoxy composite with broad glass transition temperature, T_g. The thermomechanical prop-erties of epoxy nanocomposites show dramatic improvements, particularly in the rubbery state [35]. Figure 9 illustrates how different loadings of exfoliated silicate nanolayers im-prove the modulus and strength of elastomeric matrix [35]. The dynamic storage modulus of epoxy nanocomposite containing 4 vol% clay was approximately 58% higher in the glassy region and 450% higher in the rubbery plateau region, compared to the pristine epoxy [41]. Different modifications of clay result in variations in the T_g of epoxy nanocom-posites obtained with the same curing agent [41,43]. Recently, Massam and Pinnavaia [40] demonstrated that clay nanolayers reinforce glassy epoxy matrix under compressive strain, as shown in Figure 10 [40]. The clay nanolayers also enhance the dimensional stability, thermal stability, and solvent resistance of glassy epoxy matrix.

2. Polyurethane–Clay Nanocomposites

Based on the behavior of epoxy–clay nanocomposites, Wang and Pinnavaia [66] obtained effective reinforcement of polyurethane by alkylammonium-exchanged MMT. Swelling of the organoclays by polyols commonly used in polyurethane as chain extenders or cross-linkers was a function of the surfactant chain length (carbon number ≥ 12), but indepen-dent of polyol MW or cation exchange capacity of the clay. In situ polymerization of polyol-isocyanate precursor–organoclay dispersions produced nanocomposites containing intercalated clay (~5-nm d-spacing) in cross-linked polyurethane matrix. The nanolayers

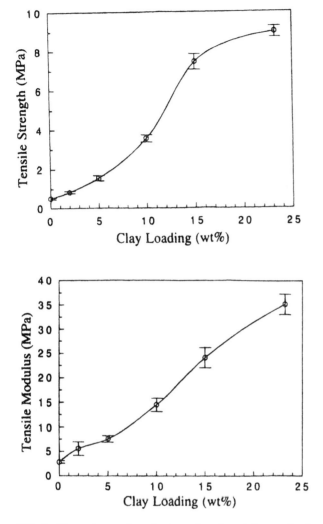

FIG. 9 Dependence of tensile strength and modulus on clay loading for epoxy-$CH_3(CH_2)_{17}NH_3^+$-montmorillonite nanocomposites. (From Ref. 35.)

increased both the strength and toughness of the elastomeric nanocomposite, as illustrated in Figure 11 [66].

3. PDMS–Clay Nanocomposites

Mark [3] reported reinforcement of poly(dimethylsiloxane) (PDMS) by precipitating silica nanoparticles as a highly dispersed phase in the elastomer. Burnside and Giannelis [67] explored an alternative method of synthesizing PDMS nanocomposites using clay. Dimethylditallow ammonium–exchanged MMT was mixed with silanol-terminated PDMS (MW = 18,000) and then cross-linked with tetraethylorthosilicate in the presence of tin octoate as catalyst. Interestingly, the XRD peak for organoclay disappeared when a small amount of water (about 0.5 wt%) was added, as shown in Figure 12 [67]. MMT treated with other organic surfactants, such as benzyldimethyloctadecylammonium, did not

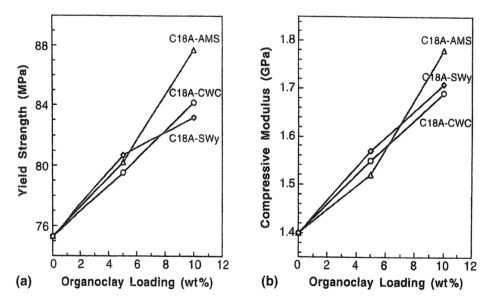

FIG. 10 Compressive (a) yield strength and (b) moduli for the pristine epoxy polymer and the exfoliated epoxy–clay nanocomposites prepared from three different kinds of organomontmorillonites. (From Ref. 40.)

result in intercalation by PDMS prepolymers. The nanocomposite obtained with dimethylditallow ammonium clay showed increased thermal stability and reduced solvent uptake compared with conventional PDMS composites containing kaolin or carbon black as filler.

4. Rubber–Clay Nanocomposites

Rubber–clay nanocomposites are particularly attractive for potential applications where enhanced barrier properties are desired. Organoclays for rubber intercalation were prepared

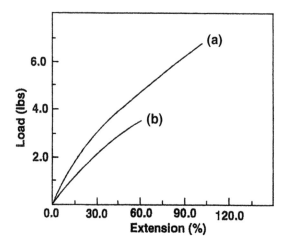

FIG. 11 Stress–strain curves for (a) a pristine polyurethane elastomer; (b) a polyurethane–clay nanocomposite prepared from organomontmorillonite (5 wt%). (From Ref. 66.)

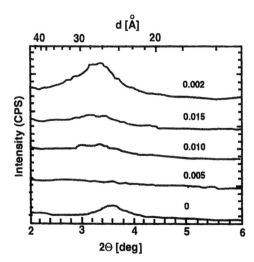

FIG. 12 X-ray diffraction patterns of poly(dimethylsiloxane)–clay nanocomposites prepared from dimethyl ditallowammonium–exchanged montmorillonite as a function of the weight ratio of water to silicate. (From Ref. 67.)

via cation exchange with a protonated form of amine-terminated butadiene acrylonitrile copolymer (ATBN) [68,69]. Nanocomposites were prepared by mixing the ATBN-intercalated clay with nitrile rubber under cross-linking (vulcanizing) conditions. The permeabilities of hydrogen and water vapor for the rubber–clay nanocomposites decreased by about one-third relative to the pristine nitrile rubber [68]. TEM results indicate that the clay nanolayers are exfoliated in the rubber matrix [69].

C. Polymer–Clay Nanocomposites Synthesized from Polymer Solution

It is well known that water-soluble polymers easily intercalate into clay galleries in aqueous suspension. A comprehensive review of the formation and properties of clay–polymer complexes is provided by Theng [14]. Recent interest has focused on the blending of organoclay with polymers dissolved in organic solvents. Ogata et al. [46] investigated poly(ethylene oxide) (PEO) mixed with distearyldimethylammonium-exchanged organoclay in chloroform. After the chloroform was evaporated, 100-μ-thick films were obtained. The organoclay accelerated the crystallization of PEO and induced a preferred orientation of the polymer. Poly(l-lactide)–clay [47] and poly(e-caprolactone)–clay [48] composite films were also prepared by solvent casting.

Polyimide–clay nanocomposites constitute another example of the synthesis of nanocomposite from polymer solution [70–76]. Polyimide–clay nanocomposite films were produced via polymerization of 4,4′-diaminodiphenyl ether and pyromellitic dianhydride in dimethylacetamide (DMAC) solvent, followed by mixing of the poly(amic acid) solution with organoclay dispersed in DMAC. Synthetic mica and MMT produced primarily exfoliated nanocomposites, while saponite and hectorite led to only monolayer intercalation in the clay galleries [71]. Dramatic improvements in barrier properties, thermal stability, and modulus were observed for these nanocomposites. Polyimide–clay nanocomposites containing only a small fraction of clay exhibited a several-fold reduction in the permeability of small gases, e.g., O_2, H_2O, He, CO_2, and ethyl acetate vapor. For example, at 2 wt% loading of synthetic mica, the permeability coefficient of water vapor decreased an order of magnitude relative to pristine polyimide [71]. At a given loading, the permeability decreased with increasing platelet aspect ratio, as shown in Figure 13. Also, the coefficient of thermal expansion decreased with increasing nanolayer aspect ratio (hectorite 46, saponite 165, montmorillonite 218, synthetic mica 123). Tyan et al. [73,74] used clay modified with *p*-phenylenediamine to synthesize polyimide–clay nanocomposites. A 2.5-fold increase in the modulus of nanocomposite film was obtained as compared to pristine polyimide. The imidization temperature and duration were reduced dramatically [73]. For instance, when 2 wt% organoclay was dispersed in the poly(amic acid), the imidization temperature was lowered by 50°C (250°C versus 300°C) for complete imidization. In addition, the imidization time at 250°C was reduced to 15 minutes with 7 wt% organoclay.

The polymer solution approach was also used to prepare polyurethane–clay nanocomposites [77]. 12-Aminolauric acid and benxidine ion–exchanged organoclays and polyurethane were mixed in dimethylformamide solvent. After degassing of the mixture and removal of solvent at 80°C, an elastic nanocomposite film was obtained. Exfoliation of polyurethane–clay nanocomposites was indicated by XRD and TEM. The tensile strength and elongation were significantly enhanced, while water absorption decreased.

D. Polymer–Clay Nanocomposites Synthesized via Melt Intercalation

Vaia and coworkers [78–81] prepared intercalated PS–clay nanocomposites via polymer melt intercalation using long-chain primary and quarternary alkylammonium–exchanged clays. The organoclay was mixed with commercially available PS at a temperature above the T_g via melt processing. The diffusion of PS into the clay galleries is slow and depends on many factors, including polymer molecular weight (MW), processing temperature, surfactant properties, and interactions between the polymer and the organoclay. Hasegawa et al. [82] used organoclay obtained via ion exchange with protonated amine–terminated PS

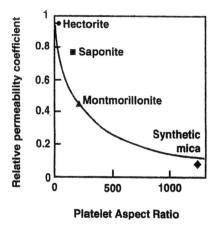

Platelet Aspect Ratio

FIG. 13 Reduction of the relative permeability coefficient is dependent on the clay platelet aspect ratio in the system of polyimide–clay hybrid with water vapor as the permeate. Each hybrid contains 2 wt% clay. The aspect ratios for hectorite, saponite, montmorillonite, and synthetic mica are 46, 165, 218, and 1230, respectively. (From Ref. 71.)

of different molecular weights (Mn = 121 and 5800 g/mol) to prepare nanocomposite by melt intercalation. Individual platelets (1 nm thick and 600 nm long) were obtained only with applied shear during melt compounding of clay modified with PS of high MW [82]. This result demonstrates that mechanical driving force is important for exfoliation of clay in a polymer melt. Exfoliated PS–clay nanocomposites were prepared by melt-blending a styrene-vinyloxazoline copolymer with organophilic clay [83]. The silicate layers were exfoliated and dispersed homogeneously at the nanometer level. The moduli of the nanocomposites were higher than that of PS copolymer.

Nylon-6–clay nanocomposites were also prepared by melt intercalation process [49]. Mechanical and thermal testing revealed that the properties of Nylon-6–clay nanocomposites are superior to Nylon. The tensile strength, flexural strength, and notched Izod impact strength are similar for both melt intercalation and in situ polymerization methods. However, the heat distortion temperature is low (112°C) for melt intercalated Nylon-6–nanocomposite, compared to 152°C for nanocomposite prepared via in situ polymerization [33].

Polypropylene (PP) is one of the most widely used polyolefin polymers. The platelet form of talc has been used as filler for PP for several decades. PP composites containing talc filler exhibit improved stiffness, dimensional stability, and heat distortion characteristics. Preparation of PP–clay nanocomposites is difficult, since PP lacks polar groups in its backbone. PP modified with polar groups was used for intercalation into clay galleries, followed by melt-compounding of organoclay with bulk PP to prepare nanocomposites [84–87]. Only a limited degree of clay exfoliation was observed by this approach. Another disadvantage of this method is the use of an organic solvent to facilitate the intercalation of modified PP into clay galleries. An alternative method was developed later by the Toyota research group [50,88,89]. A mixture of maleic anhydride–modified PP oligomer, homo-PP, and stearylammonium-exchanged MMT was melt-blended to obtain PP–clay nanocomposites. A larger fraction of clay layers was exfoliated, as shown by x-ray diffraction. The hydrolyzed maleic anhydride PP intercalated into the organoclay, expanding the galleries, and facilitating further intercalation by homo-PP. Interestingly, the density of maleic anhydride groups has a significant effect on the final morphology and properties of the nanocomposite. The nanocomposites exhibit improved storage moduli compared to pristine PP in the temperature range from T_g to 90°C. The impact of clay nanolayer reinforcement in PP nanocomposites is not as dramatic as in Nylon-6, probably due to lower degree of exfoliation, weaker interfacial adhesion between polymer matrix and clay nanolayers, and the introduction of a large amount of oligomer.

Silanol-terminated PDMS and hexadecyltrimethylammonium-exchanged clay were used to prepare PDMS–clay nanocomposites via melt intercalation [90]. The melt intercalation nanocomposites did not achieve as high a reinforcement as the aerosilica silicone hybrid, but the nanocomposite formed from solution had a nearly identical reinforcing effect on tensile strength as the aerosilica composite.

Poly(styrene-b-butadiene) copolymer–clay nanocomposites were prepared from dioctadecyldimethyl ammonium–exchanged MMT via direct melt intercalation [91]. While the identical mixing of copolymer with pristine montmorillonite showed no intercalation, the organoclay expanded from 41 to 46 Å, indicating a monolayer intercalation. The nanocomposites showed an increase in storage modulus with increasing loading. In addition, the T_g for the polystyrene block domain increased with clay content, whereas the polybutadiene block T_g remained nearly constant.

E.   Other Polymer–Clay Nanocomposite Systems

Other advancements in polymer–clay nanocomposites have included organoclay–liquid crystal nanocomposites that show a unique light-scattering effect controlled by electric field, temperature change, or shearing [92–94]. Poly(ethylene terephathalate) (PET)–clay intercalated nanocomposites showed three times faster crystallization rate than that of pristine PET, and the heat-distortion temperature (HDT) was 20–50°C higher than the pristine PET [95]. Poly(methylmethacrylate)–clay nanocomposites were prepared via the emulsion polymerization approach and showed a higher glass transition temperature [96].

VI.   THERMODYNAMICS OF INTERACTIONS BETWEEN POLYMER AND ORGANOCLAY

As described in Section V, organoclays suitable for polymer–clay nanocomposites have been reported for several systems, including ε-caprolactam [31–33], epoxies [34–36], and styrene [44,45]. The selection of an organoclay for a given polymer or the modification of a polymer for a specific organoclay is primarily empirical, as evident in this review. The surface polarities of the monomer or precursor and the clay should be similar in order to fully wet and intercalate into clay galleries. Furthermore, specific chemical reactions may play an important role. For instance, in Nylon-6 and epoxy–clay systems, protonated alkyl amine cations can catalyze reaction relative to the bulk polymer, providing a driving force for nanolayer exfoliation. In exfoliated PS nanocomposites [44,45], an important feature is that the cationic surfactant has a vinybenzyl functional group that copolymerizes with styrene monomer.

Complete exfoliation of clay nanolayers in nanocomposites yields the highest degree of reinforcement. The layers persist with a repeating stacking pattern in intercalated nanocomposites. Incomplete dispersion of the reinforcing phase reduces the interfacial contact between the polymer and clay, creating domains of pure polymer. This limits stress transfer through the composite, giving comparatively less than optimal reinforcement. Thermodynamic modeling of interactions between organoclay and polymer and the prediction of organoclay performance are reviewed next.

Thermodynamics is an important issue that can guide the design of an organoclay and predict the formation of polymer–organoclay nanocomposites. A successful thermodynamic model should be able to address questions such as why certain polymer–clay systems favor intercalated nanocomposites, some form exfoliated nanocomposites, and yet others result in immiscible conventional macro-composites. The formation of polymer–clay nanocomposites or hybrids is affected by the packing density and chain length of the surfactant in the organoclay, the charge density of the clay, specific groups on the polymer, and the type of bonding at the polymer/clay interface (i.e., hydrogen, dipole–dipole, van der Waals, or covalent). Vaia and Giannelis [97,98] developed a mean-field, lattice-based thermodynamic model to delineate the factors favoring intercalated versus exfoliated nanocomposites prepared via polymer melt intercalation. Interplay of entropic and energetic terms dictates the outcome of polymer intercalation. Free-energy values and their dependence on energetic and entropic terms determine three possible equilibrium states: immiscible, intercalated, and exfoliated. The thermodynamic calculations suggest that the entropic penalty of polymer confinement is compensated by the increased conformational freedom of the surfactant chain as the layers separate. When the total entropy change is small, small changes in the system's internal energy will determine if intercalation is thermodynamically possible, as illustrated in Figure 14 [97]. However, com-

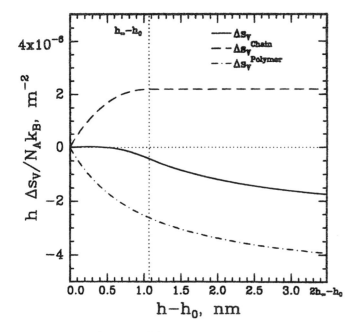

FIG. 14 $h \Delta s_v^{polymer}$, $h \Delta s_v^{chain}$, $h \Delta s_v$ as a function of the change in gallery height for an arbitrary polymer and a silicate functionalized with octadecylammonium groups. (From Ref. 97.)

plete exfoliation depends on the existence of very favorable polymer–organoclay interactions to overcome the penalty of polymer confinement.

Lyatskaya and Balazs [99] developed a free-energy expression for a mixture of polymer and solid, thin disks to model the phase behavior of polymer–clay composites. The phase diagrams for polymer–clay mixtures were constructed by minimizing the free energy and calculating the chemical potentials. The calculations indicate that favorable mixing of polymer and disks like clay is controlled through a balance of the effects of the polymerization number (N) and the Flory–Huggins interaction parameter, χ. In particular, an increase in N requires a decrease in χ to maintain thermodynamic stability of the mixture. Both numerical and analytical self-consistent field (SCF) calculations were also used to study the interactions between two closely spaced clay surfaces and the surrounding polymer [100–102]. Initially, the polymer has to intercalate into the clay galleries from an outer edge and then diffuse toward the center of the galleries. The SCF calculations and the phase diagrams suggest that the polymer and sheets are immiscible for $\chi > 0$ [99,100]. Intercalation is predicted when χ is substantially negative, i.e., the polymer and surface experience an attractive interaction, as illustrated in Figure 15 [100]. As the polymer diffuses through the energetically favorable gallery, it maximizes contact with the two confining layers. However, the real phase behavior and morphology of the mixture are affected by the kinetics of the polymers penetrating the gallery. In effect, the polymer "glues" the two surfaces together as it moves through the interlayer. This "fused" condition could represent a kinetically trapped state; consequently, increasing the attraction between the polymer and clay sheets would only lead to intercalated, rather than exfoliated, structure [100]. Further work is needed to verify the predictions of the models proposed by Vaia and Giannelis [97,98] and Balazs and coworkers [99–102].

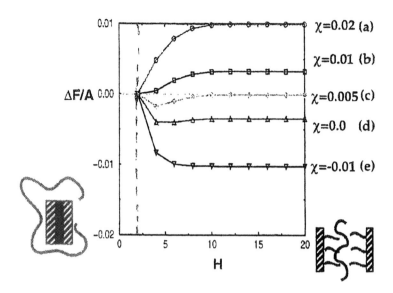

FIG. 15 Free energy per unit area as a function of surface separation for five different values of χ. The parameters are $N_{gr} = 25$, $\rho = 0.04$, $N = 100$, and $\chi_{surf} = 0$. The cartoon on the left shows the reference state, where the grafted chains form a melt between the surfaces. In the cartoon on the right, the surfaces are separated by polymers that have localized between the interfaces. (From Ref. 100.)

VII. CONCLUDING REMARKS

Clays functionalized via ion exchange with cationic surfactants have been successfully used to prepare polymer–clay nanocomposites. Organoclay nanolayers exfoliated in polymer matrix can dramatically improve mechanical reinforcement, thermal and chemical stability, barrier properties, etc. The replacement of inorganic cations in clay galleries by cationic surfactant molecules not only renders the clay organophilic but also expands the interlayer distance and lowers the cohesive energy between nanolayers. The wetting and intercalation of a monomer or a polymer is determined by the surface chemistry, in particular, the difference in polarity of the surface of organoclay and the polymer matrix. Thus, a major difficulty in the synthesis of new polymer nanocomposites is the compatibility of the organoclay surface with the polymer, particularly if it is hydrophobic. The interfacial adhesion between polymer and clay is very important for the reinforcement, due to the tremendous surface area of clay nanolayers.

Three main approaches for the synthesis of polymer–clay nanocomposites are described in this chapter. In situ polymerization of the monomer or polymer precursor seems to be most effective in achieving exfoliation of the clay nanolayers. In specific cases, cationic surfactants used for functionalization of clay can also catalyze intragallery polymerization reactions or form copolymer with a monomer that facilitates exfoliation of nanolayers in the polymer matrix. Thermodynamic calculations also predict that increasing the affinity between polymer matrix and organoclay is an effective way to prepare exfoliated nanocomposites. The melt processing of mixtures of organoclay, oligomer (compatablilizer), and hydrophobic polymers such as polypropylene is an attractive approach for the synthesis of nanocomposites.

As reflected in this chapter, most of the published literature on polymer–clay nanocomposites focuses on synthesis and characterization. Future work should address issues such as novel applications, scale-up, and industrial processing of nanocomposites, including control of rheology, morphology, and specific performance. Fundamental research is needed to improve the understanding of the mechanisms of exfoliation, interfacial adhesion, and overall reinforcement. New synthesis methods and models for the prediction of nanocomposite formation and properties are obviously desirable.

ACKNOWLEDGMENTS

The financial support from Edison Polymer Invention Corporation (EPIC) and its industrial members is gratefully acknowledged.

REFERENCES

1. Schmidt, H, J. Non-Cryst. Solids, 1985, 73, 681–691.
2. Novak, BM, Adv. Mater., 1993, 5, 422–433.
3. Mark, JE, Polym. Eng. Sci., 1996, 36, 2905–2920.
4. LeBaron, PC, Wang, Z, and Pinnavaia, TJ, Appl. Clay Sci., 1999, 15, 11–29.
5. Giannelis, EP, Adv. Mater., 1996, 8, 29–35.
6. Gilman JW, Appl. Clay Sci., 1999, 15, 31–49.
7. Jones, TR, Clay Minerals, 1983, 18, 399.
8. Grim, RE, Clay Mineralogy, McGraw-Hill, New York, 1968.
9. van Olphen H, An introduction to Clay Colloid Chemistry, 2nd ed., Wiley, New York, 1973.
10. Hunter, RJ, Foundations of Colloid Science, Vol. 1, Oxford University Press, New York, 1986.
11. Newman, ACD (ed.), Chemistry of Clays and Clay Minerals, Wiley, New York, 1987.
12. Weaver, CE, and Pollard, LD, The Chemistry of Clay Minerals, Elsevier Scientific, New York, 1973.
13. Theng, BKG, The Chemistry of Clay Organic Interactions, Wiley, New York, 1974.
14. Theng, BKG, Formation and Properties of Clay–Polymer Complexes, Elsevier, New York, 1979.
15. Parfitt, GD (ed.), Dispersion of Powders in Liquids, Applied Science Publisher, NJ, 1981.
16. Boyd, SA, Lee, JL, and Mortland, MM, Nature, 1988, 333, 345–347,
17. Bongiovanni, R, Chiarle, M, and Pelizzetti, J, J. Dispersion Sci. Tech., 1993, 14, 255.
18. Xu, S, and Boyd, SA, Langmuir, 1995, 11, 2508.
19. Xu, S, and Boyd, SA, Soil Sci. Soc. Am. J., 1994, 58, 1382.
20. Pan, J, Yang, G, Han, B, and Yan, H, J. Colloid Inter. Sci., 1998, 194, 276.
21. Chen, G, Han, B, and Yan, H, J. Colloid Interf. Sci., 1998, 201, 158.
22. Somasundaran P, and Fuerstenau, DW, In: Adsorption from aqueous solution, Adv. Chem. Ser., No. 79, ACS, Washington DC., 1968.
23. Cases, JM, and Villieras, F, Langmuir, 1992, 8, 1251.
24. Bitting, D, and Harwell, JH, Langmuir, 1987, 3, 500.
25. Bohmer, MR, and Koopal, LK, Langmuir, 1992, 8, 1908.
26. Bohmer, MR, and Koopal, LK, Langmuir, 1992, 8, 2649.
27. Lagaly, G, Clays Clay Mineral., 1982, 30, 215.
28. Lagaly, G, Solid State Ionics, 1986, 22, 43–51.
29. Vaia, RA, Teukolshy, RK, and Giannelis, EP, Chem. Mater., 1994, 6, 1017–1022.
30. Hackett, E, Manias, E, and Giannelis, EP, J. Chem. Phys., 1998, 108, 7410–7415.
31. Usuki, A, Kawasumi, M, Kojima, Y, Fujushima, A, Okada, A, and Kamigaito, O, J. Mater. Res., 1993, 8, 1174–1178.
32. Usuki, A, Kojima, Y, Kawasumi, M, Okada, A, Fujushima, A, Kurauchi, T, and Kamigaito, O, J. Mater. Res., 1993, 8, 1179–1184.

33. Kojima, Y, Usuki, A, Kawasumi, M, Okada, A, Fujushima, A, Kurauchi, T, and Kamigaito, O, J. Mater. Res., 1993, 8, 1185–1188.
34. Wang, MS, and Pinnavaia, TJ, Chem. Mater., 1994, 6, 468–474.
35. Lan, T, and Pinnavaia, TJ, Chem. Mater., 1994, 6, 2216–2219.
36. Lan, T, Kaviratna, PD, and Pinnavaia, TJ, Chem. Mater., 1995, 7, 2144–2150.
37. Shi, H, Lan, T, and Pinnavaia, TJ, Chem. Mater., 1996, 8, 1584–1587.
38. Wang, Z, Lan, T, and Pinnavaia, TJ, Chem. Mater., 1996, 8, 2200–2204.
39. Wang, Z, and Pinnavaia, TJ, Chem. Mater., 1998, 10, 1820–1826.
40. Massam, J, and Pinnavaia, Mater. Res. Soc. Symp. Proc., 1998, 520, 223–232.
41. Messersmith, PB, and Giannelis, EP, Chem. Mater., 1994, 6, 1719–1725.
42. Kelly, P, Akelah, A, Qutubuddin, S, and Moet, A, J. Mater. Sci., 1994, 29, 2274–2280.
43. Akelah, A, Kelly, P, Qutubuddin, S, and Moet A, Clay Minerals, 1994, 29, 169–178.
44. Qutubuddin, S, and Fu, X, Proceedings of SAMPE Advanced Composites Conference, Detroit Michigan, September 1999.
45. Fu, X, and Qutubuddin, S, Mater. Lett., 2000, 42, 12–15.
46. Ogata, N, Kawakage, S, and Ogihara, T, Polymer, 1997, 38, 5115–5118.
47. Ogata, N, Jimenez, G, and Ogihara, T, J. Poly. Sci. B, Polym. Phys., 1997, 35, 389–396.
48. Jimenez, G, Ogata, N, Kawai, H, and Ogihara, T, J. App. Polym. Sci., 1997, 64, 2211–2222.
49. Liu, LM, Qi, ZN, and Zhu, X, J. Appl. Polym. Sci., 1999, 71, 1133–1138.
50. Kawasumi, M, Hasegawa, N, Kato, M, Usuki, A, and Okada, A, Macromolecules, 1997, 30, 6333–6338.
51. Kojima, Y, Usuki, A, Kawasumi, M, Okada, A, Kurauchi, T, and Kamigaito, O, J. App. Polym. Sci., 1993, 49, 1259–1264.
52. Gilman, JW, Kashiwagi, T, and Lichtenhan, JD, 1997, SAMPE J. 1997, 33, 40–46.
53. Vaia, RA, Prince, G, Ruth, P, Nguyen, HT, and Lichtenhan, J, J. Appl. Clay Sci., 1999, 15, 67–92.
54. Mathias, LJ, Davis, RD, and Jarrett, WL, Macromolecules, 1999, 32, 7958–7960.
55. Okada, A, and Usuki, A, Mater. Sci. Eng., 1995, C3, 109.
56. Messersmith, PB, and Giannelis, E, Chem. Mater., 1993, 5, 1064–1066.
57. Messersmith, PB, and Giannelis, E, J. Polym. Sci. A, Polym. Chem., 1995, 33, 1047–1057; 1993, 5, 1064–1066.
58. Friedlander, HZ, and Grink, CR, J. Polym. Sci. Polym. Lett., 1964, 2, 475.
59. Blumstein, A, J. Polym. Sci. A, 1965, 3, 2653.
60. Porter, TL, Hagerman, ME, Reynolds, BP, Eastman, MP, and Parnell, RA, J. Polym. Sci. B, Polym. Phys., 1998, 36, 673–679.
61. Kato, C, Kuroda, K, and Takahara, H, Clay and Clay Minerals, 1981, 29, 294.
62. Akelah, A, and Moet, A, J. Mater. Sci., 1996, 31, 3189.
63. Doh, JG, and Cho, I, Polymer Bull., 1998, 41, 511–518.
64. Weimer, MW, Chen, H, Giannelis, EP, and Sogah, DY, J. Am. Chem. Soc., 1999, 121, 1615–1616.
65. Noh, MW, and Lee, DC, Polymer Bull., 1999, 42, 619–626.
66. Wang, Z, and Pinnavaia, TJ, Chem. Mater. 1998, 10, 3769–3771.
67. Burnside, SD, and Giannelis, EP, Chem. Mater., 1995, 7, 1597–1600.
68. Kojima, Y, Fukumori, K, Usuki, A, Okada, A, and Kurauchi, T, J. Mater. Sci. Lett., 1993, 12, 889–890.
69. Akelah, A, El-Deen, NS, Hiltner, A, Baer, E, and Moet, A, Mater. Lett., 1995, 22, 97–102.
70. Yano, K, Usuki A, Okada, A, Kurauchi, T, and Kamigaito, O, J. Polym. Sci. A, Polym. Chem., 1993, 31, 2493–2498.
71. Yano, K, Usuki A, and Okada, A, J. Polym. Sci. A, Polym. Chem., 1997, 35, 2289–2294.
72. Lan, T, Kaviratna, PD, and Pinnavaia, TJ, Chem. Mater., 1994, 6, 573–575.
73. Tyan, HL, Liu, YC, and Wei, KH, Polymer 1999, 40, 4877–4886.
74. Tyan, HL, Liu, YC, and Wei, KH, Chem. Mater., 1999, 11, 1942–1947.

75. Yang, Y, Zhu, ZK, Yin, J, Wang, XY, and Qi, ZN, Polymer, 1999, 40, 4407–4414.
76. Zhu, ZK, Yang, Y, Yin, J, Wang, XY, Ke, YC, and Qi, ZN, J. Appl. Polym. Sci., 1999, 73, 2063–2068.
77. Chen, TK, Tien, YI, and Wei, KH Polymer, 2000, 41, 1345–1353.
78. Vaia, RA, Ishii, H, and Giannelis, E, Chem. Mater., 1993, 5, 1694–1696.
79. Vaia, R, Jandt, K, Kramer, E, and Giannelis, E, Macromolecules, 1995, 28, 8080–8085.
80. Vaia, R, Jandt, K, Kramer, E, and Giannelis, E, Chem. Mater., 1996, 8, 2628–2635.
81. Krishnamoorti, R, Vaia, R, and Giannelis, E, Chem. Mater., 1996, 8, 1728–1734.
82. Hasegawa, N, Okamoto, H, Kawasumi, M, and Usuki, A, J. Appl. Polym. Sci., 1999, 74, 3359–3364.
83. Hoffmann, B, Dietrich, C, Thomann, R, Friedrich, C, and Muhaupt, R, Macromol. Rapid Commun., 2000, 21, 57–61.
84. Kurokawa, Y, Yasuda, M, and Oya, A, J. Mater. Sci. Lett., 1996, 15, 1481–1483.
85. Furuichi, N, Kurokawa, Y, Jujita K, Oya, A, Yasuda, H, and Kiso, M, J. Mater. Sci., 1996, 31, 4307–4310.
86. Kurokawa, Y, Yasuda, H, Kashiwagi, M, and Oya, A, J. Mater. Sci. Lett., 1997, 16, 1670–1672.
87. Usuki, A, Kato, M, Okada, A, and Kurauchi, T, J. Appl. Polym. Sci., 1997, 63, 137–139.
88. Kato, M, Usuki, A, and Okada, A, J. Appl. Polym. Sci., 1997, 63, 1781–1785.
89. Hsegawa, N, Kawasumi, M, Kato, M, Usuki, A, and Okada, A, J. Appl. Polym. Sci., 1998, 67, 87–92.
90. Wang SJ, Long, CF, Wang, XY, Li, Q, and Qi, ZN, J. Appl. Polym. Sci., 1998, 69, 1557–1561.
91. Laus, M, Francescangeli, O, and Sandrolini, F, J. Mater. Res., 1997, 12, 3134–3139.
92. Kawasumi, M, Hsegawa, N, Usuki, A, and Okada, A, Liq. Cryst., 1996, 21, 769–776.
93. Kawasumi, M, Usuki, A, Okada, A, and Kurauchi, T, Mol. Cryst. Liq. Cryst., 1996, 281, 91–103.
94. Kawasumi, M, Hsegawa, N, Usuki, A, and Okada, A, Appl. Clay Sci., 1999, 15, 93–108.
95. Ke, YC, Long, CF, and Qi, ZN, J. Appl. Polym. Sci., 1999, 71, 1139–1146.
96. Lee, DC, and Jang, LW, J. Appl. Polym. Sci., 1996, 61, 1117–1122.
97. Vaia, R, A, and Giannelis, EP, Macromolecules, 1997, 30, 7990–7999.
98. Vaia, R, A, and Giannelis, EP, Macromolecules, 1997, 30, 8000–8009.
99. Lyatskaya, Y, and Balazs, AC, Macromolecules, 1998, 31, 6676–6680.
100. Balazs, AC, Singh, C, and Zhulina, E, Macromolecules, 1998, 31, 8370–8381.
101. Balazs, AC, Singh, C, Zhulina, E, and Lyatskaya, Y, Acc. Chem. Res., 1999, 32, 651–657.
102. Zhulina, E, Singh, C, and Balazs, AC, Langmuir, 1999, 15, 3935–3943.

Index